Probability Theory and Stochastic Modelling

Volume 109

Probability Theory and Stochastic Modelling publishes cutting-edge research monographs in probability and its applications, as well as postgraduate-level textbooks that either introduce the reader to new developments in the field, or present a fresh perspective on fundamental topics.

Books in this series are expected to follow rigorous mathematical standards and all titles will be thoroughly peer-reviewed before being considered for publication.

Probability Theory and Stochastic Modelling covers all aspects of modern probability theory including:

- Gaussian processes
- Markov processes
- Random fields, point processes, and random sets
- Random matrices
- Statistical mechanics and random media
- Stochastic analysis
- High-dimensional probability

as well as applications that include (but are not restricted to):

- Branching processes and other models of population growth
- Communications and processing networks
- Computational methods in probability theory and stochastic processes, including simulation
- Genetics and other stochastic models in biology and the life sciences
- Information theory, signal processing, and image synthesis
- Mathematical economics and finance
- Statistical methods (e.g., empirical processes and MCMC)
- Statistics for stochastic processes
- Stochastic control and stochastic differential games
- Stochastic models in operations research and stochastic optimization
- Stochastic models in the physical sciences

Probability Theory and Stochastic Modelling is a merger and continuation of Springer's Stochastic Modelling and Applied Probability and Probability and Its Applications series.

Tianxiao Wang • Jiongmin Yong

Backward Stochastic Volterra Integral Equations

Springer

Tianxiao Wang
Department of Mathematics
Sichuan University
Chengdu, China

Jiongmin Yong
University of Central Florida
ORLANDO, FL, USA

ISSN 2199-3130 ISSN 2199-3149 (electronic)
Probability Theory and Stochastic Modelling
ISBN 978-3-032-20847-7 ISBN 978-3-032-20848-4 (eBook)
https://doi.org/10.1007/978-3-032-20848-4

Mathematics Subject Classification: 49K45, 60H20, 93E20

This Springer imprint is published by the registered company Springer Nature Switzerland AG
The registered company address is: Gewerbestrasse 11, 6330 Cham, Switzerland

To
Zheng Liu and Meifen Xu

Preface

It is well-understood that (forward) Volterra integral equations[1] (FVIE, for short) can be used to describe some dynamical state processes for which certain kind of memories exist. In other words, the current value of the state might depend on the value(s) of the state in the past. Except for some very special cases, FVIEs cannot be transformed to (forward) ordinary differential equations (FODEs, for short). Therefore, a theory of FVIE, which is a strictly more general than that of FODEs, is necessary. On the other hand, for ordinary differential equations, people can consider initial value problems as well as terminal value problems. The former is to be solved forwardly and the latter is to be solved backwardly. Since the former is called FODE, for convenience, we may call the latter the backward ordinary differential equation (BODE, for short). It is not hard to see that by making a time-reverse transformation, almost all the results for FODEs can be transformed to BODEs, except for the problem in the infinite time-horizon. Hence, having a theory for FODEs, it is almost not necessary to develop an independent theory for BODEs. With the same spirit, we have FVIE and backward Volterra integral equations (BVIEs, for short), respectively, and it almost suffices to develop a theory for FVIEs.

Next, if randomness is involved, similar to FODEs and BODEs, it is possible to consider forward stochastic differential equations (FSDEs, for short) and backward stochastic differential equations (BSDEs, for short). We point out that BSDEs are significantly different from FSDEs. The main reason is that time-reverse transformation is not allowed since it will ruin the adaptiveness of the solutions. Thus, besides a theory for FSDEs, it is necessary to establish an independent theory for BSDEs. In doing so, an adjustment process (usually denoted by $Z(s)$) was introduced. Such a process makes it possible to have the $\mathbb{F}$-adapted solution to the BSDEs. Moreover, one can prove that the introduction of this process keeps the uniqueness of $\mathbb{F}$-adapted solutions. Likewise, relevant to FVIEs and BVIEs, we could naturally consider forward stochastic Volterra integral equations (FSVIEs, for

[1] We are only talking about the second kind (forward) Volterra integral equations.

short), and backward stochastic Volterra integral equations (BSVIEs, for short). Like the SDEs, the standard theory for FSVIEs was established through many papers and monographs. In this book, we are devoted to the BSVIEs theory.

Our motivations to write the current book come from three aspects. First, it turns out that BSVIEs are significantly different from FSVIEs and BSDEs, respectively. In contrast with FSVIEs, an additional adjustment process should be introduced to keep the adaptiveness of solution. Comparing with BSDEs, this additional process has to be of form $Z(t,s)$ (with t being a parameter and $t \leqslant s$) rather than $Z(s)$. This double time variable structure brings some new phenomena and difficulties. For example, one may ask whether the other part of $Z(s,t)$ with $t \leqslant s$ should also be contained. We found that whether this $Z(s,t)$ appears will lead to quite different results in terms of theoretical results (such as regularity, comparison theorem) or applications (such as optimal control problems). For convenience, we then refer to those BSVIEs as Type I if only $Z(s,t)$ does not appear (so that the BSVIE is considered in an upper half domain of (t,s)), and Type II if $Z(s,t)$ appears (so that the BSVIE is considered in a square domain of (t,s)). Second, in the past more than 20 years, the theory of BSVIEs has been well developed, including well-posedeness, regularity of adapted solutions, comparison theorems, representation of solutions, relevant optimal control problems, and many other interesting applications. Many researchers, including the authors of the current book, have made various contributions to it. We think that it is the time to write a book summarizing the up-to-date established results for Types I and II BSVIEs. Third, we believe that the potential powerfulness of BSVIEs theory is enormous and a relevant book may help and inspire the readers for more interesting works in future. To explain this point, we take two examples from mathematical finance and optimal control theory, respectively. In recent years, more and more people are interested in financial asset problems either with rough volatilities, or non-exponential discounting functions. One way to describe these is either to let the state equation to be an FSVIE, or the cost functional to be a BSVIE. To treat the relevant pricing or optimization problems, the BSVIEs may not be avoided. On the other hand, besides SDEs, stochastic delay differential equations, stochastic fractional differential equations, stochastic integro-differential equations, and stochastic evolution equations, which include many classes of SPDEs, are also very fundamental models. The optimal control theories (including Pontryagin's maximum principle, linear quadratic control theory) and related topics (such as mean-field games) for these systems either have been established but with some open problem unsolved or are still untouched. We believe that a developed theory on BSVIEs is necessary and important in treating these problems. This is third reason for the current book to be written.

This book is intended to present a self-contained theory of BSVIEs. Some relevant applications will also be presented. Readers need only basic knowledge of stochastic analysis, stochastic differential equations, partial differential equations, and optimal control theory. For those working in the areas of stochastic optimal control, mathematical finance, financial engineering, and so on, this book will be useful.

Here are some highlights of the book. Chapter 1 presents some natural motivations of BSVIEs. Chapter 2 collects some basic probability tools that will be used in later chapters. Most well-known standard results will be stated without proofs except for those whose proofs are short or contain some different ideas than those in the standard literature. In Chap. 3, we recall the results of BSDEs and FVIEs. Chapters 4 and 5 concentrate on Type-I and Type-II BSVIEs, respectively, for which the generators are assumed to have at most linear growth in the unknown processes. In Chap. 6, we establish some comparison theorems for solutions of BSVIEs. It turns out that the situation is quite different from that of FSDEs and BSDEs. Chapter 7 gives some representations of solutions to BSVIEs by those of solutions to proper systems of partial differential equations. In Chap. 8, Type-I BSVIEs with superlinear generators are discussed with some positive and negative results. Chapter 9 considers BSVIEs containing mean-fields terms, by which we mean that the expectations of the unknown processes appear in the equations. In Chap. 10, we present a theory of optimal control for FSVIEs; Pontryagin's maximum principle is established. This is actually one of the main motivations of introducing Type-II BSVIEs. Some interesting applications are presented in Chap. 11. Finally, some technical material will be put in the last chapter.

Chengdu, China — Tianxiao Wang
Orlando, FL, USA — Jiongmin Yong
December 2025

Contents

Chapter 1
Introduction

In this chapter, we present some natural mathematical motivations for the study of backward stochastic Volterra integral equations (BSVIEs, for short) that will be presented in the current book.

1 Continuous-Time Deterministic Dynamic Equations

By a *continuous-time dynamic equation*, we mean a set of equations in continuous-time specifying how the *state variables* (describing the considered objects) change over the time. In what follows, all the involved maps are always assumed to be at least measurable.

1.1 Ordinary Differential Equations

Let us begin with the following *forward ordinary equation* (FODE, for short) or an *initial value problem* for such an equation in $\mathbb{R}^n$:

$$\begin{cases} \dot{X}(t) = b(t, X(t)), & t \in [0, T], \\ X(0) = x, \end{cases} \tag{1.1.1}$$

where $b : [0, T] \times \mathbb{R}^n \to \mathbb{R}^n$ is a given map, called the *generator*,[1] and $x \in \mathbb{R}^n$ is called an *initial state*. The above can be used to describe many dynamic processes.

[1] It is usually called the *right-hand side* of the equation. But here, we use the name *generator* to make the following theory consistent.

T. Wang, J. Yong, *Backward Stochastic Volterra Integral Equations*,
Probability Theory and Stochastic Modelling 109,
https://doi.org/10.1007/978-3-032-20848-4_1

The solution (assuming it exists) to such an equation can be regarded as the *state* or situation of the objects under consideration: position of a moving body, level of some commodity inventory, total population of certain species, remaining quantity of certain radioactive material, and spread of the virus, to mention a few.

Next, we look at the following a *backward ordinary differential equation* (BODE, for short) or a *terminal value problem* for an equation in $\mathbb{R}^m$ with a finite time horizon:

$$\begin{cases} \dot{Y}(t) = -g(t, Y(t)), & t \in [0, T], \\ Y(T) = y. \end{cases} \tag{1.1.2}$$

Here, $g : [0, T] \times \mathbb{R}^m \to \mathbb{R}^m$ is a given map and $y \in \mathbb{R}^m$, called the *terminal state*. This can be used to describe the following scenario: For a given terminal state y, is it possible to find an initial state $Y(0)$ starting from which the state $Y(t)$ will reach y exactly at $t = T$ through the equation? Since the above is to be solved backwardly, we refer to (1.1.2) as a BODE, with the generator $-g(\cdot)$. On the other hand, if we define

$$\begin{cases} X(t) \equiv Y(T-t), & t \in [0, T], \qquad x = y, \\ b(t, x) \equiv -g(T-t, x), & (t, x) \in [0, T] \times \mathbb{R}^m, \end{cases}$$

then terminal value problem (1.1.2) is transformed to an initial value problem of form (1.1.1) (with $n = m$). This tells us that under the *time-reverse* transformation $t \mapsto T - t$, initial value and terminal value problems for ordinary differential equations are equivalent. Hence, once a theory for FODEs is established, there is essentially no need to separately establish a theory for BODEs.[2]

Besides the scenario that can be described by BODEs mentioned above, it has another important feature: Linear BODE and linear FODEs are dual to each other. Let us briefly elaborate this now. Given bounded matrix-valued function $A(\cdot)$, consider linear FODE:

$$\begin{cases} \dot{X}(t) = A(t)X(t) + b(t), & t \in [0, T], \\ X(0) = x, \end{cases} \tag{1.1.3}$$

and linear BODE:

$$\begin{cases} \dot{Y}(t) = -A(t)^\top Y(t) - g(t), & t \in [0, T], \\ Y(T) = y. \end{cases} \tag{1.1.4}$$

[2] For infinite time horizon, there are some different features for BODEs. We do not discuss problems in infinite time horizon in the current book.

We call (1.1.4) the *adjoint equation* of (1.1.3). Clearly, it holds that

$$\begin{aligned}\frac{d}{dt}\langle X(t), Y(t)\rangle &= \langle A(t)X(t) + b(t), Y(t)\rangle + \langle X(t), -A(t)^\top Y(t) - g(t)\rangle \\ &= \langle b(t), Y(t)\rangle - \langle X(t), g(t)\rangle.\end{aligned}$$

Thus,

$$\langle X(T), y\rangle - \langle x, Y(0)\rangle = \int_0^T \Big(\langle b(t), Y(t)\rangle - \langle X(t), g(t)\rangle\Big)dt. \tag{1.1.5}$$

This is called the *duality relation* between (1.1.3) and (1.1.4). Such kind of relation plays a very important role in optimal control theory for FODEs and comparisons of solutions. We will explore this in more details later.

1.2 Volterra Integral Equations

An equivalent integral form of (1.1.1) reads

$$X(t) = x + \int_0^t b(s, X(s))ds, \qquad t \in [0, T]. \tag{1.1.6}$$

A natural mathematical extension of (1.1.6) is the following:

$$X(t) = \varphi(t) + \int_0^t b(t, s, X(s))ds, \qquad t \in [0, T], \tag{1.1.7}$$

where $\varphi : [0, T] \to \mathbb{R}^n$ and $b : \Delta_*[0, T] \times \mathbb{R}^n \to \mathbb{R}^n$ are given maps, called the *free term*[3] and the *generator*, respectively, with

$$\Delta_*[0.T] \overset{\Delta}{=} \big\{(t, s) \in [0, T]^2 \bigm| 0 \leqslant s < t \leqslant T\big\} \tag{1.1.8}$$

being the open lower triangle domain in the (t, s)-plane.[4]

[3] It is not proper to call it the initial condition as it is time-dependent.

[4] By considering open lower triangle domain $\Delta_*[0, T]$, we may allow the integrand to be singular along the diagonal, i.e., the map $(t, s) \mapsto b(t, s, x)$ may be singular along the line $t = s$. Similar comments apply to similar contexts later.

We call (1.1.7) a *forward Volterra integral equation* (FVIE, for short). Note that, in general, unlike (1.1.6), one could not transform FVIE (1.1.7) to an FODE, unless some compatibility conditions are satisfied, for example,

$$\varphi(t) \triangleq e^{At}x, \quad b(t,s,x) \triangleq e^{A(t-s)}b_0(s,x), \quad \forall (t,s,x) \in \Delta_*[0,T] \times \mathbb{R}^n,$$

for some $A \in \mathbb{R}^{n\times n}$ and some map $b_0 : [0,T] \times \mathbb{R}^n \to \mathbb{R}^n$. In this case, we actually have

$$X(t) = e^{At}x + \int_0^t e^{A(t-s)}b_0(s, X(s))ds, \qquad t \in [0,T],$$

which is an integral form of an FODE (through the *variation of constants formula*). It turns out that this is essentially the only case that FVIEs can be transformed to an FODE. In general, FVIEs are strictly general than FODEs. Therefore, a theory for FVIEs, independent of that for FODEs, is necessary. Also, for FODEs, the name "initial value problem" makes sense since $X(\cdot)$ has an initial value $X(0) = x$. However, for FVIEs, it is not proper to talk about the "initial value." A *free term* seems more proper.

Now, we present some examples of FVIEs in real applications, which cannot be covered by FODEs.

Example 1.1.1 Let $X(t)$ be the total population at t, in a considered region, say, a country, or a region. We neglect the migration, and want to derive an equation for $X(\cdot)$. It is known that a fraction $X(0)f(t)$ of people is surviving to age t. Thus, $f(\cdot)$, called the *surviving function*, is strictly decreasing and $f(0) = 1$. We denote the surviving population by $\widehat{X}(t)$ at time t, and

$$\widehat{X}(t) = X(0)f(t), \quad t \geqslant 0.$$

Clearly, $\widehat{X}(0) = X(0)f(0) = X(0)$.

Now, we assume that there is a continuous addition to the total population through the new births. Let the average birth rate be $r(t)$ at time t. We consider a general moment $\tau > 0$. Then the newborn children on $[\tau, \tau+\delta]$ will be of age $t-\tau$ at time t, and only a fraction $f(t-\tau)r(\tau)$ will survive to age $t-\tau$. Hence, the final addition to the total population at time t, from the children born in $[\tau, \tau+\delta]$, is roughly equal to $f(t-\tau)r(\tau)\delta$. Now, this analysis applies to each partition point $0 < \tau_1 < \tau_2 < \cdot s < \tau_k < t$; we see that the total addition will be

$$\sum_{i=1}^{k} f(t-\tau_i)r(\tau_i)\delta \to \int_0^t f(t-\tau)r(\tau)d\tau, \qquad t > 0,$$

as the partition mesh size goes to 0. Further, it is reasonable to assume that $r(t) = \alpha X(t)$, for some $\alpha \in (0, \frac{1}{2}]$. Therefore, we obtain the following equation:

$$X(t) = X(0)f(t) + \alpha \int_0^t f(t-\tau)X(\tau)d\tau, \qquad t \geqslant 0. \tag{1.1.9}$$

This gives an FVIE. Apparently, it cannot be transformed to an FODE in general.

Example 1.1.2 Let $S(t)$, $I(t)$, and $R(t)$ be the fractions of the population that are susceptible, infectious, and removed, respectively. Assume that in a time duration, the population size of certain species under consideration is a constant:

$$S(t) + I(t) + R(t) = I.$$

The population is assumed to be uniform and homogeneously mixing. A dynamic system that describes these processes is called SIR model. We now describe it. The average number of contacts (sufficient for transmission) of an infective per unit time is denoted by a constant β. Thus, the infectious are transferred at a rate equal to $\beta S(t)I(t)$ to susceptible. Let $P(t)$ be the probability of remaining infectious t units of time after becoming infectious. Assume that $P(t)$ is exponentially decay with $P(0) = 1$ (and $P(\infty) = 0$). Since the population size is constant, the birth rate must be equal to the death rate $\mu > 0$. The death rate is assumed to be the same for susceptible, infectious, and removed individuals, for simplicity. Hence, we let $Q(t) = e^{-\mu t}$ ($\mu > 0$ is actually the death rate) be the probability of being alive at time $t_0 + t$ given that an individual is alive at time t_0.

Active immunization by means of vaccine or toxoid is incorporated into the model. The fraction φ of newborns are immunized so that the flow rate of immunized newborns into the removed class is $\varphi\mu$. Immunization of susceptibles at a rate $\theta S(t)$ is also assumed.

Let the initial susceptible and removed fractions be $S_0 > 0$ and $R_0 > 0$, respectively, and let $I_0(t)e^{-\mu t}$ be the fraction of the population that was initially infectious and is still alive and infectious at time t.

The integral equation for $I(t)$ is

$$I(t) = I_0(t)e^{-\mu t} + \int_0^t \beta S(s)I(s)P(t-s)e^{-\mu(t-s)}ds,$$

where the second term is the sum of those who become infectious in the time interval $[0, t]$ and are still alive and infectious at time t. The removed fraction satisfies

$$\begin{aligned} R(t) = R_0e^{-\mu t} + [I_0(0) - I_0(t)]e^{-\mu t} + \int_0^t \beta S(s)I(s)[1 - P(t-s)]e^{-\mu(t-s)}ds \\ +\varphi(1 - e^{-\mu t}) + \int_0^t \theta S(s)e^{-\mu(t-s)}ds. \end{aligned}$$

The second term represents those initial infectives who have recovered and are still alive at time t, and the third term represents those who became infectious in the interval $[0, t]$ and are still alive at time t, but are no longer infectious. The fourth term represents the newborns in $[0, t]$ who were immunized and are still alive at time t, and the fifth term represents susceptibles immunized in $[0, t]$ who are still alive at time t. The assumption that the population is constant leads to the equation for $S(\cdot)$. The above gives a system of FVIE for the SIR model.

Example 1.1.3 Consider a problem of determining the temperature in a semi-infinite solid, whose surface can dissipate heat by nonlinear radiation. In the interior of the half-space, the heat flow is governed by a linear, one-dimensional heat equation. At the surface, energy is supplied according to the given function $f(t)$, while radiated energy escapes in proportion to $\varphi(t)^k$. Here $\varphi(t)$ denotes the temperature at the surface for all time $t \geqslant 0$. Conversion of the boundary value problem to an integral equation) yields

$$\varphi(t) = \pi^{-\frac{1}{2}} \int_0^t (t-s)^{-\frac{1}{2}} [f(s) - \varphi(s)^k] ds. \qquad t \geqslant 0.$$

Again, the above is a nonlinear FVIE with a singular kernel.

The above are some typical examples from real life. They should exhibit the usefulness of FVIEs.

Similar to FODEs, we can consider the following *backward Volterra integral equation* (BVIE, for short):

$$Y(t) = \psi(t) + \int_t^T g(t, s, Y(s)) ds, \qquad t \in [0, T], \tag{1.1.10}$$

where $\psi : [0, T] \to \mathbb{R}^m$ and $g : \Delta^*[0, T] \times \mathbb{R}^m \to \mathbb{R}^m$ are given maps with

$$\Delta^*[0, T] = \big\{(t, s) \in [0, T]^2 \bigm| 0 \leqslant t < s \leqslant T\big\} \equiv [0, T]^2 \setminus \bar{\Delta}_* \tag{1.1.11}$$

being the open upper triangle domain. We call $\psi(\cdot)$ and $g(\cdot)$ the *free term* and the *generator* of BVIE (1.1.10), respectively. Similar to the BODE case, BVIEs are quite interesting, at least from viewpoint of mathematics. In the current case, we also have the duality relation between linear FVIEs and linear BVIEs. We now illustrate that. Let us consider the following linear FVIE:

$$X(t) = \varphi(t) + \int_0^t A(t, s) X(s) ds, \qquad t \in [0, T], \tag{1.1.12}$$

and linear BVIE:

$$Y(t) = \psi(t) + \int_t^T A(s, t)^\top Y(s) ds, \qquad t \in [0, T]. \tag{1.1.13}$$

We call (1.1.13) the adjoint equation of (1.1.12). It is true that

$$\begin{aligned}\int_0^T \langle X(t), Y(t)\rangle dt &= \int_0^T \langle \varphi(t) + \int_0^t A(t,s)X(s)ds, Y(t)\rangle dt \\ &= \int_0^T \langle \varphi(t), Y(t)\rangle dt + \int_0^T \langle X(s), \int_s^T A(t,s)^\top Y(t)dt\rangle ds \\ &= \int_0^T \langle \varphi(t), Y(t)\rangle dt + \int_0^T \langle X(s), Y(s) - \psi(s)\rangle ds \\ &= \int_0^T \langle X(t), Y(t)\rangle dt + \int_0^T \Big(\langle \varphi(t), Y(t)\rangle - \langle X(t), \psi(t)\rangle\Big)dt.\end{aligned}$$

Hence,

$$\int_0^T \langle X(t), \psi(t)\rangle dt = \int_0^T \langle \varphi(t), Y(t)\rangle dt. \tag{1.1.14}$$

This is called the *duality principle* between (1.1.12) and (1.1.13).

Like FVIEs, in general, one is not able to transform a BVIE (1.1.10) to a BODE, unless some compatibility conditions are satisfied. Further, we have

$$\begin{aligned}Y(T-t) &= \psi(T-t) + \int_{T-t}^T g(T-t, s, Y(s))ds \\ &= \psi(T-t) - \int_0^t g(T-t, T-s, Y(T-s))ds.\end{aligned}$$

This shows that just like BODE case, the theory of FVIEs on finite time horizons can be carried over to that of BVIEs under the time-reverse transformation $t \mapsto T-t$. Thus, a theory for FVIEs will be sufficient.

2 Stochastic Dynamic Equations

It is easy to understand that in the real life, noises cannot be avoided. One way to describe the behavior of some noises and their effects is to use stochastic processes. In the current book, we only consider the cases when noises can be described by Brownian motions. Hence, we let $(\Omega, \mathcal{F}, \mathbb{F}, \mathbb{P})$ be a filtered complete probability space on which a d-dimensional standard Brownian motion $W(\cdot)$ is defined, with $\mathbb{F} = \{\mathcal{F}_t\}_{t\geqslant 0}$ being the natural filtration, augmented by all the $\mathbb{P}$-null sets in $\mathcal{F}$ (see Chap. 2 for more details).

2.1 Stochastic Differential Equations

We first consider the following initial value problem for Itô-type *stochastic differential equation* (SDE, for short):

$$\begin{cases} dX(t) = b(t, X(t))dt + \sigma(t, X(t))dW(t), & t \in [0, T], \\ X(0) = x, \end{cases} \tag{1.2.1}$$

where $b : [0, T] \times \mathbb{R}^n \times \Omega \to \mathbb{R}^n, \sigma : [0, T] \times \mathbb{R}^n \times \Omega \to \mathbb{R}^{n\times d}$ are given maps and $x \in \mathbb{R}^n$. They are normally called the *drift*, *diffusion* ,and *initial condition*, respectively. For convenience, we also refer to $(b(\cdot), \sigma(\cdot))$ as the *generator*. Since such an equation will be solved forwardly, we also call (1.2.1) a *forward stochastic differential equation* (FSDE, for short). The above is understood as follows:

$$X(t) = x + \int_0^t b(s, X(s))ds + \int_0^t \sigma(s, X(s))dW(s), \quad t \in [0, T], \tag{1.2.2}$$

where the first and the second integrals on the right-hand side are Lebesgue and Itô integrals, respectively. In what follows, we will not distinguish (1.2.1) and (1.2.2).

When $\sigma(\cdot)$ is identically equal to zero, (1.2.1) or (1.2.2) is reduced to an FODE. Thus, FODEs can be regarded as certain approximations of FSDEs when the noise can be neglected, and mathematically, FSDE covers FODE. On the other hand, if the noise cannot be neglected, and it can be described by Brownian motion, then by properly adding the noise to an FODE, it becomes an FSDE of Itô type.

The most important property of any solution $X(\cdot)$ to FSDE (1.2.2) is the $\mathbb{F}$-adaptiveness. This means that for each $t \in [0, T]$, $X(t)$ is required to be $\mathcal{F}_t$-measurable. This is also referred to as *non-anticipating property*, which is the most significant difference between stochastic and deterministic problems.

Next, we would like to ask the following question: Can one study terminal value problem for SDEs? Naturally, we may consider the following problem:

$$\begin{cases} dY(t) = b(t, Y(t))dt + \sigma(t, Y(t))dW(t), & t \in [0, T], \\ Y(T) = \xi, \end{cases} \tag{1.2.3}$$

with ξ being an $\mathcal{F}_T$-measurable random variable. Unfortunately, such a problem usually does not have a solution $Y(\cdot)$ which is $\mathbb{F}$-adapted. To illustrate this, let us look at the following simplest example:

$$\begin{cases} dY(t) = 0, & t \in [0, T], \\ Y(T) = W(T), \end{cases} \tag{1.2.4}$$

where $W(\cdot)$ is a one-dimensional standard Brownian motion. The above is a special case of (1.2.3) in which $n = d = 1$, $b(\cdot) = \sigma(\cdot) = 0$, and $\xi = W(T)$. For (1.2.4), as a differential equation, the only possible solution is $Y(t) \equiv W(T)$, which is not

$\mathbb{F}$-adapted. It turns out that the correct form of the problem should look like the following:

$$\begin{cases} dY(t) = -g(t, Y(t), Z(t))dt + Z(t)dW(t), & t \in [0, T], \\ Y(T) = \xi. \end{cases} \tag{1.2.5}$$

The above is called a *backward stochastic differential equation* (BSDE, for short). In the above, $g : [0, T] \times \mathbb{R}^m \times \mathbb{R}^{m\times d} \times \Omega \to \mathbb{R}^m$ is called the *generator* of the BSDE, and ξ, a given $\mathcal{F}_T$-measurable random variable valued in $\mathbb{R}^m$, is called the *terminal value*. Equation (1.2.5) is understood as the following:

$$Y(t) = \xi + \int_t^T g(s, Y(s), Z(s))ds - \int_t^T Z(s)dW(s), \quad t \in [0, T]. \tag{1.2.6}$$

By an *adapted solution*, we mean a pair $(Y(\cdot), Z(\cdot))$ of $\mathbb{F}$-adapted processes so that (1.2.5) is satisfied in the usual sense. It is interesting that process $Z(\cdot)$ is added as a part of the solution. We will establish a self-contained theory for the above BSDEs in Chap. 3. Clearly, due to the appearance of the process $Z(\cdot)$, BSDEs are significantly different from FSDEs. In particular, of course, the time-reverse transformation cannot make FSDEs and BSDEs equivalent. Consequently, a quite different theory will be needed for BSDEs. We call $Y(\cdot)$ and $Z(\cdot)$ the *backward component* and *adjustment component* of the BSDE (1.2.5), respectively. Note that it is $Z(\cdot)$ that makes the existence of adapted solution $(Y(\cdot), Z(\cdot))$ possible. Hereafter, we will not distinguish (1.2.5) and (1.2.6).

To conclude this subsection, we present the duality relation between linear FSDEs and linear BSDEs. More precisely, we consider the following linear FSDE:

$$\begin{cases} dX(t) = [A_0(t)X(t) + b(t)]dt + \displaystyle\sum_{i=1}^{d}[A_i(t)X(t) + \sigma_i(t)]dW_i(t), \\ \qquad\qquad\qquad\qquad\qquad\qquad\qquad\qquad t \in [0, T], \\ X(0) = x, \end{cases} \tag{1.2.7}$$

and corresponding linear BSDE:

$$\begin{cases} dY(t) = -\Big[A_0(t)^\top Y(t) + \displaystyle\sum_{i=1}^{d} A_i(t)^\top Z_i(t) + g(t)\Big]dt \\ \qquad\qquad\qquad + \displaystyle\sum_{i=1}^{d} Z_i(t)dW_i(t), \qquad t \in [0, T], \\ Y(T) = \xi. \end{cases} \tag{1.2.8}$$

For convenience, we assume that all the coefficients are $\mathbb{F}$-progressively measurable and bounded. We have the following duality principle, similar to (1.1.5), (1.1.14) (and the later (1.2.20)), which will be useful in the optimal control theory for FSDEs. We refer to Chap. 2, Sect. 4, for more details on the involved spaces.

Theorem 1.2.1 *Let all the coefficients of FSDE* (1.2.7) *abe* $\mathbb{F}$*-progressively measurable and bounded. Let* $b(\cdot), \sigma_i(\cdot) \in L^p_{\mathbb{F}}(0,T;\mathbb{R}^n)$ $(1 \leqslant i \leqslant d)$*, and* $g(\cdot) \in L^q_{\mathbb{F}}(0,T;\mathbb{R}^n)$*,* $\xi \in L^q_{\mathbb{F}}(\Omega;\mathbb{R}^n)$*, with* $\frac{1}{p}+\frac{1}{q}=1$*. Let* $X(\cdot) \in L^p_{\mathbb{F}}(\Omega; C([0,T];\mathbb{R}^n))$ *be the solution to FSDE* (1.2.7) *and* $(Y(\cdot), Z(\cdot)) \in \mathbf{H}^q[0,T]$ *be the adapted solution to BSDE* (1.2.8)*. Then the following duality principle holds:*

$$\begin{aligned}&\mathbb{E}_s\Big[\langle X(t),Y(t)\rangle-\langle X(s),Y(s)\rangle\Big]\\&=\mathbb{E}_s\int_s^t\Big[\langle b(\tau),Y(\tau)\rangle+\sum_{i=1}^d\langle\sigma_i(\tau),Z_i(\tau)\rangle-\langle X(\tau),g(\tau)\rangle\Big]d\tau,\\&\qquad\qquad\qquad\qquad\qquad 0\leqslant s\leqslant t\leqslant T.\end{aligned}\tag{1.2.9}$$

In particular,

$$\begin{aligned}&\mathbb{E}\Big[\langle X(T),\xi\rangle-\langle x,Y(0)\rangle\Big]\\&=\mathbb{E}\int_0^T\Big[\langle b(\tau),Y(\tau)\rangle+\sum_{i=1}^d\langle\sigma_i(\tau),Z_i(\tau)\rangle-\langle X(\tau),g(\tau)\rangle\Big]d\tau.\end{aligned}\tag{1.2.10}$$

Proof Applying Itô's formula to the process $t\mapsto\langle X(t),Y(t)\rangle$, we have

$$\begin{aligned}&d\langle X(t),Y(t)\rangle=\Big[\langle A_0(t)X(t)+b(t),Y(t)\rangle\\&-\langle X(t),A_0(t)^\top Y(t)+\sum_{i=1}^dA_i(t)^\top Z_i(t)+g(t)\rangle+\sum_{i=1}^d\langle A_i(t)X(t)+\sigma_i(t),Z_i(t)\rangle\Big]dt\\&+\sum_{i=1}^d\Big[\langle A_i(t)X(t)+\sigma_i(t),Y(t)\rangle+\langle X(t),Z_i(t)\rangle\Big]dW_i(t)\\&=\Big[\langle b(t),Y(t)\rangle+\sum_{i=1}^d\langle\sigma_i(t),Z_i(t)\rangle-\langle X(t),g(t)\rangle\Big]dt+\{\cdot s\}dW(t).\end{aligned}$$

Hence, (1.2.9) follows. □

2.2 Stochastic Volterra Integral Equations

Mimicking the FVIEs case, we can introduce the following equation:

$$X(t)=\varphi(t)+\int_0^tb(t,s,X(s))ds+\int_0^t\sigma(t,s,X(s))dW(s),\quad t\in[0,T],\tag{1.2.11}$$

Fig. 1.1 Triangle domains of Volterra system

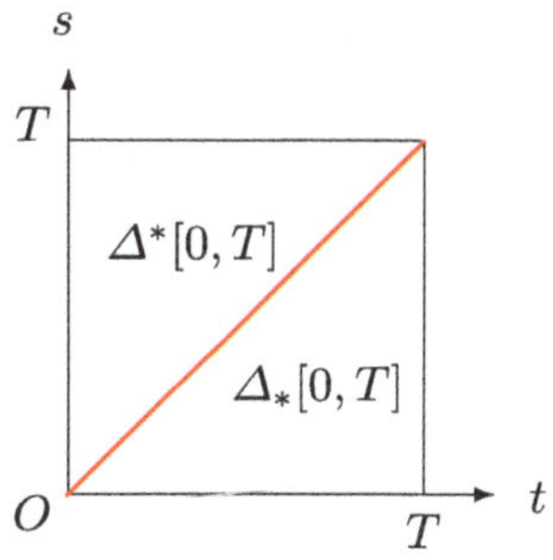

where $\varphi : [0, T] \times \Omega \to \mathbb{R}^n$, $b : \Delta_*[0, T] \times \mathbb{R}^n \times \Omega \to \mathbb{R}^n$, and $\sigma : \Delta_*[0, T] \times \mathbb{R}^n \times \Omega \to \mathbb{R}^{n\times d}$ are given maps, satisfying certain $\mathbb{F}$-adapted conditions, recalling (1.1.8) (or Fig. 1.1) for $\Delta_*[0, T]$. We call (1.2.11) a *forward stochastic Volterra integral equation* (FSVIE, for short). The term $\varphi(\cdot)$ is called the *free term* and the pair $(b(\cdot), \sigma(\cdot))$ is called the *generator* of FSVIE (1.2.11), respectively. Just like FODEs and FVIEs discussed above, FSVIE (1.2.11) is strictly more general than FSDE (1.2.1) (or its equivalent form (1.2.2)), in the sense that one cannot transform (1.2.11) to some FSDE like (1.2.2), in general. It is worthy of pointing out that the fractional Brownian motion is a special case of FSVIE:

$$W_H(t) = \int_0^t K_H(t, s)dW(s), \qquad t \geqslant 0, \tag{1.2.12}$$

where

$$K_H(t, s) = \Big[\frac{H(2H-1)}{B(2-2H, H-\frac{1}{2})}\Big]^{\frac{1}{2}} s^{\frac{1}{2}-H} \int_s^t (r-s)^{H-\frac{3}{2}} r^{H-\frac{1}{2}} dr,$$

with $B(\cdot\,,\cdot)$ being the beta function

$$B(t, s) = \int_0^1 r^{t-1}(1-r)^{s-1}dr, \qquad t, s > 0, \tag{1.2.13}$$

and $H \in (0, 1)$ is called the *Hurst index*. In Chaps. 3 and 4, we will present some more results on FSVIEs.

Next, similar to the BODE (BVIE, BSDE) being the backward version of FODE (FVIE, FSDE), we may ask what the backward version of FSVIE should look like. Naturally, inspired by (1.2.6), we could consider

$$Y(t) = \psi(t) + \int_t^T g(t, s, Y(s), Z(t, s))ds - \int_t^T Z(t, s)dW(s), \qquad t \in [0, T], \tag{1.2.14}$$

which is called a *Type I backward stochastic Volterra integral equation* (BSVIE, for short). Here $\psi : [0,T] \times \Omega \to \mathbb{R}^m$ and $g : \Delta^*[0,T] \times \mathbb{R}^m \times \mathbb{R}^{m\times d} \times \Omega \to \mathbb{R}^m$ are given maps, satisfying certain $\mathbb{F}$-adapted conditions, called the *free term* and the *generator*, respectively. Note that because of the t-dependence of $\psi(t)$ and $g(t,s,y,z)$, the adjustment component $Z(t,s)$, which is $\mathbb{F}$-adapted for $s \in [t,T]$, should naturally depend on (t,s). On the other hand, since $Z(t,s)$ is a function of two "time" variables, one could also allow $Z(s,t)$ to appear in the equation. This leads to the following more general equation:

$$\begin{aligned} Y(t) = \psi(t) + \int_t^T g(t,s,Y(s),Z(t,s),Z(s,t))ds \\ - \int_t^T Z(t,s)dW(s), \quad t \in [0,T], \end{aligned} \tag{1.2.15}$$

where now $g : \Delta^*[0,T] \times \mathbb{R}^m \times \mathbb{R}^{m\times d} \times \mathbb{R}^{m\times d} \times \Omega \to \mathbb{R}^m$. This is called a *Type II BSVIE*. We will see that Eqs. (1.2.14) and (1.2.15) are quite different. We point out that the free term $\psi(\cdot)$ is $\mathcal{F}_T$-measurable only, not necessarily $\mathbb{F}$-adapted.

In the case that the free term is independent of t, i.e., $\psi(t) \equiv \xi$, and the generator

$$g(t,s,y,z,\widehat{z}) = g(s,y,z),$$

i.e., our BSVIE reads

$$Y(t) = \xi + \int_t^T g(s,Y(s),Z(t,s))ds - \int_t^T Z(t,s)dW(s), \quad t \in [0,T]. \tag{1.2.16}$$

Now, if $(Y(\cdot), Z(\cdot))$ is an adapted solution of BSDE (1.2.6), then it is an adapted solution of BSVIE (1.2.16). By the uniqueness under proper conditions, there is no other adapted solutions of either BSDE (1.2.6) or BSVIE (1.2.16). From this sense, we see that BSVIE is a generalization of BSDEs.

To conclude this section, we present a duality principle between linear FSVIEs and BSVIEs, assuming all are well-posed, and the coefficients are bounded. They will play an interesting role in optimal control theory of FSVIEs (see Chap. 10) and comparison theorem of (adapted) solutions of BSVIEs (see Chap. 6).

Proposition 1.2.2 *Let $X(\cdot)$ be the solution of the following linear FSVIE:*

$$X(t) = \varphi(t) + \int_0^t A(t,s)X(s)ds + \int_0^t C(t,s)X(s)dW(s), \quad t \in [0,T], \tag{1.2.17}$$

and $(Y(\cdot), Z(\cdot\,,\cdot))$ *be the adapted solution of the following Type II linear BSVIE:*

$$\begin{aligned} Y(t) = \psi(t) + &\int_t^T \Big(A(s,t)^\top Y(s) + C(s,t)^\top Z(s,t)\Big)ds \\ &- \int_t^T Z(t,s)dW(s), \qquad t \in [0,T]. \end{aligned} \tag{1.2.18}$$

Furthermore, it holds that

$$Y(t) = \mathbb{E}Y(t) + \int_0^t Z(t,s)dW(s), \qquad t \in [0,T]. \tag{1.2.19}$$

Then, the following duality principle is true:

$$\mathbb{E}\int_0^T \langle\varphi(t), Y(t)\rangle dt = \mathbb{E}\int_0^T \langle X(t), \psi(t)\rangle dt. \tag{1.2.20}$$

We note that $Z(s,t)$ appears naturally because we need to change the orders of integration when we obtain the duality.

Proof We have

$$\begin{aligned} &\mathbb{E}\int_0^T \langle X(t), Y(t)\rangle dt \\ &= \mathbb{E}\int_0^T \langle\varphi(t) + \int_0^t A(t,s)X(s)ds + \int_0^t C(t,s)X(s)dW(s), Y(t)\rangle dt \\ &= \mathbb{E}\int_0^T \langle\varphi(t), Y(t)\rangle dt + \mathbb{E}\int_0^T \langle X(s), \int_s^T A(t,s)^\top Y(t)dt\rangle ds \\ &\quad + \mathbb{E}\int_0^T\int_0^t \langle C(t,s)X(s), Z(t,s)\rangle dsdt \\ &= \mathbb{E}\int_0^T \langle\varphi(t), Y(t)\rangle dt + \mathbb{E}\int_0^T \langle X(s), \int_s^T \Big(A(t,s)^\top Y(t) + C(t,s)^\top Z(t,s)\Big)dt\rangle ds \\ &= \mathbb{E}\int_0^T \langle\varphi(t), Y(t)\rangle dt + \mathbb{E}\int_0^T \langle X(s), Y(s) - \psi(s) + \int_s^T Z(s,t)dW(t)\rangle ds \\ &= \mathbb{E}\int_0^T \langle\varphi(t), Y(t)\rangle dt + \mathbb{E}\int_0^T \langle X(t), Y(t)\rangle dt - \mathbb{E}\int_0^T \langle X(t), \psi(t)\rangle dt. \end{aligned}$$

In the above, we have used relation (1.2.19). Then, (1.2.20) follows. □

We see that in the above, Itô's formula is not used. A more general form of the duality principle will be presented in Chap. 5, Sect. 4.

3 Optimal Control of FSVIEs

In this section, we will present a class of problems, some natural study of which will give rise to the BSVIEs. Consider the following controlled FSVIE:

$$X(t) = \varphi(t) + \int_0^t b(t, s, X(s), u(s))ds + \int_0^t \sigma(t, s, X(s), u(s))dW(s), \qquad t \in [0, T]. \tag{1.3.1}$$

In the above, $X(\cdot)$ is called the *state process* valued in $\mathbb{R}^n$, and $u(\cdot)$ is called the *control process*, valued in some convex subset U of a Euclidean space. Under some mild conditions, for any $u(\cdot)$ in a certain class of *admissible controls*, denoted by $\mathscr{U}[0, T]$, the above state equation admits a unique solution $X(\cdot)$. The above contains the problems involving rough volatility. We will present this in more details in Chap. 11. To measure the performance of the control $u(\cdot)$, we may introduce the following *cost functional*:[5]

$$J(u(\cdot)) = \mathbb{E}\int_0^T g(s, X(s), u(s))ds. \tag{1.3.2}$$

An optimal control problem can be stated as follows.

Problem (C) Find a $\bar{u}(\cdot) \in \mathscr{U}[0, T]$ such that

$$J(\bar{u}(\cdot)) = \inf_{u(\cdot)\in\mathscr{U}[0,T]} J(u(\cdot)).$$

Any $\bar{u}(\cdot) \in \mathscr{U}[0, T]$ satisfying the above is called an *optimal control*, and the corresponding state process $\bar{X}(\cdot)$ and the pair $(\bar{X}(\cdot), \bar{u}(\cdot))$ are called the *optimal state process* and *optimal pair*, respectively.

It is well known that for a differentiable function $f : \mathbb{R} \to \mathbb{R}$, if it attains a local minimum at some $\bar{x}$, then, we have

$$f'(\bar{x}) = 0, \tag{1.3.3}$$

which is called Férmat theorem, a first-order necessary condition of minimum. Such a condition is very useful in helping people to find a minimum. The idea of proving (1.3.3) is to compare the values of $f(x)$ for x near $\bar{x}$ and use the optimality of $\bar{x}$.

[5] We could also consider the case containing a terminal cost, which is omitted here just for simplicity. See Chap. 10 for the general case.

For Problem (C), we may use the similar natural idea. Let us assume, for the time being, the set $\mathscr{U}[0,T]$ of admissible controls is a linear space. If $\bar u(\cdot)\in\mathscr{U}[0,T]$ is optimal, then for any $u(\cdot)\in\mathscr{U}[0,T]$ and $\varepsilon>0$, the cost at the perturbed control $\bar u(\cdot)+\varepsilon u(\cdot)$ will be larger. Hence, assuming all the involved functions are smooth, we have the following *variational inequality*:

$$\begin{aligned}0&\leqslant \lim_{\varepsilon\to0}\frac{J(\bar u(\cdot)+\varepsilon u(\cdot))-J(\bar u(\cdot))}{\varepsilon}\\&=\mathbb{E}\int_0^T\Big(g_x(s,\bar X(s),\bar u(s))X_1(s)+g_u(s,\bar X(s),\bar u(s))u(s)\Big)ds\\&\equiv\mathbb{E}\int_0^T\Big(\bar g_x(s)X_1(s)+\bar g_u(s)u(s))\Big)ds,\end{aligned}\tag{1.3.4}$$

with

$$\bar g_x(s)=g_x(s,\bar X(s),\bar u(s)),\qquad \bar g_u(s)=g_u(s,\bar X(s),\bar u(s)),$$

and where $X_1(\cdot)$ is the solution of the following *variational system*:

$$\begin{aligned}X_1(t)=&\int_0^t\Big(\bar b_x(t,s)X_1(s)+\bar b_u(t,s)u(s)\Big)ds\\&+\int_0^t\Big(\bar\sigma_x(t,s)X_1(s)+\bar\sigma_u(t,s)u(s)\Big)dW(s),\end{aligned}\tag{1.3.5}$$

with

$$\begin{aligned}&\bar b_x(t,s)=b_x(t,s,\bar X(s),\bar u(s),\qquad &&\bar b_u(t,s)=b_u(t,s,\bar X(s),\bar u(s)),\\&\bar\sigma_x(t,s)=\sigma_x(t,s,\bar X(s),\bar u(s)),&&\bar\sigma_u(t,s)=\sigma_u(t,s,\bar x(s),\bar u(s)).\end{aligned}$$

Hereafter, the following convention is used.

Convention 1 For any differentiable map $F:\mathbb{R}^k\to\mathbb{R}^\ell$, $F_x(x)$ is $\mathbb{R}^{\ell\times k}$-valued. In particular, when $\ell=1$, F is scalar-valued and $F_x(x)$ is a row vector for each x.

In the variational inequality (1.3.4), $X_1(\cdot)$ is implicitly depending on the arbitrary choice of $u(\cdot)$. Therefore, it is not very convenient. Let us introduce the following Type II BSVIE:

$$\begin{aligned}Y(t)=\bar g_x(t)^\top&+\int_t^T\Big(\bar b_x(s,t)^\top Y(s)+\bar\sigma_u(s,t)^\top Z(s,t)\Big)ds\\&-\int_t^T Z(t,s)dW(s),\qquad t\in[0,T].\end{aligned}\tag{1.3.6}$$

Moreover, the additional relation (1.2.19) holds. By the general theory which will be presented later, the above admits adapted solution $(Y(\cdot), Z(\cdot\,,\cdot))$. Then, by duality relation (1.2.20), one has

$$\mathbb{E}\int_0^T \langle \bar{g}_x(t)^\top, X_1(t)\rangle dt = \mathbb{E}\int_0^T \langle \int_0^t \bar{b}_u(t,s)u(s)ds, Y(t)\rangle dt$$
$$+\mathbb{E}\int_0^T \langle \int_0^t \bar{\sigma}_u(t,s)u(s)dW(s), Y(t)\rangle dt$$

$$=\mathbb{E}\int_0^T \langle u(s), \int_s^T \bar{b}_u(t,s)^\top Y(t)dt\rangle ds + \mathbb{E}\int_0^T\int_0^t \langle \bar{\sigma}_u(t,s)u(s), Z(t,s)\rangle dsdt$$
$$=\mathbb{E}\int_0^T \langle u(s), \int_s^T \Big(\bar{b}_u(t,s)^\top Y(t) + \bar{\sigma}_u(t,s)^\top Z(t,s)\Big)dt\rangle ds.$$

Hence, (1.3.4) becomes

$$0 \leqslant \mathbb{E}\int_0^T \langle u(s), \int_s^T \Big(\bar{b}_u(t,s)^\top Y(t) + \bar{\sigma}_u(t,s)^\top Z(t,s)\Big)dt + \bar{g}_u(s)^\top\rangle ds,$$

for all $u(\cdot) \in \mathscr{U}[0,T]$ (a linear space). Consequently, if the control U is a Euclidean space, we obtain

$$\mathbb{E}_s\int_s^T \Big(\bar{b}_u(t,s)^\top Y(t) + \bar{\sigma}_u(t,s)^\top Z(t,s)\Big)dt + \bar{g}_u(s)^\top = 0, \quad s \in [0,T]. \tag{1.3.7}$$

The above says that if $(\bar{X}(\cdot), \bar{u}(\cdot))$ is an optimal pair of Problem (C), and if BSVIE (1.3.6) admits an adapted solution $(Y(\cdot), Z(\cdot\,,\cdot))$ satisfying (1.2.19), then (1.3.7) holds. Such a result is called *Pontryagin's maximum principle*. It gives a set of optimality conditions for the pair $(\bar{X}(\cdot), \bar{u}(\cdot))$. We call (1.3.1) and (1.3.6) with (1.2.19) an optimality system. Normally, one can use the above as follows: (i) for a given state–control pair $(\widetilde{X}(\cdot), \widetilde{u}(\cdot))$, if the above is not satisfied, then it is not optimal, and (ii) if a state–control pair $(\widetilde{X}(\cdot), \widetilde{u}(\cdot))$ satisfies the above, then it is a candidate of optimal pairs. Further, if a priori one has the uniqueness of the optimality system, then the pair is optimal.

4 Bibliographical Remarks

In 1823, Abel [1] proposed a generalization of the tautochrone problem whose solution involved the solution to a linear integral equation of the following form:

$$\int_0^t \frac{\phi(s)}{\sqrt{t-s}}ds = f(t),$$

with $\phi(\cdot)$ being the unknown function. The above is called the *Abel equation* which is a special case of the first kind Volterra integral equation. In 1837, while studying second-order linear differential equations, Liouville [137] introduced the following integral equation:

$$u(x) = \cos\rho(x-a) + \frac{1}{\rho}\int_a^x \sigma(\xi)\sin\rho(x-\xi)u(\xi)d\xi,$$

with $u(\cdot)$ being the unknown function (see also the book by Bôcher [31]). In 1896, Volterra [189] introduced a general version of the above which is now called Volterra integral equations. Shortly after, at the beginning of the 1900s, Fredholm introduced integral equations with the integral taken over a fixed interval, which is now called Fredholm integral equations [76]. There are many books on the topic of integral equations. Here are several standard reference books: Tricomi [183], Kondo [125], Jerri [115], Zemyan [224], and Wazwaz [212].

Example 1.1.1 is taken from [115], in which there are more details, and many more other examples. Example 1.1.2 is taken from [101] where one can find some deeper discussions, and Example 1.1.3 is taken from [96, 123, 154].

Since Itô introduced the Itô-type stochastic integrals in the early 1940s [111] and established the Itô stochastic calculus at the beginning of the 1950s [113], the study of corresponding FSDEs began. It is found that in the early years, the integral form of FSDEs were called stochastic integral equations (see Itô [112] and Protter [169], for examples), which are special cases of general FSVIEs (of the second type). Later, there were some works on *random integral equations*,[6] which are usual integral equations (of either Fredholm or Volterra type) with the generator and/or the free term being random, but there was no stochastic integral involved [184]. In 1974, Tsokos–Padgett published a book [185], where (in Chap. 8) the following equation appeared:

$$x(t) = \int_0^t k(t,\tau)f(\tau,x(\tau))d\tau + \int_0^t \Phi(\tau)d\beta(\tau),$$

with $\beta(\cdot)$ being a Brownian motion. Here, the drift term contains the kernel $k(\cdot,\cdot)$ so that, in general, such an equation cannot be transformed to an FSDE. It seems that this was the first time a (real) FSVIE appeared. The study of general theory for FSVIEs began with the works of Itô [110] in 1979 and Berger–Mizel [28] in 1980 independently.

In the early 1970s, while studying optimal controls for FSDEs, Bismut introduced linear BSDEs the first time [29]. In 1990, Pardoux–Peng extended the results of linear BSDEs to general nonlinear BSDEs [158] with standard global Lipschitz conditions. Since then the topic of BSDEs became more and more popular.

---

[6] Such a name is used to distinguish them from the stochastic integral equations for which stochastic integrals are involved.

Various further extensions have appeared later on, for example, nonlinear Feynman–Kac formula [162], forward-backward stochastic differential equation (FBSDEs, for short) [141, 143], quadratic BSDEs [124], general theory of recursive utility [133, 134], second-order BSDEs [45, 180], G-expectation [165], path-dependent partial differential equations [164, 226], and so on.

In 2002, Lin introduced a special form of Type I BSVIE for the first time, as a straightforward extension of BSDEs [135]. Motivated by the statement of Pontryagin-type maximum principle for optimal control problems of FSVIE, Yong introduced Type II BSVIEs and established a general theory for BSVIEs [218, 220]. The names of Type I and Type II BSVIEs were coined by Yong in 2013 [221] and Wang–Yong [205]. BSVIEs of other forms have been investigated in the literature as well, for example, BSVIDEs [200], BSVIEs with the drift being superlinear in Z [71], BSVIEs with jumps [168, 211], infinite-dimensional BSVIEs [11, 12], path-dependent BSVIEs [156], numerical aspects of BSVIEs [26, 93], extended BSVIEs [89, 94, 191], time-inconsistent problems involving BSVIEs [193, 194], and so on.

The duality relation between linear FSDE and BSDE is very important in stochastic optimal control theory for SDEs. Likewise, the duality relation between linear FSVIEs and BSVIEs will play an important role for optimal control of FSVIEs. Moreover, these relations will also serve as useful tools in establishing various comparison theorems for the adapted solutions of BSDEs and BSVIEs in Chap. 6.

One major motivation for studying BSVIEs comes from optimal control theory. Optimal control theory for FODEs, which appeared in 1950s, can be regarded as a further development of the classical calculus of variations. See the works of Bellman [23] and Boltyanski–Gamkrelidze–Pontryagin [32]. Optimal control problems for SDEs began in the early 1960s by the work of Kushner ([128]; see also [130, 131]). On the other hand, optimal control for FVIEs was firstly investigated by Vinokurov in the middle of the 1960s [188]. The main contribution of Pontryagin's school on maximum principle was the introduction of the adjoint equation, inspired by the derivation of the Euler–Lagrange equation in classical calculus of variations. For many years, since there was no proper object to play the role of adjoint equation for the optimal control problem of FSVIEs, no general theory was developed. To explore the optimal control for FSVIEs, in 2006, Yong introduced Type II BSVIEs [218, 220] as the (first-order) adjoint equations. To obtain maximum principle in the spirit of Pontryagin type for FSVIEs, Wang–Yong [206] introduced systems of BSVIEs, including Type II BSVIEs, as the (second-order) adjoint equations. We will present more detailed results in Chap. 10. Besides the maximum principles, it is worthy mentioning that the linear quadratic theory for FSVIEs also rely on the theory of Tye II BSVIEs or beyond. We refer to e.g. Hamaguchi–Wang [94, 95], Wang–Yong–Zhou [196], Wang–Shi [202] for more details along this line. On the other hand, these works also inspire the optimal control theory for other types of stochastic system. For example, by borrowing the ideas from [206], Meng–Shi–Wang–Zhang [147] established the Pontryagin maximum principle for stochastic delay differential system with nonconvex control domain; see also Wang–Zheng [209] for the infinite-dimensional case with convex control domain. Besides stochastic delay differential equations, FSVIEs are closely related with stochastic fractional differential equations, stochastic evolution equations (including many kinds of SPDEs), and stochastic integrodifferential equations [209]. A developed optimal control theory for FSVIEs, with the help of BSVIEs, should be helpful in establishing or improving the analogue theory (or even mean-field game theory) for these systems.

In the mathematical finance areas, we note that following the idea of Duffie–Epstein ([63]; see also [64, 68, 133, 134]), one can model the recursive utility for (time-dependent) position processes [192] via BSVIEs. In a similar spirit, extending the works [13, 58, 75, 171], the dynamic risk measure for (time-dependent) position processes can be developed as well by BSVIEs [192, 219]. In recent years, FSVIEs have been used to model rough volatility problems in finance. The number of relevant literature grows very fast. We can only mention a small portion of them here: [17, 85, 116]. Correspondingly, the optimal control or pricing problems of such FSVIEs, as well as their applications, will also be developed very fast. We will put more relevant details in Chap. 11.

Chapter 2
Basics of Stochastic Analysis

This chapter is devoted to recalling some necessary related material which will be used in later chapters. We assume that the readers have basic knowledge of stochastic analysis. Those who are familiar with basic stochastic analysis can skip most of this chapter or quickly go through it.

1 Probability, Random Variables, and Stochastic Processes

Let $\mathbb{R}^m$ be m-dimensional real Euclidean space with the usual *Euclidean inner product*:

$$\langle x, y\rangle \equiv x^\top y = \sum_{i=1}^{m} x_i y_i, \quad \forall x \equiv (x_1, \cdot s, x_m)^\top,\ y \equiv (y_1, \cdot s, y_m)^\top \in \mathbb{R}^m,$$

and its induced *Euclidean norm*:

$$|x| = \sqrt{\langle x, x\rangle} = \Big(\sum_{i=1}^{m} |x_i|^2\Big)^{\frac{1}{2}}, \qquad \forall x \equiv (x_1, \cdot s, x_m)^\top \in \mathbb{R}^m.$$

Let $\mathbb{R}^{m\times d}$ be the Euclidean space of all $(m \times d)$ real matrices with the so-called *Frobenius inner product* defined by the following:

$$\langle A, B\rangle = \operatorname{tr}[AB^\top] \equiv \sum_{i=1}^{m}\sum_{j=1}^{d} a_{ij}b_{ij}, \quad \forall A \equiv (a_{ij}),\ B \equiv (b_{ij}) \in \mathbb{R}^{m\times d}.$$

T. Wang, J. Yong, *Backward Stochastic Volterra Integral Equations*,
Probability Theory and Stochastic Modelling 109,
https://doi.org/10.1007/978-3-032-20848-4_2

The norm induced by the above is denoted by $|\cdot|$. Thus, one has

$$|A|^2 = \sum_{i=1}^{m}\sum_{j=1}^{d} a_{ij}^2, \qquad \forall A \equiv (a_{ij}) \in \mathbb{R}^{m\times d}.$$

Definition 2.1.1

(i) Let Ω be a nonempty set. The set of all subsets of Ω is denoted by 2^Ω and is called the *power set* of Ω. A set $\mathcal{F} \subseteq 2^\Omega$ is called a *σ-field* of Ω if the following three conditions hold:

$$\begin{cases} \Omega \in \mathcal{F}, \\ A, B \in \Omega \quad \Rightarrow \quad A \setminus B \equiv \{\omega \in \Omega \mid \omega \in A,\ \omega \notin B\} \in \mathcal{F}, \\ \{A_k \mid k \geqslant 1\} \subseteq \mathcal{F} \quad \Rightarrow \quad \bigcup_{k\geqslant 1} A_k \in \mathcal{F}. \end{cases}$$

In the case that $\mathcal{F}$ is a σ-field of Ω, the pair $(\Omega, \mathcal{F})$ is called a *measurable space*, and any element $A \in \mathcal{F}$ is called an *event* or a *measurable set*.

(ii) Let $\mathcal{G} \subseteq 2^\Omega$. The smallest σ-field containing $\mathcal{G}$ is called the σ-field *generated* by $\mathcal{G}$, denoted by $\sigma(\mathcal{G})$.

(iii) Let Ω be a metric space (in particular, Ω could be any open or closed set in $\mathbb{R}^m$) and $\mathscr{O}$ be the set of all open sets in Ω. Then $\sigma(\mathscr{O})$ is called the *Borel σ-field* of Ω, denoted by $\mathcal{B}(\Omega)$, and each element of $\mathcal{B}(\Omega)$ is called a *Borel set*.

(iv) A family $\mathbb{F}_T^S = \{\mathcal{F}_t^S \mid S \leqslant t \leqslant T\}$ is called a *filtration* of a measurable space $(\Omega, \mathcal{F})$ on $[S, T]$ if each $\mathcal{F}_t^S$ is a sub-σ-field of $\mathcal{F}$ and

$$\mathcal{F}_t^S \subseteq \mathcal{F}_{t'}^S, \qquad \forall S \leqslant t \leqslant t' \leqslant T.$$

In the case that $\mathbb{F}_T^S$ is a filtration of $(\Omega, \mathcal{F})$ on $[S, T]$, the triple $(\Omega, \mathcal{F}, \mathbb{F}_T^S)$ is called a *filtered measurable space* on $[S, T]$. A filtration $\mathbb{F}_T^S = \{\mathcal{F}_t^S \mid S \leqslant t \leqslant T\}$ is said to be *right-continuous* if for each $t \in [S, T)$,

$$\mathcal{F}_t^S = \bigcap_{\varepsilon>0} \mathcal{F}_{t+\varepsilon}^S,$$

and it is said to be *left-continuous* if for each $t \in (S, T]$,

$$\mathcal{F}_t^S = \sigma\Big(\bigcup_{\varepsilon>0} \mathcal{F}_{t-\varepsilon}^S\Big)$$

(v) Let $(\Omega, \mathcal{F})$ be a measurable space. A map $\mathbb{P}:\mathcal{F} \to [0,1]$ is called a *probability (measure)* on $(\Omega, \mathcal{F})$ if the following holds:

$$\begin{cases} \mathbb{P}(\varnothing) = 0, \quad \mathbb{P}(\Omega) = 1, \\ \mathbb{P}\Big(\bigcup_{k=1}^{\infty} A_k\Big) = \sum_{k=1}^{\infty} \mathbb{P}(A_k), \quad \forall \{A_k \mid k \geqslant 1\} \subseteq \mathcal{F},\ A_k \cap A_j = \varnothing,\ k \neq j. \end{cases}$$

In the case that $\mathbb{P}$ is a probability on $(\Omega, \mathcal{F})$, the triple $(\Omega, \mathcal{F}, \mathbb{P})$ is called a *probability space*. In this case, for any event $A \in \mathcal{F}$, $\mathbb{P}(A)$ is called the *probability* of the event A.

(vi) Let $(\Omega, \mathcal{F}, \mathbb{P})$ be a probability space. An event $A \in \mathcal{F}$ is called a $\mathbb{P}$-*null set* if $\mathbb{P}(A) = 0$. The σ-field $\mathcal{F}$ is said to be *complete* under $\mathbb{P}$ if for any $A \in \mathcal{F}$ with $\mathbb{P}(A) = 0$, the following holds:

$$B \in \mathcal{F},\ \mathbb{P}(B) = 0, \qquad \forall B \subseteq A.$$

When $\mathcal{F}$ is complete under $\mathbb{P}$, $(\Omega, \mathcal{F}, \mathbb{P})$ is called a *complete probability space*. A filtration $\mathbb{F}_T^S \equiv \{\mathcal{F}_t^S\}_{S \leqslant t \leqslant T}$ is said to be *augmented* if for each $t \in [S, T]$, $\mathcal{F}_t^S$ is complete under $\mathbb{P}$. In this case, $(\Omega, \mathcal{F}, \mathbb{F}_T^S, \mathbb{P})$ is called a *complete filtered probability space* on $[S, T]$. A property, denoted by $P(\omega)$, pertaining to $\omega \in \Omega$, is said to be true *almost surely* if there exists a $\mathbb{P}$-null set Ω_0 such that property $P(\omega)$ is true on $\Omega \setminus \Omega_0$. In this case, we denote $P(\omega)$, a.s.

(vii) A filtered probability space $(\Omega, \mathcal{F}, \mathbb{F}_T^S, \mathbb{P})$ is said to satisfy the *usual conditions* if it is complete and $\mathbb{F}_T^S$ is right-continuous.

The following is a useful result, called *Borel–Cantelli lemma*.

Lemma 2.1.2 (Borel–Cantelli) *Let $A_k \in \mathcal{F}$ be a sequence of events such that*

$$\sum_{k=1}^{\infty} \mathbb{P}(A_k) < \infty.$$

Then

$$\mathbb{P}\Big(\bigcap_{k=1}^{\infty} \bigcup_{\ell=k}^{\infty} A_\ell\Big) = 0.$$

In what follows, we simply denote $\mathcal{F}_t = \mathcal{F}_t^0$, and $\mathbb{F} = \mathbb{F}_\infty^0 \equiv \{\mathcal{F}_t\}_{t \geqslant 0}$. These notation will be used most of the time below.

Definition 2.1.3

(i) Let $(\Omega, \mathcal{F})$ be a measurable space. A map $\xi : \Omega \to \mathbb{R}$ is called a *random variable* if

$$(\xi \leqslant a) \equiv \{\omega \in \Omega \mid \xi(\omega) \leqslant a\} \in \mathcal{F}, \qquad \forall a \in \mathbb{R}. \tag{2.1.1}$$

In this case, we also say that ξ is $\mathcal{F}$-*measurable*. The set of all $\mathcal{F}$-measurable random variable is denoted by $L^0_{\mathcal{F}}(\Omega; \mathbb{R})$. The σ-field generated by a random variable ξ is defined to be

$$\mathcal{F}^\xi = \sigma\big\{(\xi \leqslant a) \mid a \in \mathbb{R}\big\},$$

which is the smallest sub-σ-field of $\mathcal{F}$, with respect to which ξ is measurable.

(ii) Let $(\Omega, \mathcal{F}, \mathbb{P})$ be a probability space, and $\xi \in L^0_{\mathcal{F}}(\Omega; \mathbb{R})$. Define

$$F(x) = \mathbb{P}(\xi \leqslant x), \qquad x \in \mathbb{R}. \tag{2.1.2}$$

The function $x \mapsto F(x)$ is called the *distribution* of ξ, which is monotone non-decreasing and right-continuous. In the case that $F(\cdot)$ is absolutely continuous with respect to the Lebesgue measure, i.e.,

$$F(dx) = f(x)dx,$$

we say that ξ admits a *density* $f(\cdot)$.

(iii) Let $(\Omega, \mathcal{F}, \mathbb{P})$ be a probability space. Let $\mathcal{L}^0_{\mathcal{F}}(\Omega; \mathbb{R})$ be the set of all random variables ξ of the following form:

$$\xi(\omega) = \sum_{i=1}^{k} a_i \mathbf{1}_{A_i}(\omega), \qquad \omega \in \Omega, \tag{2.1.3}$$

where, for $i = 1, 2, \cdot s, k$, $a_i \in \mathbb{R}$, $A_i \in \mathcal{F}$, are mutually disjoint, with $\bigcup_{i=1}^{k} A_i = \Omega$, and $\mathbf{1}_{A_i}$ is the characteristic function of A_i defined by

$$\mathbf{1}_{A_i}(\omega) = \begin{cases} 1, & \omega \in A_i, \\ 0, & \omega \in \Omega \setminus A_i. \end{cases}$$

Then $\mathcal{L}^0_{\mathcal{F}}(\Omega; \mathbb{R})$ is a linear space. Any $\xi \in \mathcal{L}^0_{\mathcal{F}}(\Omega; \mathbb{R})$ is called a *simple random variable*. For any $\xi \in \mathcal{L}^0_{\mathcal{F}}(\Omega; \mathbb{R})$ of form (2.1.3), its *expectation* (also called *mean*) is defined as follows:

$$\mathbb{E}\xi = \sum_{i=1}^{k} a_i \mathbb{P}(A_i) \equiv \int_\Omega \xi d\mathbb{P} = \int_{\mathbb{R}} x dF(x) = \int_{\mathbb{R}} x f(x) dx. \tag{2.1.4}$$

The last equality holds if the density exists. For any $p \geqslant 1$; and any $\xi \in \mathcal{L}^0_{\mathcal{F}}(\Omega; \mathbb{R})$ of form (2.1.3), define

$$\|\xi\|_p \equiv \Big(\sum_{i=1}^k |a_i|^p \mathbb{P}(A_i)\Big)^{\frac{1}{p}} = \Big(\int_{\mathbb{R}} |x|^p dF(x)\Big)^{\frac{1}{p}}.$$

This is a norm in $\mathcal{L}^0_{\mathcal{F}}(\Omega; \mathbb{R})$. The completion of $\mathcal{L}^0_{\mathcal{F}}(\Omega; \mathbb{R})$ under $\|\cdot\|_p$ is denoted by $L^p_{\mathcal{F}}(\Omega; \mathbb{R})$ which itself is a Banach space (complete normed space). Any $\xi \in L^p_{\mathcal{F}}(\Omega; \mathbb{R})$ is said to be *L^p-integrable*. If $\xi \in L^p_{\mathcal{F}}(\Omega; \mathbb{R})$, then there exists a sequence $\xi_k \in \mathcal{L}^0_{\mathcal{F}}(\Omega; \mathbb{R})$ converging to ξ under the norm $\|\cdot\|_p$. In this case, we define

$$\mathbb{E}\xi \equiv \int_\Omega \xi d\mathbb{P} = \lim_{k\to\infty} \mathbb{E}\xi_k \equiv \lim_{k\to\infty} \int_\Omega \xi_k d\mathbb{P},$$

which is called the expectation of ξ. Therefore, any $\xi \in L^p_{\mathcal{F}}(\Omega; \mathbb{R})$ has its expectation. Further, we define

$$\mathbb{E}[\xi \mathbf{1}_A] = \int_\Omega \xi \mathbf{1}_A d\mathbb{P} = \int_A \xi d\mathbb{P}, \qquad \forall \xi \in L^1_{\mathcal{F}}(\Omega; \mathbb{R}),\ A \in \mathcal{F}.$$

One can show that if ξ is a random variable with distribution $F(\cdot)$ and $h : \mathbb{R} \to \mathbb{R}$ such that $h(\xi) \in L^1_{\mathcal{F}}(\Omega; \mathbb{R})$, then

$$\mathbb{E}h(\xi) = \int_{\mathbb{R}} h(x) dF(x). \tag{2.1.5}$$

In the above, we may replace $\mathbb{R}$ by any Euclidean space $\mathbb{H}$, such as $\mathbb{R}^m$ and $\mathbb{R}^{m\times n}$, and denote the set of all $\mathcal{F}$-measurable, $\mathbb{H}$-valued random variables by $L^0_{\mathcal{F}}(\Omega; \mathbb{H})$, $\mathcal{L}^0_{\mathcal{F}}(\Omega; \mathbb{H})$, and $L^p_{\mathcal{F}}(\Omega; \mathbb{H})$, respectively. Next, the following result is well known and useful.

Proposition 2.1.4 *Let $p \in (1, \infty)$ and $\Xi \subseteq L^p_{\mathcal{F}}(\Omega; \mathbb{R}^m)$ such that*

$$\sup_{\xi \in \Xi} \|\xi\|_p < \infty.$$

Then Ξ is sequentially weakly compact, i.e., there exists a sequence $\xi_k \in \Xi$ such that

$$\lim_{k\to\infty} \mathbb{E}\langle \xi_k, \eta\rangle = \mathbb{E}\langle \xi_0, \eta\rangle, \qquad \forall \eta \in L^q_{\mathcal{F}}(\Omega; \mathbb{R}^m),$$

with $q = \frac{p}{p-1}$, for some $\xi_0 \in L^p_{\mathcal{F}}(\Omega; \mathbb{R}^m)$.

Unfortunately, the above is not true for the case of $p = 1$. To remedy it, we introduce the following notion.

Definition 2.1.5 A family $\Xi \subseteq L^1_{\mathcal{F}}(\Omega; \mathbb{R}^m)$ is said to be *uniformly integrable*, if

$$\lim_{k\to\infty} \sup_{\xi\in\Xi} \mathbb{E}\big[|\xi|\mathbf{1}_{\{|\xi|\geqslant k\}}\big] = 0.$$

Then, the following is true.

Proposition 2.1.6 *Let $\Xi \subseteq L^1_{\mathcal{F}}(\Omega; \mathbb{R}^m)$ be uniformly integrable. Then Ξ is sequentially weakly compact, i.e., there exists a sequence $\xi_k \in \Xi$ such that*

$$\lim_{k\to\infty} \mathbb{E}\langle \xi_k, \eta\rangle = \mathbb{E}\langle \xi_0, \eta\rangle, \qquad \forall \eta \in L^\infty_{\mathcal{F}}(\Omega; \mathbb{R}^m),$$

for some $\xi_0 \in L^1_{\mathcal{F}}(\Omega; \mathbb{R}^m)$.

Sometimes the following result is very useful.

Proposition 2.1.7 *If $\Xi \subseteq L^1_{\mathcal{F}}(\Omega; \mathbb{R}^m)$ such that*

$$\|\xi\|_{1+\varepsilon} \leqslant K, \qquad \forall \xi \in \Xi,$$

for some $\varepsilon, K > 0$, then Ξ is uniformly integrable.

Proof First of all, for any $\xi \in \Xi$, as $k \to +\infty$,

$$\mathbb{P}(|\xi| \geqslant k) \leqslant \int_{\{|\xi|\geqslant k\}} \frac{|\xi|}{k} d\mathbb{P} \leqslant \frac{1}{k}\mathbb{E}|\xi| \leqslant \frac{1}{k}\Big(\mathbb{E}|\xi|^{1+\varepsilon}\Big)^{\frac{1}{1+\varepsilon}} \leqslant \frac{K}{k} \to 0.$$

Also,

$$\mathbb{E}\big[|\xi|\mathbf{1}_{\{|\xi|\geqslant k\}}\big] \leqslant \|\xi\|_{1+\varepsilon}\Big(\mathbb{E}\big[\mathbf{1}_{\{|\xi|\geqslant k\}}\big]\Big)^{\frac{\varepsilon}{1+\varepsilon}} \leqslant K\Big(\frac{K}{k}\Big)^{\frac{\varepsilon}{1+\varepsilon}} \to 0.$$

Thus, our conclusion follows. □

Next, we introduce a new notion.

Definition 2.1.8 Let $(\Omega, \mathcal{F}, \mathbb{P})$ be a probability space and $\mathcal{G}$ be a sub-σ-field of $\mathcal{F}$. Let $\xi \in L^1_{\mathcal{F}}(\Omega; \mathbb{R}^m)$. A random variable $\eta \in L^1_{\mathcal{G}}(\Omega; \mathbb{R}^m)$ is called the *conditional expectation* of ξ given $\mathcal{G}$ if

$$\int_A \xi d\mathbb{P} = \int_A \eta d\mathbb{P}, \qquad \forall A \in \mathcal{G}.$$

We denote $\eta = \mathbb{E}[\xi \mid \mathcal{G}]$.

The following collects useful properties of conditional expectations.

Proposition 2.1.9 *Let $(\Omega, \mathcal{F}, \mathbb{P})$ be a probability space and $\mathcal{G}$ be a sub-σ-field of $\mathcal{F}$. Then:*

(i) *Map $\mathbb{E}(\cdot \mid \mathcal{G}) : L^1_{\mathcal{F}}(\Omega; \mathbb{R}^m) \to L^1_{\mathcal{G}}(\Omega; \mathbb{R}^m)$ is a well-defined linear and bounded operator.*

(ii) $\mathbb{E}(a \mid \mathcal{G}) = a, \quad$ *a.s.*, $\quad \forall a \in \mathbb{R}$.

(iii) *If $X, Y \in L^1_{\mathcal{F}}(\Omega; \mathbb{R})$ with $X \geqslant Y$, then*

$$E(X \mid \mathcal{G}) \geqslant E(Y \mid \mathcal{G}), \qquad \text{a.s.}$$

In particular,

$$X \geqslant 0, \quad \text{a.s.} \quad \Rightarrow \quad \mathbb{E}(X \mid \mathcal{G}) \geqslant 0, \quad \text{a.s.}$$

(iv) *Let $Y \in L^1_{\mathcal{G}}(\Omega; \mathbb{R}^m)$ and $Z \in L^1_{\mathcal{F}}(\Omega; \mathbb{R}^k)$. Then*

$$\mathbb{E}(YZ^\top \mid \mathcal{G}) = Y\mathbb{E}(Z \mid \mathcal{G})^\top, \qquad \text{a.s.}$$

In particular (by taking $Z \equiv 1$),

$$\mathbb{E}(Y \mid \mathcal{G}) = Y.$$

Further, if $Y = \mathbb{E}(X \mid \mathcal{G})$ for some $X \in L^1_{\mathcal{F}}(\Omega; \mathbb{R}^n)$, then

$$\mathbb{E}\big(\mathbb{E}(X \mid \mathcal{G})Z^\top \mid \mathcal{G}\big) = \mathbb{E}(X \mid \mathcal{G})\mathbb{E}(Z \mid \mathcal{G})^\top.$$

(v) *Random variable X is independent of $\mathcal{G}$ if and only if for any Borel measurable function $f : \mathbb{R} \to \mathbb{R}$ such that $\mathbb{E}f(X)$ exists, it holds*

$$\mathbb{E}\big(f(X) \mid \mathcal{G}\big) = \mathbb{E}f(X), \qquad \text{a.s.}$$

In particular, if X is independent of $\mathcal{G}$, then $\mathbb{E}(X \mid \mathcal{G}) = \mathbb{E}X$, almost surely.

(vi) *Let $\mathcal{G}_1 \subseteq \mathcal{G}_2 \subseteq \mathcal{F}$. Then*

$$\mathbb{E}\big(\mathbb{E}(X \mid \mathcal{G}_2) \mid \mathcal{G}_1\big) = \mathbb{E}\big(\mathbb{E}(X \mid \mathcal{G}_1) \mid \mathcal{G}_2\big) = \mathbb{E}(X \mid \mathcal{G}_1), \qquad \text{a.s.}$$

(vii) *(Jensen's inequality) Let $X \in L^1_{\mathcal{F}}(\Omega; \mathbb{R}^m)$ and $\varphi : \mathbb{R}^m \to \mathbb{R}$ be a convex function such that $\varphi(X) \in L^1_{\mathcal{F}}(\Omega; \mathbb{R})$. Then*

$$\varphi\big(\mathbb{E}(X \mid \mathcal{G})\big) \leqslant \mathbb{E}\big(\varphi(X) \mid \mathcal{G}\big), \qquad \text{a.s.}$$

In particular, for any $p \geqslant 1$, provided that $\mathbb{E}|X|^p$ exists, we have

$$\big|\mathbb{E}(X \mid \mathcal{G})\big|^p \leqslant \mathbb{E}\big(|X|^p \mid \mathcal{G}\big), \qquad \text{a.s.}$$

Definition 2.1.10 Let $(\Omega, \mathcal{F}, \mathbb{F}_T^S, \mathbb{P})$ be a filtered probability space on $[S, T]$.

(i) A map $X : [S, T] \times \Omega \to \mathbb{R}$ is called a *stochastic process* on $[S, T]$ if for each $t \in [S, T]$, $\omega \mapsto X(t, \omega)$ is a random variable.

(ii) Process $X(\cdot)$ is said to be $\mathbb{F}_T^S$*-adapted* if for each $t \in [S, T]$, $\omega \mapsto X(t, \omega)$ is $\mathcal{F}_t^S$-measurable, i.e.,

$$(X(t) \leqslant a) \equiv \{\omega \in \Omega \mid X(t, \omega) \leqslant a\} \in \mathcal{F}_t^S, \qquad \forall t \in [S, T],\ a \in \mathbb{R}.$$

(iii) Process $X(\cdot)$ is said to be $\mathbb{F}_T^S$*-progressively measurable* on $[S, T]$ if

$$\{(s, \omega) \in [S, t] \times \Omega \mid X(s, \omega) \leqslant a\} \in \mathcal{B}[S, t] \otimes \mathcal{F}_t^S, \quad \forall t \in [S, T],\ a \in \mathbb{R},$$

where $\mathcal{B}[S, t]$ is the Borel σ-field of $[S, t]$ and

$$\mathcal{B}[S, t] \otimes \mathcal{F}_t^S = \sigma\Big(\{A \times B \mid A \in \mathcal{B}[S, t],\ B \in \mathcal{F}_t^S\}\Big).$$

In this case, $(t, \omega) \mapsto X(t, \omega)$ is $\mathcal{B}[S, T] \otimes \mathcal{F}_T^S$-measurable.

(iv) Process $X(\cdot)$ is said to be *càdlàg* if almost surely the *path* $t \mapsto X(t, \omega)$ is right-continuous and has left-limits.

(v) Process $X(\cdot)$ is said to be *continuous* (or have *continuous paths*) if almost surely the path $t \mapsto X(t, \omega)$ is continuous.

(vi) A process $\widetilde{X}(\cdot)$ is called a *modification* of $X(\cdot)$ if

$$\mathbb{P}\big(X(t) = \widetilde{X}(t)\big) = 0, \qquad \forall t \geqslant 0. \tag{2.1.6}$$

For convenience, we identify $X(\cdot)$ with its modification $\widetilde{X}(\cdot)$.

Proposition 2.1.11 (Kolmogorov Continuity Theorem) *Suppose $X(\cdot)$ is a stochastic process on $[S, T]$ and, for some $p, \beta > 0$,*

$$\mathbb{E}|X(t) - X(s)|^p \leqslant K|t - s|^{1+\beta}, \qquad s, t \in [S, T]. \tag{2.1.7}$$

Hereafter, $K > 0$ will be a generic constant which can be different from line to line. Then $t \mapsto X(t)$ is continuous almost surely. More precisely, an $\mathcal{F}_T$-measurable random variable ξ defined by

$$\xi = \Big[\int_S^T \int_S^T \frac{|X(t) - X(s)|^p}{|t - s|^{2+\beta}\Big(\log \frac{\gamma}{|t-s|}\Big)^2} ds dt\Big]^{\frac{1}{p}}$$

is integrable, with $\gamma > e^{\frac{2}{2+\beta}}$, and for any $\varepsilon \in (0, \frac{\beta}{p})$,

$$|X(t) - X(s)| \leqslant K\xi|t - s|^{\frac{\beta}{p} - \varepsilon}, \qquad \forall t, s \in [S, T]. \tag{2.1.8}$$

Consequently,

$$\mathbb{E}\Big[\sup_{S\leqslant s\neq t\leqslant T}\frac{|X(t)-X(s)|}{|t-s|^{\frac{\beta}{p}-\varepsilon}}\Big]\leqslant K\,\mathbb{E}[\xi]. \tag{2.1.9}$$

Furthermore, it holds

$$\mathbb{E}\Big[\sup_{t\in[S,T]}|X(t)|^p\Big]\leqslant K\,\mathbb{E}\Big[1+|X(S)|^p\Big]. \tag{2.1.10}$$

In the case that $X(\cdot)$ is deterministic, (2.1.7) implies

$$\frac{|X(t)-X(t')|}{|t-t'|^{\frac{\beta}{p}+\frac{1}{p}}}\leqslant K, \tag{2.1.11}$$

which is stronger than (2.1.8). However, for the case that $X(\cdot)$ is random, we only have a weaker conclusion (2.1.8).

It is trivial that for $S\leqslant s\neq t\leqslant T$,

$$\mathbb{E}\Big[\frac{|X(t)-X(s)|}{|t-s|^{\frac{\beta}{p}-\varepsilon}}\Big]\leqslant \mathbb{E}\Big[\sup_{S\leqslant s\neq t\leqslant T}\frac{|X(t)-X(s)|}{|t-s|^{\frac{\beta}{p}-\varepsilon}}\Big].$$

Nevertheless, the right-hand side of the above is not necessarily controlled by the left-hand side of the above in general. However, when (2.1.7) holds, we do have estimate like (2.1.9), which is essentially a control of the right-hand side by the left-hand side.

For any Euclidean space $\mathbb{H}$, such as $\mathbb{R}^m$ and $\mathbb{R}^{m\times n}$, one may define $\mathbb{H}$-valued stochastic processes if each component is a stochastic process in the above sense. The corresponding results for $\mathbb{H}$-valued random variables/processes hold with some small modifications of notations. Thus, we omit them here.

Definition 2.1.12 A random variable $\tau:\Omega\to[0,\infty)$ is called an $\mathbb{F}$-*stopping time* if

$$(\tau\leqslant t)\in\mathcal{F}_t,\qquad \forall t\geqslant 0. \tag{2.1.12}$$

The set of all $\mathbb{F}$-stopping times on $[S,T]$ is denoted by $\mathscr{T}[S,T]$. The following is very important relevant to stopping times.

Proposition 2.1.13 (Debut Theorem) *Let $X(\cdot)$ be $\mathbb{F}$-progressively measurable and G be a Borel set in $\mathbb{R}^m$. Then*

$$\tau_G=\inf\{t\geqslant 0\bigm| X(t)\in G\}$$

is an $\mathbb{F}$-stopping time.

In applications, a useful case is that $X(\cdot)$ is right-continuous and G is either an open or closed set in $\mathbb{R}^m$.

2 Martingale and Brownian Motion

Let $(\Omega, \mathcal{F}, \mathbb{F}, \mathbb{P})$ be a filtered probability space with $\mathbb{F} \equiv \mathbb{F}^0_\infty \equiv \{\mathcal{F}_t\}_{t\geqslant 0}$ satisfying the usual condition.

Definition 2.2.1

(i) An $\mathbb{R}$-valued stochastic process $M(\cdot)$ is called a *martingale* on the filtered probability space $(\Omega, \mathcal{F}, \mathbb{F}, \mathbb{P})$ if $\mathbb{E}[M(t)]$ exists for all $t \in [0, T]$ and

$$\mathbb{E}_s\big[M(t)\big] = M(s), \qquad \forall\, 0 \leqslant s \leqslant t \leqslant T. \tag{2.2.1}$$

Hereafter, we denote $\mathbb{E}_s = \mathbb{E}[\,\cdot\,|\mathcal{F}_s]$. If in the above, the equality sign is replaced by "$\geqslant$" (resp. "$\leqslant$"), $M(\cdot)$ is called *sub-martingale* (resp. *super-martingale*).[1]

(ii) A martingale $M(\cdot)$ is said to be uniformly integrable if the family of random variables $\{M(t) \bigm| t \in [0, T]\}$ is uniformly integrable.

Note that martingales can be defined for any Euclidean space $\mathbb{H}$-valued processes. Here, $\mathbb{H}$ could be $\mathbb{R}^m$, $\mathbb{R}^{m\times n}$, and so on. We now look at a couple of important examples.

Example 2.2.2 Suppose $\xi \in L^1_{\mathcal{F}_T}(\Omega; \mathbb{R})$. Define

$$M(t) = \mathbb{E}_t[\xi], \qquad t \geqslant 0.$$

Then $M(\cdot)$ is a uniformly integrable martingale. The proof follows easily from the property of conditional expectation.

Example 2.2.3 Suppose $M(\cdot)$ is a martingale. Then $M(\cdot)^2$ is a sub-martingale. This follows from Jensen's inequality (see Proposition 2.1.9 (vii)).

We state the following important and standard result concerning martingales.

Proposition 2.2.4 (Doob's Inequality) *Let $p > 1$ and $M(\cdot)$ be an L^p-integrable martingale on $(\Omega, \mathcal{F}, \mathbb{F}, \mathbb{P})$. Then*

$$\mathbb{E}\Big[\sup_{t\in[0,T]} |M(t)|^p\Big] \leqslant \Big(\frac{p}{p-1}\Big)^p \mathbb{E}|M(T)|^p. \tag{2.2.2}$$

The following is the so-called Doob–Meyer decomposition theorem.

---

[1] Martingales can be defined for vector-valued stochastic processes, but not sub-martingales or super-martingales.

Proposition 2.2.5 (Doob–Meyer Decomposition) *Let $M(\cdot)$ be a square integrable martingale on $(\Omega, \mathcal{F}, \mathbb{F}, \mathbb{P})$. Then $M(\cdot)^2$ is a sub-martingale, and there exists a unique monotone nondecreasing $\mathbb{F}$-adapted process, denoted by $\langle M\rangle(\cdot)$, called the quadratic variation of M, such that $M(\cdot)^2 - \langle M\rangle(\cdot)$ is a martingale.*[2]

Next, we introduce an important martingale.

Definition 2.2.6 Let $(\Omega, \mathcal{F}, \mathbb{F}, \mathbb{P})$ be a filtered probability space. An $\mathbb{F}$-adapted process $W : [0, \infty) \to \mathbb{R}^d$ is called a d-dimensional *standard Brownian motion* if $W(0) = 0$, and for any $0 \leqslant s < t < \infty$, the *increment* $W(t) - W(s)$ is independent of $\mathcal{F}_s$, normally distributed with mean zero and covariance matrix $(t-s)I_d$. Namely, the density of $W(t) - W(s)$ exists and is given by

$$f(x, t-s) = \frac{1}{[2\pi(t-s)]^{\frac{d}{2}}} \exp\Big(-\frac{|x|^2}{2(t-s)}\Big).$$

Consequently, $W(\cdot)$ is a martingale which is L^p-integrable, for all $p \in [1, \infty)$, and further, it is uniformly integrable on any finite time interval $[0, T]$.

One can show that the above defined standard Brownian motion exists. Further, it is known that for almost all $\omega \in \Omega$, $t \mapsto W(t, \omega)$ is not of bounded variation on any interval and is merely α-Hölder continuous with $\alpha \in (0, \frac{1}{2})$.

For a d-dimensional standard Brownian motion $W(\cdot)$, the *natural filtration* $\mathbb{F}_T^S \equiv \{\mathcal{F}_t^S\}_{S \leqslant t \leqslant T}$ of $W(\cdot)$ on $[S, T]$ is defined by

$$\mathcal{F}_t^S = \sigma\Big(\{W(\tau) - W(S) \mid S \leqslant \tau \leqslant t\} \cup \mathcal{N}\Big), \qquad t \in [S, T], \tag{2.2.3}$$

with $\mathcal{N}$ being the set of all $\mathbb{P}$-null sets.

From now on, we let $(\Omega, \mathcal{F}, \mathbb{F}, \mathbb{P})$ be a fixed filtered augmented probability space on which a d-dimensional standard Brownian motion $W(\cdot)$ is defined with its natural filtration (on $[0, T]$) being $\mathbb{F}$ and its natural filtration on $[S, T]$, with $S \in (0, T)$, being $\mathbb{F}_T^S$. For convenience, we simply just use $\mathbb{F}$ instead of $\mathbb{F}_T^S$, unless S and T need to be emphasized.

3 Itô Integral

For any $p \in [1, \infty]$, let $\mathcal{L}_{\mathbb{F}}^p(S, T; \mathbb{R})$ be the set of all processes $\sigma(\cdot)$ having the following form:

[2] If $M(\cdot)$ is $\mathbb{R}^n$-valued, $M(\cdot)^2$ should be replaced by $M(\cdot)M(\cdot)^\top$ and $\langle M\rangle(\cdot)$ should be $\mathbb{R}^{n\times n}$-valued.

$$\sigma(t)=\sum_{i=0}^{k}\xi_i\mathbf{1}_{[t_i,t_{i+1})}(t), \qquad t\in[S,T], \tag{2.3.1}$$

for some partition $S=t_0<t_1<\cdot s<t_k<t_{k+1}=T$ of $[S,T]$, and for each i, ξ_i is $\mathcal{F}_{t_i}^S$-measurable, $\mathbb{R}$-valued, with $\mathbb{E}|\xi_i|^p<\infty$ (or all ξ_i are bounded uniformly in i if $p=\infty$). Clearly, $\mathcal{L}_{\mathbb{F}}^p(S,T;\mathbb{R})$ is a linear space, and any $\sigma(\cdot)\in\mathcal{L}_{\mathbb{F}}^p(S,T;\mathbb{R})$ is $\mathbb{F}$-adapted, càdlàg (see Definition 2.1.10 (iv)), piecewise constant. Any $\sigma(\cdot)\in\mathcal{L}_{\mathbb{F}}^p(S,T;\mathbb{R})$ is called a *simple L^p-integrable process*. For any $\sigma(\cdot)\in\mathcal{L}_{\mathbb{F}}^p(S,T;\mathbb{R})$ of the above form (with $p\in[1,\infty)$), we define

$$\|\sigma(\cdot)\|_{\mathcal{L}_{\mathbb{F}}^p(S,T;\mathbb{R})}=\Big[\mathbb{E}\Big(\sum_{i=0}^{k}|\xi_i|^2(t_{i+1}-t_i)\Big)^{\frac{p}{2}}\Big]^{\frac{1}{p}}=\Big[\mathbb{E}\Big(\int_S^T|\sigma(s)|^2ds\Big)^{\frac{p}{2}}\Big]^{\frac{1}{p}}. \tag{2.3.2}$$

This is a norm on the linear space $\mathcal{L}_{\mathbb{F}}^p(S,T;\mathbb{R})$. The completion of $\mathcal{L}_{\mathbb{F}}^p(S,T;\mathbb{R})$ under the norm $\|\cdot\|_{\mathcal{L}_{\mathbb{F}}^p(S,T;\mathbb{R})}$ is denoted by $L_{\mathbb{F}}^p(\Omega;L^2(S,T;\mathbb{R}))$ which is a Banach space. Because of (2.3.2), we may write

$$\|\sigma(\cdot)\|_{L_{\mathbb{F}}^p(\Omega;L^2(S,T;\mathbb{R}))}=\Big[\mathbb{E}\Big(\int_0^T|\sigma(s)|^2ds\Big)^{\frac{p}{2}}\Big]^{\frac{1}{p}}, \quad \forall\sigma(\cdot)\in L_{\mathbb{F}}^p(\Omega;L^2(S,T;\mathbb{R})).$$

The case $p=\infty$ can be treated in an obvious way. Let us make the following remark, which is relevant to the introduction of more spaces later on.

Remark 2.3.1 For any $\sigma(\cdot)\in L_{\mathbb{F}}^p(\Omega;L^2(S,T))$, let us look at the measurability of the map $t\mapsto\sigma(t)$. First, it is an almost sure limit of some simple L^p-integrable process $\sigma_k(\cdot)$ of form (2.3.1). On the other hand, for the sequence $\sigma_k(\cdot)$ of form (2.3.1), with each ξ_i in $L_{\mathcal{F}_{t_i}^S}^p(\Omega;\mathbb{R})$, there exists a sequence of $\mathcal{F}_{t_i}^S$-measurable random variables approaching ξ_i, and each of these random variables takes finitely many values. Hence, $\sigma(\cdot\,,\cdot)$ is $\mathcal{B}[S,T]\otimes\mathcal{F}_T^S$-measurable as a map from $[S,T]\times\Omega$ to $\mathbb{R}$. Next, when $p=2$, by regarding the Banach space $L_{\mathcal{F}_T^S}^p(\Omega;\mathbb{R})$ as the range space of $t\mapsto\sigma(t)$, we see that $\sigma(t)$ is the almost everywhere limit of a finitely many valued function. Hence, by definition, it is *(Bochner) strong measurable* as a map from $[S,T]$ to $L_{\mathcal{F}_T^S}^p(\Omega;\mathbb{R})$[3].

[3] A map σ from a measure space $(M,\mathscr{M},\mu)$ to a Banach space H is weakly measurable if for any $f^*\in H^*$, the dual of H, $t\mapsto\langle f^*,\sigma(t)\rangle$ is measurable, and it is strongly measurable if it is an almost everywhere limit of a sequence of at most countably many valued measurable functions. According to Pettis measurability theorem, a function is strongly measurable if and only if it is weakly measurable and essentially separably valued, which means there exists a set $N\in\mathscr{M}$ with $\mu(N)=0$ such that $\sigma(M\setminus N)$ is separable. It is known that a weakly measurable map could be not strongly measurable.

For the time being, let $d = 1$ (the one-dimensional standard Brownian motion). For any $\sigma(\cdot) \in \mathcal{L}^p_{\mathbb{F}}(S, T; \mathbb{R})$ of form (2.3.1), we introduce

$$\begin{aligned}\mathcal{I}(\sigma)(t) &= \sum_{i=0}^{k} \xi_i \big[W(t_{i+1} \wedge t) - W(t_i \wedge t)\big] \\ &= \sum_{j=0}^{k} \Big(\sum_{i=0}^{j-1} \xi_i \big[W(t_{i+1}) - W(t_i)\big] + \xi_j \big[W(t) - W(t_j)\big] \Big) I_{[t_j, t_{j+1})}(t),\end{aligned} \tag{2.3.3}$$

with the convention that $\sum_{i=0}^{-1}\{...\} = 0$. Then, $t \mapsto \mathcal{I}(\sigma)(t)$ is $\mathbb{F}^S_T$-adapted and continuous. Thus,

$$\mathcal{I}(\sigma)(T) \in L^p_{\mathcal{F}_T}(\Omega; \mathbb{R}), \qquad \forall \sigma(\cdot) \in \mathcal{L}^p_{\mathbb{F}}(S, T; \mathbb{R})$$

defines a linear operator from $\mathcal{L}^p_{\mathbb{F}}(S, T; \mathbb{R})$ to $L^p_{\mathcal{F}_T}(\Omega; \mathbb{R})$, which is called *Itô's integrator*. Now, for $p = 2$, by the independent increment property of $W(\cdot)$, we have

$$\begin{aligned}&\mathbb{E}\big[\mathcal{I}(\sigma)(t)^2\big] = \sum_{i,j=1}^{k} \mathbb{E}\Big(\xi_i \xi_j [W(t_{i+1} \wedge t) - W(t_i \wedge t)][W(t_{j+1} \wedge t) - W(t_j \wedge t)]\Big) \\ &= \sum_{i=1}^{k} \mathbb{E}\Big(\xi_i^2 \mathbb{E}_{t_i} [W(t_{i+1} \wedge t) - W(t_i \wedge t)]^2\Big) = \sum_{i=1}^{k} \mathbb{E}\xi_i^2 (t_{i+1} \wedge t - t_i \wedge t).\end{aligned}$$

In particular, taking $t = T$, one has

$$\|\mathcal{I}(\sigma)\|^2_{L^2_{\mathcal{F}_T}(\Omega)} = \mathbb{E}\big[\mathcal{I}(\sigma)(T)^2\big] = \sum_{i=1}^{k} \mathbb{E}\xi_i^2 (t_{i+1} - t_i) = \|\sigma(\cdot)\|^2_{\mathcal{L}^2_{\mathbb{F}}(S,T;\mathbb{R})}. \tag{2.3.4}$$

Hence, $\mathcal{I} : \mathcal{L}^2_{\mathbb{F}}(S, T; \mathbb{R}) \to L^2_{\mathcal{F}_T}(\Omega; \mathbb{R})$ is an *isometry*. Then $\mathcal{I}$ admits an extension $\widetilde{\mathcal{I}} : L^2_{\mathbb{F}}(\Omega; L^2(S, T; \mathbb{R})) \to L^2_{\mathcal{F}_T}(\Omega; \mathbb{R})$, which is still called Itô's integrator. Hereafter, for any $\sigma(\cdot) \in L^2_{\mathbb{F}}(\Omega; L^2(S, T; \mathbb{R}))$, we denote

$$\widetilde{\mathcal{I}}(\sigma)(T) = \int_S^T \sigma(s) dW(s), \tag{2.3.5}$$

and call it *Itô's integral* of $\sigma(\cdot)$ on $[S,T]$. From (2.3.3) to (2.3.4), we see that

$$\left.\begin{aligned}&\mathbb{E}\int_S^T \sigma(s)dW(s)=0,\\&\mathbb{E}\Big|\int_S^T \sigma(s)dW(s)\Big|^2=\mathbb{E}\int_S^T|\sigma(s)|^2ds,\end{aligned}\right\}\quad \forall\sigma(\cdot)\in L^2_{\mathbb{F}}(\Omega;L^2(S,T;\mathbb{R})). \tag{2.3.6}$$

The above can be easily extended to the case of $d>1$ (multidimensional standard Brownian motion) and $m>1$ (vector-/matrix-valued process). Further, we have the following result.

Theorem 2.3.2

(i) *For any $p>1$, there exists a $K_p>0$ (also depending on m and d) such that*

$$\mathbb{E}\Big(\int_S^T|\sigma(s)|^2ds\Big)^{\frac{p}{2}}\leqslant K_p\mathbb{E}\Big|\int_S^T\sigma(s)dW(s)\Big|^p,\quad \forall\sigma(\cdot)\in\mathcal{L}^\infty_{\mathbb{F}}(S,T;\mathbb{R}^{m\times d}). \tag{2.3.7}$$

(ii) *For any $p>0$, there are constants $0<\underline{K}_p<\overline{K}_p$ such that*

$$\begin{aligned}\underline{K}_p\mathbb{E}\Big(\int_S^T|\sigma(s)|^2ds\Big)^{\frac{p}{2}}&\leqslant\mathbb{E}\Big(\sup_{t\in[S,T]}\Big|\int_S^t\sigma(s)dW(s)\Big|^p\Big)\\&\leqslant\overline{K}_p\mathbb{E}\Big(\int_S^T|\sigma(s)|^2ds\Big)^{\frac{p}{2}},\quad \forall\sigma(\cdot)\in\mathcal{L}^\infty_{\mathbb{F}}(S,T;\mathbb{R}^{m\times d}).\end{aligned} \tag{2.3.8}$$

We call (2.3.7) the *martingale moment inequality* and (2.3.8) the *Burkholder–Davis–Gundy's inequalities*. In the above, the constants K_p, $\underline{K}_p$, and $\overline{K}_p$ are independent of T and $\sigma(\cdot)$. Inequality (2.3.7) is stronger than the first inequality in (2.3.8) as the former does not have "sup" inside the expectation $\mathbb{E}$. Moreover, it is known that

$$K_2=\underline{K}_2=1,\qquad \overline{K}_2=4;$$

and for general $p>1$, the following is a possible choice, which shows the independence of $\underline{K}_p$ and $\overline{K}_p$ on T (and $\sigma(\cdot)$):

$$\underline{K}_p=p^{-(p+1)}(7+4\sqrt{2})^{-p},\qquad \overline{K}_p=p^{p+1}(2\sqrt{6})^p. \tag{2.3.9}$$

It is also possible to have a universal choice of K_p. The case (2.3.8) with $S=0$ is standard in the literature. For general $S\in(0,T]$, by extending the function $\sigma(\cdot)$ to be zero outside of $[S, T]$, we easily see (2.3.8) is true. With the above theorem, we could extend Itô's integrator to a map from $L^p_{\mathbb{F}}(\Omega; L^2(S, T; \mathbb{R}^{m\times d}))$ with $p \geqslant 2$ to the case of $p \in [1, 2)$ and improve the integrability of Itô's integral for $\sigma(\cdot) \in L^p_{\mathbb{F}}(\Omega; L^2(0, T; \mathbb{R}^{m\times d}))$ for $p > 2$. More precisely, we have the following result.

Proposition 2.3.3 *For any $p \geqslant 1$, Itô's integral is defined for any $\sigma(\cdot) \in L^p_{\mathbb{F}}(\Omega; L^2(S, T; \mathbb{R}^{m\times d}))$. Moreover, for any $\sigma(\cdot) \in L^p_{\mathbb{F}}(\Omega; L^2(S, T; \mathbb{R}^{m\times d}))$,* (2.3.7)–(2.3.8) *hold. Further, if we define*

$$M(t) = \int_0^t \sigma(s)dW(s), \qquad t \in [0, T], \tag{2.3.10}$$

then $M(\cdot)$ is a uniformly integrable martingale.

Note that

$$\begin{aligned}
&\mathbb{E}\Big[|M(t)|\mathbf{1}_{\{|M(t)|\geqslant k\}}\Big] \leqslant \big[\mathbb{E}|M(t)|^2\big]^{\frac{1}{2}}\big[\mathbb{E}(\mathbf{1}_{\{|M(t)|\geqslant k\}})\big]^{\frac{1}{2}} \\
&\leqslant \big[\mathbb{E}|M(t)|^2\big]^{\frac{1}{2}}\Big[\frac{\mathbb{E}|M(t)|}{k}\Big]^{\frac{1}{2}} \leqslant \frac{1}{\sqrt{k}}\big[\mathbb{E}|M(t)|^2\big]^{\frac{3}{4}} \leqslant \frac{1}{\sqrt{k}}\Big[\mathbb{E}\int_0^T |\sigma(s)|^2 ds\Big]^{\frac{3}{4}}.
\end{aligned}$$

This is why $M(\cdot)$ defined by (2.3.10) is a uniformly integrable martingale on $[0, T]$ when $\sigma \in L^2_{\mathbb{F}}(S, T; \mathbb{R}^{m\times d})$. In particular, by taking $\sigma(s) \equiv I_d$, we see that the standard Brownian motion $W(\cdot)$ is a uniformly integrable martingale on $[0, T]$.

Next, we state a very useful result, called *Itô's formula*, which is essentially the chain rule for certain stochastic processes composed by smooth functions.

Theorem 2.3.4 *Let $X(\cdot)$ be an Itô process, i.e., it is defined by the following:*

$$X(t) = X(0) + \int_0^t b(s)ds + \int_0^t \sigma(s)dW(s), \qquad \forall t \geqslant 0, \tag{2.3.11}$$

for some suitable processes $b(\cdot)$ and $\sigma(\cdot)$. Let $f : [0, T] \times \mathbb{R}^n \to \mathbb{R}^n$ be a function that has required derivatives. Then

$$\begin{aligned}
f(t, X(t)) = f(0, X(0)) + \int_0^t \Big(&f_s(s, X(s)) + f_x(s, X(s))b(s) \\
&+\frac{1}{2}\, tr\,[f_{xx}(s, X(s))\sigma(s)\sigma(s)^\top]\Big)ds + \int_0^t f_x(s, X(s))\sigma(s)dW(s),
\end{aligned} \tag{2.3.12}$$

provided the right-hand side is meaningful.

In later applications, we will encounter the situation that $x \mapsto f(t, x)$ is merely convex. Thus, an extension of the above is needed. We now present that. Let $f : \mathbb{R} \to \mathbb{R}$ be convex. Then, it is known that its left-derivative

$$D^- f(x) = \lim_{\delta \to 0} \frac{f(x - \delta) - f(x)}{-\delta}$$

exists for all $x \in \mathbb{R}$ and is monotone nondecreasing. Thus, by defining

$$\mu_f([a, b)) = D^- f(b) - D^- f(a), \qquad a < b,$$

we may extend it as a Borel measure on $(\mathbb{R}, \mathcal{B}(\mathbb{R}))$. Such a measure μ_f is called the *second-order derivative measure* of $f(\cdot)$, and

$$\int_{-\infty}^{\infty} g(x) \mu_f(dx) = -\int_{-\infty}^{\infty} g'(x) D^- f(x) dx,$$

for every $g : \mathbb{R} \to \mathbb{R}$ which is piecewise C^1 and with compact support. In the case that $f(\cdot)$ is convex and C^2, one has $\mu_f(dx) = f''(x)dx$.

Let us present a simple example, from which we will get some feelings.

Example 2.3.5 Consider $\bar{f}(x) \equiv |x|$. We have

$$D^- \bar{f}(x) \equiv D^- |x| = \text{ sgn}\,(x) \equiv -\mathbf{1}_{(-\infty,0]}(x) + \mathbf{1}_{(0,\infty)}(x), \qquad x \in \mathbb{R},$$

and it holds that

$$|D^- \bar{f}(x)| = 1, \qquad \forall x \in \mathbb{R}.$$

Its second-order derivative measure is given by

$$\mu_{\bar{f}}(\{0\}) = 2, \qquad \mu_{\bar{f}}(A) = 0, \quad \forall A \in \mathcal{B}(\mathbb{R}),\ 0 \notin A.$$

Now, we consider the following process with $n = 1$ (which is more general than (2.3.11), when $n = 1$):

$$X(t) = X(0) + \int_0^t b(s)ds + \int_0^t \sigma(s)dW(s) + b_0(t), \quad \forall t \geqslant 0, \tag{2.3.13}$$

where $b_0(\cdot)$ is an adapted difference of monotone nondecreasing continuous processes (not necessarily differentiable). Any such a process is called a (continuous) *semi-martingale*. Clearly, in the case $b_0(\cdot) = 0$, the above is reduced to (2.3.11). We have the following results.

Theorem 2.3.6 *Let $X(\cdot)$ be a semi-martingale given by* (2.3.13) *and $f(\cdot)$ be convex with $\mu_f(dx)$ being its second-order derivative measure. Then for any smooth convex*

$f_k(\cdot)$ that converges to $f(\cdot)$ uniformly in any compact sets,

$$\lim_{k\to\infty}\int_0^t f_k''(X(s))\sigma(s)^2 ds \equiv L^X(t;\, f) \tag{2.3.14}$$

exists in probability, for some adapted and monotone nondecreasing process $t \mapsto L^X(t;\, f)$ (depending on $f(\cdot)$ and the given process $X(\cdot)$ satisfying (2.3.13)). Consequently, the following Itô–Tanaka formula holds:

$$\begin{aligned} f(X(t)) = f(X(0)) &+ \int_0^t D^- f(X(s))b(s)ds + \int_0^t D^- f(X(s))\sigma(s)dW(s) \\ &+ \int_0^t D^- f(X(s))db_0(s) + \frac{1}{2}L^X(t;\, f), \quad t \geqslant 0, \quad a.s. \end{aligned} \tag{2.3.15}$$

In particular, for any fixed $a \in \mathbb{R}$, with $f_\pm^a(x) = (x-a)^\pm$ and $\bar f^a(x) = |x-a|$ (which are convex), the following Tanaka–Meyer formulas hold:

$$\begin{aligned} \big(X(t)-a\big)^+ &= \big(X(0)-a\big)^+ + \int_0^t \mathbf{1}_{(a,\infty)}\big(X(s)\big)b(s)ds \\ &\quad + \int_0^t \mathbf{1}_{(a,\infty)}\big(X(s)\big)\sigma(s)dW(s) \\ &\quad + \int_0^t \mathbf{1}_{(a,\infty)}\big(X(s)\big)db_0(s) + \frac{1}{2}L^X(t;\, f_+^a), \\ \big(X(t)-a\big)^- &= \big(X(0)-a\big)^- - \int_0^t \mathbf{1}_{(-\infty,a]}\big(X(s)\big)b(s)ds \\ &\quad - \int_0^t \mathbf{1}_{(-\infty,a]}\big(X(s)\big)\sigma(s)dW(s) \\ &\quad - \int_0^t \mathbf{1}_{(-\infty,a]}\big(X(s)\big)db_0(s) + \frac{1}{2}L^X(t;\, f_-^a), \\ \big|X(t)-a\big| &= \big|X(0)-a\big| + \int_0^t sgn\,\big(X(s)-a\big)b(s)ds \\ &\quad + \int_0^t sgn\,\big(X(s)-a\big)\sigma(s)dW(s) \\ &\quad + \int_0^t sgn\,\big(X(s)-a\big)db_0(s) + \frac{1}{2}L^X(t;\, \bar f^a). \end{aligned} \tag{2.3.16}$$

Consequently,

$$L^X(t;\, \bar f^a) = L^X(t;\, f_+^a) + L^X(t;\, f_-^a). \tag{2.3.17}$$

Moreover,

$$L^X(t;\, f) = \int_{-\infty}^{\infty} L^X(t;\, \bar f^a)\mu_f(da), \qquad t \geqslant 0.$$

Note that for smooth $f(\cdot)$, one has

$$L^X(t;\,f) = \int_0^t f''(X(s))\sigma(s)^2 ds, \qquad t \geqslant 0.$$

Thus, for such a case, (2.3.15) becomes

$$\begin{aligned} f(X(t)) = f(X(0)) &+ \int_0^t f'(X(s))b(s)ds + \int_0^t f'(X(s))\sigma(s)dW(s) \\ &+ \int_0^t f'(X(s))db_0(s) + \frac{1}{2}\int_0^t f''(X(s))\sigma(s)^2 ds, \quad t \geqslant 0, \ \text{a.s.} \end{aligned} \tag{2.3.18}$$

From (2.3.14), noting the convexity of $f_k(\cdot)$, we know that $t \mapsto L^X(t;\,f)$ is nondecreasing and nonnegative valued. As to $L^X(t,\,\bar{f}^a)$, we may rewrite it

$$\begin{aligned} L^X(t,\,\bar{f}^a) = 2\Big[&\big|X(t)-a\big| - \big|X(0)-a\big| - \int_0^t \operatorname{sgn}\big(X(s)-a\big)b(s)ds \\ &- \int_0^t \operatorname{sgn}\big(X(s)-a\big)\sigma(s)dW(s) - \int_0^t \operatorname{sgn}\big(X(s)-a\big)db_0(s)\Big]. \end{aligned} \tag{2.3.19}$$

The above process is called the *local time* of the process $X(\cdot)$ at a, and we denote it simply by $L^X(t,a)$. When $a = 0$, we further denote it by $L^X(t)$.

When $X(\cdot) = W(\cdot)$, the standard Brownian motion, we have

$$L^W(t,a) = 2\Big[\big|W(t)-a\big| - |a| - \int_0^t \operatorname{sgn}\big(W(s)-a\big)dW(s)\Big], \qquad t \geqslant 0.$$

One can actually prove that

$$\int_0^t \mathbf{1}_B(W(s))ds \equiv \big|\{s \in [0,t] \mid W(s) \in B\}\big| = \frac{1}{2}\int_B L^W(t,x)dx, \qquad \forall B \in \mathcal{B}(\mathbb{R}).$$

The left-hand side represents the total time within $[0,t]$ that $W(s)$ is in B, the *occupation time* of B by the Brownian path $W(\cdot)$ up to time t, and the right-hand side says that $\frac{1}{2}L^W(t,x)$ is the density of the left-hand side with respect to the Lebesgue measure.

In the case that $X(\cdot)$ is given by (2.3.11) with $n = 1$, from the third equality in (2.3.16), by letting $a = 0$, we have

$$\big|X(t)\big| = \big|X(0)\big| + \int_0^t \operatorname{sgn}\big(X(s)\big)b(s)ds + \int_0^t \operatorname{sgn}\big(X(s)\big)\sigma(s)dW(s) + \frac{1}{2}L^X(t).$$

We now state the following corollary.

Corollary 2.3.7 *Let $X(\cdot)$ be given by (2.3.11) with $n = 1$ and $f(\cdot)$ be smooth and monotone increasing. Then it holds*

$$\begin{aligned} f(|X(t)|)=f(|X(0)|)&+\int_0^t\Big(f'(|X(s)|)b(s)\; sgn\,\big(X(s)\big)+\frac{1}{2}f''(|X(s|)\sigma(s)^2\Big)ds\\ &+\int_0^t f'(|X(s|)\sigma(s)\; sgn\,(X(s))dW(s)\\ &+\frac{1}{2}\int_0^t f'(|X(s)|)dL^X(s), \qquad t\in[0,T]. \end{aligned}$$

Further, by the nonnegativity of the last term in the above, it holds

$$\begin{aligned} f(|X(t)|)\geqslant f(|X(0)|)&+\int_0^t\Big(f'(|X(s)|)b(s)\; sgn\,\big(X(s)\big)\\ &+\frac{1}{2}f''(|X(s|)\sigma(s)^2\Big)ds\\ &+\int_0^t f'(|X(s|)\sigma(s)\; sgn\,(X(s))dW(s), \qquad t\in[0,T]. \end{aligned} \tag{2.3.20}$$

The above estimate will be very useful in a later chapter. Next, we present the following so-called *martingale representation theorem* which will play an important role in studying BSDEs and BSVIEs.

Theorem 2.3.8 *Let $M(\cdot)$ be an $\mathbb{R}^m$-valued continuous martingale on the filtered probability space $(\Omega, \mathcal{F}, \mathbb{F}_T^S, \mathbb{P})$ with $\mathbb{E}|M(t)|^p < \infty$ for all $t \in [S, T]$ with some $p \geqslant 1$. Then*

$$M(t) = M(S) + \int_S^t \sigma(s)dW(s), \qquad t\in[S,T], \;\; a.s. \tag{2.3.21}$$

for a unique $\sigma(\cdot) \in L^p_{\mathbb{F}}(\Omega; L^2(S, T; \mathbb{R}^{m\times d}))$. In this case, the quadratic variation

$$[M]([S,t]) = \lim_{\|\Pi\|\to 0}\sum_{i\geqslant 1}|M(t_i)-M(t_{i-1})|^2 = \int_S^t |\sigma(s)|^2 ds, \qquad a.s., \tag{2.3.22}$$

exists, where Π is a partition of $[S, t]$,

$$\Pi : S = t_0 < t_1 < \cdot s < t, \qquad \|\Pi\| = \max_{i\geqslant 1}|t_i - t_{i-1}|.$$

Moreover, $M(\cdot)^2 - [M]([S, \cdot\,])$ is a martingale on $[S, T]$. Therefore, in this case (recalling Proposition 2.2.5),

$$\langle M\rangle(t) = [M](t), \qquad t\in[0,T],$$

where $[M](t) = [M]([0, t])$.

Next, for any martingale $M(\cdot)$ of form (2.3.21) on $[0, T]$ (i.e., $S = 0$), with some $\sigma(\cdot) \in L^p_{\mathbb{F}}(\Omega; L^2(0, T; \mathbb{R}^{1\times d}))$, we define

$$\begin{aligned}\mathcal{E}\{M\}(t) &\equiv \exp\Big(M(t) - \frac{1}{2}\langle M\rangle(t)\Big)\\ &= \exp\Big(\int_0^t \sigma(s)dW(s) - \frac{1}{2}\int_0^t |\sigma(s)|^2 ds\Big)\\ &\equiv \widehat{\mathcal{E}}\{\sigma\}(t), \qquad t \in [0, T].\end{aligned} \tag{2.3.23}$$

This is called the *Doléans–Dade exponential* of $M(\cdot)$.[4] We call $\widehat{\mathcal{E}}\{\sigma\}$ the *exponential supermartingale* of $\sigma(\cdot)$.[5] If $M(\cdot)$ is bounded, then by martingale moment inequality (Theorem 2.3.2, (i)), we have

$$\mathbb{E}\Big(\int_0^T |\sigma(s)|^2 ds\Big)^{\frac{p}{2}} \leqslant K_p \mathbb{E}|M(T)|^p \leqslant K_p \|M(\cdot)\|_\infty < \infty. \tag{2.3.24}$$

In the current case, applying Itô's formula to $t \mapsto \mathcal{E}\{M\}(t)$, we have

$$\begin{aligned}d\mathcal{E}\{M\}(t) &= \mathcal{E}\{M\}(t)\big(\sigma(t)dW(t) - \frac{1}{2}|\sigma(t)|^2 dt\big) + \frac{1}{2}|\sigma(t)|^2 \mathcal{E}\{M\}(t)dt\\ &= \mathcal{E}\{M\}(t)\sigma(t)dW(t), \qquad t \geqslant 0.\end{aligned}$$

Consequently, by the boundedness of $\mathcal{E}\{M\}$ and (2.3.24),

$$\mathcal{E}\{M\}(t) = 1 + \int_0^t \mathcal{E}\{M\}(s)\sigma(s)dW(s), \qquad t \geqslant 0,$$

is a martingale and

$$\mathbb{E}\Big[\mathcal{E}\{M\}(t)\Big] = 1, \qquad t \in [0, T].$$

We actually have the following more general result (not just for bounded $M(\cdot)$), which is called *Girsanov's theorem*.

Theorem 2.3.9 (Girsanov) *Let $M(\cdot)$ be a martingale given by* (2.3.21) *on* $[0, T]$, *with $m = d = 1$ such that $\mathcal{E}\{M\}(\cdot)$ is a martingale. Then*

$$\widehat{\mathbb{P}}(d\omega) = \mathcal{E}\{M\}(T)\mathbb{P}(d\omega)$$

[4] The definition of Doléan–Dade exponential was the first equality, for general martingale $M(\cdot)$, the last two equalities due to the fact that the martingale $M(\cdot)$ here is given by (2.3.21) with $S = 0$ and $m = 1$. Note also that one allows $d \geqslant 1$.

[5] Such a name is used because $\widehat{\mathcal{E}}\{\sigma\}$ is always a supermartingale.

is a probability measure on $\mathcal{F}_T$, which is equivalent to $\mathbb{P}$. Moreover,

$$\widehat{W}(t) = W(t) - \int_0^t \sigma(s)ds, \qquad t \in [0, T],$$

is a Brownian motion under $\widehat{\mathbb{P}}$.

From the above, we see that it is very crucial whether $\mathcal{E}\{M\}$ (for the given martingale $M(\cdot)$, which could be defined by (2.3.21) with $S = 0$ and $m = 1$) is a (true) martingale. Here are some sufficient conditions for this.

Theorem 2.3.10 *Suppose $M(\cdot)$ is a martingale.*

(i) *For $p \in (1, \infty)$ and $q = \frac{p}{p-1}$, suppose*

$$\sup_{\tau \in \mathscr{T}[0,T]} \mathbb{E}\Big[\exp\Big(\frac{\sqrt{p}}{2(\sqrt{p}-1)} M(\tau)\Big)\Big] < \infty.$$

Then $\mathcal{E}\{M\}$ is L^q-bounded martingale, which implies the uniform integrability of $\mathcal{E}\{M\}$.

(ii) *Suppose*

$$\sup_{\tau \in \mathscr{T}[0,T]} \mathbb{E}\Big[\exp\Big(\frac{1}{2} M(\tau)\Big)\Big] < \infty. \tag{2.3.25}$$

Then $\mathcal{E}\{M\}$ is a uniformly integrable martingale.

(iii) *Suppose*

$$\mathbb{E}\Big[\exp\Big(\frac{1}{2}\langle M\rangle(T)\Big)\Big] < \infty. \tag{2.3.26}$$

Then $\mathcal{E}\{M\}$ is a uniformly integrable martingale.

Condition (2.3.25) is called the *Kazamaki condition*, and condition (2.3.26) is called *Novikov's criterion*. In the case $M(\cdot)$ is given by (2.3.21) (with $S = 0$ and $m = 1$), the above (2.3.26) reads

$$\mathbb{E}\Big[\exp\Big(\frac{1}{2}\int_0^T \sigma(s)^2 ds\Big)\Big] < \infty. \tag{2.3.27}$$

One can show that if in the above, $\frac{1}{2}$ is replaced by a smaller number, then $\widehat{\mathcal{E}}\{\sigma\}$ might not be a martingale.

To conclude this section, let us consider the following FSDE (see (1.2.1))

$$\begin{cases} dX(t) = b(t, X(t))dt + \sigma(t, X(t))dW(t), & t \geqslant S \geqslant 0, \\ X(S) = x, \end{cases} \tag{2.3.28}$$

on a complete filtered probability space $(\Omega, \mathcal{F}, \mathbb{F}, \mathbb{P})$ with $\mathbb{F} = \{\mathcal{F}_t\}_{t\geqslant 0}$ being natural filtration generated by the d-dimensional standard Brownian motion augmented by all the $\mathbb{P}$-null sets in $\mathcal{F}$. We introduce the following hypothesis.

(H2.3.1) Let $b : [0, T] \times \mathbb{R}^n \times \Omega \to \mathbb{R}^n$, $\sigma : [0, T] \times \mathbb{R}^n \times \Omega \to \mathbb{R}^{n\times d}$ be measurable and $t \mapsto (b(t, x), \sigma(t, x))$ be $\mathbb{F}$-progressively measurable, satisfying the following conditions:

$$\begin{aligned}&|b(t,0)|,\ |\sigma(t,0)| \leqslant K, \qquad t\in[0,T],\\&|b(t,x)-b(t,x')|,\ |\sigma(t,x)-\sigma(t,x')| \leqslant K|x-x'|,\\&\qquad\qquad\qquad t\in[0,T],\ x,x'\in\mathbb{R}^n,\end{aligned}\tag{2.3.29}$$

for some uniform constant $K > 0$.

The following result is standard whose proof follows from a fixed point theorem argument.

Proposition 2.3.11 *Let (H2.3.1) hold. Then, for any $x \in \mathbb{R}^n$,* (2.3.28) *admits a unique solution $X(\cdot) \equiv X(\cdot\,; S, x)$, and the following holds (with $p \geqslant 1$):*

$$\begin{aligned}\mathbb{E}\Big[\sup_{t\in[S,T]}|X(t)|^p\Big] \leqslant K\mathbb{E}\Big[|x|^p + \Big(\int_S^T |b(t,0)|dt\Big)^p\\ +\Big(\int_S^T |\sigma(t,0)|^2dt\Big)^{\frac{p}{2}}\Big] \leqslant K(1+|x|^p).\end{aligned}\tag{2.3.30}$$

If $\bar{X}(\cdot) \equiv X(\cdot\,; S, \bar{x})$ is the solution of (2.3.28) *corresponding to the similar maps $\bar{b}(\cdot), \bar{\sigma}(\cdot)$, then the following stability estimate holds:*

$$\begin{aligned}&\mathbb{E}\Big[\sup_{t\in[S,T]}|X(t)-\bar{X}(t)|^p\Big]\\&\leqslant K\mathbb{E}\Big[|x-\bar{x}|^p + \Big(\int_S^T |b(t,X(t))-\bar{b}(t,X(t))|dt\Big)^p\\&\qquad +\Big(\int_S^T |\sigma(t,X(t))-\bar{\sigma}(t,X(t))|^2dt\Big)^{\frac{p}{2}}\Big].\end{aligned}\tag{2.3.31}$$

Conditions (2.3.29) can be slightly relaxed. See $(\text{H3.4.2})_p$ in Chap. 3.

4 Some More Spaces

We now introduce some more spaces, besides $L^p_{\mathcal{F}_S}(\Omega; \mathbb{H})$, $L^p_{\mathbb{F}}(\Omega; L^2(S, T; \mathbb{H}))$ (with $\mathbb{H}$ being some Euclidean space) introduced earlier, which will be frequently used in the sequel. To this end, we denote by $\mathcal{M}$ the sub-σ-field of $\mathcal{F}$. First time readers need only roughly go through this section, skipping some delicate conclusions and examples.

Recall that $(\Omega, \mathcal{F}, \mathbb{F}, \mathbb{P})$ is a fixed complete filtered probability space, and $W(\cdot)$ is a d-dimensional standard Brownian motion with its natural filtration $\mathbb{F} \equiv \{\mathcal{F}_t\}_{t \geqslant 0}$. For any $p, q \in [1, \infty)$ and Euclidean space $\mathbb{H}$ (say, $\mathbb{R}^m$, $\mathbb{R}^{m\times d}$, etc.) whose norm is denoted by $|\cdot|$, we define (without requiring the $\mathbb{F}$-adaptiveness and taking $\mathcal{M}$ to be $\mathcal{F}_S$.)

$$
\begin{aligned}
L^p_{\mathcal{M}}(\Omega; L^q(0,T;\mathbb{H})) = \Big\{ \varphi : (0,T)\times\Omega \to \mathbb{H} \;\Big|\; & \varphi(\cdot) \text{ is } \mathcal{B}([0,T])\otimes\mathcal{M}\text{-measurable}, \\
& \mathbb{E}\Big(\int_0^T |\varphi(t)|^q dt\Big)^{\frac{p}{q}} < \infty \Big\}, \\
L^q(0,T; L^p_{\mathcal{M}}(\Omega;\mathbb{H})) = \Big\{ \varphi : (0,T)\times\Omega \to \mathbb{H} \;\Big|\; & \varphi(\cdot) \text{ is } \mathcal{B}([0,T])\otimes\mathcal{M}\text{-measurable}, \\
& \int_0^T \Big(\mathbb{E}|\varphi(t)|^p\Big)^{\frac{q}{p}} dt < \infty \Big\}.
\end{aligned}
$$

The spaces $L^\infty_{\mathcal{M}}(\Omega; L^q(0,T;\mathbb{H}))$, $L^p_{\mathcal{M}}(\Omega; L^\infty(0,T;\mathbb{H}))$, $L^\infty_{\mathcal{M}}(\Omega; L^\infty(0,T;\mathbb{H}))$, $L^\infty(0,T; L^p_{\mathcal{M}}(\Omega;\mathbb{H}))$, $L^q(0,T; L^\infty_{\mathcal{M}}(\Omega;\mathbb{H}))$, and $L^\infty(0,T; L^\infty_{\mathcal{M}}(\Omega;\mathbb{H}))$ can be defined in an obvious way. We identify

$$
\begin{aligned}
L^p_{\mathcal{M}}(\Omega; L^p(0,T;\mathbb{H})) &= L^p(0,T; L^p_{\mathcal{M}}(\Omega;\mathbb{H})) \equiv L^p_{\mathcal{M}}(0,T;\mathbb{H}) \\
&= L^p_{\mathcal{B}[0,T]\otimes\mathcal{M}}((0,T)\times\Omega;\mathbb{H}), \qquad p\in[1,\infty).
\end{aligned}
\tag{2.4.1}
$$

In the above, we have used Remark 2.3.1, concerning the measurability of the processes. Also, for convenience, we always (or only) consider those processes that the double integral exists so that the Fubini theorem is applicable when needed. Besides the above, for the case $p=\infty$, we have to be a little more careful since, in general,

$$
\begin{aligned}
L^\infty_{\mathcal{M}}(\Omega; L^\infty(0,T;\mathbb{H})) &= L^\infty(0,T; L^\infty_{\mathcal{M}}(\Omega;\mathbb{H})) \equiv L^\infty_{\mathcal{M}}(0,T;\mathbb{H}) \\
&\subsetneqq L^\infty_{\mathcal{B}[0,T]\otimes\mathcal{M}}((0,T)\times\Omega;\mathbb{H}).
\end{aligned}
\tag{2.4.2}
$$

That is, we have the last strict inclusion. This can be illustrated by the following example.

Example 2.4.1 Let $\Omega = [0,1]$ and $T=1$. Let

$$
f(t,\omega) = \mathbf{1}_{[0,t]}(\omega), \qquad t\in[0,1], \quad \omega\in[0,1].
$$

Clearly, $f \in L^\infty((0,1)\times\Omega;\mathbb{R})$. However, the range of the map $t\mapsto f(t)$ is not essentially separable. Pettis theorem says that a Banach space-valued function is strongly measurable if and only it is weakly measurable and almost separably valued. Hence, $f \notin L^\infty(0,1; L^\infty(\Omega;\mathbb{R}))$.

In what follows, to avoid technicality, we make the following convention.

Convention 2 For any map $(t,\omega)\mapsto f(t,\omega)$, which is $\mathcal{B}[0,T]\otimes\mathcal{F}_T$-measurable, we always assume that both $t\mapsto f(t,\cdot)$ and $\omega\mapsto f(\cdot,\omega)$ are (Bochner) strong measurable.

Next, for each $p\in[1,\infty]$, we define

$$\begin{aligned}
&C([0,T];L^p_{\mathcal{M}}(\Omega;\mathbb{H}))\\
&\quad=\Big\{\varphi(\cdot)\in L^\infty(0,T;L^p_{\mathcal{M}}(\Omega;\mathbb{H}))\;\Big|\;\varphi(t)\text{ is }\mathcal{M}\text{-measurable, }\forall t\in[0,T],\\
&\qquad\varphi(\cdot)\text{ is continuous from }[0,T]\text{ to }L^p_{\mathcal{M}}(\Omega;\mathbb{H}),\quad\sup_{t\in[0,T]}\mathbb{E}|\varphi(t)|^p<\infty\Big\},\\
&C^\#([0,T];L^p_{\mathcal{M}}(\Omega;\mathbb{H}))\\
&\quad=\Big\{\varphi(\cdot)\in C([0,T];L^p_{\mathcal{M}}(\Omega;\mathbb{H}))\;\Big|\;\varphi(\cdot)\text{ has continuous paths a.s.}\Big\},\\
&L^p_{\mathcal{M}}(\Omega;C([0,T];\mathbb{H}))\\
&\quad=\Big\{\varphi(\cdot)\in C^\#([0,T];L^p_{\mathcal{M}}(\Omega;\mathbb{H}))\;\Big|\;E\Big[\sup_{t\in[0,T]}|\varphi(t)|^p\Big]<\infty\Big\}.
\end{aligned}$$

We emphasize here that processes in the above spaces are not necessarily $\mathbb{F}$-adapted (merely $\mathcal{M}$-measurable).

Let us present the following lemma which is interesting by itself and will be useful below.

Lemma 2.4.2 *Let $(\Omega_1,\mathcal{F}_1,\mu_1)$ and $(\Omega_2,\mathcal{F}_2,\mu_2)$ be two finite measure spaces. Let $1\leqslant p_1<p_2<\infty$. Then*

$$L^{p_1}(\Omega_1;L^{p_2}(\Omega_2))\subseteq L^{p_2}(\Omega_2;L^{p_1}(\Omega_1)),\tag{2.4.3}$$

and for any $\mathcal{F}_1\otimes\mathcal{F}_2$-measurable bounded function $f:\Omega_1\times\Omega_2\to\mathbb{R}$, it holds

$$\begin{aligned}
&\Big[\int_{\Omega_2}\Big(\int_{\Omega_1}|f(\omega_1,\omega_2)|^{p_1}\mu_1(d\omega_1)\Big)^{\frac{p_2}{p_1}}\mu_2(d\omega_2)\Big]^{\frac{1}{p_2}}\\
&\leqslant\Big[\int_{\Omega_1}\Big(\int_{\Omega_2}|f(\omega_1,\omega_2)|^{p_2}\mu_2(d\omega_2)\Big)^{\frac{p_1}{p_2}}\mu_1(d\omega_1)\Big]^{\frac{1}{p_1}}.
\end{aligned}\tag{2.4.4}$$

In the case that $p_2=\infty$, as long as $\omega_1\mapsto f(\omega_1,\cdot)$ and $\omega_2\mapsto f(\cdot,\omega_2)$ are strongly measurable, it holds

$$\operatorname*{esssup}_{\omega_2\in\Omega_2}\Big(\int_{\Omega_1}|f(\omega_1,\omega_2)|^{p_1}\mu_1(d\omega_1)\Big)^{\frac{1}{p_1}}\leqslant\Big[\int_{\Omega}\operatorname*{esssup}_{\omega_2\in\Omega_2}|f(\omega_1,\omega_2)|^{p_1}\mu_1(d\omega_1)\Big]^{\frac{1}{p_1}}.\tag{2.4.5}$$

Proof Let us carry out the following calculation, making use of Hölder's inequality, suppressing ω_1 and ω_2.

$$
\begin{aligned}
&\int_{\Omega_2}\Big(\int_{\Omega_1}|f|^{p_1}d\mu_1\Big)^{\frac{p_2}{p_1}}d\mu_2=\int_{\Omega_2}\Big[\Big(\int_{\Omega_1}|f|^{p_1}d\mu_1\Big)^{\frac{p_2}{p_1}-1}\int_{\Omega_1}|f|^{p_1}d\mu_1\Big]d\mu_2\\
&=\int_{\Omega_2}\Big[\int_{\Omega_1}\Big(\int_{\Omega_1}|f|^{p_1}d\mu_1\Big)^{\frac{p_2-p_1}{p_1}}|f|^{p_1}d\mu_1\Big]d\mu_2\\
&=\int_{\Omega_1}\Big[\int_{\Omega_2}\Big(\int_{\Omega_1}|f|^{p_1}d\mu_1\Big)^{\frac{p_2-p_1}{p_1}}|f|^{p_1}d\mu_2\Big]d\mu_1\\
&\leqslant\int_{\Omega_1}\Big\{\Big[\int_{\Omega_2}\Big(\int_{\Omega_1}|f|^{p_1}d\mu_1\Big)^{\frac{p_2}{p_1}}d\mu_2\Big]^{\frac{p_2-p_1}{p_2}}\Big(\int_{\Omega_2}|f|^{p_2}d\mu_2\Big)^{\frac{p_1}{p_2}}\Big\}d\mu_1\\
&=\Big[\int_{\Omega_2}\Big(\int_{\Omega_1}|f|^{p_1}d\mu_1\Big)^{\frac{p_2}{p_1}}d\mu_2\Big]^{\frac{p_2-p_1}{p_2}}\int_{\Omega_1}\Big(\int_{\Omega_2}|f|^{p_2}d\mu_2\Big)^{\frac{p_1}{p_2}}d\mu_1.
\end{aligned}
$$

Then (2.4.4) follows.

The other relation is easy to obtain. □

We now present the following result which gives some relations among the spaces introduced above. We suppress the range space $\mathbb{H}$.

Proposition 2.4.3 *It holds*

$$
\begin{aligned}
&L^p_{\mathcal{M}}(\Omega;C([0,T]))\subseteq C^\#([0,T];L^p_{\mathcal{M}}(\Omega))\subseteq C([0,T];L^p_{\mathcal{M}}(\Omega))\\
&\subseteq L^\infty(0,T;L^p_{\mathcal{M}}(\Omega))\subseteq L^q(0,T;L^p_{\mathcal{M}}(\Omega))\\
&\subseteq L^p_{\mathcal{M}}(\Omega;L^q(0,T))\subseteq L^q_{\mathcal{M}}(0,T),\qquad 1\leqslant q\leqslant p\leqslant\infty,
\end{aligned}
\tag{2.4.6}
$$

and

$$
\begin{aligned}
&L^\infty_{\mathcal{M}}(\Omega;C([0,T]))=C^\#([0,T];L^\infty_{\mathcal{M}}(\Omega))=C([0,T];L^\infty_{\mathcal{M}}(\Omega))\\
&\subseteq L^\infty(0,T;L^\infty_{\mathcal{M}}(\Omega))\equiv L^\infty_{\mathcal{M}}(0,T)\equiv L^\infty_{\mathcal{M}}(\Omega;L^\infty(0,T))\\
&\subseteq L^\infty_{\mathcal{M}}(\Omega;L^q(0,T))\subseteq L^p_{\mathcal{M}}(\Omega;L^q(0,T))\\
&\subseteq L^q(0,T;L^p_{\mathcal{M}}(\Omega))\subseteq L^p_{\mathcal{M}}(0,T),\qquad 1\leqslant p\leqslant q\leqslant\infty.
\end{aligned}
\tag{2.4.7}
$$

Moreover, for $p\in[1,\infty)$,

$$
L^p_{\mathcal{M}}\big(\Omega;C([0,T])\big)\neq C^\#\big([0,T];L^p_{\mathcal{M}}(\Omega)\big)\neq C\big([0,T];L^p_{\mathcal{M}}(\Omega)\big).
\tag{2.4.8}
$$

Proof Using Lemma 2.4.2, we are able to prove the following:

$$
\begin{aligned}
&L^q(0,T;L^p_{\mathcal{M}}(\Omega))\subseteq L^p_{\mathcal{M}}(\Omega;L^q(0,T)),\qquad 1\leqslant q\leqslant p\leqslant\infty,\\
&L^p_{\mathcal{M}}(\Omega;L^q(0,T))\subseteq L^q(0,T;L^p_{\mathcal{M}}(\Omega)),\qquad 1\leqslant p\leqslant q\leqslant\infty.
\end{aligned}
$$

Other inclusions can be proved easily. For (2.4.8), we need to construct some examples. Note that the case $p=\infty$ is avoided. We only look at the case $\mathcal{M}=\mathcal{F}_T$. To show the first, we need to find a process in $C^\#([0,T];L^p_{\mathcal{F}_T}(\Omega)\big)$ that is not in $L^p_{\mathcal{F}_T}\big(\Omega;C([0,T])\big)$. To this end, let $T=1$, $\Omega=[0,1]$ with $\mathcal{F}_1$ being the Lebesgue

σ-field $\mathbb{P}$ being the Lebesgue measure, and $p \in [1, \infty)$. Let

$$f(t) = t^{p+1}(1-t)^{p+1}, \qquad t \in [0, 1].$$

Then $f : [0, 1] \to [0, 1]$, with

$$\lim_{t\to 0} \frac{f(t)}{t^p} = 0, \qquad \lim_{t\to 1} \frac{f(t)}{(1-t)^p} = 0, \tag{2.4.9}$$

$$\begin{array}{ll} t - f(t) = t - t^{p+1}(1-t)^{p+1} > t - t^{p+1} > 0, & \\ t + f(t) = t + t^{p+1}(1-t)^{p+1} < t + (1-t)^{p+1} < 1, & \end{array} \quad t \in (0, 1),$$

which leads to

$$t^2 - f(t)^2 = \big(t - f(t)\big)\big(t + f(t)\big) > 0, \qquad t \in (0, 1).$$

Also,

$$\begin{aligned} [t - f(t)]'|_{t=0} &= [1 - (p+1)t^p(1-t)^p(2t-1)]|_{t=0} = 1, \\ [t + f(t)]'|_{t=1} &= [1 + (p+1)t^p(1-t)^p(1-2t)]|_{t=1} = 1. \end{aligned}$$

Thus, near $t = 0, 1$, function $t \pm f(t)$ is increasing. Now, we define

$$\varphi(t, \omega) = \begin{cases} 0, & \omega \in [0, t - f(t)] \cup [t + f(t), 1], \\ \dfrac{\omega - [t - f(t)]}{f(t)\omega}, & \omega \in (t - f(t), t], \\ \dfrac{t + f(t) - \omega}{f(t)\omega}, & \omega \in (t, t + f(t)]. \end{cases}$$

It is seen that $t \mapsto \varphi(t, \omega)$ is continuous for $\omega \neq 0$, and

$$0 \leqslant \varphi(t, \omega) \leqslant \frac{1}{\omega} = \varphi(t, t), \qquad \omega \in (t - f(t), t + f(t)). \tag{2.4.10}$$

Thus, for $0 < \delta \leqslant s < t \leqslant 1 - \delta$, for $\delta \in (0, 1)$, small,

$$\begin{aligned} \mathbb{E}|\varphi(t) - \varphi(s)|^p &= \int_0^1 |\varphi(t, \omega) - \varphi(s, \omega)|^p d\omega \\ &\leqslant \int_{\delta - f(\delta)}^{1-\delta+f(1-\delta)} |\varphi(t, \omega) - \varphi(s, \omega)|^p d\omega \to 0, \quad \text{as } 0 < t - s \to 0. \end{aligned}$$

The convergence follows from the dominated convergence theorem as (2.4.10) holds. This implies that $t \mapsto \mathbb{E}|\varphi(t)|^p$ is continuous on $(0, 1)$. On the other hand, for $t > 0$ small, the above implies (making use of Cauchy mean value theorem and (2.4.9))

$$
\begin{aligned}
\mathbb{E}|\varphi(t)|^p &= \int_0^1 |\varphi(t,\omega)|^p d\omega \leqslant \int_{t-f(t)}^{t+f(t)} \frac{1}{\omega^p} d\omega \\
&= \frac{1}{p-1}\Big(\frac{1}{[t-f(t)]^{p-1}} - \frac{1}{[t+f(t)]^{p-1}}\Big) = \frac{[t+f(t)]^{p-1}-[t-f(t)]^{p-1}}{(p-1)[t^2-f(t)^2]^{p-1}} \\
&\leqslant \frac{2f(t)}{[t^2-f(t)^2]^{p-1}}\int_0^1\Big([t-f(t)]+2\alpha f(t)\Big)^{p-2} d\alpha \\
&= \frac{2t^{p+1}(1-t)^{p+1}t^{p-2}}{t^{2p-2}[1-o(1)]^{p-1}}\int_0^1 [1-o(1)+\alpha o(1)]^{p-2} d\alpha \leqslant Kt.
\end{aligned}
$$

Hence, $t \mapsto \mathbb{E}|\varphi(t)|^p$ is continuous on $[0, 1]$. Then

$$
\sup_{t\in[0,1]} \mathbb{E}|\varphi(t)|^p < \infty.
$$

But

$$
\sup_{t\in[0,1]} |\varphi(t,\omega)| = |\varphi(t,t)| = \frac{1}{\omega}, \qquad \forall \omega \in (0, 1].
$$

Thus,

$$
\mathbb{E}\Big[\sup_{t\in[0,1]} |\varphi(t)|^p\Big] = \int_0^1 \frac{d\omega}{\omega^p} = \infty,
$$

which means that $\varphi(\cdot) \notin L^p_{\mathcal{F}_1}\big(\Omega; C([0,T])\big)$.

Now, for the second relation in (2.4.8), we note that any $\varphi(\cdot) \in C([0,T]; L^p_{\mathcal{M}}(\Omega))$ only has the continuity of $t \mapsto \varphi(t)$ as a map from $[0, T]$ to $L^p_{\mathcal{M}}(\Omega)$, in the following sense: For any $t \in [0, T]$,

$$
\lim_{s\to t} \mathbb{E}|\varphi(s) - \varphi(t)|^p = 0.
$$

Such a process might have no continuous paths, i.e., for any $\omega \in \Omega$, $t \mapsto \varphi(t,\omega)$ could be discontinuous. To see this, again let $T = 1$ and $\Omega = [0, 1]$ with $\mathcal{F}_1$ being the Lebesgue σ-field, and $\mathbb{P}$ being the Lebesgue measure. Define

$$
\varphi(t,\omega) = \mathbf{1}_{[0,t)}(\omega) = \mathbf{1}_{[\omega,1]}(t) = \begin{cases} 0, & t < \omega, \\ 1, & t \geqslant \omega. \end{cases} \tag{2.4.11}
$$

Then for any $\omega \in \Omega$, the path $t \mapsto \varphi(t,\omega)$ has a jump at $t=\omega$. On the other hand, for any given $t \in (0,1)$, one has

$$\mathbb{E}|\varphi(s)-\varphi(t)|^p = \int_0^1 |\mathbf{1}_{[0,s)}(\omega) - \mathbf{1}_{[0,t)}(\omega)|^p d\mathbb{P}(\omega) = |s-t| \to 0,$$

as $s \to t$. Therefore, the above $\varphi(\cdot)$ satisfies

$$\varphi(\cdot) \in C([0,1]; L^p_{\mathcal{F}_1}(\Omega)) \setminus C^\#([0,1]; L^p_{\mathcal{F}_1}(\Omega)).$$

This completes the proof. □

Let us make a further observation by presenting an example which will show that for a given $T > 0$, space $C^\#([0,T]; L^p_{\mathcal{F}_T}(\Omega))$ is not complete under the norm

$$\|\varphi(\cdot)\|_{C([0,T];L^p_{\mathcal{F}_T}(\Omega))} \equiv \Big\{ \sup_{t\in[0,T]} \mathbb{E}|\varphi(t)|^p \Big\}^{\frac{1}{p}}. \tag{2.4.12}$$

Example 2.4.4 For any $k \geqslant 1$, if we define on $[0,1]^2$ the following

$$\varphi_k(t,\omega) = \begin{cases} 0, & 0 \leqslant t < \omega - \frac{1}{k}, \\ \dfrac{t-(\omega-\frac{1}{k})}{\frac{1}{k}}, & \omega - \frac{1}{k} \leqslant t < \omega, \\ 1, & \omega \leqslant t \leqslant 1, \end{cases}$$

then $\varphi_k(\cdot) \in C^\#([0,1]; L^\infty_{\mathcal{F}_1}(\Omega))$. See the Fig. 2.1: Between the blue and red lines, they are linearly continuously linked. Moreover, with $\varphi(\cdot)$ defined by (2.4.11),

$$\begin{aligned}\mathbb{E}|\varphi_k(t)-\varphi(t)|^p &= \int_t^{(t+\frac{1}{k})\wedge 1} \Big|\frac{t-\omega+\frac{1}{k}}{\frac{1}{k}}\Big|^p d\omega \\ &= k^p \int_t^{(t+\frac{1}{k})\wedge 1} \Big(t+\frac{1}{k}-\omega\Big)^p d\omega \leqslant \frac{1}{(p+1)k} \leqslant \frac{1}{k},\end{aligned}$$

which yields

$$\sup_{t\in[0,1]} \mathbb{E}|\varphi_k(t)-\varphi(t)|^p \leqslant \frac{1}{k}.$$

Thus, we find a sequence $\varphi_k(\cdot) \in C^\#([0,1]; L^\infty_{\mathcal{F}_1}(\Omega))$ which is convergent to $\varphi(\cdot) \notin C^\#([0,1]; L^\infty_{\mathcal{F}_1}(\Omega))$. This proves our assertion.

Actually, the completion of $C^\#([0,T]; L^p_{\mathcal{M}}(\Omega))$ is $C([0,T]; L^p_{\mathcal{M}}(\Omega))$.

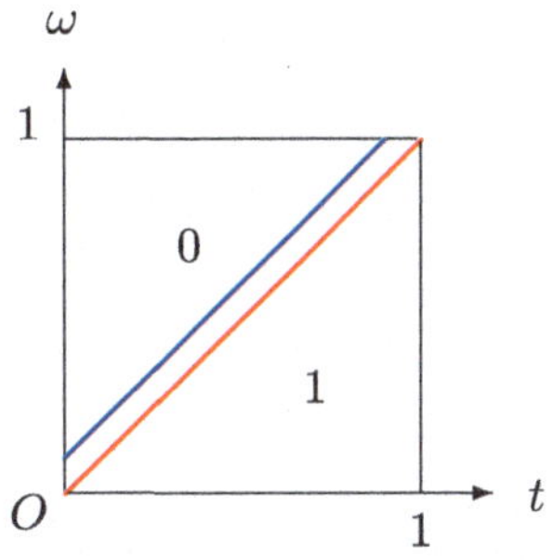

Fig. 2.1 Domain on $[0,1]^2$

We now introduce spaces involving $\mathbb{F}$-adaptiveness: For $p, q \in [1, \infty]$,

$$\begin{aligned}
&L^p_{\mathbb{F}}(\Omega; L^q(0,T)) = \big\{\varphi(\cdot) \in L^p(\Omega; L^q(0,T)) \bigm| \varphi(\cdot) \text{ is } \mathbb{F}\text{-adapted}\big\},\\
&L^q_{\mathbb{F}}(0,T; L^p(\Omega)) = \big\{\varphi(\cdot) \in L^q(0,T; L^p(\Omega)) \bigm| \varphi(\cdot) \text{ is } \mathbb{F}\text{-adapted}\big\},\\
&C_{\mathbb{F}}([0,T]; L^p(\Omega)) = \big\{\varphi(\cdot) \in C([0,T]; L^p(\Omega)) \bigm| \varphi(\cdot) \text{ is } \mathbb{F}\text{-adapted}\big\},\\
&C^{\#}_{\mathbb{F}}([0,T]; L^p(\Omega)) = \big\{\varphi(\cdot) \in C^{\#}([0,T]; L^p(\Omega)) \bigm| \varphi(\cdot) \text{ is } \mathbb{F}\text{-adapted}\big\},\\
&L^p_{\mathbb{F}}(\Omega; C([0,T])) = \big\{\varphi(\cdot) \in L^p(\Omega; C([0,T])) \bigm| \varphi(\cdot) \text{ is } \mathbb{F}\text{-adapted}\big\}.
\end{aligned}$$

Also, we identify

$$L^p_{\mathbb{F}}(\Omega; L^p(0,T)) = L^p_{\mathbb{F}}(0,T; L^p(\Omega)) \equiv L^p_{\mathbb{F}}(0,T), \qquad p \in [1,\infty].$$

Similar to the definitions of the spaces without adaptiveness, in the above, we have also suppressed the range space $\mathbb{H}$ in the notations, for simplicity. When the range space $\mathbb{H}$ needs to be emphasized, we will use the notations $L^p_{\mathbb{F}}(\Omega; L^q(0,T;\mathbb{H}))$, $C_{\mathbb{F}}([0,T]; L^p(\Omega;\mathbb{H}))$, and so on. Further, results similar to Proposition 2.4.3 hold for the above spaces.

We now look at the following process:

$$X(t) = \int_0^t \sigma(s)dW(s), \qquad t \in [0,T], \tag{2.4.13}$$

with $\sigma(\cdot) \in L^p_{\mathbb{F}}(\Omega; L^{2+\varepsilon}(0,T;\mathbb{R}^{m\times d}))$, for some $p > 2$ and $\varepsilon > \frac{4}{p-2}$ which is equivalent to $\frac{\varepsilon p}{2(2+\varepsilon)} > 1$. Then, for any $0 \leqslant r < t \leqslant T$, one has, by Burkholder-Davis–Gundy's inequalities,

$$\begin{aligned}
&\mathbb{E}|X(t) - X(r)|^p = \mathbb{E}\Big|\int_r^t \sigma(s)dW(s)\Big|^p \leqslant \mathbb{E}\Big[\sup_{\tau\in[r,t]}\Big|\int_r^\tau \sigma(s)dW(s)\Big|^p\Big]\\
&\leqslant \overline{K}_p\mathbb{E}\Big(\int_r^t |\sigma(s)|^2 ds\Big)^{\frac{p}{2}} \leqslant \overline{K}_p\mathbb{E}\Big(\int_r^t |\sigma(s)|^{2+\varepsilon} ds\Big)^{\frac{p}{2+\varepsilon}} (t-r)^{\frac{\varepsilon p}{2(2+\varepsilon)}}.
\end{aligned}$$

Thus, by Komogorov's theorem (Proposition 2.1.11), we have continuous paths of $t \mapsto X(t)$. The following result significantly improves such a result.

Proposition 2.4.5 *Let* $p \geqslant 1$ *and* $\sigma(\cdot) \in L^p_{\mathbb{F}}(\Omega; L^2(0,T;\mathbb{R}^{m\times d}))$. *Let* $X(\cdot)$ *be defined by* (2.4.13). *Then* $X(\cdot) \in L^p_{\mathbb{F}}(\Omega; C([0,T];\mathbb{R}^m))$.

Proof For any $k > 0$, let

$$\sigma_k(s) = \begin{cases} -k, & \sigma(s) < -k, \\ \sigma(s), & |\sigma(s)| \leqslant k, \\ k, & \sigma(s) > k. \end{cases}$$

Define

$$X_k(t) = \int_0^t \sigma_k(s)dW(s), \qquad t \in [0,T].$$

Since $\sigma_k(\cdot)$ is bounded, $X_k(\cdot) \in L^p_{\mathbb{F}}(\Omega; L^2(0,T;\mathbb{R}^m))$. Now,

$$\begin{aligned}&\mathbb{E}\Big[\sup_{t\in[0,T]} |X_k(t) - X(t)|^p\Big] = \mathbb{E}\Big[\sup_{t\in[0,T]} \Big|\int_0^t \Big(\sigma_k(s) - \sigma(s)\Big)dW(s)\Big|^p\Big] \\ &\leqslant \overline{K}_p \mathbb{E}\Big(\int_0^T |\sigma_k(s) - \sigma(s)|^2 ds\Big)^{\frac{p}{2}} \to 0, \qquad k \to \infty.\end{aligned}$$

Hence, our conclusion follows. □

Next, we introduce some additional spaces. First, for any $p, q \geqslant 1$, let $L^q\Big(0,T; L^p_{\mathbb{F}}(\Omega; L^2(0,T;\mathbb{H}))\Big)$ be the set of all processes $Z : [0,T]^2 \times \Omega \to \mathbb{H}$ such that for almost all $t \in [0,T]$, $Z(t,\cdot) \in L^p_{\mathbb{F}}\big(\Omega; L^2(0,T;\mathbb{H})\big)$ satisfying

$$\int_0^T \Big[\mathbb{E}\Big(\int_0^T |Z(t,s)|^2 ds\Big)^{\frac{p}{2}}\Big]^{\frac{q}{p}} dt < \infty. \tag{2.4.14}$$

The introduction of such a space is motivated by the following: For any $Z(\cdot\,,\cdot) \in L^q\big(0,T; L^p_{\mathbb{F}}(\Omega; L^2(0,T;\mathbb{R}^{m\times d}))\big)$, both of the following are well-defined:

$$\int_0^t Z(t,s)dW(s), \qquad \int_t^T Z(t,s)dW(s), \tag{2.4.15}$$

and they are in $L^q_{\mathbb{F}}\big(0,T; L^p(\Omega;\mathbb{R}^m)\big)$ and $L^q_{\mathcal{F}_T}\big(0,T; L^p(\Omega;\mathbb{R}^m)\big)$, respectively. Moreover,

$$\begin{aligned}&\int_0^T \Big[\Big(\mathbb{E}\Big|\int_0^t Z(t,s)dW(s)\Big|^p\Big)^{\frac{q}{p}} + \Big(\mathbb{E}\Big|\int_t^T Z(t,s)dW(s)\Big|^p\Big)^{\frac{q}{p}}\Big]dt \\ &\leqslant K \int_0^T \Big[\mathbb{E}\Big(\sup_{r\in[0,T]} \Big|\int_0^r Z(t,s)dW(s)\Big|^p\Big)\Big]^{\frac{q}{p}} dt \\ &\leqslant K \int_0^T \Big[\mathbb{E}\Big(\int_0^T |Z(t,s)|^2 ds\Big)^{\frac{p}{2}}\Big]^{\frac{q}{p}} dt < \infty.\end{aligned}$$

With the similar spirit, we can define the following spaces in an obvious way:

$$L^\infty\big(0,T;L^p_{\mathbb F}(\Omega;L^2(0,T;\mathbb H))\big),\qquad C\big([0,T];L^p_{\mathbb F}(\Omega;L^2(0,T;\mathbb H))\big),$$
$$L^q\big(0,T;L^\infty_{\mathbb F}(\Omega;L^2(0,T;\mathbb H))\big),\qquad C\big([0,T];L^p_{\mathbb F}(\Omega;C([0,T];\mathbb H))\big).$$

We may further refine these spaces. Let $p\geqslant 2$,

$$\begin{aligned}&L^p_{\mathbb F}\big(\Omega;\widehat L^\infty(0,T;L^2(0,T;\mathbb R^{m\times d}))\big)\\&=\Big\{Z(\cdot\,,\cdot)\in L^2((0,T)\times\Omega;\mathbb R^m)\ \big|\ Z(t,\cdot)\in L^2_{\mathbb F}(0,T;\mathbb R^{m\times d}),\ \text{a.e. } t\in[0,T]\\&\qquad\mathbb E\Big[\sup_{t\in[0,T]}\Big|\int_0^t Z(t,s)dW(s)\Big|^p\Big]\\&\qquad\quad+\mathbb E\Big[\sup_{t\in[0,T]}\Big|\int_t^T Z(t,s)dW(s)\Big|^p\Big]<\infty\Big\}.\end{aligned}$$

Also, let

$$\begin{aligned}&L^p_{\mathbb F}\big(\Omega;C([0,T];L^2(0,T;\mathbb R^{m\times d}))\big)\\&=\Big\{Z(\cdot\,,\cdot)\in L^p_{\mathbb F}\big(\Omega;\widehat L^\infty(0,T;L^2(0,T;\mathbb R^{m\times d}))\big)\ \big|\ Z(t,\cdot)\text{ being}\\&\qquad\qquad\text{continuous from }[0,T]\text{ to }L^2_{\mathcal F_T}(0,T;\mathbb R^{m\times d}),\ \text{almost surely}\Big\}.\end{aligned}$$

$$\begin{aligned}&L^p_{\mathbb F}\big(\Omega;C^\#([0,T];L^2(0,T;\mathbb R^{m\times d}))\big)\\&=\Big\{Z(\cdot\,,\cdot)\in L^p_{\mathbb F}\big(\Omega;\widehat L^\infty(0,T;L^2(0,T;\mathbb R^{m\times d}))\big)\ \big|\\&\qquad s\mapsto Z(t,s)\text{ being }\mathbb F\text{-adapted on }[0,T],\ t\in[0,T],\\&\qquad t\mapsto\int_0^T Z(t,s)dW(s)\text{ having continuous paths, almost surely}\Big\},\end{aligned}$$

and

$$\begin{aligned}&L^p_{\mathbb F}\big(\Omega;\widehat C([0,T];L^2(0,T;\mathbb R^{m\times d}))\big)\\&=\Big\{Z(\cdot\,,\cdot)\in L^p_{\mathbb F}\big(\Omega;\widehat L^\infty(0,T;L^2(0,T;\mathbb R^{m\times d}))\big)\ \big|\ \int_0^t Z(t,s)dW(s)\text{ and}\\&\qquad\qquad\int_t^T Z(t,s)dW(s)\quad\text{have continuous paths, almost surely}\Big\}.\end{aligned}$$

Clearly,

$$\begin{aligned}&L^p_{\mathbb F}\big(\Omega;\widehat C\big([0,T];L^2(0,T;\mathbb R^{m\times d})\big)\big)\subseteq L^p_{\mathbb F}\big(\Omega;C^\#([0,T];L^2(0,T;\mathbb R^{m\times d}))\big)\\&\subseteq L^p_{\mathbb F}\big(\Omega;C([0,T];L^2(0,T;\mathbb R^{m\times d}))\big)\subseteq L^p_{\mathbb F}\big(\Omega;\widehat L^\infty(0,T;L^2(0,T;\mathbb R^{m\times d}))\big)\\&\subseteq L^2((0,T)\times\Omega;\mathbb R^m).\end{aligned}$$

Note, for any $Z(\cdot\,,\cdot)\in L^p_{\mathbb{F}}(\Omega;\widehat{L}^\infty(0,T;L^2(0,T;\mathbb{R}^{m\times d})))$,

$$\begin{aligned}
&\mathbb{E}\Big[\sup_{t\in[0,T]}\Big|\int_0^t Z(t,s)dW(s)\Big|^p\Big]+\mathbb{E}\Big[\sup_{t\in[0,T]}\Big|\int_t^T Z(t,s)dW(s)\Big|^p\Big]\\
&\geqslant\mathbb{E}\Big[\sup_{t\in[0,T]}\Big(\Big|\int_0^t Z(t,s)dW(s)\Big|^p+\Big|\int_t^T Z(t,s)dW(s)\Big|^p\Big)\Big]\\
&\geqslant 2^{1-p}\mathbb{E}\Big[\sup_{t\in[0,T]}\Big|\int_0^T Z(t,s)dW(s)\Big|^p\Big]\\
&\geqslant 2^{1-p}\sup_{t\in[0,T]}\mathbb{E}\Big|\int_0^T Z(t,s)dW(s)\Big|^p\geqslant\frac{2^{1-p}}{K_p}\sup_{t\in[0,T]}\mathbb{E}\Big(\int_0^T|Z(t,s)|^2ds\Big)^{\frac{p}{2}}.
\end{aligned}$$

In the above, we have used martingale moment inequality (Theorem 2.3.2 (i)) to obtain the last inequality. The above shows that

$$L^p_{\mathbb{F}}(\Omega;\widehat{L}^\infty(0,T;L^2(0,T;\mathbb{R}^{m\times d})))\subseteq L^\infty(0,T;L^p_{\mathbb{F}}(\Omega;L^2(0,T;\mathbb{R}^{m\times d}))).$$

On the other hand, it is clear that

$$L^p_{\mathbb{F}}(\Omega;C([0,T];L^2(0,T;\mathbb{R}^{m\times d})))\subseteq L^\infty(0,T;L^p_{\mathbb{F}}(\Omega;L^2(0,T;\mathbb{R}^{m\times d}))).$$

It is not clear to us if the following two spaces are the same:

$$L^p_{\mathbb{F}}(\Omega;\widehat{C}([0,T];L^2(0,T;\mathbb{R}^{m\times d}))),\qquad L^p_{\mathbb{F}}(\Omega;C^\#([0,T];L^2(0,T;\mathbb{R}^{m\times d}))).$$

Recall the definition of upper triangle domain $\Delta^*[0,T]$ (see (1.1.11) of Chap. 1). We let

$$\begin{aligned}
L^q(0,T;L^p_{\mathbb{F}}(\Omega;L^2(\cdot\,,T;\mathbb{H})))=\Big\{&Z:\Delta^*[0,T]\times\Omega\to\mathbb{H}\;\Big|\\
&Z(t,\cdot)\in L^p_{\mathbb{F}}(\Omega;L^2(t,T;\mathbb{H})),\ \text{a.e. } t\in[0,T],\\
&\int_0^T\Big[\mathbb{E}\Big(\int_t^T|Z(t,s)|^2ds\Big)^{\frac{p}{2}}\Big]^{\frac{q}{p}}dt<\infty\Big\}.
\end{aligned}$$

The spaces $L^\infty(0,T;L^p_{\mathbb{F}}(\Omega;L^2(\cdot\,,T;\mathbb{H})))$, $C([0,T];L^p_{\mathbb{F}}(\Omega;L^2(\cdot\,,T;\mathbb{H})))$, and $L^q(0,T;L^\infty_{\mathbb{F}}(\Omega;L^2(\cdot\,,T;\mathbb{H})))$ can be defined similarly. We denote ($q=p$)

$$L^p_{\mathbb{F}}(\Omega;L^2(\Delta^*[0,T];\mathbb{H}))\equiv L^p(0,T;L^p_{\mathbb{F}}(\Omega;L^2(\cdot\,,T;\mathbb{H}))).$$

Furthermore, if $p=2$, we denote by

$$L^2_{\mathbb{F}}(\Delta^*[0,T];\mathbb{H})\equiv L^2_{\mathbb{F}}(\Omega;L^2(\Delta^*[0,T];\mathbb{H})).$$

Moreover, one can define the analogue spaces when Δ^* is replaced by Δ_* (see (1.1.8) in Chap. 1). We refer to Sect. 3, Chap. 10, for more details. Finally, we point out that in all the definitions of the relevant spaces above, $[0, T]$ can be replaced by any $[S_1, S_2]$ with $0 \leqslant S_1 < S_2 \leqslant T$.

5 Bibliographical Remarks

Most of the material presented in this chapter are standard, which can be found in, say, Karatzas–Shreve [120], Oksendal [153], and Yong–Zhou [223]. More concretely, Proposition 2.1.13 (the debut theorem) is taken from [57] (see [16] for an elementary proof). Theorem 2.3.2 is essentially taken from [120] pp. 163–166. Theorem 2.3.6 is based on [120], Theorem 7.1, and the material follows right after (see pp. 217–220 of that book). We presented a strengthened version of Kolmogorov continuity theorem which will be useful later and whose detailed proof will be presented in Chap. 12. Constants in (2.3.9) are taken from [98], p. 285. Example 2.4.1 (see Remark 3.1 also) is taken from Roubicek [173] (Example 1.42); such a situation was first noticed by Fattorini ([72], Example 5.0.10). The used Pettis theorem can be found in [102], p. 72. Lemma 4.2 is taken from [140]. Davis [53] had some very interesting discussion on $[M]$ and $\langle M \rangle$. See also [61] for the quadratic variations of martingales.

Chapter 3
BSDEs and FSVIEs

In this chapter, we will briefly present some general theory on BSDEs and FSVIEs. No previous knowledge of these equations are assumed.

1 BSDEs: A General Consideration

Let us consider the following integral form of BSDE (see Chap. 1, (1.2.6)):

$$Y(t) = \xi + \int_t^T g(s, Y(s), Z(s))ds - \int_t^T Z(s)dW(s), \qquad t \in [0, T], \tag{3.1.1}$$

where $\xi \in L^p_{\mathcal{F}_T}(\Omega; \mathbb{R}^m)$ $(p > 1)$ and $g : [0, T] \times \mathbb{R}^m \times \mathbb{R}^{m\times d} \times \Omega \to \mathbb{R}^m$ are called the *terminal condition* and the *generator*, respectively. A pair $(Y(\cdot), Z(\cdot))$ of processes is called an *adapted solution* to BSDE (3.1.1) if they are $\mathbb{F}$-adapted and (3.1.1) is satisfied in the usual Itô's sense. The key points are the following: (i) Both the processes $Y(\cdot)$ and $Z(\cdot)$ are required to be $\mathbb{F}$-adapted, and (ii) a pair of the processes consists of an adapted solution. The adaptiveness gives an extra restriction so that the existence of the "solution" becomes harder, whereas an additional process $Z(\cdot)$ makes the existence of a "solution" easier, but this makes the uniqueness problem harder. It turns out that, under reasonable conditions, our BSDEs are well-posed. Before going further, let us present the following examples.

Example 3.1.1 Consider the following BSDE:

$$Y(t) = \xi - \int_t^T Z(s)dW(s), \qquad t \in [0, T], \tag{3.1.2}$$

T. Wang, J. Yong, *Backward Stochastic Volterra Integral Equations*,
Probability Theory and Stochastic Modelling 109,
https://doi.org/10.1007/978-3-032-20848-4_3

with $\xi \in L^2_{\mathcal{F}_T}(\Omega;\mathbb{R})$. This is a special case of BSDE (3.1.1) in which $g(\cdot) = 0$. It is clear that

$$Y(t) = \mathbb{E}_t[\xi], \qquad t \in [0, T],$$

which is a martingale. Hence, by martingale representation theorem (Chap. 2, Theorem 2.3.8), there exists a unique $Z(\cdot) \in L^2_{\mathbb{F}}(0, T;\mathbb{R})$ such that

$$Y(t) \equiv \mathbb{E}_t[\xi] = \mathbb{E}\xi + \int_0^t Z(s)dW(s), \qquad t \in [0, T].$$

Thus, taking $t = T$, one has

$$\xi = \mathbb{E}[\xi] + \int_0^T Z(s)dW(s).$$

Consequently, by combining the above two, we have

$$Y(t) = \mathbb{E}_t[\xi] = \xi - \int_t^T Z(s)dW(s), \qquad t \in [0.T],$$

which means $Y(t) = \mathbb{E}_t[\xi]$, together with the above $Z(\cdot)$ is an adapted solution to BSDE (3.1.2). We can further show that such an adapted solution is unique.

From the above example, we see that solving BSDE (3.1.2) is just an application of the martingale representation theorem. The following example shows that both $Y(\cdot)$ and $Z(\cdot)$ could even be explicit.

Example 3.1.2 Consider one-dimensional BSDE:

$$Y(t) = W(T)^2 - \int_t^T Z(s)dW(s), \qquad t \in [0, T]. \tag{3.1.3}$$

Note that the terminal state is $\mathcal{F}_T$-measurable. If $(Y(\cdot), Z(\cdot))$ is an adapted solution to the above, then

$$\begin{aligned}
Y(t) &= \mathbb{E}_t[W(T)^2] = \mathbb{E}_t\Big|\int_0^T dW(s)\Big|^2 \\
&= \mathbb{E}_t\Big\{\Big|\int_0^t dW(s)\Big|^2 + \Big|\int_t^T dW(s)\Big|^2 + 2\Big(\int_0^t dW(s)\Big)\Big(\int_t^T dW(s)\Big)\Big\} \\
&= W(t)^2 + T - t, \qquad t \in [0, T],
\end{aligned}$$

which gives the explicit expression of $Y(\cdot)$. On the other hand, by Itô's formula,

$$d\Big(W(t)^2\Big) = 2W(t)dW(t) + dt.$$

Thus,

$$W(T)^2 - W(t)^2 = \int_t^T 2W(s)dW(s) + T - t.$$

Then, by BSDE (3.1.3),

$$\int_t^T Z(s)dW(s) = W(T)^2 - Y(t) = \int_t^T 2W(s)dW(s), \quad t \in [0, T].$$

Consequently,

$$Y(t) = W(t)^2 + T - t, \quad Z(t) = 2W(t), \qquad t \in [0, T], \tag{3.1.4}$$

gives the adapted solution to BSDE (3.1.3) explicitly.

Inspired by the above examples, we now look at a little more general case in $\mathbb{R}^m$. Let $\xi \in L^2_{\mathcal{F}_T}(\Omega; \mathbb{R}^m)$ and $g(\cdot) \in L^2_{\mathbb{F}}(0, T; \mathbb{R}^m)$. Consider

$$Y(t) = \xi + \int_t^T g(s)ds - \int_t^T Z(s)dW(s), \qquad t \in [0, T]. \tag{3.1.5}$$

If $(Y(\cdot), Z(\cdot))$ is an adapted solution to BSDE (3.1.5), then, by taking conditional expectation on both sides of (3.1.5), one has

$$\begin{aligned} Y(t) &= \mathbb{E}_t\Big[\xi + \int_t^T g(s)ds\Big] \\ &= \mathbb{E}_t\Big[\xi + \int_0^T g(s)ds\Big] - \int_0^t g(s)ds, \quad t \in [0, T]. \end{aligned} \tag{3.1.6}$$

We know that the first term on the most right-hand side is an L^2-martingale. Thus, by martingale representation theorem (Chap. 2, Theorem 2.3.8), we have some $\mathbb{R}^{m\times d}$-valued $\mathbb{F}$-adapted process $Z(\cdot)$ such that

$$\mathbb{E}_t\Big[\xi + \int_0^T g(s)ds\Big] = \mathbb{E}\Big[\xi + \int_0^T g(s)ds\Big] + \int_0^t Z(s)dW(s), \quad t \in [0, T]. \tag{3.1.7}$$

Hence, by taking $t = T$ in the above, we have

$$\xi + \int_0^T g(s)ds = \mathbb{E}\Big[\xi + \int_0^T g(s)ds\Big] + \int_0^T Z(s)dW(s).$$

Consequently,

$$
\begin{aligned}
Y(t) &= \mathbb{E}_t\Big[\xi + \int_0^T g(s)ds\Big] - \int_0^t g(s)ds \\
&= \mathbb{E}\Big[\xi + \int_0^T g(s)ds\Big] + \int_0^t Z(s)dW(s) - \int_0^t g(s)ds \\
&= \xi + \int_0^T g(s)ds - \int_0^T Z(s)dW(s) + \int_0^t Z(s)dW(s) - \int_0^t g(s)dW(s) \\
&= \xi + \int_t^T g(s)ds - \int_t^T Z(s)dW(s), \qquad t \in [0, T].
\end{aligned}
\tag{3.1.8}
$$

Therefore, we see that $(Y(\cdot), Z(\cdot))$ determined by (3.1.6)–(3.1.7) is an adapted solution of BSDE (3.1.5). We will establish some proper regularity of the pair $(Y(\cdot), Z(\cdot))$ below. From the above, we see that solving a BSDE like (3.1.5) essentially amounts to applying the martingale representation theorem.

Now, let us make the above situation more general and complete. Fixed $m, d \geqslant 1$ and any $p \in (1, \infty)$, $0 \leqslant S < T$, let

$$
\mathbf{H}^p[S, T] = L^p_{\mathbb{F}}(\Omega; C([S, T]; \mathbb{R}^m)) \times L^p_{\mathbb{F}}(\Omega; L^2(S, T; \mathbb{R}^{m\times d})),
$$

which is a Banach space under the norm:

$$
\|(y(\cdot), z(\cdot))\|_{\mathbf{H}^p[S,T]} = \Big\{\mathbb{E}\Big[\sup_{t\in[S,T]} |y(t)|^p + \Big(\int_S^T |z(t)|^2 dt\Big)^{\frac{p}{2}}\Big]\Big\}^{\frac{1}{p}}.
$$

For BSDE (3.1.5), we have the following result.

Proposition 3.1.3 *For any $\xi \in L^p_{\mathcal{F}_T}(\Omega; \mathbb{R}^m)$ and $g(\cdot) \in L^p_{\mathbb{F}}(\Omega; L^1(0, T; \mathbb{R}^m))$, with $p > 1$, BSDE (3.1.5) admits a unique adapted solution $(Y(\cdot), Z(\cdot)) \in \mathbf{H}^p[0, T]$ and the following holds:*

$$
\|(Y(\cdot), Z(\cdot))\|^p_{\mathbf{H}^p[0,T]} \leqslant K\mathbb{E}\Big[|\xi|^p + \Big(\int_0^T |g(t)|dt\Big)^p\Big]. \tag{3.1.9}
$$

Proof First of all, by our condition, using Doob's inequality (Chap. 2, Proposition 2.2.4), we have

$$
\begin{aligned}
\mathbb{E}\Big\{\sup_{t\in[0,T]} \mathbb{E}_t\Big[|\xi| + \int_0^T |g(s)|ds\Big]^p\Big\} &\leqslant \Big(\frac{p}{p-1}\Big)^p \mathbb{E}\Big[|\xi| + \int_0^T |g(s)|ds\Big]^p \\
&\leqslant 2^{p-1}\Big(\frac{p}{p-1}\Big)^p \mathbb{E}\Big[|\xi|^p + \Big(\int_0^T |g(s)|ds\Big)^p\Big].
\end{aligned}
$$

Then, by martingale representation theorem (Chap. 2, Theorem 2.3.8), we have some $\mathbb{R}^{m\times d}$-valued $\mathbb{F}$-adapted process $Z(\cdot)$ such that (3.1.7) holds. By defining

$$Y(t)=\mathbb{E}_t\Big[\xi+\int_0^T g(s)ds\Big],\qquad t\in[0,T],$$

and calculating (3.1.8), we see that $(Y(\cdot),Z(\cdot))$ is an adapted solution to BSDE (3.1.5). Next, by (3.1.5) itself again, together with the martingale moment inequalities (Chap. 2, Theorem 2.3.2), one has

$$\begin{aligned}&\mathbb{E}\Big(\int_t^T|Z(s)|^2ds\Big)^{\frac{p}{2}}\leqslant K_p\mathbb{E}\Big|\int_t^T Z(s)dW(s)\Big|^p\\&=K_p\mathbb{E}\Big[\Big|\xi+\int_t^T g(s)ds-Y(t)\Big|^p\Big]\\&\leqslant 3^pK_p\Big[\mathbb{E}|\xi|^p+\mathbb{E}\Big(\int_t^T|g(s)|ds\Big)^p+\mathbb{E}\Big(\sup_{\tau\in[t,T]}|Y(\tau)|^p\Big)\Big]\\&\leqslant 3^pK_p\Big[1+2^{p-1}\Big(\frac{p}{p-1}\Big)^p\Big]\Big[\mathbb{E}|\xi|^p+\mathbb{E}\Big(\int_t^T|g(s)|ds\Big)^p\Big].\end{aligned}$$

Therefore, (3.1.9) follows. Since the equation is linear, and (3.1.9) serves as a stability estimates, consequently, we obtain the uniqueness. □

It is interesting that in the above, Itô's formula is avoided. This is helpful for our later study on FSVIEs and BSVIEs since both of them generally lack Itô's formula.

Now, for nonlinear general BSDE (3.1.1), we introduce the following hypothesis.

(H3.1.1)$_p$ The map $g:[0,T]\times\mathbb{R}^m\times\mathbb{R}^{m\times d}\times\Omega\to\mathbb{R}^m$ is $\mathcal{B}([0,T]\times\mathbb{R}^m\times\mathbb{R}^{m\times d})\otimes\mathcal{F}_T$-measurable. For each $(y,z)\in\mathbb{R}^m\times\mathbb{R}^{m\times d}$, $t\mapsto g(t,y,z)$ is $\mathbb{F}$-adapted. There exist deterministic functions $L_y(\cdot)\in L^1(0,T;\mathbb{R})$, $L_z(\cdot)\in L^2(0,T;\mathbb{R})$ and a process $L_0(\cdot)\in L^p_{\mathbb{F}}(\Omega;L^1(0,T;\mathbb{R}))$ such that

$$|g(t,0,0)|\leqslant L_0(t),\qquad t\in[0,T],\ \text{a.s.},\tag{3.1.10}$$

$$\begin{aligned}|g(t,y,z)-g(t,\bar y,\bar z)|\leqslant L_y(t)|y-\bar y|+L_z(t)|z-\bar z|,&\\ t\in[0,T],\ y,\bar y\in\mathbb{R}^m,\ z,\bar z\in\mathbb{R}^{m\times d}.&\end{aligned}\tag{3.1.11}$$

The following result gives the well-posedness of BSDE (3.1.1), whose proof is a combination of Proposition 3.1.3 and the contraction mapping theorem.

Theorem 3.1.4 *Let (H3.1.1)$_p$ hold for some $p\in(1,\infty)$. Then for any $\xi\in L^p_{\mathcal{F}_T}(\Omega;\mathbb{R}^m)$, BSDE (3.1.1) admits a unique adapted solution $(Y(\cdot),Z(\cdot))\in\mathbf{H}^p[0,T]$, and the following estimate holds:*

$$\|(Y(\cdot),Z(\cdot))\|^p_{\mathbf{H}^p[0,T]}\leqslant K\mathbb{E}\big(1+|\xi|^p\big).\tag{3.1.12}$$

Further, if $\bar g$ *also satisfies* (H3.1.1)$_p$, $\bar\xi \in L^p_{\mathcal{F}_T}(\Omega;\mathbb{R}^m)$, *and* $(\bar Y(\cdot),\bar Z(\cdot)) \in \mathbf{H}^p[0,T]$ *is the adapted solution to BSDE* (3.1.1) *with* ξ *and* g *replaced by* $\bar\xi$ *and* $\bar g$, *then the following stability estimate holds:*

$$\begin{aligned}&\big\|(Y(\cdot),Z(\cdot))-(\bar Y(\cdot),\bar Z(\cdot))\big\|^p_{\mathbf{H}^p[0,T]}\\&\leqslant K\mathbb{E}\Big[|\xi-\bar\xi|^p+\Big(\int_0^T|g(t,Y(t),Z(t))-\bar g(t,Y(t),Z(t))|dt\Big)^p\Big].\end{aligned}\tag{3.1.13}$$

Proof Take any $(y(\cdot),z(\cdot)) \in \mathbf{H}^p[0,T]$, and denote

$$g(t)=g(t,y(t),z(t)),\qquad t\in[0,T].\tag{3.1.14}$$

By (H3.1.1)$_p$, we have

$$\begin{aligned}\mathbb{E}\Big|\int_0^T g(s,y(s),z(s))ds\Big|^p&\leqslant\mathbb{E}\Big[\int_0^T\Big(L_0(s)+L_y(s)|y(s)|+L_z(s)|z(s)|\Big)ds\Big]^p\\&\leqslant 3^{p-1}\mathbb{E}\Big[\Big(\int_0^T L_0(s)ds\Big)^p+\Big(\int_0^T L_y(s)ds\Big)^p\Big(\sup_{s\in[0,T]}|y(s)|^p\Big)\\&\qquad+\Big(\int_0^T L_z(s)^2ds\Big)^{\frac p2}\Big(\int_0^T|z(s)|^2ds\Big)^{\frac p2}\Big]<\infty.\end{aligned}$$

Consequently, we obtain BSDE (3.1.5) with $g(\cdot) \in L^p_{\mathbb{F}}(\Omega;L^1(0,T;\mathbb{R}^m))$, determined by (3.1.14). Thus, a map $\Phi:\mathbf{H}^p[0,T]\to\mathbf{H}^p[0,T]$ is defined as follows:

$$(Y(\cdot),Z(\cdot))=\Phi\big(y(\cdot),z(\cdot)\big),\qquad\forall(y(\cdot),z(\cdot))\in\mathbf{H}^p[0,T].\tag{3.1.15}$$

Now, let $\bar g$ be another generator also satisfying (H3.1.1)$_p$, and let $\bar\xi \in L^p_{\mathcal{F}_T}(\Omega;\mathbb{R}^m)$, $(\bar y(\cdot),\bar z(\cdot)) \in \mathbf{H}^p[0,T]$. Then, as above, there exists a unique adapted solution $(\bar Y(\cdot),\bar Z(\cdot))$ to the following BSDE:

$$\bar Y(t)=\bar\xi+\int_t^T\bar g(s,\bar y(s),\bar z(s))ds-\int_t^T\bar Z(s)dW(s),\quad t\in[0,T].\tag{3.1.16}$$

Denote

$$\begin{aligned}&\widehat y(\cdot)=y(\cdot)-\bar y(\cdot),\qquad \widehat z(\cdot)=z(\cdot)-\bar z(\cdot),\\&\widehat Y(\cdot)=Y(\cdot)-\bar Y(\cdot),\qquad \widehat Z(\cdot)=Z(\cdot)-\bar Z(\cdot),\\&\widehat\xi=\xi-\bar\xi,\qquad \widehat g(\cdot)=g(\cdot\,,y(\cdot),z(\cdot))-\bar g(\cdot\,,y(\cdot),z(\cdot)).\end{aligned}\tag{3.1.17}$$

Then

$$\widehat Y(t)=\widehat\xi+\int_t^T\Big(g(s,y(s),z(s))-\bar g(s,\bar y(s),\bar z(s))\Big)ds-\int_t^T\widehat Z(s)dW(s),$$

which leads to

$$\widehat{Y}(t) = \mathbb{E}_t\Big[\widehat{\xi} + \int_t^T \Big(g(s, y(s), z(s)) - \bar{g}(s, \bar{y}(s), \bar{z}(s))\Big)ds\Big].$$

Thus,

$$\begin{aligned}
&\mathbb{E}\Big[\sup_{\tau\in[t,T]} |\widehat{Y}(\tau)|^p\Big] \\
&= \mathbb{E}\Big\{\sup_{\tau\in[t,T]} \Big|\mathbb{E}_\tau\Big[\widehat{\xi} + \int_\tau^T \Big(g(s, y(s), z(s)) - \bar{g}(s, \bar{y}(s), \bar{z}(s))\Big)ds\Big]\Big|^p\Big\} \\
&\leqslant \mathbb{E}\Big\{\sup_{\tau\in[t,T]} \Big[\mathbb{E}_\tau\Big(|\widehat{\xi}|+\int_t^T [|\widehat{g}(s)| + L_y(s)|\widehat{y}(s)| + L_z(s)|\widehat{z}(s)|]ds\Big)\Big]^p\Big\} \\
&\leqslant \Big(\frac{p}{p-1}\Big)^p \mathbb{E}\Big[|\widehat{\xi}| + \int_t^T \Big(|\widehat{g}(s)| + L_y(s)|\widehat{y}(s)| + L_z(s)|\widehat{z}(s)|\Big)ds\Big]^p \\
&\leqslant 4^{p-1}\Big(\frac{p}{p-1}\Big)^p \mathbb{E}\Big[|\widehat{\xi}|^p + \Big(\int_t^T |\widehat{g}(s)|ds\Big)^p + \Big(\int_t^T L_y(s)|\widehat{y}(s)|ds\Big)^p \\
&\qquad\qquad + \Big(\int_t^T L_z(s)|\widehat{z}(s)|ds\Big)^p\Big].
\end{aligned}$$

By martingale moment inequalities (Chap. 2, Theorem 2.3.2), for $p > 1$, we have

$$\begin{aligned}
&\mathbb{E}\Big(\int_t^T |\widehat{Z}(s)|^2 ds\Big)^{\frac{p}{2}} \leqslant K_p \mathbb{E}\Big|\int_t^T \widehat{Z}(s)dW(s)\Big|^p \\
&= K_p \mathbb{E}\Big[\Big|\widehat{\xi} + \int_t^T \Big(g(s, y(s), z(s)) - \bar{g}(s, \bar{y}(s), \bar{z}(s)\Big)ds - \widehat{Y}(t)\Big|^p\Big] \\
&\leqslant 5^{p-1}K_p \mathbb{E}\Big[|\widehat{\xi}|^p + \Big(\int_t^T |\widehat{g}(s)|ds\Big)^p + \Big(\int_t^T L_y(s)|\widehat{y}(s)|ds\Big)^p \\
&\qquad + \Big(\int_t^T L_z(s)|\widehat{z}(s)|ds\Big)^p + \Big(\sup_{\tau\in[t,T]} |\widehat{Y}(\tau)|^p\Big)\Big] \\
&\leqslant 5^{p-1}K_p\Big[1 + 4^{p-1}\Big(\frac{p}{p-1}\Big)^p\Big]\mathbb{E}\Big[|\widehat{\xi}|^p + \Big(\int_t^T |\widehat{g}(s)|ds\Big)^p \\
&\qquad + \Big(\int_t^T L_y(s)|\widehat{y}(s)|ds\Big)^p + \Big(\int_t^T L_z(s)|\widehat{z}(s)|ds\Big)^p\Big].
\end{aligned}$$

Combining the above two, we obtain

$$\begin{aligned}
&\mathbb{E}\Big[\sup_{\tau\in[t,T]} |\widehat{Y}(\tau)|^p + \Big(\int_t^T |\widehat{Z}(s)|^2 ds\Big)^{\frac{p}{2}}\Big] \\
&\leqslant 5^{p-1}K_p\Big(\frac{4p}{p-1}\Big)^p \mathbb{E}\Big[|\widehat{\xi}|^p + \Big(\int_t^T |\widehat{g}(s)|ds\Big)^p \\
&\qquad + \Big(\int_t^T L_y(s)|\widehat{y}(s)|ds\Big)^p + \Big(\int_t^T L_z(s)|\widehat{z}(s)|ds\Big)^p\Big].
\end{aligned} \tag{3.1.18}$$

In the above, we use the fact of

$$\begin{aligned}&5^{p-1}K_p\Big[1+4^{p-1}\Big(\frac{p}{p-1}\Big)^p\Big]+4^{p-1}\Big(\frac{p}{p-1}\Big)^p\\&\leqslant 5^{p-1}2K_p4^{p-1}\Big(\frac{p}{p-1}\Big)^p+4^{p-1}\Big(\frac{p}{p-1}\Big)^p\\&=\Big[5^{p-1}2K_p+1\Big]4^{p-1}\Big(\frac{p}{p-1}\Big)^p\leqslant 5^{p-1}K_p4^p\Big(\frac{p}{p-1}\Big)^p.\end{aligned}$$

Now, we look at a special case:

$$\bar{\xi}=\xi,\qquad \bar{g}(s,y,z)=g(s,y,z).$$

Then the above estimate (3.1.18) implies

$$\begin{aligned}&\mathbb{E}\Big[\sup_{\tau\in[t,T]}|\widehat{Y}(\tau)|^p+\Big(\int_t^T|\widehat{Z}(s)|^2ds\Big)^{\frac{p}{2}}\Big]\\&\leqslant 5^{p-1}K_p\Big(\frac{4p}{p-1}\Big)^p\mathbb{E}\Big[\Big(\int_t^T L_y(s)|\widehat{y}(s)|ds\Big)^p+\Big(\int_t^T L_z(s)|\widehat{z}(s)|ds\Big)^p\Big]\\&\leqslant K_0\Big[\Big(\int_t^T L_y(s)ds\Big)^p+\Big(\int_t^T L_z(s)^2ds\Big)^{\frac{p}{2}}\Big]\\&\qquad\cdot\mathbb{E}\Big[\sup_{\tau\in[t,T]}|\widehat{y}(\tau)|^p+\Big(\int_t^T|\widehat{z}(s)|^2ds\Big)^{\frac{p}{2}}\Big].\end{aligned}$$

Note that in the above, $K_0=5^{p-1}K_p\Big(\frac{4p}{p-1}\Big)^p>0$ is an absolute constant. By (H3.1.1)$_p$, we can find $\delta\in(0,T)$, such that

$$K_0\Big[\Big(\int_{T-(i+1)\delta}^{T-i\delta}L_y(s)ds\Big)^p+\Big(\int_{T-(i+1)\delta}^{T-i\delta}L_z(s)^2ds\Big)^{\frac{p}{2}}\Big]\leqslant\alpha<1,\qquad(3.1.19)$$

for all $i=0,1,2,\cdot s,\ell$, with $\ell\delta=T$, and some $\alpha\in(0,1)$. Then the map $\Phi:\mathbf{H}^p[T-\delta,T]\to\mathbf{H}^p[T-\delta,T]$ is a contraction. Hence, Φ admits a unique fixed point $(Y(\cdot),Z(\cdot))\in\mathbf{H}^p[T-\delta,T]$, which is the unique adapted solution to BSDE (3.1.1) on $[T-\delta,T]$. Note that for any $t\in[0,T-\delta)$, BSDE (3.1.1) can be written as follows:

$$Y(t)=Y(T-\delta)+\int_t^{T-\delta}g(s,Y(s),Z(s))ds-\int_t^{T-\delta}Z(s)dW(s),\qquad(3.1.20)$$

with $Y(T-\delta)\in L^p_{\mathcal{F}_{T-\delta}}(\Omega;\mathbb{R}^m)$. Thus, the above is a BSDE on $[0,T-\delta]$. Then the above argument can be applied again to get the unique solvability of the BSDE on $[T-2\delta,T-\delta]$. By induction, we obtain a unique adapted solution $(Y(\cdot),Z(\cdot))\in\mathbf{H}^p[0,T]$.

Next, let $\bar{g}$ be another generator that also satisfies (H3.1.1)$_p$ and $\bar{\xi} \in L^p_{\mathcal{F}_T}(\Omega; \mathbb{R}^m)$. Let $(\bar{Y}(\cdot), \bar{Z}(\cdot))$ be the corresponding adapted solution to the BSDE. Denote

$$\begin{cases} \widehat{Y}(s) = Y(s) - \bar{Y}(s), & \widehat{Z}(s) = Z(s) - \bar{Z}(s), \\ \widehat{\xi} = \xi - \bar{\xi}, & \widehat{g}(s) = g(s, Y(s), Z(s)) - \bar{g}(s, Y(s), Z(s)). \end{cases}$$

Then (3.1.18) implies

$$\begin{aligned} &\mathbb{E}\Big[\sup_{\tau\in[t,T]} |\widehat{Y}(\tau)|^p + \Big(\int_t^T |\widehat{Z}(s)|^2 ds\Big)^{\frac{p}{2}}\Big] \\ &\leqslant K\mathbb{E}\Big[|\widehat{\xi}|^p + \Big(\int_t^T |\widehat{g}(s)|ds\Big)^p + \Big(\int_t^T L_y(s)|\widehat{Y}(s)|ds\Big)^p + \Big(\int_t^T L_z(s)|\widehat{Z}(s)|ds\Big)^p\Big] \\ &\leqslant K_0\mathbb{E}\Big[|\widehat{\xi}|^p + \Big(\int_t^T |\widehat{g}(s)|ds\Big)^p\Big] + K_0\Big[\Big(\int_t^T L_y(s)ds\Big)^p + \Big(\int_t^T L_z(s)^2 ds\Big)^{\frac{p}{2}}\Big] \\ &\qquad\qquad \cdot\mathbb{E}\Big[\sup_{\tau\in[t,T]} |\widehat{y}(\tau)|^p + \Big(\int_t^T |\widehat{z}(s)|^2 ds\Big)^{\frac{p}{2}}\Big]. \end{aligned}$$

By choosing $\delta > 0$ small so that (3.1.19) holds, we have

$$\mathbb{E}\Big[\sup_{\tau\in[T-\delta,T]} |\widehat{Y}(\tau)|^p + \Big(\int_{T-\delta}^T |\widehat{Z}(s)|^2 ds\Big)^{\frac{p}{2}}\Big] \leqslant K\mathbb{E}\Big[|\widehat{\xi}|^p + \Big(\int_{T-\delta}^T |\widehat{g}(s)|ds\Big)^p\Big].$$

Using the similar argument, we could get a similar estimate on $[T - 2\delta, T - \delta]$, $[T - 3\delta, T - 2\delta]$, etc. Then we obtain the stability estimate (3.1.13). $\square$

Remark 3.1.5 We note that in the above proof, Itô's formula is avoided. This is a very important point, as this formula could not be directly used for FSVIEs and BSVIEs.

The above proof is standard: extending a local solution to a global one. It is known that for ODEs, under proper conditions, the unique solution can be obtained via the so-called *Picard iteration*. A natural question is whether we can get that for BSDEs. In what follows, we would like to answer this question positively.

Let us now define the following *Picard iteration* on $\mathbf{H}^p[0, T]$ (not just on a subinterval):

$$\begin{cases} (Y^0(\cdot), Z^0(\cdot)) = 0, \\ Y^{k+1}(t) = \xi + \displaystyle\int_t^T g(s, Y^k(s), Z^k(s))ds - \int_t^T Z^{k+1}(s)dW(s), \\ \qquad\qquad\qquad\qquad\qquad\qquad t \in [0, T], \quad k \geqslant 0. \end{cases} \tag{3.1.21}$$

We present the following proposition.

Proposition 3.1.6 *Let* (H3.1.1)$_p$ *hold for some* $p > 1$. *Let the sequence* $\{(Y^k(\cdot), Z^k(\cdot))\} \subseteq \mathbf{H}^p[0,T]$ *be defined by* (3.1.21). *Then*

$$\lim_{k\to\infty} \|(Y^k(\cdot), Z^k(\cdot)) - (Y(\cdot), Z(\cdot))\|_{\mathbf{H}^p[0,T]} = 0, \tag{3.1.22}$$

where $(Y(\cdot), Z(\cdot))$ *is the unique adapted solution of* (3.1.1).

Proof First of all, from

$$Y^1(t) = \xi + \int_t^T g(s,0,0)ds - \int_t^T Z^1(s)dW(s), \qquad t \in [0,T],$$

we have

$$\big\|\big(Y^1(\cdot), Z^1(\cdot)\big)\big\|_{\mathbb{H}^p[0,T]} \leqslant K\mathbb{E}\Big[|\xi|^p + \Big(\int_0^T L_0(s)ds\Big)^p\Big] \equiv \bar\theta_0. \tag{3.1.23}$$

Next, we write (for $i = 0, 1, 2, \cdot s, \ell$ with $\ell\delta = T$)

$$\begin{aligned} Y^{k+1}(t) = Y^{k+1}(T-i\delta) + \int_t^{T-i\delta} g(s, Y^k(s), Z^k(s))ds \\ - \int_t^{T-i\delta} Z^{k+1}(s)dW(s), \quad t \in [T-(i+1)\delta, T-i\delta]. \end{aligned}$$

Then

$$\begin{aligned} Y^{k+1}(t) - Y^k(t) &= Y^{k+1}(T-i\delta) - Y^k(T-i\delta) \\ &\quad + \int_t^{T-i\delta} \Big(g(s, Y^k(s), Z^k(s)) - g(s, Y^{k-1}(s), Z^{k-1})\Big)ds \\ &\quad - \int_t^{T-i\delta} \Big(Z^{k+1}(s) - Z^k(s)\Big)dW(s), \quad t \in [T-(i+1)\delta, T-i\delta]. \end{aligned}$$

This leads to

$$\begin{aligned} \theta_i^{k+1} &\equiv \big\|\big(Y^{k+1}(\cdot), Z^{k+1}(\cdot)\big) - \big(Y^k(\cdot), Z^k(\cdot)\big)\big\|_{\mathbb{H}^p[T-(i+1)\delta, T-i\delta]} \\ &\leqslant K\mathbb{E}\Big[|Y^{k+1}(T-i\delta) - Y^k(T-i\delta)|^p \\ &\quad + \Big(\int_{T-(i+1)\delta}^{T-i\delta} |g(s, Y^k(s), Z^k(s)) - g(s, Y^{k-1}(s), Z^{k-1}(s))|ds\Big)^p\Big] \\ &\leqslant \big\|\big(Y^{k+1}(\cdot), Z^{k+1}(\cdot)\big) - \big(Y^k(\cdot), Z^k(\cdot)\big)\big\|_{\mathbb{H}^p[T-i\delta, T-(i-1)\delta]} \\ &\quad + \alpha\big\|\big(Y^k(\cdot), Z^k(\cdot)\big) - \big(Y^{k-1}(\cdot), Z^{k-1}(\cdot)\big)\big\|_{\mathbb{H}^p[T-(i+1)\delta, T-i\delta]} \\ &\equiv \theta_{i-1}^{k+1} + \alpha\theta_i^k, \qquad i, k \geqslant 1, \end{aligned} \tag{3.1.24}$$

for some $\alpha \in (0, 1)$ in (3.1.19) and

$$\begin{aligned}\theta_0^{k+1} &\equiv \big\|\big(Y^{k+1}(\cdot), Z^{k+1}(\cdot)\big) - \big(Y^k(\cdot), Z^k(\cdot)\big)\big\|_{\mathbb{H}^p[T-\delta,T]}\\ &\leqslant \alpha\big\|\big(Y^k(\cdot), Z^k(\cdot)\big) - \big(Y^{k-1}(\cdot), Z^{k-1}(\cdot)\big)\big\|_{\mathbb{H}^p[T-\delta,T]}\\ &\equiv \alpha\theta_0^k \leqslant \alpha^2\theta_0^{k-1} \leqslant \cdots \leqslant \alpha^{k+1}\theta_0^0 \leqslant \alpha^{k+1}\bar\theta_0, \qquad k \geqslant 1.\end{aligned} \tag{3.1.25}$$

Now, we claim that for each $i = 0, 1, 2, \cdots, \ell$, the series $\sum_{k=0}^{\infty}\theta_i^k$ is convergent. In fact, it is clearly true for $i = 0$. Now, suppose it is true for some i_0. Then the partial sum $S_{i_0+1,N}$ of the series $\sum_{k=1}^{N}\theta_{i_0+1}^k$ satisfies

$$\begin{aligned}S_{i_0+1,N} &\equiv \theta_{i_0+1}^1 + \sum_{k=1}^{N-1}\theta_{i_0+1}^{k+1} \leqslant \bar\theta_0 + \sum_{k=1}^{N-1}\Big(\theta_{i_0}^{k+1} + \alpha\theta_{i_0+1}^k\Big)\\ &\leqslant \bar\theta_0 + S_{i_0,N} + \alpha S_{i_0+1,N}.\end{aligned}$$

Hence,

$$\begin{aligned}\sum_{k=1}^{\infty}\theta_{i_0+1}^k &= \lim_{N\to\infty} S_{i_0+1,N} \leqslant \frac{1}{1-\alpha}\lim_{N\to\infty}\Big(\bar\theta_0 + S_{i_0,N}\Big)\\ &\leqslant \frac{1}{1-\alpha}\Big(\bar\theta_0 + \sum_{k=1}^{\infty}\theta_{i_0}^k\Big) < \infty.\end{aligned}$$

Consequently,

$$\sum_{k=0}^{\infty}\big\|\big(Y^{k+1}(\cdot) - Y^k(\cdot), Z^{k+1}(\cdot) - Z^k(\cdot)\big)\big\|_{\mathbb{H}^p[0,T]} \leqslant \sum_{k=1}^{\infty}\big(\theta_1^k + \cdots + \theta_\ell^k\big) < \infty.$$

This shows that (3.1.22) holds and

$$\big(Y(t), Z(t)\big) = \sum_{k=0}^{\infty}\big(Y^{k+1}(t) - Y^k(t), Z^{k+1}(t) - Z^k(t)\big), \qquad t \in [0, T].$$

The proof is complete. □

We see that in the above the well-posedness of BSDEs has been proved by a global Picard iteration, without using Itô's formula.

Example 3.1.2 suggests that with some particular terminal state, the adapted solution could have a nice-looking form. We now further exploit that. Consider the following BSDE:

$$Y(t) = h(W(T)) + \int_t^T g(s, Y(s), Z(s))ds - \int_t^T Z(s)dW(s), \quad t \in [0, T]. \tag{3.1.26}$$

We try the following ansatz:

$$Y(t) = \theta(t, W(t)), \quad Z(t) = \theta_x(t, W(t)), \quad t \in [0, T], \tag{3.1.27}$$

for some $\theta(\cdot\,,\cdot)$. By applying Itô's formula to $t \mapsto \theta(t, W(t))$, we have

$$\begin{aligned} &-g\big(t, \theta(t, W(t)), \theta_x(t, W(t))\big)dt = dY(t) = d\big[\theta(t, W(t))\big] \\ &= \Big(\theta_t(t, W(t)) + \frac{1}{2}\theta_{xx}(t, W(t))\Big)dt + \theta_x(t, W(t))dW(t). \end{aligned}$$

Hence, provided $\theta(\cdot\,,\cdot)$ solves the following terminal value problem of partial differential equation (PDE, for short),

$$\begin{cases} \theta_t(t,x)+\frac{1}{2}\theta_{xx}(t,x)+g(t,\theta(t,x),\theta_x(t,x))=0, & (t,x)\in[0,T]\times\mathbb{R}^n, \\ \theta(T,x) = h(x), \qquad x \in \mathbb{R}^n, \end{cases} \tag{3.1.28}$$

the adapted solution $(Y(\cdot), Z(\cdot))$ is given by (3.1.27). We call (3.1.27) a representation of adapted solution to BSDE (3.1.26). On the other hand, for any fixed $(t, x) \in [0, T] \times \mathbb{R}^n$, we can solve the following BSDE:

$$\begin{aligned} Y(r) = h(W(T) - W(t) + x) + &\int_r^T g(s, Y(s), Z(s))ds \\ - &\int_r^T Z(s)dW(s), \qquad r \in [t, T], \end{aligned} \tag{3.1.29}$$

where $W(\cdot)$ is a standard Brownian motion (starting from 0 at $t = 0$, so that $W(r) - W(t) + x$ is a Brownian motion starting from x at $r = t$). Clearly, we should denote the adapted solution by $(Y^{t,x}(\cdot), Z^{t,x}(\cdot))$, indicating the dependence on the initial pair (t, x). Then, by defining

$$\theta(t, x) = Y^{t,x}(t), \qquad (t, x) \in [0, T] \times \mathbb{R}^n,$$

and using Itô's formula, we have that $\theta(\cdot\,,\cdot)$ is the solution to the PDE (3.1.28). This is referred to as the (nonlinear) *Feyman–Kac formula*. We will come back to this point later on.

To conclude this section, we present the following corollary which will be useful in later chapters (see the proof of Proposition 4.3.4 of Chap. 4 and Proposition 5.1.3 of Chap. 5).

Corollary 3.1.7 *Let $p > 1$. Then there exists a constant $K > 0$ such that for any fixed $(S,t) \in \Delta^*[0,T]$, $\eta \in L^p_{\mathcal{F}_S}(\Omega;\mathbb{R}^m)$ and $\zeta(\cdot) \in L^p_{\mathbb{F}}(\Omega; L^2(S,t;\mathbb{R}^m))$, the following holds:*

$$\mathbb{E}\Big[|\eta|^p + \Big(\int_S^t |\zeta(s)|^2 ds\Big)^{\frac{p}{2}}\Big] \leqslant K\mathbb{E}\Big|\eta + \int_S^t \zeta(s)dW(s)\Big|^p. \tag{3.1.30}$$

Proof For fixed $(S,t) \in \Delta^*[0,T]$, let

$$\xi = \eta + \int_S^t \zeta(s)dW(s),$$

which is $\mathcal{F}_t$-measurable. Let $(Y(\cdot), Z(\cdot))$ be the adapted solution to the following BSDE:

$$Y(\tau) = \xi - \int_\tau^t Z(s)dW(s), \qquad \tau \in [S,t].$$

Then by Proposition 3.1.3, we have

$$\mathbb{E}\Big[\sup_{r\in[S,t]} |Y(r)|^p + \Big(\int_S^t |Z(s)|^2 ds\Big)^{\frac{p}{2}}\Big] \leqslant K\mathbb{E}|\xi|^p, \tag{3.1.31}$$

for an absolute constant $K > 0$ only depending on $p > 1$ and dimension d of the space. Now,

$$Y(S) + \int_S^t Z(s)dW(s) = \xi = \eta + \int_S^t \zeta(s)dW(s).$$

By taking conditional expectation $\mathbb{E}[\,\cdot\,|\,\mathcal{F}_S] \equiv E_S[\,\cdot\,]$, we see that

$$Y(S) = \eta.$$

Consequently,

$$\int_S^t \Big(Z(s) - \zeta(s)\Big)dW(s) = 0,$$

which leads to

$$Z(s) = \zeta(s), \qquad s \in [S,t], \text{ a.s.}$$

Then (3.1.30) follows from (3.1.31). □

We note that it is obvious to have

$$\mathbb{E}\Big|\eta+\int_S^t \zeta(s)dW(s)\Big|^p \leqslant K\mathbb{E}\Big[|\eta|^p+\Big(\int_S^t |\zeta(s)|^2 ds\Big)^{\frac{p}{2}}\Big], \tag{3.1.32}$$

whereas the other direction estimate, namely, estimate (3.1.30), is not trivial. This estimate will be very useful in studying BSVIEs (see Chap. 4, Sect. 3). The main point here is that Itô's formula, which is not directly available for BSVIEs, is avoided.

2 Further Regularity of Adapted Solution

In this section, we will present some further regularity results on the adapted solutions of BSDEs. For the first time readers or those who do not have knowledge of the Mallivan calculus, they can skip this section. Also, it is possible to read the appendix (Chap. 12) first.

From Theorem 3.1.4 for BSDE (3.1.1) or Proposition 3.1.3 for BSDE (3.1.5), we have seen that the process $Z(\cdot)$ is obtained through the application of martingale representation theorem. Therefore, in general, we do not expect to have the continuity of this process. However, from (3.1.4) and (3.1.27), we see that for some cases, $Z(\cdot)$ could be continuous. Thus, one expects to find conditions guaranteeing this. On the other hand, (12.1.34) of Chap. 12 suggests us that Malliavin calculus might be helpful for the study of the continuity of $Z(\cdot)$.

Before getting into the general technical details, let us first look a special case. Consider BSDE (3.1.5) which is rewritten here:

$$Y(t)=\xi+\int_t^T g(s)ds-\int_t^T Z(s)dW(s), \qquad t\in[0,T].$$

We have

$$D_\tau^{(i)}Y(t)=0, \quad D_\tau^{(i)}Z(t)=0, \qquad 0\leqslant t<\tau\leqslant T, \quad 1\leqslant i\leqslant d,$$

by the $\mathbb{F}$-adaptiveness of $(Y(\cdot),Z(\cdot))$, and the above BSDE can be rewritten as follows:

$$Y(t)=Y(r)+\int_t^r g(s)ds-\int_t^r Z(s)dW(s), \qquad t\in[0,r].$$

Then according to (12.1.34) of Chap. 12, the following holds:

$$\begin{aligned} D_\tau^{(i)}Y(t)=D_\tau^{(i)}Y(r)+&\int_{t\vee\tau}^r [D_\tau^{(i)}g](s)ds-Z_i(\tau)\mathbf{1}_{(t,r]}(\tau)\\ &-\int_{t\vee\tau}^r D_\tau^{(i)}Z(s)dW(s), \quad \text{a.e. } \tau\in[0,r],\ 1\leqslant i\leqslant d. \end{aligned} \tag{3.2.1}$$

Hence, for $t < \tau$,

$$0 = D_\tau^{(i)} Y(r) + \int_\tau^r [D_\tau^{(i)} g](s)ds - Z_i(\tau) - \int_\tau^r \sum_{j=1}^d D_\tau^{(i)} Z_j(s)dW_j(s),$$
$$\tau \leqslant r, \quad 1 \leqslant i \leqslant d. \tag{3.2.2}$$

Letting $r \to \tau$, we formally have

$$Z_i(\tau) = D_\tau^{(i)} Y(\tau), \qquad \tau \in [0, T], \text{ a.s.}, \quad 1 \leqslant i \leqslant d. \tag{3.2.3}$$

Therefore, taking $r = T$ in (3.2.2), we obtain (replacing τ by t)

$$\begin{aligned} Z_i(t) = D_t^{(i)} \xi + \int_t^T [D_t^{(i)} g](s)ds - \int_t^T D_t^{(i)} Z(s)dW(s), \\ t \in [0, T], \quad 1 \leqslant i \leqslant d. \end{aligned} \tag{3.2.4}$$

If we denote

$$\psi_i(t) = D_t^{(i)} \xi, \qquad \widetilde{g}_i(t, s) = [D_t^{(i)} g](s), \qquad \widetilde{Z}_{ij}(t, s) = D_t^{(i)} Z_j(s),$$

and note the Malliavin differentiability will not change the measurability of the random variabe, then $(Z_i(\cdot), \widetilde{Z}_i(\cdot\,, \cdot))$ is an adapted solution to the following simple BSVIE system (of Type I):

$$\begin{aligned} Z_i(t) = \psi_i(t) + \int_t^T \widetilde{g}_i(t, s)ds - \int_t^T \sum_{j=1}^d \widetilde{Z}_{ij}(t, s)dW_j(s), \\ t \in [0, T], \quad 1 \leqslant i \leqslant d. \end{aligned} \tag{3.2.5}$$

For such a BSVIE, we will show later that if $t \mapsto (\psi_i(t), \widetilde{g}_i(t, s))$ is continuous as a map from $[0, T]$ to $L^p_{\mathcal{F}_T}(\Omega; \mathbb{R}^m) \times L^p_{\mathcal{F}_T}(\Omega; \mathbb{R}^m)$, then so is the map $t \mapsto Z_i(t)$, from $[0, T]$ to $L^p_{\mathcal{F}_T}(\Omega; \mathbb{R}^m)$. Moreover, under some further conditions, $Z_i(\cdot)$ will have continuous paths. This gives the further regularity of $Z(\cdot)$.

We now look at the general case. For any $0 \leqslant r \leqslant R < S \leqslant T$, we let $\widehat{\mathbf{H}}^p_r[R, S]$ be the subspace of $\mathbf{H}^p[R, S]$ consisting of all $(y(\cdot), z(\cdot)) \in \mathbf{H}^p[R, S]$ such that

$$(D_r^{(i)} y(\cdot), D_r^{(i)} z(\cdot)) \in \mathbf{H}^p[R, S], \qquad 1 \leqslant i \leqslant d. \tag{3.2.6}$$

Clearly, $\widehat{\mathbf{H}}_r^p[S,T]$ is a Banach space under the following norm:

$$\|(y(\cdot),z(\cdot))\|_{\widehat{\mathbf{H}}_r^p[S,T]}=\Big\{\|(y(\cdot),z(\cdot))\|^p_{\mathbf{H}^p[S,T]}+\sum_{i=1}^d\|(D_r^{(i)}y(\cdot),D_r^{(i)}z(\cdot))\|^p_{\mathbf{H}^p[S,T]}\Big\}^{\frac{1}{p}}.$$

We introduce the following assumption.

(H3.2.1)$_p$ The map $g:[0,T]\times\mathbb{R}^m\times\mathbb{R}^{m\times d}\times\Omega\to\mathbb{R}^m$ is $\mathcal{B}([0,T]\times\mathbb{R}^m\times\mathbb{R}^{m\times d})\otimes\mathcal{F}_T$-measurable. For each $(y,z)\in\mathbb{R}^m\times\mathbb{R}^{m\times d}$, $t\mapsto g(t,y,z)$ is $\mathbb{F}$-adapted, and there are deterministic functions $L_y(\cdot)\in L^1(0,T;\mathbb{R})$, $L_z(\cdot)\in L^2(0,T;\mathbb{R})$, and a process $L_0(\cdot)\in L^p_{\mathbb{F}}(\Omega;L^1(0,T;\mathbb{R}))$ such that

$$\begin{aligned}&|g(t,0,0)|\leqslant L_0(t),\quad |g_y(t,y,z)|\leqslant L_y(t),\quad |g_{z_j}(t,y,z)|\leqslant L_z(t),\\&|[D_\tau^{(i)}g](t,y,z)|\leqslant L_0(t)+L_y(t)|y|+L_z(t)|z|,\\&\qquad\qquad t,\tau\in[0,T],\ y\in\mathbb{R}^m,\ z\in\mathbb{R}^{m\times d},\ 1\leqslant i,j\leqslant d.\end{aligned}\tag{3.2.7}$$

Moreover, for almost all $t,\tau\in[0,T]$, $1\leqslant i,j\leqslant d$, maps $[D_\tau^{(i)}g](t,\cdot,\cdot)$, $g_y(t,\cdot,\cdot)$, and $g_{z_j}(t,\cdot,\cdot)$ are continuous.

It is clear that (H3.2.1)$_p$ is stronger than (H3.1.1)$_p$. Therefore, under (H3.2.1)$_p$, for any $\xi\in L^p_{\mathcal{F}_T}(\Omega;\mathbb{R})$, BSDE (3.1.1) admits a unique adapted solution $(Y(\cdot),Z(\cdot))\in\mathbf{H}^p[0,T]$. The following result gives a little more regularity of the adapted solution when ξ is a little more regular, in the sense of Malliavin calculus.

Theorem 3.2.1 *Let* (H3.2.1)$_p$ *hold for some* $p>1$, *and* $\xi\in\mathbb{D}^m_{1,p}$ *with* $D_\tau^{(i)}\xi\in L^p_{\mathcal{F}_T}(\Omega;\mathbb{R}^m)$, $1\leqslant i\leqslant d$. *Let* $(Y(\cdot),Z(\cdot))$ *be the adapted solution to BSDE* (3.1.1). *Then for any* $0\leqslant\tau\leqslant S<T$, $(Y(\cdot),Z(\cdot))\in\widehat{\mathbf{H}}^p[S,T]$ *and*

$$\begin{aligned}&\mathbb{E}\Big[\sup_{t\in[S,T]}\sum_{i=1}^d|D_\tau^{(i)}Y(t)|^p+\sum_{i=1}^d\Big(\int_S^T|D_\tau^{(i)}Z(s)|^2ds\Big)^{\frac{p}{2}}\Big]\\&\leqslant K\mathbb{E}\Big(1+|\xi|^p+\sum_{i=1}^d|D_\tau^{(i)}\xi|^p\Big),\end{aligned}\tag{3.2.8}$$

and $(D_\tau^{(i)}Y(\cdot),D_\tau^{(i)}Z(\cdot))$ *is the adapted solution of the following BSDE system (parameterized by* τ*):*

$$\begin{aligned}D_\tau^{(i)}Y(t)=D_\tau^{(i)}\xi+\int_t^T\Big(&[D_\tau^{(i)}g](s,Y(s),Z(s))+g_y(s,Y(s),Z(s))D_\tau^{(i)}Y(s)\\&+\sum_{j=1}^d g_{z_j}(s,Y(s),Z(s))D_\tau^{(i)}Z_j(s)\Big)ds\\&-\sum_{j=1}^d\int_t^T D_\tau^{(i)}Z_j(s)dW_j(s),\qquad t\in[\tau,T],\ 1\leqslant i\leqslant d.\end{aligned}\tag{3.2.9}$$

Moreover,

$$Z_i(t) = D_t^{(i)} Y(t), \qquad t \in [0, T], \text{ a.s.}, \quad 1 \leqslant i \leqslant d, \tag{3.2.10}$$

which means $(Z_i(\cdot), D_\cdot^{(i)} Z(\cdot))$, $1 \leqslant i \leqslant d$ *is an adapted solution to the following BSVIE:*

$$\begin{aligned} Z_i(t) = D_t^{(i)}\xi + \int_t^T \Big([D_t^{(i)} g](s, Y(s), Z(s)) + g_y(s, Y(s), Z(s)) D_t^{(i)} Y(s) \\ + \sum_{j=1}^d g_{z_j}(s, Y(s), Z(s)) D_t^{(i)} Z_j(s) \Big) ds \\ - \sum_{j=1}^d \int_t^T D_t^{(i)} Z_j(s) dW(s), \quad t \in [0, T], \ 1 \leqslant i \leqslant d, \end{aligned} \tag{3.2.11}$$

with

$$\begin{aligned} \mathbb{E}\Big[|Z(t)|^p + \sum_{i=1}^d \Big(\int_t^T |D_t^{(i)} Z(s)|^2 ds \Big)^{\frac{p}{2}} \Big] \\ \leqslant K \mathbb{E}\Big[|\xi|^p + \sum_{i=1}^d |D_t^{(i)} \xi|^p + \Big(\int_t^T L_0(s) ds \Big)^p \Big]. \end{aligned} \tag{3.2.12}$$

Proof First of all, by Theorem 3.1.4, we see that BSDE (3.1.1) admits a unique adapted solution $(Y(\cdot), Z(\cdot)) \in \mathbf{H}^p[S, T]$ and estimate (3.1.12) holds, i.e.,

$$\begin{aligned} \|(Y(\cdot), Z(\cdot))\|_{\mathbf{H}^p[0,T]}^p \equiv \mathbb{E}\Big[\sup_{s \in [0.T]} |Y(s)|^p + \Big(\int_0^T |Z(s)|^2 ds \Big)^{\frac{p}{2}} \Big] \\ \leqslant K \mathbb{E}\Big[|\xi|^p + \Big(\int_0^T L_0(t) dt \Big)^p \Big]. \end{aligned}$$

By the $\mathbb{F}$-adaptiveness of $(Y(\cdot), Z(\cdot))$, we have

$$D_\tau^{(i)} Y(t) = 0, \quad D_\tau^{(i)} Z(t) = 0, \qquad 0 \leqslant t < \tau \leqslant T, \quad 1 \leqslant i \leqslant d.$$

Next, we consider the following BSDE (parameterized by $\tau \in [0, T)$ and $i = 1, 2, \cdot s, d$):

$$\begin{aligned} \widetilde{Y}^{\tau,i}(t) = D_\tau^{(i)}\xi + \int_t^T \Big([D_\tau^{(i)} g](s, Y(s), Z(s)) + g_y(s, Y(s), Z(s)) \widetilde{Y}^{\tau,i}(s) \\ + \sum_{j=1}^d g_{z_j}(s, Y(s), Z(s)) \widetilde{Z}_j^{\tau,i}(s) \Big) ds \\ - \int_t^T \widetilde{Z}^{\tau,i}(s) dW(s), \quad t \in [\tau, T]. \end{aligned} \tag{3.2.13}$$

By Theorem 3.1.4, under $(\mathrm{H3.1.1})_p$, the above admits a unique adapted solution $(\widetilde{Y}^{\tau,i}(\cdot), \widetilde{Z}^{\tau,i}(\cdot)) \in \mathbf{H}^p[\tau, T]$, and the following estimate holds:

$$
\begin{aligned}
&\|(\widetilde{Y}^{\tau,i}(\cdot), \widetilde{Z}^{\tau,i}(\cdot))\|^p_{\mathbf{H}^p[\tau,T]} \equiv \mathbb{E}\Big[\sup_{s\in[\tau,T]} |\widetilde{Y}^{\tau,i}(s)|^p + \Big(\int_\tau^T |\widetilde{Z}^{\tau,i}(s)|^2 ds\Big)^{\frac{p}{2}}\Big] \\
&\leqslant K\mathbb{E}\Big[| D^{(i)}_\tau \xi|^p + \Big(\int_\tau^T |[D^{(i)}_\tau g](s, Y(s), Z(s))|ds\Big)^p\Big] \\
&\leqslant K\mathbb{E}\Big[| D^{(i)}_\tau \xi|^p + \Big(\int_\tau^T \big(L_0(s) + L_y(s)|Y(s)| + L_z(s)|Z(s)|\big)ds\Big)^p\Big] \\
&\leqslant K\mathbb{E}\Big\{1 + |D^{(i)}_\tau \xi|^p + \Big(\int_0^T L_y(s)ds\Big)^p \sup_{s\in[\tau,T]} |Y(s)|^p \\
&\qquad + \Big(\int_\tau^T L_z(s)^2 ds\Big)^{\frac{p}{2}}\Big(\int_0^T |Z(s)|^2 ds\Big)^{\frac{p}{2}}\Big\} \\
&\leqslant K\mathbb{E}\big(1 + |\xi|^p + |D^{(i)}_\tau \xi|^p\big).
\end{aligned}
$$

Further, if we recall the above (3.1.21) and define

$$
\begin{cases} (Y^0(\cdot), Z^0(\cdot)) = 0, \\ (Y^{k+1}(\cdot), Z^{k+1}(\cdot)) = \Phi(Y^k(\cdot), Z^k(\cdot)), \qquad k \geqslant 0, \end{cases} \tag{3.2.14}
$$

for some mapping Φ, then by Proposition 3.1.6,

$$
\lim_{k\to\infty} \|(Y^k(\cdot), Z^k(\cdot)) - (Y(\cdot), Z(\cdot))\|_{\mathbf{H}^p[0,T]} = 0. \tag{3.2.15}
$$

On the other hand, from

$$
Y^1(t) = \xi + \int_t^T g(s, 0, 0)ds - \int_t^T Z^1(s)dW(s), \qquad t \in [0, T],
$$

we see that for fixed t, by Chap. 12, Proposition 12.1.1 (i),

$$
Y^1(t) = \mathbb{E}_t\Big[\xi + \int_t^T g(s, 0, 0)ds\Big] \in \mathbb{D}^m_{1,p}.
$$

Thus,

$$
\int_t^T Z^1(s)dW(s) = \xi + \int_t^T g(s, 0, 0)ds - Y^1(t) \in \mathbb{D}^m_{1,p}.
$$

Hence, by Chap. 12, Proposition 12.1.2 (iii), we have

$$
\begin{aligned}
&Z^1_i(\tau)\mathbf{1}_{[t,T]}(\tau) + \int_{\tau\vee t}^T D^{(i)}_\tau Z^1_i(s)dW(s) = D^{(i)}_\tau \int_t^T Z^1(s)dW(s) \\
&= D^{(i)}_\tau \xi + \int_t^T D^{(i)}_\tau g(s, 0, 0)ds - D^{(i)}_\tau Y^1(t).
\end{aligned}
$$

Now, we fix $\tau \in [0, T)$, and for $t \in (\tau, T]$, the second equality of above is equivalent to the following:

$$D_\tau^{(i)} Y^1(t) = D_\tau^{(i)}\xi + \int_t^T D_\tau^{(i)} g(s, 0, 0)ds - \int_t^T D_\tau^{(i)} Z(s)dW(s).$$

Similar to the above, from

$$Y^{k+1}(t) = \xi + \int_t^T g(s, Y^k(s), Z^k(s))ds - \int_t^T Z^{k+1}(s)dW(s),$$

inductively, one has

$$\begin{aligned} D_\tau^{(i)} Y^{k+1}(t) = D_\tau^{(i)}\xi + \int_t^T \Big\{ & [D_\tau^{(i)} g](s, Y^k(s), Z^k(s)) \\ & + g_y(s, Y^k(s), Z^k(s)) D_\tau^{(i)} Y^k(s) \\ & + \sum_{j=1}^d g_{z_j}(s, Y^k(s), Z^k(s)) D_\tau^{(i)} Z_j^k(s) \Big\} ds \\ & - \int_t^T D_\tau^{(i)} Z^{k+1}(s)dW(s), \qquad t \in [S, T], \end{aligned} \tag{3.2.16}$$

with a fixed $\tau \in [0, T)$ and any $S \in (\tau, T]$. Therefore, combining (3.2.13)–(3.2.16), we have

$$\begin{aligned} & D_\tau^{(i)} Y^{k+1}(t) - \widetilde{Y}^{\tau,i}(t) \\ & = \int_t^T \Big([D_\tau^{(i)} g](s, Y^k(s), Z^k(s)) - [D_\tau^{(i)} g](s, Y(s), Z(s)) \\ & \quad + \big[g_y(s, Y^k(s), Z^k(s)) - g_y(s, Y(s), Z(s))\big] \widetilde{Y}^{\tau,i}(s) \\ & \quad + \sum_{j=1}^d \big[g_{z_j}(s, Y^k(s), Z^k(s)) - g_{z_j}(s, Y(s), Z(s))\big] \widetilde{Z}_j^{\tau,i}(s) \\ & \quad + g_y(s, Y(s), Z(s)) \big[D_\tau^{(i)} Y^k(s) - \widetilde{Y}^{\tau,i}(s)\big] \\ & \quad + \sum_{j=1}^d g_{z_j}(s, Y(s), Z(s)) \big[D_\tau^{(i)} Z_j^k(s) - \widetilde{Z}_j^{\tau,i}(s)\big] \Big) ds \\ & \quad - \int_t^T \big[D_\tau^{(i)} Z^{k+1}(s) - \widetilde{Z}^{\tau,i}(s)\big] dW(s), \quad t \in [\tau, T]. \end{aligned}$$

Then by the standard estimate of BSDEs, with $\varepsilon > 0$ and $p - \varepsilon > 1$, we have

$$
\begin{aligned}
\theta_{k+1} &\equiv \|(D_\tau^{(i)} Y^{k+1}(\cdot), D_\tau^{(i)} Z^{k+1}(\cdot)) - (\widetilde{Y}^{\tau,i}(\cdot), \widetilde{Z}^{\tau,i}(\cdot))\|^{p-\varepsilon}_{\mathbf{H}^{p-\varepsilon}[S,T]} \\
&\leqslant K\mathbb{E}\Big\{\Big(\int_S^T |[D_\tau^{(i)} g](s, Y^k(s), Z^k(s)) - [D_\tau^{(i)} g](s, Y(s), Z(s))|ds\Big)^{p-\varepsilon} \\
&\quad + \Big(\int_S^T |g_y(s, Y^k(s), Z^k(s)) - g_y(s, Y(s), Z(s))|\,|\widetilde{Y}^{\tau,i}(s)|ds\Big)^{p-\varepsilon} \\
&\quad + \sum_{j=1}^d \Big(\int_S^T |g_{z_j}(s, Y^k(s), Z^k(s)) - g_{z_j}(s, Y(s), Z(s))|\,|\widetilde{Z}_j^{\tau,i}(s)|ds\Big)^{p-\varepsilon} \\
&\quad + \Big[\Big(\int_S^T L_y(s)ds\Big)^{p-\varepsilon} + \Big(\int_S^T L_z(s)^2 ds\Big)^{\frac{p-\varepsilon}{2}}\Big] \\
&\quad \cdot \|(D_\tau^{(i)} Y^k(\cdot), D_\tau^{(i)} Z^k(\cdot)) - (\widetilde{Y}^{\tau,i}(\cdot), \widetilde{Z}^{\tau,i}(\cdot))\|^{p-\varepsilon}_{\mathbf{H}^{p-\varepsilon}[S,T]}\Big\} \leqslant \eta_k + \alpha\theta_k,
\end{aligned}
\tag{3.2.17}
$$

where

$$
\begin{aligned}
\eta_k &\equiv K\mathbb{E}\Big\{\Big(\int_S^T |[D_\tau^{(i)} g](s, Y^k(s), Z^k(s)) - [D_\tau^{(i)} g](s, Y(s), Z(s))|ds\Big)^{p-\varepsilon} \\
&\quad + \Big(\int_S^T |g_y(s, Y^k(s), Z^k(s)) - g_y(s, Y(s), Z(s))|\,|\widetilde{Y}^{\tau,i}(s)|ds\Big)^{p-\varepsilon} \\
&\quad + \sum_{j=1}^d \Big(\int_S^T |g_{z_j}(s, Y^k(s), Z^k(s)) - g_{z_j}(s, Y(s), Z(s))|\,|\widetilde{Z}_j^{\tau,i}(s)|ds\Big)^{p-\varepsilon}\Big\} \\
&\leqslant K\mathbb{E}\Big(\int_S^T |[D_\tau^{(i)} g](s, Y^k(s), Z^k(s)) - [D_\tau^{(i)} g](s, Y(s), Z(s))|ds\Big)^{p-\varepsilon} \\
&\quad + K\Big\{\Big[\mathbb{E}\Big(\int_S^T |g_y(s, Y^k(s), Z^k(s)) - g_y(s, Y(s), Z(s))|ds\Big)^{\frac{p(p-\varepsilon)}{\varepsilon}}\Big]^{\frac{\varepsilon}{p}} \\
&\quad + \sum_{j=1}^d \Big[\mathbb{E}\Big(\int_S^T |g_{z_j}(s, Y^k(s), Z^k(s)) - g_{z_j}(s, Y(s), Z(s))|^2 ds\Big)^{\frac{p(p-\varepsilon)}{2\varepsilon}}\Big]^{\frac{\varepsilon}{p}}\Big\} \\
&\quad \cdot \Big[\mathbb{E}\big(1 + |\xi|^p + |D_\tau^{(i)}\xi|^p\big)\Big]^{\frac{p-\varepsilon}{p}},
\end{aligned}
$$

and, if necessary, we shrink $T - S$ further, so that

$$
K\Big[\Big(\int_S^T L_y(s)ds\Big)^{p-\varepsilon} + \Big(\int_S^T L_z(s)^2 ds\Big)^{\frac{p-\varepsilon}{2}}\Big] \leqslant \alpha < 1.
$$

By the convergence (3.2.15) and the dominated convergence theorem, we see that

$$
\lim_{k\to\infty} \eta_k = 0. \tag{3.2.18}
$$

Then (3.2.17) implies

$$\lim_{k\to\infty} \theta_k = 0. \tag{3.2.19}$$

Since operator $\xi \mapsto D_{\cdot}^{(i)}\xi$ is closed, we must have

$$\widetilde{Y}^{\tau,i}(t) = D_\tau^{(i)}Y(t), \qquad \widetilde{Z}^{\tau,i}(t) = D_\tau^{(i)}Z(t), \qquad t \in [S,T], \text{ a.s.} \tag{3.2.20}$$

By induction, we can obtain the result on $(\tau, T]$. Now, we look at $t = \tau$. For $t < \tau \leqslant r$, one has

$$Y(t) = Y(r) + \int_t^r g(s, Y(s), Z(s))ds - \int_t^r Z(s)dW(s).$$

Then (noting $D_\tau^{(i)}Y(t) = 0$, for $t < \tau$)

$$\begin{aligned}0 = D_\tau^{(i)}Y(r) + \int_\tau^r \Big[&[D_\tau^{(i)}g](s, Y(s), Z(s)) + g_y(s, Y(s), Z(s))D_\tau^{(i)}Y(s)\\ &+\sum_{j=1}^d g_{z_j}(s, Y(s), Z(s))D_\tau^{(i)}Z_j(s)\Big]ds - Z_i(\tau) - \int_\tau^r D_\tau^{(i)}Z(s)dW(s).\end{aligned}$$

Sending $r \searrow \tau$, we obtain

$$D_\tau^{(i)}Y(\tau) = Z_i(\tau).$$

This proves (3.2.10). Finally, (3.2.12) follows easily from (3.2.9) and (3.2.10). □

If we denote

$$\widetilde{Y}^i(\tau, t) = D_\tau^{(i)}Y(t), \qquad \widetilde{Z}^i(\tau, t) = D_\tau^{(i)}Z(t),$$

then (3.2.9) is the same as follows:

$$\begin{aligned}\widetilde{Y}^i(\tau, t) = D_\tau^{(i)}\xi + \int_t^T \Big(&[D_\tau^{(i)}g](s, Y(s), Z(s)) + g_y(s, Y(s), Z(s))\widetilde{Y}^i(\tau, s)\\ &+\sum_{j=1}^d g_{z_j}(s, Y(s), Z(s))\widetilde{Z}_j^i(\tau, s)\Big)ds\\ &-\sum_{j=1}^d \int_t^T \widetilde{Z}_j^i(\tau, s)dW_j(s), \quad t \in [\tau, T],\ 1 \leqslant i \leqslant d.\end{aligned} \tag{3.2.21}$$

In particular, with $t = \tau$, one has

$$\begin{aligned}\widetilde{Y}^i(t,t)=D_t^{(i)}\xi+\int_t^T\Big([D_t^{(i)}g](t,s,Y(s),Z(s))+g_y(s,Y(s),Z(s))\widetilde{Y}^i(t,s)\\+\sum_{j=1}^d g_{z_j}(s,Y(s),Z(s))\widetilde{Z}_j^i(t,s)\Big)ds-\int_t^T\widetilde{Z}^i(t,s)dW(s),\\ t\in[0,T],\ 1\leqslant i\leqslant d.\end{aligned}\tag{3.2.22}$$

This looks like a Type I BSVIE for $(\widetilde{Y}^i(\cdot\,,\cdot), \widetilde{Z}^i(\cdot\,,\cdot))$. Unfortunately, it is not in our sense because $\widetilde{Y}^i(t,t)$ appears on the left-hand side and $\widetilde{Y}^i(t,s)$ appears in the integral on the right-hand side. But (4.2.23) can be regarded as a BSVIE defined on the square domain $[0, T]^2$, whose general theory will not be discussed, and we only present briefly the continuity of the solutions.

We now look at the following BSDE with parameter $t \in [0, T]$, $d = 1$:

$$\begin{aligned}Y(t,r)=\psi(t)+\int_r^T g(t,s,Y(t,s),Z(t,s))ds-\int_r^T Z(t,s)dW(s),\\ r\in[0,T].\end{aligned}\tag{3.2.23}$$

Here $\psi : [0,T]\times\Omega \to \mathbb{R}^m$, $g : [0,T]^2\times\Omega\times\mathbb{R}^m\times\mathbb{R}^m \to \mathbb{R}^m$ are two measurable mappings. To show the well-posedness of (3.2.23), let us make the following assumptions.

(H3.2.2) For given $t\in[0,T]$, $\psi(t)\in L^2_{\mathcal{F}_T}(\Omega;\mathbb{R}^m)$, $g(t,\cdot,0,0)\in L^2_{\mathbb{F}}(0,T;\mathbb{R}^m)$, and there exists $K>0$ such that for $t,s\in[0,T]$, $y,z\in\mathbb{R}^m$,

$$\big|g(t,s,y_1,z_1)-g(t,s,y_2,z_2)\big|\leqslant K(|y_1-y_2|+|z_1-z_2|).$$

(H3.2.3) For each $t_0\in[0,T]$, $(Y(t_0,\cdot),Z(t_0,\cdot))$ is the solution of BSDE (3.2.23),

$$\begin{aligned}\lim_{t\to t_0}\Big[\big\|g(t,\cdot,Y(t_0,\cdot),Z(t_0,\cdot))-g(t_0,\cdot,Y(t_0,\cdot),Z(t_0,\cdot))\big\|_{L^2_{\mathbb{F}}(0,T;\mathbb{R}^m)}\\+\big\|\psi(t)-\psi(t_0)\big\|_{L^2_{\mathcal{F}_T}(\Omega;\mathbb{R}^m)}\Big]=0.\end{aligned}$$

(H3.2.4) For any $t,s\in[0,T]$, $g(t,\cdot,0,0)\in L^2_{\mathbb{F}}(0,T;\mathbb{R}^m)$, $(y,z)\mapsto g(t,s,y,z)$ is differentiable with bounded derivatives such that $(t,y,z)\mapsto g_y(t,s,y,z)$, $g_z(t,s,y,z)$ are continuous. In addition, for any $t_0\in[0,T]$ and the associated solution $(Y(t_0,\cdot),Z(t_0,\cdot))$, $t\mapsto g(t,\cdot,Y(t_0,\cdot),Z(t_0,\cdot))$, $t\mapsto\psi(t)$ are differentiable in space $L^2_{\mathbb{F}}(0,T;\mathbb{R}^m)$, $L^2_{\mathcal{F}_T}(\Omega;\mathbb{R}^m)$ at t_0, with derivative $\partial_t g(t_0,\cdot,Y(t_0,\cdot),Z(t_0,\cdot))$, $\partial_t\psi(t_0)$, respectively.

(H3.2.5) Let (H3.2.4) hold and any $t_0 \in [0, T]$,

$$\lim_{t\to t_0}\Big[\big\|\partial_t g(t,\cdot,Y(t,\cdot),Z(t,\cdot)) - \partial_t g(t_0,\cdot,Y(t_0,\cdot),Z(t_0,\cdot))\big\|_{L^2_{\mathbb{F}}(0,T;\mathbb{R}^m)} + \big\|\partial_t\psi(t) - \partial_t\psi(t_0)\big\|_{L^2_{\mathcal{F}_T}(\Omega;\mathbb{R}^m)}\Big] = 0.$$

Proposition 3.2.2 *Suppose* (H3.2.2), (H3.2.3) *hold. Then*

$$\lim_{t\to t_0}\Big[\big\|Y(t,\cdot) - Y(t_0,\cdot)\big\|_{L^2_{\mathbb{F}}(\Omega;C([0,T];\mathbb{R}^m))} + \big\|Z(t,\cdot) - Z(t_0,\cdot)\big\|_{L^2_{\mathbb{F}}(0,T;\mathbb{R}^m)}\Big] = 0. \tag{3.2.24}$$

If (H3.2.4) holds, then $t \mapsto (Y(t,\cdot), Z(t,\cdot))$, $[0,T] \to L^2_{\mathbb{F}}(\Omega; C([0,T];\mathbb{R}^m)) \times L^2_{\mathbb{F}}(0,T;\mathbb{R}^m)$ *is differentiable, with derivative given by* $(\partial_t Y(t,\cdot), \partial_t Z(t,\cdot))$ *satisfying*

$$\partial_t Y(t,r) = \partial_t\psi(t) + \int_r^T \nabla g(t,s,\partial_t Y(t,s),\partial_t Z(t,s))ds - \int_r^T \partial_t Z(t,s)dW(s), \tag{3.2.25}$$

where

$$\nabla g(t,s,y,z) \equiv g_y(t,s,Y(t,s),Z(t,s))y + g_z(t,s,Y(t,s),Z(t,s))z + \partial_t g(t,s,Y(t,s),Z(t,s)).$$

If (H3.2.5) holds, then $t \mapsto (\partial_t Y(t,\cdot), \partial_t Z(t,\cdot))$ *is continuous.*

Proof For simplicity, we assume that $m = 1$. By the standard estimate of BSDEs,

$$\begin{aligned}&\lim_{t\to t_0}\Big[\mathbb{E}\sup_{s\in[0,T]}\big|Y(t,s) - Y(t_0,s)\big|^2 + \mathbb{E}\int_0^T \big|Z(t,s) - Z(t_0,s)\big|^2 ds\Big]\\ &\leqslant K\lim_{t\to t_0}\mathbb{E}\Big[|\psi(t) - \psi(t_0)|^2 + \Big(\int_0^T |g(t,s,Y(t_0,s),Z(t_0,s)) - g(t_0,s,Y(t_0,s),Z(t_0,s))|ds\Big)^2\Big].\end{aligned}$$

Then (3.2.34) is followed by (H3.2.3) naturally.

We discuss the differentiability. To this end, define

$$\Delta_t f(t,r) \equiv \frac{f(t,r) - f(t_0,r)}{t - t_0}, \qquad f \equiv Y, Z, \quad t, r \in [0,T].$$

Then it is easy to see

$$\begin{aligned}\Delta_t Y(t,r) = &\frac{\psi(t)-\psi(t_0)}{t-t_0} + \int_r^T \Delta g(t,s,\Delta_t Y(t,s),\Delta_t Z(t,s))ds\\ &-\int_r^T \Delta_t Z(t,s)dW(s), \quad t,r\in[0,T],\end{aligned}$$

where

$$\begin{aligned}\Delta g(t,s,y,z) &\equiv A(t,s)y + B(t,s)z + \varphi(t,s),\\ A(t,s) &\equiv \int_0^1 g_y(t,s,Y(t_0,s)+\lambda(Y(t,s)-Y(t_0,s)),Z(t,s))d\lambda,\\ B(t,s) &\equiv \int_0^1 g_z(t,s,Y(t_0,s),Z_0(t,s)+\lambda(Z(t,s)-Z(t_0,s)))d\lambda,\\ \varphi(t,s) &\equiv \frac{1}{t-t_0}\big[g(t,s,Y(t_0,s),Z(t_0,s)) - g(t_0,s,Y(t_0,s),Z(t_0,s))\big].\end{aligned}$$

We claim that

$$\begin{aligned}\lim_{t\to t_0}\mathbb{E}\int_0^T &\big|\Delta g(t,s,\partial_t Y(t_0,s),\partial_t Z(t_0,s))\\ &-\nabla g(t_0,s,\partial_t Y(t_0,s),\partial_t Z(t_0,s))\big|^2 ds = 0.\end{aligned} \tag{3.2.26}$$

This together with the differentiability of $t\mapsto\psi(t)$ leads to

$$\lim_{t\to t_0}\Big[\mathbb{E}\sup_{s\in[0,T]}|\Delta_t Y(t,s)-\partial_t Y(t_0,s)|^2+\mathbb{E}\int_0^T|\Delta_t Z(t,s)-\partial_t Z(t_0,s)|^2 ds\Big]=0,$$

which implies the desired differentiability result of $t\mapsto(Y(t,\cdot),Z(t,\cdot))$. In fact, notice that

$$\begin{aligned}&\mathbb{E}\int_0^T\big|A(t,s)-g_y(t,s,Y(t_0,s),Z(t_0,s))\big|^2|\partial_t Y(t_0,s)|^2 ds\\ &\leqslant \mathbb{E}\int_0^T\int_0^1\big|g_y(t,s,Y(t_0,s)+\lambda(Y(t,s)-Y(t_0,s)),Z(t,s))\\ &\qquad -g_y(t,s,Y(t_0,s),Z(t_0,s))\big|^2|\partial_t Y(t_0,s)|^2 ds.\end{aligned} \tag{3.2.27}$$

By the previous (3.2.34), $Z(t,\cdot)$ converges to $Z(t_0,\cdot)$ in measure relative to $\mathbb{P}\otimes\lambda$, and for any $s\in[0,T]$, $Y(t,s)$ converges to $Y(t_0,s)$ in probability. Recall that $(y,z)\mapsto g_y(t,s,y,z)$ is continuous, and $g_y(t,s,y,z)$ is uniformly bounded. Then by the dominated convergence theorem, the right hand of (3.2.27) approaches to zero. Using again the continuity of $t \mapsto g_y(t,s,y,z)$,

$$\begin{aligned}\lim_{t\to t_0}\mathbb{E}\int_0^T &\big|g_y(t,s,Y(t_0,s),Z(t_0,s))\\ &-g_y(t_0,s,Y(t_0,s),Z(t_0,s))\big|^2|\partial_t Y(t_0,s)|^2ds=0.\end{aligned} \tag{3.2.28}$$

Combining (3.2.27) and (3.2.28) together, we arrive at

$$\lim_{t\to t_0}\mathbb{E}\int_0^T\big|A(t,s)-\partial_y g(t_0,s,Y(t_0,s),Z(t_0,s))\big|^2|\partial_t Y(t_0,s)|^2ds=0. \tag{3.2.29}$$

Similarly, one can obtain that

$$\lim_{t\to t_0}\mathbb{E}\int_0^T\big|B(t,s)-\partial_z g(t_0,s,Y(t_0,s),Z(t_0,s))\big|^2|\partial_t Z(t_0,s)|^2ds=0. \tag{3.2.30}$$

By the differentiability of $t\mapsto g(t,\cdot,Y(t_0,\cdot),Z(t_0,\cdot))$, we see that

$$\lim_{t\to t_0}\mathbb{E}\int_0^T\big|\varphi(t,s)-\partial_t g(t_0,s,Y(t_0,s),Z(t_0,s))\big|^2ds=0. \tag{3.2.31}$$

Then the above (3.2.26) is followed by (3.2.27), (3.2.30), and (3.2.31).

Finally, we prove the following continuity under (H3.2.5),

$$\begin{aligned}\lim_{t\to t_0}\Big[&\big\|\partial_t Y(t,\cdot)-\partial_t Y(t_0,\cdot)\big\|_{L^2_{\mathbb{F}}(\Omega;C([0,T];\mathbb{R}^m))}\\ &+\big\|\partial_t Z(t,\cdot)-\partial_t Z(t_0,\cdot)\big\|_{L^2_{\mathbb{F}}(0,T;\mathbb{R}^m)}\Big]=0.\end{aligned} \tag{3.2.32}$$

First, we have

$$\begin{aligned}&\big|\nabla g(t,s,\partial_t Y(t_0,s),\partial_t Z(t_0,s))-\nabla g(t_0,s,\partial_t Y(t_0,s),\partial_t Z(t_0,s))\big|\\ &\leqslant\big|g_y(t,s,Y(t,s),Z(t,s))-g_y(t_0,s,Y(t_0,s),Z(t_0,s))\big|\big|\partial_t Y(t_0,s)\big|\\ &\quad+\big|g_z(t,s,Y(t,s),Z(t,s))-g_z(t_0,s,Y(t_0,s),Z(t_0,s))\big|\big|\partial_t Z(t_0,s)\big|\\ &\quad+\big|\partial_t g(t,s,Y(t,s),Z(t,s))-\partial_t g(t_0,s,Y(t_0,s),Z(t_0,s))\big|.\end{aligned}$$

Then by the continuity of $(t,y,z)\mapsto g_i(t,s,y,z)$, the boundedness of $g_i(t,s,y,z)$, $i=y,z$, assumption (H3.2.5), the convergence (3.2.34), and dominated convergence arguments, we obtain the right hand converge to zero in $\mathcal{H}_2$. Using again the continuity requirement of $t\mapsto\partial_t\psi(t)$ in (H3.2.5), we then obtain the above (3.2.32).

□

As a special case, we consider a linear form of (3.2.23):

$$\begin{aligned}\widetilde{Y}(\tau,t)=\widetilde{\psi}(\tau,t)+\int_t^T\Big(g_0(s)\widetilde{Y}(\tau,s)+g(s)\widetilde{Z}(\tau,s)\Big)ds\\-\int_t^T\widetilde{Z}(\tau,s)dW(s),\qquad t\in[0,T].\end{aligned}\tag{3.2.33}$$

Note that in the case of (4.2.23), with $d=1$,

$$\widetilde{\psi}(\tau,t)=D_\tau\xi+\int_t^T[D_\tau g](s,Y(s),Z(s))ds,\qquad(\tau,t)\in[0,T]^2.$$

Corollary 3.2.3 *Suppose* $g_0(\cdot),g(\cdot)\in L^\infty_{\mathbb{F}}(0,T;\mathbb{R}^m)$, $\widetilde{\psi}(\cdot\,,\cdot)\in C([0,T];L^p_{\mathbb{F}}(\Omega;C([0,T];\mathbb{R}^m))$. *Then for any* $\tau_0\in[0,T]$,

$$\begin{aligned}\lim_{\tau\to\tau_0}\Big[\big\|\widetilde{Y}(\tau,\cdot)-\widetilde{Y}(\tau_0,\cdot)\big\|_{L^2_{\mathbb{F}}(\Omega;C([0,T];\mathbb{R}^m))}\\+\big\|\widetilde{Z}(\tau,\cdot)-\widetilde{Z}(\tau_0,\cdot)\big\|_{L^2_{\mathbb{F}}(0,T;\mathbb{R}^m))}\Big]=0.\end{aligned}\tag{3.2.34}$$

From the above corollary, we see that for any given $t_0\in[0,T]$.

$$\begin{aligned}&\mathbb{E}|\widetilde{Y}(t,t)-\widetilde{Y}(t_0,t_0)|\leqslant\mathbb{E}|\widetilde{Y}(t,t)-\widetilde{Y}(t_0,t)|+\mathbb{E}|\widetilde{Y}(t_0,t)-\widetilde{Y}(t_0,t_0)|\\&\leqslant\mathbb{E}\Big[\sup_{s\in[0,T]}|\widetilde{Y}(t,s)-\widetilde{Y}(t_0,s)|\Big]+\mathbb{E}|\widetilde{Y}(t_0,t)-\widetilde{Y}(t_0,t_0)|\to0,\end{aligned}$$

as $t\to t_0$. Hence, by the result of $Z_i(\tau)=D^{(i)}_\tau Y(\tau)$, we obtain the continuity of $t\mapsto Z_i(t)$ as a map from $[0,T]$ to $L^1_{\mathcal{F}_T}(\Omega;\mathbb{R}^{m\times d})$ under proper conditions. We leave the exact statement of such a result to the interested readers.

3 Linear BSDEs

In this section, we look at linear BSDEs of the following form:

$$\begin{cases}dY(t)=\Big[A(t)Y(t)+\sum_{i=1}^dC_i(t)Z_i(t)+g(t)\Big]dt+\sum_{i=1}^dZ_i(t)dW_i(t),\\\qquad\qquad\qquad\qquad\qquad\qquad\qquad\qquad t\in[0,T],\\Y(T)=\xi.\end{cases}\tag{3.3.1}$$

If all the coefficients are $\mathbb{F}$-progressively measurable and bounded, then for any $\xi\in L^p_{\mathcal{F}_T}(\Omega;\mathbb{R}^m)$, and $g(\cdot)\in L^p_{\mathbb{F}}(0.T;\mathbb{R}^m)$, $p>1$, the above BSDE admits a unique adapted solution $(Y(\cdot),Z(\cdot))\in\mathbf{H}^p[0,T]$. Due to the linear structure of the equation, the adapted solution should be a linear function of $(\xi, g(\cdot))$. This is usually named as a *variation of constants formula*. The main purpose of this section is to derive such a formula.

Theorem 3.3.1 *Let $A(\cdot), C_i(\cdot) \in L^\infty_{\mathbb{F}}(0,T;\mathbb{R}^{m\times m})$, $i = 1, \cdot s, d$. Then for any $\xi \in L^p_{\mathcal{F}_T}(\Omega;\mathbb{R}^m)$ and $g(\cdot) \in L^p_{\mathbb{F}}(0,T;\mathbb{R}^m)$ with some $p \in (1,\infty)$, the adapted solution $(Y(\cdot), Z(\cdot)) \in \mathbf{H}^p[0,T]$ of linear BSDE (3.3.1) admits the following representation:*

$$\begin{cases} Y(t) = \Psi(t)^{-1}\Big(\int_0^t \Psi(\tau)g(\tau)d\tau + E_t[\eta]\Big), \\ Z_i(t) = C_i(t)Y(t) + \Psi(t)^{-1}\zeta_i(t), \quad 1 \leqslant i \leqslant d, \end{cases} \quad t \in [0,T], \tag{3.3.2}$$

where $\Psi(\cdot)$ is the solution to the following SDE:

$$\begin{cases} d\Psi(t) = -\Psi(t)A(t)dt - \sum_{i=1}^d \Psi(t)C_i(t)dW_i(t), \quad t \in [0,T], \\ \Psi(0) = I, \end{cases} \tag{3.3.3}$$

and $\Psi(t)^{-1}$ must exists for all $t \in [0,T]$ almost surely; η is given by

$$\eta = \Psi(T)\xi - \int_0^T \Psi(\tau)g(\tau)d\tau, \tag{3.3.4}$$

whose martingale representation is given by the following:

$$E_t[\eta] = E\eta + \sum_{i=1}^d \int_0^t \zeta_i(\tau)dW_i(\tau), \quad \forall t \in [0,T], \ \mathbb{P}\text{-a.s.} \tag{3.3.5}$$

Proof We introduce the following linear SDEs:

$$\begin{cases} d\Psi(t;s) = -\Psi(t;s)A(t)dt - \sum_{i=1}^d \Psi(t;s)C_i(t)dW_i(t), \quad t \in [s,T], \\ \Psi(s;s) = I, \end{cases} \tag{3.3.6}$$

and

$$\begin{cases} d\widetilde{\Psi}(t;s) = \Big(A(t) + \sum_{i=1}^d C_i(t)^2\Big)\widetilde{\Psi}(t;s)dt + \sum_{i=1}^d C_i(t)\widetilde{\Psi}(t;s)dW_i(t), \\ \qquad\qquad\qquad\qquad t \in [s,T], \\ \widetilde{\Psi}(s;s) = I. \end{cases} \tag{3.3.7}$$

Under our conditions, they are well-posed. Next, by Itô's formula, we have (suppressing s)

$$\begin{aligned} d[\widetilde{\Psi}(t)\Psi(t)] &= \Big[\Big(A(t)+\sum_{i=1}^{d}C_i(t)^2\Big)[\widetilde{\Psi}(t)\Psi(t)] - [\widetilde{\Psi}(t)\Psi(t)]A(t) \\ &\quad -\sum_{i=1}^{d}C_i(t)[\widetilde{\Psi}(t)\Psi(t)]C_i(t)\Big]dt \\ &\quad +\sum_{i=1}^{d}\Big(C_i(t)[\widetilde{\Psi}(t)\Psi(t)] - [\widetilde{\Psi}(t)\Psi(t)]C_i(t)\Big)dW_i(t). \end{aligned}$$

This means that $t \mapsto [\widetilde{\Psi}(t;s)\Psi(t;s)]$ is a solution of the following FSDE:

$$\begin{cases} d\mathbf{X}(t) = \Big[\Big(A(t)+\sum_{i=1}^{d}C_i(t)^2\Big)\mathbf{X}(t) - \mathbf{X}(t)A(t) - \sum_{i=1}^{d}C_i(t)\mathbf{X}(t)C_i(t)\Big]dt \\ \qquad\qquad +\sum_{i=1}^{d}\Big(C_i(t)\mathbf{X}(t) - \mathbf{X}(t)C_i(t)\Big)dW_i(t), \qquad t\in[s,T], \\ \mathbf{X}(s) = I. \end{cases}$$

On the other hand, I is also its solution. Therefore, by the uniqueness in Chap. 2, Proposition 2.3.11, we have

$$\widetilde{\Psi}(t;s)\Psi(t;s) = I, \qquad t\in[s,T].$$

Now, let $(Y(\cdot), Z(\cdot))$ be the adapted solution of (3.3.1). Applying Itô's formula to $t \mapsto \Psi(t;s)Y(t)$ on $[s,T]$, we have

$$\begin{aligned} d[\Psi(t;s)Y(t)] &= -\Psi(t;s)A(t)Y(t)dt - \sum_{i=1}^{d}\Psi(t;s)C_i(t)Y(t)dW_i(t) \\ &\quad +\Psi(t;s)\Big[A(t)Y(t)+\sum_{i=1}^{d}C_i(t)Z_i(t)+g(t)\Big]dt \\ &\quad +\sum_{i=1}^{d}\Psi(t;s)Z_i(t)dW_i(t) - \sum_{i=1}^{d}\Psi(t;s)C_i(t)Z_i(t)dt \\ &= \Psi(t;s)g(t)dt + \sum_{i=1}^{d}\Psi(t;s)\big[Z_i(t) - C_i(t)Y(t)\big]dW_i(t). \end{aligned}$$

Thus,

$$\begin{aligned}\Psi(t;s)Y(t) &= \Psi(T;s)\xi - \int_t^T \Psi(\tau;s)g(\tau)d\tau \\ &\quad - \sum_{i=1}^d \int_t^T \Psi(\tau;s)\big[Z_i(\tau) - C_i(\tau)Y(\tau)\big]dW_i(\tau) \\ \equiv \eta(s) &+ \int_s^t \Psi(\tau;s)g(\tau)d\tau - \sum_{i=1}^d \int_t^T \Psi(\tau;s)\big[Z_i(\tau) - C_i(\tau)Y(\tau)\big]dW_i(\tau),\end{aligned}$$

where

$$\eta(s) = \Psi(T;s)\xi - \int_s^T \Psi(\tau;s)g(\tau)d\tau.$$

For fixed $s \in [0, T]$, by the integrability of Ψ and ξ, we see that there exists $1 < p_1 < p$ such that $\eta(s) \in L^{p_1}_{\mathcal{F}_T}(\Omega; \mathbb{R}^m)$. The above yields

$$\Psi(t;s)Y(t) = \int_s^t \Psi(\tau;s)g(\tau)d\tau + E_t[\eta(s)], \qquad t \in [s, T]. \tag{3.3.8}$$

Hence,

$$Y(t) = \Psi(t;s)^{-1}\Big(\int_s^t \Psi(\tau;s)g(\tau)d\tau + E_t\big[\eta(s)\big]\Big), \qquad t \in [0, T]. \tag{3.3.9}$$

This gives a representation of $Y(\cdot)$. To obtain a representation of $Z(\cdot)$, by the martingale representation theorem (Chap. 2, Theorem 2.3.8) and the integrability of η, we can find a unique $\zeta(\cdot\,;s) \equiv (\zeta_1(\cdot\ s), \cdot s, \zeta_d(\cdot\,;s)) \in L^{p_1}_{\mathbb{F}}\big(\Omega; L^2(s, T; \mathbb{R}^{m\times d})\big)$ such that

$$E_t\big[\eta(s)\big] = E_s\big[\eta(s)\big] + \sum_{i=1}^d \int_s^t \zeta_i(\tau;s)dW_i(\tau), \quad \forall t \in [s, T], \text{ a.s.} \tag{3.3.10}$$

Hence, it follows that

$$\begin{aligned}Y(t) &= \widetilde{\Psi}(t;s)\Big\{\int_s^t \Psi(\tau;s)g(\tau)d\tau + \sum_{i=1}^d \int_s^t \zeta_i(\tau;s)dW_i(\tau) + E_s[\eta(s)]\Big\}, \\ &\equiv \widetilde{\Psi}(t;s)h(t;s), \qquad \forall t \in [s, T], \text{ a.s.}\end{aligned}$$

Applying Itô's formula, we obtain

$$\begin{aligned}
dY(t) &= \Big[A(t) + \sum_{i=1}^{d} C_i(t)^2\Big]\widetilde{\Psi}(t;s)h(t,s)dt \\
&\quad + \sum_{i=1}^{d} C_i(t)\widetilde{\Psi}(t)h(t,s)dW_i(t) + \widetilde{\Psi}(t;s)\Psi(t;s)g(t)dt \\
&\quad + \sum_{i=1}^{d} \widetilde{\Psi}(t;s)\zeta_i(t;s)dW_i(t) + \sum_{i=1}^{d} C_i(t)\widetilde{\Psi}(t)\zeta_i(t;s)dt \\
&= \Big\{A(t)Y(t) + \sum_{i=1}^{d} C_i(t)[C_i(t)Y(t) + \widetilde{\Psi}(t;s)\zeta_i(t;s)] + g(t)\Big\}dt \\
&\quad + \sum_{i=1}^{d} [C_i(t)Y(t) + \widetilde{\Psi}(t;s)\zeta_i(t;s)]dW_i(t).
\end{aligned}$$

By the integrability of $Y, \widetilde{\Psi}, \zeta_i$, there exists $1 < p_2 < p_1 < p$ such that $C_i(\cdot)Y(\cdot) + \Psi(\cdot;s)^{-1}\zeta_i(\cdot;s) \in L^{p_2}_{\mathbb{F}}(\Omega; L^2(s,T;\mathbb{R}^m))$. Therefore, by the uniqueness of linear BSDE (3.3.1) in $\mathbf{H}^{p_2}[s,T]$, we must have

$$Z_i(t) = C_i(t)Y(t) + \Psi(t;s)^{-1}\zeta_i(t;s), \quad t \in [s,T], \text{ a.s., } 1 \leqslant i \leqslant d.$$

Take $s = 0$, and we complete the proof. □

Let us present an illustrative example.

Example 3.3.2 Consider the following one-dimensional BSDE:

$$\begin{cases} dY(t) = [Y(t) + Z(t) + e^{W(t)}]dt + Z(t)dW(t), & t \in [0,T], \\ Y(T) = e^{W(T)}. \end{cases}$$

Let $\Psi(\cdot)$ be the solution to the following:

$$\begin{cases} d\Psi(t) = -\Psi(t)dt - \Psi(t)dW(t), & t \in [0,T], \\ \Psi(0) = 1. \end{cases}$$

Clearly, in this case,

$$\Psi(t) = e^{-\frac{3}{2}t - W(t)}, \quad t \in [0,T].$$

Then,

$$\begin{aligned}
\eta &\equiv \Psi(T)e^{W(T)} - \int_0^T \Psi(\tau)e^{W(\tau)}d\tau \\
&= e^{-\frac{3}{2}t} - \int_0^T e^{-\frac{3}{2}\tau}d\tau = e^{-\frac{3}{2}t} + \frac{2}{3}\Big(e^{-\frac{3}{2}T} - 1\Big),
\end{aligned}$$

which is deterministic. Hence,

$$
\begin{aligned}
Y(t) &= \Psi(t)^{-1}\Big(\int_0^t \Psi(\tau)e^{W(\tau)}d\tau + E_t[\eta]\Big)\\
&= e^{\frac{3}{2}t+W(t)}\int_0^t e^{-\frac{3}{2}\tau}d\tau + e^{-\frac{3}{2}t} + \frac{2}{3}\Big(e^{-\frac{3}{2}T}-1\Big)\\
&= -\frac{2}{3}e^{\frac{3}{2}t+W(t)}\Big(e^{-\frac{3}{2}t}-1\Big) + e^{-\frac{3}{2}t} + \frac{2}{3}\Big(e^{-\frac{3}{2}T}-1\Big), \qquad t\in[0,T],
\end{aligned}
$$

and

$$
Z(t) = Y(t), \qquad t\in[0,T].
$$

The above gives a representation of adapted solution $(Y(\cdot), Z(\cdot))$.

4 General FSVIEs

In this section, we will briefly present some basic results on general *forward stochastic Volterra integral equation* (FSVIE, for short). More precisely, we consider the following equation:

$$
X(t) = \varphi(t)+\int_0^t b(t,s,X(s))ds+\int_0^t \sigma(t,s,X(s))dW(s), \qquad t\in[0,T]. \tag{3.4.1}
$$

In the above, $X(\cdot)$ is the unknown process which is supposed to be $\mathbb{F}$-progressively measurable $\mathbb{R}^n$-valued, $\varphi(\cdot)$ is called the *free term*, and the pair $(b(\cdot), \sigma(\cdot))$ is called the *generator* of the FSVIE. Also, the two integral terms on the right-hand side are called the *drift* and *diffusion*, respectively. Recall the lower triangle domain $\Delta_*[0,T]$ (see Chap. 1, (1.1.8)). Before going further, let us make a reduction. We may rewrite (3.4.1) as follows:

$$
\begin{aligned}
X(t) = \varphi(t) &+ \int_0^t b(t,s,0)ds + \int_0^t \sigma(t,s,0)dW(s)\\
&+ \int_0^t \widetilde b(t,s,X(s))ds + \int_0^t \widetilde\sigma(t,s,X(s))dW(s), \qquad t\in[0,T],
\end{aligned} \tag{3.4.2}
$$

where

$$
\widetilde b(t,s,x) = b(t,s,x) - b(t,s,0), \qquad \widetilde\sigma(t,s,x) = \sigma(t,s,x) - \sigma(t,s,0).
$$

Clearly,

$$
\widetilde b(t,s,0) = 0, \qquad \widetilde\sigma(t,s,0) = 0.
$$

We may let

$$\widetilde{\varphi}(t)=\varphi(t)+\int_0^t b(t,s,0)ds+\int_0^t \sigma(t,s,0)dW(s),\qquad t\in[0,T]$$

to be the new free term. Hence, in what follows, without loss of generality, we assume that in FSVIE (3.4.1),

$$b(t,s,0)=0,\quad \sigma(t,s,0)=0,\qquad (t,s)\in\Delta_*[0,T]. \tag{3.4.3}$$

For FSVIE (3.4.1), the most common hypothesis is the following:

(H3.4.1) Let maps $b,\sigma:\Delta_*[0,T]\times\mathbb{R}^n\times\Omega\to\mathbb{R}^n$ be measurable. For each $(t,x)\in[0,T]\times\mathbb{R}^n$, $s\mapsto(b(t,s,x),\sigma(t,s,x))$ is $\mathbb{F}$-progressively measurable on $[0,t]$. Moreover, for some constant $K>0$, it holds that

$$\begin{aligned}|b(t,s,x)-b(t,s,x')|,\ |\sigma(t,s,x)-\sigma(t,s,x')|\leqslant K|x-x'|,\\ (t,s)\in\Delta_*[0,T],\ x,x'\in\mathbb{R}^n,\ \text{a.s.}\end{aligned} \tag{3.4.4}$$

Due to (3.4.3), condition (3.4.4) implies

$$|b(t,s,x)|,\ |\sigma(t,s,x)|\leqslant K|x|,\qquad \forall(t,s,x)\in\Delta_*[0,T]\times\mathbb{R}^n. \tag{3.4.5}$$

Thus, in this case, $(t,s)\mapsto(b(t,s,x),\sigma(t,s,x))$ does not have singularities.

Theorem 3.4.1 *Let* (H3.4.1) *hold. For any* $p\geqslant 2$, $\varphi(\cdot)\in L^p_{\mathbb{F}}(0,T;\mathbb{R}^n)$, *FSVIE* (3.4.1) *admits a unique solution* $X(\cdot)\in L^p_{\mathbb{F}}(0,T;\mathbb{R}^n)$, *and the following estimate holds:*

$$\mathbb{E}|X(t)|^p\leqslant K\mathbb{E}|\varphi(t)|^p+K\int_0^t\mathbb{E}|\varphi(s)|^pds,\qquad \text{a.e. } t\in[0,T]. \tag{3.4.6}$$

Further, let $(\bar b,\bar\sigma)$ *be another generator of an FSVIE satisfying (H3.4.1) and* $\bar\varphi(\cdot)\in L^p_{\mathbb{F}}(0,T;\mathbb{R}^n)$. *Let* $\bar X(\cdot)$ *be the corresponding solution. Then the following stability estimate holds:*

$$\begin{aligned}\mathbb{E}|X(t)-\bar X(t)|^pdt\leqslant K\mathbb{E}\Big\{&|\varphi(t)-\bar\varphi(t)|^p+\int_0^t|\varphi(s)-\bar\varphi(s)|^pds\\ &+\Big(\int_0^t|b(t,s,X(s))-\bar b(t,s,X(s))|ds\Big)^p\\ &+\int_0^t\Big(\int_0^r|b(r,s,X(s))-\bar b(r,s,X(s))|ds\Big)^pdr\\ &+\Big(\int_0^t|\sigma(t,s,X(s))-\bar\sigma(t,s,X(s))|^2ds\Big)^{\frac p2}\\ &+\int_0^t\Big(\int_0^r|\sigma(r,s,X(s))-\bar\sigma(r,s,X(s))|^2ds\Big)^{\frac p2}dr\Big\}.\end{aligned} \tag{3.4.7}$$

Proof First of all, for any $X(\cdot) \in L^p_{\mathbb{F}}(0,T;\mathbb{R}^n)$, we define a map

$$\mathcal{S}[X(\cdot)](t)=\varphi(t)+\int_0^t b(t,s,X(s))ds+\int_0^t \sigma(t,s,X(s))dW(s),\quad t\in[0,T]. \tag{3.4.8}$$

By Burkholder–Davis–Gundy's inequalities (Chap. 2, Theorem 2.3.2), we have (fixing $t\in[0,T]$)

$$\begin{aligned}\mathbb{E}\Big|\int_0^t \sigma(t,s,X(s))dW(s)\Big|^p &\leqslant \mathbb{E}\Big[\sup_{r\in[0,t]}\Big|\int_0^r \sigma(t,s,X(s))dW(s)\Big|^p\Big]\\ &\leqslant \overline{K}_p\mathbb{E}\Big(\int_0^t |\sigma(t,s,X(s))|^2ds\Big)^{\frac{p}{2}},\end{aligned}$$

where $\overline{K}_p$ is the constant appearing in the Burkholder–Davis–Gundy's inequalities (Chap. 2, Theorem 2.3.2). Then it follows that

$$\begin{aligned}&\mathbb{E}\Big[|\mathcal{S}[X(\cdot)](t)|^p\Big]\\ &\leqslant K\mathbb{E}\Big\{|\varphi(t)|^p+\Big|\int_0^t b(t,s,X(s))ds\Big|^p+\Big|\int_0^t\sigma(t,s,X(s))dW(s)\Big|^p\Big\}\\ &\leqslant K\mathbb{E}\Big\{|\varphi(t)|^p+\Big(\int_0^t L|X(s)|ds\Big)^p+\sup_{r\in[0,t]}\Big|\int_0^r\sigma(t,s,X(s))dWs)\Big|^p\Big\}\\ &\leqslant K\mathbb{E}\Big\{|\varphi(t)|^p+\Big(\int_0^t|X(s)|ds\Big)^p+\overline{K}_p\Big(\int_0^t|\sigma(t,s,X(s))|^2ds\Big)^{\frac{p}{2}}\Big\}\\ &\leqslant K\mathbb{E}\Big(|\varphi(t)|^p+\int_0^t|X(s)|^pds\Big).\end{aligned} \tag{3.4.9}$$

Consequently, we have

$$\int_0^T \mathbb{E}|\mathcal{S}[X(\cdot)](t)|^pdt\leqslant K\mathbb{E}\Big(\int_0^T|\varphi(t)|^pdt+\int_0^T\mathbb{E}|X(t)|^pdt\Big),$$

with an absolute constant $K>0$. Hence, $\mathcal{S}: L^p_{\mathbb{F}}(0,T;\mathbb{R}^n)\to L^p_{\mathbb{F}}(0,T;\mathbb{R}^n)$ is well-defined. Next, for any $X(\cdot),\bar X(\cdot)\in L^p_{\mathbb{F}}(0,S;\mathbb{R}^n)$ and some undetermined parameter $\beta>0$, similar as the above (3.4.9),

$$\begin{aligned}&\int_0^T e^{-\beta t}\mathbb{E}|\mathcal{S}[X(\cdot)-\bar X(\cdot)](t)|^pdt\\ &\leqslant K\mathbb{E}\int_0^T e^{-\beta t}\int_0^t|X(s)-\bar X(s)|^pdsdt\\ &=K\mathbb{E}\int_0^T\Big[\int_s^T e^{-\beta t}dt\Big]|X(s)-\bar X(s)|^pds\\ &\leqslant\frac{K}{\beta}\mathbb{E}\int_0^T e^{-\beta s}|X(s)-\bar X(s)|^pds.\end{aligned}$$

By choosing $\beta > K$, we see that $\mathcal{S}$ is a contraction map on $L^p_{\mathbb{F}}(0,T;\mathbb{R}^n)$. Thus, it admits a unique fixed point in $L^p_{\mathbb{F}}(0,T;\mathbb{R}^n)$, which is the unique solution of FSVIE (3.4.1) on $[0,T]$.

Now by applying Gronwall's inequality to (3.4.9) (with $\mathcal{S}[X(\cdot)] = X(\cdot)$), we obtain estimate (3.4.6). Further, let $(\bar{b},\bar{\sigma})$ be the generator of another FSVIE satisfying (H3.4.1) and $\bar{\varphi}(\cdot) \in L^p_{\mathbb{F}}(0,T;\mathbb{R}^n)$. Let $\bar{X}(\cdot)$ be the corresponding solution. Then one has

$$
\begin{aligned}
\mathbb{E}|X(t)-\bar{X}(t)|^p &\leqslant K\mathbb{E}\Big\{|\varphi(t)-\bar{\varphi}(t)|^p + \Big(\int_0^t L|X(s)-\bar{X}(s)|ds\Big)^p \\
&\quad + \Big(\int_0^t |b(t,s,X(s))-\bar{b}(t,s,X(s))|ds\Big)^p \\
&\quad + \sup_{r\in[0,t]}\Big|\int_0^r \Big(\sigma(t,s,X(s))-\bar{\sigma}(t,s,\bar{X}(s))\Big)dW(s)\Big|^p \\
&\leqslant K\mathbb{E}\Big\{|\varphi(t)-\bar{\varphi}(t)|^p + \int_0^t |X(s)-\bar{X}(s)|^p ds \\
&\quad + \Big(\int_0^t |b(t,s,X(s))-\bar{b}(t,s,X(s))|ds\Big)^p \\
&\quad + \Big(\int_0^t |\sigma(t,s,X(s))-\bar{\sigma}(t,s,X(s))|^2 ds\Big)^{\frac{p}{2}}\Big\}.
\end{aligned}
$$

Then Gronwall's inequality applies to get (3.4.7). □

In the above, we allow the free term $\varphi(\cdot)$ to be just L^p-integrable. As a trade-off, the solution $X(\cdot)$ is also only L^p-integrable. We point out that, in general, $X(\cdot)$ might not have continuous paths. This is clear by considering a trivial equation with $b(\cdot)$ and $\sigma(\cdot)$ all being zero and the free term $\varphi(\cdot) \in L^p_{\mathbb{F}}(0,S;\mathbb{R}^m)$ being discontinuous.

We have seen that under (H3.4.1), the proof of well-posedness of an FSVIE is just like that for FSDEs. However, the uniform Lipschitz condition (3.4.4) is too restrictive because when one considers rough volatility involved problems, the FSVIEs encountered might have kernels that are singular along $t = s$. Fractional Brownian motion is such a special example. To get more feelings, let us present the following example.

Example 3.4.2 Consider the following FSVIE:

$$
X(t) = \frac{1}{\sqrt{t}} + \int_0^t \frac{X(s)}{\sqrt{t-s}}ds + \int_0^t \sigma dW(s), \qquad t\in[0,T]. \tag{3.4.10}
$$

Suppose $X(\cdot)$ is a solution to the above FSVIE. Then,

$$
X(t) = \frac{1}{\sqrt{t}} + \int_0^t \frac{1}{\sqrt{t-s}}\Big(\frac{1}{\sqrt{s}} + \int_0^s \frac{X(r)}{\sqrt{s-r}}dr + \int_0^s \sigma dW(r)\Big)ds + \int_0^t \sigma dW(s)
$$

$$
\begin{aligned}
&= \frac{1}{\sqrt{t}} + \int_0^t \frac{ds}{\sqrt{t-s}\sqrt{s}} + \int_0^t \Big(\int_r^t \frac{ds}{\sqrt{t-s}\sqrt{s-r}}\Big) X(r)dr \\
&\quad + \int_0^t \Big(\int_r^t \frac{\sigma}{\sqrt{t-s}} ds\Big) dW(r) + \int_0^t \sigma dW(s) \\
&= \frac{1}{\sqrt{t}} + B + B\int_0^t X(r)dr + \int_0^t \sigma(1+2\sqrt{t-r})dW(r),
\end{aligned}
$$

where $B = B\big(\frac{1}{2}, \frac{1}{2}\big)$ is the value of the beta function at $(\frac{1}{2}, \frac{1}{2})$ (see (1.2.13) in Chap. 1). Note that

$$
B\Big(\frac{1}{2}, \frac{1}{2}\Big) = \frac{\Gamma(\frac{1}{2})^2}{\Gamma(1)} = \pi .
$$

Here, $\Gamma(\cdot)$ is the gamma function. Denote

$$
\bar{X}(t) = \int_0^t X(s)ds, \quad \varphi(t) = \frac{1}{\sqrt{t}} + \pi + \int_0^t \sigma(1+2\sqrt{t-r})dW(r), \qquad t \in [0, T].
$$

Then, the above becomes

$$
\dot{\bar{X}}(t) = \pi \bar{X}(t) + \varphi(t), \qquad t \in [0, T].
$$

Hence,

$$
\bar{X}(t) = \int_0^t e^{\pi(t-s)} \varphi(s)ds, \qquad t \in [0, T].
$$

Consequently,

$$
\begin{aligned}
X(t) &= \frac{1}{\sqrt{t}} + \pi + \pi \int_0^t e^{\pi(t-s)} \Big(\frac{1}{\sqrt{s}} + \pi + \int_0^s \sigma(1+2\sqrt{s-r})dW(r) \Big) ds \\
&\quad + \int_0^t \sigma(1+2\sqrt{t-r})dW(r), \qquad t \in [0, T].
\end{aligned}
$$

This gives the solution of FSVIE (3.4.10).

The above example shows that the uniform Lipschitz conditions assumed in (H3.4.1) are not necessary for the well-posedness of FSVIEs. We now introduce the following hypothesis. Let $p \in (2, \infty)$.

(H3.4.2)$_p$ Let maps $b, \sigma : \Delta_*[0,T] \times \mathbb{R}^n \times \Omega \to \mathbb{R}^n$ be measurable. For each $(t,x) \in [0,T] \times \mathbb{R}^n$, $s \mapsto (b(t,s,x), \sigma(t,s,x))$ is $\mathbb{F}$-progressively measurable on $[0,t]$. Moreover,

$$\begin{aligned} &|b(t,s,x) - b(t,s,x')| \leqslant L_b(t,s)|x - x'|, \\ &|\sigma(t,s,x) - \sigma(t,s,x')| \leqslant L_\sigma(t,s)|x - x'|, \\ &\qquad\qquad (t,s) \in \Delta_*[0,T],\ x, x' \in \mathbb{R}^n,\ \text{a.s.}, \end{aligned} \tag{3.4.11}$$

for some deterministic measurable functions $L_b, L_\sigma : \Delta_*[0,T] \to \mathbb{R}$ satisfying

$$\int_0^T \Big[\Big(\int_0^t L_b(t,s)^{\frac{p}{p-1}} ds\Big)^{p-1} + \Big(\int_0^t L_\sigma(t,s)^{\frac{2p}{p-2}} ds\Big)^{\frac{p-2}{2}}\Big] dt < \infty. \tag{3.4.12}$$

We see that $b(t,s,x)$ and $\sigma(t,s,x)$ could have singularity along $t = s$. Similar to (3.4.5), due to (3.4.3), we have

$$\begin{aligned} &|b(t,s,x)| \leqslant L_b(t,s)|x|, \\ &|\sigma(t,s,x)| \leqslant L_\sigma(t,s|x|, \end{aligned} \quad \forall (t,s,x) \in \Delta_*[0,T] \times \mathbb{R}^n. \tag{3.4.13}$$

Note that the coefficients of Eq. (3.4.10) in Example 3.4.2 are given by

$$L_b(t,s) = (t-s)^{-\frac{1}{2}}, \quad L_\sigma(t,s) = 0, \qquad (t,s) \in \Delta_*[0,T].$$

Clearly, (3.4.12) is satisfied for $p > 2$. In general, if we take

$$L_b(t,s) = \frac{K}{(t-s)^\alpha}, \quad L_\sigma(t,s) = \frac{K}{(t-s)^\beta}, \qquad (t,s) \in \Delta_*[0,T],$$

for some $K > 0$ and $\alpha \in (0,1)$, $\beta \in (0, \frac{1}{2})$, then (H3.4.2)$_p$ holds if

$$p > \frac{1}{1-\alpha} \vee \frac{2}{1-2\beta}.$$

We have the following result concerning the well-posedness of FSVIEs.

Theorem 3.4.3 *Let* (H3.4.2)$_p$ *hold for some* $p > 2$. *Then for any* $\varphi(\cdot) \in L^p_{\mathbb{F}}(0,T;\mathbb{R}^n)$, *FSVIE* (3.4.1) *admits a unique solution* $X(\cdot) \in L^p_{\mathbb{F}}(0,T;\mathbb{R}^n)$, *and the following estimate holds:*

$$\int_0^T \mathbb{E}|X(t)|^p dt \leqslant K\mathbb{E}\int_0^T |\varphi(t)|^p dt. \tag{3.4.14}$$

Further, let $(\bar b,\bar\sigma)$ be the generator of another FSVIE satisfying $(H3.4.2)_p$ and $\bar\varphi(\cdot)\in L^p_{\mathbb{F}}(0,T;\mathbb{R}^n)$. Let $\bar X(\cdot)$ be the corresponding solution. Then the following stability estimate holds:

$$\begin{aligned}\int_0^T \mathbb{E}|X(t)-\bar X(t)|^p dt \leqslant K\mathbb{E}\Big\{\int_0^T |\varphi(t)-\bar\varphi(t)|^p dt\\ +\int_0^T\Big(\int_0^t |b(t,s,X(s))-\bar b(t,s,X(s))|ds\Big)^p dt\\ +\int_0^T\Big(\int_0^t |\sigma(t,s,X(s))-\bar\sigma(t,s,X(s))|^2 ds\Big)^{\frac p2} dt\Big\}.\end{aligned} \tag{3.4.15}$$

Proof First of all, for any $X(\cdot)\in L^p_{\mathbb{F}}(0,T;\mathbb{R}^n)$, we define

$$\mathcal{S}[X(\cdot)](t)=\varphi(t)+\int_0^t b(t,s,X(s))ds+\int_0^t\sigma(t,s,X(s))dW(s),\quad t\in[0,T]. \tag{3.4.16}$$

By Burkholder–Davis–Gundy's inequalities (Chap. 2, Theorem 2.3.2), we have (fixing $t\in[0,T]$)

$$\begin{aligned}\mathbb{E}\Big|\int_0^t \sigma(t,s,X(s))dW(s)\Big|^p &\leqslant \mathbb{E}\Big[\sup_{r\in[0,t]}\Big|\int_0^r \sigma(t,s,X(s))dW(s)\Big|^p\Big]\\ &\leqslant \overline{K}_p\mathbb{E}\Big(\int_0^t |\sigma(t,s,X(s))|^2 ds\Big)^{\frac p2},\end{aligned}$$

where $\overline{K}_p$ is the constant appearing in Burkholder–Davis–Gundy's inequalities. Then it follows that

$$\begin{aligned}\int_0^T \mathbb{E}|\mathcal{S}[X(\cdot)](t)|^p dt &\leqslant K\Big[\int_0^T \mathbb{E}|\varphi(t)|^p dt+\int_0^T\mathbb{E}\Big(\int_0^t L_b(t,s)|X(s)|ds\Big)^p dt\\ &\quad+\int_0^T \mathbb{E}\Big(\int_0^t L_\sigma(t,s)^2|X(s)|^2 ds\Big)^{\frac p2} dt\Big]\\ &\leqslant K\Big[\int_0^T \mathbb{E}|\varphi(t)|^p dt+\int_0^T\Big(\int_0^t L_b(t,s)^{\frac{p}{p-1}}ds\Big)^{p-1}\Big(\int_0^t \mathbb{E}|X(s)|^p ds\Big)dt\\ &\quad+\int_0^T\Big(\int_0^t L_\sigma(t,s)^{\frac{2p}{p-2}}ds\Big)^{\frac{p-2}{2}}\Big(\int_0^t \mathbb{E}|X(s)|^p ds\Big)dt\Big]\\ &= K\Big[\int_0^T \mathbb{E}|\varphi(t)|^p dt+\int_0^T \Lambda_0(t)\int_0^t \mathbb{E}|X(s)|^p ds dt\Big],\end{aligned}$$

where

$$\Lambda_0(t)=\Big(\int_0^t L_b(t,s)^{\frac{p}{p-1}}ds\Big)^{p-1}+\Big(\int_0^t L_\sigma(t,s)^{\frac{2p}{p-2}}ds\Big)^{\frac{p-2}{2}},\quad t\in[0,T]. \tag{3.4.17}$$

Hence, $\mathcal{S} : L^p_{\mathbb{F}}(0,T;\mathbb{R}^n) \to L^p_{\mathbb{F}}(0,T;\mathbb{R}^n)$. Likewise, for any $X(\cdot),\bar X(\cdot)\in L^p_{\mathbb{F}}(0,T;\mathbb{R}^n)$, we have

$$\begin{aligned}
&\int_0^T \mathbb{E}|\mathcal{S}[X(\cdot)](t)-\mathcal{S}[\bar X(t)]|^p dt\\
&\leqslant K_0\Big[\int_0^T\Big(\mathbb{E}\int_0^t L_b(t,s)|X(s)-\bar X(s)|ds\Big)^p dt\\
&\qquad+\int_0^T\Big(\mathbb{E}\int_0^t L_\sigma(t,s)^2|X(s)-\bar X(s)|^2ds\Big)^{\frac{p}{2}}dt\Big]\\
&\leqslant K_0\Big[\int_0^T\Big(\int_0^t L_b(t,s)^{\frac{p}{p-1}}ds\Big)^{p-1}\Big(\int_0^t \mathbb{E}|X(s)-\bar X(s)|^pds\Big)dt\\
&\qquad+\int_0^T\Big(\int_0^t L_\sigma(t,s)^{\frac{2p}{p-2}}ds\Big)^{\frac{p-2}{2}}\Big(\int_0^t \mathbb{E}|X(s)-\bar X(s)|^pds\Big)dt\Big]\\
&=K_0\int_0^T\Lambda_0(t)\int_0^t\mathbb{E}|X(s)-\bar X(s)|^pdsdt,
\end{aligned} \tag{3.4.18}$$

for some specific absolute constant $K_0>0$. We now introduce the following Picard iteration sequence:

$$\begin{cases} X^0(t)=0,\\ X^{k+1}(t)=\varphi(t)+\displaystyle\int_0^t b(t,s,X^k(s))ds+\int_0^t\sigma(t,s,X^k(s))dW(s),\\ \qquad\qquad\qquad\qquad\qquad\qquad t\in[0,T],\quad k\geqslant 0.\end{cases} \tag{3.4.19}$$

Then, for $k=1$, $X^1(\cdot)=\varphi(\cdot)$. Thus,

$$\int_0^T\mathbb{E}|X^1(t)|^pdt=\int_0^T\mathbb{E}|\varphi(t)|^pdt.$$

Further, for $X^2(\cdot)-X^1(\cdot)$, by (3.4.18),

$$\begin{aligned}
&\int_0^T\mathbb{E}|X^2(t)-X^1(t)|^pdt\leqslant K_0\int_0^T\Lambda_0(t)\int_0^t\mathbb{E}|X^1(s)|^pdsdt\\
&=K_0\int_0^T\Lambda_0(t)\int_0^t\mathbb{E}|\varphi(s)|^pdsdt=K_0\int_0^T\Big(\int_s^T\Lambda_0(t)dt\Big)\mathbb{E}|\varphi(s)|^pds,
\end{aligned}$$

for some absolute constant $K_0 > 0$. For $X^3(\cdot) - X^2(\cdot)$, by (3.4.18) again,

$$\begin{aligned}
&\int_0^T \mathbb{E}|X^3(t) - X^2(t)|^p dt \leqslant K_0 \int_0^T \Lambda_0(t) \int_0^t \mathbb{E}|X^2(s) - X^1(s)|^p ds dt \\
&\leqslant K_0^2 \int_0^T \Lambda_0(t) \int_0^t \Big(\int_s^t \Lambda_0(\tau)d\tau\Big)\mathbb{E}|\varphi(s)|^p ds dt \\
&= K_0^2 \int_0^T \Big[\int_s^T \Lambda_0(t)\Big(\int_s^t \Lambda_0(\tau)d\tau\Big)dt\Big]\mathbb{E}|\varphi(s)|^p ds \\
&= \frac{K_0^2}{2!}\int_0^T \Big(\int_s^T \Lambda_0(\tau)d\tau\Big)^2 \mathbb{E}|\varphi(s)|^p ds.
\end{aligned}$$

Suppose

$$\int_0^T \mathbb{E}|X^{k+1}(t) - X^k(t)|^p dt \leqslant \frac{K_0^k}{k!}\int_0^T \Big(\int_s^T \Lambda_0(\tau)d\tau\Big)^k \mathbb{E}|\varphi(s)|^p ds. \qquad (3.4.20)$$

Then

$$\begin{aligned}
&\int_0^T \mathbb{E}|X^{k+2}(t) - X^{k+1}(t)|^p dt \leqslant K_0 \int_0^T \Lambda_0(t)\int_0^t \mathbb{E}|X^{k+1}(s) - X^k(s)|^p ds dt \\
&\leqslant \frac{K_0^{k+1}}{k!}\int_0^T \Lambda_0(t)\int_0^t \Big(\int_s^t \Lambda_0(\tau)d\tau\Big)^k \mathbb{E}|\varphi(s)|^p ds dt \\
&= \frac{K_0^{k+1}}{k!}\int_0^T \Big[\int_s^T \Lambda_0(t)\Big(\int_s^t \Lambda_0(\tau)d\tau\Big)^k dt\Big]\mathbb{E}|\varphi(s)|^p ds \\
&= \frac{K_0^{k+1}}{(k+1)!}\int_0^T \Big(\int_s^T \Lambda_0(\tau)d\tau\Big)^{k+1}\mathbb{E}|\varphi(s)|^p ds.
\end{aligned}$$

By induction, (3.4.20) is true for all $k \geqslant 0$. Then,

$$\|X^{k+1}(\cdot) - X^k(\cdot)\|_{L^p_{\mathbb{F}}(0,T;\mathbb{R}^n)} \leqslant \Big[\frac{K_0^k}{k!}\Big(\int_0^T \Lambda_0(\tau)d\tau\Big)^k\Big]^{\frac{1}{p}} \|\varphi(\cdot)\|_{L^p_{\mathbb{F}}(0,T;\mathbb{R}^n)}.$$

Since the series

$$\begin{aligned}
&\sum_{k=0}^\infty \|X^{k+1}(\cdot) - X^k(\cdot)\|_{L^p_{\mathbb{F}}(0,T;\mathbb{R}^n)} \\
&\leqslant \sum_{k=0}^\infty \Big[\frac{K_0^k}{k!}\Big(\int_0^T \Lambda_0(\tau)d\tau\Big)^k\Big]^{\frac{1}{p}} \|\varphi(\cdot)\|_{L^p_{\mathbb{F}}(0,T;\mathbb{R}^n)}
\end{aligned} \qquad (3.4.21)$$

is convergent, we have

$$\lim_{k\to\infty} X^k(\cdot) = X(\cdot) = \sum_{k=0}^{\infty}\big[X^{k+1}(\cdot) - X^k(\cdot)\big], \quad \text{in } L^p_{\mathbb{F}}(0,T;\mathbb{R}^n), \tag{3.4.22}$$

for some $X(\cdot) \in L^p_{\mathbb{F}}(0,T;\mathbb{R}^n)$, which is clearly a solution of FSVIE (3.4.1). Moreover, by (3.4.22) and (3.4.21), we obtain (3.4.14).

Now, suppose $\bar X(\cdot)$ is a solution of (3.4.1) with b,σ replaced by $\bar b,\bar\sigma$, for which $(\text{H3.4.2})_p$ (with the same $p>2$) is also satisfied and $\bar\varphi(\cdot)\in L^p_{\mathbb{F}}(0,T;\mathbb{R}^n)$. Then with $S\in[0,T]$,

$$\begin{aligned}\int_0^S \mathbb{E}|X(t)-\bar X(t)|^p dt \leqslant K_0\Big[&\int_0^S |\varphi(t)-\bar\varphi(t)|^p dt\\&+\int_0^S \mathbb{E}\Big(\int_0^t |b(t,s,X(s))-\bar b(t,s,X(s))|ds\Big)^p dt\\&+\int_0^S \mathbb{E}\Big(\int_0^t |\sigma(t,s,X(s))-\bar\sigma(t,s,X(s))|^2 ds\Big)^{\frac p2} dt\\&+\int_0^S \Lambda_0(t)\int_0^t \mathbb{E}|X(s)-\bar X(s)|^p ds dt\Big].\end{aligned}$$

Thus, (3.4.15) follows from Gronwall's inequality. □

The solution $X(\cdot)$ obtained above is just L^p-integrable. We do not have continuity of $t\mapsto X(t)$ in any sense. To explore some kind of continuity, we introduce the following stronger hypothesis.

(H3.4.3)$_p$ Let $(3.4.2)_p$ hold with given $p\in(2,\infty)$. In addition, for some modulus of continuity $\rho:[0,\infty)\to[0,\infty)$, i.e., $\rho(\cdot)$ is deterministic strictly increasing continuous with $\rho(0)=0$, it holds that

$$\begin{aligned}|b(t,s,x)-b(t',s,x)|+|\sigma(t,s,x)-\sigma(t',s,x)| \leqslant \rho(|t-t'|)|x|,\\ (t,s),(t',s)\in\Delta_*[0,T],\ x\in\mathbb{R}^n.\end{aligned} \tag{3.4.23}$$

Moreover, for any fixed $\tau\in[0,T)$, $t\mapsto\mathcal{L}(t,\tau)$ is continuous on $[\tau,T]$, and for any fixed $t\in[0,T]$, $\tau\mapsto\mathcal{L}(t,\tau)$ is continuous on $[0,t]$, where

$$\mathcal{L}(t,\tau)\equiv\Big(\int_\tau^t L_b(t,s)^{\frac{p}{p-1}}ds\Big)^{p-1}+\Big(\int_\tau^t L_\sigma(t,s)^{\frac{2p}{p-2}}ds\Big)^{\frac{p-2}{2}}.$$

We have the following result concerning the solution $X(\cdot)$ of FSVIE (3.4.1).

Proposition 3.4.4 *Let* (H3.4.3)$_p$ *hold with any given* $p > 2$. *Then for any free term* $\varphi(\cdot) \in C_{\mathbb{F}}([0,T]; L^p(\Omega;\mathbb{R}^n))$, *FSVIE* (3.4.1) *admits a unique solution* $X(\cdot) \in C_{\mathbb{F}}([0,T]; L^p(\Omega;\mathbb{R}^n))$, *and it holds that*

$$\sup_{t\in[0,T]} \mathbb{E}|X(t)|^p \leqslant K \sup_{t\in[0,T]} \mathbb{E}|\varphi(t)|^p. \tag{3.4.24}$$

Let $(\bar{b},\bar{\sigma})$ *be the generator of another FSVIE satisfying (H3.4.3)*$_p$ *and* $\bar{\varphi}(\cdot) \in C_{\mathbb{F}}([0,T]; L^p(\Omega;\mathbb{R}^n))$. *Let* $\bar{X}(\cdot)$ *be the corresponding solution. Then the following stability estimate holds:*

$$\begin{aligned}\sup_{t\in[0,T]} \mathbb{E}|X(t)-\bar{X}(t)|^p \leqslant K \sup_{t\in[0,T]} \Big[&\mathbb{E}|\varphi(t)-\bar{\varphi}(t)|^p \\ &+\Big(\int_0^t \mathbb{E}|b(t,s,X(s))-\bar{b}(t,s,X(s))|ds\Big)^p \\ &+\Big(\int_0^t \mathbb{E}|\sigma(t,s,X(s))-\bar{\sigma}(t,s,X(s))|^2 ds\Big)^{\frac{p}{2}}\Big].\end{aligned} \tag{3.4.25}$$

Proof First of all, by Theorem 3.4.3 for any $\varphi(\cdot) \in C_{\mathbb{F}}([0,T]; L^p(\Omega;\mathbb{R}^n))$, FSVIE (3.4.1) admits a unique solution $X(\cdot) \in L^p_{\mathbb{F}}(0,T;\mathbb{R}^n)$. We now prove the continuity of $X(\cdot)$. Let $0 \leqslant t < t' \leqslant T$. We look at the following:

$$\begin{aligned}\mathbb{E}|X(t')-X(t)|^p &= \mathbb{E}\Big|\varphi(t')-\varphi(t)+\int_0^t \Big(b(t',s,X(s))-b(t,s,X(s))\Big)ds \\ &\quad+\int_t^{t'} b(t',s,X(s))ds+\int_0^t \Big(\sigma(t',s,X(s))-\sigma(t,s,X(s))\Big)dW(s) \\ &\quad+\int_t^{t'} \sigma(t',s,X(s))dW(s)\Big|^p \\ &\leqslant K\mathbb{E}\Big[|\varphi(t')-\varphi(t)|^p+[\rho(t'-t)]^p\int_0^t |X(s)|^p ds+\Big(\int_t^{t'} L_b(t',s)|X(s)|ds\Big)^p \\ &\qquad+\Big(\int_t^{t'} L_\sigma(t',s)^2|X(s)|^2 ds\Big)^{\frac{p}{2}}\Big] \\ &\leqslant K\mathbb{E}\Big[|\varphi(t')-\varphi(t)|^p+[\rho(t'-t)]^p\int_0^T |X(s)|^p ds \\ &\quad+\Big[\Big(\int_t^{t'} L_b(t',s)^{\frac{p}{p-1}}ds\Big)^{p-1}+\Big(\int_t^{t'} L_\sigma(t',s)^{\frac{2p}{p-2}}ds\Big)^{\frac{p-2}{2}}\Big]\Big(\int_0^T |X(s)|^p ds\Big)\Big] \to 0,\end{aligned}$$

as $t'-t \to 0$. In the above, we have used the following estimation:

$$\begin{aligned}\mathbb{E}\Big|\int_t^{t'} \sigma(t',s,X(s))dW(s)\Big|^p &\leqslant \mathbb{E}\Big[\sup_{r\in[t,t']}\Big|\int_t^r \sigma(t',s,X(s))dW(s)\Big|^p\Big] \\ &\leqslant K\mathbb{E}\Big(\int_t^{t'} |\sigma(t',s,X(s))|^2 ds\Big)^{\frac{p}{2}} \leqslant K\mathbb{E}\Big(\int_t^{t'} L_\sigma(t',s)^2|X(s)|^2 ds\Big)^{\frac{p}{2}}.\end{aligned} \tag{3.4.26}$$

Likewise,

$$\begin{aligned}
&\mathbb{E}\Big|\int_0^t \Big(\sigma(t',s,X(s))-\sigma(t,s,X(s))\Big)dW(s)\Big|^p\\
&\leqslant \mathbb{E}\Big[\sup_{r\in[0,t]}\Big|\int_0^r \Big(\sigma(t',s,X(s))-\sigma(t,s,X(s))\Big)dW(s)\Big|^p\Big]\\
&\leqslant K\mathbb{E}\Big(\int_0^t |\sigma(t',s,X(s))-\sigma(t,s,X(s))|^2ds\Big)^{\frac{p}{2}}\\
&\leqslant K\mathbb{E}\Big(\int_0^t \big[\rho(t'-t)\big]^2|X(s)|^2ds\Big)^{\frac{p}{2}}.
\end{aligned}$$

By switching the above t, t', we obtain a similar result when $0\leqslant t'<t\leqslant T$. Thus, $X(\cdot)\in C_{\mathbb{F}}([0,T];L^p(\Omega;\mathbb{R}^n))$. Next,

$$\begin{aligned}
&\mathbb{E}|X(t)|^p\\
&\leqslant K\Big[|\varphi(t)|^p+\Big(\int_0^t|b(t,s,X(s))|ds\Big)^p+\Big|\int_0^t\sigma(t,s,X(s))dW(s)\Big|^p\Big]\\
&\leqslant K\Big[|\varphi(t)|^p+\Big(\int_0^t L_b(t,s)|X(s)|ds\Big)^p+\Big(\int_0^t L_\sigma(t,s)^2|X(s)|^2ds\Big)^{\frac{p}{2}}\Big]\\
&\leqslant K\Big[|\varphi(t)|^p+\int_0^t \Lambda_0(t)\mathbb{E}|X(s)|^pds\Big].
\end{aligned} \tag{3.4.27}$$

By our assumption we see that $\Lambda_0(\cdot)$ in (3.4.17) is bounded, and then by Gronwall's inequality, we obtain (3.4.24). In the same way, one obtains (3.4.25). This competes the proof. □

Note that in the above, the solution $X(\cdot)\in C_{\mathbb{F}}([0,T];L^p(\Omega;\mathbb{R}^n))$, and it might not have continuous paths, since

$$t\mapsto\int_0^t\sigma(t,s,X(s))dW(s)$$

might not be in $L^p_{\mathbb{F}}(\Omega;C([0,T];\mathbb{R}^n))$[1]. We now strengthen the condition further in the following.

(H3.4.4)$_p$ Let (H3.4.3)$_p$ hold for given $p>4$. In addition, suppose there exist $K>0$, $\alpha>\frac{1}{p}$, and $\beta\in(0,\alpha p-1)$ such that

$$\begin{aligned}
&|\sigma(t,s,x)-\sigma(t',s,x)|\leqslant K|t-t'|^\alpha|x|,\\
&\qquad\qquad\forall(t,s,x),(t',s,x)\in\Delta_*[0,T]\times\mathbb{R}^n,
\end{aligned}\tag{3.4.28}$$

[1] We believe that there should be an example for which (H3.4.3)$_p$ holds, but the solution does not have continuous paths. However, at the moment, we do not have a proper example.

and

$$\Big(\int_t^{t'} L_\sigma(t',s)^{\frac{2p}{p-2}}ds\Big)^{\frac{p-2}{2}} \leqslant K(t'-t)^{1+\beta}, \qquad \forall\, 0 \leqslant t < t' \leqslant T. \tag{3.4.29}$$

Notice that if $p \in (2,4)$, then (3.4.29) even excludes the case of $L_\sigma \equiv K$ with constant K. Therefore, in the above, we choose $p > 4$.

(H3.4.5)$_p$ Let (H3.4.4)$_p$ hold with given $p > 4$. Suppose that for some $K > 0$, $\alpha \in (\frac{1}{p}, 1)$, and $\beta \in (0, \alpha p - 1)$,

$$\begin{aligned} &|b(t,s,x) - b(t',s,x)| \leqslant K|t-t'|^\alpha |x|, \\ &\qquad \forall (t,s,x), (t',s,x) \in \Delta_*[0,T] \times \mathbb{R}^n, \end{aligned} \tag{3.4.30}$$

and

$$\Big(\int_t^{t'} L_b(t',s)^{\frac{p}{p-1}}ds\Big)^{p-1} \leqslant K(t'-t)^{1+\beta}, \qquad 0 \leqslant t < t' \leqslant T. \tag{3.4.31}$$

Notice that if $p = 2$, then (3.4.30) even excludes the case of $L_b \equiv K$ with constant K. Therefore, we choose at least $p > 2$ in (H3.4.5)$_p$.

We have the following result concerning the path continuity of the solution.

Proposition 3.4.5 *Let* (H3.4.4)$_p$ *hold. Then for any* $\varphi(\cdot) \in L^p_{\mathbb{F}}(\Omega; C([0,T];\mathbb{R}^n))$, *the solution* $X(\cdot)$ *to FSVIE* (3.4.1) *is in* $L^p_{\mathbb{F}}(\Omega; C([0,T];\mathbb{R}^n))$. *Further, if (H3.4.5)*$_p$ *holds, then for any* $\varphi(\cdot) \in L^p_{\mathbb{F}}(\Omega; C([0,T];\mathbb{R}^n))$ *with* ($\varepsilon \in (0, \frac{\beta}{p})$)

$$\mathbb{E}\Big[\sup_{0\leqslant t\neq t'\leqslant T} \frac{|\varphi(t)-\varphi(t')|}{|t-t'|^{\frac{\beta}{p}-\varepsilon}}\Big] \leqslant K, \tag{3.4.32}$$

the solution $X(\cdot)$ *of FSVIE* (3.4.1) *satisfies*

$$\mathbb{E}\Big[\sup_{0\leqslant t\neq t'\leqslant T} \frac{|X(t)-X(t')|}{|t-t'|^{\frac{\beta}{p}-\varepsilon}}\Big] \leqslant K, \tag{3.4.33}$$

and

$$\mathbb{E}\Big[\sup_{t\in[0,T]} |X(t)|^p\Big] \leqslant K\mathbb{E}\Big[1 + \sup_{\tau\in[0,T]} |\varphi(\tau)|^p\Big]. \tag{3.4.34}$$

Proof Since (H3.4.4)$_p$ implies (H3.4.3)$_p$, for $\varphi(\cdot) \in L^p_{\mathbb{F}}(\Omega; C([0,T];\mathbb{R}^n)) \subseteq C_{\mathbb{F}}([0,T]; L^p(\Omega;\mathbb{R}^n))$, by Proposition 3.4.4, the solution $X(\cdot)$ of FSVIE (3.4.1) is in $C_{\mathbb{F}}([0,T]; L^p(\Omega;\mathbb{R}^n))$, and (3.4.24) holds.

First, let us prove that $X(\cdot) \in L^p_{\mathbb{F}}(\Omega; C([0, T]; \mathbb{R}^n))$. We first look at the drift term. Notice that for $0 \leqslant t_0 < t \leqslant T$, by $(\mathrm{H3.4.3})_p$,

$$\begin{aligned}
&\Big|\int_0^t b(t,s,X(s))ds - \int_0^{t_0} b(t_0,s,X(s))ds\Big| \\
&\leqslant \Big|\int_0^{t_0} \big|b(t,s,X(s)) - b(t_0,s,X(s))\big|ds\Big| + \Big|\int_{t_0}^t \big|b(t_0,s,X(s))\big|ds\Big| \\
&\leqslant \rho(|t-t_0|)\int_0^{t_0} |X(s)|ds + \Big[\int_{t_0}^t L_b(t,s)^{\frac{p}{p-1}}ds\Big]^{\frac{p-1}{p}}\Big[\int_{t_0}^T |X(s)|^p ds\Big]^{\frac{1}{p}} \to 0,
\end{aligned}$$

as $t \downarrow t_0$. The case of $t < t_0$ can also be obtained similarly.

For the diffusion term, we have

$$\begin{aligned}
&\mathbb{E}\Big|\int_0^t \sigma(t,s,X(s))dW(s) - \int_0^{t'} \sigma(t',s,X(s))dW(s)\Big|^p \\
&\leqslant K\mathbb{E}\Big\{\Big(\int_0^t |t'-t|^{2\alpha}|X(s)|^2 ds\Big)^{\frac{p}{2}} + \sup_{r\in[t,t']}\Big|\int_t^r \sigma(t',s,X(s))dW(s)\Big|^p\Big\} \\
&\leqslant K\Big\{|t'-t|^{\alpha p}\mathbb{E}\int_0^T |X(s)|^p ds + \mathbb{E}\Big(\int_t^{t'} L_\sigma(t',s)^2|X(s)|^2 ds\Big)^{\frac{p}{2}}\Big\} \\
&\leqslant K\Big[|t'-t|^{\alpha p} + \Big(\int_t^{t'} L_\sigma(t',s)^{\frac{2p}{p-2}}ds\Big)^{\frac{p-2}{2}}\Big] \leqslant K|t'-t|^{1+\beta}.
\end{aligned}$$

We point out that the above estimate also holds in case of $0 \leqslant t' < t \leqslant T$. Thus, applying the strengthened version of Kolmogorov theorem (Chap. 2, Theorem 2.1.11), we have continuous paths for the process

$$t \mapsto \int_0^t \sigma(t,s,X(s))dW(s).$$

This shows that $X(\cdot) \in L^p_{\mathbb{F}}(\Omega; C([0, T]; \mathbb{R}^n))$.

Next, we prove some Hölder continuity in the sense of (3.4.33) under $(\mathrm{H3.4.5})_p$. To this end, for $0 \leqslant r < r' \leqslant T$, we have

$$\begin{aligned}
&\mathbb{E}\Big[\Big|\int_0^r b(r,s,X(s))ds - \int_0^{r'} b(r',s,X(s))ds\Big|^p\Big] \\
&\leqslant K\mathbb{E}\Big\{\Big(\int_0^r |r'-r|^\alpha|X(s)|ds\Big)^p + \Big(\int_r^{r'} L_b(r',s)|X(s)|ds\Big)^p\Big\} \\
&\leqslant K\Big[|r'-r|^{\alpha p} + \Big(\int_r^{r'} L_b(r',s)^{\frac{p}{p-1}}ds\Big)^{p-1}\Big] \\
&\leqslant K\Big[|r'-r|^{\alpha p} + |r'-r|^{1+\beta}\Big].
\end{aligned}$$

We point out that the above estimate also holds in case of $0 \leqslant r' < r \leqslant T$.

Notice that $\varphi(\cdot) \in L^p_{\mathbb{F}}(\Omega; C([0,T]; \mathbb{R}^n))$ with (3.4.32), one has

$$\mathbb{E}|X(t) - X(t')|^p \leqslant K|t - t|^{1+\beta}, \qquad 0 \leqslant t, t' \leqslant T.$$

Consequently, our conclusions (3.4.33) and (3.4.34) follow from Chap. 2, Theorem 2.1.11 (Kolmogorov theorem).

□

As to the previous conclusions, let us point out two interesting facts: (i) If L_σ reduces to constant, then Proposition 3.4.4 holds true with $p = 2$. However, to obtain the path continuity in such a case, according to Proposition 3.4.5, especially (H3.4.4)$_p$, we need $p > 4$, and (ii) if the diffusion term disappears, according to the proof of Proposition 3.4.5, we can obtain the continuity of $X(\cdot)$ under (H3.4.4)$_p$ with $p \in (1, \infty)$.

Let us now look at a special case of FSVIE (3.4.1). Suppose

$$b(t,s,x) = b(s,x), \qquad \sigma(t,s,x) = \sigma(s,x), \qquad \varphi(t) = X_0.$$

Then (3.4.1) reads

$$X(t) = X_0 + \int_0^t b(s, X(s))ds + \int_0^t \sigma(s, X(s))dW(s), \quad t \in [0,T]. \tag{3.4.35}$$

This is the integral form of the following *(forward) stochastic differential equation* (FSDE, for short):

$$\begin{cases} dX(t) = b(t, X(t))dt + \sigma(t, X(t))dW(t), & t \in [0,T], \\ X(0) = X_0. \end{cases} \tag{3.4.36}$$

By Theorem 3.4.3, we have the following result concerning (3.4.36).

Proposition 3.4.6 *Suppose for some* $p \geqslant 1$, (H3.4.2)$_p$ *holds with measurable functions* $L_b, L_\sigma : [0,T] \to \mathbb{R}$ *satisfying*

$$\int_0^T L_b(t)dt + \int_0^T L_\sigma(t)^2 dt < \infty. \tag{3.4.37}$$

Then FSDE (3.4.36) *admits a unique solution* $X(\cdot) \in L^p_{\mathbb{F}}(\Omega; C([0,T]; \mathbb{R}^n))$.

The above recovers a classical result for FSDEs.

5 Linear FSVIEs

We now look at the following linear FSVIE:

$$X(t) = \varphi(t) + \int_0^t A_0(t,s)X(s)ds + \sum_{i=1}^d \int_0^t A_i(t,s)X(s)dW_i(s), \qquad t \in [0,T]. \tag{3.5.1}$$

Because the equation is linear, it defines a linear operator mapping the free term to the solution. This is the so-called *variation of constants formula*. To derive this formula, we need to use Skorohod integral and the resulting basics of Malliavin calculus below. Therefore, those readers who do not have those could either skip this section or read the Appendices first. Let us now introduce the following standing assumption.

(H3.5.1) Suppose $A_i : \Delta_*[0,T] \times \Omega \to \mathbb{R}^{n\times n}$ $(0 \leqslant i \leqslant d)$ are measurable and bounded. Suppose for each $t \in [0,T]$, $s \mapsto A_i(t,s)$ are $\mathbb{F}$-progressively measurable on $[0,t]$. Moreover, for $0 \leqslant i \leqslant d$, the Malliavin derivatives $D_s[A_i(t,r)]$ exist and are all bounded $(r,s,t \in [0,T])$.

The conditions assumed in (H3.5.1) are more than enough. Assuming the above is for the simplicity of presentation. According to Theorem 3.4.3, under (H3.5.1), for any suitable free term $\varphi(\cdot)$, linear FSVIE (3.5.1) admits a unique solution $X(\cdot)$. We have the following further result.

Theorem 3.5.1 *Let* (H3.5.1) *hold. Then for any* $\varphi(\cdot) \in L^\infty_{\mathbb{F}}(0,T;\mathbb{R}^n)$ *satisfying* $D.\varphi(\cdot)$ *being bounded, linear FSVIE* (3.5.1) *admits a unique solution* $X(\cdot) \in L^2_{\mathbb{F}}(0,T;\mathbb{R}^n)$*, and it has the following representation, involving Lebesgue and Skorohod integrals:*

$$X(t) = \varphi(t) + \int_0^t \Lambda_0(t,s)\varphi(s)ds + \sum_{i=1}^d \int_0^t \Lambda_i(t,s)\varphi(s)dW_i(s), \qquad t \in [0,T], \tag{3.5.2}$$

where for all $0 \leqslant i \leqslant d$*, processes* $\Lambda_i : \Delta_*[0,T]\times\Omega \to \mathbb{R}^{n\times n}$ *are measurable such that for each* $s \in [0,T)$*, the maps* $t \mapsto \Lambda_i(t,s)$ *are* $\mathbb{F}$*-progressively measurable and the following are satisfied, which only involve Lebesgue and Itô's integrals:*

$$\begin{cases} \Lambda_0(t,s) = A_0(t,s) + \int_s^t \Big(A_0(t,r)\Lambda_0(r,s) + \sum_{j=1}^d D_s^{(j)}[A_0(t,r)]\Lambda_j(r,s)\Big)dr \\ \qquad + \sum_{i=1}^d \int_s^t \Big(A_i(t,r)\Lambda_0(r,s) + \sum_{j=1}^d D_s^{(j)}[A_i(t,r)]\Lambda_j(r,s)\Big)dW_i(r), \\ \Lambda_j(t,s) = A_j(t,s) + \int_s^t A_0(t,r)\Lambda_j(r,s)dr \\ \qquad + \sum_{i=1}^d \int_s^t A_i(t,r)\Lambda_j(r,s)dW_i(r), \quad 0 \leqslant s \leqslant t \leqslant T,\ 1 \leqslant j \leqslant d. \end{cases} \tag{3.5.3}$$

Proof For notational simplicity, we assume that $d = n = 1$. We separate the proof into several steps. In the following, we always assume that $p \geqslant 2$.

Step 1. First, under (H3.5.1), FSVIE (3.5.1) admits a unique solution $X(\cdot) \in L^2_{\mathbb{F}}(0, T; \mathbb{R})$. Next, let us look at the solvability of Λ_0 and Λ_1. Notice

$$\Lambda_1(t,s) = A_1(t,s) + \int_s^t A_0(t,r)\Lambda_1(r,s)dr + \int_s^t A_1(t,r)\Lambda_1(r,s)dW_1(r), \quad s \leqslant t \leqslant T.$$

Since $A_1(\cdot,\cdot)$, $A_0(\cdot,\cdot)$ are bounded, there exists a unique $\Lambda_1(\cdot,\cdot)$ satisfying the above equation such that for any $p \geqslant 2$,

$$\sup_{s\in[0,T]} \mathbb{E}\int_s^T |\Lambda_1(t,s)|^p dt < \infty. \tag{3.5.4}$$

Given the above integrability of Λ_1, let us turn to look at Λ_0, the equation of which can be rewritten as follows:

$$\Lambda_0(t,s) = \varphi_0(t,s) + \int_s^t A_0(t,r)\Lambda_0(r,s)dr + \int_s^t A_1(t,r)\Lambda_0(r,s)dW_1(r),$$

where

$$\varphi_0(t,s) \equiv A_0(t,s) + \int_s^t \big[D_s A_0(t,r)\big]\Lambda_1(r,s)dr + \int_s^t \big[D_s A_1(t,r)\big]\Lambda_1(r,s)dW_1(r).$$

As to $\varphi_0(\cdot,\cdot)$, one has

$$\begin{aligned}
&\mathbb{E}\int_s^T |\varphi_0(t,s)|^p dt \\
&\leqslant K + \mathbb{E}\int_s^T \Big|\int_s^t |\Lambda_1(r,s)| dr\Big|^p dt + \mathbb{E}\int_s^T \Big|\int_s^t \big[D_s A_1(t,r)\big]\Lambda_1(r,s)dW_1(r)\Big|^p dt \\
&\leqslant K + K\mathbb{E}\int_s^T |\Lambda_1(r,s)|^p dr < \infty.
\end{aligned}$$

Using the above (3.5.12), we see that

$$\sup_{s\in[0,T]} \mathbb{E}\int_s^T |\varphi_0(t,s)|^p dt < \infty,$$

which then implies that

$$\sup_{s\in[0,T]} \mathbb{E}\int_s^T |\Lambda_0(t,s)|^p dt < \infty. \tag{3.5.5}$$

Step 2. In this step, we look at the integrals in (3.5.2). By the above integrability of Λ_0 and the boundedness of $\varphi(\cdot)$, we see the Lebesgue integral is well-defined. Next, we show the Skorohod integral involving Λ_1 is well-defined. To this end, let us make some preparations.

Under (H3.5.1), we observe that for $\tau \in [0, s]$,

$$D_\tau \Lambda_1(t,s) = \varphi(t,\tau,s) + \int_s^t A_0(t,r) D_\tau \Lambda_1(r,s) dr + \int_s^t A_1(t,r) D_\tau \Lambda_1(r,s) dW_1(r),$$

where

$$\begin{aligned}\varphi(t,\tau,s) \equiv D_\tau A_1(t,s) &+ \int_s^t \big[D_\tau A_0(t,r)\big]\Lambda_1(r,s)dr\\ &+ \int_s^t \big[D_\tau A_1(t,r)\big]\Lambda_1(r,s)dW_1(r),\end{aligned}$$

while for $\tau \in (s, t]$,

$$D_\tau \Lambda_1(t,s) = \bar\varphi(t,\tau,s) + \int_\tau^t A_0(t,r) D_\tau \Lambda_1(r,s) dr + \int_\tau^t A_1(t,r) D_\tau \Lambda_1(r,s) dW_1(r),$$

where

$$\begin{aligned}\bar\varphi(t,\tau,s) \equiv \int_\tau^t \big[D_\tau A_0(t,r)\big]\Lambda_1(r,s)dr &+ A_1(t,\tau)\Lambda_1(\tau,s)\\ &+ \int_\tau^t \big[D_\tau A_1(t,r)\big]\Lambda_1(r,s)dW_1(r).\end{aligned}$$

We discuss the solvability and integrability of $D_\cdot \Lambda_1(\cdot,\cdot)$. For $\tau \in [0, s]$, we have

$$\begin{aligned}&\mathbb{E}\int_0^s\int_s^T |\varphi(t,\tau,s)|^p dt d\tau\\ &\leqslant K + \mathbb{E}\int_0^s\int_s^T \Big|\int_s^t |\Lambda_1(r,s)|dr\Big|^p dt d\tau\\ &\quad + \mathbb{E}\int_0^s\int_s^T \Big|\int_s^t \big[D_\tau A_1(t,r)\big]\Lambda_1(r,s)dW_1(r)\Big|^p dt d\tau\\ &\leqslant K + K\mathbb{E}\int_s^T |\Lambda_1(r,s)|^p dr < \infty.\end{aligned}$$

Then by (3.5.12), we see that

$$\sup_{s\in[0,T]} \mathbb{E}\int_0^s\int_s^T |\varphi(t,\tau,s)|^p dt d\tau < \infty,$$

which implies that

$$\sup_{s\in[0,T]} \mathbb{E}\int_0^s \int_s^T |D_\tau \Lambda_1(t,s)|^p dt d\tau < \infty. \tag{3.5.6}$$

For the case when $\tau > s$, for any $p \geqslant 2$, we have

$$\begin{aligned}
&\mathbb{E}\int_s^T \int_s^T |\bar{\varphi}(t,\tau,s)|^p dt d\tau \\
&\leqslant K + \mathbb{E}\int_s^T \int_s^T \Big|\int_\tau^t |\Lambda_1(r,s)| dr\Big|^p dt d\tau + \mathbb{E}\int_s^T \int_s^T |\Lambda_1(\tau,s)|^p dt d\tau \\
&\quad + \mathbb{E}\int_s^T \int_s^T \Big|\int_\tau^t \big[D_\tau A_1(t,r)\big]\Lambda_1(r,s) dW_1(r)\Big|^p dt d\tau \\
&\leqslant K + K\mathbb{E}\int_s^T |\Lambda_1(r,s)|^p dr < \infty.
\end{aligned}$$

Using again (3.5.12), we see that

$$\sup_{s\in[0,T]} \mathbb{E}\int_s^T \int_s^T |\bar{\varphi}(t,\tau,s)|^p dt d\tau < \infty,$$

which then implies that

$$\sup_{s\in[0,T]} \mathbb{E}\int_s^T \int_s^T |D_\tau \Lambda_1(t,s)|^p dt d\tau < \infty. \tag{3.5.7}$$

From the above (3.5.6) and (3.5.7), we then have

$$\begin{aligned}
&\sup_{s\in[0,T]} \mathbb{E}\int_s^T \int_0^T |D_\tau \Lambda_1(t,s)|^p d\tau dt \\
&\leqslant \sup_{s\in[0,T]} \mathbb{E}\int_s^T \int_0^s |D_\tau \Lambda_1(t,s)|^p d\tau dt \\
&\quad + \sup_{s\in[0,T]} \mathbb{E}\int_s^T \int_s^T |D_\tau \Lambda_1(t,s)|^p d\tau dt < \infty,
\end{aligned} \tag{3.5.8}$$

which implies that for almost $s \in [0,T]$, and almost $t \in [s,T]$, and any $p \geqslant 2$,

$$\mathbb{E}\int_0^T |D_\tau \Lambda_1(t,s)|^p d\tau < \infty. \tag{3.5.9}$$

Similarly,

$$\begin{aligned}
&\mathbb{E}\int_0^T\int_0^t\int_0^T |D_\tau \Lambda_1(t,s)|^p d\tau ds dt = \mathbb{E}\int_0^T\int_s^T\int_0^T |D_\tau \Lambda_1(t,s)|^p d\tau dt ds \\
&\leqslant \sup_{s\in[0,T]} \mathbb{E}\int_s^T\int_0^T |D_\tau \Lambda_1(t,s)|^p d\tau dt < \infty,
\end{aligned} \tag{3.5.10}$$

which then implies that for almost $t \in [0, T]$,

$$\mathbb{E}\int_0^t\int_0^T |D_\tau \Lambda_1(t,s)|^p d\tau ds < \infty. \tag{3.5.11}$$

For later usefulness, we define $\mathscr{L}(\cdot)$ as follows:

$$\mathscr{L}_1(t) \equiv \int_0^t \Lambda_1(t,s)\varphi(s)dW_1(s), \quad t \in [0, T].$$

According to Proposition 12.1.3 in the Appendices, we need to verify three facts. First, by (3.5.12) we have

$$\begin{aligned}
&\mathbb{E}\int_0^T\int_0^t |\Lambda_1(t,s)|^p ds dt = \mathbb{E}\int_0^T\int_s^T |\Lambda_1(t,s)|^p dt ds \\
&\leqslant \sup_{s\in[0,T]} \mathbb{E}\int_s^T |\Lambda_1(t,s)|^p dt < \infty, \\
&\Longrightarrow \mathbb{E}\int_0^t |\Lambda_1(t,s)|^p ds < \infty, \quad t \in [0, T]. \text{ a.e.}
\end{aligned} \tag{3.5.12}$$

Recall the boundedness for $\varphi(\cdot)$, for almost $t \in [0, T]$, we have

$$\mathbb{E}\int_0^t |\Lambda_1(t,s)\varphi(s)|^2 ds < \infty.$$

To show $\mathscr{L}_1(t)$ is well-defined, for almost $s \in [0, t]$, the second fact we need to show is $\Lambda_1(t,s)\varphi(s) \in \mathbb{D}^{1,2}$, i.e.,

$$\mathbb{E}\big|\Lambda_1(t,s)\varphi(s)\big|^2 + \mathbb{E}\int_0^t \big|D_\tau\big[\Lambda_1(t,s)\varphi(s)\big]\big|^2 d\tau < \infty.$$

In fact, by Proposition 12.1.6 in the appendix, we first have

$$\mathbb{E}\Big|D_\tau\big[\Lambda_1(t,s)\varphi(s)\big]\Big|^2 = \mathbb{E}\Big|D_\tau\big[\Lambda_1(t,s)\big]\varphi(s) + \Lambda_1(t,s)D_\tau\varphi(s)\Big|^2.$$

Hence by the bounded requirement of $\varphi(\cdot)$ and $D_\cdot\varphi(\cdot)$, as well as (3.5.9), for almost $t \in [0,T]$ and almost $s \in [0,t]$,

$$\begin{aligned}&\mathbb{E}\int_0^t \big|D_\tau\big[\Lambda_1(t,s)\varphi(s)\big]\big|^2 d\tau \\ &\leqslant K\mathbb{E}\int_0^t \big|D_\tau\big[\Lambda_1(t,s)\big]\big|^2 d\tau + T\mathbb{E}\big|\Lambda_1(t,s)\big|^2 < \infty.\end{aligned}$$

Then the above second fact is easy to verify. Third, we need to show that there exists a measurable version of $D_\tau\big[\Lambda_1(t,s)\varphi(s)\big]$ such that

$$\mathbb{E}\int_0^t\int_0^t \big|D_\tau\big[\Lambda_1(t,s)\varphi(s)\big]\big|^2 ds d\tau < \infty.$$

Of course, this can be implied by the boundedness of $\varphi(\cdot)$ and its Malliavin derivative, (3.5.11) and (3.5.12).

Eventually, by Proposition 12.1.3 in the Appendices,

$$\begin{aligned}\mathbb{E}\big|\mathscr{L}_1(t)\big|^2 &\leqslant K\mathbb{E}\int_0^t \big|\Lambda_1(t,s)\varphi(s)\big|^2 ds \\ &+\mathbb{E}\int_0^t\int_0^t \Big|D_r\big[\Lambda_1(t,s)\varphi(s)\big]\Big|^2 dr ds < \infty.\end{aligned} \tag{3.5.13}$$

Step 3. In this step, for $i = 0, 1$, we show that

$$\begin{aligned}A_i(t,s)\int_0^s \Lambda_1(s,\tau)\varphi(\tau)dW_1(\tau) &= \int_0^s A_i(t,s)\Lambda_1(s,\tau)\varphi(\tau)dW_1(\tau) \\ &+ \int_0^s D_\tau\big[A_i(t,s)\big]\Lambda_1(s,\tau)\varphi(\tau)d\tau.\end{aligned} \tag{3.5.14}$$

To begin with, we first look at the second Lebesgue term on the right hand. By the boundedness of $D_\cdot A_i$, $\varphi(\cdot)$ and (3.5.12), it is well-defined and square integrable. Hence for almost $t \in [0,T]$, the following term is well-defined:

$$\int_0^t\int_0^s D_\tau\big[A_i(t,s)\big]\Lambda_1(s,\tau)\varphi(\tau)d\tau ds.$$

Next, by Step 2 and the boundedness of A_i, the left-hand term is well-defined and square integrable. Hence for almost $t \in [0, T]$, the following term is well-defined:

$$\int_0^t A_i(t,s) \int_0^s \Lambda_1(s,\tau)\varphi(\tau)dW_1(\tau)ds.$$

Eventually, according to Proposition 12.1.5 in the Appendices, to finish the aim of this step, we only need to show that the first term on the right hand, i.e.,

$$\mathscr{L}_2(t,s) \equiv \int_0^s A_i(t,s)\Lambda_1(s,\tau)\varphi(\tau)dW_1(\tau), \quad t \geqslant s,$$

is well-defined in the Skorohod integral sense. According to Proposition 12.1.3 in the Appendices, we need to verify three facts. The first one is for almost $0 \leqslant s \leqslant t \leqslant T$,

$$\mathbb{E}\int_0^s \left|A_i(t,s)\Lambda_1(s,\tau)\varphi(\tau)\right|^2 d\tau < \infty.$$

This is easy to see by the boundedness of A_i, φ and (3.5.12). Second, for almost $0 \leqslant s \leqslant t \leqslant T$, to show $\mathscr{L}_2(t,s)$ is well-defined, we need that

$$\mathbb{E}\left|A_i(t,s)\Lambda_1(s,\tau)\varphi(\tau)\right|^2 + \mathbb{E}\int_0^s \left|D_r\big[A_i(t,s)\Lambda_1(s,\tau)\varphi(\tau)\big]\right|^2 dr < \infty, \ s \in [0,\tau].$$

The result for the first term is easy to see. As to the second one, by Proposition 12.1.6 in the Appendices,

$$\begin{aligned}
&\mathbb{E}\left|D_r\big[A_i(t,s)\Lambda_1(s,\tau)\varphi(\tau)\big]\right|^2 \\
&= \mathbb{E}\Big|D_r\big[A_i(t,s)\big]\Lambda_1(s,\tau)\varphi(\tau) + A_i(t,s)D_r\big[\Lambda_1(s,\tau)\big]\varphi(\tau) \\
&\qquad + A_i(t,s)\Lambda_1(s,\tau)\big[D_r\varphi(\tau)\big]\Big|^2.
\end{aligned}$$

Then the above second fact is easy to verify by the bounded conditions on $\varphi(\cdot)$, $D_\cdot A_i$, (3.5.12) and (3.5.9). Third, we need to show that there exists a measurable version of $D_r\big[A_i(t,s)\Lambda_1(s,\tau)\varphi(\tau)\big]$ such that

$$\mathbb{E}\int_0^s\int_0^s \left|D_r\big[A_i(t,s)\Lambda_1(s,\tau)\varphi(\tau)\big]\right|^2 d\tau dr < \infty.$$

Of course, this can be implied by the bounded conditions on $\varphi(\cdot)$, $D.A_i$, (3.5.12) and (3.5.11). By the above arguments in this step, we can also derive

$$\begin{aligned}
&\int_0^t A_i(t,s)\int_0^s \Lambda_1(s,\tau)\varphi(\tau)dW_1(\tau)ds \\
&=\int_0^t\int_0^s A_i(t,s)\Lambda_1(s,\tau)\varphi(\tau)dW_1(\tau)ds \\
&\quad+\int_0^t\int_0^s D_\tau\big[A_i(t,s)\big]\Lambda_1(s,\tau)\varphi(\tau)d\tau ds, \\
&\int_0^t A_i(t,s)\int_0^s \Lambda_1(s,\tau)\varphi(\tau)dW_1(\tau)dW_1(s) \\
&=\int_0^t\int_0^s A_i(t,s)\Lambda_1(s,\tau)\varphi(\tau)dW_1(\tau)dW_1(s) \\
&\quad+\int_0^t\int_0^s D_\tau\big[A_i(t,s)\big]\Lambda_1(s,\tau)\varphi(\tau)d\tau dW_1(s).
\end{aligned} \tag{3.5.15}$$

Step 4. In this step, we show that

$$\begin{aligned}
&\int_0^t\int_0^s A_1(t,s)\Lambda_1(s,\tau)\varphi(\tau)dW_1(\tau)dW_1(s) \\
&=\int_0^t\Big[\int_\tau^t A_1(t,s)\Lambda_1(s,\tau)dW_1(s)\Big]\varphi(\tau)dW_1(\tau).
\end{aligned}$$

According to Theorem 12.1.7 in the Appendices and the previous discussion, we only verify the right-hand term is well-defined. For notational simplicity, let

$$\begin{aligned}
&\mathscr{L}_4(t)\equiv\int_0^t\Big[\int_\tau^t A_1(t,s)\Lambda_1(s,\tau)dW_1(s)\Big]\varphi(\tau)dW_1(\tau)=\int_0^t \mathscr{L}_3(t,\tau)\varphi(\tau)dW_1(\tau), \\
&\mathscr{L}_3(t,\tau)\equiv\Big[\int_\tau^t A_1(t,s)\Lambda_1(s,\tau)dW_1(s)\Big].
\end{aligned}$$

To show the above $\mathscr{L}_4(t)$ is well-defined in Skorohod integral sense, by Proposition 12.1.3 in the Appendices we need to show

$$\mathbb{E}\int_0^t|\mathscr{L}_3(t,\tau)\varphi(\tau)|^2d\tau+\mathbb{E}\int_0^t\Big|D_r\big[\mathscr{L}_3(t,\tau)\varphi(\tau)\big]\Big|^2dr<\infty,$$

and there exists a measurable version such that

$$\mathbb{E}\int_0^t\int_0^t\Big|D_s\big[\mathscr{L}_3(t,\tau)\varphi(\tau)\big]\Big|^2dsd\tau<\infty.$$

In fact, by the measurability and integrability of A_1 and Λ_1, $\mathscr{L}_3(t,\tau)$ is a well-defined Itô integral such that

$$\mathbb{E}\int_0^T\int_0^t\big|\mathscr{L}_3(t,\tau)\big|^2d\tau dt=\mathbb{E}\int_0^T\int_0^t\int_\tau^t\big|A_1(t,s)\Lambda_1(s,\tau)\big|^2dsd\tau dt<\infty.$$

According to (iii) of Proposition 12.1.2 in the Appendices, for $r\in[0,t]$,

$$D_r\mathscr{L}_3(t,\tau)=A_1(t,r)\Lambda_1(r,\tau)\mathbf{1}_{(\tau,t]}(r)+\int_{r\vee\tau}^t D_r\big[A_1(t,s)\Lambda_1(s,\tau)\big]dW_1(s).$$

From the above (3.5.12) and the boundedness of A_1,

$$\mathbb{E}\int_0^t\int_\tau^t\big|A_1(t,r)\Lambda_1(r,\tau)\big|^2drd\tau<\infty.$$

Also by the Fubini theorem and the above results of (3.5.12), (3.5.10), as well as the boundedness A_1, $D.A_1$,

$$\begin{aligned}&\mathbb{E}\int_0^t\int_0^t\int_{r\vee\tau}^t\Big|D_r\big[A_1(t,s)\Lambda_1(s,\tau)\big]\Big|^2dsdrd\tau\\&=\mathbb{E}\int_0^t\int_0^t\int_\tau^t\Big|D_r\big[A_1(t,s)\Lambda_1(s,\tau)\big]\Big|^2dsdrd\tau\\&\quad+\mathbb{E}\int_0^t\int_0^t\int_r^t\Big|D_r\big[A_1(t,s)\Lambda_1(s,\tau)\big]\Big|^2dsdrd\tau\\&=\mathbb{E}\int_0^t\int_0^s\int_0^s\Big|D_r\big[A_1(t,s)\Lambda_1(s,\tau)\big]\Big|^2drd\tau ds<\infty.\end{aligned}$$

Therefore, we must have

$$\begin{aligned}&\mathbb{E}\int_0^t\int_0^t\Big|D_r\big[\mathscr{L}_3(t,\tau)\big]\Big|^2drd\tau\\&\leqslant\mathbb{E}\int_0^t\int_\tau^t\big|A_1(t,r)\Lambda_1(r,\tau)\big|^2drd\tau\\&\quad+\mathbb{E}\int_0^t\int_0^t\int_{r\vee\tau}^t\Big|D_r\big[A_1(t,s)\Lambda_1(s,\tau)\big]\Big|^2dsdrd\tau<\infty.\end{aligned}$$

This finishes the proof of the current step.

Step 5. In this step, we show that

$$\int_0^t\int_0^s A_0(t,s)\Lambda_1(s,\tau)\varphi(\tau)dW_1(\tau)ds=\int_0^t\Big[\int_\tau^t A_0(t,s)\Lambda_1(s,\tau)ds\Big]\varphi(\tau)dW_1(\tau).$$

According to Theorem 12.1.7 in the appendix, we need to verify the above two integrals on both sides are well-defined.

By means of (3.5.15), the term on the left hand is well-defined. Next, we look at the term on the right hand. We define

$$\mathscr{L}_5(t,\tau) \equiv \int_\tau^t A_0(t,s)\Lambda_1(s,\tau)ds,\ t \geqslant \tau.$$

According to Proposition 12.1.3 in the Appendices, it suffices to show that

$$\mathbb{E}\int_0^t \left|\mathscr{L}_5(t,\tau)\varphi(\tau)\right|^2 d\tau + \mathbb{E}\int_0^t\int_0^t \left|D_r\left[\mathscr{L}_5(t,\tau)\varphi(\tau)\right]\right|^2 dr d\tau < \infty.$$

By (3.5.12) and the boundedness of A_0, φ, we immediately obtain that

$$\mathbb{E}\int_0^t \left|\mathscr{L}_5(t,\tau)\varphi(\tau)\right|^2 d\tau \leqslant K\mathbb{E}\int_0^t\int_\tau^t \left|\Lambda_1(s,\tau)\right|^2 ds d\tau < \infty.$$

Using again the boundedness of φ, $D_\cdot\varphi$, A_0, $D_\cdot A_0$,

$$\begin{aligned}
&\left|D_r\left[\mathscr{L}_5(t,\tau)\varphi(\tau)\right]\right|^2 \leqslant K\left|D_r\mathscr{L}_5(t,\tau)\right|^2 + K\left|\mathscr{L}_5(t,\tau)\right|^2,\\
&\left|D_r\mathscr{L}_5(t,\tau)\right|^2 \leqslant K\int_\tau^t \left|D_r\left[A_0(t,s)\Lambda_1(s,\tau)\right]\right|^2 ds\\
&\leqslant K\int_\tau^t \left|D_r\Lambda_1(s,\tau)\right|^2 ds + \int_\tau^t \left|\Lambda_1(s,\tau)\right|^2 ds.
\end{aligned}$$

Then by (3.5.10), (3.5.12) and the Fubini theorem, we obtain that

$$\begin{aligned}
&\mathbb{E}\int_0^t\int_0^t \left|D_r\left[\mathscr{L}_5(t,\tau)\varphi(\tau)\right]\right|^2 dr d\tau\\
&\leqslant K\mathbb{E}\int_0^t\int_0^t\int_\tau^t \left|D_r\Lambda_1(s,\tau)\right|^2 ds dr d\tau + K\mathbb{E}\int_0^t\int_0^t\int_\tau^t \left|\Lambda_1(s,\tau)\right|^2 ds dr d\tau\\
&\quad + K\mathbb{E}\int_0^t\int_0^t \left|\mathscr{L}_5(t,\tau)\right|^2 dr d\tau\\
&\leqslant K\mathbb{E}\int_0^t\int_0^s\int_0^t \left|D_r\Lambda_1(s,\tau)\right|^2 dr d\tau ds + K\mathbb{E}\int_0^t\int_\tau^t \left|\Lambda_1(s,\tau)\right|^2 ds d\tau\\
&\quad + K\mathbb{E}\int_0^t \left|\mathscr{L}_5(t,\tau)\right|^2 d\tau < \infty.
\end{aligned}$$

This finishes the proof of the current step.

Step 6. Based on the above results, in this step we prove (3.5.2). More precisely, by defining $X(\cdot)$ as (3.5.2), we show that it is the unique solution to (3.5.1). To this end, we look at the following:

$$
\begin{aligned}
&\int_0^t A_0(t,s)X(s)ds + \int_0^t A_1(t,s)X(s)dW_1(s)\\
&= \int_0^t A_0(t,s)\Big[\varphi(s) + \int_0^s \Lambda_0(s,\tau)\varphi(\tau)d\tau + \int_0^s \Lambda_1(s,\tau)\varphi(\tau)dW_j(\tau)\Big]ds\\
&\quad+\int_0^t A_1(t,s)\Big[\varphi(s)+\int_0^s \Lambda_0(s,\tau)\varphi(\tau)d\tau+\int_0^s \Lambda_1(s,\tau)\varphi(\tau)dW_1(\tau)\Big]dW_1(s)\\
&= \int_0^t A_0(t,s)\varphi(s)ds + \int_0^t\int_0^s A_0(t,s)\Lambda_0(s,\tau)\varphi(\tau)d\tau ds\\
&\quad+\int_0^t A_0(t,s)\int_0^s \Lambda_1(s,\tau)\varphi(\tau)dW_1(\tau)ds + \int_0^t A_1(t,s)\varphi(s)dW_1(s)\\
&\quad+\int_0^t\int_0^s A_1(t,s)\Lambda_0(s,\tau)\varphi(\tau)d\tau dW_1(s)\\
&\quad+\int_0^t A_1(t,s)\int_0^s \Lambda_1(s,\tau)\varphi(\tau)dW_1(\tau)dW_i(s).
\end{aligned}
$$

In terms of Step 3 and Step 5, we have

$$
\begin{aligned}
&\int_0^t A_0(t,s)\int_0^s \Lambda_1(s,\tau)\varphi(\tau)dW_1(\tau)ds\\
&=\int_0^t\Big[\int_0^s A_0(t,s)\Lambda_1(s,\tau)\varphi(\tau)dW_1(\tau) + \int_0^s D_\tau[A_0(t,s)]\Lambda_1(s,\tau)\varphi(\tau)d\tau\Big]ds\\
&=\int_0^t\int_r^t A_0(t,s)\Lambda_1(s,\tau)\varphi(\tau)dsdW_1(\tau)+\int_0^t\int_r^t D_\tau[A_0(t,s)]\Lambda_1(s,\tau)\varphi(\tau)dsd\tau\\
&=\int_0^t\Big[\int_s^t A_0(t,\tau)\Lambda_1(\tau,s)d\tau\Big]\varphi(s)dW_1(s)+\int_0^t\Big[\int_s^t D_s[A_0(t,\tau)]\Lambda_1(\tau,s)d\tau\Big]\varphi(s)ds.
\end{aligned}
$$

The first integral on the right-hand side of the above is a Skorohod integral. Since $\varphi(\cdot)$ is $\mathbb{F}$-adapted, we have

$$
\begin{aligned}
&\int_0^t\int_0^s A_1(t,s)\Lambda_0(s,\tau)\varphi(\tau)d\tau dW_1(s) = \int_0^t\int_\tau^t A_1(t,s)\Lambda_0(s,\tau)\varphi(\tau)dW_1(s)d\tau\\
&= \int_0^t\Big[\int_s^t A_1(t,\tau)\Lambda_0(\tau,s)dW_1(\tau)\Big]\varphi(s)ds.
\end{aligned}
$$

Here the stochastic integrals hold in Itô's sense. Using the results in Step 3 and Step 4,

$$
\begin{aligned}
&\int_0^t A_1(t,s)\int_0^s \Lambda_1(s,\tau)\varphi(\tau)dW_1(\tau)dW_1(s)\\
&=\int_0^t\Big[\int_0^s A_1(t,s)\Lambda_1(s,\tau)\varphi(\tau)dW_1(\tau)+\int_0^s D_\tau[A_1(t,s)]\Lambda_1(s,\tau)\varphi(\tau)d\tau\Big]dW_1(s)\\
&=\int_0^t\int_\tau^t A_1(t,s)\Lambda_1(s,\tau)\varphi(\tau)dW_1(s)dW_1(\tau)\\
&\qquad\qquad+\int_0^t\int_\tau^t D_\tau[A_1(t,s)]\Lambda_1(s,\tau)\varphi(\tau)dW_1(s)d\tau\\
&=\int_0^t\Big[\int_s^t A_1(t,\tau)\Lambda_1(\tau,s)dW_1(\tau)\Big]\varphi(s)dW_1(s)\\
&\qquad\qquad+\int_0^t\Big[\int_s^t D_s[A_1(t,\tau)]\Lambda_1(\tau,s)dW_1(\tau)\Big]\varphi(s)ds.
\end{aligned}
$$

Hence, we obtain

$$
\begin{aligned}
&\int_0^t A_0(t,s)X(s)ds+\int_0^t A_1(t,s)X(s)dW_1(s)\\
&=\int_0^t A_0(t,s)\varphi(s)ds+\int_0^t\Big[\int_s^t A_0(t,\tau)\Lambda_0(\tau,s)d\tau\Big]\varphi(s)ds\\
&\quad+\int_0^t\Big[\int_s^t A_0(t,\tau)\Lambda_1(\tau,s)d\tau\Big]\varphi(s)dW_1(s)\\
&\quad+\int_0^t\Big[\int_s^t D_s[A_0(t,\tau)]\Lambda_1(\tau,s)d\tau\Big]\varphi(s)ds\\
&\quad+\int_0^t A_1(t,s)\varphi(s)dW_1(s)+\int_0^t\Big[\int_s^t A_1(t,\tau)\Lambda_0(\tau,s)dW_1(\tau)\Big]\varphi(s)ds\\
&\quad+\int_0^t\Big[\int_s^t A_1(t,\tau)\Lambda_1(\tau,s)dW_1(\tau)\Big]\varphi(s)dW_1(s)\\
&\quad+\int_0^t\Big[\int_s^t D_s[A_1(t,\tau)]\Lambda_1(\tau,s)dW_1(\tau)\Big]\varphi(s)ds\\
&=\int_0^t\Big[A_0(t,s)+\int_s^t A_0(t,\tau)\Lambda_0(\tau,s)d\tau+\int_s^t A_1(t,\tau)\Lambda_0(\tau,s)dW_1(\tau)\\
&\quad+\int_s^t D_s[A_0(t,\tau)]\Lambda_1(\tau,s)d\tau+\int_s^t D_s[A_1(t,\tau)]\Lambda_1(\tau,s)dW_1(\tau)\Big]\varphi(s)ds\\
&\quad+\int_0^t\Big[A_1(t,s)+\int_s^t A_0(t,s)\Lambda_1(\tau,s)d\tau\\
&\quad+\int_s^t A_1(t,\tau)\Lambda_1(\tau,s)dW_1(\tau)\Big]\varphi(s)dW_1(s)\\
&=\int_0^t \Lambda_0(t,s)\varphi(s)ds+\int_0^t \Lambda_1(t,s)\varphi(s)dW_1(s)=X(t)-\varphi(t).
\end{aligned}
$$

Consequently, $X(\cdot)$ defined by (3.5.2) is the unique solution to (3.5.1). □

In the case that all the coefficients are deterministic, (3.5.3) becomes

$$\begin{cases} \Lambda_j(t,s) = A_j(t,s) + \int_s^t A_0(t,\tau)\Lambda_j(\tau,s)d\tau \\ \qquad + \sum_{i=1}^{d} \int_s^t A_i(t,\tau)\Lambda_j(\tau,s)dW_i(\tau), \\ \qquad\qquad 0 \leqslant s \leqslant t \leqslant T,\ 0 \leqslant j \leqslant d. \end{cases} \tag{3.5.16}$$

Now, we consider linear FSDE, which is a special case of FSVIE:

$$\begin{cases} dX(t) = [A_0(t)X(t) + b(t)]dt + \sum_{i=1}^{d}[A_i(t)X(t) + \sigma_i(t)]dW_i(t), \\ X(0) = x. \end{cases} \tag{3.5.17}$$

The case of the above with $b(\cdot) = \sigma_1(\cdot) = \cdot s = \sigma_d(\cdot) \equiv 0$ is called *homogeneous linear FSDE*, denoted by $[A_0(\cdot); A_1(\cdot). \cdot s, A_d(\cdot)]$. We have the following proposition.

Proposition 3.5.2 *Let* $A_0(\cdot), A_1(\cdot), \cdot s, A_d(\cdot) \in L^\infty_{\mathbb{F}}(0,T;\mathbb{R}^{n\times n})$ *and* $b(\cdot)$, $\sigma_1(\cdot), \cdot s, \sigma_d(\cdot) \in L^2_{\mathbb{F}}(0,T;\mathbb{R}^n)$. *Then*

$$\begin{aligned} X(t) = \Phi(t;0)x + \int_0^t \Phi(t;\tau)b(\tau)d\tau - \int_0^t \sum_{i=1}^{d} \Phi(t;\tau)A_i(\tau)\sigma_i(\tau)d\tau \\ + \Phi(t;0)\int_0^t \sum_{i=1}^{d} \Phi(\tau;0)^{-1}\sigma_i(\tau)dW_i(\tau), \quad t \in [0,T]. \end{aligned} \tag{3.5.18}$$

Here the process $\Phi(\cdot;\cdot)$ *that is reversible and satisfies*

$$\begin{cases} d\Phi(t;s) = A_0(t)\Phi(t;s)dt + \sum_{i=1}^{d} A_i(t)\Phi(t;s)dW_i(t), \qquad t \in [s,T], \\ \Phi(s;s) = I, \end{cases} \tag{3.5.19}$$

is called the fundamental matrix of linear FSDE $[A_0(\cdot); A_1(\cdot), \cdot s, A_d(\cdot)]$.

Proof For any $s \in [0, T)$, let $\Phi(\cdot\,; s)$ be the solution to (3.5.19). We claim that for any $0 \leqslant s \leqslant t \leqslant T$, $\Phi(t; s)$ is invertible. To see this, let $\widetilde{\Phi}(\cdot\,; s)$ be the solution to the following linear SDE:

$$\begin{cases} d\widetilde{\Phi}(t; s) = \widetilde{\Phi}(t; s)\Big[\sum_{i=1}^{d} A_i(t)^2 - A_0(t)\Big]dt - \sum_{i=1}^{d} \widetilde{\Phi}(t; s)A_i(t)dW_i(t), \\ \qquad\qquad\qquad\qquad\qquad\qquad\qquad\qquad\qquad\qquad t \in [s, T], \\ \widetilde{\Phi}(s; s) = I. \end{cases} \tag{3.5.20}$$

Then applying Itô's formula to the process $t \mapsto \Phi(t; s)\widetilde{\Phi}(t; s)$, we have

$$\begin{aligned} d\Big[\Phi(t; s)\widetilde{\Phi}(t; s)\Big] = &\Big\{A_0(t)\Phi(t; s)\widetilde{\Phi}(t; s) + \Phi(t; s)\widetilde{\Phi}(t; s)\Big[\sum_{i=1}^{d} A_i(t)^2 - A_0(t)\Big] \\ &- \sum_{i=1}^{d} A_i(t)\Phi(t; s)\widetilde{\Phi}(t; s)A_i(t)\Big\}dt \\ &+ \sum_{i=1}^{d}\Big[A_i(t)\Phi(t; s)\widetilde{\Phi}(t; s) - \Phi(t; s)\widetilde{\Phi}(t; s)A_i(t)\Big]dW_i(t). \end{aligned}$$

The above is a linear equation for $\Phi(\cdot\,; s)\widetilde{\Phi}(\cdot\,; s)$, and I is a solution. Thus, by uniqueness,

$$\widetilde{\Phi}(t; s) = \Phi(t; s)^{-1}, \qquad 0 \leqslant s \leqslant t \leqslant T.$$

This means that $\Phi(t; s)$ is invertible. Now, let

$$X(t) = \Phi(t; s)\widetilde{X}(t), \qquad t \in [s, T],$$

be the solution of (3.5.17), with

$$d\widetilde{X}(t) = \beta_0(t)dt + \sum_{i=1}^{d} \beta_i(t)dW_i(t),$$

for some $\beta_i(\cdot)$, $0 \leqslant i \leqslant d$. Then

$$X(s) = \Phi(s; s)\widetilde{X}(s) = \widetilde{X}(s),$$

and

$$\begin{aligned}
&\Big[A_0(t)\Phi(t;s)\widetilde{X}(t)+b(t)\Big]dt+\sum_{i=1}^{d}\Big[A_i(t)\Phi(t;s)\widetilde{X}(t)+\sigma_i(t)\Big]dW_i(t)\\
&=d\Big[\Phi(t;s)\widetilde{X}(t)\Big]\\
&=\Big[A_0(t)\Phi(t;s)\widetilde{X}(t)+\Phi(t;s)\beta_0(t)+\sum_{i=1}^{d}A_i(t)\Phi(t;s)\beta_i(t)\Big]dt\\
&\quad+\sum_{i=1}^{d}\Big[A_i(t)\Phi(t;s)\widetilde{X}(t)+\Phi(t;s)\beta_i(t)\Big]dW_i(t).
\end{aligned}$$

Hence,

$$\begin{cases} b(t)=\Phi(t;s)\beta_0(t)+\sum_{i=1}^{d}A_i(t)\Phi(t;s)\beta_i(t),\\ \sigma_i(t)=\Phi(t;s)\beta_i(t), \qquad 1\leqslant i\leqslant d,\end{cases}$$

which leads to

$$\begin{aligned}
&\beta_i(t)=\Phi(t;s)^{-1}\sigma_i(t), \qquad 1\leqslant i\leqslant d,\\
&\beta_0(t)=\Phi(t;s)^{-1}\Big[b(t)-\sum_{i=1}^{d}A_i(t)\sigma_i(t)\Big].
\end{aligned}$$

Consequently,

$$\begin{aligned}
\widetilde{X}(t)=X(s)&+\int_s^t\Big\{\Phi(\tau;s)^{-1}\Big[b(\tau)-\sum_{i=1}^{d}A_i(\tau)\sigma_i(\tau)\Big]\Big\}d\tau\\
&+\int_s^t\sum_{i=1}^{d}\Phi(\tau;s)^{-1}\sigma_i(\tau)dW_i(\tau), \qquad t\in[s,T].
\end{aligned}$$

Thus, we obtain the variation of constants formula for linear FSDE:

$$\begin{aligned}
X(t)=\Phi(t;s)X(s)&+\Phi(t;s)\int_s^t\Phi(\tau;s)^{-1}\Big[b(\tau)-\sum_{i=1}^{d}A_i(\tau)\sigma_i(\tau)\Big]d\tau\\
&+\Phi(t;s)\int_s^t\sum_{i=1}^{d}\Phi(\tau;s)^{-1}\sigma_i(\tau)dW_i(\tau), \quad t\in[s,T].
\end{aligned}$$

In particular, by setting $s=0$, we obtain (3.5.18). □

At this moment, let us make some comments on the above two variation of constants formulas. First, the ideas of the two proofs are different. In fact, Theorem 3.5.1 relies on the Fubini theorem (see Step 6), (anticipated) Skorohod integral, and the resulting Malliavin calculus (see H3.5.1), while Proposition 3.5.2 makes full use of Itô's formula and bypasses the involvement of Malliavin calculus. Second, when (3.5.1) reduces to linear SDE (3.5.17), these two results also have some tight connections. To this end, using the same ideas as in Theorem 3.5.1, we give one variation of constants formula for

$$X(t) = \varphi(t) + \int_\theta^t A_0(t,s)X(s)ds + \sum_{i=1}^d \int_\theta^t A_i(t,s)X(s)dW_i(s), \quad t \in [\theta, T]. \tag{3.5.21}$$

More precisely, under (H3.5.1), for any $\varphi(\cdot) \in L^\infty_{\mathbb{F}}(0,T;\mathbb{R}^n)$ satisfying $D_\cdot\varphi(\cdot)$ being bounded, the solution to linear FSVIE (3.5.21) has the following representation, involving Lebesgue and Skorohod integrals:

$$X(t) = \varphi(t) + \int_\theta^t \Lambda_0(t,s)\varphi(s)ds + \sum_{i=1}^d \int_\theta^t \Lambda_i(t,s)\varphi(s)dW_i(s), \quad t \in [0, T], \tag{3.5.22}$$

where for all $0 \leqslant i \leqslant d$, processes $\Lambda_i : \Delta_*[0,T] \times \Omega \to \mathbb{R}^{n\times n}$ satisfy system (3.5.3). On the other hand, for the following SDE,

$$\begin{cases} dX(t) = [A_0(t)X(t) + b(t)]dt + \sum_{i=1}^d [A_i(t)X(t) + \sigma_i(t)]dW_i(t), & t \in [\theta, T], \\ X(\theta) = x, \end{cases} \tag{3.5.23}$$

by Proposition 3.5.2, its solution can be represented as

$$\begin{aligned} X(t) &= \Phi(t;\theta)x + \Phi(t;\theta)\int_\theta^t \Phi(\tau;\theta)^{-1}\Big[b(\tau) - \sum_{i=1}^d A_i(\tau)\sigma_i(\tau)\Big]d\tau \\ &\quad + \Phi(t;\theta)\int_\theta^t \sum_{i=1}^d \Phi(\tau;\theta)^{-1}\sigma_i(\tau)dW_i(\tau), \quad t \in [\theta, T]. \end{aligned}$$

Note that linear FSDE (3.5.23) can be written as follows:

$$X(t) = \varphi(t) + \int_\theta^t A_0(s)X(s)ds + \sum_{i=1}^d \int_\theta^t A_i(s)X(s)dW_i(s), \quad t \in [\theta, T],$$

with

$$\varphi(t) = x + \int_\theta^t b(s)ds + \int_\theta^t \sum_{i=1}^d \sigma_i(s)dW_i(s). \tag{3.5.24}$$

To see the relation between Φ and Λ_i, let $A_i(t,s)$ be independent of t, $\varphi(\cdot) \equiv x$, and $b = \sigma = 0$. Then (3.5.21) reduces to (3.5.23), and

$$X(t) = \Big[I + \int_\theta^t \Lambda_0(t,s)ds + \sum_{i=1}^d \int_\theta^t \Lambda_i(t,s)dW_i(s)\Big]x = \Phi(t;\theta)x.$$

For any given $0 \leqslant \theta \leqslant t \leqslant T$, by the arbitrariness of $x \in \mathbb{R}^n$, we see that

$$\Phi(t;\theta) = I + \int_\theta^t \Lambda_0(t,\tau)d\tau + \sum_{i=1}^d \int_\theta^t \Lambda_i(t,\tau)dW_i(\tau). \tag{3.5.25}$$

Of course, to keep the anticipated integral on the right hand to be well-defined, we still need (H3.5.1).

Next, let us look at the case when $\sigma = 0$, $b(\cdot)$ and its Malliavin derivative $D.b(\cdot)$ are bounded. Then by (3.5.22),

$$\begin{aligned} X(t) &= \Big(I + \int_\theta^t \Lambda_0(t,s)ds + \sum_{i=1}^d \int_\theta^t \Lambda_i(t,s)dW_i(s)\Big)x + \int_\theta^t b(s)ds \\ &\quad + \int_\theta^t \Lambda_0(t,s)\Big(\int_\theta^s b(\tau)d\tau\Big)ds + \sum_{i=1}^d \int_\theta^t \Lambda_i(t,s)\Big(\int_\theta^s b(\tau)d\tau\Big)dW_i(s). \end{aligned} \tag{3.5.26}$$

Similar to Step 5 in the proof of Theorem 3.5.1, we have

$$\sum_{i=1}^d \int_\theta^t \Lambda_i(t,s)\Big(\int_\theta^s b(\tau)d\tau\Big)dW_i(s) = \sum_{i=1}^d \int_\theta^t \int_\tau^t \Lambda_i(t,s)b(\tau)dW_i(s)d\tau.$$

Also we have

$$\int_\theta^t \Lambda_0(t,s)\Big(\int_\theta^s b(\tau)d\tau\Big)ds = \int_\theta^t \int_\tau^t \Lambda_0(t,s)b(\tau)dsd\tau.$$

Then to sum up the above arguments, using again (3.5.25), we rewrite (3.5.26) as

$$X(t) = \Phi(t;\theta)x + \int_\theta^t \Phi(t;s)b(s)ds, \quad t \in [\theta, T]. \tag{3.5.27}$$

This is just the result in Proposition 3.5.2.

Eventually, we consider the general case of $\varphi(\cdot)$ in (3.5.24). According to the above discussion, to show the coincidence between Theorem 3.5.1 and Proposition 3.5.2 under this setting, we only need to verify that

$$\begin{aligned}
&\sum_{i=1}^d \Big[-\int_\theta^t \Phi(t;\tau)A_i(\tau)\sigma_i(\tau)d\tau + \Phi(t;\theta)\int_\theta^t \Phi(\tau;\theta)^{-1}\sigma_i(\tau)dW_i(\tau)\Big] \\
&= \int_\theta^t \sum_{i=1}^d \sigma_i(s)dW_i(s) + \int_\theta^t \Lambda_0(t,s)\Big(\sum_{i=1}^d \int_\theta^s \sigma_i(\tau)dW_i(\tau)\Big)ds \\
&\quad + \sum_{i=1}^d \int_\theta^t \Lambda_i(t,s)\Big(\int_\theta^s \sum_{j=1}^d \sigma_j(\tau)dW_j(\tau)\Big)dW_i(s), \qquad t \in [\theta, T],
\end{aligned} \tag{3.5.28}$$

provided the anticipated integrals on the right hand are well-defined with proper choice of σ.

Next, we look at (3.5.28). First of all, for the left term of (3.5.28), by proper assumption on σ, one has

$$\begin{aligned}
&-\int_\theta^t \Phi(t;\tau)A_i(\tau)\sigma_i(\tau)d\tau + \Phi(t;\theta)\int_\theta^t \Phi(\tau;\theta)^{-1}\sigma_i(\tau)dW_i(\tau) \\
&= \int_\theta^t \Big(\{D_\tau^{(i)}[\Phi(t;\theta)]\}\Phi(\tau;\theta)^{-1} - \Phi(t;\tau)A_i(\tau)\Big)\sigma_i(\tau)d\tau + \int_\theta^t \Phi(t;\tau)\sigma_i(\tau)dW_i(\tau).
\end{aligned}$$

Recalling $\Phi(\tau;\theta)^{-1} = \widetilde{\Phi}(\tau;\theta)$, we have

$$\begin{aligned}
\Phi(\tau;\theta)^{-1} &= I + \int_\theta^\tau \Phi(r;\theta)^{-1}\Big[\sum_{i=1}^d A_i(r)^2 - A_0(r)\Big]dr \\
&\quad - \int_\theta^\tau \sum_{i=1}^d \Phi(r;\theta)^{-1}A_i(r)dW_i(r), \qquad \tau \in [\theta, T].
\end{aligned}$$

Since the integrands in the above are all $\mathbb{F}$-adapted, one has

$$D_\tau^{(i)}[\Phi(\tau;\theta)^{-1}] = -\Phi(\tau;\theta)^{-1}A_i(\tau), \qquad \text{a.e. } \tau \in [\theta, T].$$

Suppose the following is true:

$$
\begin{aligned}
&D_\tau^{(i)}\big[\Phi(t;\tau)\big] = D_\tau^{(i)}\big[\Phi(t;\theta)\Phi(\tau;\theta)^{-1}\big]\\
&= \Big\{D_\tau^{(i)}\big[\Phi(t;\theta)\big]\Big\}\Phi(\tau;\theta)^{-1} + \Phi(t;\theta)\Big\{D_\tau^{(i)}\big[\Phi(\tau;\theta)^{-1}\big]\Big\}\\
&= \Big\{D_\tau^{(i)}\big[\Phi(t;\theta)\big]\Big\}\Phi(\tau;\theta)^{-1} - \Phi(t;\tau)A_i(\tau).
\end{aligned}
$$

On the other hand, by (3.5.25) with θ replaced by τ, one has (using (12.1.34) with $S=\tau$ and $T=t$)

$$
D_\tau^{(i)}\Phi(t;\tau) = \int_\tau^t \{D_\tau^{(i)}\Lambda_0(t,s)\}ds + \sum_{j=1}^d \int_\tau^t D_\tau^{(i)}\Lambda_j(t,s)dW_j(s).
$$

Thus,

$$
\begin{aligned}
&-\int_\theta^t \Phi(t;\tau)A_i(\tau)\sigma_i(\tau)d\tau + \Phi(t;\theta)\int_\theta^t \Phi(\tau;\theta)^{-1}\sigma_i(\tau)dW_i(\tau)\\
&= \int_\theta^t \{D_\tau^{(i)}[\Phi(t;\tau)]\}\sigma_i(\tau)d\tau + \int_\theta^t \Phi(t;\tau)\sigma_i(\tau)dW_i(\tau)\\
&= \int_\theta^t \Big(\int_\tau^t \{D_\tau^{(i)}\Lambda_0(t,s)\}ds + \sum_{j=1}^d \int_\tau^t D_\tau^{(i)}\Lambda_j(t,s)dW_j(s)\Big)\sigma_i(\tau)d\tau\\
&\quad + \int_\theta^t \Phi(t;\tau)\sigma_i(\tau)dW_i(\tau).
\end{aligned}
$$

Eventually, one has

$$
\begin{aligned}
&\sum_{i=1}^d \Big[\int_\theta^t \sigma_i(s)dW_i(s) + \int_\theta^t \Lambda_0(t,s)\Big(\int_\theta^s \sigma_i(\tau)dW_i(\tau)\Big)ds\Big]\\
&\qquad + \sum_{j=1}^d \int_\theta^t \Lambda_j(t,s)\Big(\int_\theta^s \sum_{i=1}^d \sigma_i(\tau)dW_i(\tau)\Big)dW_j(s)\\
&= \sum_{i=1}^d \Big\{\int_\theta^t \sigma_i(s)dW_i(s)\\
&\qquad + \int_\theta^t \Big[\int_\theta^s \Lambda_0(t,s)\sigma_i(\tau)dW_i(\tau) + \int_\theta^s \{D_\tau^{(i)}\Lambda_0(t,s)\}\sigma_i(\tau)d\tau\Big]ds\Big\}\\
&\quad + \sum_{i,j=1}^d \int_\theta^t \Big[\int_\theta^s \Lambda_j(t,s)\sigma_i(\tau)dW_i(\tau) + \int_\theta^s \{D_\tau^{(i)}\Lambda_j(t,s)\}\sigma_i(\tau)d\tau\Big]dW_j(s)
\end{aligned}
$$

$$
\begin{aligned}
&= \sum_{i=1}^d \Big\{ \int_\theta^t \sigma_i(\tau) dW_i(\tau) + \int_\theta^t \Big(\int_\tau^t \Lambda_0(t,s)ds \Big) \sigma_i(\tau) dW_i(\tau) \\
&\quad + \int_\theta^t \Big(\int_\tau^t \{ D_\tau^{(i)} \Lambda_0(t,s) \} ds \Big) \sigma_i(\tau) d\tau + \sum_{j=1}^d \Big[\int_\theta^t \Big(\int_\tau^t \Lambda_j(t,s) dW_j(s) \Big) \sigma_i(\tau) dW_i(\tau) \\
&\quad + \int_\theta^t \Big(\int_\tau^t \{ D_\tau^{(i)} \Lambda_j(t,s) \} dW_j(s) \Big) \sigma_i(\tau) d\tau \Big] \Big\} \\
&= \sum_{i=1}^d \Big\{ \int_\theta^t \Big(I + \int_\tau^t \Lambda_0(t,s) ds + \sum_{j=1}^d \int_\tau^t \Lambda_j(t,s) dW_j(s) \Big) \sigma_i(\tau) dW_i(\tau) \\
&\quad + \int_\theta^t \Big(\int_\tau^t \{ D_\tau^{(i)} \Lambda_0(t,s) \} ds + \sum_{j=1}^d \int_\tau^t \{ D_\tau^{(i)} \Lambda_j(t,s) \} dW_j(s) \Big) \sigma_i(\tau) d\tau \Big\}.
\end{aligned}
$$

Of course, one has to impose proper assumption on σ to keep the above anticipated integral and Fubini theorem true. Thus, (3.5.28) is true. At last, as to the appropriate assumption on σ to keep the above arguments go through, we leave the details to the interested readers.

6 Bibliographical Remarks

Results presented in Sect. 1 on general (nonlinear) BSDEs are essentially due to Pardoux–Peng [158] and El Karoui–Peng–Quenez [68]. However, our proof is a little different from that in the literature; in particular, we do not use Itô's formula which might be hard to apply for BSVIEs later on, and we provide a detailed Picard iteration scheme over the whole interval $[0, T]$. Corollary 3.1.7 was proved in Shi–Wang–Yong [176], which will be useful in proving the path continuity of adapted solution to BSVIEs later. The Malliavin differentiability part in Sect. 2 is based on the work by El Karoui–Peng–Quenez [68] with little modifications. For linear BSDEs in Sect. 3, we derived a variation of constants formula, essentially following the lines of Yong–Zhou [223], Chapter 7. See Bismut [30] and Bensoussan [27] for early works. In Sect. 4, we briefly presented a theory of FSVIEs, partially in the spirit of Berger–Mizel [28] (see also Pardoux–Protter [160], Wang–Zheng [209]). However, we established the well-posedness of the equation by means of global Picard iteration, not only for continuous free terms, but also for L^p-integrable free terms. For the path continuity of the solution, we used the strengthened version of Kolmogorov continuity theorem which can be proved by the Garsia–Rodemith–Rumsey inequality. For readers' convenience, we will provide a detailed proof in the appendix. Similar results of the path continuity of the solution to the FSVIE with more results were presented by Zhang [227]. In Sect. 5, some efforts are made to derive the variation of constants formula for linear FSVIEs. To our best knowledge, at the moment when this book was started to be written, the variation of constants formula for linear FSVIEs seems to be the first time to be derived, at least in this way. Shortly after, Hamaguchi [92] found a different approach.

Chapter 4
Type I BSVIEs

In this chapter, we will establish the well-posedness of Type I BSVIEs, by which we mean to study the following type BSVIE:

$$Y(t)=\psi(t)+\int_t^T g(t,s,Y(s),Z(t,s))ds-\int_t^T Z(t,s)dW(s),\qquad t\in[0,T]. \tag{4.0.1}$$

We note that only $Z(t,s)$ appear in the above. In (4.0.1), $g:\Delta^*[0,T]\times\mathbb{R}^m\times\mathbb{R}^m\times\Omega\to\mathbb{R}^m$ is a *generator* and $\psi:[0,T]\times\Omega\to\mathbb{R}^m$ is a *free term*. For simplicity, the Brownian motion in this chapter is assumed to be one-dimensional ($d=1$).

1 Adapted Solutions

We recall that $\Delta^*[0,T]$ is the upper triangle domain given by the following:

$$\Delta^*[0,T]=\big\{(t,s)\in[0,T]^2\bigm|0\leqslant t<s\leqslant T\big\}.$$

We may rewrite (4.0.1) as follows:

$$Y(t)=\widetilde{\psi}(t)+\int_t^T\widetilde{g}(t,s,Y(s),Z(t,s))ds-\int_t^T Z(t,s)dW(s),\qquad t\in[0,T], \tag{4.1.1}$$

T. Wang, J. Yong, *Backward Stochastic Volterra Integral Equations*,
Probability Theory and Stochastic Modelling 109,
https://doi.org/10.1007/978-3-032-20848-4_4

with

$$\begin{cases} \widetilde{\psi}(t) = \psi(t) + \int_t^T g(t,s,0,0)ds, \quad t \in [0,T], \\ \widetilde{g}(t,s,y,z) = g(t,s,y,z) - g(t,s,0,0), \quad \forall (t,s,y,z) \in \Delta^*[0,T] \times \mathbb{R}^m \times \mathbb{R}^m. \end{cases}$$

Clearly, $\widetilde{g}(t,s,0,0) = 0$, for all $(t,s) \in \Delta^*[0,T]$. Because of the above reduction, from now on, we assume that in BSVIE (4.0.1),

$$g(t,s,0,0) = 0, \qquad \forall (t,s) \in \Delta^*[0,T]. \tag{4.1.2}$$

We now introduce the following definition.

Definition 4.1.1 A pair $(Y(\cdot), Z(\cdot\,,\cdot))$ is called an *adapted solution* to the Type I BSVIE (4.0.1) if for almost all $t \in [0,T]$, $s \mapsto (Y(s), Z(t,s))$ is $\mathbb{F}$-adapted on $[t,T]$ and (4.0.1) is satisfied almost surely.

Note that $t \mapsto (\psi(t), g(t,s,y,z))$ is not necessarily continuous. Thus, we only require "for almost all $t \in [0,T]$" in the definition of adapted solution and we do not expect to have the continuity of $t \mapsto Y(t)$ in any sense. Actually, in general $Y(\cdot)$ could be discontinuous. Here is a simple example.

Example 4.1.2 Consider the following one-dimensional Type I BSVIE:

$$Y(t) = \psi_0(t)W(T)^2 - \int_t^T Z(t,s)dW(s), \qquad t \in [0,T],$$

where $\psi_0(\cdot)$ is deterministic, measurable, and bounded. Then

$$Y(t) = \mathbb{E}_t[\psi_0(t)W(T)^2] = \psi_0(t)(T-t), \qquad t \in [0,T].$$

Hence, if $\psi_0(\cdot)$ is discontinuous, so is $Y(\cdot)$.

Let us compare the above Type I BSVIE (4.0.1) with the following family of BSDEs parameterized by $t \in [0,T]$:

$$\eta(t,r) = \psi(t) + \int_r^T g(t,s,\eta(t,s),\zeta(t,s))ds - \int_r^T \zeta(t,s)dW(s), \quad r \in [t,T]. \tag{4.1.3}$$

According to the BSDE theory presented in the previous chapter, we know that for almost all $t \in [0,T]$, under some mild conditions, BSDE (4.1.3) admits a unique adapted solution $(\eta(t,\cdot), \zeta(t,\cdot))$. Setting $r = t$ in the above, we obtain

$$\eta(t,t) = \psi(t) + \int_t^T g(t,s,\eta(t,s),\zeta(t,s))ds - \int_t^T \zeta(t,s)dW(s). \tag{4.1.4}$$

A careful comparison of (4.0.1) with (4.1.4) shows that we could not simply set

$$Y(t) = \eta(t,t), \qquad Z(t,s) = \zeta(t,s), \qquad 0 \leqslant t \leqslant s \leqslant T$$

to get an adapted solution to BSVIE (4.0.1) since on the right-hand side of (4.1.4), $\eta(t,s)$ still appears, unless the generator is independent of y:

$$g(t,s,y,z) \equiv g(t,s,z), \qquad \forall (t,s,y,z) \in \Delta^*[0,T] \times \mathbb{R}^m \times \mathbb{R}^{m\times d}.$$

Next, we make an observation from another angle. Consider BSDE:

$$Y(t) = \xi + \int_t^T g(s, Y(s), Z(s))ds - \int_t^T Z(s)dW(s), \qquad t \in [0,T]. \tag{4.1.5}$$

Let us recall a way of solving the above BSDE presented in Chap. 3, Sect. 1. Suppose (4.1.5) is solved on $[T-\delta, T]$ for some $\delta \in (0,T)$, which can be proved. Then the values $(Y(t), Z(t))$ of $(Y(\cdot), Z(\cdot))$ are determined for all $t \in [T-\delta, T]$. Thus, for $t \in [0, T-\delta]$, we may rewrite (4.1.5) as follows:

$$Y(t) = Y(T-\delta) + \int_t^{T-\delta} g(s, Y(s), Z(s))ds - \int_t^{T-\delta} Z(s)dW(s), \tag{4.1.6}$$

where

$$Y(T-\delta) = \xi + \int_{T-\delta}^T g(s, Y(s), Z(s))ds - \int_{T-\delta}^T Z(s)dW(s), \tag{4.1.7}$$

is $\mathcal{F}_{T-\delta}$-measurable (and completely determined by $(Y(s), Z(s))$, $s \in [T-\delta, T]$). Hence, (4.1.6) is a BSDE on $[0, T-\delta]$. Because of this feature, one can establish the well-posedness of BSDE (4.1.5) first on $[T-\delta, T]$, then use the "similar argument" to prove the solvability of (4.1.5) on $[T-2\delta, T-\delta]$, and so on, and eventually obtain the solvability of (4.1.5) over $[0,T]$.

Now, let us look at Type I BSVIE (4.0.1). Suppose it admits an adapted solution $(Y(\cdot), Z(\cdot\,,\cdot))$ on $[T-\delta, T]$ for some $\delta \in (0,T)$, which is to be proved. From the equation, we see that at this moment, the following are determined:

$$\begin{cases} Y(s), & s \in [T-\delta, T], \\ Z(t,s), & (t,s) \in \Delta \cap [T-\delta, T]^2. \end{cases} \tag{4.1.8}$$

Namely, $Y(s)$ is known in the regions marked ① and ②, whereas $Z(t,s)$ is known only in the region marked ①. See the figure below.

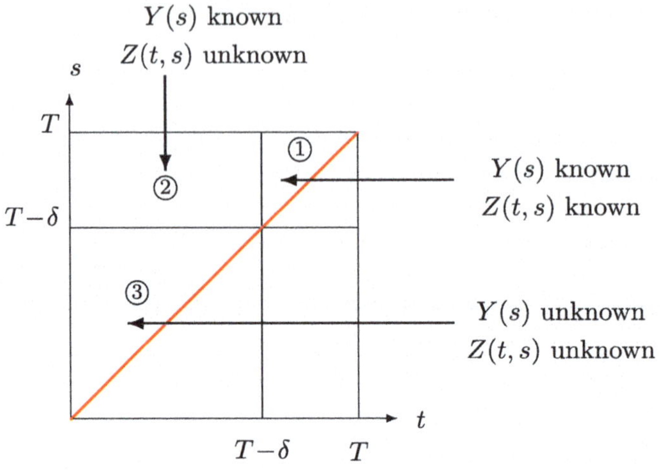

Mimicking BSDEs, for $t \in [0, T-\delta]$, we rewrite Eq. (4.0.1) as follows:

$$Y(t)=\psi^{T-\delta}(t)+\int_t^{T-\delta} g(t,s,Y(s),Z(t,s))ds-\int_t^{T-\delta} Z(t,s)dW(s),\quad t\in[0,T-\delta], \tag{4.1.9}$$

where (compare with (4.1.7))

$$\psi^{T-\delta}(t)=\psi(t)+\int_{T-\delta}^{T} g(t,s,Y(s),Z(t,s))ds-\int_{T-\delta}^{T} Z(t,s)dW(s),\quad t\in[0,T-\delta]. \tag{4.1.10}$$

In order (4.1.9) to be a Type I BSVIE over $[0, T-\delta]$, we need $\psi^{T-\delta}(t)$ to be $\mathcal{F}_{T-\delta}$-measurable for almost all $t \in [0, T-\delta]$. Note that we only have the values $Y(s)$ and $Z(t,s)$ given in (4.1.8) (see the region labeled ① in the figure), whereas, in (4.1.10), one needs the values $Z(t,s)$ for $(t,s) \in [0, T-\delta]\times[T-\delta, T]$ as well (see the region labeled ② in the figure). However, these values $Z(t,s)$ of $Z(\cdot\,,\cdot)$ are not available at the moment. Thus, when (4.0.1) is solvable on $[T-\delta, T]$, one could not simply use the "similar arguments" (as BSDE case) to obtain the solvability of (4.0.1) over $[T-2\delta, T-\delta]$, etc.

Note that (4.1.10) is a *stochastic Fredholm integral equation*(SFDE, for short), with the unknown $(\psi^{T-\delta}(\cdot), Z(\cdot\,,\cdot))$, for which the process $\psi^{T-\delta}(\cdot)$ is required to be $\mathcal{F}_{T-\delta}$-measurable. To solve (4.1.10), we could consider the following family of BSDEs, parameterized by $t \in [0, T-\delta]$:

$$\widetilde{Y}^t(r)=\psi(t)+\int_r^{T} g(t,s,Y(s),\widetilde{Z}^t(s))ds-\int_r^{T} \widetilde{Z}^t(s)dW(s),\quad r\in[T-\delta,T]. \tag{4.1.11}$$

Then we set

$$\psi^{T-\delta}(t) = \widetilde{Y}^t(T-\delta), \quad Z(t,s) = \widetilde{Z}^t(s).$$

By such a way, we obtain that $\psi^{T-\delta}(t)$ is $\mathcal{F}_{T-\delta}$-measurable. Hence, (4.1.9) is really a Type I BSVIE on $[0, T-\delta]$. Then a "similar argument" can be applied.[1] In the following section, however, we will use a different approach which gives the solvability of the BSVIE on the whole interval $[0, T]$ with one shot, without extending the global adapted solution from a local one. The assumed conditions will also be much more general.

2 Well-Posedness

In this section, we will establish well-posedness of Type I BSVIE (4.0.1).

First, for $\beta \geqslant 0$, let $L^{p,\beta}_{\mathbb{F}}(0,T;\mathbb{R}^m)$ be the set of all adapted process $Y : [0,T] \times \Omega \to \mathbb{R}^m$ with

$$\|Y(\cdot)\|_{L^{p,\beta}_{\mathbb{F}}(0,T;\mathbb{R}^m)} = \Big(\mathbb{E}\int_0^T e^{\beta pt}|Y(t)|^p dt\Big)^{\frac{1}{p}} < \infty,$$

which is equivalent to $\|\cdot\|_{L^p_{\mathbb{F}}(0,T;\mathbb{R}^m)}$. Thus, all $L^{p,\beta}_{\mathbb{F}}(0,T;\mathbb{R}^m)$ $(\beta \geqslant 0)$ are the same as $L^{p,0}_{\mathbb{F}}(0,T;\mathbb{R}^m) \equiv L^p_{\mathbb{F}}(0,T;\mathbb{R}^m)$. Next, let $L^{p,\beta}_{\mathbb{F}}(\Omega;L^2(\Delta^*[0,T];\mathbb{R}^m))$ be the set of all $Z : \Delta^*[0,T] \times \Omega \to \mathbb{R}^m$ such that for almost all $t \in [0,T]$, $s \mapsto Z(t,s)$ is $\mathbb{F}$-adapted, and

$$\|Z(\cdot\,,\cdot)\|_{L^{p,\beta}_{\mathbb{F}}(\Omega;L^2(\Delta^*[0,T];\mathbb{R}^m))} = \Big[\mathbb{E}\int_0^T e^{\beta pt}\Big(\int_t^T |Z(t,s)|^2 ds\Big)^{\frac{p}{2}} dt\Big]^{\frac{1}{p}} < \infty,$$

which is equivalent to $\|\cdot\|_{L^p_{\mathbb{F}}(\Omega;L^2(\Delta^*[0,T];\mathbb{R}^m))}$ and thus $L^{p,\beta}_{\mathbb{F}}(\Omega;L^2(\Delta^*[0,T];\mathbb{R}^m))$ is equal to $L^{p,0}_{\mathbb{F}}(\Omega;L^2(\Delta^*[0,T];\mathbb{R}^m)) \equiv L^p_{\mathbb{F}}(\Omega;L^2(\Delta^*[0,T];\mathbb{R}^m))$ (whose definition is given in Chap. 2, Sect. 4), for all $\beta \geqslant 0$. We let

$$\mathcal{H}^{p,\beta}_{\Delta^*}[0,T] = L^{p,\beta}_{\mathbb{F}}(0,T;\mathbb{R}^m) \times L^{p,\beta}_{\mathbb{F}}(\Omega;L^2(\Delta^*[0,T];\mathbb{R}^m)), \qquad \beta \geqslant 0,$$

[1] This approach of solving BSVIEs was taught to us by Professor Jianfeng Zhang, University of Southern California. We are going to use a different method starting from the next section.

be the Banach space with the norm

$$\begin{aligned}&\|(Y(\cdot), Z(\cdot\,,\cdot))\|_{\mathcal{H}^{p,\beta}_{\Delta^*}[0,T]}\\&\equiv\Big[\mathbb{E}\int_0^T e^{\beta pt}|Y(t)|^p dt+\mathbb{E}\int_0^T e^{\beta pt}\Big(\int_t^T |Z(t,s)|^2 ds\Big)^{\frac{p}{2}}dt\Big]^{\frac{1}{p}}<\infty,\end{aligned}$$

and simply denote

$$\mathcal{H}^{p,0}_{\Delta^*}[0,T]=\mathcal{H}^{p}_{\Delta^*}[0,T]\equiv L^p_{\mathbb{F}}(0,T;\mathbb{R}^m)\times L^p_{\mathbb{F}}(\Omega;L^2(\Delta^*[0,T];\mathbb{R}^m)),$$

which is actually equal to $\mathcal{H}^{p,\beta}_{\Delta^*}[0,T]$ for any $\beta\geqslant 0$.

For $p\in(1,\infty)$, we introduce the following hypothesis for the generator $g(\cdot)$ of Type I BSVIE (4.0.1).

(H4.2.1)$_p$ Let $g:\Delta^*[0,T]\times\mathbb{R}^m\times\mathbb{R}^m\times\Omega\to\mathbb{R}^m$ be measurable such that $s\mapsto g(t,s,y,z)$ is $\mathbb{F}$-adapted and (4.1.2) holds. There are deterministic measurable functions $L_y, L_z:\Delta^*[0,T]\to[0,\infty)$ with the property that

$$\int_0^T\Big(\int_t^T L_y(t,s)^{\frac{p(1+\varepsilon)}{p-1}}ds\Big)^{\frac{p-1}{1+\varepsilon}}dt+\sup_{t\in[0,T]}\int_t^T L_z(t,s)^2 ds<\infty, \tag{4.2.1}$$

for some $\varepsilon>0$ such that

$$\begin{aligned}&|g(t,s,y,z)-g(t,s,\bar y,\bar z)|\leqslant L_y(t,s)|y-\bar y|+L_z(t,s)|z-\bar z|,\\&\qquad\qquad (t,s)\in\Delta^*[0,T],\ y,\bar y,z,\bar z\in\mathbb{R}^m,\ \text{a.s.}\end{aligned} \tag{4.2.2}$$

Note that the above will automatically be true if $L_y(\cdot)$ and $L_z(\cdot)$ are positive constants. Namely, if $(y,z)\mapsto g(t,s,y,z)$ is globally Lipschitz (together with (4.1.2)), then (H4.2.1)$_p$ holds for all $p\geqslant 1$.

If we let

$$\begin{aligned}L^p(0,T;L^q(\cdot\,;\mathbb{R}_+))=\Big\{L:\Delta^*[0,T]\to[0,\infty)\ \big|\ L(\cdot\,,\cdot)\text{ is measurable,}\\ \int_0^T\Big(\int_t^T L(t,s)^q ds\Big)^{\frac{p}{q}}dt<\infty\Big\},\end{aligned} \tag{4.2.3}$$

and let $L^\infty(0,T;L^q(\cdot\,,T;\mathbb{R}_+))$ be defined similarly. Then condition (4.2.1) can be equivalently written as

$$\begin{aligned}&L_y(\cdot\,,\cdot)\in L^p(0,T;L^{\frac{p(1+\varepsilon)}{p-1}}(\cdot\,,T;\mathbb{R}_+)),\\&L_z(\cdot\,,\cdot)\in L^\infty(0,T;L^2(\cdot\,,T;\mathbb{R}_+)).\end{aligned} \tag{4.2.4}$$

Condition (4.2.1) allows the generator $g(t, s, y, z)$ to have some singularities along $t = s$, and/or at $t = 0, T$. As a matter of fact, we may, say, suppose the following:

$$\begin{cases} L_y(t,s) = \dfrac{K}{t^{\beta_1}(T-t)^{\beta_2}(s-t)^{\beta_3}(T-s)^{\beta_4}}, & 0 \leqslant \beta_1, \beta_2, \beta_3, \beta_4 < 1, \\ L_z(t,s) = \dfrac{K}{(s-t)^{\beta_5}(T-t)^{\beta_6}}, & 0 \leqslant \beta_5, \beta_6 < 1, \end{cases} \tag{4.2.5}$$

for some constant $K > 0$ and with

$$\begin{aligned} &\beta_1, \beta_2, \beta_3, \beta_4, \beta_5, \beta_6 \geqslant 0, \quad \beta_1 + \beta_2 + \beta_3 + \beta_4 < 1, \\ &\beta_1 < \frac{1}{p}, \qquad \beta_3, \beta_4 < \frac{p-1}{p}, \quad \beta_5 + \beta_6 < \frac{1}{2}. \end{aligned} \tag{4.2.6}$$

In this case,

$$\begin{aligned} &\int_0^T \Big(\int_t^T L_y(t,s)^{\frac{p(1+\varepsilon)}{p-1}} ds \Big)^{\frac{p-1}{1+\varepsilon}} dt \\ &= \int_0^T \frac{K}{t^{\beta_1 p}(T-t)^{\beta_2 p}} \Big(\int_t^T \frac{ds}{(s-t)^{\beta_3 \frac{p(1+\varepsilon)}{p-1}} (T-s)^{\beta_4 \frac{p(1+\varepsilon)}{p-1}}} \Big)^{\frac{p-1}{1+\varepsilon}} dt \\ &= \int_0^T \frac{K}{t^{\beta_1 p}(T-t)^{\beta_2 p}} \Big(\int_0^1 \frac{(T-t)d\tau}{[(T-t)\tau]^{\beta_3 \frac{p(1+\varepsilon)}{p-1}} [(T-t)(1-\tau)]^{\beta_4 \frac{p(1+\varepsilon)}{p-1}}} \Big)^{\frac{p-1}{1+\varepsilon}} dt \\ &= \int_0^T \frac{1}{t^{\beta_1 p}(T-t)^{(\beta_2+\beta_3+\beta_4)p - \frac{p-1}{1+\varepsilon}}} \Big(\int_0^1 \tau^{-\beta_3 \frac{p(1+\varepsilon)}{p-1}} (1-\tau)^{-\beta_4 \frac{p(1+\varepsilon)}{p-1}} d\tau \Big)^{\frac{p-1}{1+\varepsilon}} dt \\ &= K B\Big(1 - \beta_3 \frac{p(1+\varepsilon)}{p-1}, 1 - \beta_4 \frac{p(1+\varepsilon)}{p-1}\Big)^{\frac{p-1}{1+\varepsilon}} \\ &\qquad \cdot \int_0^1 \frac{T d\tau}{(T\tau)^{\beta_1 p}[T(1-\tau)]^{(\beta_2+\beta_3+\beta_4)p - \frac{p-1}{1+\varepsilon}}} \\ &= K T^{1-(\beta_1+\beta_2+\beta_3+\beta_4)p + \frac{p-1}{1+\varepsilon}} B\Big(1 - \beta_3 \frac{p(1+\varepsilon)}{p-1}, 1 - \beta_4 \frac{p(1+\varepsilon)}{p-1}\Big)^{\frac{p-1}{1+\varepsilon}} \\ &\qquad \cdot B\Big(1 - \beta_1 p, 1 - (\beta_2 + \beta_3 + \beta_4)p + \frac{p-1}{1+\varepsilon}\Big), \end{aligned}$$

where $B(r, s)$ is the beta function (see (1.2.13) in Chap. 1). The above is meaningful provided

$$\begin{aligned} &\beta_1 + \beta_2 + \beta_3 + \beta_4 \leqslant \frac{1}{p} + \frac{p-1}{p(1+\varepsilon)}, \quad \beta_3, \beta_4 < \frac{p-1}{p(1+\varepsilon)}, \\ &\beta_1 < \frac{1}{p}, \qquad \beta_2 + \beta_3 + \beta_4 < \frac{1}{p} + \frac{p-1}{p(1+\varepsilon)}, \end{aligned}$$

for some $\varepsilon > 0$, which are implied by (4.2.6) (on $\beta_1, \beta_2, \beta_3, \beta_4$). Also,

$$\begin{aligned}&\int_t^T L_z(t,s)^2 ds = \int_t^T \frac{K ds}{(s-t)^{2\beta_5}(T-s)^{2\beta_6}}\\ =&\int_0^1 \frac{K(T-t)d\tau}{[(T-t)\tau]^{2\beta_5}[(T-t)(1-\tau)]^{2\beta_6}} = KT^{1-2(\beta_5+\beta_6)}B\Big(1-2\beta_5, 1-2\beta_6\Big),\end{aligned}$$

which will be finite if (4.2.6) (on β_5, β_6) holds.

The following technical lemma will be frequently used below.

Lemma 4.2.1 *Let*

$$\int_0^T \Big(\int_t^T L_y(t,s)^{\frac{p(1+\varepsilon)}{p-1}} ds\Big)^{\frac{p-1}{1+\varepsilon}} dt < \infty, \tag{4.2.7}$$

for some $\varepsilon > 0$. *Then, there exists a constant* $K = K(\varepsilon, p) > 0$, *such that for any* $\beta > 0$ *and* $y(\cdot) \in L^p_{\mathbb{F}}(0,T;\mathbb{R}^n)$,

$$\begin{aligned}&\mathbb{E}\Big(\int_t^T e^{\beta t} L_y(t,s)|y(s)|ds\Big)^p\\ \leqslant&\frac{K}{\beta^{\frac{(p-1)\varepsilon}{1+\varepsilon}}}\Big(\int_t^T L_y(t,s)^{\frac{p(1+\varepsilon)}{p-1}} ds\Big)^{\frac{p-1}{1+\varepsilon}} \mathbb{E}\Big(\int_t^T e^{\beta ps}|y(s)|^p ds\Big), \quad a.e.\ t\in[0,T],\end{aligned} \tag{4.2.8}$$

and

$$\int_0^T \mathbb{E}\Big(\int_t^T e^{\beta t} L_y(t,s)|y(s)|ds\Big)^p dt \leqslant \frac{K}{\beta^{\frac{(p-1)\varepsilon}{1+\varepsilon}}}\mathbb{E}\Big(\int_0^T e^{\beta ps}|y(s)|^p ds\Big). \tag{4.2.9}$$

Proof First, for almost every $t \in [0, T]$,

$$\begin{aligned}&\mathbb{E}\Big(\int_t^T e^{\beta t} L_y(t,s)|y(s)|ds\Big)^p = \mathbb{E}\Big(\int_t^T e^{\beta(t-s)} e^{\beta s} L_y(t,s)|y(s)|ds\Big)^p\\ \leqslant&\,\mathbb{E}\Big(\int_t^T e^{\frac{\beta p}{p-1}(t-s)} L_y(t,s)^{\frac{p}{p-1}} ds\Big)^{p-1}\Big(\int_t^T e^{\beta ps}|y(s)|^p ds\Big)\\ \leqslant&\Big(\int_t^T e^{\frac{\beta p(1+\varepsilon)}{(p-1)\varepsilon}(t-s)} ds\Big)^{\frac{(p-1)\varepsilon}{1+\varepsilon}}\Big(\int_t^T L_y(t,s)^{\frac{p(1+\varepsilon)}{p-1}} ds\Big)^{\frac{p-1}{1+\varepsilon}}\mathbb{E}\Big(\int_t^T e^{\beta ps}|y(s)|^p ds\Big)\\ \leqslant&\Big(\frac{(p-1)\varepsilon}{\beta p(1+\varepsilon)}\Big)^{\frac{(p-1)\varepsilon}{1+\varepsilon}}\Big(\int_t^T L_y(t,s)^{\frac{p(1+\varepsilon)}{p-1}} ds\Big)^{\frac{p-1}{1+\varepsilon}}\mathbb{E}\Big(\int_t^T e^{\beta ps}|y(s)|^p ds\Big)\\ \leqslant&\frac{K}{\beta^{\frac{(p-1)\varepsilon}{1+\varepsilon}}}\Big(\int_t^T L_y(t,s)^{\frac{p(1+\varepsilon)}{p-1}} ds\Big)^{\frac{p-1}{1+\varepsilon}}\mathbb{E}\Big(\int_t^T e^{\beta ps}|y(s)|^p ds\Big).\end{aligned}$$

□

This leads to (4.2.8). Integrating the above over $[0, T]$, one obtains (4.2.9).

We are now ready to present the well-posedness of Type I BSVIE (4.0.1).

Theorem 4.2.2 *Let $(H4.2.1)_p$ hold for some $p \in (1, \infty)$. Then for any free term $\psi(\cdot) \in L^p_{\mathcal{F}_T}(0, T; \mathbb{R}^m)$, Type I BSVIE* (4.0.1) *admits a unique adapted solution $(Y(\cdot), Z(\cdot\,,\cdot)) \in \mathcal{H}^p_{\Delta^*}[0, T]$, and the following estimate holds for some constant $K > 0$:*

$$\begin{aligned}&\|(Y(\cdot), Z(\cdot\,,\cdot))\|^p_{\mathcal{H}^p_{\Delta^*}[0,T]}\\&\equiv \mathbb{E}\int_0^T |Y(t)|^p dt + \mathbb{E}\int_0^T \Big(\int_t^T |Z(t,s)|^2 ds\Big)^{\frac{p}{2}} dt \leqslant K\mathbb{E}\int_0^T |\psi(t)|^p dt.\end{aligned} \tag{4.2.10}$$

Further, let $\bar g : \Delta^*[0, T] \times \mathbb{R}^m \times \mathbb{R}^m \times \Omega \to \mathbb{R}^m$ satisfy the same conditions as $g(\cdot)$ and $\bar\psi(\cdot) \in L^p_{\mathcal{F}_T}(0, T; \mathbb{R}^m)$. Let $(\bar Y(\cdot), \bar Z(\cdot\,,\cdot))$ be the adapted solution of the Type I BSVIE corresponding to $(\bar g(\cdot), \bar\psi(\cdot))$. Then the following stability estimate holds for some constant $K > 0$:

$$\begin{aligned}&\|(Y(\cdot), Z(\cdot\,,\cdot)) - (\bar Y(\cdot), \bar Z(\cdot\,,\cdot))\|^p_{\mathcal{H}^p_{\Delta^*}[0,T]}\\&\leqslant K\mathbb{E}\Big[\int_0^T |\psi(t) - \bar\psi(t)|^p dt \\&\quad+\int_0^T\Big(\int_t^T |g(t,s,Y(s),Z(t,s)) - \bar g(t,s,Y(s),Z(t,s))| ds\Big)^p dt\Big].\end{aligned} \tag{4.2.11}$$

Proof Since $\psi(\cdot) \in L^p_{\mathcal{F}_T}(0, T; \mathbb{R}^m)$, $\psi(t)$ is defined for almost all $t \in [0, T]$. For any $y(\cdot) \in L^{p,\beta}_{\mathbb{F}}(0, T; \mathbb{R}^m)$ with $\beta \geqslant 0$ and almost all fixed $t \in [0, T)$ at which $\psi(t)$ is defined, consider the following BSDE for $(\eta(t,\cdot), \zeta(t,\cdot))$:

$$\eta(t,\tau) = \psi(t) + \int_\tau^T g(t,s,y(s),\zeta(t,s)) ds - \int_\tau^T \zeta(t,s) dW(s), \qquad \tau \in [t, T]. \tag{4.2.12}$$

Note

$$\begin{aligned}&\mathbb{E}\int_0^T\Big(\int_t^T |g(t,s,y(s),0)| ds\Big)^p dt \leqslant \mathbb{E}\int_0^T\Big(\int_t^T L_y(t,s)|y(s)| ds\Big)^p dt\\&\leqslant \Big[\int_0^T\Big(\int_t^T L_y(t,s)^{\frac{p}{p-1}} ds\Big)^{p-1} dt\Big]\mathbb{E}\int_0^T |y(s)|^p ds < \infty.\end{aligned}$$

Hence, for almost all $t \in [0, T]$,

$$\mathbb{E}\Big(\int_t^T |g(t,s,y(s),0)| ds\Big)^p < \infty.$$

Also, by (4.2.2), we have

$$|g(t,s,y(s),\zeta)-g(t,s,y(s),\bar{\zeta})|\leqslant L_z(t,s)|\zeta-\bar{\zeta}|,$$

with (4.2.1) for $L_z(t,\cdot)$. Thus, by Chap. 3, Sect. 1, BSDE (4.2.12) has a unique adapted solution $(\eta(t,\cdot),\zeta(t,\cdot))$ on $[t,T]$, parameterized by $t\in[0,T]$, and the following holds:

$$\begin{aligned}&\mathbb{E}\Big[\sup_{\tau\in[t,T]}|\eta(t,\tau)|^p+\Big(\int_t^T|\zeta(t,s)|^2ds\Big)^{\frac{p}{2}}\Big]\\&\leqslant K\mathbb{E}\Big[|\psi(t)|^p+\Big(\int_t^T|g(t,s,y(s),0)|ds\Big)^p\Big],\qquad t\in[0,T].\end{aligned}\tag{4.2.13}$$

Taking $\tau=t$ in (4.2.12) and defining

$$Y(t)=\eta(t,t),\quad Z(t,s)=\zeta(t,s),\qquad (t,s)\in\Delta^*[0,T],\tag{4.2.14}$$

we obtain

$$Y(t)=\psi(t)+\int_t^Tg(t,s,y(s),Z(t,s))ds-\int_t^TZ(t,s)dW(s),\quad t\in[0,T].\tag{4.2.15}$$

This means that $(Y(\cdot),Z(\cdot\,,\cdot))$ is an adapted solution to BSVIE (4.2.15) whose generator is $(t,s,Y,Z)\mapsto g(t,s,y(s),Z)$, independent of Y. In addition, (4.2.13) leads to

$$\begin{aligned}&\mathbb{E}\Big[|Y(t)|^p+\Big(\int_t^T|Z(t,s)|^2ds\Big)^{\frac{p}{2}}\Big]\leqslant K\mathbb{E}\Big[|\psi(t)|^p+\Big(\int_t^T|g(t,s,y(s),0)|ds\Big)^p\Big]\\&\leqslant K\mathbb{E}\Big[|\psi(t)|^p+\Big(\int_t^TL_y(t,s)|y(s)|ds\Big)^p\Big],\qquad \text{a.e. } t\in[0,T].\end{aligned}$$

Integrating the above over $[0,T]$, by (4.2.1)–(4.2.2) and Lemma 4.2.1, one has

$$\begin{aligned}&\mathbb{E}\Big[\int_0^Te^{\beta pt}|Y(t)|^pdt+\int_0^Te^{\beta pt}\Big(\int_t^T|Z(t,s)|^2ds\Big)^{\frac{p}{2}}dt\Big]\\&\leqslant K\mathbb{E}\Big[\int_0^Te^{\beta pt}|\psi(t)|^pdt+\int_0^Te^{\beta pt}\Big(\int_t^TL_y(t,s)|y(s)|ds\Big)^pdt\Big]\\&\leqslant K\mathbb{E}\Big\{\int_0^Te^{\beta pt}|\psi(s)|^pds+\frac{1}{\beta^{\frac{(p-1)\varepsilon}{1+\varepsilon}}}\Big(\int_0^Te^{\beta ps}|y(s)|^pds\Big)\Big\}.\end{aligned}$$

Therefore, the above constructed $(Y(\cdot),Z(\cdot\,,\cdot))\in\mathcal{H}^{p,\beta}_{\Delta^*}[0,T]$ is an adapted solution to (4.2.15), a special case of Type I BSVIE. This defines a map $y(\cdot)\mapsto Y(\cdot)$ from $L^{p,\beta}_{\mathbb{F}}(0,T;\mathbb{R}^m)$ to itself through (4.2.14).

Now, let $y_i(\cdot)\in L^{p,\beta}_{\mathbb{F}}(0,T;\mathbb{R}^m)$ $(i=1,2)$ and let $(Y_i(\cdot),Z_i(\cdot\,,\cdot))\in \mathcal{H}^{p,\beta}_{\Delta^*}[0,T]$ be an adapted solution of the following BSVIEs:

$$Y_i(t)=\psi(t)+\int_t^T g(t,s,y_i(s),Z_i(t,s))ds-\int_t^T Z_i(t,s)dW(s),\qquad t\in[0,T],\quad i=1,2, \tag{4.2.16}$$

constructed as (4.2.14). More precisely, for almost all fixed $t\in[0,T]$, we let $(\eta_i(t,\cdot),\zeta_i(t,\cdot))$ be the adapted solution to the following BSDE:

$$\eta_i(t,\tau)=\psi(t)+\int_\tau^T g(t,s,y_i(s),\zeta_i(t,s))ds-\int_\tau^T \zeta_i(t,s)dW(s),\ \ \tau\in[t,T]. \tag{4.2.17}$$

Then by the stability estimate of BSDEs, one has the following estimate:

$$\begin{aligned}&\mathbb{E}\Big[\sup_{\tau\in[t,T]}|\eta_1(t,\tau)-\eta_2(t,\tau)|^p+\Big(\int_t^T|\zeta_1(t,s)-\zeta_2(t,s)|^2ds\Big)^{\frac{p}{2}}\Big]\\&\leqslant K\mathbb{E}\Big(\int_t^T|g(t,s,y_1(s),\zeta_1(t,s))-g(t,s,y_2(s),\zeta_1(t,s))|ds\Big)^p.\end{aligned} \tag{4.2.18}$$

Recall that $(Y_i(\cdot),Z_i(\cdot\,,\cdot))$ is defined as follows:

$$Y_i(t)=\eta_i(t,t),\quad Z_i(t,s)=\zeta_i(t,s),\qquad (t,s)\in\Delta^*[0,T]. \tag{4.2.19}$$

Thus, (4.2.18) implies

$$\begin{aligned}&\mathbb{E}|Y_1(t)-Y_2(t)|^p\\&\leqslant K\mathbb{E}\Big(\int_t^T|g(t,s,y_1(s),Z_1(t,s))-g(t,s,y_2(s),Z_1(t,s))|ds\Big)^p\\&\leqslant K\mathbb{E}\Big(\int_t^T L_y(t,s)|y_1(s)-y_2(s)|ds\Big)^p,\qquad \text{a.e. } t\in[0,T].\end{aligned} \tag{4.2.20}$$

Hence, for $\beta>0$ and small $\varepsilon>0$, by Lemma 4.2.1,

$$\mathbb{E}\int_0^T e^{\beta pt}|Y_1(t)-Y_2(t)|^pdt\leqslant\frac{K}{\beta^{\frac{(p-1)\varepsilon}{1+\varepsilon}}}\mathbb{E}\int_0^T e^{\beta pt}|y_1(t)-y_2(t)|^pdt, \tag{4.2.21}$$

with some constant $K>0$ independent of $\beta>0$. Therefore, by letting $\beta>0$ large enough, we obtain that the map $y(\cdot)\mapsto Y(\cdot)$ is a contraction mapping on $L^p_{\mathbb{F}}(0,T;\mathbb{R}^m)$ (under the equivalent norm $\|\cdot\|_{L^{p,\beta}_{\mathbb{F}}(0,T;\mathbb{R}^m)}$). Hence, it admits a unique fixed point. Then by solving BSDE (4.2.15), we obtain a unique adapted solution $(Y(\cdot), Z(\cdot\,,\cdot)) \in \mathcal{H}^p_{\Delta^*}[0,T]$ to (4.0.1). By (4.2.13) (with $\tau = t$), we obtain

$$\begin{aligned}
&\mathbb{E}\Big[|Y(t)|^p + \Big(\int_t^T |Z(t,s)|^2 ds\Big)^{\frac{p}{2}}\Big] \\
&\leqslant K\mathbb{E}\Big[|\psi(t)|^p + \Big(\int_t^T |g(t,s,Y(s),0)|ds\Big)^p\Big] \\
&\leqslant K\mathbb{E}\Big[|\psi(t)|^p + \Big(\int_t^T L_y(t,s)|Y(s)|ds\Big)^p\Big], \qquad \text{a.e. } t \in [0,T].
\end{aligned} \tag{4.2.22}$$

By Lemma 4.2.1, one has

$$\begin{aligned}
&\mathbb{E}\Big[\int_0^T e^{\beta pt}|Y(t)|^p dt + \int_0^T e^{\beta pt}\Big(\int_t^T |Z(t,s)|^2 ds\Big)^{\frac{p}{2}} dt\Big] \\
&\leqslant K\mathbb{E}\Big[\int_0^T e^{\beta pt}|\psi(t)|^p dt + \int_0^T \Big(\int_t^T e^{\beta t} L_y(t,s)|Y(s)|ds\Big)^p\Big] \\
&\leqslant K\mathbb{E}\int_0^T e^{\beta pt}|\psi(t)|^p dt + \frac{K}{\beta^{\frac{(p-1)\varepsilon}{1+\varepsilon}}}\mathbb{E}\Big(\int_0^T e^{\beta ps}|Y(s)|^p ds\Big).
\end{aligned} \tag{4.2.23}$$

In the above, the constant $K > 0$ is independent of β. Hence, by choosing $\beta > 0$ large enough (and fixed), we may absorb the last term to the left to obtain

$$\mathbb{E}\Big[\int_0^T e^{\beta pt}|Y(t)|^p dt + \int_0^T e^{\beta pt}\Big(\int_t^T |Z(t,s)|^2 ds\Big)^{\frac{p}{2}} dt\Big] \leqslant K\mathbb{E}\int_0^T e^{\beta pt}|\psi(t)|^p dt.$$

Then (4.2.10) follows by changing $K > 0$ properly.

Next, let $\bar{g} : \Delta^*[0,T] \times \mathbb{R}^m \times \mathbb{R}^m \times \Omega \to \mathbb{R}^m$ satisfy the same conditions as $g(\cdot)$ and $\bar{\psi}(\cdot) \in L^p_{\mathcal{F}_T}(0,T;\mathbb{R}^m)$. Let $(\bar{Y}(\cdot), \bar{Z}(\cdot\,,\cdot)) \in \mathcal{H}^{p,\beta}_{\Delta^*}[0,T]$ be the adapted solution of corresponding Type I BSVIE. Then similar to (4.2.18)–(4.2.20), we obtain

$$\begin{aligned}
&\mathbb{E}\Big[|Y(t)-\bar{Y}(t)|^p + \Big(\int_t^T |Z(t,s)-\bar{Z}(t,s)|^2 ds\Big)^{\frac{p}{2}}\Big] \\
&\leqslant K\mathbb{E}\Big[|\psi(t)-\bar{\psi}(t)|^p + \Big(\int_t^T |g(t,s,Y(s),Z(t,s)) \\
&\qquad\qquad -\bar{g}(t,s,\bar{Y}(s),Z(t,s))|ds\Big)^p\Big] \\
&\leqslant K\mathbb{E}\Big[|\psi(t)-\bar{\psi}(t)|^p + \Big(\int_t^T |g(t,s,Y(s),Z(t,s)) \\
&\qquad\qquad -\bar{g}(t,s,Y(s),Z(t,s))|ds\Big)^p \\
&\qquad + \Big(\int_t^T L_y(t,s)|Y(s)-\bar{Y}(s)|ds\Big)^p\Big], \qquad \text{a.e. } t \in [0,T].
\end{aligned} \tag{4.2.24}$$

Hence, similar to the proof of (4.2.10) (see (4.2.22)–(4.2.23)), we obtain (4.2.11). □

Remark 4.2.3 In the above approach, we bypass the reliance of Itô's formula and obtain the contraction property of the relevant solution map on the whole interval $[0, T]$ with the help of the introduced factor $e^{\beta pt}$. On the other hand, since BSDEs are a special case of BSVIEs, one may wonder if it can be used to seek adapted solution $(Y(\cdot), Z(\cdot))$ in $L^p_{\mathbb{F}}(0, T; \mathbb{R}^m) \times L^p_{\mathbb{F}}(\Omega; L^2(0, T; \mathbb{R}^{m\times d}))$. At this moment, we point out that there is no double integral for this Z on $[0, T]$, a difference between BSDEs and BSVIEs. Even so, the above technique can be adjusted to obtain the global solution. However, if we want to get an adapted solution in $\mathbf{H}^p[0, T] \equiv L^p_{\mathbb{F}}(\Omega; C([0, T]; \mathbb{R}^n)) \times L^p_{\mathbb{F}}(\Omega; L^2(0, T; \mathbb{R}^{m\times d}))$ with continuous process $Y(\cdot)$, then the above technique seems not to be applicable. Hence, in Chap. 3, Sect. 1, for BSDEs, the classical approach is used: extending a local solution to a global one.

Remark 4.2.4 From the above proof, we see the following. If the generator $g(t, s, y, z) \equiv g(t, s, z)$ is independent of y, then by solving the following family of BSDEs:

$$\eta(t, \tau) = \psi(t) + \int_\tau^T g(t, s, \zeta(t, s))ds - \int_\tau^T \zeta(t, s)dW(s), \qquad \tau \in [t, T], \tag{4.2.25}$$

we can obtain the adapted solution $(Y(\cdot), Z(\cdot\,, \cdot)) \in \mathcal{H}^p_{\Delta^*}[0, T]$ of Type I BSVIE (4.0.1) by setting $(Y(\cdot), Z(\cdot\,, \cdot))$ as in (4.2.14). On the other hand, in the general case, if $(Y(\cdot), Z(\cdot\,, \cdot)) \in \mathcal{H}^p_{\Delta^*}[0, T]$ is the unique adapted solution of Type iI BSVIE (4.0.1), and for almost all $t \in [0, T]$, $(\eta(t, \cdot), \zeta(t, \cdot))$ is the adapted solution of the following family of BSDEs:

$$\eta(t, \tau) = \psi(t) + \int_\tau^T g(t, s, Y(s), \zeta(t, s))ds - \int_\tau^T \zeta(t, s)dW(s), \qquad \tau \in [t, T], \tag{4.2.26}$$

then (4.2.14) must be true.

Now, in the above, with the aid of the factor $e^{\beta pt}$, we obtained global contraction of a certain solution map. Then it is natural to ask whether we have convergence of the corresponding global *Picard iteration scheme*. More precisely, we introduce the following iteration scheme:

$$\begin{cases} Y^0(t) = \psi(t) - \displaystyle\int_t^T Z^0(t, s)dW(s), \qquad t \in [0, T], \\ Y^k(t) = \psi(t) + \displaystyle\int_t^T g(t, s, Y^{k-1}(s), Z^k(t, s))ds \\ \qquad\qquad - \displaystyle\int_t^T Z^k(t, s)dW(s), \qquad t \in [0, T], \quad k \geqslant 1. \end{cases} \tag{4.2.27}$$

We claim that the following convergence holds:

$$(Y^k(\cdot), Z^k(\cdot\,,\cdot)) \to (Y(\cdot), Z(\cdot\,,\cdot)), \qquad \text{in } \mathcal{H}^{p,\beta}_{\Delta^*}[0,T], \quad k \to \infty.$$

In fact, from (4.2.21), we have

$$\mathbb{E}\int_0^T e^{\beta pt}|Y^{k+1}(t) - Y^k(t)|^p dt \leqslant \frac{K}{\beta^{\frac{(p-1)\varepsilon}{1+\varepsilon}}}\mathbb{E}\int_0^T e^{\beta pt}|Y^k(t) - Y^{k-1}(t)|^p dt,$$

which implies that $Y^k(\cdot)$ is Cauchy in $L^{p,\beta}_{\mathbb{F}}(0,T;\mathbb{R}^m)$. Also (note (4.2.19)),

$$\begin{aligned}
&\mathbb{E}\Big(\int_t^T |Z^{k+1}(t,s) - Z^k(t,s)|^2 ds\Big)^{\frac{p}{2}}\\
&\leqslant K\mathbb{E}\Big(\int_t^T |g(t,s,Y^k(s),Z^{k-1}(t,s)) - g(t,s,Y^{k-1}(s),Z^{k-1}(t,s))|ds\Big)^p\\
&\leqslant K\mathbb{E}\Big(\int_t^T L_y(t,s)|Y^k(s) - Y^{k-1}(s)|ds\Big)^p.
\end{aligned}$$

Now, by first multiplying the factor $e^{\beta pt}$ in the above and then using Lemma 4.2.1, we have

$$\begin{aligned}
&\mathbb{E}\int_0^T e^{\beta pt}\Big(\int_t^T |Z^{k+1}(t,s) - Z^k(t,s)|^2 ds\Big)^{\frac{p}{2}} dt\\
&\leqslant K\int_0^T \mathbb{E}\Big(\int_t^T e^{\beta(t-s)}L_y(t,s)e^{\beta s}|Y^k(s) - Y^{k-1}(s)|ds\Big)^p dt\\
&\leqslant \frac{K}{\beta^{\frac{(p-1)\varepsilon}{1+\varepsilon}}}\mathbb{E}\Big(\int_0^T e^{\beta ps}|Y^k(s) - Y^{k-1}(s)|^p ds\Big).
\end{aligned}$$

Thus, $Z^k(\cdot\,,\cdot)$ is Cauchy in $L^{p,\beta}_{\mathbb{F}}(\Omega; L^2(\Delta^*[0,T];\mathbb{R}^m))$. We can check that the limit $(Y(\cdot), Z(\cdot\,,\cdot))$ of the sequence $(Y^k(\cdot), Z^k(\cdot\,,\cdot))$ is the unique adapted solution to the BSVIE (4.0.1).

In principle, the above global Picard iteration scheme gives a computational algorithm of Type I BSVIEs.

3 Continuity of Adapted Solutions

In the previous sections, we have obtained the existence and uniqueness of the adapted solution $(Y(\cdot), Z(\cdot\,,\cdot))$ to the Type I BSVIE (4.0.1), in the space $\mathcal{H}^p_{\Delta^*}[0,T] \equiv L^p_{\mathbb{F}}(0,T;\mathbb{R}^m) \times L^p_{\mathbb{F}}(\Omega; L^2(\Delta^*[0,T];\mathbb{R}^m))$, under $(\text{H4.2.1})_p$. It is not hard to see that we only have L^p-integrability of $Y(\cdot)$ and it does not have any continuity in general. In this section, we will discuss certain types of continuity for the adapted solution $(Y(\cdot), Z(\cdot\,,\cdot))$.

As a bridge, we first introduce the following hypothesis which is slightly more than $(H4.2.1)_p$.

(H4.3.1)$_p$ Let $(H4.2.1)_p$ hold and, in addition, the following be satisfied

$$\sup_{t\in[0,T]}\int_t^T L_y(t,s)^{\frac{p}{p-1}}ds<\infty. \tag{4.3.1}$$

Note (4.3.1) amounts to saying that

$$L_y(\cdot\,,\cdot)\in L^\infty(0,T;L^{\frac{p}{p-1}}(\cdot\,,T;\mathbb{R}_+)). \tag{4.3.2}$$

Comparing the above with the first in (4.2.4), we see that none of the one implies the other, in general.

The following proposition contains some parts of the proof of Theorem 4.2.2. We state them separately below, since they will be useful and frequently used right below.

Proposition 4.3.1

(i) Let $(H4.2.1)_p$ hold for some $p>1$. Then for any $\psi(\cdot)\in L^p_{\mathcal{F}_T}(0,T;\mathbb{R}^m)$, the unique adapted solution $(Y(\cdot),Z(\cdot\,,\cdot))\in\mathcal{H}^p_{\Delta^*}[0,T]$ of BSVIE (4.0.1) satisfies the following:

$$\begin{aligned}&\mathbb{E}\Big[|Y(t)|^p+\Big(\int_t^T|Z(t,s)|^2ds\Big)^{\frac{p}{2}}\Big]\\&\leqslant K\mathbb{E}\Big[|\psi(t)|^p+\Big(\int_t^T L_y(t,s)|Y(s)|ds\Big)^p\Big],\qquad \text{a.e. } t\in[0,T].\end{aligned} \tag{4.3.3}$$

If $\bar g:\Delta^*[0,T]\times\mathbb{R}^m\times\mathbb{R}^m\times\Omega\to\mathbb{R}^m$ also satisfies $(H4.2.1)_p$, $\bar\psi(\cdot)\in L^p_{\mathcal{F}_T}(0,T;\mathbb{R}^m)$, and $(\bar Y(\cdot),\bar Z(\cdot\,,\cdot))$ is the adapted solution of the Type I BSVIE corresponding to $(\bar g(\cdot),\bar\psi(\cdot))$, then the following stability estimate holds:

$$\begin{aligned}&\mathbb{E}\Big[|Y(t)-\bar Y(t)|^p+\Big(\int_t^T|Z(t,s)-\bar Z(t,s)|^2ds\Big)^{\frac{p}{2}}\Big]\\&\leqslant K\mathbb{E}\Big[|\psi(t)-\bar\psi(t)|^p\\&\quad+\Big(\int_t^T|g(t,s,Y(s),Z(t,s))-\bar g(t,s,Y(s),Z(t,s))|ds\Big)^p\\&\quad+\Big(\int_t^T L_y(t,s)|Y(s)-\bar Y(s)|ds\Big)^p\Big],\quad \text{a.e. } t\in[0,T].\end{aligned} \tag{4.3.4}$$

(ii) Let $(H4.3.1)_p$ hold for some $p > 1$. Then (4.3.3) *can be strengthened to the following, respectively:*

$$\begin{aligned}&\mathbb{E}\Big[|Y(t)|^p+\Big(\int_t^T|Z(t,s)|^2ds\Big)^{\frac{p}{2}}\Big]\\&\leqslant K\mathbb{E}|\psi(t)|^p+K\mathbb{E}\int_t^T|\psi(s)|^p ds,\quad \text{a.e. } t\in[0,T],\end{aligned}\tag{4.3.5}$$

and (4.3.4) *can be strengthened to the following:*

$$\begin{aligned}&\mathbb{E}\Big[|Y(t)-\bar{Y}(t)|^p+\Big(\int_t^T|Z(t,s)-\bar{Z}(t,s)|^2ds\Big)^{\frac{p}{2}}\Big]\\&\leqslant K\mathbb{E}\Big[|\psi(t)-\bar{\psi}(t)|^p+\Big(\int_t^T|g(t,s,Y(s),Z(t,s))\\&\qquad\qquad -\bar{g}(t,s,Y(s),Z(t,s))|ds\Big)^p\Big]\\&\quad+K\mathbb{E}\Big[\int_t^T|\psi(s)-\bar{\psi}(s)|^p ds+\int_t^T\Big(\int_s^T|g(s,r,Y(r),Z(s,r))\\&\qquad\qquad -\bar{g}(s,r,Y(r),Z(s,r))|dr\Big)^p ds\Big],\quad \text{a.e. } t\in[0,T].\end{aligned}\tag{4.3.6}$$

Consequently, if $\psi(\cdot)\in L^\infty_{\mathcal{F}_T}(0,T;L^p(\Omega;\mathbb{R}^m))$, then

$$\sup_{t\in[0,T]}\mathbb{E}|Y(t)|^p+\sup_{t\in[0,T]}\mathbb{E}\Big(\int_t^T|Z(t,s)|^2ds\Big)^{\frac{p}{2}}\leqslant K\Big[\sup_{t\in[0,T]}\mathbb{E}|\psi(t)|^p\Big].\tag{4.3.7}$$

Proof

(i) Under $(H4.2.1)_p$, it is seen that (4.3.3) and (4.3.4) follow from (4.2.22) and (4.2.24).

(ii) Next, if $(H4.3.1)_p$ holds, then (4.3.3) implies that

$$\begin{aligned}\mathbb{E}|Y(t)|^p&\leqslant K\mathbb{E}\Big[|\psi(t)|^p+\Big(\int_t^T L_y(t,s)|Y(s)|ds\Big)^p\Big]\\&\leqslant K\mathbb{E}\Big[|\psi(t)|^p+\Big(\int_t^T L_y(t,s)^{\frac{p}{p-1}}ds\Big)^{p-1}\int_t^T|Y(s)|^p ds\Big]\\&\leqslant K\mathbb{E}\Big[|\psi(t)|^p+\int_t^T\mathbb{E}|Y(s)|^p ds\Big].\end{aligned}$$

By Gronwall-type inequality, we obtain (4.3.5).

By similar arguments, we can obtain (4.3.6) from (4.3.4). Finally, (4.3.7) follows from (4.3.5) immediately. □

The main point in part (i) of the above proposition is that the term containing $|Y(s)|$ on the right-hand side cannot be absorbed by the corresponding term on the left-hand side, unless condition (4.3.1) is assumed so that the Gronwall-type inequality is applicable to get (4.3.5). The same comments apply to (4.3.4) and (4.3.6).

We now make an observation. Under $(\text{H4.3.1})_p$, which includes (H4.2.1), for any $\psi(\cdot) \in L^{\infty}_{\mathcal{F}_T}(0, T; L^p(\Omega; \mathbb{R}^m))$, by Theorem 4.2.2, we may let $(Y(\cdot), Z(\cdot\,,\cdot)) \in \mathcal{H}^p_{\Delta^*}[0, T]$ be an adapted solution of (4.0.1). Then BSVIE (4.0.1) can be written as

$$\begin{aligned} Y(t) = \psi(t) &+ \int_t^T g(t, s, Y(s), 0)ds \\ &+ \int_y^T \big[g(t, s, Y(s), Z(t, s)) - g(t, s, Y(s), 0)\big]ds \\ &- \int_t^T Z(t, s)dW(s). \end{aligned} \tag{4.3.8}$$

Denote

$$\begin{cases} \widehat{\psi}(t) = \psi(t) + \displaystyle\int_t^T g(t, s, Y(s), 0)ds, \\ \widehat{g}(t, s, z) = g(t, s, Y(s), z) - g(t, s, Y(s), 0). \end{cases} \tag{4.3.9}$$

Then the above BSVIE (4.3.8) can be written as

$$Y(t) = \widehat{\psi}(t) + \int_t^T \widehat{g}(t, s, Z(t, s))ds - \int_t^T Z(t, s)dW(s), \qquad t \in [0, T], \tag{4.3.10}$$

with

$$\widehat{g}(t, s, 0) = 0, \qquad (t, s) \in \Delta^*[0, T], \tag{4.3.11}$$

and

$$\begin{aligned} &\sup_{t\in[0,T]} \mathbb{E}|\widehat{\psi}(t)|^p \leqslant \sup_{t\in[0,T]} \mathbb{E}|\psi(t)|^p + \sup_{t\in[0,T]} \mathbb{E}\Big[\Big(\int_t^T |g(t, s, Y(s), 0)|ds\Big)^p\Big] \\ &\leqslant \sup_{t\in[0,T]} \mathbb{E}\Big[|\psi(t)|^p + K\Big(\int_t^T L_y(t, s)^{\frac{p}{p-1}}ds\Big)^{p-1}\Big(\int_0^T |Y(s)|^p ds\Big)\Big] < \infty, \end{aligned}$$

giving $\widehat{\psi}(\cdot) \in L^{\infty}_{\mathcal{F}_T}(0, T; L^p(\Omega; \mathbb{R}^m))$. The main feature of BSVIE (4.3.10) is that the generator is independent of $Y(\cdot)$. To obtain the continuity of $Y(\cdot)$, it suffices to look at such a BSVIE.

In the rest of this section, we will concentrate on the above BSVIE (4.3.10), which does not have to be reduced from (4.3.8), with the generator $\widehat{g}$ satisfying the following.

(H4.3.2) Let $\widehat{g} : \Delta^*[0, T] \times \mathbb{R}^m \times \Omega \to \mathbb{R}^m$ be measurable such that $s \mapsto \widehat{g}(t, s, z)$ is $\mathbb{F}$-progressively measurable satisfying

$$\widehat{g}(t, s, 0) = 0, \qquad (t, s) \in \Delta^*[0, T]. \tag{4.3.12}$$

Moreover, there is a deterministic measurable function $L_z : \Delta^*[0, T] \to [0, \infty)$ with

$$\sup_{t\in[0,T]} \int_t^T L_z(t, s)^2 ds < \infty, \tag{4.3.13}$$

such that

$$\begin{aligned} |\widehat{g}(t, s, z_1) - \widehat{g}(t, s, z_2)| &\leqslant L_z(t, s)|z_1 - z_2|, \\ &\qquad \forall (t, s) \in \Delta^*[0, T],\ z_1, z_2 \in \mathbb{R}^m. \end{aligned} \tag{4.3.14}$$

In particular, if $\widehat{g}(t, s, z)$ is defined by (4.3.9) with (H4.3.1)$_p$ $(p > 1)$ holding, then (H4.3.2) is true. On the other hand, by Proposition 4.3.1, under (H4.3.2), for any $\widehat{\psi}(\cdot) \in L^\infty_{\mathcal{F}_T}(0, T; L^p(\Omega; \mathbb{R}^m))$, with $p > 1$, BSVIE (4.3.10) admits a unique adapted solution $(Y(\cdot), Z(\cdot\,, \cdot)) \in \mathcal{H}^p_{\Delta^*}[0, T]$, and the following is satisfied (take supremum with respect to $t \in [0, T]$ on both sides of (4.3.3)):

$$\sup_{t\in[0,T]} \mathbb{E}|Y(t)|^p + \sup_{t\in[0,T]} \mathbb{E}\Big(\int_t^T |Z(t, s)|^2 ds\Big)^{\frac{p}{2}} \leqslant K\Big[\sup_{t\in[0,T]} \mathbb{E}|\widehat{\psi}(t)|^p\Big]. \tag{4.3.15}$$

By applying the conditional expectation operator $\mathbb{E}_t$ to the BSVIE (4.3.10), we have

$$Y(t) = \mathbb{E}_t[\widehat{\psi}(t)] + \int_t^T \mathbb{E}_t\big[\widehat{g}(t, s, Z(t, s))\big]ds, \qquad t \in [0, T]. \tag{4.3.16}$$

Thus, if we are able to prove the certain continuity of the two terms on the right-hand side of (4.3.16), under reasonable conditions, then the same continuity of $Y(\cdot)$ will follow. We now present the following lemma which takes care of the first term on the right hand of the above.

Lemma 4.3.2 *Let $\widehat{\psi}(\cdot) \in L^1_{\mathcal{F}_T}(0, T; \mathbb{R}^m)$, and let*

$$Y(t) = \mathbb{E}_t\big[\widehat{\psi}(t)\big], \qquad t \in [0, T]. \tag{4.3.17}$$

Then the following hold with $p \geqslant 1$*:*

$$\widehat{\psi}(\cdot) \in C([0,T]; L^p_{\mathcal{F}_T}(\Omega;\mathbb{R}^m)) \quad \Rightarrow \quad Y(\cdot) \in C_{\mathbb{F}}([0,T]; L^p(\Omega;\mathbb{R}^m)), \tag{4.3.18}$$

and with $p > 1$,

$$\widehat{\psi}(\cdot) \in L^p_{\mathcal{F}_T}(\Omega; C([0,T];\mathbb{R}^m)) \quad \Rightarrow \quad Y(\cdot) \in L^p_{\mathbb{F}}(\Omega; C([0,T];\mathbb{R}^m)). \tag{4.3.19}$$

Proof First, let $\widehat{\psi}(\cdot) \in C([0,T]; L^p_{\mathcal{F}_T}(\Omega;\mathbb{R}^m))$ for some $p \geqslant 1$. For any $t \in (0,T)$, there exists a unique $\widehat{\zeta}(t,\cdot) \in L^p_{\mathbb{F}}(\Omega; L^2(0,T;\mathbb{R}^{m\times d}))$ such that

$$\widehat{\psi}(t) = \mathbb{E}_\tau[\widehat{\psi}(t)] + \int_\tau^T \widehat{\zeta}(t,s)dW(s), \qquad \tau \in [0,T].$$

Then for any $\delta \in (0, T-t)$,

$$\begin{aligned}
&\mathbb{E}|Y(t+\delta) - Y(t)|^p = \mathbb{E}\big|\mathbb{E}_{t+\delta}[\widehat{\psi}(t+\delta)] - \mathbb{E}_t[\widehat{\psi}(t)]\big|^p \\
&\leqslant 2^{p-1}\mathbb{E}\Big(\big|\mathbb{E}_{t+\delta}[\widehat{\psi}(t+\delta) - \widehat{\psi}(t)]\big|^p + \big|\mathbb{E}_{t+\delta}[\widehat{\psi}(t)] - \mathbb{E}_t[\widehat{\psi}(t)]\big|^p\Big) \\
&\leqslant 2^{p-1}\mathbb{E}\Big(|\widehat{\psi}(t+\delta) - \widehat{\psi}(t)|^p + \Big|\int_t^{t+\delta} \widehat{\zeta}(t,s)dW(s)\Big|^p\Big) \\
&\leqslant K\mathbb{E}\Big[|\widehat{\psi}(t+\delta) - \widehat{\psi}(t)|^p + \Big(\int_t^{t+\delta} |\widehat{\zeta}(t,s)|^2 ds\Big)^{\frac{p}{2}}\Big],
\end{aligned}$$

which goes to zero as $\delta \to 0$. Likewise,

$$\begin{aligned}
&\mathbb{E}|Y(t-\delta) - Y(t)|^p = \mathbb{E}\big|\mathbb{E}_{t-\delta}[\widehat{\psi}(t-\delta)] - \mathbb{E}_t[\widehat{\psi}(t)]\big|^p \\
&\leqslant 2^{p-1}\mathbb{E}\Big(\big|\mathbb{E}_{t-\delta}[\widehat{\psi}(t-\delta) - \widehat{\psi}(t)]\big|^p + \big|\mathbb{E}_{t-\delta}[\widehat{\psi}(t)] - \mathbb{E}_t[\widehat{\psi}(t)]\big|^p\Big) \\
&\leqslant 2^{p-1}\mathbb{E}\Big(|\widehat{\psi}(t-\delta) - \widehat{\psi}(t)|^p + \Big|\int_{t-\delta}^{t} \widehat{\zeta}(t,s)dW(s)\Big|^p\Big) \\
&\leqslant K\mathbb{E}\Big[|\widehat{\psi}(t-\delta) - \widehat{\psi}(t)|^p + \Big(\int_{t-\delta}^{t} |\widehat{\zeta}(t,s)|^2 ds\Big)^{\frac{p}{2}}\Big],
\end{aligned}$$

which also goes to zero as $\delta \to 0$. The cases $t = 0, T$ need only a one-side estimate. Hence, we obtain $Y(\cdot) \in C_{\mathbb{F}}([0,T]; L^p(\Omega;\mathbb{R}^m))$.

Next, let $\widehat{\psi}(\cdot) \in L^p_{\mathcal{F}_T}(\Omega; C([0,T];\mathbb{R}^m))$ for some $p > 1$. Since

$$\widehat{\xi} \equiv \sup_{s\in[0,T]} |\widehat{\psi}(s)| \in L^p_{\mathcal{F}_T}(\Omega;\mathbb{R}),$$

and $t \mapsto \mathbb{E}_t[\widehat{\xi}]$ is an L^p-integrable martingale, one can apply Doob's martingale inequality to obtain

$$\begin{aligned}
&\mathbb{E}\Big[\sup_{t\in[0,T]}|Y(t)|^p\Big]\leqslant\mathbb{E}\Big[\sup_{t\in[0,T]}\Big(\mathbb{E}_t|\widehat{\psi}(t)|\Big)^p\Big]\leqslant\mathbb{E}\Big[\sup_{t\in[0,T]}\Big(\mathbb{E}_t[\widehat{\xi}]\Big)^p\Big]\\
&\leqslant\Big(\frac{p}{p-1}\Big)^p\mathbb{E}\Big(\mathbb{E}_T[\widehat{\xi}]\Big)^p=\Big(\frac{p}{p-1}\Big)^p\mathbb{E}\Big[\sup_{s\in[0,T]}|\widehat{\psi}(s)|^p\Big].
\end{aligned}$$

The above implies that

$$Y(\cdot)\in L^p_{\mathbb{F}}(\Omega;L^\infty(0,T;\mathbb{R}^m)). \tag{4.3.20}$$

We now try to prove that $t\mapsto Y(t)$ has continuous paths, almost surely. To this end, for any $\ell>1$, we define random times τ_i^ℓ $(i\geqslant 0)$ as follows:

$$\begin{cases}\tau_0^\ell=0,\\ \tau_i^\ell=\inf\big\{t\in(\tau_{i-1}^\ell,T]\bigm||\widehat{\psi}(t)-\widehat{\psi}(\tau_{i-1}^\ell)|=\frac{1}{\ell}\big\}, \qquad i\geqslant 1,\end{cases}$$

with the convention that $\inf\varnothing=T$. Since there exists an $\Omega_0\subseteq\Omega$ with $\mathbb{P}(\Omega_0)=0$ such that for any $\omega\in\Omega\setminus\Omega_0$, $t\mapsto\widehat{\psi}(t,\omega)$ is continuous on $[0,T]$, we must have the following:

$$\big\{\omega\in\Omega\bigm|\tau_i^\ell(\omega)=T,\quad\text{for some } i\geqslant 1\big\}\supseteq\Omega\setminus\Omega_0.$$

Now, we define

$$\widehat{\psi}^\ell(t)=\sum_{i\geqslant 1}\widehat{\psi}(\tau_{i-1}^\ell)\mathbf{1}_{[\tau_{i-1}^\ell,\tau_i^\ell)}(t)\mathbf{1}_{\Omega\setminus\Omega_0},\qquad t\in[0,T),$$

which, almost surely, is a random $\mathcal{F}_T$-measurable step function with jump size being no more than $\frac{1}{\ell}$, and let

$$\widetilde{\psi}^\ell(t)=\widehat{\psi}(\tau_{i-2}^\ell)+\frac{\widehat{\psi}(\tau_{i-1}^\ell)-\widehat{\psi}(\tau_{i-2}^\ell)}{\tau_i^\ell-\tau_{i-1}^\ell}(t-\tau_{i-1}^\ell),\quad t\in[\tau_{i-1}^\ell,\tau_i^\ell),\ i\geqslant 1,$$

with the convention that $\widehat{\psi}(\tau_{-1}^\ell)=\widehat{\psi}(0)$, which is an $\mathcal{F}_T$-measurable, piecewise linear continuous function. Define

$$\begin{aligned}
Y^\ell(t)&=\mathbb{E}_t\big[\widehat{\psi}^\ell(t)\big],\qquad t\in[0,T],\\
\widetilde{Y}^\ell(t)&=\mathbb{E}_t\big[\widetilde{\psi}^\ell(t)\big]=\mathbb{E}_t\big[\widehat{\psi}(\tau_{i-2}^\ell)\big]+\mathbb{E}_t\Big[\frac{\widehat{\psi}(\tau_{i-1}^\ell)-\widehat{\psi}(\tau_{i-2}^\ell)}{\tau_i^\ell-\tau_{i-1}^\ell}\Big]t\\
&\quad-\mathbb{E}_t\Big[\frac{\widehat{\psi}(\tau_{i-1}^\ell)-\widehat{\psi}(\tau_{i-2}^\ell)}{\tau_i^\ell-\tau_{i-1}^\ell}\tau_{i-1}^\ell\Big],\qquad t\in[\tau_{i-1}^\ell,\tau_i^\ell].
\end{aligned}$$

Then by martingale inequality,

$$\begin{aligned}&\mathbb{E}\Big[\sup_{t\in[0,T]}|Y^\ell(t)|^p\Big]=\mathbb{E}\Big[\sup_{t\in[0,T]}|\mathbb{E}_t[\widehat\psi^\ell(t)]|^p\Big]\leqslant\mathbb{E}\Big[\sup_{t\in[0,T]}\big[\mathbb{E}_t|\widehat\psi^\ell(t)|\big]^p\Big]\\&\leqslant\mathbb{E}\Big[\sup_{t\in[0,T]}\big[\mathbb{E}_t\sup_{t\in[0,T]}|\widehat\psi^\ell(t)|\big]^p\Big]\leqslant K\mathbb{E}\Big[\sup_{t\in[0,T]}|\widehat\psi^\ell(t)|^p\Big]\leqslant K\mathbb{E}\Big[\sup_{t\in[0,T]}|\widehat\psi(t)|^p\Big].\end{aligned}$$

Similarly,

$$\mathbb{E}\Big[\sup_{t\in[0,T]}|\widetilde Y^\ell(t)|^p\Big]\leqslant K\mathbb{E}\Big[\sup_{t\in[0,T]}|\widetilde\psi^\ell(t)|^p\Big]\leqslant K\mathbb{E}\Big[\sup_{t\in[0,T]}|\widehat\psi(t)|^p\Big].$$

Moreover,

$$\begin{aligned}&Y^\ell(\cdot)\in L^p_{\mathbb{F}}(\Omega;L^\infty(0,T;\mathbb{R}^m)),\qquad \widehat\psi^\ell(\cdot)\in L^p_{\mathcal{F}_T}(\Omega;L^\infty(0,T;\mathbb{R}^m)),\\&\widetilde Y^\ell(\cdot)\in L^p_{\mathbb{F}}(\Omega;C([0,T];\mathbb{R}^m)),\qquad \widetilde\psi^\ell(\cdot)\in L^p_{\mathcal{F}_T}(\Omega;C([0,T];\mathbb{R}^m)).\end{aligned}$$

Also,

$$\begin{aligned}&|\widehat\psi^\ell(t)-\widehat\psi(t)|\leqslant\frac{1}{\ell},\qquad t\in[0,T],\ \text{a.s.}\\&|\widetilde\psi^\ell(t)-\widehat\psi(t)|\leqslant|\widetilde\psi^\ell(t)-\widehat\psi^\ell(t)|+|\widehat\psi^\ell(t)-\widehat\psi(t)|\leqslant\frac{2}{\ell},\quad \forall t\in[0,T],\ \text{a.s.}\end{aligned}$$

Hence,

$$|Y^\ell(t,\omega)-Y(t,\omega)|=\big|\mathbb{E}_t\big[\widehat\psi^\ell(t)\big]-\mathbb{E}_t\big[\widehat\psi(t)\big]\big|\leqslant\frac{1}{\ell},\quad \forall t\in[0,T),\omega\in\Omega\setminus\Omega_0.$$

Consequently,

$$\mathbb{E}\Big[\sup_{t\in[0,T]}|Y^\ell(t)-Y(t)|^p\Big]\leqslant\frac{1}{\ell^p}.$$

Further, from

$$\begin{aligned}|Y^\ell(t)-\widetilde Y^\ell(t)|&=\big|\mathbb{E}_t\big[\widehat\psi^\ell(t)\big]-\mathbb{E}_t\big[\widetilde\psi^\ell(t)\big]\big|\\&\leqslant\mathbb{E}_t|\widehat\psi(\tau^\ell_{i-1})-\widehat\psi(\tau^\ell_{i-2})|\leqslant\frac{1}{\ell},\quad t\in[\tau^\ell_{i-1},\tau^\ell_i],\quad \text{a.s.}\end{aligned}$$

one has

$$\begin{aligned}&\mathbb{E}\Big[\sup_{t\in[0,T]}|Y(t)-\widetilde Y^\ell(t)|^p\Big]\\&\leqslant 2^{p-1}\mathbb{E}\Big[\sup_{t\in[0,T]}|Y(t)-Y^\ell(t)|^p+\sup_{t\in[0,T]}|Y^\ell(t)-\widetilde Y^\ell(t)|^p\Big]\leqslant\frac{2^p}{\ell^p}.\end{aligned}$$

Then letting $\ell \to \infty$, we obtain

$$\begin{aligned} Y(\cdot) &\in L^p_{\mathbb{F}}(\Omega; L^\infty(0,T;\mathbb{R}^m)) \bigcap L^p_{\mathcal{F}_T}(\Omega; C([0,T];\mathbb{R}^m)) \\ &\subseteq L^p_{\mathbb{F}}(\Omega; C([0,T];\mathbb{R}^m)). \end{aligned}$$

This completes the proof. □

The above result seems to suggest that $t \mapsto \mathbb{E}_t[\widehat{\psi}(t)]$ should keep the Hölder continuity if $\widehat{\psi}(\cdot)$ is so. Unfortunately, this is unknown to us. To handle the second term on the right-hand side of (4.3.16), we introduce a stronger assumption than (H4.3.2), which naturally imposes proper continuity of the map $t \mapsto \widehat{g}(t,s,z)$. Let $\alpha > 0$.

(H4.3.3)$_\alpha$ Let (H4.3.2) hold with

$$\sup_{t\in[0,T]} \Big[\int_t^T L_z(t,s)^{2(1+\alpha)} ds\Big] < \infty. \tag{4.3.21}$$

Moreover, there exists a nonnegative valued function $L(\cdot) \in L^2(0,T)$, an adapted (nonnegative) process $\kappa(\cdot)$, and some modulus of continuity $\rho : [0,\infty) \to [0,\infty)$ such that

$$\begin{aligned} |\widehat{g}(t_1,s,z) - \widehat{g}(t_2,s,z)| &\leqslant \rho(|t_1 - t_2|) L(s)\big(1 + |z| + \kappa(s)\big), \\ &\forall (t_1,s),(t_2,s) \in \Delta^*[0,T],\ z \in \mathbb{R}^m, \text{ a.s.} \end{aligned} \tag{4.3.22}$$

Note that $\alpha > 0$, and (4.3.21) means $L_z(\cdot\,,\cdot) \in L^\infty(0,T; L^{2(1+\alpha)}(\cdot\,,T;\mathbb{R}_+))$ which is stronger than the second relation in (4.2.4). Clearly, the larger the α is, the more restrictive condition (4.3.21) will be. In particular, $L_z(\cdot\,,\cdot)$ is bounded if and only if $\alpha = \infty$. The following is a further strengthened assumption. Let $\alpha > 0$ and $\delta \in (0,1]$.

(H4.3.4)$_{\alpha,\delta}$ Let (H4.3.3)$_\alpha$ hold with modulus of continuity $\rho(r)$ given by

$$\rho(r) = r^\delta, \qquad r \geqslant 0. \tag{4.3.23}$$

We have the following result.

Theorem 4.3.3

(i) Let (H4.3.3)$_\alpha$ hold for some $\alpha > 0$. Then for any given $p > 1$, $\widehat{\psi}(\cdot) \in C_{\mathcal{F}_T}([0,T]; L^p(\Omega;\mathbb{R}^m))$, $\kappa(\cdot) \in L^p_{\mathbb{F}}(\Omega; L^2(0,T;\mathbb{R}^+))$, the adapted solution $(Y(\cdot), Z(\cdot\,,\cdot))$ of (4.3.10) satisfies

$$\sup_{t\in[0,T]} \mathbb{E}\Big[|Y(t)|^p + \Big(\int_t^T |Z(t,s)|^2 ds\Big)^{\frac{p}{2}}\Big] \leqslant K \sup_{t\in[0,T]} \mathbb{E}|\widehat{\psi}(t)|^p, \tag{4.3.24}$$

and

$$\begin{cases} Y(\cdot) \in C_{\mathbb{F}}([0,T]; L^p(\Omega;\mathbb{R}^m)), \\ \displaystyle\int_\cdot^T Z(\cdot\,,s)dW(s) \in C_{\mathcal{F}_T}([0,T]; L^p(\Omega;\mathbb{R}^m)). \end{cases} \tag{4.3.25}$$

(*ii*) *Let* (H4.3.4)$_{\alpha,\delta}$ *hold for some* $\alpha > 0$, $\delta \in (0,1]$. *Then for given* $\gamma \in (0,1)$ *and any* $p > \frac{2(1+\alpha)}{\alpha} \vee \frac{1}{\gamma} \vee \frac{1}{\delta}$, *any* $\kappa(\cdot) \in L^p_{\mathbb{F}}(\Omega; L^2(0,T;\mathbb{R}^+))$, *any* $\widehat{\psi}(\cdot) \in L^p_{\mathcal{F}_T}(\Omega; C([0,T];\mathbb{R}^m))$ *satisfying*

$$\Big(\mathbb{E}|\widehat{\psi}(t_1) - \widehat{\psi}(t_2)|^p\Big)^{\frac{1}{p}} \leqslant K|t_1 - t_2|^\gamma, \qquad \forall t_1, t_2 \in [0,T], \tag{4.3.26}$$

the adapted solution $(Y(\cdot), Z(\cdot\,,\cdot))$ *of the Type I BSVIE* (4.3.10) *satisfies the following:*

$$\mathbb{E}\Big[\sup_{t\in[0,T]} |Y(t)|^p\Big] \leqslant K\mathbb{E}\Big[\sup_{t\in[0,T]} |\widehat{\psi}(t)|^p\Big], \tag{4.3.27}$$

and

$$\begin{cases} Y(\cdot) \in L^p_{\mathbb{F}}(\Omega; C([0,T];\mathbb{R}^m)), \\ \displaystyle\int_\cdot^T Z(\cdot\,,s)dW(s) \in L^p_{\mathcal{F}_T}(\Omega; C([0,T];\mathbb{R}^m)). \end{cases} \tag{4.3.28}$$

Proof

(i) Since $\widehat{\psi}(\cdot) \in C_{\mathcal{F}_T}([0,T]; L^p(\Omega;\mathbb{R}^m))$, for any $t \in [0,T]$, $\widehat{\psi}(t) \in L^p(\Omega;\mathbb{R}^m)$ is well-defined. Hence, we may let $(\eta(t,\cdot), \zeta(t,\cdot))$ be the adapted solution to the following auxiliary BSDE:

$$\eta(t,\tau) = \widehat{\psi}(t) + \int_\tau^T \widehat{g}(t,s,\zeta(t,s))ds - \int_\tau^T \zeta(t,s)dW(s), \qquad \tau \in [t,T]. \tag{4.3.29}$$

By the result in Chap. 3, Sect. 1, under (H4.3.3)$_\alpha$, one has the following estimate for the BSDEs:

$$\mathbb{E}\Big[\sup_{\tau\in[t,T]} |\eta(t,\tau)|^p + \Big(\int_t^T |\zeta(t,s)|^2 ds\Big)^{\frac{p}{2}}\Big] \leqslant K\mathbb{E}|\widehat{\psi}(t)|^p. \tag{4.3.30}$$

Moreover, we know that the unique adapted solution $(Y(\cdot), Z(\cdot\,,\cdot))$ to BSVIE (4.3.10) is given by

$$Y(t) = \eta(t,t), \quad t \in [0,T]; \qquad Z(t,s) = \zeta(t,s), \quad (t,s) \in \Delta^*[0,T]. \tag{4.3.31}$$

Thus, (4.3.30) implies

$$\mathbb{E}\Big[|Y(t)|^p + \Big(\int_t^T |Z(t,s)|^2 ds\Big)^{\frac{p}{2}}\Big] \leqslant K\mathbb{E}|\widehat{\psi}(t)|^p. \tag{4.3.32}$$

Then (4.3.24) follows. Now, let $0 \leqslant t_1 < t_2 \leqslant T$. By looking at the BSDE (4.3.29) on $[t_2, T]$ with the parameter $t = t_1, t_2$, using the stability estimate for BSDEs, we have (noting (4.3.30))

$$\begin{aligned}
&\mathbb{E}\Big[\sup_{\tau\in[t_2,T]} |\eta(t_1,\tau) - \eta(t_2,\tau)|^p + \Big(\int_{t_2}^T |\zeta(t_1,s) - \zeta(t_2,s)|^2 ds\Big)^{\frac{p}{2}}\Big]\\
&\leqslant K\mathbb{E}\Big[|\widehat{\psi}(t_1) - \widehat{\psi}(t_2)|^p + \Big(\int_{t_2}^T |\widehat{g}(t_1,s,\zeta(t_2,s)) - \widehat{g}(t_2,s,\zeta(t_2,s))|ds\Big)^p\Big]\\
&\leqslant K\mathbb{E}\Big[|\widehat{\psi}(t_1) - \widehat{\psi}(t_2)|^p + \Big(\int_{t_2}^T \rho(|t_1-t_2|)L(s)(1+|\zeta(t_2,s)| + |\kappa(s)|)ds\Big)^p\Big]\\
&\leqslant K\mathbb{E}\Big\{|\widehat{\psi}(t_1) - \widehat{\psi}(t_2)|^p + [\rho(|t_1-t_2|)]^p\Big[\Big(\int_{t_2}^T L(s)ds\Big)^p\\
&\qquad + \Big(\int_{t_2}^T L(s)^2 ds\Big)^{\frac{p}{2}}\Big\{\Big(\int_{t_2}^T |\kappa(s)|^2 ds\Big)^{\frac{p}{2}} + \Big(\int_{t_2}^T |Z(t_2,s)|^2 ds\Big)^{\frac{p}{2}}\Big\}\Big]\Big\}\\
&\leqslant K\mathbb{E}\Big(|\widehat{\psi}(t_1) - \widehat{\psi}(t_2)|^p + [\rho(|t_1-t_2|)]^p\Big),
\end{aligned}$$

where (4.3.31) and (4.3.32) are used. By (4.3.31) again, we obtain from the above

$$\mathbb{E}\Big(\int_{t_2}^T |Z(t_1,s) - Z(t_2,s)|^2 ds\Big)^{\frac{p}{2}} \leqslant K\mathbb{E}\Big(|\widehat{\psi}(t_1) - \widehat{\psi}(t_2)|^p + [\rho(|t_1-t_2)]^p\Big). \tag{4.3.33}$$

Now, we estimate

$$\begin{aligned}
&\mathbb{E}\Big|\int_{t_1}^T \widehat{g}(t_1,s,Z(t_1,s))ds - \int_{t_2}^T \widehat{g}(t_2,s,Z(t_2,s))ds\Big|^p\\
&\leqslant K\mathbb{E}\Big[\Big(\int_{t_1}^{t_2} |\widehat{g}(t_1,s,Z(t_1,s))|ds\Big)^p\\
&\qquad + \Big(\int_{t_2}^T |\widehat{g}(t_1,s,Z(t_1,s)) - \widehat{g}(t_2,s,Z(t_2,s))|ds\Big)^p\Big].
\end{aligned} \tag{4.3.34}$$

Note that

$$
\begin{aligned}
&\mathbb{E}\Big(\int_{t_1}^{t_2}|\widehat g(t_1,s,Z(t_1,s))|ds\Big)^p\leqslant\mathbb{E}\Big[\int_{t_1}^{t_2}L_z(t_1,s)|Z(t_1,s)|ds\Big]^p\\
&\leqslant K\mathbb{E}\Big[\Big(\int_{t_1}^{t_2}L_z(t_1,s)^2ds\Big)^{\frac p2}\Big(\int_{t_1}^{t_2}|Z(t_1,s)|^2ds\Big)^{\frac p2}\Big]\\
&\leqslant K\mathbb{E}\Big[|t_1-t_2|^{\frac{p\alpha}{2(1+\alpha)}}\Big(\int_{t_1}^{T}L_z(t_1,s)^{2(1+\alpha)}ds\Big)^{\frac{p}{2(1+\alpha)}}\Big(\int_{t_1}^{T}|Z(t_1,s)|^2ds\Big)^{\frac p2}\Big]\\
&\leqslant K|t_1-t_2|^{\frac{p\alpha}{2(1+\alpha)}}.
\end{aligned}
$$

Again, we used (4.3.32). Also, by (4.3.33),

$$
\begin{aligned}
&\mathbb{E}\Big(\int_{t_2}^{T}|\widehat g(t_1,s,Z(t_1,s))-\widehat g(t_2,s,Z(t_2,s))|ds\Big)^p\\
&\leqslant\mathbb{E}\Big[\Big(\int_{t_2}^{T}\rho\big(|t_1-t_2|\big)L(s)\big(1+|Z(t_1,s)|+|\kappa(s)|\big)ds\Big)^p\\
&\qquad+\Big(\int_{t_2}^{T}L_z(t_2,s)|Z(t_1,s)-Z(t_2,s)|ds\Big)^p\Big]\\
&\leqslant K\mathbb{E}\Big\{\big[\rho\big(|t_1-t_2|\big)\big]^p\Big[\Big(\int_0^TL(s)ds\Big)^p+\Big(\int_0^TL(s)^2ds\Big)^{\frac p2}\Big(\int_{t_1}^{T}|Z(t_1,s)|^2ds\Big)^{\frac p2}\Big]\\
&\qquad+\Big(\int_{t_2}^{T}L(s)^2ds\Big)^{\frac p2}\Big(\int_{t_2}^{T}|\kappa(s)|^2ds\Big)^{\frac p2}\\
&\qquad+\Big(\int_{t_2}^{T}L_z(t_2,s)^2ds\Big)^{\frac p2}\Big(\int_{t_2}^{T}|Z(t_1,s)-Z(t_2,s)|^2ds\Big)^{\frac p2}\Big\}\\
&\leqslant K\mathbb{E}\Big\{\big[\rho\big(|t_1-t_2|\big)\big]^p+\Big(\int_{t_2}^{T}L_z(t_2,s)^2ds\Big)^{\frac p2}\Big(|\widehat\psi(t_1)-\widehat\psi(t_2)|^p+\big[\rho\big(|t_1-t_2|\big)\big]^p\Big)\Big\}\\
&\leqslant K\mathbb{E}\Big[|\widehat\psi(t_1)-\widehat\psi(t_2)|^p+\big[\rho\big(|t_1-t_2|\big)\big]^p\Big].
\end{aligned}
$$

Hence, we obtain from (4.3.34)

$$
\begin{aligned}
&\mathbb{E}\Big|\int_{t_1}^{T}\widehat g(t_1,s,Z(t_1,s))ds-\int_{t_2}^{T}\widehat g(t_2,s,Z(t_2,s))ds\Big|^p\\
&\leqslant K\mathbb{E}\Big[|t_1-t_2|^{\frac{p\alpha}{2(1+\alpha)}}+|\widehat\psi(t_1)-\widehat\psi(t_2)|^p+\big[\rho\big(|t_1-t_2|\big)\big]^p\Big].
\end{aligned}
\tag{4.3.35}
$$

Consequently, when $\widehat\psi(\cdot)\in C_{\mathcal{F}_T}([0,T];L^p(\Omega;\mathbb{R}^m))$, we have

$$
\widehat\psi(\cdot)+\int_{\cdot}^{T}\widehat g(\cdot\,,s,Z(\cdot\,,s))ds\in C_{\mathcal{F}_T}([0,T];L^p(\Omega;\mathbb{R}^m)).
$$

Then by Lemma 4.3.2 and noting (4.3.16), we obtain that $Y(\cdot)\in C_{\mathbb{F}}([0,T];L^p(\Omega;\mathbb{R}^m))$. Further, from

$$
\int_t^TZ(t,s)dW(s)=\widehat\psi(t)-Y(t)+\int_t^T\widehat g(t,s,Z(t,s))ds,\qquad t\in[0,T],
$$

we obtain the second relation in (4.3.25).

(ii) We still consider the following auxiliary BSDE on $[t, T]$:

$$\eta(t,\tau) = \widehat{\psi}(t) + \int_\tau^T \widehat{g}(t,s,\zeta(t,s))ds - \int_\tau^T \zeta(t,s)dW(s), \qquad \tau \in [t,T].$$

Under $(\mathrm{H4.3.4})_{\alpha,\delta}$, for any $p > 1$, the above admits a unique adapted solution $(\eta(t,\cdot),\zeta(t,\cdot)) \in \mathbf{H}^p[0,T] \equiv L^p_{\mathbb{F}}(\Omega; C([t,T];\mathbb{R}^m)) \times L^p_{\mathbb{F}}(\Omega; L^2(t,T;\mathbb{R}^m))$. Applying Itô's formula to the process $\tau \mapsto e^{-\int_\tau^T \beta(s)ds}|\eta(t,\tau)|^2$ with $\beta(\cdot)$ being undetermined deterministic function, one has

$$\begin{aligned}
&|\widehat{\psi}(t)|^2 - e^{-\int_\tau^T \beta(s)ds}|\eta(t,\tau)|^2 \\
&= \int_\tau^T e^{-\int_s^T \beta(\theta)d\theta}\Big(\beta(s)|\eta(t,s)|^2 - 2\langle \eta(t,s), \widehat{g}(t,s,\zeta(t,s))\rangle + |\zeta(t,s)|^2\Big)ds \\
&\quad + \int_\tau^T 2e^{-\int_s^T \beta(\theta)d\theta}\langle \eta(t,s),\zeta(t,s)\rangle dW(s).
\end{aligned}$$

This implies

$$\begin{aligned}
&e^{-\int_\tau^T \beta(\theta)d\theta}|\eta(t,\tau)|^2 + \int_\tau^T e^{-\int_s^T \beta(\theta)d\theta}\Big(\beta(s)|\eta(t,s)|^2 + |\zeta(t,s)|^2\Big)ds \\
&\quad + \int_\tau^T 2e^{-\int_s^T \beta(\theta)d\theta}\langle \eta(t,s),\zeta(t,s)\rangle dW(s) \\
&= |\widehat{\psi}(t)|^2 + 2\int_\tau^T e^{-\int_s^T \beta(\theta)d\theta}\langle \eta(t,s), \widehat{g}(t,s,\zeta(t,s))\rangle ds \\
&\leqslant |\widehat{\psi}(t)|^2 + 2\int_\tau^T e^{-\int_s^T \beta(\theta)d\theta}|\eta(t,s)|L_z(t,s)|\zeta(t,s)|ds \\
&\leqslant |\widehat{\psi}(t)|^2 + \int_\tau^T e^{-\int_s^T \beta(\theta)d\theta}\Big[L_z(t,s)^2|\eta(t,s)|^2 + |\zeta(t,s)|^2\Big]ds.
\end{aligned}$$

Hence,

$$\begin{aligned}
&e^{-\int_\tau^T \beta(\theta)d\theta}|\eta(t,\tau)|^2 + \int_\tau^T e^{-\int_s^T \beta(\theta)d\theta}\Big(\beta(s) - L_z(t,s)^2\Big)|\eta(t,s)|^2 ds \\
&\quad + \int_\tau^T 2e^{-\int_s^T \beta(\theta)d\theta}\langle \eta(t,s),\zeta(t,s)\rangle dW(s) \leqslant |\widehat{\psi}(t)|^2.
\end{aligned}$$

Taking $\beta(s) = L_z(t,s)^2$, we get

$$|\eta(t,\tau)|^2 + \int_\tau^T 2e^{\int_\tau^s L_z(t,\theta)^2 d\theta}\langle \eta(t,s),\zeta(t,s)\rangle dW(s) \leqslant e^{\int_\tau^T L_z(t,\theta)^2 d\theta}|\widehat{\psi}(t)|^2.$$

Thus,

$$|\eta(t,\tau)|^2 \leqslant \mathbb{E}_\tau\Big[e^{\int_\tau^T L_z(t,\theta)^2 d\theta}|\widehat{\psi}(t)|^2\Big] \leqslant K\mathbb{E}_\tau|\widehat{\psi}(t)|^2.$$

Then

$$|Y(t)|^2 \leqslant K\mathbb{E}_t|\widehat{\psi}(t)|^2, \quad t \in [0, T].$$

Consequently, when $p > 2$, making use of Doob's inequality, one has

$$\begin{aligned}
&\mathbb{E}\Big[\sup_{t\in[0,T]} |Y(t)|^p\Big] \leqslant K\mathbb{E}\Big[\sup_{t\in[0,T]} \Big(\mathbb{E}_t\big[|\widehat{\psi}(t)|^2\big]\Big)^{\frac{p}{2}}\Big] \\
&\leqslant K\Big(\frac{p}{p-2}\Big)^{\frac{p}{2}}\mathbb{E}\Big[\Big(\mathbb{E}_T|\widehat{\psi}(t)|^2\Big)^{\frac{p}{2}}\Big] \leqslant K\mathbb{E}\Big[\sup_{t\in[0,T]} |\widehat{\psi}(t)|^p\Big].
\end{aligned}$$

This gives (4.3.27). Next, for $0 \leqslant t_1 < t_2 \leqslant T$, similar to (4.3.33), we have

$$\mathbb{E}\Big(\int_{t_2}^T |Z(t_1, s) - Z(t_2, s)|^2 ds\Big)^{\frac{p}{2}} \leqslant K\mathbb{E}\Big(|\widehat{\psi}(t_1) - \widehat{\psi}(t_2)|^p + |t_1 - t_2|^{\delta p}\Big). \qquad (4.3.36)$$

Then, similar to (4.3.35), replacing $\kappa(|t_1 - t_2|)$ by $|t_1 - t_2|^{\delta}$, one has

$$\begin{aligned}
&\mathbb{E}\Big|\int_{t_1}^T \widehat{g}(t_1, s, Z(t_1, s))ds - \int_{t_2}^T \widehat{g}(t_2, s, Z(t_2, s))ds\Big|^p \\
&\leqslant K\mathbb{E}\Big(|t_1 - t_2|^{\frac{p\alpha}{2(1+\alpha)}} + |t_1 - t_2|^{\delta p} + |\widehat{\psi}(t_1) - \widehat{\psi}(t_2)|^p\Big) \\
&\leqslant K|t_1 - t_2|^{p(\frac{\alpha}{2(1+\alpha)} \wedge \delta \wedge \gamma)}.
\end{aligned}$$

This implies

$$\begin{aligned}
&\mathbb{E}\Big|\Big(\widehat{\psi}(t_1) + \int_{t_1}^T \widehat{g}(t_1, s, Z(t_1, s))ds\Big) - \Big(\widehat{\psi}(t_2) + \int_{t_2}^T \widehat{g}(t_2, s, Z(t_2, s))ds\Big)\Big|^p \\
&\leqslant K|t_1 - t_2|^{p(\frac{\alpha}{2(1+\alpha)} \wedge \delta \wedge \gamma)} \equiv K|t_1 - t_2|^{p\delta_0},
\end{aligned}$$

where

$$\delta_0 = \frac{\alpha}{2(1+\alpha)} \wedge \delta \wedge \gamma.$$

Hence, for any $p > \frac{1}{\delta_0} = \frac{2(1+\alpha)}{\alpha} \vee \frac{1}{\delta} \vee \frac{1}{\gamma}$, by Kolmogorov theorem (Chap. 2, Proposition 2.1.11), for any $\varepsilon \in (0, \delta_0 - \frac{1}{p})$, one has

$$\begin{aligned}
&\frac{1}{|t_1 - t_2|^{\delta_0 - \frac{1}{p} - \varepsilon}}\Big|\Big(\widehat{\psi}(t_1) + \int_{t_1}^T \widehat{g}(t_1, s, Z(t_1, s))ds\Big) \\
&\quad -\Big(\widehat{\psi}(t_2) + \int_{t_2}^T \widehat{g}(t_2, s, Z(t_2, s))ds\Big)\Big|^p \leqslant K\xi, \quad a.s.
\end{aligned}$$

for some integrable random variable ξ. This means that the process

$$t \mapsto \widehat{\psi}(t) + \int_t^T \widehat{g}(t,s,Z(t,s))ds$$

is Hölder continuous almost surely, which implies

$$\widehat{\psi}(\cdot) + \int_\cdot^T \widehat{g}(\cdot\,,s,Z(\cdot\,,s))ds \in L^p_{\mathcal{F}_T}(\Omega; C([0,T];\mathbb{R}^m)). \tag{4.3.37}$$

Making use of Lemma 4.3.2, we obtain the first relation in (4.3.28). Then, by the form of BSVIE (4.3.10), we obtain the second relation in (4.3.28). □

To conclude this section, we present the following result.

Proposition 4.3.4 *Let $(H4.3.4)_{\alpha,\delta}$ hold for some $\alpha > 0$ and $\delta \in (0,1]$. Let $(Y(\cdot),Z(\cdot\,,\cdot)) \in \mathcal{H}^p_{\Delta^*}[0,T]$ be an adapted solution of BSVIE* (4.3.10), *for some $p > 1$, with $\widehat{\psi}(\cdot)$ satisfying* (4.3.26), *and $Z(\cdot\,,\cdot) \in C([0,T];L^p_{\mathbb{F}}(\Omega;L^2(\cdot\,,T)))$. Then the following holds:*

$$\begin{aligned}&\Big|\Big[\mathbb{E}\Big(\int_{t_1}^T |Z(t_1,s)|^2ds\Big)^{\frac{p}{2}}\Big]^{\frac{1}{p}} - \Big[\mathbb{E}\Big(\int_{t_2}^T |Z(t_2,s)|^2ds\Big)^{\frac{p}{2}}\Big]^{\frac{1}{p}}\Big| \\ &\leqslant K\Big\{\mathbb{E}\Big|\int_{t_1}^T Z(t_1,s)dW(s) - \int_{t_2}^T Z(t_2,s)dW(s)\Big|^p\Big\}^{\frac{1}{p}}.\end{aligned} \tag{4.3.38}$$

Consequently, the map

$$t \mapsto \mathbb{E}\Big(\int_t^T |Z(t,s)|^2ds\Big)^{\frac{p}{2}}$$

is continuous.

Proof Let $0 \leqslant t_1 < t_2 \leqslant T$. Then by Corollary 3.1.7 of Chap. 3 and the martingale moment inequality (Theorem 2.3.2 of Chap. 2, (i)),

$$\begin{aligned}&\mathbb{E}\Big|\int_{t_2}^T Z(t_2,s)dW(s) - \int_{t_1}^T Z(t_1,s)dW(s)\Big|^p \\ &= \mathbb{E}\Big|\int_{t_2}^T [Z(t_2,s) - Z(t_1,s)]dW(s) - \int_{t_1}^{t_2} Z(t_1,s)dW(s)\Big|^p \\ &\geqslant K\mathbb{E}\Big[\Big(\int_{t_2}^T |Z(t_2,s) - Z(t_1,s)|^2ds\Big)^{\frac{p}{2}} + \Big|\int_{t_1}^{t_2} Z(t_1,s)dW(s)\Big|^p\Big] \\ &\geqslant K\Big[\mathbb{E}\Big(\int_{t_2}^T |Z(t_2,s) - Z(t_1,s)|^2ds\Big)^{\frac{p}{2}} + \Big(\int_{t_1}^{t_2} |Z(t_1,s)|^2ds\Big)^{\frac{p}{2}}\Big].\end{aligned}$$

By the triangle inequality, we have

$$
\begin{aligned}
&\Big[\mathbb{E}\Big(\int_{t_2}^{T}|Z(t_2,s)|^2ds\Big)^{\frac{p}{2}}\Big]^{\frac{1}{p}}-\Big[\mathbb{E}\Big(\int_{t_1}^{T}|Z(t_1,s)|^2ds\Big)^{\frac{p}{2}}\Big]^{\frac{1}{p}}\\
&\leqslant\Big[\mathbb{E}\Big(\int_{t_2}^{T}|Z(t_2,s)|^2ds\Big)^{\frac{p}{2}}\Big]^{\frac{1}{p}}-\Big[\mathbb{E}\Big(\int_{t_2}^{T}|Z(t_1,s)|^2ds\Big)^{\frac{p}{2}}\Big]^{\frac{1}{p}}\\
&\leqslant\Big[\mathbb{E}\Big(\int_{t_2}^{T}|Z(t_2,s)-Z(t_1,s)|^2ds\Big)^{\frac{p}{2}}\Big]^{\frac{1}{p}}\\
&\leqslant\Big[\mathbb{E}\Big(\int_{t_2}^{T}|Z(t_2,s)-Z(t_1,s)|^2ds\Big)^{\frac{p}{2}}\Big]^{\frac{1}{p}}+\Big[\mathbb{E}\Big(\int_{t_1}^{t_2}\mathbb{E}|Z(t_1,s)|^2ds\Big)^{\frac{p}{2}}\Big]^{\frac{1}{p}}.
\end{aligned}
$$

Also,

$$
\begin{aligned}
&\Big[\mathbb{E}\Big(\int_{t_1}^{T}|Z(t_1,s)|^2ds\Big)^{\frac{p}{2}}\Big]^{\frac{1}{p}}-\Big[\mathbb{E}\Big(\int_{t_2}^{T}|Z(t_2,s)|^2ds\Big)^{\frac{p}{2}}\Big]^{\frac{1}{p}}\\
&\leqslant\Big[\mathbb{E}\int_{t_1}^{t_2}|Z(t_1,s)|^2ds\Big)^{\frac{p}{2}}\Big]^{\frac{1}{p}}\\
&\qquad+\Big[\mathbb{E}\Big(\int_{t_2}^{T}|Z(t_1,s)|^2ds\Big)^{\frac{p}{2}}\Big]^{\frac{1}{p}}-\Big[\mathbb{E}\int_{t_2}^{T}|Z(t_2,s)|^2ds\Big)^{\frac{p}{2}}\Big]^{\frac{1}{p}}\\
&\leqslant\Big[\mathbb{E}\Big(\int_{t_1}^{t_2}|Z(t_1,s)|^2ds\Big)^{\frac{p}{2}}\Big]^{\frac{1}{p}}+\Big[\mathbb{E}\Big(\int_{t_2}^{T}|Z(t_2,s)-Z(t_1,s))|^2ds\Big)^{\frac{p}{2}}\Big]^{\frac{1}{p}}.
\end{aligned}
$$

As a result,

$$
\begin{aligned}
&\Big|\Big[\mathbb{E}\Big(\int_{t_1}^{T}|Z(t_1,s)|^2ds\Big)^{\frac{p}{2}}\Big]^{\frac{1}{p}}-\Big[\mathbb{E}\Big(\int_{t_2}^{T}|Z(t_2,s)|^2ds\Big)^{\frac{1}{2}}\Big]^{\frac{1}{p}}\Big|\\
&\leqslant\Big[\mathbb{E}\Big(\int_{t_2}^{T}|Z(t_2,s)-Z(t_1,s)|^2ds\Big)^{\frac{p}{2}}\Big]^{\frac{1}{p}}+\Big[\mathbb{E}\Big(\int_{t_1}^{t_2}|Z(t_1,s)|^2ds\Big)^{\frac{p}{2}}\Big]^{\frac{1}{p}}\\
&\leqslant K\Big(\mathbb{E}\Big|\int_{t_2}^{T}Z(t_2,s)dW(s)-\int_{t_1}^{T}Z(t_1,s)dW(s)\Big|^{p}\Big)^{\frac{1}{p}}.
\end{aligned}
$$

Then our conclusion follows from (4.3.28). □

We have seen again that Corollary 3.1.7 of Chap. 3 plays a crucial role here.

4 Linear Type I BSVIEs

For linear Type I BSVIEs, besides the well-posedness, one expects a little more: the variation of constants formula. The purpose of this section is to present that. We will use some Malliavin calculus. Therefore, those who do not know these preliminary knowledge could skip this section, or read the Appendices first.

We consider the following linear Type I BSVIE:

$$\begin{aligned} Y(t) = \psi(t) &+ \int_t^T \Big(A(t,s)Y(s) + \sum_{i=1}^d C_i(t,s)Z_i(t,s)\Big)ds \\ &- \int_t^T \sum_{i=1}^d Z_i(t,s)dW_i(s), \quad t \in [0,T]. \end{aligned} \tag{4.4.1}$$

Note that each $Z_i(t,s)$ is valued in $\mathbb{R}^m$. In the following, we assume that the coefficients $A(\cdot\,,\cdot)$ and $C_i(\cdot\,,\cdot)$ are bounded, and the free term $\psi(\cdot) \in L^2_{\mathcal{F}_T}(0,T;\mathbb{R}^n)$. Therefore, BSVIE (4.4.1) admits a unique adapted solution $(Y(\cdot), Z(\cdot\,,\cdot))$. From the theory of ODEs and SDEs, as well as FSVIEs and BSDEs, for linear equations, one expects to have the variation of constants formula for the adapted solution $(Y(\cdot), Z(\cdot\,,\cdot))$ in terms of the free term $\psi(\cdot)$. Such a result is stated as follows.

Theorem 4.4.1 *Suppose $A(\cdot\,,\cdot)$ and $C_i(\cdot\,,\cdot)$ are bounded, $\psi(\cdot) \in L^2_{\mathcal{F}_T}(0,T;\mathbb{R}^n)$. Let $(Y(\cdot), Z(\cdot\,,\cdot))$ be the adapted solution to the linear BSVIE* (4.4.1).

(i) Let $M(\cdot\,,\cdot)$ be the solution to the following linear SDE:

$$\begin{cases} dM(t,s) = \sum_{i=1}^d M(t,s)C_i(t,s)dW_i(s), & s \in [t,T], \\ M(t,t) = I_m, \end{cases} \tag{4.4.2}$$

and $\Gamma(\cdot,\cdot)$ be the solution to the following system:

$$\Gamma(t,\tau) = \Phi_1(t,\tau) + \int_t^\tau \Phi_1(t,s)\Gamma(s,\tau)ds. \tag{4.4.3}$$

Then $Y(\cdot)$ admits the following representation:

$$Y(t) = \mathbb{E}_t\Big[M(t,T)\psi(t) + \int_t^T \Gamma(t,s)M(s,T)\psi(s)ds\Big]. \tag{4.4.4}$$

(ii) Suppose the coefficients $A(\cdot\,,\cdot)$ and $C_i(\cdot\,,\cdot)$ as well as the free term $\psi(\cdot)$ have bounded first-order Malliavin derivatives such that the following defined $U(\cdot)$ satisfies $D_\cdot U(t) \in L^2_{\mathcal{F}_t}(0,t;\mathbb{R}^m)$:

$$\begin{aligned} U(t) = \mathbb{E}_t\Big[M(t,T)\psi(t) &+ \int_t^T \Gamma(t,s)M(s,T)\psi(s)ds\Big] - \psi(t) \\ &- \int_t^T A(t,s)\mathbb{E}_s\Big[M(s,T)\psi(s) + \int_s^T \Gamma(s,r)M(r,T)\psi(r)dr\Big]ds, \\ &\qquad t \in [0,T]. \end{aligned} \tag{4.4.5}$$

For $0 \leqslant t \leqslant r \leqslant s \leqslant T$, $1 \leqslant i \leqslant d$, *denote*

$$\mathbf{Z}(t,r) = \begin{pmatrix} Z_1(t,r) \\ Z_2(t,r) \\ \vdots \\ Z_d(t,r) \end{pmatrix}_{md\times 1}, \qquad \widetilde{\mathbf{U}}(t,r) = \begin{pmatrix} D_r^{(1)}U(t) \\ D_r^{(2)}U(t) \\ \vdots \\ D_r^{(d)}U(t) \end{pmatrix}_{md\times 1},$$

$$\mathbf{C}_i(t,s) = \begin{pmatrix} C_i(t,s) & 0 & \cdots & 0 \\ 0 & C_i(t,s) & \cdots & 0 \\ \vdots & \vdots & \ddots & \vdots \\ 0 & 0 & \cdots & C_i(t,s) \end{pmatrix}_{md\times md},$$

$$\widetilde{\mathbf{C}}(t,r,s) = \begin{pmatrix} D_r^{(1)}C_1(t,s) & D_r^{(1)}C_2(t,s) & \cdots & D_r^{(1)}C_d(t,s) \\ D_r^{(2)}C_1(t,s) & D_r^{(2)}C_2(t,s) & \cdots & D_r^{(2)}C_d(t,s) \\ \vdots & \vdots & \ddots & \vdots \\ D_r^{(d)}C_1(t,s) & D_r^{(d)}C_2(t,s) & \cdots & D_r^{(d)}C_d(t,s) \end{pmatrix}_{md\times md}.$$

Let $\mathbf{M}(t,r,\cdot)$ *be the* $\mathbb{R}^{md\times md}$*-valued solution to the following SDE parameterized by* (t,r)*:*

$$\begin{cases} d\mathbf{M}(t,r,s) = \displaystyle\sum_{i=1}^{d} \mathbf{M}(t,r,s)\mathbf{C}_i(t,s)dW_i(s), & s \in [r,T], \\ \mathbf{M}(t,r,r) = I_{md}. \end{cases} \tag{4.4.6}$$

Define the sequence

$$\begin{aligned} &\mathbf{\Phi}_1(t,r,s) = \mathbf{M}(t,r,s)\widetilde{\mathbf{C}}(t,r,s), \\ &\mathbf{\Phi}_{k+1}(t,r,s) = \int_t^s \mathbf{\Phi}_1(t,r,\theta)\mathbf{\Phi}_k(t,\theta,s)d\theta, \quad k \geqslant 2. \end{aligned}$$

Then the following function defined by the series is well-defined:

$$\mathbf{\Gamma}(t,r,s) = \sum_{k=1}^{\infty} \mathbf{\Phi}_k(t,r,s), \qquad s \in [r,T],$$

and $\mathbf{Z}(\cdot\,,\cdot)$ *admits the following representation:*

$$\mathbf{Z}(t,r) = \mathbb{E}_r\Big[\mathbf{M}(t,r,T)\widetilde{\mathbf{U}}(t,r) + \int_r^T \mathbf{\Gamma}(t,r,s)\mathbf{M}(t,s,T)\widetilde{\mathbf{U}}(t,s)ds\Big], \quad 0 \leqslant t \leqslant r \leqslant T. \tag{4.4.7}$$

Proof

(i) *Step 1*. In this step, we discuss the solvability of (4.4.2) and (4.4.3).

In the similar spirit of Proposition 2.3.11 in Chap. 2, for bounded coefficient $C(\cdot,\cdot)$, we see that (4.4.2) admits a unique solution $M(t,\cdot)$. To study the case of (4.4.3), let us first define the sequence

$$\begin{aligned}&\Phi_1(t,s)=M(t,s)A(t,s),\\&\Phi_{k+1}(t,s)=\int_t^s\Phi_1(t,r)\Phi_k(r,s)dr,\quad k\geqslant 1.\end{aligned}\tag{4.4.8}$$

We claim that the following function is well-defined and a solution of (4.4.3) such that for any $0\leqslant t\leqslant s\leqslant T$, $\mathbb{E}_t\Big[\big|\Gamma(t,s)\big|^2\Big]\leqslant K$ for some constant K,

$$\Gamma(t,s)=\sum_{k=1}^{\infty}\Phi_k(t,s),\qquad s\in[t,T].\tag{4.4.9}$$

In fact, by the boundedness of $C_i(\cdot,\cdot)$, for any $p\geqslant 2$, we first obtain that there exists a constant K_0 such that

$$\operatorname*{esssup}_{\omega\in\Omega}\ \sup_{0\leqslant t\leqslant\tau\leqslant T}\ E_t\big[|\Phi_1(t,\tau)|^p\big]\leqslant K_0<\infty.$$

This then implies that

$$\begin{aligned}&\operatorname*{esssup}_{\omega\in\Omega}\ \sup_{0\leqslant t\leqslant\tau\leqslant T}\ \mathbb{E}_t\Big[|\Phi_{k+1}(t,\tau)|^2\Big]\\&\leqslant\operatorname*{esssup}_{\omega\in\Omega}\ \sup_{0\leqslant t\leqslant\tau\leqslant T}\ \mathbb{E}_t\Big|\int_t^\tau\big|\Phi_1(t,s)\Phi_k(s,\tau)\big|ds\Big|^2\\&\leqslant(\tau-t)\operatorname*{esssup}_{\omega\in\Omega}\ \sup_{0\leqslant t\leqslant\tau\leqslant T}\int_t^\tau\mathbb{E}_t\Big[\big|\Phi_1(t,s)\big|^2\big|\Phi_k(s,\tau)\big|^2\Big]ds\\&=(\tau-t)\operatorname*{esssup}_{\omega\in\Omega}\ \sup_{0\leqslant t\leqslant\tau\leqslant T}\int_t^\tau\mathbb{E}_t\Big[\big|\Phi_1(t,s)\big|^2\Big\{\mathbb{E}_s\big|\Phi_k(s,\tau)\big|^2\Big\}\Big]ds.\end{aligned}\tag{4.4.10}$$

By induction, we can show from the above that

$$\mathbb{E}_t\Big[|\Phi_{k+1}(t,\tau)|^2\Big]\leqslant\frac{K_0^{k+1}(\tau-t)^{2k}}{(2k-1)!!},\qquad s\in[t,T].$$

Actually, for $k = 1$, we have

$$\begin{aligned}\mathbb{E}_t\Big[|\Phi_2(t,\tau)|^2\Big] &\leqslant \mathbb{E}_t\Big|\int_t^\tau |\Phi_1(t,s)\Phi_1(s,\tau)|ds\Big|^2\\ &\leqslant (\tau-t)\int_t^\tau \mathbb{E}_t\Big[\big|\Phi_1(t,s)\big|^2\Big\{\mathbb{E}_s\big|\Phi_1(s,\tau)\big|^2\Big\}\Big]ds\\ &\leqslant K_0(\tau-t)\int_t^\tau \mathbb{E}_t\Big[\big|\Phi_1(t,s)\big|^2\Big]ds \leqslant K_0^2(\tau-t)^2.\end{aligned}$$

Suppose the above holds for k. Then by (4.4.10),

$$\begin{aligned}&\mathbb{E}_t\Big[|\Phi_{k+2}(t,\tau)|^2\Big]\\ &\leqslant (\tau-t)\operatorname*{esssup}_{\omega\in\Omega}\sup_{0\leqslant t\leqslant\tau\leqslant T}\int_t^\tau \mathbb{E}_t\Big[\big|\Phi_1(t,s)\big|^2\Big\{\mathbb{E}_s\big|\Phi_{k+1}(s,\tau)\big|^2\Big\}\Big]ds\\ &\leqslant (\tau-t)\frac{K_0^{k+1}}{(2k-1)!!}\operatorname*{esssup}_{\omega\in\Omega}\sup_{0\leqslant t\leqslant\tau\leqslant T}\int_t^\tau \mathbb{E}_t\Big[\big|\Phi_1(t,s)\big|^2(\tau-s)^{2k}\Big]ds\\ &\leqslant \frac{K_0^{k+2}(\tau-t)^{2k+2}}{(2k+1)!!}.\end{aligned}$$

Notice that

$$\begin{aligned}&(2k+1)!! = 1\times 3\times 5\times(2k+1) > (k+1)!,\\ &K_0^{k+2}(\tau-t)^{2k+2} \leqslant T^2K_0^2\big[K_0T^2\big]^k.\end{aligned}$$

Hence,

$$\sum_{k\geqslant 1}\frac{K_0^{k+2}(\tau-t)^{2k+2}}{(2k+1)!!} \leqslant \mathscr{K} \equiv \sum_{k\geqslant 1} T^2K_0^2\frac{(K_0T^2)^{k+1}}{(k+1)!} < \infty.$$

For any square-integrable random variable ξ, we have

$$\begin{aligned}&\mathbb{E}_t\big|\Gamma(t,s)\xi\big| = \mathbb{E}_t\Big|\sum_{k=1}^\infty \Phi_k(t,s)\xi\Big| \leqslant \mathbb{E}_t\sum_{k=1}^\infty\Big|\Phi_k(t,s)\xi\Big| = \lim_{N\to\infty}\sum_{k=1}^N \mathbb{E}_t\Big|\Phi_k(t,s)\xi\Big|\\ &\leqslant \lim_{N\to\infty}\sum_{k=1}^N\Big[\mathbb{E}_t|\Phi_k(t,s)|^2\Big]^{\frac12}\big[\mathbb{E}_t|\xi|^2\big]^{\frac12} \leqslant \mathscr{K}\big[\mathbb{E}_t|\xi|^2\big]^{\frac12}.\end{aligned}$$

For any given $l\in\mathbb{N}$, we choose $\xi = \big|\Gamma(t,s)\big|\mathbf{1}_{|\Gamma(t,s)|\leqslant l}$ which is obviously square-integrable. Then plugging it into the above, we have

$$\mathbb{E}_t\big|\Gamma(t,s)\xi\big| \equiv \mathbb{E}_t\Big[\big|\Gamma(t,s)\big|^2\mathbf{1}_{|\Gamma(t,s)|\leqslant l}\Big] \leqslant \mathscr{K}\Big\{\mathbb{E}_t\Big[\big|\Gamma(t,s)\big|^2\mathbf{1}_{|\Gamma(t,s)|\leqslant l}\Big]\Big\}^{\frac12},$$

from which we see that

$$\mathbb{E}_t\Big[\big|\Gamma(t,s)\big|^2\mathbf{1}_{|\Gamma(t,s)|\leqslant l}\Big]\leqslant\mathscr{K}^2.$$

By monotone convergence theorem, we have

$$\mathbb{E}_t\Big[\big|\Gamma(t,s)\big|^2\Big]=\lim_{l\to\infty}\mathbb{E}_t\Big[\big|\Gamma(t,s)\big|^2\mathbf{1}_{|\Gamma(t,s)|\leqslant l}\Big]\leqslant\mathscr{K}^2.$$

Taking expectation on both sides, we have $\mathbb{E}\Big[\big|\Gamma(t,s)\big|^2\Big]\leqslant\mathscr{K}^2$ which implies that for given (t,s) and almost all $\omega\in\Omega$, $\Gamma(t,s)$ is well-defined.

Next, let us show that the above Γ is the solution of (4.4.3). Notice that

$$\begin{aligned}&\Big|\sum_{k=1}^N\Phi_1(t,r)\Phi_k(r,\tau)\Big|\leqslant\sum_{k=1}^N\big|\Phi_1(t,r)\Phi_k(r,\tau)\big|\\&=\big|\Phi_1(t,r)\big|\sum_{k=1}^N\big|\Phi_k(r,\tau)\big|\leqslant\big|\Phi_1(t,r)\big|\sum_{k=1}^\infty\big|\Phi_k(r,\tau)\big|\equiv\eta(t,r,\tau),\end{aligned}$$

and

$$\begin{aligned}&\mathbb{E}_t\int_t^\tau\eta(t,r,\tau)dr=\sum_{k=1}^\infty\mathbb{E}_t\int_t^\tau\big|\Phi_1(t,r)\big|\big|\Phi_k(r,\tau)\big|dr\\&\leqslant\sum_{k=1}^\infty\int_t^\tau\Big[\mathbb{E}_t\big|\Phi_1(t,r)\big|^2\Big]^{\frac12}\Big[\mathbb{E}_t\big|\Phi_k(r,\tau)\big|^2\Big]^{\frac12}dr\\&=K_0^{\frac12}\sum_{k=1}^\infty\int_t^\tau\frac{K_0^{\frac{k}{2}}(\tau-r)^{(k-1)}}{\sqrt{(2k-3)!!}}dr=K_0^{\frac12}\sum_{k=1}^\infty\frac{K_0^{\frac{k}{2}}(\tau-t)^k}{k\sqrt{(2k-3)!!}}\equiv\mathscr{K}_1,\end{aligned}$$

for some constant $\mathscr{K}_1>0$. Therefore, it follows from convergence theorem that

$$\begin{aligned}&\sum_{k=2}^\infty\Phi_k(t,\tau)=\sum_{k=1}^\infty\Phi_{k+1}(t,\tau)=\sum_{k=1}^\infty\int_t^\tau\Phi_1(t,r)\Phi_k(r,\tau)dr\\&=\int_t^\tau\Phi_1(t,r)\sum_{k=1}^\infty\Phi_k(r,\tau)dr=\int_t^\tau\Phi_1(t,r)\Gamma(r,\tau)dr,\end{aligned}$$

from which we finish the existence proof.

Step 2. In this step, we prove the above representation (4.4.4) under given conditions. To this end, we need some preparations.

Given $\Gamma(\cdot,\cdot)$ in the above, we define

$$\bar Y(t)=\mathbb{E}_t\Big[M(t,T)\psi(t)+\int_t^T\Gamma(t,s)M(s,T)\psi(s)ds\Big].\tag{4.4.11}$$

For any given $p \in (1,2)$, we first have

$$\begin{aligned}
&\mathbb{E}\int_0^T \Big|\mathbb{E}_t\big[M(t,T)\psi(t)\big]\Big|^p dt\\
&\leqslant \mathbb{E}\int_0^T \big|M(t,T)\psi(t)\big|^p dt\\
&\leqslant \Big[\mathbb{E}\int_0^T |M(t,T)|^{\frac{2p}{2-p}}dt\Big]^{\frac{2-p}{2}}\Big[\mathbb{E}\int_0^T |\psi(t)|^2 dt\Big]^{\frac{p}{2}} < \infty.
\end{aligned}$$

Next,

$$\begin{aligned}
&\mathbb{E}\int_0^T \Big|\mathbb{E}_t\Big[\int_t^T \Gamma(t,s)M(s,T)\psi(s)ds\Big]\Big|^2 dt\\
&\leqslant \mathbb{E}\int_0^T \Big|\mathbb{E}_t\int_t^T \big|\Gamma(t,s)M(s,T)\psi(s)\big|ds\Big|^2 dt\\
&= \mathbb{E}\int_0^T \Big|\mathbb{E}_t\int_t^T \big[\mathbb{E}_s\big|\Gamma(t,s)M(s,T)\psi(s)\big]\big|ds\Big|^2 dt\\
&\leqslant \mathbb{E}\int_0^T \Big|\mathbb{E}_t\int_t^T \big[\mathbb{E}_s\big|\Gamma(t,s)M(s,T)\big|^2\big]^{\frac{1}{2}}\big[\mathbb{E}_s\big|\psi(s)\big|^2\big]^{\frac{1}{2}}\big|ds\Big|^2 dt\\
&\leqslant \mathbb{E}\int_0^T \Big[\mathbb{E}_t\int_t^T \mathbb{E}_s\big|\Gamma(t,s)M(s,T)\big|^2 ds\Big]\Big[\mathbb{E}_t\int_t^T \mathbb{E}_s\big|\psi(s)\big|^2 ds\Big]dt\\
&= \mathbb{E}\int_0^T \Big[\mathbb{E}_t\int_t^T \big[|\Gamma(t,s)|^2\mathbb{E}_s\big|M(s,T)\big|^2\big]ds\Big]\Big[\mathbb{E}_t\int_t^T \big|\psi(s)\big|^2 ds\Big]dt\\
&\leqslant K\mathbb{E}\int_0^T \big|\psi(s)\big|^2 ds.
\end{aligned}$$

Therefore, we see that $\bar{Y}(\cdot) \in L^p(0,T;\mathbb{R}^n)$ with any $p \in (1,2)$.

Next, we show that $\bar{Y}(\cdot)$, together with another proper process, constructs an adapted solution to BSVIE (4.4.1). To this end, by Fubini theorem and (4.4.3), we first obtain the following:

$$\begin{aligned}
&\mathbb{E}_t\Big\{M(t,T)\psi(t) + \int_t^T M(t,s)A(t,s)\bar{Y}(s)ds\Big\}\\
&= \mathbb{E}_t\Big\{M(t,T)\psi(t) + \int_t^T \Phi_1(t,s)M(s,T)\psi(s)ds\\
&\qquad\qquad + \int_t^T \Phi_1(t,s)\int_s^T \Gamma(s,\tau)M(\tau,T)\psi(\tau)d\tau ds\Big\}\\
&= \mathbb{E}_t\Big\{M(t,T)\psi(t) + \int_t^T \Phi_1(t,s)M(s,T)\psi(s)ds\\
&\qquad\qquad + \int_t^T\int_t^\tau \Phi_1(t,s)\Gamma(s,\tau)M(\tau,T)\psi(\tau)dsd\tau\Big\}\\
&= \mathbb{E}_t\Big[M(t,T)\psi(t) + \int_t^T \Gamma(t,s)M(s,T)\psi(s)ds\Big] = \bar{Y}(t).
\end{aligned} \tag{4.4.12}$$

On the other hand, let $(\bar{\eta}(t,\cdot),\bar{\zeta}(t,\cdot))$ be the adapted solution to the following BSDE:

$$\begin{aligned}\bar{\eta}(t,\tau)=\psi(t)+\int_\tau^T\Big(A(t,s)\bar{Y}(s)+\sum_{i=1}^d C_i(t,s)\bar{\zeta}_i(t,s)\Big)ds\\ -\int_\tau^T\sum_{i=1}^d\bar{\zeta}_i(t,s)dW_i(s),\qquad \tau\in[t,T].\end{aligned}$$

The relevant solvability can be obtained from Theorem 3.1.4 in Chap. 3. Now, applying Itô's formula to $\tau\mapsto M(t,\tau)\bar{\eta}(t,\tau)$, we have

$$\begin{aligned}d\big[M(t,\tau)\bar{\eta}(t,\tau)\big]&=-M(t,\tau)A(t,\tau)Y(\tau)d\tau\\&\quad+\sum_{i=1}^d M(t,\tau)\Big(C_i(t,\tau)\bar{\eta}(t,\tau)+\bar{\zeta}_i(t,\tau)\Big)dW_i(\tau).\end{aligned}$$

Hence,

$$\begin{aligned}M(t,\tau)\bar{\eta}(t,\tau)&=M(t,T)\psi(t)+\int_\tau^T M(t,\tau)A(t,\tau)\bar{Y}(\tau)d\tau\\&\quad-\sum_{i=1}^d\int_\tau^T M(t,\tau)\Big(C_i(t,\tau)\bar{\eta}(t,\tau)+\bar{\zeta}_i(t,\tau)\Big)dW_i(\tau),\\&\quad\tau\in[t,T].\end{aligned}$$

Taking conditional expectation and letting $\tau=t$, we obtain

$$\bar{\eta}(t,t)=\mathbb{E}_t\big[M(t,T)\psi(t)\big]+\int_t^T\mathbb{E}_t\big[M(t,s)A(t,s)\bar{Y}(s)\big]ds. \tag{4.4.13}$$

Considering the above obtained result (4.4.12), we see that $\bar{\eta}(t,t)=\bar{Y}(t)$, which then implies that $(\bar{Y}(\cdot),\bar{\zeta}(\cdot,\cdot))$ is an adapted solution to BSVIE (4.4.1). By uniqueness, we have $Y(t)=\bar{Y}(t)$ which implies representation (4.4.4).

Step 3. In this step, let us show the uniqueness of (4.4.3). Suppose (4.4.3) admits another $\widetilde{\Gamma}(\cdot,\cdot)$ such that there exists some constant K satisfying $\mathbb{E}_t\Big[\big|\widetilde{\Gamma}(t,s)\big|^2\Big]\leqslant K$ with any $t\leqslant s$. By the above arguments in Step 1 and Step 2, for any $\psi(\cdot)\in L^2_{\mathcal{F}_T}(0,T;\mathbb{R}^n)$, the solution $Y(\cdot)$ to BSVIE (4.4.1) can be represented as

$$Y(t)=\mathbb{E}_t\Big[M(t,T)\psi(t)+\int_t^T\widetilde{\Gamma}(t,s)M(s,T)\psi(s)ds\Big].$$

Hence, by defining $\widehat{\Gamma}\equiv\Gamma-\widetilde{\Gamma}$, we have

$$\mathbb{E}_t\Big[\int_t^T\widehat{\Gamma}(t,s)M(s,T)\psi(s)ds\Big]=0,\quad\forall\psi(\cdot)\in L^2_{\mathcal{F}_T}(0,T;\mathbb{R}^n).$$

Given $t \in (0, T)$, we see that

$$\mathbb{E}\int_t^T \big|\widehat{\Gamma}(t,s)M(s,T)\big|^2 ds = \mathbb{E}\int_t^T \Big[\big|\widehat{\Gamma}(t,s)\big|^2 \mathbb{E}_s\big|M(s,T)\big|^2\Big] ds < \infty.$$

Then by the arbitrariness of $\psi(\cdot)$, we have

$$\mathbb{E}_t\Big[\int_t^T \big|\widehat{\Gamma}(t,s)M(s,T)\big|^2 ds\Big] = 0, \quad t \in [0,T].$$

By the existence of $M(\cdot,T)^{-1}$, we obtain the uniqueness of Γ. This finishes the proof of (i).

(ii) Note that the process $U(\cdot)$ defined by (4.4.5) is actually the following:

$$U(t) = Y(t) - \psi(t) - \int_t^T A(t,s)Y(s)ds, \qquad t \in [0,T].$$

Therefore, by BSVIE (4.4.1), we have

$$U(t) = \int_t^T \sum_{i=1}^d C_i(t,s)Z_i(t,s)ds - \int_t^T \sum_{i=1}^d Z_i(t,s)dW_i(s).$$

It is known that

$$D_\tau^{(j)} \int_t^T Z_i(t,s)dW_i(s) = \int_\tau^T D_\tau^{(j)} Z_i(t,s)dW_i(s) + Z_i(t,\tau)\delta_{ij}.$$

Hence,

$$\begin{aligned} D_\tau^{(j)}U(t) = &\int_t^T \sum_{i=1}^d \Big(\big[D_\tau^{(j)}C_i(t,s)\big]Z_i(t,s) + C_i(t,s)\big[D_\tau^{(j)}Z_i(t,s)\big]\Big)ds \\ &+ \int_\tau^T \sum_{i=1}^d D_\tau^{(j)} Z_i(t,s)dW_i(s) + Z_j(t,\tau). \end{aligned}$$

This leads to

$$\begin{aligned} Z_j(t,\tau) = D_\tau^{(j)}U(t) - &\int_t^T \sum_{i=1}^d \Big(\big[D_\tau^{(j)}C_i(t,s)\big]Z_i(t,s) + C_i(t,s)\big[D_\tau^{(j)}Z_i(t,s)\big]\Big)ds \\ - &\int_\tau^T \sum_{i=1}^d D_\tau^{(j)} Z_i(t,s)dW_i(s). \end{aligned}$$

Then, with the notations $\mathbf{Z}(t,\tau)$, $\widetilde{\mathbf{U}}(t,\tau)$, $\mathbf{C}_i(t,s)$, $\widetilde{\mathbf{C}}(t,\tau,s)$, as well as

$$\widetilde{\mathbf{Z}}_i(t,\tau,s)=\begin{pmatrix}D_\tau^{(1)}Z_i(t,s)\\ D_\tau^{(2)}Z_i(t,s)\\ \vdots\\ D_\tau^{(d)}Z_i(t,s)\end{pmatrix},\qquad 0\leqslant t\leqslant\tau\leqslant s\leqslant T,\ 1\leqslant i\leqslant d,$$

we have

$$\begin{aligned}\mathbf{Z}(t,\tau)=\widetilde{\mathbf{U}}(t,\tau)&-\int_\tau^T\Big(\widetilde{\mathbf{C}}(t,\tau,s)\mathbf{Z}(t,s)+\sum_{i=1}^d\mathbf{C}_i(t,s)\widetilde{\mathbf{Z}}_i(t,\tau,s)\Big)ds\\&-\int_\tau^T\sum_{i=1}^d\widetilde{\mathbf{Z}}_i(t,\tau,s)dW_i(s),\qquad \tau\in[t,T].\end{aligned}$$

For given $t\in[0,T)$, the above is a BSVIE for $(\mathbf{Z}(t,\cdot),\widetilde{\mathbf{Z}}(t,\cdot,\cdot))$ on $[t,T]$. Hence, similar to (i), we can obtain the conclusion of (ii). □

To prove the result for $\mathbf{Z}(\cdot,\cdot)$, we need some requirements involving Malliavin calculus, such that the above $D.U(t)\in L^2_{\mathcal{F}_t}(0,t;\mathbb{R}^m)$. We will give more relevant details in Chap. 5.

5 Bibliographical Remarks

For the case that $\psi(\cdot)=\xi$, a given $\mathcal{F}_T$-measurable random variable, Type I BSVIEs were first studied by Lin in 2002 [135], under Lipschitz condition of the generator $g(t,s,y,z)$ in (y,z). Such a result was extended by Aman–N'Zi [9] when the global Lipschitz continuity requirement weakened to a local version. However, it was pointed by Wang–Zhang [211] (p. 480) that the proofs in both [135] and [9] had a gap. With some proper arguments, including the application of Itô's formula, Wang–Zhang rigorously established the existence and uniqueness of the adapted solution to Type I BSVIEs [211]. We point out that all the aforementioned papers assumed that $\psi(\cdot)=\xi$ and there was no singularity in the sense of (4.2.1). To this end, Shi–Wang [175] overcame these limitations by introducing some (equivalent) weighted norm for the solution and bypassed the tool of Itô's formula in establishing the well-posedness. In Sect. 2, we have presented a refinement of [175, 211]. In contrast with BSDEs, BSVIEs do not have the flow property, which makes Itô's formula hard to apply directly. However, it turns out that BSVIEs are closely related to a family of BSDEs, and the well-posedness of the former can be obtained by that of the latter, combined with a contraction mapping theorem. In establishing the relevant map to be a contraction, we have derived concise weighted norm estimates for the relevant processes. This is mainly a proper improvement/modification of those in [175, 211]. Further more results on Type I BSVIEs can be found in, e.g., [4, 52, 59, 60, 100, 191, 195, 197, 201, 213].

One expects the existing adapted solution $(Y(\cdot), Z(\cdot\,,\cdot))$ to a Type I BSVIE has certain continuity, under proper conditions. It is not hard to see that one needs only to look at the case that the generator is independent of $Y(\cdot)$. Section 4 follows such an idea to establish the continuity of the solution. This is in some way inspired by [210, 211]. Further regularities can be discussed beyond the continuity. We will present that in the next chapter.

It is well-accepted that linear evolution equations should have a variation of constants formula. We presented relevant results for SDEs, BSDEs, and FSVIEs in Chap. 3. For one-dimensional linear BSVIE with deterministic coefficients, Hu–Oksendal [103] obtained a variation of constants formula, and it was extended to the case of random coefficient case by Wang–Yong–Zhang [194]. The result presented in Sect. 5 is a full extension of the abovementioned works to multidimensional cases. In addition, the representation of correction term Z (see (ii) in Theorem 10.3.2) was first given by means of Malliavin calculus. Eventually we mention that Hamaguchi [92] also studied the variation of constants formula for BSVIEs with a different approach.

Chapter 5
Type II BSVIEs

In this chapter, we will discuss Type II BSVIEs which is referred to the following type BSVIE:

$$Y(t) = \psi(t) + \int_t^T g(t, s, Y(s), Z(t, s), Z(s, t))ds - \int_t^T Z(t, s)dW(s), \qquad t \in [0, T]. \tag{5.1}$$

Note that in (5.1), both $Z(t, s)$ and $Z(s, t)$ appear.

1 Adapted M-Solution

Similar to the previous chapter, without loss of generality, we may (and hereafter) assume

$$g(t, s, 0, 0, 0) = 0, \qquad (t, s) \in \Delta^*[0, T]. \tag{5.1.1}$$

It is natural to call a pair $(Y(\cdot), Z(\cdot\,, \cdot))$ an adapted solution to the above Type II BSVIE if for each $t \in [0, T]$, $s \mapsto (Y(s), Z(t, s))$ is $\mathbb{F}$-adapted and (5.1) is satisfied in the usual Itô's sense. It is seen that different from Type I BSVIE, if $(Y(\cdot), Z(\cdot\,, \cdot))$ is an adapted solution to the above Type II BSVIE, then $Z(t, s)$ has to be defined for $(t, s) \in [0, T]^2$ which is a square domain, instead of an upper triangle domain $\Delta^*[0, T]$. Because of this, a new issue arises. To illustrate this, let us look at the following example.

T. Wang, J. Yong, *Backward Stochastic Volterra Integral Equations*,
Probability Theory and Stochastic Modelling 109,
https://doi.org/10.1007/978-3-032-20848-4_5

Example 5.1.1 Let $m = d = 1$. Consider the following Type II BSVIE:

$$Y(t) = \int_t^T Z(s,t)ds - \int_t^T Z(t,s)dW(s), \qquad t \in [0,T]. \tag{5.1.2}$$

We claim that

$$\begin{cases} Y(t) = (T-t)\zeta(t), & t \in [0,T], \\ Z(t,s) = \mathbf{1}_{[0,t]}(s)\zeta(s), & (t,s) \in [0,T]^2, \end{cases} \tag{5.1.3}$$

is an adapted solution of Type II BSVIE (5.1.2) for any $\zeta(\cdot) \in L^2_{\mathbb{F}}(0,T;\mathbb{R})$. In fact, with the above definition (5.1.3),

$$\begin{aligned} &\int_t^T Z(s,t)ds - \int_t^T Z(t,s)dW(s) \\ &= \int_t^T \mathbf{1}_{[0,s]}(t)\zeta(t)ds - \int_t^T \mathbf{1}_{[0,t]}(s)\zeta(s)dW(s) \\ &= \int_t^T \mathbf{1}_{[t,T]}(s)\zeta(t)ds - 0 = (T-t)\zeta(t) \equiv Y(t), \qquad t \in [0,T]. \end{aligned}$$

Hence, the adapted solution defined in an obvious or a "natural" way is not unique since different $\zeta(\cdot) \in L^2_{\mathbb{F}}(0,T;\mathbb{R})$ will give different adapted solutions. To avoid this, we need a new notion of solutions.

Similar to $\Delta^*[0,T]$ and $\Delta_*[0,T]$, for any $0 \leqslant S_1 < S_2 \leqslant T$, we define

$$\begin{cases} \Delta_*[S_1,S_2] = \{(t,s) \in [S_1,S_2]^2 \mid S_1 \leqslant s < t \leqslant S_2\}, \\ \Delta^*[S_1,S_2] = \{(t,s) \in [S_1,S_2]^2 \mid S_1 \leqslant t < s \leqslant S_2\}. \end{cases} \tag{5.1.4}$$

See the picture below for an illustration.

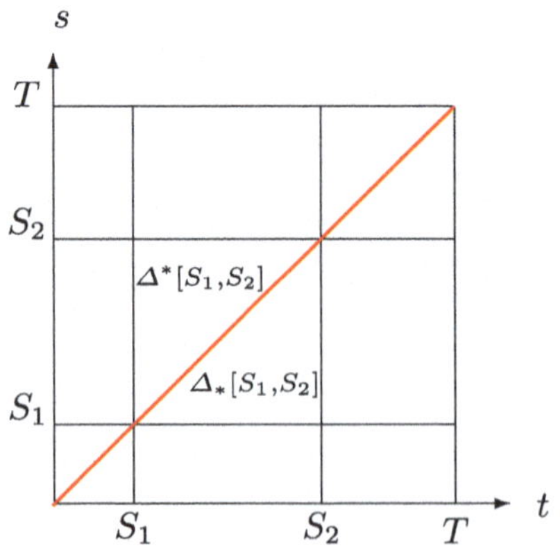

Next, we let

$$\mathcal{H}^p[S_1, S_2] = L^p_{\mathbb{F}}(S_1, S_2; \mathbb{R}^m) \times L^p\big(S_1, S_2; L^p_{\mathbb{F}}\big(\Omega; L^2(S_1, S_2; \mathbb{R}^{m\times d})\big)\big),$$
$$p \in (1, \infty),$$

with the naturally induced norm (for $p \in (1, \infty)$):

$$\|(Y(\cdot), Z(\cdot\,,\cdot)\|_{\mathcal{H}^p[S_1,S_2]} = \Big[\mathbb{E}\int_{S_1}^{S_2} |Y(t)|^p dt + \int_{S_1}^{S_2} \mathbb{E}\Big(\int_{S_1}^{S_2} |Z(t,s)|^2 ds\Big)^{\frac{p}{2}} dt\Big]^{\frac{1}{p}},$$
$$\forall (Y(\cdot), Z(\cdot\,,\cdot)) \in \mathcal{H}^p[S_1, S_2].$$

Note that for $Z(\cdot\,,\cdot) \in L^p(S_1, S_2; L^2_{\mathbb{F}}(S_1, S_2; \mathbb{R}^m)))$, the map $s \mapsto Z(t,s)$ is $\mathbb{F}$-adapted on $[S_1, S_2]$.

Let us introduce the following definition.

Definition 5.1.2 Let $S \in [0, T)$. A pair $(Y(\cdot), Z(\cdot\,,\cdot)) \in \mathcal{H}^p[S, T]$ is called an *adapted M-solution* of Type II BSVIE (5.1) on $[S, T]$ if (5.1) holds in the usual Itô's sense for almost all $t \in [S, T]$ and, in addition, the following holds:

$$Y(t) = E_S[Y(t)] + \int_S^t Z(t,s)dW(s), \qquad \text{a.e. } t \in [S, T]. \tag{5.1.5}$$

In the above, the letter "M" in "M-solution" stands for "a martingale representation" (for $Y(t)$ to determine $Z(\cdot\,,\cdot)$ on $\Delta_*[S, T]$). We will see that additional constraint (5.1.5) guarantees the uniqueness of adapted M-solutions.

It is natural to require that if $(Y(\cdot), Z(\cdot\,,\cdot))$ is an adapted M-solution of (5.1) on $[S, T]$, it should be an adapted M-solution of (5.1) on any $[\bar{S}, T]$ with $\bar{S} \in (S, T)$. In fact, if (5.1.5) holds, then for any $\bar{S} \in (S, T)$ and almost all $t \in [\bar{S}, T]$, one has

$$\mathbb{E}_{\bar{S}}[Y(t)] = \mathbb{E}_S[Y(t)] + \int_S^{\bar{S}} Z(t,s)dW(s).$$

Consequently, for almost all $t \in [\bar{S}, T]$,

$$Y(t) = \mathbb{E}_S[Y(t)] + \int_S^t Z(t,s)dW(s) = \mathbb{E}_{\bar{S}}[Y(t)] + \int_{\bar{S}}^t Z(t,s)dW(s).$$

In particular, if $(Y(\cdot), Z(\cdot\,,\cdot))$ is an adapted M-solution of (5.1) on $[0, T]$, then

$$Y(t) = \mathbb{E}Y(t) + \int_0^t Z(t,s)dW(s), \qquad t \in [0, T].$$

For any $S \in [0, T)$, $p \in (1, \infty)$, and $\beta \geqslant 0$, let $\mathcal{M}^{p,\beta}[S, T]$ be the space of all processes $(y(\cdot), z(\cdot\,,\cdot)) \in \mathcal{H}^p[S, T]$ such that

$$y(t) = \mathbb{E}_S[y(t)] + \int_S^t z(t,s)dW(s), \qquad \text{a.e. } t \in [S, T], \ \text{a.s.} \tag{5.1.6}$$

and with the norm

$$\begin{aligned}&\|(y(\cdot), z(\cdot\,,\cdot))\|_{\mathcal{M}^{p,\beta}[S,T]} = \mathbb{E}\Big[\int_S^T e^{\beta pt}|y(t)|^p dt + \int_S^T e^{\beta pt}\Big(\int_t^T |z(t,s)|^2 ds\Big)^{\frac{p}{2}} dt\Big]^{\frac{1}{p}}\\ &\leqslant K\mathbb{E}\Big[\int_S^T |y(t)|^p dt + \int_S^T \mathbb{E}\Big(\int_S^T |z(t,s)|^2 ds\Big)^{\frac{p}{2}} dt\Big]^{\frac{1}{p}} \equiv K\|(y(\cdot), z(\cdot\,,\cdot))\|_{\mathcal{H}^p[S,T]}.\end{aligned}$$

This means $\|\cdot\|_{\mathcal{M}^{p,\beta}[S,T]}$ is weaker than $\|\cdot\|_{\mathcal{H}^p[S,T]}$. Note that in the above definition of the norm $\|\cdot\|_{\mathcal{M}^{p,\beta}[S,T]}$, although $z(t,s)$ is defined on $[S,T]^2$, only the values of $z(t,s)$ with $(t,s) \in \Delta^*[S,T]$ are used. We now state and prove an important result below.

Proposition 5.1.3 *For all $\beta \geqslant 0$, $p > 1$, and the elements in $\mathcal{M}^{p,\beta}[S,T]$, the norms $\|\cdot\|_{\mathcal{M}^{p,\beta}[S,T]}$ are equivalent to $\|\cdot\|_{\mathcal{H}^p[S,T]}$.*

Proof For any $(y(\cdot), z(\cdot\,,\cdot)) \in \mathcal{M}^{p,\beta}[S,T]$, from Corollary 3.1.7 of Chap. 3, we have

$$\begin{aligned}\mathbb{E}\Big(\int_S^t |z(t,s)|^2 ds\Big)^{\frac{p}{2}} &\leqslant \mathbb{E}\Big[\big|\mathbb{E}_S[y(t)]\big|^p + \Big(\int_S^t |z(t,s)|^2 ds\Big)^{\frac{p}{2}}\Big]\\ &\leqslant K\mathbb{E}\Big|\mathbb{E}_S[y(t)] + \int_S^t z(t,s)dW(s)\Big|^p = K\mathbb{E}|y(t)|^p.\end{aligned} \tag{5.1.7}$$

Consequently,

$$\mathbb{E}\int_S^T e^{\beta pt}\Big(\int_S^t |z(t,s)|^2 ds\Big)^{\frac{p}{2}} dt \leqslant K\mathbb{E}\int_S^T e^{\beta pt}|y(t)|^p dt. \tag{5.1.8}$$

Thus, for any $(y(\cdot), z(\cdot\,,\cdot)) \in \mathcal{M}^{p,\beta}[S,T] \subseteq \mathcal{H}^p[S,T]$,

$$\begin{aligned}
&\|(y(\cdot), z(\cdot\,,\cdot))\|^p_{\mathcal{H}^p[S,T]} \equiv \mathbb{E}\Big[\int_S^T |y(t)|^p dt + \int_S^T \Big(\int_S^T |z(t,s)|^2 ds\Big)^{\frac{p}{2}} dt\Big] \\
&\leqslant \mathbb{E}\Big[\int_S^T e^{\beta pt}|y(t)|^p dt + \int_S^T e^{\beta pt}\Big(\int_S^T |z(t,s)|^2 ds\Big)^{\frac{p}{2}} dt\Big] \\
&\leqslant K\mathbb{E}\Big[\int_S^T e^{\beta pt}|y(t)|^p dt + \int_S^T e^{\beta pt}\Big(\int_S^t |z(t,s)|^2 ds\Big)^{\frac{p}{2}} dt \\
&\qquad\qquad + \int_S^T e^{\beta pt}\Big(\int_t^T |z(t,s)|^2 ds\Big)^{\frac{p}{2}} dt\Big] \\
&\leqslant K\mathbb{E}\Big[\int_S^T e^{\beta pt}|y(t)|^p dt + \int_S^T e^{\beta pt}\Big(\int_t^T |z(t,s)|^2 ds\Big)^{\frac{p}{2}} dt\Big] \\
&= K\|(y(\cdot), z(\cdot\,,\cdot))\|^p_{\mathcal{M}^{p,\beta}[S,T]} \leqslant K\|(y(\cdot), z(\cdot\,,\cdot))\|^p_{\mathcal{H}^p[S,T]}.
\end{aligned} \tag{5.1.9}$$

This shows the equivalence between $\|\cdot\|_{\mathcal{H}^p[S,T]}$ and any $\|\cdot\|_{\mathcal{M}^{p,\beta}[S,T]}$, $\beta \geqslant 0$. □

We have seen that Corollary 3.1.7 of Chap. 3 plays a crucial role here. In what follows, we denote $\mathcal{M}^{p,0}[S,T] = \mathcal{M}^p[S,T]$.

2 Well-Posedness

We now establish the well-posedness of Type II BSVIEs in the sense of adapted M-solutions. To begin with, we first introduce the following hypothesis. Let $p > 1$.

(H5.2.1)$_p$ The map $g : \Delta^*[0,T] \times \Omega \times \mathbb{R}^m \times \mathbb{R}^{m\times d} \times \mathbb{R}^{m\times d} \to \mathbb{R}^m$ is measurable, and for all $(t, y, z, \widehat{z}) \in [0,T] \times \mathbb{R}^m \times \mathbb{R}^{m\times d} \times \mathbb{R}^{m\times d}$, the map $(s,\omega) \mapsto g(t,s,\omega,y,z,\widehat{z})$ is $\mathbb{F}$-progressively measurable satisfying (5.1.1). There exist deterministic functions $L_y, L_z, L_{\widehat{z}} : \Delta^*[0,T] \to (0,\infty)$ satisfying

$$\begin{aligned}
&\mathbb{E}\int_0^T \Big(\int_t^T L_y(t,s)^{\frac{(p\wedge 2)(1+\varepsilon)}{(p\wedge 2)-1}} ds\Big)^{\frac{(p\wedge 2)-1}{1+\varepsilon}} dt \\
&\qquad + \sup_{t\in[0,T]} \int_t^T \Big(L_z(t,s)^2 + L_{\widehat{z}}(t,s)^{\frac{(p\wedge 2)(1+\varepsilon)}{(p\wedge 2)-1}}\Big) ds < \infty,
\end{aligned} \tag{5.2.1}$$

for some $\varepsilon > 0$ such that

$$\begin{aligned}
&|g(t,s,y_1,z_1,\widehat{z}_1) - g(t,s,y_2,z_2,\widehat{z}_2)| \\
&\leqslant L_y(t,s)|y_1 - y_2| + L_z(t,s)|z_1 - z_2| + L_{\widehat{z}}(t,s)|\widehat{z}_1 - \widehat{z}_2|, \\
&\qquad (t,s) \in \Delta^*[0,T],\ y_i \in \mathbb{R}^m,\ z_i, \widehat{z}_i \in \mathbb{R}^{m\times d},\ i = 1,2, \quad \text{a.s.}
\end{aligned} \tag{5.2.2}$$

and

$$|g(t,s,0,0,\widehat{z})| \leqslant L_{\widehat{z}}(t,s)|\widehat{z}|^{\frac{2}{p}\wedge 1}, \quad (t,s)\in\Delta^*[0,T],\ \widehat{z}\in\mathbb{R}^{m\times d}, \ \text{a.s.} \tag{5.2.3}$$

Let us make some comments to the above assumptions. First of all, recall the space defined in Chap. 4, (4.2.3), we see that (5.2.1) is equivalent to the following:

$$\begin{aligned} &L_y(\cdot\,,\cdot)\in L^{p\wedge 2}(0,T;L^{\frac{(p\wedge 2)(1+\varepsilon)}{(p\wedge 2)-1}}(\cdot\,,T;\mathbb{R})),\\ &L_z(\cdot\,,\cdot)\in L^{\infty}(0,T;L^2(\cdot\,,T;\mathbb{R})),\\ &L_{\widehat{z}}(\cdot\,,\cdot)\in L^{\infty}(0,T;L^{\frac{(p\wedge 2)(1+\varepsilon)}{(p\wedge 2)-1}}(\cdot,T;\mathbb{R})). \end{aligned} \tag{5.2.4}$$

Since $\frac{(p\wedge 2)(1+\varepsilon)}{(p\wedge 2)-1} > 2$ (equivalent to $(p\wedge 2) < \frac{2}{1-\varepsilon}$, which is always true), we see that (5.2.4) is stronger than

$$\begin{aligned} &L_y(\cdot\,,\cdot)\in L^{p\wedge 2}(0,T;L^2(\cdot\,,T;\mathbb{R})),\\ &L_z(\cdot\,,\cdot),L_{\widehat{z}}(\cdot\,,\cdot)\in L^{\infty}(0,T;L^2(\cdot\,,T;\mathbb{R})), \end{aligned} \tag{5.2.5}$$

or

$$\mathbb{E}\int_0^T\Big(\int_t^T L_y(t,s)^2ds\Big)^{\frac{p\wedge 2}{2}}dt+\sup_{t\in[0,T]}\int_t^T\Big(L_z(t,s)^2+L_{\widehat{z}}(t,s)^2\Big)ds<\infty. \tag{5.2.6}$$

Next, for $p=\infty$, (5.2.3) stands for the following:

$$|g(t,s,0,0,\widehat{z})| \leqslant L_{\widehat{z}}(t,s), \quad (t,s)\in\Delta^*[0,T],\ \widehat{z}\in\mathbb{R}^{m\times d}, \ \text{a.s.} \tag{5.2.7}$$

On the other hand, by the previous convention (5.1.1) and the above (5.2.2), we obtain that

$$|g(t,s,0,0,\widehat{z})| \leqslant L_{\widehat{z}}(t,s)|\widehat{z}|, \qquad (t,s)\in\Delta^*[0,T], \quad \widehat{z}\in\mathbb{R}^{m\times d}, \ \text{a.s.} \tag{5.2.8}$$

In other words, in the case of $p\in(1,2]$, (5.2.3) is reductant, while in the case of $p\in(2,\infty)$, (5.2.3) actually is a stronger assumption.

Similar to Lemma 4.2.1 of Chap. 4, we have the following lemma which will be used several times below.

Lemma 5.2.1 *Let*

$$\sup_{t\in[0,T]}\Big(\int_t^T L_{\widehat{z}}(t,s)^{\frac{p(1+\varepsilon)}{p-1}}ds\Big)^{\frac{p-1}{1+\varepsilon}}<\infty, \tag{5.2.9}$$

for some $p > 1$ and $\varepsilon > 0$. Then there exists a constant $K > 0$ only depending on p and ε such that for any $z(\cdot\,,\cdot) \in L^2_{\mathbb{F}}(\Delta_*[0,T];\mathbb{R}^m)$,

$$\begin{aligned}&\mathbb{E}\int_0^T e^{\beta pt}\Big(\int_t^T L_{\widehat{z}}(t,s)|z(s,t)|^{\frac{2}{p}\wedge 1}ds\Big)^p dt\\&\leqslant \frac{K}{\beta^{\frac{(p-1)\varepsilon}{1+\varepsilon}}}\mathbb{E}\int_0^T e^{\beta pt}\Big(\int_0^t |z(t,s)|^2 ds\Big)^{\frac{p}{2}\wedge 1}dt.\end{aligned}\tag{5.2.10}$$

Proof First, we look at the case of $p \geqslant 2$. It follows that

$$\begin{aligned}&\mathbb{E}\int_0^T e^{\beta pt}\Big(\int_t^T L_{\widehat{z}}(t,s)|z(s,t)|^{\frac{2}{p}}ds\Big)^p dt\\&=\mathbb{E}\int_0^T\Big(\int_t^T e^{-\beta(s-t)}e^{\beta s}L_{\widehat{z}}(t,s)|z(s,t)|^{\frac{2}{p}}ds\Big)^p dt\\&\leqslant \mathbb{E}\int_0^T\Big(\int_t^T e^{-\frac{\beta p}{p-1}(s-t)}L_{\widehat{z}}(t,s)^{\frac{p}{p-1}}ds\Big)^{p-1}\Big(\int_t^T e^{\beta ps}|z(s,t)|^2ds\Big)dt\\&\leqslant \mathbb{E}\int_0^T\Big[\Big(\int_t^T e^{-\frac{\beta p(1+\varepsilon)}{(p-1)\varepsilon}(s-t)}ds\Big)^{\frac{(p-1)\varepsilon}{1+\varepsilon}}\Big(\int_t^T L_{\widehat{z}}(t,s)^{\frac{p(1+\varepsilon)}{p-1}}ds\Big)^{\frac{p-1}{1+\varepsilon}}\\&\qquad\qquad\cdot\Big(\int_t^T e^{\beta ps}|z(s,t)|^2ds\Big)\Big]dt\\&\leqslant\Big(\frac{(p-1)\varepsilon}{\beta p(1+\varepsilon)}\Big)^{\frac{(p-1)\varepsilon}{1+\varepsilon}}\sup_{t\in[0,T]}\Big(\int_t^T L_{\widehat{z}}(t,s)^{\frac{p(1+\varepsilon)}{p-1}}ds\Big)^{\frac{p-1}{1+\varepsilon}}\\&\qquad\qquad\cdot\mathbb{E}\int_0^T\Big(\int_t^T e^{\beta ps}|z(s,t)|^2ds\Big)dt\\&\leqslant\frac{K}{\beta^{\frac{(p-1)\varepsilon}{1+\varepsilon}}}\mathbb{E}\int_0^T e^{\beta pt}\int_0^t|z(t,s)|^2dsdt.\end{aligned}$$

Now, in the case $p \in (1,2]$, we have

$$\begin{aligned}&\mathbb{E}\int_0^T e^{\beta pt}\Big(\int_t^T L_{\widehat{z}}(t,s)|z(s,t)|ds\Big)^p dt\\&=\mathbb{E}\int_0^T\Big(\int_t^T e^{-\beta(s-t)}e^{\beta s}L_{\widehat{z}}(t,s)|z(s,t)|ds\Big)^p dt\\&\leqslant\mathbb{E}\Big[\int_0^T\Big(\int_t^T e^{-\frac{\beta p}{p-1}(s-t)}L_{\widehat{z}}(t,s)^{\frac{p}{p-1}}ds\Big)^{p-1}\Big(\int_t^T e^{\beta ps}|z(s,t)|^pds\Big)dt\Big]\\&\leqslant\mathbb{E}\int_0^T\Big(\int_t^T e^{-\frac{\beta p(1+\varepsilon)}{(p-1)\varepsilon}(s-t)}ds\Big)^{\frac{(p-1)\varepsilon}{1+\varepsilon}}\Big(\int_t^T L_{\widehat{z}}(t,s)^{\frac{p(1+\varepsilon)}{p-1}}ds\Big)^{\frac{p-1}{1+\varepsilon}}\\&\qquad\qquad\cdot\Big(\int_t^T e^{\beta ps}|z(s,t)|^pds\Big)dt\\&\leqslant\frac{K}{\beta^{\frac{(p-1)\varepsilon}{1+\varepsilon}}}\mathbb{E}\int_0^T\int_0^s e^{\beta ps}|z(s,t)|^pdtds\leqslant\frac{K}{\beta^{\frac{(p-1)\varepsilon}{1+\varepsilon}}}\mathbb{E}\int_0^T e^{\beta pt}\Big(\int_0^t|z(t,s)|^2ds\Big)^{\frac{p}{2}}dt.\end{aligned}$$

This proves the lemma. □

We now state the well-posedness theorem for Type II BSVIEs.

Theorem 5.2.2 *Let (H5.2.1)$_p$ hold with $1 < p < \infty$. Then for any $\psi(\cdot) \in L^p_{\mathcal{F}_T}(0,T;\mathbb{R}^m)$, Type II BSVIE* (5.1) *admits a unique adapted M-solution $(Y(\cdot), Z(\cdot\,,\cdot)) \in \mathcal{M}^p[0,T]$, and the following estimate holds:*

$$\begin{aligned}&\|(Y(\cdot), Z(\cdot\,,\cdot))\|^p_{\mathcal{M}^p[0,T]}\\&\equiv \mathbb{E}\Big[\int_0^T |Y(t)|^p dt + \int_0^T \Big(\int_0^T |Z(t,s)|^2 ds\Big)^{\frac{p}{2}} dt\Big] \leqslant K\mathbb{E}\int_0^T |\psi(t)|^p dt.\end{aligned} \tag{5.2.11}$$

Moreover, if, for $i = 1, 2$, $g_i(\cdot)$ satisfy (H5.2.1)$_p$ with, additionally, either $g_1(\cdot)$ or $g_2(\cdot)$ satisfying the following:

$$\begin{aligned}&|g(t,s,y_1,z_1,\widehat{z}_1) - g(t,s,y_2,z_2,\widehat{z}_2)|\\&\leqslant L_y(t,s)|y_1 - y_2| + L_z(t,s)|z_1 - z_2| + L_{\widehat{z}}(t,s)|\widehat{z}_1 - \widehat{z}_2|^{\frac{2}{p}\wedge 1},\\&\qquad\qquad (t,s) \in \Delta^*[0,T],\ (y_i, z_i, \widehat{z}_i) \in \mathbb{R}^{m(1+2d)},\ i = 1,2,\ \ a.s.\end{aligned} \tag{5.2.12}$$

then, for any $\psi_i(\cdot) \in L^p_{\mathcal{F}_T}(0,T;\mathbb{R}^n)$, the corresponding adapted M-solutions $(Y_i(\cdot), Z_i(\cdot\,,\cdot)) \in \mathcal{M}^p[0,T]$ satisfy the following stability estimate:

$$\begin{aligned}&\|(Y_1(\cdot), Z_1(\cdot\,,\cdot)) - (Y_2(\cdot), Z_2(\cdot\,,\cdot))\|^p_{\mathcal{M}^p[0,T]}\\&\leqslant K\mathbb{E}\Big\{\int_0^T |\psi_1(t) - \psi_2(t)|^p dt + \int_0^T \Big(\int_t^T |(g_1 - g_2)(t,s)|^2 ds\Big)^{\frac{p}{2}} dt\Big\},\end{aligned} \tag{5.2.13}$$

where

$$\begin{aligned}(g_1 - g_2)(t,s) = g_1\big(t,s,Y_1(s),Z_1(t,s),Z_1(s,t)\big)\\ -g_2\big(t,s,Y_1(s),Z_1(t,s),Z_1(s,t)\big).\end{aligned}$$

Note that (5.2.12) is different from (5.2.2) only if $p > 2$.

Proof We split the proof into several steps.
Step 1. Existence and uniqueness of an adapted M-solution in $\mathcal{M}^p[0,T]$ with $p \in (1,2]$ under (H5.2.1)$_p$.

Let $\psi(\cdot) \in L^p_{\mathcal{F}_T}(0,T;\mathbb{R}^m)$ be given. For any $(y(\cdot), z(\cdot\,,\cdot)) \in \mathcal{M}^p[0,T]$, we consider the following Type I BSVIE

$$Y(t) = \widetilde{\psi}(t) + \int_t^T \widetilde{g}\big(t,s,Y(s),Z(t,s)\big)ds - \int_t^T Z(t,s)dW(s), \quad t \in [0,T], \tag{5.2.14}$$

with

$$\begin{aligned}&\widetilde{\psi}(t) = \psi(t) + \int_t^T g(t,s,0,0,z(s,t))ds,\\&\widetilde{g}(t,s,Y,Z) = g(t,s,Y,Z,z(s,t)) - g(t,s,0,0,z(s,t)).\end{aligned}$$

Let us estimate the following (making use of (5.1.7)):

$$
\begin{aligned}
&\mathbb{E}\int_0^T\Big(\int_t^T|g(t,s,0,0,z(s,t))|ds\Big)^p dt \leqslant K\mathbb{E}\Big[\int_0^T\Big(\int_t^T L_{\widehat{z}}(t,s)|z(s,t)|ds\Big)^p dt\Big]\\
&\leqslant K\mathbb{E}\int_0^T\Big(\int_t^T L_{\widehat{z}}(t,s)^{\frac{p}{p-1}}ds\Big)^{p-1}\Big(\int_t^T|z(s,t)|^p ds\Big)dt\\
&\leqslant K\sup_{t\in[0,T]}\Big(\int_t^T L_{\widehat{z}}(t,s)^{\frac{p}{p-1}}ds\Big)^{p-1}\mathbb{E}\int_0^T\int_0^t|z(t,s)|^p dsdt\\
&\leqslant K\int_0^T\mathbb{E}\Big(\int_0^t|z(t,s)|^2ds\Big)^{\frac{p}{2}}dt \leqslant K\mathbb{E}\int_0^T|y(t)|^p dt<\infty.
\end{aligned}
$$

Hence,

$$
\begin{aligned}
&\mathbb{E}\int_0^T|\widetilde{\psi}(t)|^p dt \leqslant 2^{p-1}\mathbb{E}\int_0^T\Big(|\psi(t)|^p+\int_t^T|g(t,s,0,0,z(s,t))|ds\Big)^p dt\\
&\leqslant K\mathbb{E}\int_0^T\Big(|\psi(t)|^p+|y(t)|^p\Big)dt<\infty.
\end{aligned}
$$

According to Theorem 4.2.2 in Chap. 4, there exists a unique adapted solution $(Y(\cdot),Z(\cdot\,,\cdot))\in\mathcal{H}^p_{\Delta^*}[0,T]$ to Type I BSVIE (5.2.14). In addition, by Proposition 4.3.1 of Chap. 4, for almost $t\in[0,T]$, the following holds:

$$
\begin{aligned}
&\mathbb{E}\Big\{|Y(t)|^p+\Big(\int_t^T|Z(t,s)|^2ds\Big)^{\frac{p}{2}}\Big\}\\
&\leqslant K\mathbb{E}\Big\{|\widetilde{\psi}(t)|^p+\Big(\int_t^T L_y(t,s)|Y(s)|ds\Big)^p\Big\}\\
&\leqslant K\mathbb{E}\Big\{|\psi(t)|^p+\Big(\int_t^T L_y(t,s)|Y(s)|ds\Big)^p+\Big(\int_t^T L_{\widehat{z}}(t,s)|z(s,t)|ds\Big)^p\Big\}.
\end{aligned}
\tag{5.2.15}
$$

Then for any $\beta>0$,

$$
\begin{aligned}
&\mathbb{E}\Big[\int_0^T e^{\beta pt}|Y(t)|^p dt+\int_0^T e^{\beta pt}\Big(\int_t^T|Z(t,s)|^2ds\Big)^{\frac{p}{2}}dt\Big]\\
&\leqslant K\mathbb{E}\Big[\int_0^T e^{\beta pt}|\psi(t)|^p dt+\int_0^T\Big(\int_t^T e^{\beta t}L_y(t,s)|Y(s)|ds\Big)^p dt\\
&\qquad+\int_0^T\Big(\int_t^T e^{\beta t}L_{\widehat{z}}(t,s)|z(s,t)|ds\Big)^p dt\Big].
\end{aligned}
$$

By Lemma 4.2.1 of Chap. 4, we have

$$
\int_0^T\mathbb{E}\Big(\int_t^T e^{\beta t}L_y(t,s)|Y(s)|ds\Big)^p dt\leqslant\frac{K}{\beta^{\frac{(p-1)\varepsilon}{1+\varepsilon}}}\mathbb{E}\Big(\int_0^T e^{\beta ps}|Y(s)|^p ds\Big),
\tag{5.2.16}
$$

and by Lemma 5.2.1 (of this chapter), we have

$$\begin{aligned}&\mathbb{E}\int_0^T e^{\beta pt}\Big(\int_t^T L_{\widehat{z}}(t,s)|z(s,t)|ds\Big)^p dt\\&\leqslant \frac{K}{\beta^{\frac{(p-1)\varepsilon}{1+\varepsilon}}}\mathbb{E}\int_0^T e^{\beta pt}\Big(\int_0^t |z(t,s)|^2 ds\Big)^{\frac{p}{2}} dt.\end{aligned}\tag{5.2.17}$$

Thus,

$$\begin{aligned}&\mathbb{E}\Big[\int_0^T e^{\beta pt}|Y(t)|^p dt+\int_0^T e^{\beta pt}\Big(\int_t^T |Z(t,s)|^2 ds\Big)^{\frac{p}{2}} dt\Big]\\&\leqslant K\mathbb{E}\int_0^T e^{\beta pt}|\psi(t)|^p dt\\&\quad+\frac{K}{\beta^{\frac{(p-1)\varepsilon}{1+\varepsilon}}}\mathbb{E}\Big[\int_0^T e^{\beta pt}|Y(t)|^p dt+\int_0^T e^{\beta pt}\Big(\int_0^t |z(t,s)|^2 ds\Big)^{\frac{p}{2}} dt\Big].\end{aligned}\tag{5.2.18}$$

Choosing $\beta > 0$ large enough so that $\beta^{\frac{(p-1)\varepsilon}{1+\varepsilon}} \geqslant 3K$, then the above leads to

$$\begin{aligned}&\mathbb{E}\Big[\int_0^T e^{\beta pt}|Y(t)|^p dt+\int_0^T e^{\beta pt}\Big(\int_t^T |Z(t,s)|^2 ds\Big)^{\frac{p}{2}} dt\Big]\\&\leqslant \frac{3K}{2}\mathbb{E}\Big[\int_0^T e^{\beta pt}|\psi(t)|^p dt\Big]+\frac{1}{2}\mathbb{E}\int_0^T e^{\beta pt}\Big(\int_0^t |z(t,s)|^2 ds\Big)^{\frac{p}{2}} dt\Big].\end{aligned}$$

Next, define $Z(t,s)$ with $(t,s)\in\Delta_*[0,T]$ by the following:

$$Y(t)=\mathbb{E}Y(t)+\int_0^t Z(t,s)dW(s),\qquad t\in[0,T].\tag{5.2.19}$$

Then, for the given fixed $\beta > 0$, we obtain (using (5.1.9))

$$\begin{aligned}&\|(Y(\cdot),Z(\cdot\,,\cdot))\|^p_{\mathcal{M}^p_\beta[0,T]}\\&=\mathbb{E}\Big[\int_0^T e^{\beta pt}|Y(t)|^p dt+\int_0^T e^{\beta pt}\Big(\int_t^T |Z(t,s)|^2 ds\Big)^{\frac{p}{2}} dt\Big]\\&\leqslant K\mathbb{E}\Big[\int_0^T e^{\beta pt}|\psi(t)|^p dt+\int_0^T e^{\beta pt}\Big(\int_0^t |z(t,s)|^2 ds\Big)^{\frac{p}{2}} dt\Big].\\&\leqslant K\mathbb{E}\Big[\int_0^T |\psi(t)|^p dt+\int_0^T \Big(\int_0^t |z(t,s)|^2 ds\Big)^{\frac{p}{2}} dt\Big]\\&\leqslant K\mathbb{E}\Big[\int_0^T |\psi(t)|^p dt+\int_0^T |y(t)|^p dt\Big]\\&\leqslant K\Big[\|\psi(\cdot)\|^p_{L^p_{\mathcal{F}_T}(0,T;\mathbb{R}^m)}+\|(y(\cdot),z(\cdot\,,\cdot))\|_{\mathcal{M}^p[0,T]}\Big].\end{aligned}\tag{5.2.20}$$

Here, the constant K on the right-hand side of the above is allowed to depend on β. Hence, if we define $\Theta(y(\cdot), z(\cdot\,,\cdot)) = (Y(\cdot), Z(\cdot\,,\cdot))$, then Θ maps from $\mathcal{M}^p[0, T]$ to itself. We now show that the map Θ is a contraction. To this end, let us take any $(y_i(\cdot), z_i(\cdot\,,\cdot)) \in \mathcal{M}^p[0, T]$ $(i = 1, 2)$, and let

$$(Y_i(\cdot), Z_i(\cdot\,,\cdot)) = \Theta(y_i(\cdot), z_i(\cdot\,,\cdot)).$$

By Proposition 4.3.1 in Chap. 4, we have that for almost all $t \in [0, T]$,

$$\begin{aligned}
&\mathbb{E}\Big[|Y_1(t) - Y_2(t)|^p + \Big(\int_t^T |Z_1(t,s) - Z_2(t,s)|^2 ds\Big)^{\frac{p}{2}}\Big]\\
&\leqslant K\mathbb{E}\Big[\Big(\int_t^T |g\big(t,s,Y_1(s),Z_1(t,s),z_1(s,t)\big)\\
&\qquad\qquad -g\big(t,s,Y_1(s),Z_1(t,s),z_2(s,t)\big)|ds\Big)^p\\
&\qquad +\Big(\int_t^T L_y(t,s)|Y_1(s) - Y_2(s)|ds\Big)^p\Big]\\
&\leqslant K\mathbb{E}\Big[\int_t^T L_y(t,s)|Y_1(s) - Y_2(s)|ds\Big)^p + \Big(\int_t^T L_z(t,s)|z_1(s,t) - z_2(s,t)|ds\Big)^p\Big].
\end{aligned}$$

Then, using the same arguments in proving (5.2.18), we obtain

$$\begin{aligned}
&\mathbb{E}\Big[\int_0^T e^{\beta pt}|Y_1(t) - Y_2|^p dt + \int_0^T e^{\beta pt}\Big(\int_t^T |Z_1(t,s) - Z_2(t,s)|^2 ds\Big)^{\frac{p}{2}} dt\Big]\\
&\leqslant \frac{K}{\beta^{\frac{(p-1)\varepsilon}{1+\varepsilon}}}\mathbb{E}\Big[\int_0^T e^{\beta pt}|Y_1(t) - Y_2(t)|^p dt + \int_0^T e^{\beta pt}\Big(\int_0^t |z_1(t,s) - z_2(t,s)|^2 ds\Big)^{\frac{p}{2}} dt\Big].
\end{aligned}$$

Choosing $\beta > 0$ large so that $\beta^{\frac{(p-1)\varepsilon}{1+\varepsilon}} \geqslant 3K$, then

$$\begin{aligned}
&\mathbb{E}\Big[\int_0^T e^{\beta pt}|Y_1(t) - Y_2|^p dt + \int_0^T e^{\beta pt}\Big(\int_t^T |Z_1(t,s) - Z_2(t,s)|^2 ds\Big)^{\frac{p}{2}} dt\Big]\\
&\leqslant \frac{1}{2}\mathbb{E}\int_0^T e^{\beta pt}\Big(\int_0^t |z_1(t,s) - z_2(t,s)|^2 ds\Big)^{\frac{p}{2}} dt.
\end{aligned}$$

Hence,

$$\begin{aligned}
&\|\Theta(y_1(\cdot), z_1(\cdot\,,\cdot)) - \Theta(y_2(\cdot), z_2(\cdot\,,\cdot))\|^p_{\mathcal{M}^{p,\beta}[0,T]}\\
&\leqslant \frac{1}{2}\|(y_1(\cdot), z_1(\cdot\,,\cdot)) - (y_2(\cdot), z_2(\cdot\,,\cdot))\|^p_{\mathcal{M}^{p,\beta}[0,T]}.
\end{aligned} \tag{5.2.21}$$

This means that Θ is a contraction. Hence, there exists a unique fixed point $(Y(\cdot), Z(\cdot\,,\cdot)) \in \mathcal{M}^p[0, T]$, which is the unique adapted M-solution of (5.1).
Step 2. Existence and uniqueness of adapted M-solutions in $\mathcal{M}^p[0, T]$ with $p \in [2, \infty)$.

Let $p \in [2, \infty)$ and $\psi(\cdot) \in L^p_{\mathcal{F}_T}(0, T; \mathbb{R}^m) \subseteq L^2_{\mathcal{F}_T}(0, T; \mathbb{R}^m)$. According to Step 1, there exists a unique adapted M-solution $(Y(\cdot), Z(\cdot\,,\cdot)) \in \mathcal{M}^2[0, T]$. We want to show that in the current case, $(Y(\cdot), Z(\cdot\,,\cdot))$ is actually in $\mathcal{M}^p[0, T]$. To show this, for the obtained adapted M-solution $(Y(\cdot), Z(\cdot\,,\cdot))$, let us consider the following Type I BSVIE for $(\widetilde{Y}(\cdot), \widetilde{Z}(\cdot\,,\cdot))$:

$$\widetilde{Y}(t) = \widetilde{\psi}(t) + \int_t^T \widetilde{g}\big(t, s, \widetilde{Y}(s), \widetilde{Z}(t, s), Z(s, t)\big)ds - \int_t^T \widetilde{Z}(t, s)dW(s), \qquad t \in [0, T], \tag{5.2.22}$$

with

$$\begin{aligned}
&\widetilde{\psi}(t) = \psi(t) + \int_t^T g(t, s, 0, 0, Z(s, t))ds,\\
&\widetilde{g}(t, s, \widetilde{Y}, \widetilde{Z}) = g(t, s, \widetilde{Y}, \widetilde{Z}, Z(s, t)) - g(t, s, 0, 0, Z(s, t)).
\end{aligned}$$

Note that (making use of (5.2.3))

$$\begin{aligned}
&\mathbb{E}\int_0^T\Big(\int_t^T |g(t, s, 0, 0, Z(s, t))|ds\Big)^p dt \leqslant K\mathbb{E}\int_0^T\Big(\int_t^T L_{\bar z}(t, s)|Z(s, t)|^{\frac{2}{p}} ds\Big)^p dt\\
&\leqslant K\mathbb{E}\Big\{\int_0^T\Big(\int_t^T L_{\bar z}(t, s)^{\frac{p}{p-1}} ds\Big)^{p-1}\Big(\int_t^T |Z(s, t)|^2 ds\Big)dt\Big\}\\
&\leqslant K\mathbb{E}\Big\{\int_0^T\int_0^t |Z(t, s)|^2 ds dt\Big\} < \infty.
\end{aligned}$$

Thus, $\widetilde{\psi}(\cdot) \in L^p_{\mathcal{F}_T}(0, T; \mathbb{R}^m)$. Hence, by Theorem 4.2.2 of Chap. 4, Type I BSVIE (5.2.22) admits a unique adapted solution $(\widetilde{Y}(\cdot), \widetilde{Z}(\cdot\,,\cdot)) \in \mathcal{H}^p_{\Delta^*}[0, T]$. Now, define $\widetilde{Z}(t, s)$ for $(t, s) \in \Delta_*[0, T]$ as follows (similar to (5.2.19)):

$$\widetilde{Y}(t) = \mathbb{E}\widetilde{Y}(t) + \int_0^t \widetilde{Z}(t, s)dW(s), \qquad t \in [0, T], \tag{5.2.23}$$

which leads to $(\widetilde{Y}(\cdot), \widetilde{Z}(\cdot\,,\cdot)) \in \mathcal{M}^p[0, T]$ because (see (5.1.9))

$$\begin{aligned}
&\mathbb{E}\Big[\int_0^T |\widetilde{Y}(t)|^p dt + \int_0^T\Big(\int_0^T |\widetilde{Z}(t, s)|^2 ds\Big)^{\frac{p}{2}} dt\Big]\\
&\leqslant K\mathbb{E}\Big[\int_0^T |\widetilde{Y}(t)|^p dt + \int_0^T\Big(\int_t^T |\widetilde{Z}(t, s)|^2 ds\Big)^{\frac{p}{2}} dt\Big].
\end{aligned} \tag{5.2.24}$$

Thus, $(\widetilde{Y}(\cdot), \widetilde{Z}(\cdot\,,\cdot)) \in \mathcal{M}^p[0, T] \subseteq \mathcal{M}^2[0, T]$ is an adapted M-solution to (5.2.22). But $(Y(\cdot), Z(\cdot\,,\cdot))$ is already the unique adapted M-solution to (5.2.22) in $\mathcal{M}^2[0,T]$. Hence, it is necessary that

$$(Y(\cdot), Z(\cdot\,,\cdot)) = (\widetilde{Y}(\cdot), \widetilde{Z}(\cdot\,,\cdot)) \in \mathcal{M}^p[0,T].$$

Step 3. Some estimates.

First, we consider the case $p > 2$. By Proposition 4.3.1 in Chap. 4, we have

$$\begin{aligned}&\mathbb{E}\Big[|Y(t)|^p + \Big(\int_t^T |Z(t,s)|^2 ds\Big)^{\frac{p}{2}}\Big]\\&\leqslant K\mathbb{E}\Big[|\psi(t)|^p + \Big(\int_t^T L_y(t,s)|Y(s)|ds\Big)^p + \Big(\int_t^T L_{\widehat{z}}(t,s)|Z(s,t)|^{\frac{2}{p}}ds\Big)^p\Big].\end{aligned}$$

Then, for any $\beta > 0$, by Lemma 5.2.1 of this chapter, we have

$$\mathbb{E}\int_0^T e^{\beta pt}\Big(\int_t^T L_{\widehat{z}}(t,s)|Z(s,t)|^{\frac{2}{p}}ds\Big)^p dt \leqslant \frac{K}{\beta^{\frac{(p-1)\varepsilon}{1+\varepsilon}}}\mathbb{E}\int_0^T e^{\beta pt}\Big(\int_0^t |Z(t,s)|^2 ds\Big)dt.$$

Hence, combining the above with (5.2.16), for $p > 2$, we get

$$\begin{aligned}&\mathbb{E}\Big[\int_0^T e^{\beta pt}|Y(t)|^p dt + \int_0^T e^{\beta pt}\Big(\int_t^T |Z(t,s)|^2 ds\Big)^{\frac{p}{2}} dt\Big]\\&\leqslant K\mathbb{E}\Big[\int_0^T e^{\beta pt}|\psi(t)|^p dt + \int_0^T e^{\beta pt}\Big(\int_t^T L_y(t,s)|Y(s)|ds\Big)^p dt\\&\qquad+\int_0^T e^{\beta pt}\Big(\int_t^T L_{\widehat{z}}(t,s)|Z(s,t)|^{\frac{2}{p}}ds\Big)^p dt\Big]\\&\leqslant K\mathbb{E}\Big[\int_0^T e^{\beta pt}|\psi(t)|^p dt\Big]\\&\qquad+\frac{K}{\beta^{\frac{(p-1)\varepsilon}{1+\varepsilon}}}\mathbb{E}\Big[\int_0^T e^{\beta pt}|Y(t)|^p dt + \int_0^T e^{\beta pt}\Big(\int_0^t |Z(t,s)|^2 ds\Big)^{\frac{p}{2}} dt\Big].\end{aligned}$$

Then by choosing $\beta > 0$ large, one has (5.2.11).

Next, we let $g_i(\cdot)$ satisfy $(\text{H5.2.1})_p$, and $\psi_i(\cdot) \in L^p_{\mathcal{F}_T}(0,T;\mathbb{R}^m)$. Let $(Y_i(\cdot), Z_i(\cdot\,,\cdot))$ be the adapted M-solution of the corresponding Type II BSVIEs. Then (assuming $g_2(\cdot)$ satisfies (5.2.12))

$$\begin{aligned}Y_1(t) - Y_2(t) = \psi_1(t) - \psi_2(t) + \int_t^T &\Big(g_1(t,s,Y_1(s),Z_1(t,s),Z_1(s,t))\\&-g_2(t,s,Y_2(s),Z_2(t,s),Z_2(s,t))\Big)ds\\&-\int_t^T \Big(Z_1(t,s) - Z_2(t,s)\Big)dW(s).\end{aligned}$$

By Proposition 4.3.1 in Chap. 4, we obtain that for almost all $t \in [0, T]$,

$$
\begin{aligned}
&\mathbb{E}\Big[|Y_1(t) - Y_2(t)|^p + \Big(\int_t^T |Z_1(t,s) - Z_2(t,s)|^2 ds\Big)^{\frac{p}{2}}\Big] \\
&\leqslant K\mathbb{E}\Big[|\psi_1(t) - \psi_2(t)|^p + \Big(\int_t^T L_y(t,s)|Y_1(s) - Y_2(s)|ds\Big)^p \\
&+\Big(\int_t^T |g_1(t,s,Y_1(s),Z_1(t,s),Z_1(s,t)) - g_2(t,s,Y_1(s),Z_1(t,s),Z_1(s,t))|ds\Big)^p \\
&+\Big(\int_t^T |g_2(t,s,Y_1(s),Z_1(t,s),Z_1(s,t)) - g_2(t,s,Y_1(s),Z_1(t,s),Z_2(s,t))|ds\Big)^p\Big] \\
&\leqslant K\mathbb{E}\Big[|\psi_1(t) - \psi_2(t)|^p + \Big(\int_t^T |(g_1 - g_2)(t,s)|ds\Big)^p \\
&+\Big(\int_t^T L_y(t,s)|Y_1(s,t) - Y_2(s,t)|ds\Big)^p + \Big(\int_t^T L_{\widehat{z}}(t,s)|Z_1(s,t) - Z_2(s,t)|^{\frac{2}{p}} ds\Big)^p\Big].
\end{aligned}
$$

By Lemma 5.2.1 in the current chapter, for any $\beta > 0$, we have

$$
\begin{aligned}
&\mathbb{E}\int_0^T e^{\beta pt}\Big(\int_t^T L_{\widehat{z}}(t,s)|Z_1(s,t) - Z_2(s,t)|^{\frac{2}{p}} ds\Big)^p dt \\
&\leqslant \frac{K}{\beta^{\frac{(p-1)\varepsilon}{1+\varepsilon}}}\mathbb{E}\int_0^T e^{\beta pt}\Big(\int_0^t |Z_1(t,s) - Z_2(t,s)|^2 ds\Big)dt.
\end{aligned}
$$

Hence, we obtain

$$
\begin{aligned}
&\mathbb{E}\int_0^T e^{\beta pt}|Y_1(t) - Y_2(t)|^p dt + \int_0^T e^{\beta pt}\Big(\int_t^T |Z_1(t,s) - Z_2(t,s)|^2 ds\Big)^{\frac{p}{2}} dt\Big] \\
&\leqslant K\mathbb{E}\int_0^T e^{\beta pt}\Big[|\psi_1(t) - \psi_2(t)|^p + \Big(\int_t^T |(g_1 - g_2)(t,s)|ds\Big)^p dt \\
&\quad + \frac{K}{\beta^{\frac{(p-1)\varepsilon}{1+\varepsilon}}}\mathbb{E}\Big[\int_0^T e^{\beta pt}|Y_1(t) - Y_2(t)|^p dt + \int_0^T e^{\beta pt}\Big(\int_0^t |Z_1(t,s) - Z_2(t,s)|^2 ds\Big)dt\Big].
\end{aligned}
$$

Since $p > 2$, by choosing β large enough, we obtain stability estimate (5.2.13).

The case $1 < p < 2$ can be proved similarly. □

Let us make a remark on the above result, together with its proof.

We have seen that the growth of the map

$$
\widehat{z} \mapsto g(t,s,y,z,\widehat{z}) \tag{5.2.25}
$$

plays an important role in proving the well-posedness of Type II BSVIEs, especially for the case of $p > 2$. If (H5.2.1)$_\infty$ holds, then for any $p > 1$, as long as $\psi(\cdot) \in L^p_{\mathcal{F}_T}(0,T;\mathbb{R}^m)$, (5.1) admits a unique adapted M-solution $(Y(\cdot), Z(\cdot\,,\cdot)) \in \mathcal{M}^p[0,T]$. On the other hand, if the map in (5.2.25) grows linearly, the adapted

M-solution $(Y(\cdot), Z(\cdot\,,\cdot))$ of (5.1) may not be in $\mathcal{M}^p[0,T]$ for $p > 2$, even if $\psi(\cdot) \in L^p_{\mathcal{F}_T}(0,T;\mathbb{R}^m)$. This can be seen from the following example.

Example 5.2.3 Consider BSVIE:

$$Y(t) = \psi(t) + \int_t^T Z(s,t)ds - \int_t^T Z(t,s)dW(s), \quad t \in [0,T]. \tag{5.2.26}$$

Let

$$\psi(t) \equiv \int_0^T \psi_1(s)dW(s), \qquad \forall t \in [0,T],$$

with $\psi_1(\cdot)$ being deterministic and

$$\psi_1(\cdot) \in L^2(0,T;\mathbb{R}) \setminus \bigcup_{p>2} L^p(0,T;\mathbb{R}). \tag{5.2.27}$$

We may take, say, $\psi_1(s) = \frac{1}{\sqrt{T-s}\,\log(T-s)}$. Thus, $\psi(t)$ is a t independent $\mathcal{F}_T$-measurable random variable. Hence, for any $p > 1$,

$$\mathbb{E}\int_0^T |\psi(t)|^p ds = T\mathbb{E}\Big|\int_0^T \psi_1(s)dW(s)\Big|^p \leqslant K\Big(\int_0^T |\psi_1(s)|^2 ds\Big)^{\frac{p}{2}},$$

which means $\psi(\cdot) \in L^p_{\mathcal{F}_T}(0,T;\mathbb{R})$ for any $p > 1$. If we define

$$\begin{cases} Y(t) = \displaystyle\int_0^t \psi_1(s)dW(s) + \psi_1(t)(T-t), & t \in [0,T], \\ Z(t,s) = \psi_1(s), \qquad (s,t) \in [0,T]^2, \end{cases} \tag{5.2.28}$$

then

$$\begin{aligned} Y(t) &= \int_0^t \psi_1(s)dW(s) + \psi_1(t)(T-t) \\ &= \psi(t) - \int_t^T \psi_1(s)dW(s) + \int_t^T \psi_1(t)ds \\ &= \psi(t) - \int_t^T Z(s,t)ds + \int_t^T Z(t,s)dW(s), \end{aligned}$$

and

$$Y(t) = \mathbb{E}Y(t) + \int_0^t Z(t,s)dW(s), \qquad t \in [0,T].$$

Thus, $(Y(\cdot), Z(\cdot\,,\cdot))$ is the adapted M-solution of (5.2.26). Since

$$\begin{aligned}&\mathbb{E}\int_0^T\Big|\int_0^t \psi_1(s)dW(s)\Big|^p dt \leqslant \overline{K}_p\int_0^T\Big(\int_0^t|\psi_1(s)|^2ds\Big)^{\frac{p}{2}}dt\\&\leqslant \overline{K}_pT\Big(\int_0^T|\psi_1(s)|^2ds\Big)^{\frac{p}{2}}<\infty,\end{aligned}$$

then, from (5.2.28), we must have that $Y(\cdot)\notin L^p_{\mathbb{F}}(0,T;\mathbb{R})$ for any $p>2$. In fact, if $Y(\cdot)\in L^p_{\mathbb{F}}(0,T;\mathbb{R})$ for some $p>2$, then

$$\begin{aligned}&\delta^p\mathbb{E}\int_0^{T-\delta}|\psi_1(s)|^pds\leqslant\int_0^T(T-s)^p|\psi_1(s)|^pds\\&\leqslant 2^{p-1}\mathbb{E}\int_0^T\Big(|Y(t)|^p+\Big|\int_0^t\psi_1(r)dW(r)\Big|^p\Big)dt\\&\leqslant K\Big\{\int_0^T\mathbb{E}|Y(t)|^pdt+\Big(\int_0^T|\psi(s)|^2ds\Big)^{\frac{p}{2}}\Big\}<\infty.\end{aligned}$$

This leads to a contradiction.

From Theorem 5.2.2, we see that the Type II BSVIE (5.1) is well-posed in $\mathcal{M}^p[0,T]$; in particular, (5.2.11) holds. It is interesting that the estimates for the adapted M-solution can be improved if the assumptions on the generator and the free term are strengthened. More precisely, we have the following result.

Theorem 5.2.4 *Let $(H5.2.1)_p$ hold for some $p>1$, with*

$$\sup_{t\in[0,T]}\int_t^T L_y(t,s)^{\frac{p}{p-1}}ds<\infty. \tag{5.2.29}$$

Let $(Y(\cdot), Z(\cdot\,,\cdot))$ be the adapted M-solution of BSVIE (5.1).

(i) *Suppose*

$$|g(t,s,0,0,\widehat{z})|\leqslant L_{\widehat{z}}(t,s),\qquad \forall(t,s,\widehat{z})\in\Delta^*[0,T]\times\mathbb{R}^{m\times d},\quad a.s. \tag{5.2.30}$$

Then for any $p\in[2,\infty)$,

$$\begin{aligned}&\sup_{t\in[0,T]}\mathbb{E}|Y(t)|^p+\sup_{t\in[0,T]}\mathbb{E}\Big(\int_0^T|Z(t,s)|^2ds\Big)^{\frac{p}{2}}\\&\leqslant K\Big\{\sup_{t\in[0,T]}\mathbb{E}|\psi(t)|^p+\sup_{t\in[0,T]}\mathbb{E}\Big(\int_t^T L_{\widehat{z}}(t,s)ds\Big)^p\Big\}.\end{aligned} \tag{5.2.31}$$

(ii) *Suppose*

$$|g(t,s,0,z,\widehat{z})|\leqslant L_{\widehat{z}}(t,s),\ \forall(t,s,z,\widehat{z})\in\Delta^*[0,T]\times\mathbb{R}^{m\times d}\times\mathbb{R}^{m\times d},\quad a.s. \tag{5.2.32}$$

Then for any $p \in (1, \infty)$,

$$\begin{aligned}&\mathbb{E}\Big[\sup_{t\in[0,T]}|Y(t)|^p+\sup_{t\in[0,T]}\Big|\int_0^t Z(t,s)dW(s)\Big|^p+\sup_{t\in[0,T]}\Big|\int_t^T Z(t,s)dW(s)\Big|^p\Big]\\&\leqslant K\mathbb{E}\Big[\sup_{t\in[0,T]}|\psi(t)|^p+\sup_{t\in[0,T]}\Big(\int_t^T L_{\widehat{z}}(t,s)ds\Big)^p\Big].\end{aligned}\tag{5.2.33}$$

Before proving the above result, let us make some comments. First of all, according to (5.2.1), one has for some $\varepsilon > 0$, $p \in (1, 2)$,

$$\int_0^T \Big(\int_t^T L_y(t,s)^{\frac{p(1+\varepsilon)}{p-1}}ds\Big)^{\frac{p-1}{1+\varepsilon}}dt < \infty.$$

This is different from (5.2.29), and none of them implies the other. Second, under $(H5.2.1)_p$, we have

$$|g(t,s,y,z,\widehat{z})| \leqslant L_y(t,s)|y| + L_z(t,s)|z| + L_{\widehat{z}}(t,s)|\widehat{z}|^{\frac{2}{p}\wedge 1}.$$

If, in addition, (5.2.30) is assumed, then

$$|g(t,s,y,z,\widehat{z})| \leqslant L_y(t,s)|y| + L_z(t,s)|z| + L_{\widehat{z}}(t,s).$$

Further, if (5.2.32) is assumed, then

$$|g(t,s,y,z,\widehat{z})| \leqslant L_y(t,s)|y| + L_{\widehat{z}}(t,s).$$

Hence, they give the different growth rates for $(z,\widehat{z}) \mapsto g(t,s,y,z,\widehat{z})$. From the conclusions of Theorems 5.2.2 and 5.2.4, one can see that estimates (5.2.11), (5.2.31), and (5.2.33) show the role played by such a growth rate: The lower the growth rate, the better the integrability of the adapted M-solutions. Third, even by assuming (5.2.32), the generator is still allowed to depend on both z and $\widehat{z}$. Hence, the BSVIE will still be Type II. Fourth, by the martingale moment inequality and Burkholder-Davis–Gundy's inequalities (see Chap. 2, Theorem 2.3.2), we know that

$$\begin{aligned}&\mathbb{E}\Big(\int_0^T |Z(t,s)|^2ds\Big)^{\frac{p}{2}} \leqslant K\mathbb{E}\Big|\int_0^T Z(t,s)dW(s)\Big|^p\\&\leqslant K\Big(\mathbb{E}\Big|\int_0^t Z(t,s)dW(s)\Big|^p+\mathbb{E}\Big|\int_t^T Z(t,s)dW(s)\Big|^p\Big)\\&\leqslant K\Big\{\mathbb{E}\Big(\int_0^t |Z(t,s)|^2ds\Big)^{\frac{p}{2}}+\mathbb{E}\Big(\int_t^T |Z(t,s)|^2ds\Big)^{\frac{p}{2}}\Big\}\\&\leqslant K\mathbb{E}\Big(\int_0^T |Z(t,s)|^2ds\Big)^{\frac{p}{2}}.\end{aligned}\tag{5.2.34}$$

Thus, in (5.2.31), we can replace $\mathbb{E}\Big(\int_0^T |Z(t,s)|^2 ds\Big)^{\frac{p}{2}}$ by

$$\mathbb{E}\Big|\int_0^t Z(t,s)dW(s)\Big|^p + E\Big|\int_t^T Z(t,s)dW(s)\Big|^p.$$

Proof Since we had the well-posedness in L^2-spaces, it suffices to establish proper *a priori* estimates.

(i) Let $(Y(\cdot), Z(\cdot\,,\cdot)) \in \mathcal{M}^2[0,T]$ be the (unique) adapted M-solution of BSVIE (1.1). Using auxiliary BSDE (with t being the parameter),

$$\eta(t,\tau) = \psi(t) + \int_\tau^T g(t,s,Y(s),\zeta(t,s),Z(s,t))ds - \int_\tau^T \zeta(t,s)dW(s),$$

with $\tau \in [t,T]$, we have

$$\begin{aligned}&\mathbb{E}\Big[\sup_{\tau\in[t,T]} |\eta(t,\tau)|^p + \Big(\int_t^T |\zeta(t,s)|^2 ds\Big)^{\frac{p}{2}}\Big]\\ &\leqslant K\mathbb{E}\Big[|\psi(t)|^p + \Big(\int_t^T |g(t,s,Y(s),0,Z(s,t))|ds\Big)^p\Big].\end{aligned}$$

Then, by the facts $Y(t) = \eta(t,t)$ and $Z(t,s) = \zeta(t,s)$, one has

$$\begin{aligned}&\mathbb{E}\Big\{|Y(t)|^p + \Big(\int_t^T |Z(t,s)|^2 ds\Big)^{\frac{p}{2}}\Big\}\\ &\leqslant K\mathbb{E}\Big\{|\psi(t)|^p + \Big[\int_t^T |g(t,s,Y(s),0,Z(s,t))|ds\Big]^p\Big\}\\ &\leqslant K\mathbb{E}\Big\{|\psi(t)|^p + \Big[\int_t^T \big(L_{\widehat z}(t,s) + L_y(t,s)|Y(s)|\big)ds\Big]^p\Big\}\\ &\leqslant K\mathbb{E}\Big\{|\psi(t)|^p + \Big(\int_0^T L_{\widehat z}(t,s)ds\Big)^p\\ &\qquad + \Big(\int_t^T L_y(t,s)^{\frac{p}{p-1}} ds\Big)^{p-1}\Big(\int_t^T |Y(s)|^p ds\Big)\Big\}.\end{aligned} \tag{5.2.35}$$

By Gronwall's inequality, we obtain

$$\mathbb{E}\Big[|Y(t)|^p + \Big(\int_t^T |Z(t,s)|^2 ds\Big)^{\frac{p}{2}}\Big] \leqslant K\mathbb{E}\Big[|\psi(t)|^p + \Big(\int_t^T L_{\widehat z}(t,s)ds\Big)^p\Big]. \tag{5.2.36}$$

On the other hand, by martingale moment inequality, we have

$$\begin{aligned}&\mathbb{E}\Big(\int_0^t|Z(t,s)|^2ds\Big)^{\frac{p}{2}}\leqslant K\mathbb{E}\Big|\int_0^t Z(t,s)dW(s)\Big|^p\\&=K\mathbb{E}\big|Y(t)-\mathbb{E}Y(t)\big|^p\leqslant K\mathbb{E}|Y(t)|^p,\qquad \text{a.e. } t\in[0,T].\end{aligned}\tag{5.2.37}$$

Combining the above, we obtain (5.2.31).

(ii) By BSVIE, one has

$$Y(t)=\mathbb{E}_t\Big[\psi(t)+\int_t^T g(t,s,Y(s),Z(t,s),Z(s,t))ds\Big],\quad t\in[0,T].$$

Thus, for any $S\in[0,T)$, we have

$$\begin{aligned}&\mathbb{E}\Big[\sup_{t\in[S,T]}|Y(t)|^p\Big]\leqslant K(p)\mathbb{E}\Big[\sup_{t\in[S,T]}\Big|\psi(t)+\int_t^T g(t,s,Y(s),Z(t,s),Z(s,t))ds\Big|^p\Big]\\&\leqslant K(p)\mathbb{E}\Big\{\sup_{t\in[S,T]}\Big[|\psi(t)|+\int_t^T\Big(L_{\widehat z}(t,s)+L_y(t,s)|Y(s)|\Big)ds\Big]^p\Big\}\\&\leqslant 3^{p-1}K(p)\mathbb{E}\Big[\sup_{t\in[0,T]}|\psi(t)|^p+\sup_{t\in[0,T]}\Big(\int_t^T L_{\widehat z}(t,s)ds\Big)^p\Big]\\&\qquad+3^{p-1}K(p)(T-S)\sup_{t\in[S,T]}\Big(\int_t^T L_y(t,s)^{\frac{p}{p-1}}ds\Big)^{p-1}\mathbb{E}\Big[\sup_{t\in[S,T]}|Y(t)|^p\Big].\end{aligned}$$

Now, by choosing S_1 such that $T-S_1>0$ is small, we obtain

$$\mathbb{E}\Big[\sup_{t\in[S_1,T]}|Y(t)|^p\Big]\leqslant K\mathbb{E}\Big[\sup_{t\in[0,T]}|\psi(t)|^p+\sup_{t\in[0,T]}\Big(\int_t^T L_{\widehat z}(t,s)ds\Big)^p\Big].$$

For $S\in[0,S_1]$, we have

$$Y(t)=\mathbb{E}_t\Big[\psi^{S_1}(t)+\int_t^{S_1} g(t,s,Y(s),Z(t,s),Z(s,t))ds\Big],\quad t\in[0,S_1],$$

where

$$\psi^{S_1}(t)=\mathbb{E}_t\Big[\psi(t)+\int_{S_1}^T g(t,s,Y(s),Z(t,s),Z(s,t))ds\Big],\quad t\in[0,S_1].$$

We have that

$$
\begin{aligned}
&\mathbb{E}\Big[\sup_{t\in[0,S_1]}|\psi^{S_1}(t)|^p\Big]\\
&\leqslant K(p)\mathbb{E}\Big[\sup_{t\in[0,S_1]}\Big|\psi(t)+\int_{S_1}^T g(t,s,Y(s),Z(t,s),Z(s,t))ds\Big|^p\Big]\\
&\leqslant K(p)\mathbb{E}\Big\{\sup_{t\in[0,S_1]}\Big[|\psi(t)|+\int_{S_1}^T\Big(L_{\hat{z}}(t,s)+L_y(t,s)|Y(s)|\Big)ds\Big]^p\Big\}\\
&\leqslant 3^{p-1}K(p)\mathbb{E}\Big[\sup_{t\in[0,T]}|\psi(t)|^p+\sup_{t\in[0,T]}\Big(\int_t^T L_{\hat{z}}(t,s)ds\Big)^p\Big]\\
&\quad+3^{p-1}K(p)(T-S_1)\sup_{t\in[0,T]}\Big(\int_t^T L_y(t,s)^{\frac{p}{p-1}}ds\Big)^{p-1}\mathbb{E}\Big[\sup_{t\in[S_1,T]}|Y(t)|^p\Big]\\
&\leqslant K\mathbb{E}\Big[\sup_{t\in[0,T]}|\psi(t)|^p+\sup_{t\in[0,T]}\Big(\int_t^T L_{\hat{z}}(t,s)ds\Big)^p\Big].
\end{aligned}
$$

Similar to the study on $[S_1,T]$, we can choose S_2 such that S_1-S_2 is small and obtain

$$
\mathbb{E}\Big[\sup_{t\in[S_2,S_1]}|Y(t)|^p\Big]\leqslant K\mathbb{E}\Big[\sup_{t\in[0,T]}|\psi(t)|^p+\sup_{t\in[0,T]}\Big(\int_t^T L_{\hat{z}}(t,s)ds\Big)^p\Big].
$$

By induction, it follows that

$$
\mathbb{E}\Big[\sup_{t\in[0,T]}|Y(t)|^p\Big]\leqslant K\mathbb{E}\Big[\sup_{t\in[0,T]}|\psi(t)|^p+\sup_{t\in[0,T]}\Big(\int_t^T L_{\hat{z}}(t,s)ds\Big)^p\Big].
$$

Having this, one obtains

$$
\begin{aligned}
&\mathbb{E}\Big[\sup_{t\in[0,T]}\Big|\int_0^t Z(t,s)dW(s)\Big|^p\Big]=\mathbb{E}\Big[\sup_{t\in[0,T]}|Y(t)-\mathbb{E}Y(t)|^p\Big]\\
&\leqslant K\mathbb{E}\Big[\sup_{t\in[0,T]}|Y(t)|^p\Big]\leqslant K\mathbb{E}\Big[\sup_{t\in[0,T]}|\psi(t)|^p+\sup_{t\in[0,T]}\Big(\int_t^T L_{\hat{z}}(t,s)ds\Big)^p\Big],
\end{aligned}
$$

and

$$
\begin{aligned}
&\mathbb{E}\Big[\sup_{t\in[0,T]}\Big|\int_t^T Z(t,s)dW(s)\Big|^p\Big]\\
&=E\Big[\sup_{t\in[0,T]}\Big|Y(t)-\psi(t)-\int_t^T g(t,s,Y(s),Z(t,s),Z(s,t))ds\Big|^p\Big]\\
&\leqslant K\mathbb{E}\Big[\sup_{t\in[0,T]}|Y(t)|^p+\sup_{t\in[0,T]}|\psi(t)|^p+\sup_{t\in[0,T]}\Big(\int_t^T L_{\hat{z}}(t,s)ds\Big)^p\Big]\\
&\leqslant K\mathbb{E}\Big[\sup_{t\in[0,T]}|\psi(t)|^p+\sup_{t\in[0,T]}\Big(\int_t^T L_{\hat{z}}(t,s)ds\Big)^p\Big].
\end{aligned}
$$

This proves estimate (5.2.33). □

The above result gives some better estimates for the adapted M-solution to the Type II BSVIEs. However, we do not expect to have the continuity of $(Y(\cdot), Z(\cdot\,,\cdot))$. In fact, let us recall Example 5.2.3. Choose $\psi_1(\cdot)$ to be discontinuous, bounded by 1, and choose the generator g to be smooth and satisfy the following:

$$g(t,s,y,z,\widehat{z}) = \begin{cases} \widehat{z}, & |\widehat{z}| \leqslant 2, \\ 0, & |\widehat{z}| \geqslant 3. \end{cases}$$

Then (5.2.28) is still the adapted M-solution. Clearly, $Y(\cdot)$ is discontinuous, even if the generator and the free term are smooth.

3 Regularity of Adapted M-Solutions

In the previous section, we have seen that the growth rate of $\widehat{z} \mapsto g(t,s,y,z,\widehat{z})$ plays an important role in getting integrability estimates for the adapted M-solution $(Y(\cdot), Z(\cdot\,,\cdot))$. However, these have nothing to do with the continuity of the solution. By carefully looking at Example 5.2.3, it seems that in order to obtain the continuity of $Y(\cdot)$, in general, one at least needs the continuity of the map $t \mapsto Z(s,t)$. To get more feeling about this, let us make an observation. Consider an adapted M-solution $(Y(\cdot), Z(\cdot\,,\cdot))$ of some Type II BSVIE. Suppose both $Y(\cdot)$ and $Z(\cdot\,,\cdot)$ are Malliavin differentiable. Then from

$$Y(t) = \mathbb{E}Y(t) + \int_0^t Z(t,s)dW(s), \qquad t \in [0,T], \tag{5.3.1}$$

we have

$$D_\tau^{(i)} Y(t) = Z_i(t,\tau) + \int_\tau^t D_\tau^{(i)} Z(t,s)dW(s), \qquad (t,\tau) \in \Delta^*[0,T], \quad 1 \leqslant i \leqslant d. \tag{5.3.2}$$

By writing the above as follows,

$$Z_i(t,\tau) = D_\tau^{(i)} Y(t) - \int_\tau^t D_\tau^{(i)} Z(t,s)dW(s), \qquad (t,\tau) \in \Delta^*[0,T], \ 1 \leqslant i \leqslant d,$$

we see that the behavior of $\tau \mapsto Z(t,\tau)$ can be investigated through those of the right-hand side terms which involve the Malliavin derivatives of the adapted M-solution $(Y(\cdot), Z(\cdot\,,\cdot))$. Inspired by this, we are going to establish some Malliavin regularity of the adapted M-solution $(Y(\cdot), Z(\cdot\,,\cdot))$ and then obtain its continuity in some sense. For the simplicity of the presentation, we restrict ourselves to the L^2-framework. Similar results can be expected under the L^p-framework, for $p \neq 2$. We leave it to the interested readers.

To begin with, for any $0 \leqslant S < T$, let $\Psi[S,T] = L^{\infty}_{\mathcal{F}_T}(S,T;\mathbb{D}^m_{1,2})$ (see Chap. 12, Sect. 1, for details), which is the space consisting of all processes $\psi(\cdot) \in L^{\infty}(S,T;L^2_{\mathcal{F}_T}(\Omega))$ such that $D^{(i)}_t\psi(s)$ $(1 \leqslant i \leqslant d)$ exists and

$$\|\psi(\cdot)\|^2_{\Psi[S,T]} \equiv \sup_{t\in[S,T]} \mathbb{E}\Big[|\psi(t)|^2 + \int_t^T \sum_{i=1}^d |D^{(i)}_t\psi(s)|^2 ds\Big] < \infty.$$

We let $\mathcal{Y}[S,T]$ be the space of all processes $y(\cdot) \in \Psi[S,T]$, which are $\mathbb{F}$-adapted. Likewise, let $\mathcal{Z}[S,T]$ be the space consisting of all processes $z : [S,T]^2 \times \Omega \to \mathbb{R}^{m\times d}$ such that $s \mapsto z(t,s)$ is $\mathbb{F}$-adapted for almost all $t \in [S,T]$, $D^{(i)}_t z(\tau,s)$ exists for almost $t \leqslant s$, and the following holds:

$$\begin{aligned}\|z(\cdot\,,\cdot)\|^2_{\mathcal{Z}[S,T]} \equiv \Big\{ \sup_{t\in[S,T]} \mathbb{E}\Big[\int_S^T |z(t,s)|^2 ds + \int_t^T |z(s,t)|^2 ds\\ + \int_t^T\int_t^T \sum_{i=1}^d |D^{(i)}_t z(\tau,s)|^2 ds d\tau\Big]\Big\} < \infty.\end{aligned} \tag{5.3.3}$$

Clearly, $\|\cdot\|_{\Psi[S,T]}$ and $\|\cdot\|_{\mathcal{Z}[S,T]}$ are norms under which $\Psi[S,T]$ and $\mathcal{Z}[S,T]$ are Banach spaces, respectively, and $\mathcal{Y}[S,T]$ is a closed subspace of $\Psi[S,T]$. For notational better-looking, when $y(\cdot) \in \mathcal{Y}[S,T]$, we use $\|y(\cdot)\|_{\mathcal{Y}[S,T]}$ instead of $\|y(\cdot)\|_{\Psi[S,T]}$ for its norm. We note that in $\Psi[S,T]$, $\mathcal{Y}[S,T]$ and $\mathcal{Z}[S,T]$, no continuity in the time variable is assumed.

Now, we introduce the following assumption which is based on $(\text{H5.2.1})_2$.

$(\text{H5.3.1})_2$ Let $(\text{H5.2.1})_2$ hold. Let $(y,z,\widehat{z}) \mapsto g(t,s,y,z,\widehat{z})$ be continuously differentiable. Moreover, there are deterministic functions $\widehat{L}_y, \widehat{L}_z : \Delta^*[0,T] \to [0,\infty)$ satisfying

$$\int_0^T\int_t^T \widehat{L}_y(t,s)^2 ds dt + \sup_{t\in[0,T]}\int_t^T \widehat{L}_z(t,s)^2 ds < \infty, \tag{5.3.4}$$

such that

$$\begin{aligned}&\sum_{i=1}^d \Big|[D^{(i)}_\tau g](t,s,y,z,\widehat{z})\Big| \leqslant \widehat{L}_y(t,s)|y| + \widehat{L}_z(t,s)\big(|z| + |\widehat{z}|\big),\\ &\qquad\forall (t,s) \in \Delta^*[0,T],\ (y,z,\widehat{z}) \in \mathbb{R}^m \times \mathbb{R}^{m\times d} \times \mathbb{R}^{m\times d},\ \tau \in [0,T], \text{ a.s.}\end{aligned} \tag{5.3.5}$$

Note that under $(\text{H5.3.1})_2$, the Lipschitz condition (5.2.2) is equivalent to the following:

$$\begin{aligned}&|g_y(t,s,y,z,\widehat{z})| \leqslant L_y(t,s),\\ &|g_{z_j}(t,s,y,z,\widehat{z})|, |g_{\widehat{z}_j}(t,s,y,z,\widehat{z})| \leqslant L_z(t,s),\\ &\qquad\forall (t,s,y,z,\widehat{z}) \in \Delta^*[0,T] \times \mathbb{R}^m \times \mathbb{R}^{m\times d} \times \mathbb{R}^{m\times d}, \text{ a.s.}\end{aligned}$$

From (5.3.5), we see that only the linear growth of the map

$$(y, z, \widehat{z}) \mapsto [D_\tau^{(i)} g](t, s, y, z, \widehat{z})$$

is assumed, instead of the Lipschitz continuity. Also, (5.1.1) implies that

$$[D_\tau^{(i)} g](t, s, 0, 0, 0) = 0, \qquad (t, s) \in \Delta^*[0, T], \ \tau \in [0, T].$$

The first main result of this section is the following theorem.

Theorem 5.3.1 *Let (H5.3.1)*$_2$ *hold,* $\psi(\cdot) \in \Psi[0, T]$*, and* $(Y(\cdot), Z(\cdot\,, \cdot)) \in \mathcal{M}^2[0, T]$ *be the adapted M-solution of (1.1). Then for almost* $\tau \in [0, T]$*,* $(D_\tau^{(i)} Y(\cdot), D_\tau^{(i)} Z(\cdot\,, \cdot))$ *exists (*$1 \leqslant i \leqslant d$*) and is the adapted M-solution of the following linear BSVIE:*

$$\begin{aligned} D_\tau^{(i)} Y(t) = D_\tau^{(i)} \psi(t) &+ \int_t^T [D_\tau^{(i)} g](t, s, Y(s), Z(t, s), Z(s, t)) ds \\ &+ \int_t^T \Big\{ g_y(t, s, Y(s), Z(t, s), Z(s, t)) D_\tau^{(i)} Y(s) \\ &+ \sum_{j=1}^d \Big[g_{z_j}(t, s, Y(s), Z(t, s), Z(s, t)) D_\tau^{(i)} Z_j(t, s) \\ &+ g_{\widehat{z}_j}(t, s, Y(s), Z(t, s), Z(s, t)) D_\tau^{(i)} Z_j(s, t) \Big] \Big\} ds \\ &- \int_t^T D_\tau^{(i)} Z(t, s) dW(s), \qquad t \in [\tau, T], \end{aligned} \tag{5.3.6}$$

and for any $S \in [\tau, T)$*,*

$$\begin{aligned} &\mathbb{E} \sum_{i=1}^d \Big(\int_S^T |D_\tau^{(i)} Y(t)|^2 dt + \int_S^T \int_S^T |D_\tau^{(i)} Z(t, s)|^2 ds dt \Big) \\ &\leqslant K \mathbb{E} \Big[\int_S^T |\psi(t)|^2 dt + \sum_{i=1}^d \int_S^T | D_\tau^{(i)} \psi(t)|^2 dt \Big]. \end{aligned} \tag{5.3.7}$$

Moreover,

$$D_\tau^{(i)} Y(t) = Z_i(t, \tau) + \int_\tau^t D_\tau^{(i)} Z(t, s) dW(s), \qquad (t, \tau) \in \Delta^*[0, T], \tag{5.3.8}$$

which implies that a measurable version of Z_i *can be shown as*

$$Z_i(t, \tau) = \mathbb{E}_\tau \big[D_\tau^{(i)} Y(t) \big], \qquad (t, \tau) \in \Delta^*[0, T], \ \ a.s., \quad 1 \leqslant i \leqslant d. \tag{5.3.9}$$

Further, if, in addition, (5.2.30) *and* (5.2.29) *hold for* $p = 2$, *then*

$$\begin{aligned}&\|Y(\cdot)\|^2_{\mathcal{Y}[0,T]}+\|Z(\cdot\,,\cdot)\|^2_{\mathcal{Z}[0,T]}\equiv \sup_{t\in[0,T]}\mathbb{E}\Big[|Y(t)|^2+\int_t^T\sum_{i=1}^d|D_t^{(i)}Y(s)|^2ds\Big]\\&+\mathbb{E}\Big[\int_0^T|Z(t,s)|^2ds+\int_t^T|Z(s,t)|^2ds+\int_t^T\int_t^T\sum_{i=1}^d|D_t^{(i)}Z(\tau,s)|^2dsd\tau\Big]\\&\leqslant K\Big(1+\|\psi(\cdot)\|^2_{\Psi[0,T]}\Big).\end{aligned} \tag{5.3.10}$$

Proof Since $(\text{H5.3.1})_2$ is on top of $(\text{H5.2.1})_2$, by Theorem 5.2.2 for any $\psi(\cdot)\in\Psi[0,T]$, BSVIE (5.1) admits a unique adapted M-solution $(Y(\cdot),Z(\cdot))\in\mathcal{M}^2[0,T]$ and estimate (5.2.11) holds (with $p=2$). Next, we introduce the following linear BSVIE for $(\widetilde{Y}^{\tau,i}(\cdot),\widetilde{Z}^{\tau,i}(\cdot\,,\cdot))$ (which is a formal Malliavin differentiation of (5.1)):

$$\begin{aligned}\widetilde{Y}^{\tau,i}(t)=\boldsymbol{\psi}(t)+\int_t^T\mathbf{g}(t,s,\widetilde{Y}^{\tau,i}(s),\widetilde{Z}^{\tau,i}(t,s),\widetilde{Z}^{\tau,i}(s,t))ds\\-\int_t^T\widetilde{Z}^{\tau,i}(t,s)dW(s),\qquad t\in[\tau,T],\end{aligned}\tag{5.3.11}$$

where

$$\boldsymbol{\psi}(t)=D_\tau^{(i)}\psi(t)+\int_t^T[D_\tau^{(i)}g](t,s,Y(s),Z(t,s),Z(s,t))ds,$$

and

$$\begin{aligned}\mathbf{g}(t,s,\widetilde{y},\widetilde{z},\widetilde{\zeta})=\;&g_y(t,s,Y(s),Z(t,s),Z(s,t))\widetilde{y}\\&+\sum_{j=1}^d\Big(g_{z_j}(t,s,Y(s),Z(t,s),Z(s,t))\widetilde{z}_j\\&\qquad+g_{\widehat{z}_j}(t,s,Y(s),Z(t,s),Z(s,t))\widetilde{\zeta}_j\Big).\end{aligned}$$

Let us estimate the following:

$$\begin{aligned}&\mathbb{E}\int_0^T\Big|\int_t^T[D_\tau^{(i)}g](t,s,Y(s),Z(t,s),Z(s,t))ds\Big|^2dt\\&\leqslant\mathbb{E}\int_0^T\Big[\int_t^T\Big(\widehat{L}_y(t,s)|Y(s)|+\widehat{L}_z(t,s)|Z(t,s)|+\widehat{L}_z(t,s)|Z(s,t)|\Big)ds\Big]^2dt\\&\leqslant K\Big\{\Big(\int_0^T\int_t^T\widehat{L}_y(t,s)^2dsdt\Big)\mathbb{E}\Big(\int_0^T|Y(s)|^2ds\Big)\\&\qquad+\Big[\sup_{t\in[0,T]}\int_t^T\widehat{L}_z(t,s)^2ds\Big]\mathbb{E}\int_0^T\int_0^T|Z(t,s)|^2dsdt\Big\}<\infty.\end{aligned}$$

Thus, the free term $\psi(\cdot) \in L^p_{\mathcal{F}_T}(0, T; \mathbb{R}^m)$ and the generator $\mathbf{g}(t, s, \widetilde{y}, \widetilde{z}, \widetilde{\zeta})$ satisfy (H5.2.1)$_2$. Hence, by Theorem 5.2.2, the linear BSVIE (5.3.11) admits a unique adapted M-solution $(\widetilde{Y}^{\tau,i}(\cdot), \widetilde{Z}^{\tau,i}(\cdot\,,\cdot)) \in \mathcal{M}^2[\tau, T]$, and for any $S \in [\tau, T]$,

$$
\begin{aligned}
&\|(\widetilde{Y}^{\tau,i}(\cdot), \widetilde{Z}^{\tau,i}(\cdot\,,\cdot))\|^2_{\mathcal{M}^2[S,T]} \\
&\equiv \mathbb{E}\Big\{\int_S^T |\widetilde{Y}^{\tau,i}(t)|^2 dt + \int_S^T\int_S^T |\widetilde{Z}^{\tau,i}(t,s)|^2 ds dt\Big\} \leqslant K\mathbb{E}\int_S^T |\psi(t)|^2 dt \\
&\leqslant K\mathbb{E}\Big\{\int_S^T |D_\tau^{(i)}\psi(t)|^2 dt + \Big(\int_S^T\int_t^T \widehat{L}_y(t,s)^2 ds dt\Big)\int_S^T |Y(s)|^2 ds \\
&\qquad + \Big(\sup_{t\in[S,T]}\int_t^T \widehat{L}_z(t,s)^2 ds\Big)\int_S^T\int_S^T |Z(t,s)|^2 ds dt\Big\} \\
&\leqslant K\mathbb{E}\Big(\int_S^T |\psi(t)|^2 dt + \int_S^T |D_\tau^{(i)}\psi(t)|^2 dt\Big).
\end{aligned}
\tag{5.3.12}
$$

We now claim that the following holds so that (5.3.6) holds, and (5.3.7) will follow from (5.3.12):

$$
\widetilde{Y}^{\tau,i}(t) = D_\tau^{(i)}Y(t), \quad \widetilde{Z}^{\tau,i}(t,s) = D_\tau^{(i)}Z(t,s), \qquad t,s \in [\tau, T], \text{ a.s.} \tag{5.3.13}
$$

To show this, let us define a sequence:

$$
\begin{cases} (Y^0(\cdot), Z^0(\cdot)) = 0, \\ (Y^{k+1}(\cdot), Z^{k+1}(\cdot)) = \Theta(Y^k(\cdot), Z^k(\cdot)), \qquad k \geqslant 0, \end{cases} \tag{5.3.14}
$$

where we recall that $\Theta : \mathcal{M}^{2,\beta}[0, T] \to \mathcal{M}^{2,\beta}[0, T]$ is defined by the solution map of the following:

$$
\begin{aligned}
Y^{k+1}(t) = \psi(t) + &\int_t^T g(t, s, Y^k(s), Z^{k+1}(t,s), Z^k(s,t)) ds \\
&- \int_t^T Z^{k+1}(t,s) dW(s), \qquad t \in [0, T].
\end{aligned}
\tag{5.3.15}
$$

Under (H5.2.1)$_2$, by the proof of Theorem 5.2.2, one has that $\Theta : \mathcal{M}^{2,\beta}[0, T] \to \mathcal{M}^{2,\beta}[0, T]$ is a contraction, for $\beta > 0$ large. Hence,

$$
\lim_{k\to\infty} \|(Y^k(\cdot), Z^k(\cdot)) - (Y(\cdot), Z(\cdot))\|_{\mathcal{M}^{2,\beta}[0,T]} = 0. \tag{5.3.16}
$$

On the other hand, for each $k \geqslant 1$, we may consider the following auxiliary BSDE:

$$\begin{aligned}\eta^{k+1}(t,\tau)=\psi(t)+&\int_\tau^T g(t,s,Y^k(s),\zeta^{k+1}(t,s),Z^k(s,t))ds\\&-\int_\tau^T\zeta^{k+1}(t,s)dW(s),\qquad \tau\in[t,T],\end{aligned}$$

so that

$$\begin{aligned}&Y^{k+1}(t)=\eta^{k+1}(t,t),\quad Z^{k+1}(t,s)=\zeta^{k+1}(t,s),\qquad (t,s)\in\Delta^*[0,T],\\&Y^{k+1}(t)=\mathbb{E}Y^{k+1}(t)+\int_0^t Z^{k+1}(t,s)dW(s),\qquad t\in[0,T].\end{aligned}$$

By Theorem 3.2.1 of Chap. 3, we have that $(D_\theta^{(i)}\eta^{k+1}(t,\tau),D_\theta^{(i)}\zeta^{k+1}(t,\tau))$ exists for any $\theta\in[0,T]$, satisfying BSDE (parameterized by (θ,i,t)):

$$\begin{aligned}D_\theta^{(i)}\eta^{k+1}(t,\tau)=&D_\theta^{(i)}\psi(t)+\int_\tau^T\Big[[D_\theta^{(i)}g](t,s,Y^k(s),\zeta^{k+1}(t,s),Z^k(s,t))\\&+g_y(t,s,Y^k(s),\zeta^{k+1}(t,s),Z^k(s,t))D_\theta^{(i)}Y^k(s)\\&+\sum_{j=1}^d\Big(g_{\widehat z_j}(t,s,Y^k(s),\zeta^{k+1}(t,s),Z^k(s,t))D_\theta^{(i)}Z_j^k(s,t)\\&+g_{z_j}(t,s,Y^k(s),\zeta^{k+1}(t,s),Z^k(s,t))D_\theta^{(i)}\zeta_j^{k+1}(t,s)\Big)\Big]ds\\&-\int_\tau^T D_\theta^{(i)}\zeta^{k+1}(t,s)dW(s),\quad \tau\in[t,T],\ 1\leqslant i\leqslant d.\end{aligned}\tag{5.3.17}$$

Corresponding to (5.3.11), we also introduce the following auxiliary BSDE:

$$\begin{aligned}\widetilde\eta^{\theta,i}(t,\tau)=D_\theta^{(i)}\psi(t)+&\int_\tau^T\Big[[D_\theta^{(i)}g](t,s,Y(s),Z(t,s),Z(s,t))\\&+g_y(t,s,Y(s),Z(t,s),Z(s,t))\widetilde Y^{\theta,i}(s)\\&+\sum_{j=1}^d\Big(g_{\widehat z_j}(t,s,Y(s),Z(t,s),Z(s,t))\widetilde Z_j^{\theta,i}(s,t)\\&+g_{z_j}(t,s,Y(s),Z(t,s),Z(s,t))\widetilde\zeta_j^{\theta,i}(t,s)\Big)\Big]ds\\&-\int_\tau^T\widetilde\zeta^{\theta,i}(t,s)dW(s),\qquad\tau\in[t,T].\end{aligned}\tag{5.3.18}$$

Here $\widetilde Y^{\theta,i}(\cdot)$ satisfies the above (5.3.11). Then, applying the stability estimate to BSDEs (5.3.17) and (5.3.18), we have

$$
\begin{aligned}
&\mathbb{E}\Big[\sup_{\tau\in[t,T]}|D_\theta^{(i)}\eta^{k+1}(t,\tau)-\widetilde{\eta}^{\theta,i}(t,\tau)|^2 \\
&\qquad\qquad +\int_t^T|D_\theta^{(i)}\zeta^{k+1}(t,s)-\widetilde{\zeta}^{\theta,i}(t,s)|^2ds\Big] \\
&\leqslant K\mathbb{E}\Big[\int_t^T\Big(\big|[D_\theta^{(i)}g](t,s,Y^k(s),Z^{k+1}(t,s),Z^k(s,t)) \\
&\quad -[D_\theta^{(i)}g](t,s,Y(s),Z(t,s),Z(s,t))\big| \\
&\quad +\big|g_y(t,s,Y(s),Z(t,s),Z(s,t))\big(D_\theta^{(i)}Y^k(s)-\widetilde{Y}^{\theta,i}(s)\big)\big| \\
&\quad +\sum_{j=1}^d\big|g_{\widehat{z}_j}(t,s,Y(s),Z(t,s),Z(s,t))\big(D_\theta^{(i)}Z(s,t)-\widetilde{Z}^{\theta,i}(s,t)\big)\big|\Big)ds\Big]^2 \\
&\leqslant K\mathbb{E}\Big[\int_t^T\Big(\delta_k^{(i)}(t,s)+L_y(t,s)\big|D_\theta^{(i)}Y^k(s)-\widetilde{Y}^{\theta,i}(s)\big| \\
&\qquad\qquad +L_{\widehat{z}}(t,s)\big|D_\theta^{(i)}Z^k(s,t)-\widetilde{Z}^{\theta,i}(s,t)\big|\Big)ds\Big]^2,
\end{aligned}
$$

where

$$
\begin{aligned}
\delta_k^{(i)}(t,s)=\big|&[D_\theta^{(i)}g](t,s,Y^k(s),Z^{k+1}(t,s),Z^k(s,t)) \\
&-[D_\theta^{(i)}g](t,s,Y(s),Z(t,s),Z(s,t))\big|.
\end{aligned}
$$

By the dominated convergence theorem, one has

$$
\lim_{k\to\infty}\mathbb{E}\int_0^T\int_t^T\delta_k^{(i)}(t,s)^2dsdt=0.
$$

For $0\leqslant s\leqslant t\leqslant T$, denote

$$
\begin{aligned}
y^k(t)&=|D_\theta^{(i)}Y^k(t)-\widetilde{Y}^{\theta,i}(t)|, \\
z^k(t,s)&=|D_\theta^{(i)}Z^k(t,s)-\widetilde{Z}^{\theta,i}(t,s)|.
\end{aligned}
$$

Then we have

$$
\begin{aligned}
&\mathbb{E}\Big[y^{k+1}(t)^2+\int_t^T z^{k+1}(t,s)^2ds\Big] \\
&\leqslant K\mathbb{E}\Big[\int_t^T\Big(\delta_k^{(i)}(t,s)+L_y(t,s)y^k(s)+L_{\widehat{z}}(t,s)z^k(s,t)\Big)ds\Big]^2.
\end{aligned}
$$

Now, let $\beta>0$; we estimate the following:

$$
\begin{aligned}
&\mathbb{E}\int_0^T e^{2\beta t}y^{k+1}(t)^2dt+\mathbb{E}\int_0^T e^{2\beta t}\int_t^T z^{k+1}(t,s)^2dsdt \\
&\leqslant K\mathbb{E}\int_0^T e^{2\beta t}\Big[\int_t^T\Big(\delta_k^{(i)}(t,s)+L_y(t,s)y^k(s)+L_{\widehat{z}}(t,s)z^k(s,t)\Big)ds\Big]^2dt
\end{aligned}
$$

$$
\begin{aligned}
&\leqslant K\mathbb{E}\Big\{\int_0^T e^{2\beta t}\Big(\int_t^T \delta_k^{(i)}(t,s)ds\Big)^2 dt+\int_0^T\Big(\int_t^T e^{\beta(t-s)}L_y(t,s)[e^{\beta s}y^k(s)]ds\Big)^2 dt\\
&\qquad+\int_0^T\Big(\int_t^T e^{\beta(t-s)}L_{\widehat{z}}(s,t)[e^{\beta s}z^k(s,t)]ds\Big)^2 dt\Big\}\equiv I_1+I_2+I_3.
\end{aligned}
$$

It follows that

$$
\begin{aligned}
I_2&\equiv\mathbb{E}\int_0^T\Big(\int_t^T e^{\beta(t-s)}L_y(t,s)[e^{\beta s}y^k(s)]ds\Big)^2 dt\\
&\leqslant\int_0^T\Big(\int_t^T e^{2\beta(t-s)}L_y(t,s)^2 ds\Big)\mathbb{E}\Big(\int_t^T e^{2\beta s}y^k(s)^2 ds\Big)dt\\
&\leqslant\int_0^T\Big(\int_t^T e^{\frac{2\beta(1+\varepsilon)}{\varepsilon}(t-s)}ds\Big)^{\frac{\varepsilon}{1+\varepsilon}}\Big(\int_t^T L_y(t,s)^{2(1+\varepsilon)}ds\Big)^{\frac{1}{1+\varepsilon}}\mathbb{E}\Big(\int_0^T e^{2\beta s}y^k(s)^2 ds\Big)dt\\
&\leqslant\Big(\frac{\varepsilon}{2\beta(1+\varepsilon)}\Big)^{\frac{\varepsilon}{1+\varepsilon}}\Big[\int_0^T\Big(\int_t^T L_y(t,s)^{2(1+\varepsilon)}ds\Big)^{\frac{1}{1+\varepsilon}}dt\Big]\mathbb{E}\Big(\int_0^T e^{2\beta s}y^k(s)^2 ds\Big)\\
&\leqslant\frac{K}{\beta^{\frac{\varepsilon}{1+\varepsilon}}}\mathbb{E}\Big(\int_0^T e^{2\beta s}y^k(s)^2 ds\Big),
\end{aligned}
$$

and

$$
\begin{aligned}
I_3&\equiv\mathbb{E}\int_0^T\Big(\int_t^T e^{\beta(t-s)}L_{\widehat{z}}(s,t)[e^{\beta s}z^k(s,t)]ds\Big)^2 dt\\
&\leqslant\int_0^T\Big[\Big(\int_t^T e^{2\beta(t-s)}L_{\widehat{z}}(s,t)^2 ds\Big)\mathbb{E}\Big(\int_t^T e^{2\beta s}z^k(s,t)^2 ds\Big)\Big]dt\\
&\leqslant\int_0^T\Big(\int_t^T e^{\frac{2\beta(1+\varepsilon)}{\varepsilon}(t-s)}ds\Big)^{\frac{\varepsilon}{1+\varepsilon}}\Big(\int_t^T L_{\widehat{z}}(t,s)^{2(1+\varepsilon)}ds\Big)^{\frac{1}{1+\varepsilon}}\mathbb{E}\Big(\int_t^T e^{2\beta s}z^k(s,t)^2 ds\Big)dt\\
&\leqslant\Big(\frac{\varepsilon}{2\beta(1+\varepsilon)}\Big)^{\frac{\varepsilon}{1+\varepsilon}}\sup_{t\in[0,T]}\Big(\int_t^T L_{\widehat{z}}(t,s)^{2(1+\varepsilon)}ds\Big)^{\frac{1}{1+\varepsilon}}\\
&\qquad\qquad\cdot\mathbb{E}\int_0^T e^{2\beta s}\Big(\int_0^s z^k(s,t)^2 dt\Big)ds\\
&\leqslant\frac{K}{\beta^{\frac{\varepsilon}{1+\varepsilon}}}\mathbb{E}\int_0^T\Big(\int_0^s e^{2\beta s}z^k(s,t)^2 dt\Big)ds.
\end{aligned}
$$

Hence,

$$
\begin{aligned}
&\mathbb{E}\int_0^T e^{2\beta t}y^{k+1}(t)^2 dt+\mathbb{E}\int_0^T e^{2\beta t}\int_t^T z^{k+1}(t,s)^2 dsdt\\
&\leqslant K\int_0^T e^{2\beta t}\mathbb{E}\Big(\int_t^T \delta_k^{(i)}(t,s)ds\Big)^2 dt\\
&\qquad+\frac{K}{\beta^{\frac{\varepsilon}{1+\varepsilon}}}\Big[\mathbb{E}\int_0^T e^{2\beta s}y^k(s)^2 ds+\mathbb{E}\int_0^T e^{2\beta t}\int_0^t z^k(t,s)^2 dsdt\Big].
\end{aligned}
\tag{5.3.19}
$$

Since

$$Y^k(t) = \mathbb{E}Y^k(t) + \int_0^t Z^k(t,s)dW(s),$$

which leads to

$$D_\theta^{(i)}Y^k(t) = Z_i^k(t,\theta) + \int_\theta^t D_\theta^{(i)}Z^k(t,s)dW(s),$$

with $Z_i^k(t,\theta) = \mathbb{E}_\theta[D_\theta^{(i)}Y^k(t)]$, and

$$\widetilde{Y}^{\theta,i}(t) = \mathbb{E}_\theta\widetilde{Y}^{\theta,i}(t) + \int_\theta^t \widetilde{Z}^{\theta,i}(t,s)dW(s),$$

we have

$$\begin{aligned} D_\theta^{(i)}Y^k(t) - \widetilde{Y}^{\theta,i}(t) &= \mathbb{E}_\theta[D_\theta^{(i)}Y^k(t) - Y^{\theta,i}(t)] \\ &\quad + \int_\theta^t \Big(D_\theta^{(i)}Z^k(t,s) - \widetilde{Z}^{\theta,i}(t,s)\Big)dW(s). \end{aligned}$$

Thus, by the definitions of $y^k(t)$ and $z^k(t,s)$, one has

$$\mathbb{E}_\theta \int_\theta^t z^k(t,s)^2 ds \leqslant K\mathbb{E}_\theta[y^k(t)^2].$$

Consequently,

$$\mathbb{E}\int_0^T e^{2\beta t}\int_0^t z^k(t,s)^2 dsdt \leqslant K\mathbb{E}\int_0^T e^{2\beta t}y^k(t)^2 dt.$$

Therefore, (5.3.19) implies

$$\begin{aligned} \boldsymbol{\theta}_{k+1} &\equiv \mathbb{E}\int_0^T e^{2\beta t}y^{k+1}(t)^2 dt + \mathbb{E}\int_0^T e^{2\beta t}\int_t^T z^{k+1}(t,s)^2 dsdt \\ &\leqslant K\int_0^T e^{2\beta t}\mathbb{E}\Big(\int_t^T \delta_k^{(i)}(t,s)ds\Big)^2 dt + \frac{K}{\beta^{\frac{\varepsilon}{1+\varepsilon}}}\mathbb{E}\int_0^T e^{2\beta s}y^k(s)^2 ds \\ &\leqslant K\int_0^T e^{2\beta t}\mathbb{E}\Big(\int_t^T \delta_k^{(i)}(t,s)ds\Big)^2 dt + \frac{K}{\beta^{\frac{\varepsilon}{1+\varepsilon}}}\boldsymbol{\theta}_k. \end{aligned}$$

Now, choosing $\beta > 0$ large so that $\beta^{\frac{\varepsilon}{1+\varepsilon}} > 2K$, we see that $\{\boldsymbol{\theta}_k\}_{k\geqslant 1}$ is uniformly bounded. Then there exists a subsequence, $\{\boldsymbol{\theta}_{k_j}\}_{j\geqslant 1}$, such that

$$\lim_{j\to\infty}\boldsymbol{\theta}_{k_j} = 0.$$

Since operator $\xi \mapsto D_r^i \xi$ is closed, we must have (5.3.13). This proves (5.3.6) and (5.3.7). Then (5.3.8) and (5.3.9) follow easily.

Next, in the case that (5.2.30) holds and (5.2.29) holds for $p = 2$, from (5.3.9) and (5.3.7), we have

$$\begin{aligned} &\mathbb{E}\int_t^T |Z(s,t)|^2 ds \leqslant \sum_{i=1}^d \mathbb{E}\int_t^T |D_t^{(i)} Y(s)|^2 ds \\ &\leqslant K\mathbb{E}\Big\{\int_t^T |\psi(s)|^2 ds + \sum_{i=1}^d \int_t^T |D_t^{(i)}\psi(s)|^2 ds\Big\} \leqslant K\|\psi(\cdot)\|^2_{\Psi[0,T]}. \end{aligned} \tag{5.3.20}$$

Combining the above with (5.2.31) in Theorem 5.2.4 (i), we obtain

$$\mathbb{E}\Big\{|Y(t)|^2 + \int_0^T |Z(t,s)|^2 ds + \int_t^T |Z(s,t)|^2 ds\Big\} \leqslant K\big(1 + \|\psi(\cdot)\|^2_{\Psi[0,T]}\big). \tag{5.3.21}$$

Then, combining the above with (5.3.7), we obtain (5.3.10). □

Note that the above theorem does not give the continuity of $t \mapsto Y(t)$ in any sense. As a matter of fact, due to the dependence of the generator $g(t,s,y,z,\widehat{z})$ on $\widehat{z}$ (namely, the term $Z(s,t)$ appears in the drift of the BSVIE), the continuity of $Y(\cdot)$ becomes very subtle. The following example shows that the continuity of $t \mapsto \psi(t)$ alone (even with smooth $g(t,s,y,z,\widehat{z})$) is not enough to guarantee the continuity of $t \mapsto Y(t)$.

Example 5.3.2 Let $m = d = 1$. Consider the following BSVIE (compare with Example 1.1):

$$Y(t) = \psi(t) + \int_t^T Z(s,t)ds - \int_t^T Z(t,s)dW(s), \qquad t \in [0,T]. \tag{5.3.22}$$

Let

$$\psi(t) = \psi_0(t)\int_0^T \psi_1(s)dW(s), \qquad t \in [0,T], \tag{5.3.23}$$

where $\psi_0(\cdot)$ and $\psi_1(\cdot)$ are two deterministic square integrable functions. We assume that $\psi_0(\cdot)$ is continuous and $\psi_1(\cdot)$ is discontinuous. Then $t \mapsto \psi(t)$ is continuous and $\mathcal{F}_T$-measurable. If we define

$$\begin{cases} Y(t) = \psi_0(t)\displaystyle\int_0^t \psi_1(s)dW(s) + \psi_1(t)\int_t^T \psi_0(s)ds, & t \in [0,T], \\ Z(t,s) = \psi_0(t)\psi_1(s), \qquad (s,t) \in [0,T]^2, \end{cases} \tag{5.3.24}$$

then

$$\begin{aligned}&\psi(t)+\int_t^T Z(s,t)ds-\int_t^T Z(t,s)dW(s)\\&=\psi_0(t)\int_0^T\psi_1(s)dW(s)+\psi_1(t)\int_t^T\psi_0(s)ds-\psi_0(t)\int_t^T\psi_1(s)dW(s)\\&=\psi_0(t)\int_0^t\psi_1(s)dW(s)+\psi_1(t)\int_t^T\psi_0(s)ds=Y(t),\end{aligned}$$

and

$$\begin{aligned}Y(t)-\mathbb{E}Y(t)&=\psi_0(t)\int_0^t\psi_1(s)dW(s)+\psi_1(t)\int_t^T\psi_0(s)ds-\psi_1(t)\int_t^T\psi_0(s)ds\\&=\psi_0(t)\int_0^t\psi_1(s)dW(s)=\int_0^t Z(t,s)dW(s).\end{aligned}$$

Thus, $(Y(\cdot),Z(\cdot\,,\cdot))$ defined by (5.3.24) is the adapted M-solution of (5.3.22). Clearly, $t\mapsto Y(t)$ is not necessarily continuous. On the other hand, from (5.3.24), we see that $Y(\cdot)$ is continuous if both $\psi_0(\cdot)$ and $\psi_1(\cdot)$ are continuous. We also note that

$$D_r\psi(t)=\psi_0(t)\psi_1(r),\qquad (t,r)\in[0,T]^2.$$

Thus, in order to obtain the continuity of $Y(\cdot)$, some continuity conditions on the Malliavin derivative $D_r\psi(\cdot)$ of $\psi(\cdot)$, etc. should be imposed.

Now, for convenience, let us introduce some new notations. Let $\Psi_c[S,T]$ be the space consisting of all $\psi(\cdot)\in\Psi[S,T]$ such that

$$\lim_{\substack{t,\bar t\in[S,T]\\|t-\bar t|\to 0}}\mathbb{E}\Big\{|\psi(t)-\psi(\bar t)|^2+\int_{t\vee\bar t}^T\sum_{i=1}^d|D_t^{(i)}\psi(s)-D_{\bar t}^{(i)}\psi(s)|^2ds\Big\}=0,\qquad(5.3.25)$$

and let $\mathcal{Y}_c[S,T]$ be the space of all process $y(\cdot)\in\Psi_c[S,T]$ that are $\mathbb{F}$-adapted. For any $\psi(\cdot)\in\Psi_c[S,T]$, we have (with $i=1,2,\cdot s,d$, $S\leqslant t<\bar t\leqslant T$)

$$\begin{aligned}&\mathbb{E}\Big|\int_t^T D_t^{(i)}\psi(s)ds-\int_{\bar t}^T D_{\bar t}^{(i)}\psi(s)ds\Big|^2\\&\leqslant 2\mathbb{E}\Big(\int_{\bar t}^T|D_t^{(i)}\psi(s)-D_{\bar t}^{(i)}\psi(s)|ds\Big)^2+2\mathbb{E}\Big(\int_t^{\bar t}|D_t^{(i)}\psi(s)|ds\Big)^2\\&\leqslant 2(T-\bar t)\mathbb{E}\int_{\bar t}^T|D_t^{(i)}\psi(s)-D_{\bar t}^{(i)}\psi(s)|^2ds+2(\bar t-t)\mathbb{E}\int_t^T|D_t^{(i)}\psi(s)|^2ds.\end{aligned}$$

Thus, for each $i = 1, 2, \cdot s, d$,

$$t \mapsto \int_t^T D_t^{(i)} \psi(s) ds \quad \text{is continuous from } [S, T] \text{ to } L^2_{\mathcal{F}_T}(\Omega; \mathbb{R}^m).$$

Likewise, let $\mathcal{Z}_c[S, T]$ be the space of all $z(\cdot\,,\cdot) \in \mathcal{Z}[S, T]$ such that

$$\begin{aligned}\lim_{\substack{t,\bar{t}\in[S,T]\\ |t-\bar{t}|\to 0}} \mathbb{E}\Big\{ &\int_{t\vee\bar{t}}^T |z(t,s) - z(\bar{t},s)|^2 ds + \int_{t\vee\bar{t}}^T |z(s,t) - z(s,\bar{t})|^2 ds \\ &+ \int_{t\vee\bar{t}}^T \int_{t\vee\bar{t}}^T \sum_{i=1}^d |D_t^{(i)} z(s,\tau) - D_{\bar{t}}^{(i)} z(s,\tau)|^2 d\tau ds \Big\} = 0.\end{aligned} \tag{5.3.26}$$

For any $z(\cdot\,,\cdot) \in \mathcal{Z}_c[S, T]$ given, for each $i = 1, 2, \cdot s, d$ and $S \leqslant t < \bar{t} \leqslant T$, we have

$$\begin{aligned}&\mathbb{E}\Big\{ \Big| \int_t^T z(t,s)ds - \int_{\bar{t}}^T z(\bar{t},s)ds \Big|^2 + \Big| \int_t^T z(s,t)ds - \int_{\bar{t}}^T z(s,\bar{t})ds \Big|^2 \\ &\quad + \Big| \int_t^T \int_t^T D_t^{(i)} z(s,\tau) d\tau ds - \int_{\bar{t}}^T \int_{\bar{t}}^T D_{\bar{t}}^{(i)} z(s,\tau) d\tau ds \Big|^2 \Big\} \\ &\leqslant K\mathbb{E}\Big\{ \int_{\bar{t}}^T |z(t,s) - z(\bar{t},s)|^2 ds + (\bar{t} - t) \int_S^T |z(t,s)|^2 ds \\ &\quad + \int_{\bar{t}}^T |z(s,t) - z(s,\bar{t})|^2 ds + (\bar{t} - t) \int_t^T |z(s,t)|^2 ds \\ &\quad + \int_{\bar{t}}^T \int_{\bar{t}}^T |D_t^{(i)} z(s,\tau) - D_{\bar{t}}^{(i)} z(s,\tau)|^2 d\tau ds + (\bar{t} - t) \int_t^T \int_t^T |D_t^{(i)} z(s,\tau)|^2 d\tau ds \Big\}.\end{aligned}$$

This together with (5.3.3) indicates that the following maps

$$t \mapsto \int_t^T z(t,s)ds, \quad t \mapsto \int_t^T z(s,t)ds, \quad t \mapsto \int_t^T \int_t^T D_t z(s,\tau) d\tau ds,$$

are continuous from $[S, T]$ to $L^2_{\mathcal{F}_T}(\Omega; \mathbb{R}^{m\times d})$. It is easy to see that $\Psi_c[S, T]$, $\mathcal{Y}_c[S, T]$, and $\mathcal{Z}_c[S, T]$ are closed subspaces of $\Psi[S, T]$, $\mathcal{Y}[S, T]$, and $\mathcal{Z}[S, T]$, respectively, and $\mathcal{Y}_c[S, T]$ is a closed subspace of $\Psi_c[S, T]$. Also, we note that

$$\mathcal{Y}_c[S, T] \subseteq C_{\mathbb{F}}([S, T]; L^2(\Omega; \mathbb{R}^m)). \tag{5.3.27}$$

Now, we further strengthen $(\text{H5.3.1})_2$.

(H5.3.2)$_2$ Let $(\text{H5.3.1})_2$ hold with $(L_z(t,s) = L_{\hat{z}}(t,s))$

$$\sup_{t\in[0,T]} \int_t^T \Big(L_y(t,s)^{2(1+\varepsilon)} + L_z(t,s)^{2(1+\varepsilon)} \Big) ds < \infty, \tag{5.3.28}$$

for some $\varepsilon > 0$, and the map $\tau \mapsto [D_\tau^{(i)} g](t, s, y, z, \widehat{z})$ is continuously uniform in $(t, s, y, z, \widehat{z})$. There exists a modulus of continuity $\kappa : [0, \infty) \to [0, \infty)$ and functions $L(\cdot) \in L^2(0, T)$, $\widehat{L}(\cdot\,,\cdot) \in L^2(\Delta^*[0, T])$ (i.e., square integrable in $\Delta^*[0, T]$) such that

$$
\begin{aligned}
&|g(t_1, s, y, z, \widehat{z}) - g(t_2, s, y, z, \widehat{z})| \leqslant \kappa(|t_1 - t_2|)L(s)(1 + |y| + |z| + |\widehat{z}|),\\
&\qquad (t_1, s), (t_2, s) \in \Delta^*[0, T],\ (y, z, \widehat{z}) \in \mathbb{R}^m \times \mathbb{R}^{m\times d} \times \mathbb{R}^{m\times d},
\end{aligned}
\tag{5.3.29}
$$

and

$$
\begin{aligned}
&\big|[D_{t_1}^{(i)} g](t, s, y, z, \widehat{z}) - [D_{t_1}^{(i)} g](t, s, y, z, \widehat{z})\big|\\
&\leqslant \kappa(|t_1 - t_2|)\widehat{L}(t, s)(1 + |y| + |z| + |\widehat{z}|),\\
&\qquad t_1, t_2 \in [0, T],\ (t, s, y, z, \widehat{z}) \in \Delta^*[0, T] \times \mathbb{R}^m \times \mathbb{R}^{m\times d} \times \mathbb{R}^{m\times d}.
\end{aligned}
\tag{5.3.30}
$$

The next result is about the continuity of the adapted M-solution.

Theorem 5.3.3 *Let (H5.3.2)$_2$ hold. Then for any $\psi(\cdot) \in \Psi_c[0, T]$, the unique M-solution $(Y(\cdot), Z(\cdot\,,\cdot)) \in \mathcal{H}^2[0, T]$ admits a measurable version $Y(\cdot) \in \mathcal{Y}_c[0, T]$ and $Z(\cdot\,,\cdot) \in \mathcal{Z}_c[0, T]$, namely,*

$$
\begin{aligned}
\lim_{|t-\bar{t}|\to 0} \mathbb{E}\Big\{&|Y(t) - Y(\bar{t})|^2 + \sum_{i=1}^d \int_{t\vee\bar{t}}^T |D_t^{(i)} Y(s) - D_{\bar{t}}^{(i)} Y(s)|^2 ds\\
&+ \int_{t\vee\bar{t}}^T \Big(|Z(t, s) - Z(\bar{t}, s)|^2 + |Z(s, t) - Z(s, \bar{t})|^2\Big) ds\\
&+ \sum_{i=1}^d \int_{t\vee\bar{t}}^T \int_{t\vee\bar{t}}^T |D_t^{(i)} Z(s, \tau) - D_{\bar{t}}^{(i)} Z(s, \tau)|^2 d\tau ds\Big\} = 0,
\end{aligned}
\tag{5.3.31}
$$

and

$$
\int_\cdot^T Z(\cdot\,, s) dW(s) \in C_{\mathcal{F}_T}([0, T]; L^2(\Omega; \mathbb{R}^m)).
\tag{5.3.32}
$$

Proof For given adapted M-solution $(Y(\cdot), Z(\cdot\,,\cdot)) \in \mathcal{H}^2[0, T]$ of BSVIE (5.1), consider the following auxiliary BSDE on $[t, T]$ (parameterized by $t \in [0, T)$):

$$
\eta(t, \tau) = \psi(t) + \int_\tau^T g(t, s, Y(s), \zeta(t, s), Z(s, t)) ds - \int_\tau^T \zeta(t, s) dW(s).
$$

Let $0 \leqslant t_1 < t_2 \leqslant T$. Then by the stability estimate in Sect. 1 of Chap. 3, with $t = t_1, t_2$, one has

$$
\begin{aligned}
&\mathbb{E}\Big[\sup_{\tau\in[t_2,T]}|\eta(t_1,\tau)-\eta(t_2,\tau)|^2+\int_{t_2}^T|\zeta(t_1,s)-\zeta(t_2,s)|^2ds\Big]\\
&\leqslant K\mathbb{E}\Big[|\psi(t_1)-\psi(t_2)|^2+\Big(\int_{t_2}^T|g(t_1,s,Y(s),\zeta(t_2,s),Z(s,t_1))\\
&\qquad\qquad -g(t_2,s,Y(s),\zeta(t_2,s),Z(s,t_2))|ds\Big)^2\Big]\\
&\leqslant K\mathbb{E}\Big\{|\psi(t_1)-\psi(t_2)|^2+\Big[\int_{t_2}^T\Big(\kappa(|t_1-t_2|)L(s)\big(1+|Y(s)|+|\zeta(t_2,s)|\\
&\qquad\qquad +|Z(s,t_2)|\big)+\widehat{L}_z(t,s)|Z(s,t_1)-Z(s,t_2)|\Big)ds\Big]^2\Big\}\\
&\leqslant K\mathbb{E}\Big[|\psi(t_1)-\psi(t_2)|^2+\big[\kappa(|t_1-t_2|)\big]^2+\int_{t_2}^T|Z(s,t_1)-Z(s,t_2)|^2ds\Big].
\end{aligned}
$$

Since

$$
Y(t)=\eta(t,t),\qquad Z(t,s)=\zeta(t,s),
$$

one has, from the above, that

$$
\begin{aligned}
\mathbb{E}\int_{t_2}^T|Z(t_1,s)-Z(t_2,s)|^2ds\leqslant K\mathbb{E}\Big[&|\psi(t_1)-\psi(t_2)|^2+\big[\kappa(|t_1-t_2|)\big]^2\\
&+\int_{t_2}^T|Z(s,t_1)-Z(s,t_2)|^2ds\Big].
\end{aligned}
\tag{5.3.33}
$$

Next, we estimate

$$
\begin{aligned}
&\mathbb{E}\Big|\int_{t_1}^T g(t_1,s,Y(s),Z(t_1,s),Z(s,t_1))ds-\int_{t_2}^T g(t_2,s,Y(s),Z(t_2,s),Z(s,t_2))ds\Big|^2\\
&\leqslant K\mathbb{E}\Big[\Big(\int_{t_1}^{t_2}|g(t_1,s,Y(s),Z(t_1,s),Z(s,t_1))|ds\Big)^p\\
&\quad+\Big(\int_{t_2}^T|g(t_1,s,Y(s),Z(t_1,s),Z(s,t_1))-g(t_2,s,Y(s),Z(t_2,s),Z(s,t_2))|ds\Big)^p\Big].
\end{aligned}
$$

Note that

$$
\begin{aligned}
&\mathbb{E}\Big(\int_{t_2}^T|g(t_1,s,Y(s),Z(t_1,s),Z(s,t_1))-g(t_2,s,Y(s),Z(t_2,s),Z(s,t_2))|ds\Big)^2\\
&\leqslant\mathbb{E}\Big[\Big(\int_{t_2}^T\kappa\big(|t_1-t_2|\big)L(s)\big(1+|Z(t_1,s)|\big)ds\Big)^2\\
&\quad+\Big(\int_{t_2}^T L_z(t_2,s)|Z(t_1,s)-Z(t_2,s)|ds\Big)^2\Big]\\
&\quad+\Big(\int_{t_2}^T L_{\widehat{z}}(t_2,s)Z(s,t_1)-Z(s,t_2)|ds\Big)^2\Big]
\end{aligned}
$$

$$
\begin{aligned}
&\leqslant K\mathbb{E}\Big\{\big[\kappa(|t_1-t_2|)\big]^2\Big[\Big(\int_0^T L(s)ds\Big)^2+\Big(\int_0^T L(s)^2ds\Big)\Big(\int_{t_1}^T |Z(t_1,s)|^2ds\Big)\Big]\\
&\quad+\Big(\int_{t_2}^T L_z(t_2,s)^2ds\Big)\Big(\int_{t_2}^T |Z(t_1,s)-Z(t_2,s)|^2ds\Big)\\
&\quad+\Big(\int_{t_2}^T L_{\widehat{z}}(t_2,s)^2ds\Big)\Big(\int_{t_2}^T |Z(s,t_1)-Z(s,t_2)|^2ds\Big)\Big\}\\
&\leqslant K\mathbb{E}\Big[|\psi(t_1)-\psi(t_2)|^2+\big[\kappa(|t_1-t_2|)\big]^2+\Big(\int_{t_2}^T |Z(s,t_1)-Z(s,t_2)|^2ds\Big)\Big].
\end{aligned}
$$

Also, we have (recall that $g(t,s,0,0,0)=0$)

$$
\begin{aligned}
&\mathbb{E}\Big(\int_{t_1}^{t_2}|g(t_1,s,Y(s),Z(t_1,s),Z(s,t_1))|ds\Big)^2\\
&\leqslant \mathbb{E}\Big[\int_{t_1}^{t_2}\Big(L_y(t_1,s)|Y(s)|+L_z(t_1,s)|Z(t_1,s)|+L_{\widehat{z}}(t_1,s)|Z(s,t_1)|\Big)ds\Big]^2\\
&\leqslant K\mathbb{E}\Big[\Big(\int_{t_1}^{t_2} L_y(t_1,s)^2ds\Big)\Big(\int_{t_1}^{t_2} |Y(s)|^2ds\Big)\\
&\quad+\Big(\int_{t_1}^{t_2} L_z(t_1,s)^2ds\Big)\Big(\int_{t_1}^{t_2} |Z(t_1,s)|^2ds\Big)\\
&\quad+\Big(\int_{t_1}^{t_2} L_{\widehat{z}}(t_1,s)^2ds\Big)\Big(\int_{t_1}^{t_2} |Z(s,t_1)|^2ds\Big)\Big]\\
&\leqslant K\mathbb{E}\Big[|t_1-t_2|^{\frac{\varepsilon}{1+\varepsilon}}\Big(\int_{t_1}^T L_y(t_1,s)^{2(1+\varepsilon)}ds\Big)^{\frac{1}{1+\varepsilon}}\Big(\int_{t_1}^{t_2} |Y(s)|^2ds\Big)\\
&\quad+|t_1-t_2|^{\frac{\varepsilon}{1+\varepsilon}}\Big(\int_{t_1}^T L_z(t_1,s)^{2(1+\varepsilon)}ds\Big)^{\frac{1}{1+\varepsilon}}\Big(\int_{t_1}^T |Z(t_1,s)|^2ds\Big)\\
&\quad+|t_1-t_2|^{\frac{\varepsilon}{1+\varepsilon}}\Big(\int_{t_1}^T L_{\widehat{z}}(t_1,s)^{2(1+\varepsilon)}ds\Big)^{\frac{1}{1+\varepsilon}}\Big(\int_{t_1}^T |Z(s,t_1)|^2ds\Big)\Big]\\
&\leqslant K|t_1-t_2|^{\frac{\varepsilon}{1+\varepsilon}}.
\end{aligned}
$$

Hence, we obtain

$$
\begin{aligned}
&\mathbb{E}\Big|\int_{t_1}^T g(t_1,s,Y(s),Z(t_1,s),Z(s,t_1))ds\\
&\qquad-\int_{t_2}^T g(t_2,s,Y(s),Z(t_2,s),Z(s,t_2))ds\Big|^2\\
&\leqslant K\mathbb{E}\Big[|t_1-t_2|^{\frac{\varepsilon}{1+\varepsilon}}+|\psi(t_1)-\psi(t_2)|^2+\big[\kappa(|t_1-t_2|)\big]^2\\
&\qquad+\int_{t_2}^T |Z(s,t_1)-Z(s,t_2)|^2ds\Big].
\end{aligned}
\tag{5.3.34}
$$

We now estimate the last term on the right-hand side of the above. Since

$$\begin{aligned}
D^{(i)}_{t_1}Y(s) - D^{(i)}_{t_2}Y(s) &= D^{(i)}_{t_1}\psi(s) - D^{(i)}_{t_2}\psi(s)\\
&+ \int_s^T \Big\{[D^{(i)}_{t_1}g](s,\tau,Y(\tau),Z(s,\tau),Z(\tau,s))\\
&\qquad -[D^{(i)}_{t_2}g](s,\tau,Y(\tau),Z(s,\tau),Z(\tau,s))\\
&\qquad +g_y(s,\tau,Y(\tau),Z(s,\tau),Z(\tau,s))[D^{(i)}_{t_1}Y(\tau) - D^{(i)}_{t_2}Y(\tau)]\\
&\qquad +\sum_{i=1}^d \Big[g_{z_j}(s,\tau,Y(\tau),Z(s,\tau),Z(\tau,s))[D^{(i)}_{t_1}Z_j(s,\tau) - D^{(i)}_{t_2}Z_j(s,\tau)]\\
&\qquad +g_{\zeta_j}(s,\tau,Y(\tau),Z(s,\tau),Z(\tau,s))[D^{(i)}_{t_1}Z_j(\tau,s) - D^{(i)}_{t_2}Z_j(\tau,s)]\Big]\Big\}d\tau\\
&- \int_s^T [D^{(i)}_{t_1}Z(s,\tau) - D^{(i)}_{t_2}Z(s,\tau)]dW(\tau), \qquad s \in [t_2,T].
\end{aligned}$$

It is a Type II BSVIE for which

$$(D^{(i)}_{t_1}Y(\cdot) - D^{(i)}_{t_2}Y(\cdot), D^{(i)}_{t_1}Z(\cdot\,,\cdot) - D^{(i)}_{t_2}Z(\cdot\,,\cdot))$$

is an adapted M-solution. Consequently,

$$\begin{aligned}
&\mathbb{E}\Big\{\int_{t_2}^T |D^{(i)}_{t_1}Y(s) - D^{(i)}_{t_2}Y(s)|^2 ds\\
&\qquad + \int_{t_2}^T\int_{t_2}^T |D^{(i)}_{t_1}Z(s,\tau) - D^{(i)}_{t_2}Z(s,\tau)|^2 d\tau ds\Big\}\\
&\leqslant K\mathbb{E}\Big\{\int_{t_2}^T |D^{(i)}_{t_1}\psi(s) - D^{(i)}_{t_2}\psi(s)|^2 ds\\
&\quad + \int_{t_2}^T\Big(\int_{t_2}^T |[D^{(i)}_{t_1}g](s,\tau,Y(\tau),Z(s,\tau),Z(\tau,s))\\
&\quad -[D^{(i)}_{t_2}g](s,\tau,Y(\tau),Z(s,\tau),Z(\tau,s))|d\tau\Big)^2 ds\Big\}\\
&\leqslant K\mathbb{E}\Big\{\int_{t_2}^T |D^{(i)}_{t_1}\psi(s) - D^{(i)}_{t_2}\psi(s)|^2 ds\\
&\quad +\int_{t_2}^T\Big(\int_{t_2}^T \kappa(|t_1-t_2|)\widehat{L}(s,\tau)(1+|Y(\tau)|+|Z(s,\tau)|+|Z(\tau,s)|)d\tau\Big)^2 ds\Big\}\\
&\leqslant K\mathbb{E}\Big[\int_{t_2}^T |D^{(i)}_{t_1}\psi(s) - D^{(i)}_{t_2}\psi(s)|^2 ds + \big[\kappa(|t_1-t_2|)\big]^2\Big].
\end{aligned} \tag{5.3.35}$$

Next, since $Z_i(s,t) = \mathbb{E}_t[D_t^{(i)} Y(s)]$, we have

$$
\begin{aligned}
&\mathbb{E}\int_{t_2}^{T} |Z_i(s,t_1) - Z_i(s,t_2)|^2 ds = \mathbb{E}\int_{t_2}^{T} |\mathbb{E}_{t_1}[D_{t_1}^{(i)} Y(s)] - \mathbb{E}_{t_2}[D_{t_2}^{(i)} Y(s)]|^2 ds \\
&\leqslant 2\mathbb{E}\int_{t_2}^{T} \Big\{ |\mathbb{E}_{t_1}[D_{t_1}^{(i)} Y(s)] - \mathbb{E}_{t_2}[D_{t_1}^{(i)} Y(s)]|^2 + |\mathbb{E}_{t_2}[D_{t_1}^{(i)} Y(s) - D_{t_2}^{(i)} Y(s)]|^2 \Big\} ds \\
&\leqslant 2\mathbb{E}\int_{t_2}^{T} \Big\{ \Big| \int_{t_1}^{t_2} D_{t_1}^{(i)} Z(s,\tau) dW(\tau) \Big|^2 + |D_{t_1}^{(i)} Y(s) - D_{t_2}^{(i)} Y(s)|^2 \Big\} ds \\
&\leqslant K\mathbb{E}\Big[\int_{t_2}^{T} \Big(\int_{t_1}^{t_2} |D_{t_1}^{(i)} Z(s,\tau)|^2 d\tau + |D_{t_1}^{(i)} \psi(s) - D_{t_2}^{(i)} \psi(s)|^2 \Big) ds + \big[\kappa(|t_1 - t_2|)\big]^2 \Big].
\end{aligned}
$$

Hence,

$$
\lim_{t_2 \downarrow t_1} \mathbb{E}\int_{t_2}^{T} \big|Z_i(s,t_1) - Z_i(s,t_2)\big|^2 ds = 0.
$$

Next, we show the analogue result when $t_1 \leqslant t_2$ with fixed t_2. Notice that for fixed (s,t_2) with $s \geqslant t_2$, by Doob's inequality in Proposition 2.2.4 of Chap. 2, we have

$$
\int_{t_2}^{T} \mathbb{E}\Big[\sup_{r \in [0,t_2]} \big|\mathbb{E}_r Z_i(s,t_2)\big|^2 \Big] ds \leqslant 4 \int_{t_2}^{T} \mathbb{E}|Z_i(s,t_2)|^2 ds < \infty.
$$

It then follows from dominated convergence theorem that

$$
\lim_{t_1 \uparrow t_2} \mathbb{E}\int_{t_2}^{T} \big|\mathbb{E}_{t_1} Z_i(s,t_2) - Z_i(s,t_2)\big|^2 ds = 0.
$$

On the other hand,

$$
\begin{aligned}
&\mathbb{E}\int_{t_2}^{T} \big|Z_i(s,t_1) - \mathbb{E}_{t_1} Z_i(s,t_2)\big|^2 ds \\
&\leqslant \mathbb{E}\int_{t_2}^{T} \big|\mathbb{E}_{t_1}\big[D_{t_1}^{(i)} Y(s)\big] - \mathbb{E}_{t_1}\big[D_{t_2}^{(i)} Y(s)\big]\big|^2 ds \\
&\leqslant K\mathbb{E}\Big[\int_{t_2}^{T} |D_{t_1}^{(i)} \psi(s) - D_{t_2}^{(i)} \psi(s)|^2 ds + \big[\kappa(|t_1 - t_2|)\big]^2 \Big] \to 0, \quad t_1 \uparrow t_2.
\end{aligned}
$$

Since

$$
\begin{aligned}
&\mathbb{E}\int_{t_2}^{T} \big|Z_i(s,t_1) - Z_i(s,t_2)\big|^2 ds \\
&\leqslant 2\mathbb{E}\int_{t_2}^{T} \Big\{ \big|D_{t_1}^{(i)} Y(s) - D_{t_2}^{(i)} Y(s)\big|^2 + \big|\mathbb{E}_{t_1} Z_i(s,t_2) - Z_i(s,t_2)\big|^2 \Big\} ds,
\end{aligned}
$$

we then have

$$\lim_{t_1 \uparrow t_2} \mathbb{E} \int_{t_2}^{T} \big|Z_i(s, t_1) - Z_i(s, t_2)\big|^2 ds = 0.$$

Combining the above, we obtain

$$\lim_{|t_1 - t_2| \to 0} \mathbb{E} \int_{t_2}^{T} |Z_i(s, t_1) - Z_i(s, t_2)|^2 ds = 0. \tag{5.3.36}$$

Thus, from (5.3.34), we have

$$\psi(\cdot) + \int_{\cdot}^{T} g(\cdot\,, s, Y(s), Z(\cdot\,, s), Z(s, \cdot))ds \in C_{\mathcal{F}_T}([0, T]; L^2(\Omega; \mathbb{R}^m)).$$

Then by Lemma 4.3.2 in Chap. 4 and noting

$$Y(t) = \mathbb{E}_t\Big[\psi(t) + \int_t^T g(t, s, Y(s), Z(t, s), Z(s, t))ds\Big],$$

we obtain that $Y(\cdot) \in C_{\mathbb{F}}([0, T]; L^2(\Omega; \mathbb{R}^m))$. Hence, combining (5.3.33), (5.3.35), and (5.3.36), we obtain (5.3.31). Further, from

$$\int_t^T Z(t, s)dW(s) = \psi(t) - Y(t) + \int_t^T g(t, s, Y(s), Z(t, s), Z(s, t))ds, \qquad t \in [0, T],$$

we obtain

$$\int_{\cdot}^{T} Z(\cdot, s)dW(s) \in C_{\mathcal{F}_T}([0, T]; L^2(\Omega; \mathbb{R}^m)).$$

This completes the proof. □

To conclude this section, let us make some remarks.

- It is possible to discuss the above in the L^p-framework. But we expect to have lengthier proofs. We therefore prefer not to present that here.
- It is possible to discuss when $Y(\cdot)$ is in $L^p_{\mathbb{F}}(\Omega; C([0, T]; \mathbb{R}^m))$, in the spirit of Theorem 4.3.3 (ii) of Chap. 4.
- We have not presented linear Type II BSVIEs here. We expect that it should also have a variational constants formula. We leave it to the interested readers to derive it using our method.

4 Duality Principle

Recall that in Sect. 2.2 of Chap. 1, we have established a duality principle for linear FSVIEs and linear Type II BSVIEs in a special case. In this section, we want to establish a much more general version.

Let us first look at the duality of a linear FSVIE and a linear BSVIE of Type II, with the generator independent of $Z(t,s)$.

Theorem 5.4.1 *Suppose $A(\cdot\,,\cdot), C(\cdot\,,\cdot) \in L^\infty(0,T;L^\infty_{\mathbb{F}}(0,T;\mathbb{R}^{m\times m}))$, $\varphi(\cdot) \in L^2_{\mathbb{F}}(0,T;\mathbb{R}^m)$, and $\psi(\cdot) \in L^2_{\mathcal{F}_T}(0,T;\mathbb{R}^m)$. Let $X(\cdot) \in L^2_{\mathbb{F}}(0,T;\mathbb{R}^m)$ be the solution of FSVIE:*

$$X(t) = \varphi(t)+\int_0^t A(t,s)X(s)ds+\int_0^t C(t,s)X(s)dW(s), \quad t \in [0,T], \tag{5.4.1}$$

and $(Y(\cdot), Z(\cdot\,,\cdot)) \in \mathcal{H}^2(0,T)$ be the adapted M-solution to the following BSVIE:

$$Y(t)=\psi(t)+\int_t^T \big[A(s,t)^\top Y(s)+C(s,t)^\top Z(s,t)\big]ds-\int_t^T Z(t,s)dW(s), \quad t \in [0,T]. \tag{5.4.2}$$

Then the following duality principle holds:

$$\mathbb{E}\int_0^T \langle X(t), \psi(t)\rangle dt = \mathbb{E}\int_0^T \langle \varphi(t), Y(t)\rangle dt. \tag{5.4.3}$$

Further, suppose the map $s \mapsto (A(s,t), C(s,t), \varphi(s))$ is continuous at $s = T$ with value $(A(T,t), C(T,t), \varphi(T))$, and $\eta \in L^2_{\mathcal{F}_T}(\Omega;\mathbb{R}^n)$. Let $(Y(\cdot), Z(\cdot\,,\cdot)) \in \mathcal{H}^2(0,T)$ be the adapted M-solution to the following BSVIE:

$$\begin{aligned} Y(t) = \psi(t) + A(T,t)^\top\eta + C(T,t)^\top\zeta(t) + \int_t^T \big[A(s,t)^\top Y(s) \\ +C(s,t)^\top Z(s,t)\big]ds - \int_t^T Z(t,s)dW(s), \quad t \in [0,T], \end{aligned} \tag{5.4.4}$$

where $\zeta(\cdot) \in L^2_{\mathcal{F}}(0,T;\mathbb{R}^m)$ is the unique process that satisfies

$$\eta = \mathbb{E}\eta + \int_0^T \zeta(t)dW(t). \tag{5.4.5}$$

Then the following holds:

$$\mathbb{E}\Big\{\langle X(T),\eta\rangle+\int_0^T\langle X(t),\psi(t)\rangle dt\Big\}=\mathbb{E}\Big\{\langle\varphi(T),\eta\rangle+\int_0^T\langle\varphi(t),Y(t)\rangle dt\Big\}. \qquad (5.4.6)$$

Proof Let us first look at (5.4.3). We observe the following:

$$\begin{aligned}
&\mathbb{E}\int_0^T\langle\varphi(t),Y(t)\rangle dt\\
&=\mathbb{E}\int_0^T\langle X(t)-\int_0^t A(t,s)X(s)ds-\int_0^t C(t,s)X(s)dW(s),Y(t)\rangle dt\\
&=\mathbb{E}\int_0^T\langle X(t),Y(t)\rangle dt-\mathbb{E}\int_0^T\int_s^T\langle X(s),A(t,s)^\top Y(t)\rangle dtds\\
&\quad-\mathbb{E}\int_0^T\langle\int_0^t C(t,s)X(s)dW(s),\mathbb{E}Y(t)+\int_0^t Z(t,s)dW(s)\rangle dt\\
&=\mathbb{E}\int_0^T\langle X(t),Y(t)\rangle dt-\mathbb{E}\int_0^T\int_t^T\langle X(t),A(s,t)^\top Y(s)\rangle dsdt\\
&\quad-\mathbb{E}\int_0^T\int_0^t\langle C(t,s)X(s),Z(t,s)\rangle dsdt\\
&=\mathbb{E}\int_0^T\langle X(t),Y(t)-\int_t^T\Big[A(s,t)^\top Y(s)+C(s,t)^\top Z(s,t)\Big]ds\rangle dt\\
&=\mathbb{E}\int_0^T\langle X(t),\psi(t)-\int_t^T Z(t,s)dW(s)\rangle dt=\mathbb{E}\int_0^T\langle X(t),\psi(t)\rangle dt.
\end{aligned}$$

Thus, (5.4.3) holds. Now, for (5.4.9), we denote

$$\widehat{\psi}(t)=\psi(t)+A(T,t)^\top\eta+C(T,t)^\top\zeta(t).$$

Then by (5.4.3), one has

$$\begin{aligned}
&\mathbb{E}\int_0^T\langle\varphi(t),Y(t)\rangle dt=\mathbb{E}\int_0^T\langle X(t),\widehat{\psi}(t)\rangle dt\\
&=\mathbb{E}\int_0^T\langle X(t),\psi(t)+A(T,t)^\top\eta+C(T,t)^\top\zeta(t)\rangle dt\\
&=\mathbb{E}\int_0^T\Big(\langle X(t),\psi(t)\rangle+\langle A(T,t)X(t),\eta\rangle+\langle C(T,t)X(t),\zeta(t)\rangle\Big)dt\\
&=\mathbb{E}\int_0^T\langle X(t),\psi(t)\rangle dt+\mathbb{E}\langle X(T)-\varphi(T),\eta\rangle.
\end{aligned}$$

This proves (5.4.9). □

BSVIE (5.4.2) is called the adjoint equation of (5.4.1), and vice versa, i.e., the latter is the adjoint equation of the former. Note that BSVIE (5.4.4) is essentially the same as (5.4.2) (with a different free term). A natural question is for the linear Type

II BSVIE with generator containing $Z(t,s)$, what the adjoint equation will look like. We now consider the following Type II BSVIE:

$$
\begin{aligned}
Y(t) = \psi(t) + \int_t^T \big[A(s,t)^\top Y(s) + B(s,t)^\top Z(t,s) + C(s,t)^\top Z(s,t)\big] ds \\
- \int_t^T Z(t,s)dW(s), \qquad t \in [0,T];
\end{aligned}
\tag{5.4.7}
$$

then what does its adjoint equation look like? We point out that $Z(t,s)$ appears in the generator of the above BSVIE. In the rest of this section, we are going to investigate such an equation. First, let us introduce the following hypothesis.

(H5.4.1) The maps $A, B, C : [0,T]^2 \times \Omega \to \mathbb{R}^{m\times m}$ are measurable and bounded such that $s \mapsto (A(s,t), B(s,t), C(s,t))$ is $\mathbb{F}$-adapted on $[t,T]$ for every $t \in [0,T]$.

By Theorem 5.2.2, we know that under (H5.4.1), for any $\psi(\cdot) \in L^2_{\mathcal{F}_T}(0,T;\mathbb{R}^m)$, linear BSVIE (5.4.7) admits a unique adapted M-solution $(Y(\cdot), Z(\cdot\,,\cdot)) \in \mathcal{M}^2[0,T]$. For any $(\alpha(\cdot), \beta(\cdot\,,\cdot)) \in L^2_{\mathbb{F}}(0,T;\mathbb{R}^m) \times L^2(0,T;L^2_{\mathbb{F}}(0,T;\mathbb{R}^m))$, consider the following stochastic integral equation:

$$
\begin{aligned}
\xi(t) = \alpha(t) + \int_0^T \beta(t,s)dW(s) + \int_0^t A(t,s)\mathbb{E}_t[\xi(s)]ds \\
+ \int_0^t \mathbb{E}_s[C(t,s)\xi(s)]dW(s) + \int_t^T B(t,s)\mathbb{E}_s[\xi(t)]dW(s), \\
\text{a.s., a.e. } t \in [0,T].
\end{aligned}
\tag{5.4.8}
$$

We will show that the above is the adjoint equation of (5.4.7). Note that each Itô's integral in the above is well-defined. However, process $\xi(\cdot)$ is not $\mathbb{F}$-adapted. In the case $B(\cdot\,,\cdot) = 0$, $\beta(\cdot\,,\cdot) = 0$, with $X(\cdot) = \xi(\cdot)$ and $\alpha(\cdot) = \varphi(\cdot)$, (5.4.8) is reduced to (5.4.1). We call the above (5.4.8) a *mean-field stochastic Fredholm–Volterra integral equation* (MF-SFVIEs, for short). We are not planning to study general MF-SFVIEs in this book. However, for the completeness of duality, we will establish well-posedness of the above special form of MF-SFVIE.

Theorem 5.4.2 *Let (H5.4.1) hold. Then for any* $(\alpha(\cdot), \beta(\cdot\,,\cdot)) \in L^2_{\mathbb{F}}(0,T;\mathbb{R}^m) \times L^2(0,T;L^2_{\mathbb{F}}(0,T;\mathbb{R}^m))$, *equation* (5.4.8) *admits a unique solution* $\xi(\cdot) \in L^2_{\mathcal{F}_T}(0,T;\mathbb{R}^m)$. *Further, if* $(Y(\cdot), Z(\cdot,\cdot))$ *is the unique M-solution for* (5.4.7) *with* $\psi(\cdot) \in L^2_{\mathcal{F}_T}(0,T;\mathbb{R}^m)$, *then*

$$
\mathbb{E}\int_0^T \langle \psi(t), \xi(t)\rangle dt = \mathbb{E}\Big\{\int_0^T \langle Y(t), \alpha(t)\rangle dt + \int_0^T\int_0^T \langle Z(t,s), \beta(t,s)\rangle ds dt\Big\}.
\tag{5.4.9}
$$

Proof We first prove the well-posedness of (5.4.8). Let $(\alpha(\cdot), \beta(\cdot\,,\cdot)) \in L^2_{\mathbb{F}}(0,T;\mathbb{R}^m) \times L^2(0,T;L^2_{\mathbb{F}}(0,T;\mathbb{R}^m))$ be given. Suppose $\xi(\cdot)$ is a solution to (5.4.8). For almost all $t \in [0,T]$ and any $r \in [t,T]$, applying $\mathbb{E}_r$ on the both sides of (5.4.8), one gets

$$
\begin{aligned}
\mathbb{E}_r[\xi(t)] &= \alpha(t) + \int_0^r \beta(t,s)dW(s) + \int_t^r B(t,s)\mathbb{E}_s[\xi(t)]dW(s) \\
&\quad + \int_0^t A(t,s)\mathbb{E}_t[\xi(s)]ds + \int_0^t \mathbb{E}_s[C(t,s)\xi(s)]dW(s) \\
&= \alpha(t)+\int_0^t \beta(t,s)dW(s)+\int_t^r \Big(B(t,s)\mathbb{E}_s[\xi(t)]+\beta(t,s)\Big)dW(s) \\
&\quad + \int_0^t A(t,s,t)\mathbb{E}_t[\xi(s)]ds+\int_0^t \mathbb{E}_s[C(t,s)\xi(s)]dW(s), \\
&\qquad\qquad \forall r \in [t,T],\ t \in [0,T].
\end{aligned}
\tag{5.4.10}
$$

On the other hand, if $\xi(\cdot)$ satisfies (5.4.10) for any $r \in [t,T]$, then taking $r = T$, we recover (5.4.8). Thus, to solve (5.4.8), it suffices to solve (5.4.10).

For any $\bar\xi(\cdot) \in L^2_{\mathcal{F}_T}(0,T;\mathbb{R}^m)$ and almost every $t \in [0,T]$, consider the following family of SDE with parameter t:

$$
\begin{aligned}
\lambda(t,r) &= \alpha(t)+\int_0^t \beta(t,s)dW(s)+\int_t^r \Big(B(t,s)\lambda(t,s)+\beta(t,s)\Big)dW(s) \\
&\quad +\int_0^t A(t,s)\mathbb{E}_t[\bar\xi(s)]ds]+\int_0^t \mathbb{E}_s[C(t,s)\bar\xi(s)]dW(s), \qquad r \in [t,T].
\end{aligned}
\tag{5.4.11}
$$

It admits a unique solution $\lambda(t,\cdot)$ which is $\mathbb{F}$-adapted on $[t,T]$, and

$$
\begin{aligned}
\mathbb{E}|\lambda(t,r)|^2 \leqslant K\mathbb{E}\Big\{|\alpha(t)|^2 + \int_0^r |\beta(t,s)|^2 ds + \int_t^r |\lambda(t,s)|^2 ds \\
+ \int_0^t |\bar\xi(s)|^2 ds\Big\}, \qquad r \in [t,T].
\end{aligned}
$$

Hence, it follows from Gronwall's inequality that

$$
\mathbb{E}|\lambda(t,r)|^2 \leqslant K\mathbb{E}\Big\{|\alpha(t)|^2+\int_0^T |\beta(t,s)|^2 ds+\int_0^t |\bar\xi(s)|^2 ds\Big\}, \quad r \in [t,T]. \tag{5.4.12}
$$

We define

$$
\xi(t) \stackrel{\Delta}{=} \lambda(t,T), \qquad \text{a.e. } t \in [0,T].
$$

Then for any $r \in [t,T]$, from (5.4.11), one has

$$
\mathbb{E}_r[\xi(t)] = \mathbb{E}_r[\lambda(t,T)] = \lambda(t,r), \tag{5.4.13}
$$

and it follows from (5.4.12) that

$$
\mathbb{E}|\xi(t)|^2 \leqslant K\mathbb{E}\Big\{|\alpha(t)|^2+\int_0^T |\beta(t,s)|^2 ds+\int_0^t |\bar\xi(s)|^2 ds\Big\}, \qquad \text{a.e. } t \in [0,T].
$$

Thus, $\xi(\cdot) \in L^2_{\mathcal{F}_T}(0, T; \mathbb{R}^m)$, and we have defined a map $\bar{\xi}(\cdot) \mapsto \xi(\cdot)$ from $L^2_{\mathcal{F}_T}(0, T; \mathbb{R}^m)$ to itself. We now show that this map is a contraction. To this end, we take any $\bar{\xi}_i(\cdot) \in L^2_{\mathcal{F}_T}(0, T; \mathbb{R}^m)$, $i = 1, 2$, and let $\lambda_i(\cdot\,, \cdot)$ solve (5.4.11). Then for almost every $t \in [0, T]$ and any $r \in [t, T]$,

$$\begin{aligned}\lambda_1(t,r) - \lambda_2(t,r) &= \int_t^r B(t,s)\Big(\lambda_1(t,s) - \lambda_2(t,s)\Big)dW(s)\\&+ \int_0^t A(t,s)\mathbb{E}_t[\bar{\xi}_1(s) - \bar{\xi}_2(s)]ds + \int_0^t \mathbb{E}_s\Big[C(t,s)\Big(\bar{\xi}_1(s) - \bar{\xi}_2(s)\Big)\Big]dW(s).\end{aligned}$$

Hence,

$$\mathbb{E}|\lambda_1(t,r)-\lambda_2(t,r)|^2 \leqslant K\mathbb{E}\Big(\int_t^r|\lambda_1(t,s)-\lambda_2(t,s)|^2ds+\int_0^t|\bar{\xi}_1(s)-\bar{\xi}_2(s)|^2ds\Big).$$

By Gronwall's inequality, we obtain

$$\mathbb{E}|\lambda_1(t,r) - \lambda_2(t,r)|^2 \leqslant K\mathbb{E}\int_0^t |\bar{\xi}_1(s) - \bar{\xi}_2(s)|^2ds, \qquad \forall r \in [t, T].$$

With the definition

$$\xi_i(t) = \lambda_i(t, T), \qquad t \in [0, T], \qquad \text{a.e.},$$

one obtains

$$\mathbb{E}|\xi_1(t) - \xi_2(t)|^2 \leqslant K\mathbb{E}\int_0^t |\bar{\xi}_1(s) - \bar{\xi}_2(s)|^2ds, \qquad \text{a.e. } t \in [0, T].$$

Now, for any $\mu > 0$, it follows that

$$\begin{aligned}&\mathbb{E}\int_0^T e^{-\mu t}|\xi_1(t) - \xi_2(t)|^2dt \leqslant K\mathbb{E}\int_0^T e^{-\mu t}\int_0^t |\bar{\xi}_1(s) - \bar{\xi}_2(s)|^2dsdt\\&= K\mathbb{E}\int_0^T\Big(\int_s^T e^{-\mu t}dt\Big)|\bar{\xi}_1(s) - \bar{\xi}_2(s)|^2ds \leqslant \frac{K}{\mu}\int_0^T e^{-\mu s}|\bar{\xi}_1(s) - \bar{\xi}_2(s)|^2ds.\end{aligned}$$

Since K in the above is an absolute constant, by choosing $\mu > 0$ large, we get that the map $\bar{\xi}(\cdot) \mapsto \xi(\cdot)$ is a contraction in $L^2_{\mathcal{F}_T}(0, T; \mathbb{R}^m)$ with a weighted norm. Hence, it admits a unique fixed point which is the unique solution of (5.4.10).

Next, we prove the duality relation (5.4.9). For any given $\psi(\cdot) \in L^2_{\mathcal{F}_T}(0, T; \mathbb{R}^m)$, let $(Y(\cdot), Z(\cdot\,, \cdot)) \in \mathcal{M}^2[0, T]$ be the unique adapted M-solution of (5.4.4). Then we have

$$Y(t) = \mathbb{E}Y(t) + \int_0^t Z(t,s)dW(s), \qquad \text{a.e. } t \in [0, T]. \tag{5.4.14}$$

Hence, for any $(\alpha(\cdot), \beta(\cdot\,,\cdot)) \in L^2_{\mathbb{F}}(0, T; \mathbb{R}^m) \times L^2(0, T; L^2_{\mathbb{F}}(0, T; \mathbb{R}^m))$, we have

$$\begin{aligned}&\mathbb{E}\Big\{\int_0^T\langle Y(t), \alpha(t)+\int_0^t\beta(t,s)dW(s)\rangle dt+\int_0^T\int_t^T\langle Z(t,s), \beta(t,s)\rangle dsdt\Big\}\\&=\mathbb{E}\Big\{\int_0^T\langle Y(t), \alpha(t)\rangle dt+\int_0^T\int_0^T\langle Z(t,s), \beta(t,s)\rangle dsdt\Big\}.\end{aligned} \tag{5.4.15}$$

On the other hand, let $\xi(\cdot) \in L^2_{\mathcal{F}_T}(0, T; \mathbb{R}^m)$ be the solution to the adjoint equation (5.4.8) corresponding to $(\alpha(\cdot), \beta(\cdot\,,\cdot))$. Then

$$\begin{aligned}\xi(t)&=\alpha(t)+\int_0^T\beta(t,s)dW(s)+\int_t^T B(t,s)\mathbb{E}_s[\xi(t)]dW(s)\\&\quad+\int_0^t A(t,s)\mathbb{E}_t[\xi(s)]ds+\int_0^t\mathbb{E}_s[C(t,s)\xi(s)]dW(s)\\&=\alpha(t)+\int_0^t A(t,s)\mathbb{E}_t[\xi(s)]ds+\int_0^t\Big(\mathbb{E}_s[C(t,s)\xi(s)]+\beta(t,s)\Big)dW(s)\\&\quad+\int_t^T\Big(B(t,s)\mathbb{E}_s[\xi(t)]+\beta(t,s)\Big)dW(s)\\&=\mathbb{E}_t[\xi(t)]+\int_t^T\zeta(t,s)dW(s),\end{aligned} \tag{5.4.16}$$

with

$$\zeta(t,s)=B(t,s)\mathbb{E}_s[\xi(t)\big]+\beta(t,s),\qquad s\in[t,T],\ \text{ a.e. } t\in[0,T]. \tag{5.4.17}$$

Consequently,

$$\begin{aligned}&\mathbb{E}\int_0^T\langle\psi(t), \xi(t)\rangle dt\\&=\mathbb{E}\Big\{\int_0^T\langle Y(t)-\int_t^T\big[A(s,t)^\top Y(s)+B(t,s)^\top Z(t,s)+C(s,t)^\top Z(s,t)\big]ds, \xi(t)\rangle dt\\&\quad+\int_0^T\int_t^T\langle Z(t,s), \zeta(t,s)\rangle dsdt\Big\}\\&=\mathbb{E}\Big\{\int_0^T\langle Y(t), \xi(t)-\int_0^t A(t,s,t)\xi(s)ds\rangle dt\\&\quad+\int_0^T\int_t^T\langle Z(t,s), \zeta(t,s)-B(t,s)\xi(t)\rangle dsdt-\int_0^T\int_0^t\langle Z(t,s), C(t,s)\xi(s)\rangle dsdt\Big\}\\&=\mathbb{E}\Big\{\int_0^T\langle Y(t), \xi(t)-\int_0^t A(t,s)\xi(s)ds-\int_0^t\mathbb{E}_s\big[C(t,s)\xi(s)\big]dW(s)\rangle dt\\&\quad+\int_0^T\int_t^T\langle Z(t,s), \zeta(t,s)-B(t,s)\xi(t)\rangle dsdt\\&=\mathbb{E}\Big\{\int_0^T\langle Y(t), \int_0^t\beta(t,s)dW(s)+\alpha(t)\rangle dt+\int_0^T\int_t^T\langle Z(t,s), \beta(t,s)\rangle dsdt,\end{aligned}$$

where the last equality in the above follows from (5.4.16) and (5.4.17). Thanks to (5.4.15), our conclusion follows. □

5 Bibliographical Remarks

For an optimal control problem of an FSVIE, when stating a Pontryagin-type maximum principle (see, e.g., Chap. 10), the adjoint equation is needed, which is a BSVIE different from Type I BSVIE in the previous chapter. Motivated by this fact, Yong [218] introduced a new BSVIEs in 2006, without knowing the work of Lin [135]. Shortly, it was found that for this kind of BSVIEs, adapted solutions are not unique, in general. The notion of adapted M-solution (with "M" meaning martingale representation) was introduced by Yong in 2008 [220]; see also Definition 5.1.2 in the current chapter. In Yong [221] and Wang-Yong [205], the names of Type I and Type II BSVIEs were introduced and used to distinguish these two different types of BSVIEs. The theory on Type II BSVIEs was extended to infinite dimensions by Anh-Grecksch-Yong [12] and Wang-Zheng [209]. For the non-Lipschitz case, we refer to Ren [170] which used the method proposed in Anh-Yong [11]. However, it was pointed by Shi-Wang [175] that the method used in [170] was questionable. They bridged the gap by specifying new weaker assumptions. There are some other extension of Type II BSVIEs; see Ahmadova-Mahmudov [7] and Wang-Zheng [209] for the extension to singular framework, Shi-Wen-Xiong [178] for the extension to the double stochastic integral framework, Hamaguchi [90] for the extension to the infinite horizon framework, Wang [197] for the extension to the L^p with $p \in (1, 2]$, and Wang [200] for the extension to the integrodifferential framework. For the contents from Sects. 1 to 3, we point out some remarks. First, the equivalent weighted norm in Proposition 5.1.3, which was partially inspired by Shi-Wang [175], plays an important role in establishing the well-posedness in Sect. 2. Second, for the proof of Theorem 5.2.2, it is different from yet simpler than the four-step arguments proposed in Yong [220]. Moreover, the conclusions in Theorem 5.2.2 also extend those in [220] from $p = 2$ to $p > 1$. Third, the regularity results in Sect. 3 also improve/simplify/modify the analogue arguments in [220].

Duality principle of linear FSDE and linear BSDE (Theorem 1.2.1 of Chap. 1) is standard and plays an important role in the optimal control of FSDEs. Proposition 1.2.2 (of Chap. 1) and Theorem 5.4.1 are natural extensions of Theorem 1.2.1 (of Chap. 1) to linear FSVIEs and linear BSVIEs. One extension to the forward–backward system is Theorem 5.4.2, which was proved in Shi-Wang-Yong [177] and later improved by Wang-Zhang [207]. Another extension is given in Hamaguchi-Wang [94, 95], which is quite useful in obtaining feedback control/strategy for linear quadratic problems of forward SVIEs.

For Type II BSVIEs, there are some other interesting topics. Using a different method involving some different tools, Hamaguchi [92] derived the variation of constants formula. Besides the adapted M-solution, another kind of solutions, called symmetrical solutions of BSVIEs, was introduced in Wang-Shi [201]. This idea was extended into the backward double stochastic Volterra integral equations; see Wen-Shi [215]. To study the linear quadratic problem of forward SVIEs, there are two interesting notions proposed in the literature. The first one is the extended BSVIEs, an extension of linear Type II BSVIEs, which was introduced and studied in Hamaguchi-Wang [94]. Another interesting one, called Type III BSVIE, was introduced in Wang-Yong-Zhou [196]. More related results on Type II BSVIEs can be found in [90, 168, 178, 197, 214, 215], and so on.

Chapter 6
Comparison Theorems

In this chapter, we will present various comparison theorems. The main results are that under certain conditions, various dynamic equations, including BSVIEs, preserve the monotonicity of the initial or terminal conditions. For simplicity of the presentation, in this chapter, we let $d = 1$, i.e., the Brownian motion is one dimensional. In the case $m = n = 1$, comparison theorems are well-understood. In this chapter, we will study the case $m = n \geqslant 1$.

1 Comparison Theorems for FSDEs, FSVIEs, and BSDEs

In this section, we are going to present comparison theorems for FSDEs, FSVIEs, and BSDEs, allowing the dimension $n > 1$. Some of them are known, but our proofs are a little different from those in the literature.

Let us first make some preparations. Denote

$$\mathbb{R}^n_+ \stackrel{\Delta}{=} \Big\{(x_1, \cdots, x_n) \in \mathbb{R}^n \bigm| x_i \geqslant 0,\ 1 \leqslant i \leqslant n\Big\}.$$

When $x \in \mathbb{R}^n_+$, we also denote it by $x \geqslant 0$ and say that x is *nonnegative*. By $x \leqslant 0$ and $x \geqslant y$ (for $x, y \in \mathbb{R}^n$), we mean $-x \geqslant 0$ and $x - y \geqslant 0$, respectively. In what follows, we let $e_i \in \mathbb{R}^n_+$ be the vector that the i-th entry is 1 and all other entries are zero. Let

$$\begin{cases} \mathbb{R}^{n\times n}_+ \stackrel{\Delta}{=} \Big\{A = (a_{ij}) \in \mathbb{R}^{n\times n} \bigm| a_{ij} \geqslant 0,\ 1 \leqslant i, j \leqslant n\Big\}, \\ \mathbb{R}^{n\times n}_{*+} \stackrel{\Delta}{=} \Big\{A = (a_{ij}) \in \mathbb{R}^{n\times n} \bigm| a_{ij} \geqslant 0,\ i \neq j\Big\}, \\ \mathbb{R}^{n\times n}_{d} \stackrel{\Delta}{=} \Big\{A = (a_{ij}) \in \mathbb{R}^{n\times n} \bigm| a_{ij} = 0,\ i \neq j\Big\}. \end{cases}$$

T. Wang, J. Yong, *Backward Stochastic Volterra Integral Equations*,
Probability Theory and Stochastic Modelling 109,
https://doi.org/10.1007/978-3-032-20848-4_6

Note that $\mathbb{R}_+^{n\times n}$ is the set of all $(n\times n)$ matrices with all the entries being nonnegative, $\mathbb{R}_{*+}^{n\times n}$ is the set of all $(n \times n)$ square matrices with all the off-diagonal entries being nonnegative (no conditions are imposed on the diagonal entries), and $\mathbb{R}_d^{n\times n}$ is the set of all $(n \times n)$ diagonal matrices, with the diagonal entries allowing to be any real numbers. Clearly, $\mathbb{R}_+^{n\times n}$ and $\mathbb{R}_{*+}^{n\times n}$ are closed convex cones of $\mathbb{R}^{n\times m}$ and $\mathbb{R}^{n\times n}$, respectively; $\mathbb{R}_+^{n\times n}$ is a proper subset of $\mathbb{R}_{*+}^{n\times n}$; and $\mathbb{R}_d^{n\times n}$ is a proper subspace of $\mathbb{R}^{n\times n}$, contained in $\mathbb{R}_{*+}^{n\times n}$. Also, we have

$$\mathbb{R}_{*+}^{n\times n} = \mathbb{R}_+^{n\times n} + \mathbb{R}_d^{n\times n} \equiv \Big\{A+B \;\Big|\; A\in\mathbb{R}_+^{n\times n},\ B\in\mathbb{R}_d^{n\times n}\Big\}.$$

The definitions of $\mathbb{R}_+^{m\times m}$, $\mathbb{R}_{*+}^{m\times m}$, and $\mathbb{R}_d^{m\times m}$ are similar. Further, for $n = m = 1$, one has

$$\mathbb{R}_{*+}^{1\times 1} = \mathbb{R}_d^{1\times 1} = \mathbb{R}, \qquad \mathbb{R}_+^{1\times 1} = \mathbb{R}_+ \equiv [0,\infty). \tag{6.1.1}$$

We now discuss several cases.

1.1 FSDEs

In this subsection, we are going to establish comparisons for the following nonlinear FSDEs:

$$X^i(t) = x^i + \int_s^t b^i(r, X^i(r))dr + \int_s^t \sigma(r, X^i(r))dW(r), \quad t\in[s,T]. \tag{6.1.2}$$

Note that unlike the drift $b^i(r,x)$, the diffusion $\sigma(r,x)$ is independent of $i = 0, 1$. This condition is necessary for the consideration of comparison theorem for FSDEs. Here is a simple example.

Example 6.1.1 Let $\sigma^0 \equiv 0 < 1 \equiv \sigma^1$ and $b^0 = b^1 = 0$. Then, the solutions of the corresponding SDEs with 0 initial condition are $X^0(t) = 0$ and $X^1(t) = W(t)$. Clearly, it holds neither $X^0(t) \leqslant X^1(t)$, for $t\in[0,T]$ almost everywhere, nor $X^0(t) \geqslant X^1(t)$, for $t\in[0,T]$ almost everywhere.

As we know, the Lebesgue integral has the nonnegativity property in the sense that $f(\cdot)\geqslant 0$ implies $\int_A f(x)dx \geqslant 0$ for some non-empty set A. However, such a property does not hold for Itô integral. In our opinion, this gives the explanation of the above example. Therefore, the comparison for (6.1.2) can be roughly explained as follows: under the same random perturbation, the solution corresponding to a larger initial state and a larger drift (a larger return rate) will be larger.

We now make this precise. To this end, we introduce the following assumption, which is comparable with (H2.3.1) of Chap. 2.

(H6.1.1) The maps $b^0, b^1, \sigma : [0, T] \times \mathbb{R}^n \times \Omega \to \mathbb{R}^n$ are measurable such that $t \mapsto (b^0(t, x), b^1(t, x), \sigma(t, x))$ is $\mathbb{F}$-progressively measurable satisfying

$$\begin{aligned}&|b^i(t,0)|,\ |\sigma(t,0)| \leqslant K, \qquad t \in [0,T],\ i = 0,1,\\ &|b^i(t,x) - b^i(t,x')|,\ |\sigma(t,x) - \sigma(t,x')| \leqslant K|x - x'|,\\ &\qquad\qquad t \in [0,T],\ x, x' \in \mathbb{R}^n,\ i = 0,1.\end{aligned}$$

From the standard theory of SDEs (see Proposition 2.3.11 of Chap. 2, e.g.,), we know that under (H6.1.1), for any $(s, x^i) \in [0, T) \times \mathbb{R}^n$, (6.1.2) admits a unique strong solution:

$$X^i(\cdot) \equiv X^i(\cdot\,; s, x^i) \in L^2_{\mathbb{F}}(\Omega; C([s,T]; \mathbb{R}^n)),$$

and the following estimate holds: for $p > 1$,

$$\begin{aligned}\mathbb{E}\Big[\sup_{t\in[s,T]} |X^i(t)|^p\Big] \leqslant K\Big[|x^i|^p + \mathbb{E}\Big(\int_s^T |b^i(t,0)|dt\Big)^p\\ +\Big(\int_s^T |\sigma(t,0)|^2 dt\Big)^{\frac{p}{2}}\Big] \leqslant K(1 + |x^i|^p).\end{aligned} \tag{6.1.3}$$

The main result of this subsection is the following comparison theorem.

Theorem 6.1.2 *Let* (H6.1.1) *hold and* $x \mapsto \sigma(t, x)$ *be differentiable. Suppose* $\bar b : [0, T] \times \mathbb{R}^n \times \Omega \to \mathbb{R}^n$ *is measurable,* $t \mapsto \bar b(t, x)$ *is* $\mathbb{F}$*-progressively measurable, and* $\bar b_x(t, x)$ *exists and is uniformly bounded. Then, the following holds:*

(i) *Suppose*

$$\bar b_x(t,x) \in \mathbb{R}^{n\times n}_{*+}, \quad \sigma_x(t,x) \in \mathbb{R}^{n\times n}_d, \qquad \forall x \in \mathbb{R}^n,\ \text{a.s.}\ \forall t \in [0,T], \tag{6.1.4}$$

and

$$b^0(t,x) \leqslant \bar b(t,x) \leqslant b^1(t,x), \qquad \forall x \in \mathbb{R}^n,\ \text{a.s.}\ \forall t \in [0,T]. \tag{6.1.5}$$

Then, for any $(s, x^i) \in [0, T) \times \mathbb{R}^n$ *with* $x^0 \leqslant x^1$*, the unique solutions* $X^i(\cdot) \equiv X^i(\cdot\,; s, x^i)$ *of* (6.1.2) *satisfy the following comparison:*

$$X^0(t) \leqslant X^1(t), \qquad t \in [s,T],\ \text{a.s.} \tag{6.1.6}$$

(ii) *Suppose*

$$b^0(t,x) = \bar b(t,x) = b^1(t,x), \qquad \forall (t,x) \in [0,T] \times \mathbb{R}^n,\ \text{a.s.},$$

and $(t, x) \mapsto (\bar b(t, x), \sigma(t, x), \bar b_x(t, x), \sigma_x(t, x))$ *is continuous, uniformly almost surely. Then,* (6.1.4) *is necessary for the conclusion of* (i) *to hold.*

Remark 6.1.3 When $m = n = 1$, the above (6.1.4)–(6.1.5) reduces to

$$\begin{aligned} &\bar{b}_x(t,x) \in \mathbb{R}, \quad \sigma_x(t,x) \in \mathbb{R}, && \forall x \in \mathbb{R}, \text{ a.s. } \forall t \in [0,T], \\ &b^0(t,x) \leqslant \bar{b}(t,x) \leqslant b^1(t,x), && \forall x \in \mathbb{R}, \text{ a.s. } \forall t \in [0,T]. \end{aligned}$$

Then, one can easily see the comparison theorem. On the other hand, even for this case, when diffusions are different, no comparisons will hold; see Example 6.1.1.

To present a proof of Theorem 6.1.2, and also for our later purpose, we first make some preparations. The following simple lemma, seemingly obvious, will play an interesting role below.

Lemma 6.1.4 *Suppose* $\varphi(\cdot) \in L^1_{\mathcal{F}_T}(0,T;\mathbb{R})$ *and*

$$\varphi(t) \geqslant 0, \qquad \text{a.s., a.e. } t \in [0,T]. \tag{6.1.7}$$

Then

$$\int_0^t \varphi(s)ds \geqslant 0, \qquad \forall t \in [0,T], \text{ a.s.} \tag{6.1.8}$$

According to condition (6.1.7), there is a set $E \subseteq [0,T]$ with Lebesgue measure $|E| = 0$ such that for each $t \in [0,T] \setminus E$ one can find an $\Omega_t \in \mathcal{F}$ with $\mathbb{P}(\Omega_t) = 0$ satisfying

$$\varphi(t,\omega) \geqslant 0, \qquad \forall \omega \in \Omega \setminus \Omega_t \equiv \Omega_t^c, \ t \in [0,T] \setminus E \equiv E^c.$$

Thus, one can only guarantee that

$$\int_0^t \varphi(s,\omega)ds \geqslant 0, \qquad \omega \in \bigcap_{t \in E^c} \Omega_t^c = \Big(\bigcup_{t \in E^c} \Omega_t\Big)^c.$$

But since E^c is uncountable, the set $\bigcup_{t \in E^c} \Omega_t$ might not be a $\mathbb{P}$-null set. Hence, (6.1.8) is not obvious.

Proof For any $\xi \in L^\infty_{\mathcal{F}_T}(\Omega;\mathbb{R})$ satisfying $\xi \geqslant 0$, a.s., we have

$$\mathbb{E}[\xi\varphi(s)] \geqslant 0, \qquad \text{a.s. } s \in [0,T].$$

Then, for any $t \in [0,T]$,

$$0 \leqslant \int_0^t \mathbb{E}[\xi\varphi(s)]ds = \mathbb{E}\Big[\xi \int_0^t \varphi(s)ds\Big].$$

Since ξ is arbitrary, one obtains

$$\int_0^t \varphi(s)ds \geqslant 0, \qquad \text{a.s., } \forall t \in [0, T].$$

Now, let $\mathbb{Q}[0, T]$ be the set of rational numbers in $[0, T]$. Then, for any $r \in \mathbb{Q}[0, T]$, we have a $\mathbb{P}$-null set Ω_r such that

$$\int_0^r \varphi(s)ds \geqslant 0, \qquad \omega \in \Omega \setminus \Omega_r.$$

Let $\widetilde{\Omega} = \bigcup_{r\in\mathbb{Q}[0,T]} \Omega_r$. Then, $\mathbb{P}(\widetilde{\Omega}) = 0$. For any $\omega \in \Omega \setminus \widetilde{\Omega}$ and any $t \in [0, T]$, there exists a sequence $r_k \in \mathbb{Q}[0, T]$ such that $r_k \to t$. Then, by the continuity of Lebesgue integral, we have

$$\int_0^t \varphi(s)ds = \lim_{k\to\infty} \int_0^{r_k} \varphi(s)ds \geqslant 0,$$

proving (6.1.8). □

Now, we look at the following linear nonhomogeneous FSDE:

$$\begin{cases} dX(t) = \big[A_0(t)X(t) + b(t)\big]dt + A_1(t)X(t)dW(t), & t \in [s, T], \\ X(s) = x, \end{cases} \tag{6.1.9}$$

with $A_0(\cdot), A_1(\cdot)$ being some suitable matrix-valued processes, satisfying the following hypothesis:

(H6.1.2) The maps $A_0(\cdot), A_1(\cdot) \in L^\infty_{\mathbb{F}}(0, T; \mathbb{R}^{n\times n})$.

In the case of (H6.1.2), (6.1.9) is well-posed and by (6.1.3), for $p > 1$,

$$\mathbb{E}\Big[\sup_{t\in[s,T]} |X(t)|^p\Big] \leqslant K(1 + |x|^p). \tag{6.1.10}$$

Now, for a (nonhomogeneous) linear equation, roughly speaking, the comparison is equivalent to the fact that the nonnegativity of the initial state and nonhomogeneous term leads to that of the solution. Before the comparison theorem, let $\Phi(\cdot\,,\cdot)$ be the stochastic fundamental matrix $\Phi(\cdot\,,\cdot)$ of $[A_0(\cdot); A_1(\cdot)]$, i.e.,

$$\begin{cases} d\Phi(t, s) = A_0(t)\Phi(t, s)dt + A_1(t)\Phi(t, s)dW(t), & t \in [s, T], \\ \Phi(s, s) = I. \end{cases} \tag{6.1.11}$$

Proposition 6.1.5

(i) *Let* (H6.1.2) *hold. Suppose*

$$A_0(t) \in \mathbb{R}^{n\times n}_{*+}, \qquad \text{a.s., a.e. } t \in [0, T], \tag{6.1.12}$$

and

$$A_1(t) \in \mathbb{R}_d^{n\times n}, \qquad \text{a.s., a.e. } t \in [0, T]. \tag{6.1.13}$$

Then, $\Phi(t, s) \in \mathbb{R}_+^{n\times n}$ *for any* $t \in [s, T]$*, and* $s \in [0, T)$*. In addition, for any* $(s, x) \in [0, T) \times \mathbb{R}^n$ *and* $b(\cdot) \in L^2_{\mathbb{F}}(\Omega; L^1(s, T;\mathbb{R}^n))$ *with*

$$x \geqslant 0, \quad b(t) \geqslant 0, \qquad \text{a.s., a.e. } t \in [s, T], \tag{6.1.14}$$

the unique solution $X(\cdot) \equiv X(\cdot\,; s, x, b(\cdot)) \in L^2_{\mathbb{F}}(\Omega; C([s, T]; \mathbb{R}^n))$ *of linear FSDE* (6.1.9) *corresponding to* $(x, b(\cdot))$ *on* $[s, T]$ *satisfies the following:*

$$X(t) \geqslant 0, \qquad \forall t \in [s, T], \quad \text{a.s.} \tag{6.1.15}$$

(ii) *If* $\Phi(t, s) \in \mathbb{R}_+^{n\times n}$ *for any* $t \in [s, T]$*, and* $s \in [0, T)$*, then* (6.1.12) *and* (6.1.13) *hold true.*

Proof Let $X^0(\cdot) \equiv X(\cdot\,; s, x, 0)$ be the solution to the linear SDE (6.1.9) with $b(\cdot) = 0$. Then, (6.1.15) is implied by

$$X^0(t) = \Phi(t, s)x \geqslant 0, \qquad \forall x \geqslant 0,\ t \in [s, T],\ \text{a.s. } s \in [0, T), \tag{6.1.16}$$

as well as the fact that

$$X(t) = \Phi(t, s)x + \int_s^t \Phi(t, r)b(r)dr, \qquad t \in [s, T].$$

Sufficiency. We now prove (6.1.16) under conditions (6.1.12) and (6.1.13). To this end, we first let $x = (x_1, \cdots, x_n)$ with $x_k > 0\ (1 \leqslant k \leqslant n)$, and let

$$\tau \triangleq \inf\Big\{t \in [s, T] \;\Big|\; \min_{1\leqslant k\leqslant n} \langle X^0(t), e_k\rangle \leqslant 0\Big\}, \qquad \inf\varnothing \triangleq T.$$

Then, we have

$$X_k^0(t) \equiv \langle X^0(t), e_k\rangle > 0, \qquad t \in [s, \tau),\ \text{a.s.,} \quad k = 1, 2, \cdots, n,$$

and

$$\min_{1\leqslant k\leqslant n} X_k^0(\tau) = 0, \qquad \text{a.s. } \omega \in (\tau < T).$$

By conditions (6.1.12)–(6.1.13), we

$$\begin{aligned}\langle A_0(r)X^0(r), e_k\rangle &= \sum_{\ell=1}^{n}\langle A_0(r)X^0_\ell(r)e_\ell, e_k\rangle\\ &= \sum_{\ell\neq k}^{n} X^0_\ell(r)\langle A_0(r)e_\ell, e_k\rangle + X^0_k(r)\langle A_0(r)e_k, e_k\rangle \geqslant X^0_k(r)\langle A_0(r)e_k, e_k\rangle,\end{aligned}$$

and

$$\big(\langle A_1(r)X^0(r), e_k\rangle\big)^2 = \Big[\sum_{\ell=1}^{n}\langle A_1(r)X^0_\ell(r)e_\ell, e_k\rangle\Big]^2 = |X^0_k(r)|^2\Big[\langle A_1(r)e_k, e_k\rangle\Big]^2.$$

On the other hand, by Itô's formula, and Lemma 6.1.4,

$$\begin{aligned}\mathbb{E}\Big[\frac{1}{X^0_k(t\wedge\tau)+\varepsilon}\Big] &= \frac{1}{x_k+\varepsilon} + \mathbb{E}\int_s^{t\wedge\tau}\Big(-\frac{1}{[X^0_k(r)+\varepsilon]^2}\langle A_0(r)X^0(r), e_k\rangle\\ &\qquad + \frac{1}{4}\frac{1}{[X^0_k(r)+\varepsilon]^3}\big(\langle A_1(r)X^0(r), e_k\rangle\big)^2\Big)dr\\ &\leqslant \frac{1}{x_k+\varepsilon} + \mathbb{E}\int_s^{t\wedge\tau}\Big(-\frac{1}{[X^0_k(r)+\varepsilon]^2}\langle A_0(r)e_k, e_k\rangle X^0_k(r)\\ &\qquad + \frac{1}{4}\frac{1}{[X^0_k(r)+\varepsilon]^3}\big(\langle A_1(r)e_k, e_k\rangle\big)^2 X^0_k(r)^2\Big)dr\\ &\leqslant \frac{1}{x_k+\varepsilon} + K\mathbb{E}\int_s^{t\wedge\tau}\frac{1}{X^0_k(r)+\varepsilon}dr.\end{aligned}$$

Then

$$\mathbb{E}\sum_{k=1}^{n}\frac{1}{X^0_k(t\wedge\tau)+\varepsilon} \leqslant \sum_{k=1}^{n}\frac{1}{x_k+\varepsilon} + K\mathbb{E}\int_s^{\tau\wedge t}\sum_{k=1}^{n}\frac{1}{X^0_k(r)+\varepsilon}dr.$$

By Gronwall's inequality,

$$\mathbb{E}\sum_{k=1}^{n}\frac{1}{X^0_k(\tau)+\varepsilon} \leqslant K\sum_{k=1}^{n}\frac{1}{x_k+\varepsilon}.$$

Sending $\varepsilon \to 0$, by Fatou's lemma, we obtain

$$\mathbb{E}\sum_{k=1}^{n}\frac{1}{X^0_k(\tau)} \leqslant K\sum_{k=1}^{n}\frac{1}{x_k} < \infty.$$

Thus, it is necessary that $\tau = T$, and for any $s \in [0, T]$, $\min_{1 \leqslant k \leqslant n} X_k^0(s) > 0$. Therefore, $\Phi(t, s) \in \mathbb{R}_+^{n\times n}$ for any $t \in [s, T]$ and $s \in [0, T]$ and (6.1.16) holds naturally. Now, in general, if we just have $x \geqslant 0$, from the above, we have

$$X(t; s, x + \delta\mathbb{1}, 0) \geqslant 0, \qquad t \in [s, T], \text{ a.s.,}$$

where $\mathbb{1} = (1, 1, \cdots, 1)^\top \in \mathbb{R}^n$. Then, sending $\delta \to 0$, we can obtain our conclusion (6.1.16).

Necessity. To begin with, we claim that there exists measurable set $\Gamma \subset [0, T]$ satisfying $|\Gamma| = T$ such that for any $t \in \Gamma$,

$$\begin{aligned}
&\lim_{\varepsilon\to 0}\frac{1}{\varepsilon}\mathbb{E}\int_t^{t+\varepsilon}\big|A_0(s) - A_0(t)\big|^2 ds = 0,\\
&\lim_{\varepsilon\to 0}\frac{1}{\varepsilon}\mathbb{E}\int_t^{t+\varepsilon}\big|A_1(s) - A_1(t)\big|^2 ds = 0.
\end{aligned}$$

In fact, since $L^2_{\mathcal{F}_T}(\Omega; \mathbb{R}^{n\times n})$ is separable, there exists a countable dense subset, denoted by $\mathscr{C}$. By Lebesgue differentiation theorem, for any $\xi \in \mathscr{C}$, there exists full measure $\Gamma_\xi \subset [0, T]$ such that

$$\lim_{\varepsilon\to 0}\frac{1}{\varepsilon}\mathbb{E}\int_t^{t+\varepsilon}\big|A_0(s) - \xi\big|^2 ds = \mathbb{E}|A_0(t) - \xi|^2, \quad t \in \Gamma_\xi.$$

We define by $\Gamma_0 \triangleq \bigcap\limits_{\xi\in\mathscr{C}} \Gamma_\xi$. It is easy to see that Γ_0 is also full measure in $[0, T]$, and for any $t \in \Gamma_0$,

$$\lim_{\varepsilon\to 0}\frac{1}{\varepsilon}\mathbb{E}\int_t^{t+\varepsilon}\big|A_0(s) - \xi\big|^2 ds = \mathbb{E}|A_0(t) - \xi|^2, \quad \forall \xi \in \mathscr{C}.$$

On the other hand, for any small $\eta > 0$, and any $t \in \Gamma_0$, there exists $\bar\xi \in \mathscr{C}$ such that $\mathbb{E}\big|A_0(t) - \bar\xi\big|^2 < \eta$. As a result,

$$\begin{aligned}
&\lim_{\varepsilon\to 0}\frac{1}{\varepsilon}\mathbb{E}\int_t^{t+\varepsilon}\big|A_0(s) - A_0(t)\big|^2 ds\\
&\leqslant 2\lim_{\varepsilon\to 0}\frac{1}{\varepsilon}\mathbb{E}\int_t^{t+\varepsilon}\big|A_0(s) - \bar\xi\big|^2 ds + 2\mathbb{E}\big|A_0(t) - \bar\xi\big|^2 < 2\eta.
\end{aligned}$$

By the arbitrariness of η, we see that for any $t \in \Gamma_0$,

$$\lim_{\varepsilon\to 0}\frac{1}{\varepsilon}\mathbb{E}\int_t^{t+\varepsilon}\big|A_0(s) - A_0(t)\big|^2 ds = 0.$$

In the similar spirit, there exists full measure $\Gamma_1 \subset [0, T]$ such that

$$\lim_{\varepsilon \to 0} \frac{1}{\varepsilon} \mathbb{E} \int_t^{t+\varepsilon} \big|A_1(s) - A_1(t)\big|^2 ds = 0.$$

Then, our conclusion follows by letting $\Gamma = \Gamma_0 \bigcap \Gamma_1$.

To prove (6.1.12) holds, it is sufficient to prove that there exists a full measure subset of Γ, say, Υ, such that for any $t \in \Upsilon$, $A_0(t) \in \mathbb{R}^{n\times n}_{*+}$. a.s. Suppose it fails. Then, there exists some $s \in \Gamma \subset [0, T]$ and some $k \neq \ell$,

$$\mathbb{P}\big(\Omega^0_{k\ell}(s)\big) > 0, \quad \Omega^0_{k\ell}(s) = \big\{\omega \in \Omega \bigm| \langle A_0(s)e_k, e_\ell\rangle < 0\big\}. \tag{6.1.17}$$

In other words, the (ℓ, k)-th (off-diagonal) entry of $A_0(s)$ is not almost surely nonnegative. Let $\Phi_0(\cdot\,,\cdot)$ be the fundamental matrix of $A_0(\cdot)$, i.e.,

$$\Phi_0(t,s) = I + \int_s^t A_0(r)\Phi_0(r,s)dr, \qquad 0 \leqslant s \leqslant t \leqslant T.$$

With bounded A_0, it is known that $\Phi_0(\cdot,\cdot)^{-1}$ exists and satisfies

$$\Phi_0(t,s)^{-1} = I - \int_s^t \Phi_0(r,s)^{-1}A_0(r)dr, \qquad 0 \leqslant s \leqslant t \leqslant T,$$

and

$$\Phi_0(t,s)\Phi_0(r,s)^{-1} = \Phi_0(t,r), \qquad 0 \leqslant s \leqslant r \leqslant t \leqslant T.$$

Hence, for some constant $K > 0$,

$$|\Phi_0(t,s) - I| + |\Phi_0(t,s)^{-1} - I| \leqslant K(t-s), \quad 0 \leqslant s \leqslant t \leqslant T.$$

Now, let $X^{0,k}(\cdot) \equiv X(\cdot\,; s, e_k, 0)$ be the solution to linear FSDE (6.1.9) with $(s, x, b(\cdot)) = (s, e_k, 0)$. Then, by (6.1.10), for $p > 1$,

$$\mathbb{E}\Big[\sup_{t\in[s,T]} |X^{0,k}(t)|^p\Big] \leqslant K(1 + |e_k|^p) \leqslant K. \tag{6.1.18}$$

Also

$$X^{0,k}(t) = \Phi_0(t,s)\Big[e_k + \int_s^t \Phi_0(r,s)^{-1}A_1(r)X^{0,k}(r)dW(r)\Big], \quad 0 \leqslant s \leqslant t \leqslant T,$$

and (note (6.1.18))

$$
\begin{aligned}
\mathbb{E}|X^{0,k}(t)-e_k|^p &\leqslant K\Big[\mathbb{E}|\Phi_0(t,s)-I|^p+\mathbb{E}\Big|\int_s^t \Phi_0(r,s)^{-1}A_1(r)X^{0,k}(r)dW(r)\Big|^p\Big] \\
&\leqslant K\Big[|t-s|^p+\Big(\int_s^t \mathbb{E}|X^{0,k}(r)|^2dr\Big)^{\frac{p}{2}}\Big]\leqslant K|t-s|^{\frac{p}{2}}.
\end{aligned}
$$

Next

$$
\begin{aligned}
\mathbb{E}\big[\langle X^{0,k}(t),e_\ell\rangle \mathbf{1}_{\Omega^0_{k\ell}(s)}\big] &= \int_s^t \mathbb{E}[\mathbf{1}_{\Omega^0_{k\ell}(s)}\langle A_0(r)X^{0,k}(r),e_\ell\rangle\big]dr \\
&= \mathbb{E}[\langle A_0(s)e_k,e_\ell\rangle \mathbf{1}_{\Omega^0_{k\ell}(s)}\big](t-s)+\eta_{k\ell}(t),
\end{aligned}
$$

where

$$
\eta_{k\ell}(t)=\int_s^t \mathbb{E}\big[\mathbf{1}_{\Omega^0_{k\ell}(s)}\langle A_0(r)X^{0,k}(r)-A_0(s)e_k,e_\ell\rangle\big]dr.
$$

Then, by the choice of $s\in\Gamma$,

$$
\begin{aligned}
|\eta_{k\ell}(t)| &\leqslant \int_s^t \mathbb{E}\Big(|A_0(r)-A_0(s)||X^{0,k}(r)|+|X^{0,k}(r)-e_k|\Big)dr \\
&\leqslant \Big[\mathbb{E}\sup_{r\in[s,T]}|X^{0,k}(r)|^2\Big]^{\frac12}\Big\{\mathbb{E}\int_s^t |A_0(r)-A_0(s)|^2dr\Big\}^{\frac12}(t-s)^{\frac12} \\
&\quad +(t-s)^{\frac12}\Big[\int_s^t \mathbb{E}|X^{0,k}(r)-e_k|^2dr\Big]^{\frac12}\leqslant o(|t-s|).
\end{aligned}
$$

Hence

$$
\mathbb{E}\big[\langle X^{0,k}(t),e_\ell\rangle \mathbf{1}_{\Omega^0_{k\ell}(s)}\big]=\mathbb{E}[\langle A_0(s)e_k,e_\ell\rangle \mathbf{1}_{\Omega^0_{k\ell}(s)}\big](t-s)+o(t-s),
$$

for $t-s>0$ small. Thus, $X^{0,k}(t)=\Phi(t,s)e_k\geqslant 0$ fails for some $t\in[s,T]$ that is close to s. This is a contradiction, which shows that (6.1.12) is necessary.

Next, to prove (6.1.13) holds, it is sufficient to prove that there exists a full measure subset of Γ, say, Υ, such that for any $t\in\Upsilon$, $A_1(t)\in\mathbb{R}^{n\times n}_d$. a.s. Suppose it fails. Then, there exists some $s\in\Gamma\subset[0,T]$ and some $k\neq\ell$,

$$
\mathbb{P}\Big(\langle A_1(s)e_k,e_\ell\rangle\neq 0\Big)>0.
$$

That is, the (ℓ,k)-th (off-diagonal) entry of $A_1(s)$ is not identically equal to zero (almost surely). Hence, there exist $\delta>0$ such that

$$
\mathbb{P}\big(\bar{\Omega}^+(s)\cup\bar{\Omega}^-(s)\big)>\delta,
$$

where

$$\bar{\Omega}^+(s) \stackrel{\triangle}{=} \{\omega \in \Omega \mid \langle A_1(s)e_k, e_\ell\rangle > \delta\} \subset \Omega,$$
$$\bar{\Omega}^-(s) \stackrel{\triangle}{=} \{\omega \in \Omega \mid \langle A_1(s)e_k, e_\ell\rangle < -\delta\} \subset \Omega.$$

Without loss of generality, we suppose (the other case will be similar.)

$$\mathbb{P}\big(\bar{\Omega}^+(s)\big) > \delta.$$

Thus, for $\ell \neq k$,

$$\begin{aligned}\langle X^{0,k}(t), e_\ell\rangle &= \langle \Phi_0(t,s)\Big[e_k + \int_s^t \Phi_0(r,s)^{-1}A_1(r)X^{0,k}(r)dW(r)\Big], e_\ell\rangle \\ &= \langle A_1(s)e_k, e_\ell\rangle[W(t) - W(s)] + \xi_{k\ell}(t),\end{aligned} \tag{6.1.19}$$

where (note $\langle e_k, e_\ell\rangle = 0$)

$$\begin{aligned}\xi_{k\ell}(t) = &\langle[\Phi_0(t,s) - I]e_k, e_\ell\rangle \\ &+\langle[\Phi_0(t,s) - I]\int_s^t \Phi_0(r,s)^{-1}A_1(r)X^{0,k}(r)dW(r), e_\ell\rangle \\ &+\langle\int_s^t [\Phi_0(r,s)^{-1} - I]A_1(r)X^{0,k}(r)dW(r), e_\ell\rangle \\ &+\langle\int_s^t [A_1(r) - A_1(s)]X^{0,k}(r)dW(r), e_\ell\rangle \\ &+\langle\int_s^t A_1(s)[X^{0,k}(r) - e_k]dW(r), e_\ell\rangle.\end{aligned}$$

Hence, due to the choice of $s \in \Gamma$,

$$\begin{aligned}\mathbb{E}|\xi_{k\ell}(t)| \leqslant &K\Big((t-s) + (t-s)\mathbb{E}\Big|\int_s^t \Phi_0(r,s)^{-1}A_1(r)X_k^0(r)dW(r)\Big| \\ &+\mathbb{E}\Big|\int_s^t [\Phi_0(r,s)^{-1} - I]A_1(r)X_k^0(r)dW(r)\Big| \\ &+\mathbb{E}\Big|\int_s^t [A_1(r) - A_1(s)]X_k^0(r)dW(r)\Big| \\ &+\mathbb{E}\Big|\int_s^t A_1(s)[X_k^0(r) - e_k]dW(r)\Big|\Big) \\ \leqslant &K\Big[(t-s)+(t-s)\Big(\int_s^t \mathbb{E}|X_k^0(r)|^2dr\Big)^{\frac{1}{2}}+\Big(\int_s^t (r-s)^2\mathbb{E}|X_k^0(r)|^2dr\Big)^{\frac{1}{2}} \\ &+\Big[\mathbb{E}\sup_{r\in[s,T]}|X^{0,k}(r)|^2\Big]^{\frac{1}{2}}\Big\{\mathbb{E}\int_s^t |A_1(r) - A_1(s)|^2dr\Big\}^{\frac{1}{2}} \\ &+\Big(\int_s^t \mathbb{E}|X^{0,k}(r) - e_k|^2dr\Big)^{\frac{1}{2}}\Big] \\ \leqslant &K\Big[(t-s) + (t-s)^{\frac{3}{2}} + o(\sqrt{t-s})\Big] = o(\sqrt{t-s}).\end{aligned}$$

On the other hand, recalling the definition of $\bar{\Omega}^+(s)$ above, and dividing (6.1.19) by $\sqrt{t-s}\langle A_1(s)e_k, e_\ell\rangle \mathbf{1}_{\bar{\Omega}^+(s)}$, we have

$$\frac{\langle X^{0,k}(t), e_\ell\rangle \mathbf{1}_{\bar{\Omega}^+(s)}}{\sqrt{t-s}\langle A_1(s)e_k, e_\ell\rangle} = \theta_1(t) + \theta_2(t), \tag{6.1.20}$$

where

$$\theta_1(t) \triangleq \frac{W(t) - W(s)}{\sqrt{t-s}} \mathbf{1}_{\bar{\Omega}^+(s)}, \quad \theta_2(t) \triangleq \frac{\xi_{k\ell}(t) \cdot \mathbf{1}_{\bar{\Omega}^+(s)}}{\sqrt{t-s}\langle A_1(s)e_k, e_\ell\rangle}.$$

For $\theta_2(t)$ in (6.1.20), we have

$$\mathbb{E}|\theta_2(t)| \leqslant \frac{\mathbb{E}\big[|\xi_{k\ell}(t)| \cdot \mathbf{1}_{\bar{\Omega}^+(s)}(\omega)\big]}{\delta\sqrt{t-s}} \leqslant \frac{\mathbb{E}\big[|\xi_{k\ell}(t)|\big]}{\delta\sqrt{t-s}} \leqslant o(1).$$

Next, let $\zeta(t) = \frac{W(t)-W(s)}{\sqrt{t-s}} \sim N(0,1)$ and

$$\theta_1(t) = \zeta(t)\mathbf{1}_{\bar{\Omega}^+(s)}.$$

Since $\mathbb{E}\zeta(t) = 0$, and

$$\mathbb{E}|\zeta(t)| = \frac{1}{\sqrt{2\pi}} \int_{\mathbb{R}} |x| e^{\frac{-x^2}{2}} dx = \sqrt{\frac{2}{\pi}},$$

we have

$$\mathbb{E}\zeta(t)^+ = \mathbb{E}\zeta(t)^- = \frac{1}{\sqrt{2\pi}}.$$

Also, note that $\bar{\Omega}^+(s) \in \mathcal{F}_s$, and $\frac{W(t)-W(s)}{\sqrt{t-s}}$ is independent of $\mathcal{F}_s$, we see that $\zeta(t)$ and $\mathbf{1}_{\bar{\Omega}^+(s)}$ are independent. Therefore,

$$\mathbb{E}\big[\zeta(t)^- \cdot \mathbf{1}_{\bar{\Omega}^+(s)}\big] = \mathbb{E}\zeta(t)^- \cdot \mathbb{E}\mathbf{1}_{\bar{\Omega}^+(s)} \geqslant \frac{\delta}{\sqrt{2\pi}}. \tag{6.1.21}$$

By the definition of $X^{0,k}(\cdot)$, one has $X^{0,k}(\cdot) \geqslant 0$ and $\langle X^{0,k}(\cdot), e_\ell\rangle \geqslant 0$ on $[s, T]$. Then

$$0 \leqslant \theta_1(t) + \theta_2(t) = [\zeta(t)^+ - \zeta(t)^-] \cdot \mathbf{1}_{\bar{\Omega}^+(s)} + \theta_2(t).$$

Therefore,

$$\zeta(t)^- \cdot \mathbf{1}_{\bar{\Omega}^+(s)} \leqslant \zeta(t)^+ \mathbf{1}_{\bar{\Omega}^+(s)} + \theta_2(t).$$

Multiplying the above by $\mathbf{1}_{\Omega_0^-(t)}$, where $\Omega_0^-(t) \triangleq \{\omega \in \Omega \mid \zeta(t) \leqslant 0\}$, then

$$\zeta(t)^- \cdot \mathbf{1}_{\bar{\Omega}^+(s)} \leqslant \theta_2(t) \cdot \mathbf{1}_{\Omega_0^-(t)},$$

since $\zeta(t)^- \mathbf{1}_{\Omega_0^-(t)} = \zeta(t)^-$. Therefore, by (6.1.21),

$$\frac{\delta}{\sqrt{2\pi}} \leqslant \mathbb{E}\big[\zeta(t)^- \cdot \mathbf{1}_{\bar{\Omega}^+(s)}\big] \leqslant \mathbb{E}|\theta_2(t)| \leqslant o(1),$$

which is impossible if $t - s$ is small. Hence

$$\mathbb{P}\Big(\langle A_1(s)e_k, e_\ell\rangle \neq 0\Big) > 0$$

leads to a contradiction, proving the necessity. □

We point out that in the above, the dimension $n \geqslant 1$; and if $n = 1$, conditions (6.1.12)–(6.1.13) are automatically true.

Proof of Theorem 6.1.2

(i) We may use our Proposition 6.1.5 to prove. More precisely, let $X^i(\cdot)$ be solution of (6.1.2). Then

$$\begin{aligned}
\widehat{X}(t) \equiv X^1(t) - X^0(t) = x^1 - x^0 &+ \int_0^t \Big(b^1(s, X^1(s)) - b^0(s, X^0(s))\Big)ds \\
&+ \int_0^t \Big(\sigma(s, X^1(s)) - \sigma(s, X^0(s))\Big)dW(s) \\
= x^1 - x^0 + \int_0^t &\Big(b^1(s, X^1(s)) - \bar{b}(s, X^1(s)) + \bar{b}(s, X^0(s)) - b^0(s, X^0(s))\Big)ds \\
&+ \int_0^t \Big[\int_0^1 \bar{b}_x(s, X^0(s) + \alpha\widehat{X}(s))d\alpha\Big]\widehat{X}(s)ds \\
&+ \int_0^t \Big[\int_0^1 \bar{\sigma}_x(s, X^0(s) + \alpha\widehat{X}(s))d\alpha\Big]\widehat{X}(s)dW(s) \\
\equiv \widehat{x} + \int_0^t &\Big(\widehat{A}_0(s)\widehat{X}(s) + \widehat{b}(s)\Big)ds + \int_0^t \widehat{A}_1(s)\widehat{X}(s)dW(s),
\end{aligned}$$

where

$$\begin{cases}
x = x^1 - x^0 \geqslant 0, \\
\widehat{b}(t) = b^1(t, X^1(t)) - \bar{b}(t, X^1(t)) + \bar{b}(t, X^0(t)) - b^0(t, X^0(t)) \geqslant 0, \\
\widehat{A}_0(t) = \displaystyle\int_0^1 \bar{b}_x(t, X^0(t) + \alpha\widehat{X}(t))d\alpha, \\
\widehat{A}_1(t) = \displaystyle\int_0^1 \bar{\sigma}_x(t, X^0(t) + \alpha\widehat{X}(t))d\alpha.
\end{cases}$$

Then, Proposition 6.1.5, part (i) applies.

(ii) For any $x, \widetilde{x} \in \mathbb{R}^n$ and $\widetilde{x} \geqslant 0$, $\delta \geqslant 0$, let $X^\delta(\cdot)$ be the solution to the following:

$$X^\delta(t) = x + \delta\widetilde{x} + \int_s^t \bar{b}(r, X^\delta(r))dr + \int_s^t \sigma(r, X^\delta(r))dW(r), \quad t \in [s, T],$$

and $\widetilde{X}(\cdot)$ be the solution to the following: for $t \in [s, T]$,

$$\widetilde{X}(t) = \widetilde{x} + \int_s^t \bar{b}_x(r, X^0(r))\widetilde{X}(r)dr + \int_s^t \sigma_x(r, X^0(r))\widetilde{X}(r)dW(r).$$

Then, it is straightforward that

$$\lim_{\delta \to 0} \sup_{t \in [s,T]} \mathbb{E}\left|\frac{X^\delta(t) - X^0(t)}{\delta} - \widetilde{X}(t)\right|^2 = 0, \quad s \in [0, T].$$

Since $X^\delta(t) - X^0(t) \geqslant 0$ for $\delta > 0$, we then have

$$\widetilde{X}(t) \geqslant 0, \qquad \forall t \in [s, T], \text{ a.s.}$$

Then, by Proposition 6.1.5, part (ii), we must have

$$\bar{b}_x(r, X^0(r)) \in \mathbb{R}^{n\times n}_{*+}, \quad \sigma_x(r, X^0(r)) \in \mathbb{R}^{n\times n}_d, \text{ a.s. a.e., } r \in [s, T].$$

Setting $r = s$, we obtain (6.1.4).

□

1.2 BSDEs

We now look at nonlinear m-dimensional BSDEs: for $i = 0, 1$,

$$Y^i(t) = \xi^i + \int_t^T g^i(s, Y^i(s), Z^i(s))ds - \int_t^T Z^i(s)dW(s), \quad t \in [0, T], \qquad (6.1.22)$$

where ξ^i is $\mathcal{F}_T$-measurable, and $g^i(\cdot)$ is a suitable map. Let us introduce the following standard assumption.

(H6.1.4) For $i = 0, 1$, the map $g^i : [0, T] \times \mathbb{R}^m \times \mathbb{R}^m \times \Omega \to \mathbb{R}^m$ is measurable, $s \mapsto g^i(s, y, z)$ is $\mathbb{F}$-progressively measurable, $(y, z) \mapsto g^i(s, y, z)$ is uniformly Lipschitz, and $s \mapsto g^i(s, 0, 0)$ is uniformly bounded.

Under (H6.1.4), for any $\xi^i \in L^p_{\mathcal{F}_T}(\Omega;\mathbb{R}^m)$ (with $p > 1$), BSDE (6.1.22) admits a unique adapted solution:

$$(Y^i(\cdot), Z^i(\cdot)) \in L^p_{\mathbb{F}}(\Omega; C([0,T];\mathbb{R}^m)) \times L^p_{\mathbb{F}}(\Omega; L^2(0,T;\mathbb{R}^m)) \equiv \mathbf{H}^p[0,T],$$

and the following estimate holds:

$$\mathbb{E}\Big[\sup_{t\in[0,T]}|Y^i(t)|^p+\Big(\int_0^T|Z^i(s)|^2ds\Big)^{\frac{p}{2}}\Big] \leqslant K\mathbb{E}\Big[|\xi^i|^p+\Big(\int_0^T|g^i(s,0,0)|ds\Big)^p\Big].$$

We have the following comparison theorem.

Theorem 6.1.6 *Let* (H6.1.4) *hold. Suppose* $\bar g : [0,T]\times\mathbb{R}^m\times\mathbb{R}^m\times\Omega\to\mathbb{R}^m$ *is measurable,* $s\mapsto \bar g(s,y,z)$ *is* $\mathbb{F}$*-progressively measurable, and* $\bar g_y(s,y,z)$ *and* $\bar g_z(s,y,z)$ *exist and are uniformly bounded.*

(i) *Suppose for almost all* $s\in[0,T]$,

$$\bar g_y(s,y,z)\in\mathbb{R}^{m\times m}_{*+},\quad \bar g_z(s,y,z)\in\mathbb{R}^{m\times m}_d,\qquad \forall (y,z)\in\mathbb{R}^m\times\mathbb{R}^m,\ \text{a.s.,} \tag{6.1.23}$$

which is automatically true if $m=1$, *and*

$$g^0(s,y,z)\leqslant \bar g(s,y,z)\leqslant g^1(s,y,z),\qquad \forall (y,z)\in\mathbb{R}^m\times\mathbb{R}^m,\ \text{a.s..} \tag{6.1.24}$$

Then, for any $\xi^0,\xi^1\in L^2_{\mathcal{F}_T}(\Omega;\mathbb{R}^m)$ *with* $\xi^0\leqslant\xi^1$, *almost surely, the corresponding adapted solutions* $(Y^i(\cdot),Z^i(\cdot))$ *of BSDEs* (6.1.22) *satisfy*

$$Y^0(t)\leqslant Y^1(t),\qquad t\in[0,T],\ \text{a.s.} \tag{6.1.25}$$

(ii) *Suppose for almost all* $s\in[0,T]$,

$$g^0(s,y,z)=\bar g(s,y,z)=g^1(s,y,z),\qquad \forall (y,z)\in\mathbb{R}^m\times\mathbb{R}^m,\ \text{a.s,}$$

and $(s,y,z)\mapsto(\bar g(s,y,z),\bar g_y(s,y,z),\bar g_z(s,y,z))$ *is continuous. Then,* (6.1.23) *is necessary for the conclusion of* (i) *to be true.*

We make the following remarks:

- In the case that $m=n=1$, the above comparison is well-understood with (6.1.23)–(6.1.24) becoming

$$\begin{aligned}&\bar g_y(s,y,z)\in\mathbb{R},\ \ \bar g_z(s,y,z)\in\mathbb{R},\quad \forall (y,z)\in\mathbb{R}\times\mathbb{R},\ \text{a.s.,}\\ &g^0(s,y,z)\leqslant\bar g(s,y,z)\leqslant g^1(s,y,z),\qquad \forall (y,z)\in\mathbb{R}\times\mathbb{R},\ \text{a.s..}\end{aligned}$$

- For the case that $g^0(\cdot)$ and $g^1(\cdot)$ are different, our sufficient condition seems to be easier to check. For example, suppose there exists a $\bar{g}(\cdot)$ satisfies (6.1.23). Then, the comparison theorem holds for the case, say,

$$g^0(s,y,z)=\bar{g}(s,y,z)-|z|,\qquad g^1(s,y,z)=\bar{g}(s,y,z)+|z|,$$
$$(s,y,z)\in[0,T]\times\mathbb{R}^m\times\mathbb{R}^m.$$

  Here, $z\in\mathbb{R}^m$ is a vector. So, our result is different from FSDEs.
- We will see below that our proof is based on the duality and a corresponding result for linear FSDEs (Proposition 6.1.5).
- For the pricing European contingent claim, roughly speaking, comparison of BSDEs means a larger initial endowment, and a larger return rate leads to a larger contingent price.

To prove our theorem, we first look at the following n-dimensional linear BSDE:

$$\begin{cases} dY(t)=\big[A(t)Y(t)+C(t)Z(t)-g(t)\big]dt+Z(t)dW(t), & t\in[0,T],\\ Y(T)=\xi, \end{cases} \tag{6.1.26}$$

where $\xi\in L^2_{\mathcal{F}_T}(\Omega;\mathbb{R}^m)$. Similar to (H6.1.2)–(H6.1.3), we introduce the following hypotheses.

(H6.1.5) The maps $A(\cdot),C(\cdot)\in L^\infty_{\mathbb{F}}\big(0,T;\mathbb{R}^{m\times m}\big)$.

The following is comparable with Proposition 6.1.5.

Proposition 6.1.7 *Let* (H6.1.5) *hold. Then, for any* $g(\cdot)\in L^2_{\mathbb{F}}(0,T;\mathbb{R}^m)$ *and* $\xi\in L^2_{\mathcal{F}_T}(\Omega;\mathbb{R}^m)$ *with*

$$\xi\geqslant 0,\qquad g(t)\geqslant 0,\quad \text{a.s. a.e.}\quad t\in[0,T], \tag{6.1.27}$$

the adapted solution $(Y(\cdot),Z(\cdot))$ *to BSDE* (6.1.26) *satisfies*

$$Y(t)\geqslant 0,\qquad t\in[0,T],\ \text{a.s.}, \tag{6.1.28}$$

if

$$-A(t)\in\mathbb{R}^{m\times m}_{*+},\quad C(t)\in\mathbb{R}^{m\times m}_d,\ \text{a.s., a.e. } t\in[0,T]. \tag{6.1.29}$$

Conversely, (6.1.29) *is necessary for* (6.1.28) *to be true.*

Proof *Sufficiency.* Let τ be any $\mathbb{F}$-stopping times such that $0\leqslant\tau\leqslant T$, almost surely. For any $x\in\mathbb{R}^m$, let $X(\cdot)$ be the strong solution to the following FSDE:

$$\begin{cases} dX(t)=-A(t)^\top X(t)dt-C(t)^\top X(t)dW(t), & t\in[\tau,T],\\ X(\tau)=x. \end{cases} \tag{6.1.30}$$

By Theorem 1.2.1 of Chap. 1, we have the following duality principle:

$$\langle x, Y(\tau)\rangle = \mathbb{E}_\tau\Big[\langle X(T), \xi\rangle + \int_\tau^T \langle X(r), g(r)\rangle dr\Big]. \tag{6.1.31}$$

Now, for any $x \in \mathbb{R}^m_+$, under our conditions, by Proposition 6.1.5, the solution $X(\cdot)$ of (6.1.30) satisfies

$$X(t) \geqslant 0, \qquad t \in [\tau, T], \text{ a.s.}$$

Therefore, by duality principle (6.1.31),

$$\langle x, Y(\tau)\rangle = \mathbb{E}_\tau\Big[\langle X(T), \xi\rangle + \int_\tau^T \langle X(r), g(r)\rangle dr\Big] \geqslant 0,$$

for any $\mathbb{F}$-stopping time $\tau \leqslant T$, proving our conclusion.

Necessity. Suppose (6.1.29) fails. Then, by Proposition 6.1.5, for some $1 \leqslant k, \ell \leqslant m$ and $\tau \in [0, T]$, the solution $X(\cdot) \equiv X(\cdot\,; \tau, e_k)$ of (6.1.30) satisfies

$$\mathbb{P}\big(\langle X(\tau'), e_\ell\rangle < 0\big) > 0,$$

for some time $\tau' > \tau$. For such a τ', choosing $\xi = e_\ell \mathbf{1}_{\{\langle X(\tau'), e_\ell\rangle < 0\}}$, and $g(\cdot) = 0$, we have

$$\langle e_k, Y(\tau)\rangle = \mathbb{E}_\tau\Big[\langle X(\tau'), e_\ell\rangle \mathbf{1}_{\{\langle X(\tau'), e_\ell\rangle < 0\}}\Big] < 0,$$

a contradiction. □

We now prove Theorem 6.1.6.

Proof of Theorem 6.1.6

(i) Let $\bar\xi \in L^2_{\mathcal{F}_\tau}(\Omega; \mathbb{R}^m)$ such that

$$\xi^0 \leqslant \bar\xi \leqslant \xi^1, \qquad \text{a.s.}$$

Let $(\bar Y(\cdot), \bar Z(\cdot))$ be the adapted solution to the following BSDE:

$$\bar Y(t) = \bar\xi + \int_t^T \bar g(s, \bar Y(s), \bar Z(s))ds - \int_t^T \bar Z(s)dW(s), \qquad t \in [0, T].$$

Observe

$$\begin{aligned}\bar{Y}(t)-Y^0(t)=\bar{\xi}-\xi^0&+\int_t^T\Big[\bar{g}(s,Y^0(s),Z^0(s))-g^0(s,Y^0(s),Z^0(s))\Big]ds\\&+\int_t^T\Big[A(s)\Big(\bar{Y}(s)-Y^0(s)\Big)+C(s)\Big(\bar{Z}(s)-Z^0(s)\Big)\Big]ds\\&-\int_t^T\Big(\bar{Z}(s)-Z^0(s)\Big)dW(s),\end{aligned}$$

where

$$\begin{aligned}A(s)&\overset{\Delta}{=}\int_0^1\bar{g}_y\big(s,Y^0(s)+\beta[\bar{Y}(s)-Y^0(s)],Z^0(s)+\beta[\bar{Z}(s)-Z^0(s)]\big)d\beta\in\mathbb{R}^{m\times m}_{*+},\\C(s)&\overset{\Delta}{=}\int_0^1\bar{g}_z\big(s,Y^0(s)+\beta[\bar{Y}(s)-Y^0(s)],Z^0(s)+\beta[\bar{Z}(s)-Z^0(s)]\big)d\beta\in\mathbb{R}^{m\times m}_{d}.\end{aligned}$$

Hence, by Proposition 6.1.7, we obtain our conclusion.

(ii) For any $\xi,\widetilde{\xi}\in L^2_{\mathcal{F}_T}(\Omega;\mathbb{R}^m)$ and $\widetilde{\xi}\geqslant 0$, $\delta\geqslant 0$, let $(Y^\delta(\cdot),Z^\delta(\cdot))$ be the adapted solution to the following BSDE:

$$Y^\delta(t)=\xi+\delta\widetilde{\xi}+\int_t^T\bar{g}(s,Y^\delta(s),Z^\delta(s))ds-\int_t^T Z^\delta(s)dW(s),\qquad t\in[0,T].$$

In particular

$$Y^0(t)=\xi+\int_t^T\bar{g}(s,Y^0(s),Z^0(s))ds-\int_t^T Z^0(s)dW(s),\quad t\in[0,T].\tag{6.1.32}$$

If $(\widetilde{Y}(\cdot),\widetilde{Z}(\cdot))$ is the adapted solution to the following BSDE:

$$\begin{aligned}\widetilde{Y}(t)=\widetilde{\xi}&+\int_t^T\Big(\bar{g}_y(s,Y^0(s),Z^0(s))\widetilde{Y}(s)+\bar{g}_z(s,Y^0(s),Z^0(s))\widetilde{Z}(s)\Big)ds\\&-\int_t^T\widetilde{Z}(s)dW(s),\qquad t\in[0,T],\end{aligned}$$

then, it is straightforward to show that

$$\lim_{\delta\to 0}\frac{Y^\delta(t)-Y^0(t)}{\delta}=\widetilde{Y}(t),\quad\forall t\in[0,T],\ \text{a.s.}$$

Hence, conclusion of (i) implies that

$$\widetilde{Y}(t)\geqslant 0,\qquad\forall t\in[0,T],\ \text{a.s.}$$

Consequently, by Proposition 6.1.7, we obtain

$$\bar g_y(s,Y^0(s),Z^0(s))\in\mathbb{R}^{m\times m}_{*+},\quad \bar g_z(s,Y^0(s),Z^0(s))\in\mathbb{R}^{m\times m}_d,\quad \text{a.s., a.e., } s\in[0,T], \tag{6.1.33}$$

for the adapted solution $(Y^0(\cdot),Z^0(\cdot))$ of BSDE (6.1.32) with any $\xi\in L^2_{\mathcal{F}_T}(\Omega;\mathbb{R}^m)$. Now, for any $s\in[0,T)$ and $y,z\in\mathbb{R}^m$, let

$$\xi=y+z[W(T)-W(s)]-\int_s^T\bar g(r,\bar Y^0(r),z)dr,$$

where $\bar Y^0(\cdot)$ is the unique solution of (forward) Volterra integral equation

$$\bar Y^0(t)=y+z[W(t)-W(s)]-\int_s^t\bar g(r,\bar Y^0(r),z)dr,\qquad t\in[s,T].$$

Then, it is easy to show that $(\bar Y^0(\cdot),z)$ is the unique adapted solution to the following BSDE:

$$Y^0(t)=\xi+\int_t^T\bar g(r,Y^0(r),Z^0(r))dr-\int_t^T Z^0(r)dW(r),\qquad t\in[s,T].$$

Clearly, $Y^0(s)=y$, and from (6.1.33), as well as the continuity of $(s,y,z)\mapsto(\bar g(s,y,z),\bar g_y(s,y,z),\bar g_z(s,y,z))$, we have

$$\bar g_y(s,y,z)\in\mathbb{R}^{m\times m}_{*+},\qquad \bar g_z(s,y,z)\in\mathbb{R}^{m\times m}_d,\qquad \text{a.s.}$$

Hence, (6.1.23) follows.

□

In Theorem 6.1.6, we assumed that $\bar g_y,\bar g_z$ exist and are bounded. However, when $m=1$, we can slightly relax this condition into $(y,z)\mapsto g(s,y,z)$ being Lispchitz. Hence, we can take $\bar g=g^1$ or $\bar g=g^2$. In fact, the above proof also works if we replace $g_y(s,y,z)$, $g_z(s,y,z)$, respectively by the following form:

$$\frac{[g(s,y_1,z)-g(s,y_0,z)]}{|y_1-y_0|}\mathbf{1}_{\{y_1-y_0\neq0\}},$$
$$\frac{[g(s,y,z_1)-g(s,y,z_0)]}{|z_1-z_0|}\mathbf{1}_{\{z_1-z_0\neq0\}}.$$

Then, we have the following corollary, which is useful in the sequel.

Corollary 6.1.8 *Let* (H6.1.4) *hold and* $m=1$ *and*

$$g^0(s,y,z)\leqslant g^1(s,y,z),\qquad \forall(y,z)\in\mathbb{R}^m\times\mathbb{R}^m,\ \text{a.s.}.$$

Then, for any $\xi^0, \xi^1 \in L^2_{\mathcal{F}_T}(\Omega;\mathbb{R}^m)$ *with* $\xi^0 \leqslant \xi^1$*, almost surely, the corresponding adapted solutions* $(Y^i(\cdot), Z^i(\cdot))$ *of BSDEs* (6.1.22) *satisfy*

$$Y^0(t) \leqslant Y^1(t), \qquad t \in [0, T], \text{ a.s.}$$

1.3 FSVIEs

Let us now turn to FSVIEs. Thanks to the dependence of the integrand on (t, s), it is very different from the case of FSDEs. Now, we first consider the following linear FSVIE:

$$\begin{aligned} X(t) = \varphi(t) &+ \int_0^t \Big(A_0(t,s)X(s)+b(s)\Big)ds \\ &+ \int_0^t \Big(A_1(t,s)X(s)+\sigma(s)\Big)dW(s), \quad t \in [0, T]. \end{aligned} \tag{6.1.34}$$

Replacing $\varphi(\cdot)$ by

$$\varphi(\cdot) + \int_0^{\cdot} b(s)ds + \int_0^{\cdot} \sigma(s)dW(s),$$

we see that without loss of generality, it suffices to consider the following linear FSVIE:

$$X(t) = \varphi(t) + \int_0^t A_0(t,s)X(s)ds + \int_0^t A_1(t,s)X(s)dW(s), \quad t \in [0, T], \tag{6.1.35}$$

namely, we may assume $b(\cdot) = \sigma(\cdot) = 0$ in (10.1.5). We now present several examples which will address the following points:

(i) Even for a (deterministic) linear FVIEs, the comparison of the solutions may fail (see Example 6.1.9 below).
(ii) When the diffusion is not identically zero, nonnegativity of the free term $\varphi(\cdot)$ is not enough to ensure the nonnegativity of the solution $X(\cdot)$ to linear FSVIE (6.1.35) (see Example 6.1.10 below).
(iii) If $A_1(t, s)$ is not independent of t, even if the free term $\varphi(\cdot)$ is a constant, comparison theorem could fail in general (see Example 6.1.11 below).

Example 6.1.9 Consider the following one-dimensional (deterministic) FVIE:

$$X(t) = 1 - \int_0^t 2e^{t-s}X(s)ds, \qquad t \in [0, T].$$

In this case, we have

$$\varphi(t)=1,\quad A_0(t,s)=-2e^{t-s},\quad A_1(t,s)=0,\qquad \forall(t,s)\in\Delta^*[0,T].$$

It is straightforward that the solution is given by

$$X(t)=-1+2e^{-t},\qquad t\in[0,T].$$

Consequently, for $T>\ln 2$, we have

$$X(T)=-1+2e^{-T}<0.$$

A main reason that makes the comparison theorem fails in the above example is due to the fact that $A_0(\cdot\,,\cdot)$ is negative and $t\mapsto A_0(t,s)\equiv-2e^{t-s}$ is decreasing.

Example 6.1.10 Consider the following one-dimensional FSVIE:

$$X(t)=2T-t+\int_0^t X(s)dW(s),\qquad t\in[0,T].\tag{6.1.36}$$

Clearly, (6.1.36) is a special case of (6.1.35) with $n=1$:

$$\varphi(t)=2T-t>0,\quad A_0(t,s)=0,\quad A_1(t,s)=1,\qquad t\in[0,T].$$

Thus, $\varphi(\cdot)$ is (strictly) positive, and both $A_0(\cdot\,,\cdot)$ and $A_1(\cdot\,,\cdot)$ are constants. Note that (6.1.36) is equivalent to the following FSDE:

$$\begin{cases}dX(t)=-dt+X(t)dW(t),\qquad t\in[0,T],\\ X(0)=2T.\end{cases}$$

By letting

$$d\Phi(t)=\Phi(t)dW(t),\qquad \Phi(0)=1,$$

we have

$$\Phi(t)=e^{-\frac{t}{2}+W(t)},\qquad t\geqslant 0.$$

This $\Phi(\cdot)$ is the fundamental matrix of the system $[A_0;A_1]$ with $A_0=0$ and $A_1=1$. Therefore, by variation of constants formula for linear SDEs (Proposition 3.5.2 of Chap. 3), the solution $X(\cdot)$ of the above is given by the following:

$$
\begin{aligned}
X(t) &= e^{-\frac{1}{2}t+W(t)}\Big[2T - \int_0^t e^{\frac{1}{2}s-W(s)}ds\Big] \\
&\leqslant e^{-\frac{1}{2}t+W(t)}\Big[2T - \int_0^t e^{-W(s)}ds\Big] \\
&= te^{-\frac{1}{2}t+W(t)}\Big[\frac{2T}{t} - \frac{1}{t}\int_0^t e^{-W(s)}ds\Big] \\
&\leqslant te^{-\frac{1}{2}t+W(t)}\Big[\frac{2T}{t} - e^{-\frac{1}{t}\int_0^t W(s)ds}\Big],
\end{aligned}
\tag{6.1.37}
$$

where we have applied Jensen's inequality and the convexity of $\lambda \mapsto e^{\lambda}$. Thus, for any $t > 0$, $X(t) < 0$ is implied by

$$
e^{-\frac{1}{t}\int_0^t W(s)ds} \geqslant \frac{2T}{t},
$$

which is equivalent to the following:

$$
-\frac{1}{t}\int_0^t W(s)ds \geqslant \log\frac{2T}{t}.
$$

Since the left-hand side of the above is a normal random variable, we therefore obtain

$$
\mathbb{P}(X(t) < 0) \geqslant \mathbb{P}\Big(-\frac{1}{t}\int_0^t W(s)ds \geqslant \log\frac{2T}{t}\Big) > 0. \tag{6.1.38}
$$

This means that the comparison theorem fails for this example.

The main reason that the comparison fails in the above example is due to the fact that $t \mapsto \varphi(t)$ is decreasing.

Example 6.1.11 Consider

$$
X(t) = 1 + \int_0^t \frac{2T-s}{2T-t}X(s)dW(s), \qquad t \in [0, T]. \tag{6.1.39}
$$

We see that the above is a special case of (6.1.35) with

$$
\varphi(t) = 1, \quad A_0(t,s) = 0, \quad A_1(t,s) = \frac{2T-s}{2T-t} > 0, \quad s \in [0, T].
$$

The main feature of the above is that the diffusion coefficient $A_1(t,s)$ depends on (t,s). Clearly, the process $\widetilde{X}(t) \equiv (2T-t)X(t)$ satisfies the following FSVIE:

$$
\widetilde{X}(t) = 2T - t + \int_0^t \widetilde{X}(s)dW(s), \qquad t \in [0, T],
$$

which coincides with (6.1.36). Hence, by Example 6.1.9, although the free term $\varphi(t) = 1 > 0$ in (6.1.39), we have

$$\mathbb{P}\Big(X(t) < 0\Big) > 0,$$

comparison theorem fails for (6.1.39).

Take into account the above examples, to consider a general comparison theorem for linear FSVIEs, we had better restrict ourselves to the following type:

$$X(t) = \varphi(t) + \int_0^t A_0(t,s)X(s)ds + \int_0^t A_1(s)X(s)dW(s), \qquad t \in [0,T]. \tag{6.1.40}$$

In the case that $A_1(s) = 0$, almost surely and almost everywhere, Eq. (6.1.40) reads

$$X(t) = \varphi(t) + \int_0^t A_0(t,s)X(s)ds, \qquad t \in [0,T]. \tag{6.1.41}$$

Thus, Itô integral is not involved, and the sample point ω is treated as a parameter. We refer to the above as a *random Volterra integral equation*. To present positive results, we introduce the following assumption.

(H6.1.7) The maps $A_0 : \Delta^*[0,T] \times \Omega \to \mathbb{R}^{n\times n}$ and $A_1 : [0,T] \times \Omega \to \mathbb{R}^{n\times n}$ are measurable, uniformly bounded. For any $t \in [0,T]$, $s \mapsto (A_0(t,s), A_1(s))$ is $\mathbb{F}$-progressively measurable on $[0,t]$.

We present the following result.

Proposition 6.1.12 *Let* (H6.1.7) *hold.*

(i) *Suppose for almost* $(t,s) \in \Delta_*[0,T]$,

$$A_0(t,s) \in \mathbb{R}^{n\times n}_+, \quad A_1(s) = 0, \quad \text{a.s.} \tag{6.1.42}$$

Then, for any $\varphi(\cdot) \in L^2_{\mathbb{F}}(0,T;\mathbb{R}^n)$ *with*

$$\varphi(t) \geqslant 0, \qquad \text{a.s., a.e. } t \in [0,T],$$

FSVIE (6.1.40) *admits a unique solution* $X(\cdot) \in L^2_{\mathbb{F}}(0,T;\mathbb{R}^n)$, *and it satisfies*

$$X(t) \geqslant \varphi(t) \geqslant 0, \qquad \text{a.e. } t \in [0,T], \text{ a.s.} \tag{6.1.43}$$

In addition, if for almost all (s,ω), $t \mapsto A_0(t,s)$ *is continuous, and* $\varphi(\cdot) \in L^2_{\mathbb{F}}(\Omega; C([0,T];\mathbb{R}^n))$, *then the above can be strengthened to*

$$X(t) \geqslant \varphi(t) \geqslant 0, \qquad \forall t \in [0,T], \text{ a.s.} \tag{6.1.44}$$

(ii) *Suppose*

$$A_0(t,s)\in\mathbb{R}^{n\times n}_{*+},\quad A_1(s)\in\mathbb{R}^{n\times n}_d,\quad \text{a.s., a.e. } (t,s)\in\Delta_*[0,T]. \tag{6.1.45}$$

Moreover, there exists a modulus of continuity $\rho(\cdot)$ *such that*

$$|A_0(t,s)-A_0(t',s)|\leqslant\rho(|t-t'|),\qquad \text{a.s., } t,t'\in[0,T],\ s\in[0,t\wedge t'], \tag{6.1.46}$$

and for almost $s\in[0,T)$,

$$A_0(t',s)-A_0(t,s)\in\mathbb{R}^{n\times n}_{+},\quad \text{a.s., } \forall t,t'\in[s,T],\ t\leqslant t'. \tag{6.1.47}$$

Then, for any $\varphi(\cdot)\in C_{\mathbb{F}}([0,T];L^2(\Omega,\mathbb{R}^n))$ *satisfying*

$$\varphi(t')\geqslant\varphi(t)\geqslant 0,\quad \text{a.s.,}\quad 0\leqslant t\leqslant t'\leqslant T, \tag{6.1.48}$$

FSVIE (6.1.40) *admits a unique solution* $X(\cdot)\in C_{\mathbb{F}}([0,T];L^2(\Omega;\mathbb{R}^n))$, *and it satisfies*

$$X(t)\geqslant 0,\qquad \text{a.s., } \forall t\in[0,T]. \tag{6.1.49}$$

(iii) *Suppose* (6.1.45) *holds, and there exists a modulus of continuity* $\rho:[0,T]\to[0,\infty)$ *such that*

$$|A_0(t,s)-A_0(t',s)|\leqslant\rho(|t-t'|),\qquad t,t'\in[0,T],\ s\in[0,t\wedge t'],\ \text{a.s.}, \tag{6.1.50}$$

and for almost every $s\in[0,T)$,

$$A_0(t',s)-A_0(t,s)\in\mathbb{R}^{n\times n}_{+},\quad \forall t,t'\in[s,T],\ t\leqslant t',\ \text{a.s.} \tag{6.1.51}$$

Then, for any $\varphi(\cdot)\in L^2_{\mathbb{F}}(\Omega;C([0,T];\mathbb{R}^n))$ *satisfying*

$$\varphi(t')\geqslant\varphi(t)\geqslant 0,\qquad 0\leqslant t\leqslant t'\leqslant T,\ \text{a.s.} \tag{6.1.52}$$

FSVIE (6.1.40) *admits a unique solution* $X(\cdot)\in L^2_{\mathbb{F}}(\Omega;C([0,T];\mathbb{R}^n))$, *and it satisfies*

$$X(t)\geqslant 0,\qquad \forall t\in[0,T],\ \text{a.s.} \tag{6.1.53}$$

Remark 6.1.13 As to the above results, we make the following comments.

First, Part (i) is about random Volterra integral equations where we assumed that all entries of $A_0(\cdot,\cdot)$ (including the diagonal ones) are nonnegative, and no monotonicity condition is assumed for $\varphi(\cdot)$. Because of Example 6.1.10, $\mathbb{R}^{n\times n}_{+}$ in

(6.1.42) cannot be replaced by $\mathbb{R}^{n\times n}_{*+}$ (unless some other monotonicity conditions are assumed).

Second, the last two parts are about linear FSVIEs where condition on $A_0(\cdot\,,\cdot)$ is relaxed and $A_1(\cdot)$ is allowed to be nonzero. As a tradeoff, some monotonicity conditions are assumed which will play essential roles.

Third, if we compare (ii) and (iii), we point out that (ii) is a weaker comparison under a set of weaker conditions.

Proof

(i) For any $X(\cdot) \in L^2_{\mathbb{F}}(0,T;\mathbb{R}^n)$, we define

$$(\mathcal{A}X)(t) = \int_0^t A_0(t,s)X(s)ds, \qquad t \in [0,T].$$

By condition (6.1.42) and Lemma 6.1.4, we see that for any $X(\cdot) \in L^2_{\mathbb{F}}(0,T;\mathbb{R}^n)$ satisfying

$$X(t) \geqslant 0, \qquad \text{a.s., a.e. } t \in [0,T],$$

one has

$$(\mathcal{A}X)(t) \geqslant 0, \qquad \text{a.s., a.e. } t \in [0,T].$$

Now, we define the following Picard iteration sequence:

$$X^0(\cdot) = \varphi(\cdot), \qquad X^k(\cdot) = \varphi(\cdot) + (\mathcal{A}X^{k-1})(\cdot), \qquad k \geqslant 1.$$

By induction, it is easy to see that

$$X^k(t) \geqslant \varphi(t), \qquad \text{a.s., a.e. } t \in [0,T],\ \forall k \geqslant 0.$$

$$\lim_{k\to\infty} \mathbb{E}\int_0^T |X^k(t) - X(t)|^2 dt = 0,$$

with $X(\cdot)$ being the solution to (6.1.40). Then, it is easy to see that (6.1.43) holds.

Next, let us discuss the case when $t \mapsto (A_0(t,s),\varphi(t))$ is continuous for almost all (s,ω). First, for any $0 \leqslant t_0 < t \leqslant T$, by the boundedness of A_0, the integrability of $X(\cdot)$, we have

$$\begin{aligned}
&\Big|\int_0^t A_0(t,s)X(s)ds - \int_0^{t_0} A_0(t_0,s)X(s)ds\Big| \\
&\leqslant \Big|\int_0^{t_0} \big|A_0(t,s) - A_0(t_0,s)\big||X(s)|ds\Big| + \Big|\int_{t_0}^t |A_0(t_0,s)||X(s)|ds\Big| \to 0,
\end{aligned}$$

as $t \to t_0$. The case of $0 \leqslant t < t_0 \leqslant T$ can be obtained similarly. Then, one has $X(\cdot) \in L^2_{\mathbb{F}}(\Omega; C([0,T]; \mathbb{R}^n))$.

By condition (6.1.42) and Lemma 6.1.4, for any $X(\cdot) \in L^2_{\mathbb{F}}(\Omega; C([0,T]; \mathbb{R}^n))$ satisfying

$$X(t) \geqslant 0, \qquad , \forall t \in [0,T], \ \text{a.s.}$$

by the continuity of A_0, one has

$$(\mathcal{A}X)(t) \geqslant 0, \qquad \forall t \in [0,T], \quad \text{a.s.}$$

Now, we define the following Picard iteration sequence:

$$X^0(\cdot) = \varphi(\cdot), \qquad X^k(\cdot) = \varphi(\cdot) + (\mathcal{A}X^{k-1})(\cdot), \qquad k \geqslant 1.$$

By induction, it is easy to see that

$$X^k(t) \geqslant \varphi(t), \qquad \forall t \in [0,T], \quad \text{a.s.}, \ \ \forall k \geqslant 0.$$

Further

$$\lim_{k \to \infty} \mathbb{E} \sup_{t \in [0,T]} |X^k(t) - X(t)|^2 = 0,$$

with $X(\cdot)$ being the solution to (6.1.40). Then, it is easy to see that (6.1.44) holds.

(ii) First, under the given assumptions, (6.1.40) admits a unique solution $X(\cdot) \in C_{\mathbb{F}}([0,T]; L^2(\Omega; \mathbb{R}^n))$. Let $\Pi = \{t_k, 0 \leqslant k \leqslant N\}$ be a finite sequence of real numbers with $0 = t_0 < t_1 < \cdots < t_N = T$, and we define its mesh size by

$$\|\Pi\| = \max_{1 \leqslant k \leqslant N} |t_k - t_{k-1}|.$$

Define

$$A_0^{\Pi}(t,s) = \sum_{k=0}^{N-1} A_0(t_{k+1}, s)\mathbf{1}_{[t_k, t_{k+1})}(t), \qquad \varphi^{\Pi}(t) = \sum_{k=0}^{N-1} \varphi(t_k)\mathbf{1}_{[t_k, t_{k+1})}(t).$$

Clearly, for each $k \geqslant 0$,

$$A_0(t_{k+1}, s) \in \mathbb{R}^{n \times n}_{*+}, \qquad \text{a.s., a.e. } s \in [t_k, t_{k+1}),$$

and

$$0 \leqslant \varphi(t_k) \leqslant \varphi(t_{k+1}), \qquad \text{a.s.} \tag{6.1.54}$$

For any $0 \leqslant s \leqslant t \leqslant T$, we note that

$$
\begin{aligned}
|A_0^{\Pi}(t,s) - A_0(t,s)| &= \sum_{k=0}^{N-1} |A_0(t_{k+1},s) - A_0(t,s)| \mathbf{1}_{[t_k,t_{k+1})}(t) \\
\leqslant \sum_{k=0}^{N-1} \rho\big(t_{k+1} - t\big) \mathbf{1}_{[t_k,t_{k+1})}(t) \leqslant \rho\big(\|\Pi\|\big),
\end{aligned}
$$

and since $\varphi(\cdot) \in C_{\mathbb{F}}([0,T]; L^2(\Omega;\mathbb{R}^n))$, by the uniform continuity, one has

$$
\begin{aligned}
&\sup_{t\in[0,T]} \mathbb{E}|\varphi^{\Pi}(t) - \varphi(t)|^2 \\
&\leqslant \sup_{t\in[0,T]} \mathbb{E}\Big(\sum_{k=0}^{N-1} |\varphi(t_k) - \varphi(t)| \cdot \mathbf{1}_{[t_k,t_{k+1})}(t)\Big)^2 \\
&= \sup_{t\in[0,T]} \sum_{k=0}^{N-1} \mathbb{E}|\varphi(t_k) - \varphi(t)|^2 \cdot \mathbf{1}_{[t_k,t_{k+1})}(t) \\
&\leqslant \sup_{|t-t'|\leqslant\|\Pi\|} \mathbb{E}|\varphi(t) - \varphi(t')|^2 \leqslant \widetilde{\rho}(\|\Pi\|),
\end{aligned}
$$

with $\widetilde{\rho} : [0,\infty) \to [0,\infty)$ being a modulus of continuity. Now, we let $X^{\Pi}(\cdot)$ be the solution to the following FSVIE:

$$
\begin{aligned}
X^{\Pi}(t) = \varphi^{\Pi}(t) &+ \int_0^t A_0^{\Pi}(t,s) X^{\Pi}(s) ds \\
&+ \int_0^t A_1(s) X^{\Pi}(s) dW(s), \qquad t \in [0,T].
\end{aligned}
$$

Then

$$
\begin{aligned}
&\mathbb{E}|X^{\Pi}(t) - X(t)|^2 \\
&\leqslant K\mathbb{E}\Big[|\varphi^{\Pi}(t) - \varphi(t)|^2 + \Big(\int_0^t |A_0^{\Pi}(t,s)X^{\Pi}(s) - A_0(t,s)X(s)| ds\Big)^2 \\
&\qquad + \int_0^t |A_1(s)|^2 |X^{\Pi}(s) - X(s)|^2 ds\Big] \\
&\leqslant K\mathbb{E}\Big[\widetilde{\rho}(\|\Pi\|) + \Big(\int_0^t \big\{\rho\big(\|\Pi\|\big)|X(s)| + K|X^{\Pi}(s) - X(s)|\big\} ds\Big)^2 \\
&\qquad + K\int_0^t |X^{\Pi}(s) - X(s)|^2 ds\Big] \\
&\leqslant K\mathbb{E}\Big[\widetilde{\rho}(\|\Pi\|) + \rho\big(\|\Pi\|\big)^2 + \int_0^t |X^{\Pi}(s) - X(s)|^2 ds\Big].
\end{aligned}
$$

By Gronwall's inequality, we get

$$\sup_{t\in[0,T]} \mathbb{E}|X^{\Pi}(t) - X(t)|^2 \leqslant K\Big(\widetilde{\rho}(\|\Pi\|) + \rho\big(\|\Pi\|\big)^2\Big).$$

Hence

$$\lim_{\|\Pi\|\to 0}\Big[\sup_{t\in[0,T]} \mathbb{E}|X^{\Pi}(t) - X(t)|^2\Big] = 0. \tag{6.1.55}$$

We claim that

$$X^{\Pi}(t) \geqslant 0, \quad \text{a.s.,} \ t \in [0,T], \tag{6.1.56}$$

which, together with (6.1.55) will lead to (6.1.49).

First, on interval $[0, t_1)$, we have

$$X^{\Pi}(t) = \varphi(0) + \int_0^t A_0(t_1, s)X^{\Pi}(s)ds + \int_0^t A_1(s)X^{\Pi}(s)dW(s),$$

which is an FSDE, and $X^{\Pi}(\cdot)$ has continuous paths (on $[0, t_1)$). From Proposition 6.1.5, we have

$$X^{\Pi}(t) \geqslant 0, \qquad t \in [0, t_1), \ \text{a.s.}$$

In particular

$$\begin{aligned} X^{\Pi}(t_1 - 0) = \varphi(0) &+ \int_0^{t_1} A_0(t_1, s)X^{\Pi}(s)ds \\ &+ \int_0^{t_1} A_1(s)X^{\Pi}(s)dW(s) \geqslant 0. \end{aligned} \tag{6.1.57}$$

Next, on $[t_1, t_2)$, we have (making use of (6.1.57))

$$\begin{aligned} X^{\Pi}(t) &= \varphi(t_1) + \int_0^{t_1} A_0(t_2, s)X^{\Pi}(s)ds + \int_0^{t_1} A_1(s)X^{\Pi}(s)dW(s) \\ &\quad + \int_{t_1}^t A_0(t_2, s)X^{\Pi}(s)ds + \int_{t_1}^t A_1(s)X^{\Pi}(s)dW(s) \\ &= \varphi(t_1) - \varphi(0) + X^{\Pi}(t_1 - 0) + \int_0^{t_1} \Big(A_0(t_2, s) - A_0(t_1, s)\Big)X^{\Pi}(s)ds \\ &\quad + \int_{t_1}^t A_0(t_2, s)X^{\Pi}(s)ds + \int_{t_1}^t A_1(s)X^{\Pi}(s)dW(s) \\ &\equiv \widetilde{X}(t_1) + \int_{t_1}^t A_0(t_2, s)X^{\Pi}(s)ds + \int_{t_0}^t A_1(s)X^{\Pi}(s)dW(s), \end{aligned}$$

where, by (6.1.48) and (6.1.57),

$$\widetilde{X}(t_1) \equiv \varphi(t_1) - \varphi(0) + X^{\Pi}(t_1 - 0) \\ + \int_0^{t_1} \Big(A_0(t_2, s) - A_0(t_1, s)\Big) X^{\Pi}(s) ds \geqslant 0.$$

Hence, by Proposition 6.1.5, one obtains

$$X^{\Pi}(t) \geqslant 0, \qquad t \in [t_1, t_2).$$

By induction, we obtain (6.1.56). Hence, (6.1.49) follows.

(iii) First, under the given assumptions, (6.1.40) admits a unique solution $X(\cdot) \in L^2_{\mathbb{F}}(\Omega; C([0, T]; \mathbb{R}^n))$. In the current case, we can make use of the path-wise continuity of the coefficients. Let $\tau_0^{\delta} = 0$, and for $k \geqslant 1$,

$$\tau_k^{\delta} \stackrel{\triangle}{=} \inf\big\{t \in (\tau_{k-1}^{\delta}, (\tau_{k-1}^{\delta} + \delta) \wedge T] \bigm| |\varphi(t) - \varphi(\tau_{k-1}^{\delta})| \geqslant \delta\big\}, \qquad \text{a.s.}$$

or $\tau_k^{\delta} \stackrel{\triangle}{=} \inf \phi \equiv \tau_{k-1}^{\delta} + \delta$. Then, we have

$$\mathbb{P}\Big(\tau_k^{\delta} = T, \quad \text{for some } k \geqslant 1\Big) = 1,$$

and

$$0 < \tau_k^{\delta} - \tau_{k-1}^{\delta} \leqslant \delta, \qquad k \geqslant 1.$$

Now, we define

$$A_0^{\delta}(t, s) \stackrel{\triangle}{=} \sum_{k \geqslant 0} A_0(\tau_k^{\delta} \vee s, s) \mathbf{1}_{[\tau_k^{\delta}, \tau_{k+1}^{\delta})}(t), \quad \varphi^{\delta}(t) \stackrel{\triangle}{=} \sum_{k \geqslant 0} \varphi(\tau_k^{\delta}) \mathbf{1}_{[\tau_k^{\delta}, \tau_{k+1}^{\delta})}(t).$$

Clearly, for each $k \geqslant 0$,

$$A_0(\tau_k^{\delta} \vee s, s) \in \mathbb{R}^{n \times n}_{*+}, \quad \text{a.s., a.e. } s \in [\tau_k^{\delta}, \tau_{k+1}^{\delta}), \tag{6.1.58}$$

and

$$0 \leqslant \varphi(\tau_k^{\delta}) \leqslant \varphi(\tau_{k+1}^{\delta}), \qquad \text{a.s.} \tag{6.1.59}$$

Further, for $0 \leqslant s \leqslant t \leqslant T$,

$$|A_0^{\delta}(t, s) - A_0(t, s)| = \sum_{k \geqslant 0} |A_0(\tau_k^{\delta} \vee s, s) - A_0(t, s)| \mathbf{1}_{[\tau_k^{\delta}, \tau_{k+1}^{\delta})}(t) \\ \leqslant \sum_{k \geqslant 0} \rho\big(|\tau_k^{\delta} \vee s - t|\big) \mathbf{1}_{[\tau_k^{\delta}, \tau_{k+1}^{\delta})}(t) \leqslant \rho(\delta),$$

and

$$|\varphi^\delta(t)-\varphi(t)|\leqslant\sum_{k\geq 0}|\varphi(\tau_k^\delta)-\varphi(t)|\cdot\mathbf{1}_{[\tau_k^\delta,\tau_{k+1}^\delta)}(t)\leqslant\delta.$$

Thus

$$\mathbb{E}\Big[\sup_{t\in[0,T]}|\varphi^\delta(t)-\varphi(t)|^2\Big]\leqslant\delta^2.$$

Now, let $X^\delta(\cdot)$ be the solution to the following:

$$X^\delta(t)=\varphi^\delta(t)+\int_0^t A_0^\delta(t,s)X^\delta(s)ds+\int_0^t A_1^\delta(s)X^\delta(s)dW(s),\quad t\in[0,T].$$

Then, we can show that

$$\lim_{\delta\to 0}\mathbb{E}\Big[\sup_{t\in[0,T]}|X^\delta(t)-X(t)|^2\Big]=0.$$

The rest of the proof is similar to that for (ii).

□

Based on the above result, it is not very hard for us to present comparison theorems for nonlinear FSVIEs. We prefer not to give the details here. One can cook them up by following the relevant details for BSVIEs presented in the following section.

To conclude this subsection, we present an example, which says if the monotonicity condition $t\mapsto A_0(t,s)$ is removed, comparison of solutions might fail. Here is an example.

Example 6.1.14 Consider the following FSVIE:

$$X(t)=1+\int_0^t \mathbf{1}_{[0,\tau]}(t)X(s)ds+\int_0^t X(s)dW(s),\qquad t\in[0,T],$$

where $\tau\in(0,T)$. Clearly, the above is a special case of (6.1.35) with

$$\varphi(t)=1,\quad A_0(t,s)=I_{[0,\tau]}(t),\quad A_1(s)=1.$$

Thus, $t\mapsto A_0(t,s)$ is not nondecreasing. We can obtain that the solution to the above FSVIE is given by (noting the variation of constants formula, Proposition 3.5.2, Chap. 3)

$$X(t)=\begin{cases} e^{\frac{t}{2}+W(t)}, & t\in[0,\tau),\\ e^{-\frac{t-\tau}{2}+W(t)-W(\tau)}\Big[e^{\frac{\tau}{2}+W(\tau)}-\int_0^\tau e^{\frac{s}{2}+W(s)}ds\Big], & t\in[\tau,T].\end{cases}$$

By convexity of $\lambda \mapsto e^{\lambda}$ and Jensen's inequality, we have that

$$e^{\frac{\tau}{2}+W(\tau)} - \int_0^\tau e^{\frac{s}{2}+W(s)}ds \leqslant e^{\frac{\tau}{2}+W(\tau)} - \tau e^{\frac{1}{\tau}\int_0^\tau [\frac{s}{2}+W(s)]ds}.$$

Hence, $X(t) < 0$ for some $t \in [\tau, T]$ is implied by

$$\begin{aligned}\frac{\tau}{2} + W(\tau) < \ln\tau + \frac{1}{\tau}\int_0^\tau \Big(\frac{s}{2} + W(s)\Big)ds &= \ln\tau + \frac{\tau}{4} + \frac{1}{\tau}\int_0^\tau W(s)ds \\ &= \ln\tau + \frac{\tau}{4} + \frac{1}{\tau}sW(s)\Big|_0^\tau - \int_0^\tau s dW(s) \\ &= \ln\tau + \frac{\tau}{4} + W(\tau) - \frac{1}{\tau}\int_0^\tau s dW(s).\end{aligned}$$

This is equivalent to the following:

$$\int_0^\tau s dW(s) < \tau\ln\tau - \frac{\tau^2}{4}.$$

The left-hand side of the above is a normal random variable. Hence

$$\mathbb{P}\Big(\int_0^\tau s dW(s) < \tau\ln\tau - \frac{\tau^2}{4}\Big) > 0,$$

which implies

$$\mathbb{P}\Big(X(t) < 0\Big) > 0, \qquad t \in (\tau, T].$$

Although, in the above, $A_0(\cdot\,,\cdot)$ is discontinuous, it is not hard for us to replace it by a continuous one and still have the same conclusion.

2 Comparison Theorems for Type-I BSVIEs

We now consider BSVIEs for which we will establish various comparison theorems.

2.1 *Case I*

We consider the following Type-I BSVIEs: for $i = 0, 1$,

$$Y^i(t) = \psi^i(t) + \int_t^T g^i(t, s, Y^i(s), Z^i(t, s))ds - \int_t^T Z^i(t, s)dW(s), \qquad t \in [0, T]. \tag{6.2.1}$$

The key feature here is that the generator $g^i(\cdot)$ is independent of $Z^i(s,t)$. We note that for any adapted solution $(Y^i(\cdot), Z^i(\cdot\,,\cdot)) \in \mathcal{H}^p_{\Delta^*}[0,T]$ of the above, $Z^i(t,s)$ is only (uniquely) defined for $(t,s) \in \Delta^*[0,T]$ (the upper triangle domain), and we do not need the values $Z^i(t,s)$ of $Z^i(\cdot\,,\cdot)$ for $(t,s) \in \Delta_*[0,T]$ (the lower triangle domain). For the generator $g^i(\cdot)$ of BSVIE (6.2.1), we adopt the following assumption.

(H6.2.1) Let $g^i : \Delta^*[0,T] \times \mathbb{R}^m \times \mathbb{R}^m \times \Omega \to \mathbb{R}^m$ be measurable such that $s \mapsto g^i(t,s,y,z)$ is $\mathbb{F}$-progressively measurable, $(y,z) \mapsto g^i(t,s,y,z)$ is uniformly Lipschitz, $(t,s) \mapsto g^i(t,s,0,0)$ is uniformly bounded, and there exists a modulus of continuity $\rho_i : [0,\infty) \to [0,\infty)$ such that

$$
\begin{aligned}
&|g^i(t,s,y,z) - g^i(t',s,y,z)| \leqslant \rho_i(|t-t'|)\big(1+|y|+|z|\big),\\
&\qquad\qquad\qquad\qquad \forall (t,s),(t',s) \in \Delta^*[0,T],\ (y,z) \in \mathbb{R}^m \times \mathbb{R}^m.
\end{aligned}
$$

Note here that we do not assume $g^i(t,s,0,0) = 0$. However, by using a simple reduction argument, we can easily reduce the general case to the case presented in Chap. 4 and have the following result concerning BSVIE (6.2.1).

Proposition 6.2.1 *Let* (H6.2.1) *hold. Then, for any* $\psi^i(\cdot) \in L^2_{\mathcal{F}_T}(0,T;\mathbb{R}^m)$, *BSVIE* (6.2.1) *admits a unique adapted solution* $(Y^i(\cdot), Z^i(\cdot\,,\cdot)) \in \mathcal{H}^2_{\Delta^*}[0,T]$, *and the following estimate holds:*

$$
\begin{aligned}
&\mathbb{E}\Big\{\int_0^T |Y^i(t)|^2 dt + \int_0^T\int_t^T |Z^i(t,s)|^2 ds dt\Big\}\\
&\leqslant K\mathbb{E}\Big\{\int_0^T |\psi^i(t)|^2 dt + \int_0^T\Big(\int_t^T |g^i(t,s,0,0)|ds\Big)^2 dt\Big\}.
\end{aligned}
\tag{6.2.2}
$$

Further, if $\psi^i(\cdot) \in C_{\mathcal{F}_T}([0,T];L^2(\Omega;\mathbb{R}^m))$, *then* $Y^i(\cdot) \in C_{\mathbb{F}}([0,T];L^2(\Omega;\mathbb{R}^m))$ *such that*

$$
\begin{aligned}
&\sup_{t\in[0,T]} \mathbb{E}\Big\{|Y^i(t)|^2 + \int_t^T |Z^i(t,s)|^2 ds\Big\}\\
&\leqslant K \sup_{t\in[0,T]} \mathbb{E}\Big\{|\psi^i(t)|^2 + \Big(\int_t^T |g^i(t,s,0,0)|ds\Big)^2\Big\}.
\end{aligned}
\tag{6.2.3}
$$

Eventually, suppose the modulus ρ_i *in* (H6.2.1) *is given by* $\rho_i(r) = r^\delta$ *for some* $\delta \in (0,1]$. *Then, for given* $\gamma \in (0,1)$, *constant* $K > 0$, *and any* $p > 2 \vee \frac{1}{\gamma} \vee \frac{1}{\delta}$, $\psi^i(\cdot) \in L^p_{\mathcal{F}_T}(\Omega;C([0,T];\mathbb{R}^m))$ *satisfying*

$$
\Big(\mathbb{E}|\psi^i(t_1) - \psi^i(t_2)|^p\Big)^{\frac{1}{p}} \leqslant K|t_1-t_2|^\gamma, \qquad \forall t_1,t_2 \in [0,T], \tag{6.2.4}
$$

we have $Y^i(\cdot) \in L^p_{\mathbb{F}}\big(\Omega;C([0,T];\mathbb{R}^m)\big)$.

Proof For reader's convenience, let us give a sketch of the proof. We can rewrite the above (6.2.1):

$$
\begin{aligned}
Y^i(t) &= \widetilde{\psi}^i(t) + \int_t^T \widetilde{g}^i(t,s,Y^i(s),Z^i(t,s))ds - \int_t^T Z^i(t,s)dW(s) \\
&= \widehat{\psi}^i(t) + \int_t^T \widehat{g}^i(t,s,Z^i(t,s))ds - \int_t^T Z^i(t,s)dW(s), \\
&\qquad\qquad t \in [0,T],
\end{aligned} \tag{6.2.5}
$$

where

$$
\begin{aligned}
&\widetilde{\psi}^i(t) \equiv \psi^i(t) + \int_t^T g^i(t,s,0,0)ds, \\
&\widetilde{g}^i(t,s,y,z) \equiv g^i(t,s,y,z) - g^i(t,s,0,0), \\
&\widehat{\psi}^i(t) \equiv \psi^i(t) + \int_t^T g^i(t,s,Y^i(s),0)ds, \\
&\widehat{g}^i(t,s,z) \equiv g^i(t,s,Y^i(s),z) - g^i(t,s,Y^i(s),0).
\end{aligned}
$$

Then, the existence and uniqueness of $(Y^i(\cdot), Z^i(\cdot\,,\cdot)) \in \mathcal{H}^2_{\Delta *}[0,T]$ and (6.2.2) are implied by Theorem 4.2.2 of Chap. 4 with $\psi^i(\cdot)$ replaced by $\widetilde{\psi}^i(\cdot)$.

Notice that

$$
\begin{aligned}
\big|\widehat{g}^i(t_1,s,z) - \widehat{g}^i(t_2,s,z)\big| &\leqslant \big|g^i(t_1,s,Y^i(s),z) - g^i(t_2,s,Y^i(s),z)\big| \\
&\quad + \big|g^i(t_1,s,Y^i(s),0) - g^i(t_2,s,Y^i(s),0)\big| \\
&\leqslant \rho(|t_1-t_2|)\big(1+|Y^i(s)|+|z|\big).
\end{aligned}
$$

Therefore, the counterpart of $(\text{H4.3.3})_\alpha$ holds in our setting. To show $Y^i(\cdot) \in C_{\mathbb{F}}([0,T];L^2(\Omega;\mathbb{R}^m))$ and (11.3.10), we verify that $\widehat{\psi}^i(\cdot) \in C_{\mathcal{F}_T}([0,T];L^2(\Omega;\mathbb{R}^m))$. For any given $t_0 \in [0,T]$, $t \geqslant t_0$, and $p > 1$,

$$
\begin{aligned}
&\mathbb{E}\Big|\int_t^T g^i(t,s,Y^i(s),0)ds - \int_{t_0}^T g^i(t_0,s,Y^i(s),0)ds\Big|^p \\
&\leqslant K\mathbb{E}\Big|\int_t^T |g^i(t,s,Y^i(s),0) - g^i(t_0,s,Y^i(s),0)|ds\Big|^p \\
&\quad + K\mathbb{E}\Big|\int_{t_0}^t |g(t_0,s,Y^i(s),0)|ds\Big|^p \\
&\leqslant K\big[\rho_i(|t-t_0|)\big]^p\mathbb{E}\int_t^T (1+|Y^i(s)|)^p ds \\
&\quad + K(t-t_0)^{p-1}\mathbb{E}\int_{t_0}^t \big(1+|Y^i(s)|\big)^p ds \to 0, \quad t \downarrow t_0.
\end{aligned}
$$

The case of $t \leqslant t_0$ can be obtained similarly. This, together with $\psi^i(\cdot) \in C_{\mathcal{F}_T}([0,T];L^2(\Omega;\mathbb{R}^m))$, implies our conclusion on $Y^i(\cdot)$.

For the case of $Y^i(\cdot) \in L^p_{\mathbb{F}}\big(\Omega; C([0,T];\mathbb{R}^m)\big)$, it is a direct conclusion of the above arguments. □

Next, we want to look at a proper comparison between $Y^0(\cdot)$ and $Y^1(\cdot)$ under certain additional conditions on $g^i(\cdot)$ and $\psi^i(\cdot)$. To begin with, let us first look at the following simple Type-I BSVIEs: for $i = 0, 1$,

$$Y^i(t) = \psi^i(t) + \int_t^T g^i(t,s,Z^i(t,s))ds - \int_t^T Z^i(t,s)dW(s), \qquad t \in [0,T], \tag{6.2.6}$$

with the generators $g^i(\cdot,\cdot,\cdot)$ $(i = 0, 1)$ being independent of $Y^i(\cdot)$. We have the following result.

Proposition 6.2.2 *For $i = 0, 1$, let $g^i : \Delta^*[0,T] \times \mathbb{R}^m \times \Omega \to \mathbb{R}^m$ satisfy* (H6.2.1). *Moreover*

$$g^0(t,s,z) \leqslant g^1(t,s,z), \quad \forall (t,z) \in [0,s] \times \mathbb{R}^m, \text{ a.s., a.e. } s \in [0,T], \tag{6.2.7}$$

and for either $i = 0$ or $i = 1$, $g^i_z(t,s,z)$ exists and is bounded, as well as

$$g^i_z(t,s,z) \in \mathbb{R}^{m\times m}_d, \quad (t,z) \in [0,s] \times \mathbb{R}^m, \text{ a.s., a.e. } s \in [0,T]. \tag{6.2.8}$$

Then, for any $\psi^i(\cdot) \in C_{\mathcal{F}_T}([0,T]; L^2(\Omega;\mathbb{R}^m))$ with

$$\psi^0(t) \leqslant \psi^1(t), \qquad \text{a.s., } \forall t \in [0,T], \tag{6.2.9}$$

the adapted solutions $(Y^i(\cdot), Z^i(\cdot\,,\cdot))$ of BSVIE (6.2.6) *satisfy*

$$Y^0(t) \leqslant Y^1(t), \qquad \text{a.s., } \forall t \in [0,T]. \tag{6.2.10}$$

In addition, suppose the modulus ρ_i in (H6.2.1) *is given by $\rho_i(r) = r^\delta$ for some $\delta \in (0,1]$. Then, for given $\gamma \in (0,1)$, constant $K > 0$, and any $p > 2 \vee \frac{1}{\gamma} \vee \frac{1}{\delta}$, $\psi^i(\cdot) \in L^p_{\mathcal{F}_T}(\Omega; C([0,T];\mathbb{R}^m))$ satisfying*

$$\begin{aligned} &\Big(\mathbb{E}|\psi^i(t_1) - \psi^i(t_2)|^p\Big)^{\frac{1}{p}} \leqslant K|t_1 - t_2|^\gamma, \qquad \forall t_1, t_2 \in [0,T], \\ &\psi^0(t) \leqslant \psi^1(t), \qquad \forall t \in [0,T], \quad \text{a.s.,} \end{aligned} \tag{6.2.11}$$

the adapted solutions $(Y^i(\cdot), Z^i(\cdot\,,\cdot))$ of BSVIE (6.2.6) *satisfy*

$$Y^0(t) \leqslant Y^1(t), \qquad \forall t \in [0,T], \quad \text{a.s.} \tag{6.2.12}$$

Proof Fixed $t \in [0, T]$. For $i = 0, 1$, let $(\lambda^i(t, \cdot), \mu^i(t, \cdot))$ be the adapted solution to the following BSDE:

$$\lambda^i(t, r) = \psi^i(t) + \int_r^T g^i(t, s, \mu^i(t, s))ds - \int_r^T \mu^i(t, s)dW(s), \qquad r \in [t, T].$$

By (H6.2.1), $z \mapsto g^i(t, s, z)$ is uniformly Lipschitz. Then, by applying Theorem 6.1.6, we have

$$\lambda^0(t, r) \leqslant \lambda^1(t, r), \qquad r \in [t, T], \text{ a.s.} \tag{6.2.13}$$

By setting

$$Y^i(t) \triangleq \lambda^i(t, t), \qquad Z^i(t, s) \triangleq \mu^i(t, s), \qquad \forall (t, s) \in \Delta^*[0, T], \tag{6.2.14}$$

we see that $(Y^i(\cdot), Z^i(\cdot\,, \cdot))$ is the adapted solution to the BSVIE (6.2.6). Also, by letting $r = t$ in (6.2.13), we get comparison (6.2.10). The conclusion of (6.2.12) follows from Proposition 6.2.1. □

For general type-I BSVIE (6.2.1), we have the following result, making use of the above proposition.

Theorem 6.2.3 *Let* (H6.2.1) *hold. Suppose* $\bar{g} : \Delta^*[0, T] \times \mathbb{R}^m \times \mathbb{R}^m \times \Omega \to \mathbb{R}^m$ *is measurable such that:*

(i) (Regular conditions) $s \mapsto \bar{g}(t, s, y, z)$ *is* $\mathbb{F}$*-progressively measurable,* $(y, z) \mapsto \bar{g}(t, s, y, z)$ *is uniformly Lipschitz, and* $(t, s) \mapsto \bar{g}(t, s, 0, 0)$ *is uniformly bounded.*

(ii) (Monotonicity condition) $y \mapsto \bar{g}(t, s, y, z)$ *is nondecreasing, and*

$$\begin{aligned} g^0(t, s, y, z) \leqslant \bar{g}(t, s, y, z) \leqslant g^1(t, s, y, z),& \\ \forall (t, y, z) \in [0, s] \times \mathbb{R}^m \times \mathbb{R}^m, \text{ a.s., a.e. } s \in [0, T].& \end{aligned} \tag{6.2.15}$$

(iii) (Structural condition) $\bar{g}_z(t, s, y, z)$ *exists and*

$$\bar{g}_z(t, s, y, z) \in \mathbb{R}^{m\times m}_d, \ (t, y, z) \in [0, s] \times \mathbb{R}^m \times \mathbb{R}^m, \text{ a.s., a.e. } s \in [0, T]. \tag{6.2.16}$$

Then, for any $\psi^i(\cdot) \in C_{\mathcal{F}_T}([0, T]; L^2(\Omega; \mathbb{R}^m))$ *satisfying*

$$\psi^0(t) \leqslant \psi^1(t), \qquad \text{a.s. } t \in [0, T], \tag{6.2.17}$$

the corresponding adapted solution $(Y^i(\cdot), Z^i(\cdot\,, \cdot))$ *of BSVIE* (6.2.1) *satisfies*

$$Y^0(t) \leqslant Y^1(t), \qquad \text{a.s. } \forall t \in [0, T]. \tag{6.2.18}$$

In addition, suppose the modulus ρ_i in (H6.2.1) *is given by* $\rho_i(r) = r^\delta$ *for some* $\delta \in (0,1]$, *and*

(iv) *(Special* ψ*) for given* $\gamma \in (0,1)$, *constant* $K > 0$, *and any* $p > 2 \vee \frac{1}{\gamma} \vee \frac{1}{\delta}$, $\psi^i(\cdot) \in L^p_{\mathcal{F}_T}(\Omega; C([0,T];\mathbb{R}^m))$ *satisfying*

$$\begin{aligned} &\Big(\mathbb{E}|\psi^i(t_1) - \psi^i(t_2)|^p\Big)^{\frac{1}{p}} \leqslant K|t_1 - t_2|^\gamma, \qquad \forall t_1, t_2 \in [0,T], \\ &\psi^0(t) \leqslant \psi^1(t), \qquad \forall t \in [0,T], \ \text{a.s.} \end{aligned} \tag{6.2.19}$$

Then, the adapted solutions $(Y^i(\cdot), Z^i(\cdot\,,\cdot))$ *of BSVIE* (6.2.6) *satisfy*

$$Y^0(t) \leqslant Y^1(t), \qquad \forall t \in [0,T], \ \textit{a.s.} \tag{6.2.20}$$

Proof Let $\bar\psi(\cdot) \in C_{\mathcal{F}_T}([0,T]; L^2(\Omega;\mathbb{R}^m))$ such that

$$\psi^0(t) \leqslant \bar\psi(t) \leqslant \psi^1(t), \qquad \text{a.s. } t \in [0,T].$$

By Proposition 6.2.1, the following Type-I BSVIE

$$\bar Y(t) = \bar\psi(t) + \int_t^T \bar g(t,s,\bar Y(s),\bar Z(t,s))ds - \int_t^T \bar Z(t,s)dW(s), \qquad t \in [0,T],$$

admits a unique adapted solution such that $(\bar Y, \bar Z) \in \mathcal{H}^2_{\Delta^*}[0,T]$. Set $\widetilde Y_0(\cdot) \stackrel{\Delta}{=} Y^1(\cdot)$ and consider

$$\widetilde Y_1(t) = \bar\psi(t) + \int_t^T \bar g(t,s,\widetilde Y_0(s),\widetilde Z_1(t,s))ds - \int_t^T \widetilde Z_1(t,s)dW(s), \qquad t \in [0,T].$$

Let $(\widetilde Y_1(\cdot), \widetilde Z_1(\cdot\,,\cdot))$ be the unique adapted solution to the above. Since

$$\begin{cases} \bar g(t,s,\widetilde Y_0(s),z) \leqslant g^1(t,s,\widetilde Y_0(s),z), \\ \qquad\qquad (t,z) \in [0,s] \times \mathbb{R}^m, \ \text{a.s., a.e. } s \in [0,T], \\ \bar g_z(t,s,\widetilde Y_0(s),z) \in \mathbb{R}^{m\times m}_d, \\ \qquad\qquad (t,z) \in [0,s] \times \mathbb{R}^m, \ \text{a.s., a.e. } s \in [0,T], \\ \bar\psi(t) \leqslant \psi^1(t), \qquad \text{a.s., } t \in [0,T], \end{cases}$$

by Proposition 6.2.2, there exists full measure $\Gamma_1 \subset [0,T]$, and for any $t \in \Gamma_1$, there exists a measurable set Ω^1_t satisfying $\mathbb{P}(\Omega^1_t) = 0$ such that

$$\widetilde Y_1(t,\omega) \leqslant \widetilde Y_0(t,\omega), \qquad \omega \in \Omega \setminus \Omega^1_t, \ t \in \Gamma_1.$$

Next, we consider the following BSVIE:

$$\begin{aligned}\widetilde{Y}_2(t) = \bar{\psi}(t) + \int_t^T \bar{g}(t, s, \widetilde{Y}_1(s), \widetilde{Z}_2(t, s))ds \\ - \int_t^T \widetilde{Z}_2(t, s)dW(s), \quad t \in [0, T],\end{aligned}$$

and let $(\widetilde{Y}_2(\cdot), \widetilde{Z}_2(\cdot\,,\cdot))$ be the adapted solution in $\mathcal{H}^2_{\Delta^*}[0, T]$ to the above. Now, since $y \mapsto \bar{g}(t, s, y, z)$ is nondecreasing, we have

$$\bar{g}(t, s, \widetilde{Y}_1(s), z) \leqslant \bar{g}(t, s, \widetilde{Y}_0(s), z), \ \ \forall (t, z) \in [0, s] \times \mathbb{R}^n, \text{ a.s., a.e. } s \in [0, T].$$

Hence, similar to the above, there exists Γ_2, and for any $t \in \Gamma_2$, there exists a measurable set Ω_t^2 satisfying $\mathbb{P}(\Omega_t^2) = 0$ such that

$$\widetilde{Y}_2(t, \omega) \leqslant \widetilde{Y}_1(t, \omega), \quad \omega \in \Omega \setminus \Omega_t^2, \ t \in \Gamma_2.$$

By induction, we can construct a sequence $\{(\widetilde{Y}_k(\cdot), \widetilde{Z}_k(\cdot\,,\cdot))\}_{k\geqslant 1} \subseteq \mathcal{H}^2_{\Delta^*}[0, T]$ satisfying

$$\begin{aligned}\widetilde{Y}_k(t) = \bar{\psi}(t) + \int_t^T \bar{g}(t, s, \widetilde{Y}_{k-1}(s), \widetilde{Z}_k(t, s))ds \\ - \int_t^T \widetilde{Z}_k(t, s)dW(s), \quad t \in [0, T].\end{aligned}$$

In addition, for any $t \in \bigcap_{k\geqslant 1} \Gamma_k$,

$$\begin{aligned}Y^1(t, \omega) = \widetilde{Y}_0(t, \omega) \geqslant \widetilde{Y}_1(t, \omega) \geqslant \widetilde{Y}_2(t, \omega) \cdots, \\ \omega \in \Omega \setminus \Big(\bigcup_{k\geq 1} \Omega_t^k\Big).\end{aligned} \tag{6.2.21}$$

Note that $\bigcap_{k\geqslant 1} \Gamma_k$ is full measure in $[0, T]$ and $\mathbb{P}\Big(\bigcup_{k\geqslant 1} \Omega_t^k\Big) = 0$. Next, we will show that the sequence $\{(\widetilde{Y}_k(\cdot), \widetilde{Z}_k(\cdot\,,\cdot))\}_{k\geqslant 1}$ is Cauchy in $\mathcal{H}^2_{\Delta^*}[0, T]$. To show this, we introduce an equivalent norm of $\mathcal{H}^2_{\Delta^*}[0, T]$ as

$$\|(y(\cdot), z(\cdot\,,\cdot))\|^2_{\mathcal{H}^2_{\Delta^*}[0,T]} \triangleq \mathbb{E}\int_0^T e^{\beta t}|y(t)|^2 dt + \mathbb{E}\int_0^T e^{\beta t}\int_t^T |z(t, s)|^2 ds dt,$$

with $(y(\cdot), z(\cdot\,,\cdot)) \in \mathcal{H}^2_{\Delta*}[0,T]$, and β being a constant undetermined. By utilizing a stability estimate (e.g., Proposition 4.3.1 in Chap. 4), we have

$$\begin{aligned}&\mathbb{E}|\widetilde{Y}_k(t)-\widetilde{Y}_\ell(t)|^2+\mathbb{E}\int_t^T|\widetilde{Z}_k(t,s)-\widetilde{Z}_\ell(t,s)|^2ds\\&\leqslant K\mathbb{E}\Big(\int_t^T|\bar{g}(t,s,\widetilde{Y}_{k-1}(s),\widetilde{Z}_k(t,s))-\bar{g}(t,s,\widetilde{Y}_{\ell-1}(s),\widetilde{Z}_k(t,s))|ds\Big)^2.\end{aligned}\tag{6.2.22}$$

Consequently, we arrive at

$$\begin{aligned}&\mathbb{E}\int_0^T e^{\beta t}|\widetilde{Y}_k(t)-\widetilde{Y}_\ell(t)|^2dt+\mathbb{E}\int_0^T e^{\beta t}\Big(\int_t^T|\widetilde{Z}_k(t,s)-\widetilde{Z}_\ell(t,s)|^2ds\Big)dt\\&\leqslant K\mathbb{E}\int_0^T e^{\beta t}\Big(\int_t^T|\bar{g}(t,s,\widetilde{Y}_{k-1}(s),\widetilde{Z}_k(t,s))\\&\qquad\qquad-\bar{g}(t,s,\widetilde{Y}_{\ell-1}(s),\widetilde{Z}_k(t,s))|ds\Big)^2dt\\&\leqslant K\mathbb{E}\int_0^T e^{\beta t}\Big(\int_t^T|\widetilde{Y}_{k-1}(s)-\widetilde{Y}_{\ell-1}(s)|ds\Big)^2dt\\&\leqslant K\mathbb{E}\int_0^T|\widetilde{Y}_{k-1}(s)-\widetilde{Y}_{\ell-1}(s)|^2ds\int_0^s e^{\beta t}dt\\&\leqslant\frac{K}{\beta}\mathbb{E}\int_0^T e^{\beta s}|\widetilde{Y}_{k-1}(s)-\widetilde{Y}_{\ell-1}(s)|^2ds.\end{aligned}\tag{6.2.23}$$

Note that the constant $K > 0$ in the above can be chosen independent of $\beta > 0$. Thus, by choosing a β such that $\frac{K}{\beta} < 1$, we obtain immediately that $\{(\widetilde{Y}_k(\cdot), \widetilde{Z}_k(\cdot\,,\cdot))\}_{k\geqslant 1}$ is Cauchy in $\mathcal{H}^2_{\Delta*}[0,T]$. Hence, there exists a $(\widetilde{Y}(\cdot), \widetilde{Z}(\cdot\,,\cdot)) \in \mathcal{H}^2_{\Delta*}[0,T]$ such that

$$\begin{aligned}\lim_{k\to\infty}\Big[\mathbb{E}\int_0^T e^{\beta t}|\widetilde{Y}_k(t)-\widetilde{Y}(t)|^2dt&\\+\mathbb{E}\int_0^T e^{\beta t}\Big(\int_t^T|\widetilde{Z}_k(t,s)-\widetilde{Z}(t,s)|^2ds\Big)dt\Big]&=0,\end{aligned}$$

and

$$\widetilde{Y}(t)=\bar{\psi}(t)+\int_t^T\bar{g}(t,s,\widetilde{Y}(s),\widetilde{Z}(t,s))ds-\int_t^T\widetilde{Z}(t,s)dW(s),\qquad t\in[0,T].$$

By uniqueness in $\mathcal{H}^2_{\Delta*}[0,T]$, we have

$$\bar{Y}(t)=\widetilde{Y}(t)\leqslant\widetilde{Y}_0(t)=Y^1(t),\qquad\text{a.s., } t\in[0,T].\ \text{a.e.}$$

Similarly, we can prove that

$$Y^0(t) \leqslant \bar{Y}(t), \qquad \text{a.s., } t \in [0, T]. \text{ a.e.}$$

Therefore, we end up with

$$Y^0(t) \leqslant Y^1(t), \qquad \text{a.s., } t \in [0, T]. \text{ a.e.}$$

This implies that for any $\tau \in [0, T)$, and $\tau + \varepsilon \leqslant T$, the following holds:

$$\int_\tau^{\tau+\varepsilon} \mathbb{E}\big(Y^1(s) - Y^0(s)\big)ds \geqslant 0.$$

By the continuity of $s \mapsto \mathbb{E}\big(Y^1(s) - Y^0(s)\big)$ and Lebesgue differential theorem, we obtain the above (6.2.18).

In addition, under (6.2.19), we know that $Y^i(\cdot) \in L^2_{\mathbb{F}}(\Omega; C([0, T]; \mathbb{R}^m))$. For any $t \in \mathbb{Q}[0, T]$, the set of rational number in $[0, T]$, there exists full measure Ω_t such that for any $\omega \in \Omega_t$, we have $Y^0(t, \omega) \leqslant Y^1(t, \omega)$. For any $\omega \in \bar{\Omega} \triangleq \bigcap_{t\in\mathbb{Q}[0,T]} \Omega_t \subset \Omega$, one has

$$Y^0(t, \omega) \leqslant Y^1(t, \omega), \quad \forall t \in \mathbb{Q}[0, T].$$

By the continuity of $Y^0(\cdot)$, $Y^1(\cdot)$, we obtain the desired conclusion. □

If $m = 1$, similar to the above Corollary 6.1.8, we can slightly relax the differentiability assumption of $z \mapsto \bar{g}(t, s, y, z)$ into the Lipschitz case. Since the proof is essentially the same as that in Theorem 6.2.3, we only present the statement.

Corollary 6.2.4 *Let* (H6.2.1) *hold with* $m = 1$, *and*
(i) (monotonicity) either $y \mapsto g^0(t, s, y, z)$ *or* $y \mapsto g^1(t, s, y, z)$ *is nondecreasing such that* $\forall (t, y, z) \in [0, s] \times \mathbb{R} \times \mathbb{R}$, a.s., a.e. $s \in [0, T]$, $g^0(t, s, y, z) \leqslant g^1(t, s, y, z)$, *and* $\psi^i(\cdot) \in C_{\mathcal{F}_T}([0, T]; L^2(\Omega; \mathbb{R}))$ *such that* $\psi^0(t) \leqslant \psi^1(t)$, a.s., $t \in [0, T]$.

Then, the adapted solution $(Y^i(\cdot), Z^i(\cdot\,,\cdot))$ *of BSVIE* (6.2.1) *satisfies*

$$Y^0(t) \leqslant Y^1(t), \qquad \text{a.s. } \forall t \in [0, T].$$

In addition, suppose the modulus ρ_i *in* (H6.2.1) *is given by* $\rho_i(r) = r^\delta$ *for some* $\delta \in (0, 1]$ *and* ψ^i *satisfies the above (iv) in Theorem 6.2.3. Then, the adapted solutions* $(Y^i(\cdot), Z^i(\cdot\,,\cdot))$ *of BSVIE* (6.2.6) *satisfy*

$$Y^0(t) \leqslant Y^1(t), \qquad \forall t \in [0, T], \quad \text{a.s.}$$

It is easy to cook up an example for which $y \mapsto g^i(t,s,y,z)$ is not nondecreasing for $i = 0, 1$, but a $\bar{g}(\cdot)$ satisfying conditions of Theorem 6.2.3 can be constructed. For example,

$$g^0(t,s,y,z) \equiv \sin y \leqslant 1 \equiv \bar{g}(t,s,y,z) \leqslant 2 + \cos y \equiv g^1(t,s,y,z).$$

Condition (6.2.15) means that in the tube

$$\Big\{\Big[g^0(t,s,y,z), g^1(t,s,y,z)\Big] \mid (t,s,y,z) \in \Delta^*[0,T] \times \mathbb{R}^n \times \mathbb{R}^n\Big\},$$

there exists a selection $\bar{g}(t,s,y,z)$ which is nondecreasing in y, and (6.2.16) is satisfied. Therefore, the condition assumed in Theorem 6.2.3 is a kind of *generalized nondecreasing condition* for the maps $y \mapsto g^i(t,s,y,z)$, although these maps themselves are not necessarily nondecreasing.

The above Theorem 6.2.3 provides very general comparison theorem for Type-I BSVIEs. The following examples show that the comparison may fail if the imposed conditions are not satisfied.

Example 6.2.5 Consider one-dimensional linear BSVIE:

$$Y(t) = t - \int_t^T Y(s)ds - \int_t^T Z(t,s)dW(s), \qquad t \in [0,T].$$

It is clear that if $(Y(\cdot), Z(\cdot\,,\cdot))$ is the adapted solution, then $Z(\cdot\,,\cdot) = 0$ and

$$Y(t) = e^{t-T}(T+1) - 1, \qquad t \in [0,T].$$

Consequently

$$Y(t) < 0, \qquad t \in [0, T - \ln(T+1)].$$

Therefore, comparison theorem fails for this example. This example corresponds to the case:

$$g^i(t,s,y,z) = -y, \quad i = 0,1, \qquad \psi^1(t) = t, \quad \psi^0(t) = 0.$$

Apparently, $\bar{g}(\cdot)$ satisfying the conditions in Theorem 6.2.3 does not exist.

Example 6.2.6 Consider

$$Y(t) = 1 + \int_t^T (t-1)Y(s)ds - \int_t^T Z(t,s)dW(s), \qquad t \in [0,T].$$

Again, if $(Y(\cdot), Z(\cdot\,,\cdot))$ the adapted solution, then $Z(\cdot\,,\cdot) = 0$. Now, we denote

$$y(t) \stackrel{\Delta}{=} \int_t^T Y(s)ds, \quad t \in [0, T].$$

Then

$$\dot{y}(t) = -Y(t) = -1 - (t-1)\int_t^T Y(s)ds = -1 - (t-1)y(t).$$

Hence

$$0 = y(T) = e^{-\int_t^T (s-1)ds} y(t) - \int_t^T e^{-\int_\tau^T (s-1)ds} d\tau.$$

This yields

$$y(t) = \int_t^T e^{\int_t^\tau (s-1)ds} d\tau = \int_t^T e^{\frac{1}{2}[(\tau-1)^2 - (t-1)^2]} d\tau.$$

Therefore

$$Y(t) = 1 + (t-1)y(t) = 1 + (t-1)\int_t^T e^{\frac{1}{2}[\tau^2 - t^2 - 2(\tau - t)]} d\tau.$$

Consequently

$$Y(0) = 1 - \int_0^T e^{\frac{1}{2}\tau^2 - \tau} d\tau < 0,$$

provided $T > 0$ is large. Thus, comparison theorem fails for this example as well. This example corresponds to the case:

$$g^i(t, s, y, z) = (t-1)y, \quad i = 0, 1, \qquad \psi^1(t) = 1, \quad \psi^0(t) = 0.$$

Again, for this example, the generator $\bar{g}(\cdot)$ satisfying the conditions in Theorem 6.2.3 does not exist.

2.2 *Case II*

In this subsection, we are going to establish some more comparison theorems that are not covered by those presented in the last subsection.

Let us take a closer look at the above two examples. In Example 6.2.5, $t \mapsto \psi^1(t)$ is (strictly) increasing, and in Example 6.2.6, $t \mapsto g^i(t, s, y, z)$ is (strictly) increasing for $y > 0$. In a certain sense, these conditions actually prevent the comparison theorem from being true for these examples. On the other hand, we keep in mind that when $\psi(t)$ and $g^i(t,s,y,z)$ are independent of t, the above two situations do not appear. Hence, it is natural to ask if comparison theorem remains when $\psi(t)$ and $g^i(t,s,y,z)$ do depend on t, and the generalized nondecreasing condition (6.2.15) is not assumed. The answer is positive. Before we state and prove a general positive result, let us look at the following example first.

Example 6.2.7 Consider the following BSVIE:

$$Y(t)=\int_t^T\Big[s-t-Y(s)\Big]ds-\int_t^T Z(t,s)dW(s),\qquad t\in[0,T].$$

In this case, we have

$$\psi(t)\equiv 0,\quad g(t,s,y,z)=s-t-y.$$

Thus, conditions of Theorem 6.2.3 fail. However, it is easy to check that the unique adapted solution $(Y(\cdot),Z(\cdot\,,\cdot))$ is given by

$$Y(s)=e^{s-T}+T-s-1,\quad Z(t,s)=0.$$

Clearly

$$Y(s)\geqslant 0,\qquad s\in[0,T],$$

comparison theorem holds. Note that in this case, $t\mapsto g(t,s,y,z)$ is decreasing. On the other hand, the BSVIE is equivalent to the following:

$$Y(t)=\frac{(T-t)^2}{2}-\int_t^T Y(s)ds-\int_t^T Z(t,s)dW(s),\qquad t\in[0,T],$$

with

$$\psi(t)=\frac{(T-t)^2}{2},\qquad g(t,s,y)=-y.$$

For this, we have that $t\mapsto\psi(t)$ is nonincreasing.

Inspired by the above example, we see that without condition (6.2.15), one might still have comparison theorem. We now establish such a result. Let us begin with a result for linear Type-I BSVIEs. More precisely, we consider the following:

$$\begin{aligned}Y(t)=\psi(t)+&\int_t^T\Big[A(t,s)Y(s)+B(s)Z(t,s)\Big]ds\\&-\int_t^T Z(t,s)dW(s),\quad t\in[0,T].\end{aligned}\tag{6.2.24}$$

Note that the coefficient $B(s)$ of $Z(t,s)$ is independent of t. We have the following theorem.

Proposition 6.2.8 *Let $A:\Delta^*[0,T]\times\Omega\to\mathbb{R}^{n\times n}$ and $B:[0,T]\times\Omega\to\mathbb{R}^{n\times n}$ be uniformly bounded such that:*

(i) (Regular condition) $B(\cdot)$ is $\mathbb{F}$-progressively measurable, for each $t\in[0,T]$, $s\mapsto A(t,s)$ is $\mathbb{F}$-progressively measurable on $[t,T]$, and there exists a modulus of continuity $\rho:[0,T]\to[0,\infty)$ such that

$$|A(t,s)-A(t',s)|\leqslant\rho(|t-t'|),\qquad t,t'\in[0,s],\ \text{a.s., a.e. } s\in[0,T]. \tag{6.2.25}$$

(ii) (Structural conditions)

$$A(t,s)\in\mathbb{R}^{n\times n}_{*+},\qquad t\in[0,s],\ \text{a.s., a.e. } s\in[0,T], \tag{6.2.26}$$

$$\begin{aligned}A(t,s)-A(\tau,s)\in\mathbb{R}^{n\times n}_{+},&\qquad 0\leqslant t\leqslant\tau\leqslant s,\ \text{a.s., a.e. } s\in[0,T],\\ B(s)\in\mathbb{R}^{n\times n}_{d},&\qquad \text{a.s., a.e. } s\in[0,T].\end{aligned} \tag{6.2.27}$$

Then, for any $\psi(\cdot)\in C_{\mathcal{F}_T}([0,T];L^2(\Omega;\mathbb{R}^n))$ with

$$\psi(t)\geqslant\psi(s)\geqslant0,\qquad \text{a.s., } 0\leqslant t\leqslant s\leqslant T, \tag{6.2.28}$$

the adapted solution $(Y(\cdot),Z(\cdot\,,\cdot))$ of linear BSVIE (6.2.24) satisfies

$$Y(t)\geqslant0,\qquad \text{a.s., } \forall t\in[0,T]. \tag{6.2.29}$$

We point out that $A(t,s)$ satisfying (6.2.26) (which is always true if $n=1$) is not necessarily in $\mathbb{R}^{n\times n}_{+}$ (since the diagonal elements are not required to be nonnegative). Therefore, the map $y\mapsto A(t,s)y$ is not necessarily nondecreasing. Also, when $A(t,s)$ is independent of t, (6.2.27) is automatically true.

Proof Let $\Pi=\{t_k,0\leqslant k\leqslant N\}$ be a finite sequence of real numbers with $0=t_0<t_1<\cdots<t_N=T$. We define its mesh size by

$$\|\Pi\|\triangleq\max_{1\leqslant k\leqslant N}|t_k-t_{k-1}|.$$

Let us first consider the simple case of

$$A^\Pi(t,s)\triangleq\sum_{k=1}^N A(t_{k-1},s)\mathbf{1}_{(t_{k-1},t_k\wedge s]}(t),\qquad \psi^\Pi(t)\triangleq\sum_{k=1}^N\psi_k\mathbf{1}_{(t_{k-1},t_k]}(t),$$

where

$$
\begin{aligned}
&A^{\Pi}(t,s)\in\mathbb{R}^{n\times n}_{*+}, \qquad t\in[0,s],\ \text{a.s., a.e. } s\in[0,T],\\
&A^{\Pi}(t,s)-A^{\Pi}(\tau,s)\in\mathbb{R}^{n\times n}_{+}, \qquad 0\leqslant t\leqslant\tau\leqslant s,\ \text{a.s., a.e. } s\in[0,T],
\end{aligned}
$$

and each ψ_k is an $\mathcal{F}_T$-measurable random variable valued in $\mathbb{R}^n$ such that

$$
\psi_1\geqslant\psi_2\geqslant\cdots\geqslant\psi_{N-1}\geqslant\psi_N\geqslant 0, \qquad \text{a.s.}
$$

Let $(Y^{\Pi}(\cdot), Z^{\Pi}(\cdot\,,\cdot))$ be the adapted solution to the BSVIE with (A^{Π}, ψ^{Π}). On $(t_{N-1}, t_N]$, we have

$$
\begin{aligned}
Y^{\Pi}(t)=\psi_N+\int_t^T\Big(A(t_{N-1},s)Y^{\Pi}(s)+B(s)Z^{\Pi}(t,s)\Big)ds\\
-\int_t^T Z^{\Pi}(t,s)dW(s).
\end{aligned}
$$

By uniqueness of BSDEs, we see that

$$
(Y^{\Pi}(s), Z^{\Pi}(t,s))\equiv(Y_N(s), Z_N(s)), \qquad \forall t_{N-1}<t\leqslant s\leqslant T,
$$

with $(Y_N(\cdot), Z_N(\cdot))$ being the adapted solution to the following BSDE:

$$
\begin{aligned}
Y_N(t)=\psi_N+\int_t^T\Big(A(t_{N-1},s)Y_N(s)+B(s)Z_N(s)\Big)ds\\
-\int_t^T Z_N(s)dW(s), \qquad t\in(t_{N-1},t_N].
\end{aligned}
$$

Further, under our condition, by Proposition 6.1.7, we have

$$
Y^{\Pi}(t)\equiv Y_N(t)\geqslant 0, \qquad t\in(t_{N-1},t_N],\ \text{a.s.}
$$

In particular

$$
\begin{aligned}
Y^{\Pi}(t_{N-1}+0)=\psi_N+\int_{t_{N-1}}^T\Big(A(t_{N-1},s)Y^{\Pi}(s)+B(s)Z_N(s)\Big)ds\\
-\int_{t_{N-1}}^T Z_N(s)dW(s)\geqslant 0, \qquad \text{a.s.}
\end{aligned}
$$

Next, for $t \in (t_{N-2}, t_{N-1}]$, we have

$$\begin{aligned} Y^\Pi(t) &= \psi_{N-1} - \psi_N + Y^\Pi(t_{N-1}+0) + \int_{t_{N-1}}^T \big[A(t_{N-2}, s) - A(t_{N-1}, s)\big] Y^\Pi(s) ds \\ &\quad + \int_{t_{N-1}}^T B(s)\big[Z^\Pi(t,s) - Z_N(s)\big] ds - \int_{t_{N-1}}^T \big[Z^\Pi(t,s) - Z_N(s)\big] dW(s) \\ &\quad + \int_t^{t_{N-1}} \Big(A(t_{N-2}, s) Y^\Pi(s) + B(s) Z^\Pi(t,s)\Big) ds - \int_t^{t_{N-1}} Z^\Pi(t,s) dW(s). \end{aligned}$$

Let $(\widetilde{Y}_N(\cdot), \widetilde{Z}_N(\cdot))$ be the adapted solution to the following BSDE:

$$\begin{aligned} \widetilde{Y}_N(\tau) &= \psi_{N-1} - \psi_N + Y^\Pi(t_{N-1}+0) + \int_{t_{N-1}}^T \big[A(t_{N-2}, s) - A(t_{N-1}, s)\big] Y^\Pi(s) ds \\ &\quad + \int_\tau^T B(s) \widetilde{Z}_N(s) ds - \int_\tau^T \widetilde{Z}_N(s) dW(s), \qquad \tau \in (t_{N-1}, T]. \end{aligned}$$

Since v

$$\psi_{N-1} - \psi_N + Y(t_{N-1}+0) + \int_{t_{N-1}}^T \big[A(t_{N-2}, s) - A(t_{N-1}, s)\big] Y^\Pi(s) ds \geqslant 0,$$

by our conditions, using Proposition 6.1.7, we have

$$\widetilde{Y}_N(\tau) \geqslant 0, \qquad \tau \in (t_{N-1}, T], \text{ a.s.}$$

In particular

$$\begin{aligned} &\widetilde{Y}_N(t_{N-1}+0) \\ &= \psi_{N-1} - \psi_N + Y^\Pi(t_{N-1}+0) + \int_{t_{N-1}}^T \big[A(t_{N-2}, s) - A(t_{N-1}, s)\big] Y^\Pi(s) ds \\ &\quad + \int_{t_{N-1}}^T B(s) \widetilde{Z}_N(s) ds - \int_{t_{N-1}}^T \widetilde{Z}_N(s) dW(s) \geqslant 0, \qquad \text{a.s.} \end{aligned}$$

On the other hand, by the uniqueness of adapted solutions to the above BSVIEs, it is necessary that

$$Z^\Pi(t,s) = Z_N(s) + \widetilde{Z}_N(s), \qquad (t,s) \in (t_{N-2}, t_{N-1}] \times (t_{N-1}, t_N].$$

Then, $\widetilde{Y}_N(t_{N-1})$ is $\mathcal{F}_{t_{N-1}}$-measurable, $s \mapsto Z^\Pi(t,s)$ is $\mathbb{F}$-adapted, and for $t \in (t_{N-2}, t_{N-1}]$, and

$$\begin{aligned} Y^\Pi(t) &= \widetilde{Y}_N(t_{N-1}+0) + \int_t^{t_{N-1}} \Big(A(t_{N-2}, s) Y^\Pi(s) + B(s) Z^\Pi(t,s)\Big) ds \\ &\quad - \int_t^{t_{N-1}} Z^\Pi(t,s) dW(s). \end{aligned}$$

Next, let $(Y_{N-1}(\cdot), Z_{N-1}(\cdot))$ be the adapted solution to the following BSDE:

$$\begin{aligned} Y_{N-1}(t) = \widetilde{Y}(t_{N-1}+0) + \int_t^{t_{N-1}} \Big(A(t_{N-2}, s) Y_{N-1}(s) + B(s) Z_{N-1}(s) \Big) ds \\ - \int_t^{t_{N-1}} Z_{N-1}(s) dW(s), \qquad t \in [t_{N-2}, t_{N-1}]. \end{aligned}$$

By uniqueness of adapted solutions to BSDEs, we must have

$$(Y^{\Pi}(s), Z^{\Pi}(t,s)) = (Y_{N-1}(s), Z_{N-1}(s)), \qquad t \in (t_{N-2}, t_{N-1}],\ s \in [t, t_{N-1}].$$

Also, by $\widetilde{Y}(t_{N-1}) \geqslant 0$, we obtain

$$Y^{\Pi}(t) \equiv Y_{N-1}(t) \geqslant 0, \qquad t \in (t_{N-2}, t_{N-1}].$$

Therefore

$$Y^{\Pi}(t) \geqslant 0, \qquad t \in (t_{N-2}, t_N], \text{ a.s.}$$

Then, by induction, we obtain

$$Y^{\Pi}(t) \geqslant 0, \qquad t \in [0, T].$$

Finally, as to the general case, by the above conditions on A, we can obtain

$$\lim_{\|\Pi\| \to 0} \sup_{t \in [0,T]} \mathbb{E}|Y^{\Pi}(t) - Y(t)|^2 = 0,$$

with $Y(\cdot)$ satisfying (6.2.24). Then, the above conclusion holds naturally. □

Having the above result, we now state a result for nonlinear case.

Theorem 6.2.9 *Let* $g^i : \Delta^*[0,T] \times \mathbb{R}^n \times \mathbb{R}^n \times \Omega \to \mathbb{R}^n$ *be*

$$g^i(t,s,y,z) = h^i(t,s,y) + B(s)z, \quad (t,y,z) \in [0,s] \times \mathbb{R}^n \times \mathbb{R}^n, \text{ a.s.,} \tag{6.2.30}$$

for measurable h^i *and bounded process* $B(\cdot)$ *satisfying:*

(i) *(Regular condition)* $y \mapsto h^i(t,s,y)$ *is Lipschitz,* $(t,s) \mapsto h^i(t,s,0)$ *is bounded.*
(ii) *(Monotonicity) for* $y \in \mathbb{R}^n$, $0 \leqslant t \leqslant \tau \leqslant s$, $s \in [0,T]$,

$$h^1(t,s,y) - h^0(t,s,y) \geqslant h^1(\tau,s,y) - h^0(\tau,s,y) \geqslant 0 \tag{6.2.31}$$

and for either $i = 0$ *or* $i = 1$,

$$\begin{aligned} h^i_y(t,s,y) \in \mathbb{R}^{n\times n}_{*+}, \quad h^i_y(t,s,y) - h^i_y(\tau,s,y) \in \mathbb{R}^{n\times n}_{+}, \\ 0 \leqslant t \leqslant \tau \leqslant s,\ y \in \mathbb{R}^n, \end{aligned} \tag{6.2.32}$$

(iii) (Continuity) there exists a modulus of continuity ρ *such that*

$$|h_y^0(t,s,y)-h_y^0(t',s,y)|\leqslant\rho(|t-t'|),\qquad t,t'\in[0,s],\quad y\in\mathbb{R}^n,\ \text{a.s., a.e. } s\in[0,T]. \tag{6.2.33}$$

Then, for any $\psi^i(\cdot)\in C_{\mathcal{F}_T}([0,T];L^2(\Omega;\mathbb{R}^n))$ *with*

$$\psi^1(t)-\psi^0(t)\geqslant\psi^1(\tau)-\psi^0(\tau)\geqslant 0,\qquad 0\leqslant t\leqslant\tau\leqslant T,\ \text{a.s.},$$

the corresponding adapted solutions $(Y^i(\cdot),Z^i(\cdot\,,\cdot))$ *of BSVIEs* (6.2.1) *satisfy*

$$Y^1(t)\geqslant Y^0(t),\qquad a.s.,\ t\in[0,T].$$

Proof Suppose that $y\mapsto h^0(t,s,y)$ is differentiable and (6.2.32) holds for $i=0$. Then, we have

$$\begin{aligned}Y^1(t)-Y^0(t)&=\psi^1(t)-\psi^0(t)+\int_t^T\Big[h^1(t,s,Y^1(s))-h^0(t,s,Y^1(s))\Big]ds\\&\quad+\int_t^T\Big[A(t,s)\Big(Y^1(s)-Y^0(s)\Big)+B(s)\Big(Z^1(t,s)-Z^0(t,s)\Big)\Big]ds\\&\quad-\int_t^T\Big(Z^1(t,s)-Z^0(t,s)\Big)dW(s),\end{aligned}$$

where

$$A(t,s)=\int_0^1 h_y^0(t,s,Y^0(s)+\beta[Y^1(s)-Y^0(s)])d\beta,\quad (t,s)\in\Delta^*[0,T].$$

Then, our conclusion follows from Proposition 6.2.8. □

Note that in the above theorem, we have not assumed any sort of nondecreasing conditions on $y\mapsto g^i(t,s,y,z)$. Also, when $h^i(t,s,y)$ are independent of t, condition (6.2.31) is reduced to

$$h^1(s,y)\geqslant h^0(s,y),\qquad (s,y)\in[0,T]\times\mathbb{R}^n,\ \text{a.s.},$$

and condition (6.2.32) is automatically true. Finally, if $y\mapsto h^i(t,s,y)$ is just Lipschitz and not necessarily differentiable, we may modify condition (6.2.32) in the following way: for any $0\leqslant t\leqslant\tau\leqslant s\leqslant T$,

$$\begin{cases}h_k^i(t,s,y)-h_k^i(t,s,y')\geqslant 0,\quad y_k=y_k',\ y_\ell\geqslant y_\ell',\ \ell\neq k,\ \text{a.s.},\\ h^i(t,s,y)-h^i(t,s,y')\geqslant h^i(\tau,s,y)-h^i(\tau,s,y'),\quad y\geqslant y'.\end{cases} \tag{6.2.34}$$

The same conclusion will remain.

2.3 Case III

In the last subsection, the condition that the coefficient $B(s)$ of $Z(t,s)$ is independent of t is crucial. It is desired if there are comparison theorems that can recover that for general nonlinear n-dimensional BSDEs. We give a confirmative answer as follows.

For any $\tau \in [0,T]$, let $L^2_{\mathbb{F},2}([0,\tau]\times[0,T]\times\Omega;\mathbb{R}^n)$ be the set of $\mathcal{B}([0,\tau])\otimes\mathcal{B}([0,T])\otimes\mathcal{F}$-measurable process $X:[0,\tau]\times[0,T]\times\Omega\to\mathbb{R}^n$ such that $s\to\bar{X}(t,s)$ is $\mathbb{F}$-adapted with $t\in[0,\tau]$, and

$$\mathbb{E}\int_0^\tau\int_0^T |X(t,s)|^2dsdt<\infty.$$

Before showing the main comparison theorem, we give two preparation lemmas.

Lemma 6.2.10 *Suppose $X\in C([0,T];L^2_{\mathbb{F}}(0,T;\mathbb{R}^m))$. Then, there exists a unique $\bar{X}\in L^2_{\mathbb{F},2}([0,T]\times[0,T]\times\Omega;\mathbb{R}^m)$ such that*

$$\begin{aligned}&\int_0^T \|\big(X(t)\big)(\cdot)-\bar{X}(t,\cdot)\|^2_{L^2_{\mathbb{F}}(0,T;\mathbb{R}^m)}dt\\&=\int_0^T\mathbb{E}\int_0^T|\bar{X}(t,s)-\big(X(t)\big)(s)|^2dsdt=0,\end{aligned}\tag{6.2.35}$$

and for any $\tau\in[0,T]$, $\int_0^\tau\bar{X}(t,\cdot)dt\in L^2_{\mathbb{F}}(0,T,\mathbb{R}^m)$,

$$\mathbb{E}\int_0^T\Big|\Big[\int_0^\tau X(t)dt\Big](s)-\int_0^\tau\bar{X}(t,s)dt\Big|^2ds=0.\tag{6.2.36}$$

The above conclusion is standard in the infinite-dimensional case and should appear somewhere else in the literature. For reader's convenience, we give a proof here.

Proof In the following, for notational simplicity, let $\mathcal{H}_2\equiv L^2_{\mathbb{F}}(0,T;\mathbb{R}^m)$. For any $X\in C([0,T];\mathcal{H}_2)$, there exists a sequence of step process $\{X^n\}_{n\geq1}$ such that

$$\begin{aligned}&\sup_{t\in[0,T]}\Big\|X(t)-X^n(t)\Big\|_{\mathcal{H}_2}\to0,\ \ n\to\infty,\\&X^n(t)\equiv\sum_{k=1}^{l_n}\xi_k^{(n)}(\cdot)I_{E_k^{(n)}}(t),\ \ t\in[0,T],\end{aligned}$$

for some $l_n\in\mathbb{N}$, $\xi_k^{(n)}\in\mathcal{H}_2$, and mutually disjoint sets $E_1^{(n)},\cdots,E_k^{(n)}\in\mathcal{B}([0,T])$ satisfying $\bigcup_{i=1}^k E_i^{(n)}=[0,T]$. Given X^n, there exists a full measure $\Gamma^n\subset[0,T]\times\Omega$

such that for $(s,\omega)\in\Gamma^n$, $\xi_k^{(n)}(s,\omega)$ is well-defined. We define $\bar X^n$ as

$$\bar X^n(t,s)\equiv\sum_{k=1}^{l_n}\xi_k^{(n)}(s)\chi_{E_k^{(n)}}(t),\qquad t\in[0,T],\ \ (s,\omega)\in\Gamma^n,$$

and $\bar X^n(t,s)=0$ if $(s,\omega)\in(\Gamma^n)^c$. Obviously, $\bar X^n$ is $\mathcal{B}([0,T])\otimes\mathcal{B}([0,T])\otimes\mathcal{F}_T$-measurable, and

$$\begin{aligned}&\sup_{t\in[0,T]}\|X^n(t)-\bar X^n(t)\|_{\mathcal{H}_2}^2\\&=\sup_{t\in[0,T]}\mathbb{E}\int_0^T|\big(X^n(t)\big)(s)-\bar X^n(t,s)|^2ds=0.\end{aligned}\tag{6.2.37}$$

Since

$$\begin{aligned}&\int_0^T\mathbb{E}\int_0^T\big|\bar X^n(t,s)-\big(X(t)\big)(s)\big|^2dsdt\\&\leqslant T\sup_{t\in[0,T]}\mathbb{E}\int_0^T\big|\bar X^n(t,s)-\big(X(t)\big)(s)\big|^2ds=T\sup_{t\in[0,T]}\big\|\bar X^n(t,\cdot)-\big(X(t)\big)(\cdot)\big\|_{\mathcal{H}_2}^2\\&\leqslant T\Big[\sup_{t\in[0,T]}\|X^n(t)-\bar X^n(t)\|_{\mathcal{H}_2}^2+\sup_{t\in[0,T]}\big\|X^n(t)-X(t)\big\|_{\mathcal{H}_2}^2\Big]\to 0,\quad\text{as } n\to\infty.\end{aligned}$$

Therefore, there exists a subsequence, still denoted by $\{\bar X^n\}$ and a full measure $\Upsilon\subset[0,T]\times[0,T]\times\Omega$ such that for any $(t,s,\omega)\in\Upsilon$, $\bar X^n(t,s,\omega)\to\big(X(t)\big)(s,\omega)$. Define $\bar X(t,s,\omega)\equiv\big(X(t)\big)(s,\omega)$ if $(t,s,\omega)\in\Upsilon$ and $\bar X(t,s,\omega)\equiv 0$ if $(t,s,\omega)\in(\Upsilon)^c$. Then, $\bar X$ is obviously $\mathcal{B}([0,T])\otimes\mathcal{B}([0,T])\otimes\mathcal{F}_T$-measurable. The conclusion of $s\to\bar X(t_0,s)$ being $\mathbb{F}$-adapted with $t_0\in[0,T]$ is easy to see. Moreover

$$\begin{aligned}&\mathbb{E}\int_0^T\int_0^T|\bar X(t,s)|^2dsdt=\int_0^T\mathbb{E}\int_0^T|\bar X(t,s)|^2dsdt\\&=\int_\Upsilon|\bar X(t,s)|^2dsd\mathbb{P}(\omega)dt=\int_\Upsilon|\big(X(t)\big)(s)|^2dsd\mathbb{P}(\omega)dt\\&=\int_0^T\mathbb{E}\int_0^T|\big(X(t)\big)(s)|^2dsdt<\infty.\end{aligned}$$

We look at the uniqueness. Suppose there is $\widetilde X\in L^2_{\mathbb{F},2}([0,T]\times[0,T]\times\Omega;\mathbb{R}^m)$ such that the corresponding (6.2.35) holds. Then

$$\int_0^T\|\widetilde X(t,\cdot)-\bar X(t,\cdot)\|_{\mathcal{H}_2}^2dt=\int_0^T\mathbb{E}\int_0^T|\bar X(t,s)-\widetilde X(t,s)|^2dsdt=0,$$

which proves the uniqueness.

Let us look at the integral $\int_0^\tau \bar{X}(t,\cdot)dt$ with $\tau \in [0,T]$ with $\bar{X} \in L^2_{\mathbb{F},2}([0,T] \times [0,T] \times \Omega; \mathbb{R}^m)$. By Hölder inequality and Fubini theorem,

$$\begin{aligned}\mathbb{E}\int_0^\tau \Big|\int_0^\tau \bar{X}(t,s)dt\Big|^2 ds &\le T\mathbb{E}\int_0^\tau\int_0^\tau |\bar{X}(t,s)|^2 dtds\\ = T\mathbb{E}\int_0^\tau\int_0^\tau |\bar{X}(t,s)|^2 dsdt &< \infty.\end{aligned}$$

As a result, for almost all (s,ω), the Lebesgue integral $\bar{X}(\cdot,s)$ is Lebesgue integrable on $[0,\tau]$. In addition, one has $s \to \int_0^\tau \bar{X}(t,s)dt$ is $\mathbb{F}$-adapted and measurable.

Next, we prove (6.2.36). By the above relation between X^n and X, for any $\varepsilon > 0$, there exists $N \in \mathbb{N}$, such that for any $n \ge N$, and $\tau \in [0,T]$,

$$\sup_{t\in[0,\tau]} \Big\|X(t) - X^n(t)\Big\|_{\mathcal{H}_2} = \sup_{t\in[0,\tau]} \Big[\mathbb{E}\int_0^T \big|X(t,s) - X^n(t,s)\big|^2 ds\Big]^{\frac{1}{2}} < \varepsilon.$$

Thanks to the property of Bochner integral, for given $\tau \in [0,T]$, we have

$$\Big\|\int_0^\tau [X(t) - X^N(t)]dt\Big\|_{\mathcal{H}_2} \le \int_0^\tau \|X(t) - X^N(t)\|_{\mathcal{H}_2} dt \le T\varepsilon. \tag{6.2.38}$$

On the one hand, by definition, for the above fixed N and $(s,\omega) \in \Gamma^N$, where $\Gamma^N \subset [0,T] \times \Omega$ is a full measure defined in above, we have

$$\Big[\int_0^\tau X^N(t)dt\Big](s,\omega) = \Big[\sum_{k=1}^{l_N} \xi_k^{(N)}(\cdot)\lambda(E_k^{(N)})\Big](s,\omega) = \int_0^\tau \bar{X}^N(t,s,\omega)dt. \tag{6.2.39}$$

Then, by (6.2.35), (6.2.37), we have

$$\begin{aligned}&\Big\|\int_0^\tau [\bar{X}^N(t,\cdot) - \bar{X}(t,\cdot)]dt\Big\|^2_{\mathcal{H}_2} \equiv \mathbb{E}\int_0^T \Big|\int_0^\tau [\bar{X}^N(t,s) - \bar{X}(t,s)]dt\Big|^2 ds\\
&\le T\mathbb{E}\int_0^T\int_0^\tau \big|\bar{X}^N(t,s) - \bar{X}(t,s)\big|^2 dtds\\
&\le T\int_0^\tau \mathbb{E}\int_0^T \Big\{\big|\bar{X}^N(t,s) - \big(X^N(t)\big)(s)\big|^2 + \big|\big(X^N(t)\big)(s) - \big(X(t)\big)(s)\big|^2\\
&\qquad\qquad + \big|\bar{X}(t,s) - \big(X(t)\big)(s)\big|^2\Big\} dsdt\\
&\le T\sup_{t\in[0,\tau]} \|X^N(t) - X(t)\|^2_{\mathcal{H}_2}.\end{aligned}$$

Therefore, we have

$$\Big\|\int_0^\tau [\bar{X}^N(t,\cdot) - \bar{X}(t,\cdot)]dt\Big\|_{\mathcal{H}_2} \le T\varepsilon. \tag{6.2.40}$$

From the above (6.2.38), (6.2.39), and (6.2.40), we then see that

$$\Big\|\Big(\int_0^\tau X(t)dt\Big)(\cdot)-\int_0^\tau \bar X(t,\cdot)dt\Big\|_{\mathcal{H}_2}<K\varepsilon.$$

Therefore, our conclusion is easy to see. □

Remark 6.2.11 We point out two important facts.

First, by (6.2.36), there exists a full measure $\Gamma\subset[0,T]\times\Omega$ such that for any $(s,\omega)\in\Gamma$, we have

$$H_X(\tau,s,\omega)\equiv\Big[\int_0^\tau X(t)dt\Big](s,\omega)=R_X(\tau,s,\omega)\equiv\int_0^\tau \bar X(t,s,\omega)dt. \tag{6.2.41}$$

Rough speaking, the above equality (6.2.41) shows that (s,ω) and the integral $\int_0^\tau$ is exchangeable.

Second, for any fixed $\tau\in[0,T]$, the variable s in $\Big[\int_0^\tau X(t)dt\Big](s)$ depends on the given τ. To drop such a dependence, we introduce the term $\bar X(\cdot,\cdot)$ and use $\int_0^\tau \bar X(t,s)dt$ instead.

For later usefulness, we introduce two more spaces. Let $L^2_{\mathbb{F}}(\Omega;C([0,T]^2;\mathbb{R}^m))$ be the set of measurable process $X:[0,T]^2\times\Omega\to\mathbb{R}^m$ such that $(t,s)\mapsto X(t,s)$ is continuous, for any $t\in[0,T]$, $s\mapsto X(t,s)$ is $\mathbb{F}$-adapted, and

$$\mathbb{E}\sup_{s,t\in[0,T]}|X(t,s)|^2<\infty.$$

Let $L^2_{\mathbb{F}}(0,T;C([0,T];\mathbb{R}^m))$ be the set of measurable process $X:[0,T]^2\times\Omega\to\mathbb{R}^m$ such that for almost all (s,ω), $t\mapsto X(t,s)$ is continuous in $[0,T]$, and $s\mapsto X(t,s)$ is $\mathbb{F}$-adapted,

$$\mathbb{E}\int_0^T\sup_{t\in[0,T]}|X(t,s)|^2ds<\infty.$$

Before proving the comparison theorem, let us consider the following parameterized BSDE:

$$Y(t,r)=\psi(t)+\int_r^T g(t,s,Y(t,s),Z(t,s))ds-\int_r^T Z(t,s)dW(s),\qquad r\in[0,T]. \tag{6.2.42}$$

Such kind of equation was studied in Sect. 2, Chap. 3. With the help of previous Lemma 6.2.10, we will give (6.2.42) a careful revisit. More precisely, we prove a new well-posedness for (6.2.42) and then apply it to comparison theorems for

BSVIEs in the sequel. For reader's convenience, we rewrite (H3.2.5) in Chap. 3 as follows.

(H6.2.2) Suppose g is measurable function such that:

(i) (Regular condition) for any $t, s \in [0, T]$, $g(t, \cdot, 0, 0) \in L^2_{\mathbb{F}}(0, T; \mathbb{R}^m)$, $(y, z) \mapsto g(t, s, y, z)$ is differentiable with bounded derivatives such that $(t, y, z) \mapsto g_y(t, s, y, z), g_z(t, s, y, z)$ are continuous.

(ii) (Differentiability) for any $t_0 \in [0, T]$ and the associated solution $(Y(t_0, \cdot), Z(t_0, \cdot))$, $t \mapsto g(t, \cdot, Y(t_0, \cdot), Z(t_0, \cdot))$, $t \mapsto \psi(t)$ are differentiable in space $L^2_{\mathbb{F}}(0, T; \mathbb{R}^m)$, $L^2_{\mathcal{F}_T}(\Omega; \mathbb{R}^m)$ at t_0, with derivative $\partial_t g(t_0, \cdot, Y(t_0, \cdot), Z(t_0, \cdot)), \partial_t \psi(t_0)$, respectively.

(iii) (Continuity) for any $t_0 \in [0, T]$,

$$\lim_{t \to t_0} \Big[\big\| \partial_t g(t, \cdot, Y(t, \cdot), Z(t, \cdot)) - \partial_t g(t_0, \cdot, Y(t_0, \cdot), Z(t_0, \cdot)) \big\|_{L^2_{\mathbb{F}}(0,T;\mathbb{R}^m)} + \big\| \partial_t \psi(t) - \partial_t \psi(t_0) \big\|_{L^2_{\mathcal{F}_T}(\Omega;\mathbb{R}^m)} \Big] = 0.$$

According to Proposition 3.2.2 of Chap. 3, under (H6.2.2), we have $t \mapsto (\partial_t Y(t, \cdot), \partial_t Z(t, \cdot))$ is continuous in $L^2_{\mathbb{F}}(0, T; \mathbb{R}^m)$. Then, according to Lemma 6.2.10, there exists $(\bar{Y}(\cdot, \cdot), \bar{Z}(\cdot, \cdot)) \in L^2_{\mathbb{F},2}([0, \tau] \times [0, T] \times \Omega; \mathbb{R}^{2m})$ such that

$$\int_0^\tau \mathbb{E} \int_0^T |\bar{f}(t, s) - \partial_t f(t, s)|^2 ds dt = 0, \quad f \equiv Y, Z.$$

Suppose $(Y(0, \cdot), Z(0, \cdot))$ satisfies BSDE (6.2.42) with $t = 0$. For almost all $(r, \omega) \in [0, T] \times \Omega$, we define

$$\begin{aligned} \mathcal{Z}(t, r) &\equiv Z(0, r) + \int_0^t \bar{Z}(s, r) ds, \quad \forall t \in [0, T], \\ \mathcal{Y}(t, r) &\equiv Y(0, r) + \int_0^t \bar{Y}(s, r) ds, \quad \forall t \in [0, T]. \end{aligned} \tag{6.2.43}$$

In terms of Lemma 6.2.10, the right-hand Lebesgue integral is well-defined for given (r, ω).

Lemma 6.2.12 *Suppose (H6.2.2) holds. Then, the above pair of $(\mathcal{Y}, \mathcal{Z})$ is the unique pair of solution to BSDE (6.2.42) in the following spaces:*

$$(\mathcal{Y}, \mathcal{Z}) \in L^2_{\mathbb{F}}(\Omega; C([0, T]^2; \mathbb{R}^m)) \times L^2_{\mathbb{F}}(0, T; C([0, T]; \mathbb{R}^m)). \tag{6.2.44}$$

In particular, $\mathcal{Y}^d(\cdot) \in L^2_{\mathbb{F}}(\Omega; C([0, T]; \mathbb{R}^m))$ with $\mathcal{Y}^d(t) \equiv \mathcal{Y}(t, t)$, $t \in [0, T]$.

Proof We first show the pair of $(\mathcal{Y}, \mathcal{Z})$ in (6.2.43) satisfies (6.2.44).

By the definition of $\mathcal{Z}$, for almost all $(s, \omega) \in [0, T] \times \Omega$, $\mathcal{Z}^d(s) \equiv \mathcal{Z}(s, s)$ is well-defined, $t \mapsto \mathcal{Z}(t, s)$ is continuous, and

$$\begin{aligned}&\mathbb{E}\int_0^T \sup_{t\in[0,T]} |\mathcal{Z}(t,r)|^2 dr\\&\le 2\mathbb{E}\int_0^T |Z(0,r)|^2 dr + 2T\mathbb{E}\int_0^T\int_0^T |\bar{Z}(s,r)|^2 drds < \infty.\end{aligned}$$

This implies that $\mathcal{Z}^d(\cdot) \in L^2_{\mathbb{F}}(0, T; \mathbb{R}^m)$, and $\mathcal{Z} \in L^2_{\mathbb{F}}(0, T; C([0, T]; \mathbb{R}^m))$.

Similarly, for almost all (s, ω), $\mathcal{Y}^d(s) \equiv \mathcal{Y}(s, s)$ is well-defined, and

$$\begin{aligned}&\mathbb{E}\Big[\sup_{t,r\in[0,T]} |\mathcal{Y}(t,r)|^2\Big]\\&\leqslant 2\mathbb{E}\Big[\sup_{r\in[0,T]} |Y(0,r)|^2\Big] + 2T\mathbb{E}\Big[\sup_{r\in[0,T]}\int_0^T |\bar{Y}(s,r)|^2 ds\Big],\end{aligned}$$

and

$$\begin{aligned}&\mathbb{E}\Big[\sup_{r\in[0,T]}\int_0^T |\bar{Y}(s,r)|^2 ds\Big]\\&\leqslant \mathbb{E}\Big[\int_0^T \sup_{r\in[0,T]} |\bar{Y}(s,r)|^2 ds\Big] \leqslant \sup_{s\in[0,T]} \mathbb{E} \sup_{r\in[0,T]} |\bar{Y}(s,r)|^2 < \infty.\end{aligned}$$

The last equality also implies that there exists a full measure Ω_1 such that for any $\omega \in \Omega_1$,

$$\begin{aligned}&Y(0,r) < \infty, \quad \int_0^T \bar{Y}(s,r,\omega) ds < \infty, \quad \forall r \in [0,T],\\&\int_0^T \sup_{r\in[0,T]} \big|\bar{Y}(s,r,\omega)\big| ds < \infty.\end{aligned}$$

Therefore, by dominated convergence theorem, for such $\omega \in \Omega_1$,

$$\lim_{r\to r_0}\int_0^T \big|\bar{Y}(s,r,\omega) - \bar{Y}(s,r_0,\omega)\big| ds = 0.$$

Therefore, for any $\omega \in \Omega_1$,

$$\begin{aligned}&\sup_{t\in[0,T]} \big|\mathcal{Y}(t,r,\omega) - \mathcal{Y}(t,r_0,\omega)\big|\\&\leqslant 2\big|Y(0,r) - Y(0,r_0)\big| + 2\int_0^T \big|\bar{Y}(s,r) - \bar{Y}(s,r_0)\big| ds \to 0, \quad r \to r_0.\end{aligned} \tag{6.2.45}$$

On the other hand, for any $\omega \in \Omega_1$,

$$\big|\mathcal{Y}(t,r_0)-\mathcal{Y}(t_0,r_0)\big| \leqslant \int_{t_0\wedge t}^{t\vee t_0} |\bar{Y}(s,r_0)|ds \to 0, \quad t\to t_0.$$

Therefore, for any $\omega \in \Omega_1$, as $(t,r)\to(t_0,r_0)$,

$$\big|\mathcal{Y}(t,r)-\mathcal{Y}(t_0,r_0)\big| \leqslant \sup_{t\in[0,T]} \big|\mathcal{Y}(t,r)-\mathcal{Y}(t,r_0)\big| + \big|\mathcal{Y}(t,r_0)-\mathcal{Y}(t_0,r_0)\big| \to 0.$$

This implies that $\mathcal{Y} \in L^2_{\mathbb{F}}(\Omega; C([0,T]^2;\mathbb{R}^m))$. The conclusion of $\mathcal{Y}^d(\cdot)$ is easy to see.

Next, we show that $(\mathcal{Y},\mathcal{Z})$ satisfies BSDE (6.2.42). To begin with, let us recall the Newton-Leibniz formula in Banach space, say, $\mathscr{X}$ in terms of Bochner integral. Suppose $x:[0,T]\to\mathscr{X}$ is strong differentiable on $[0,T]$ with derivative $x'(\cdot)$ satisfying

$$\lim_{t\to t_0}\|x'(t)-x'(t_0)\|_{\mathscr{X}}=0, \quad \forall t_0\in[0,T].$$

Then, we have

$$\int_0^T x'(t)dt = x(T)-x(0), \quad \text{in } \mathscr{X}, \tag{6.2.46}$$

where the left integral is well-defined in the sense of Bochner integral. By means of Proposition 3.2.2 and the previous Newton-Leibniz formula, under (H6.2.2), for any $t\in[0,T]$, one has

$$\begin{aligned}
&\Big\|Y(t,\cdot)-Y(0,\cdot)-\int_0^t \partial_s Y(s,\cdot)ds\Big\|_{L^2_{\mathbb{F}}(0,T;\mathbb{R}^m)}\\
&\equiv \mathbb{E}\int_0^T \Big|Y(t,r)-Y(0,r)-\Big(\int_0^t \partial_s Y(s,\cdot)ds\Big)(r)\Big|^2 dr = 0,\\
&\Big\|Z(t,\cdot)-Z(0,\cdot)-\int_0^t \partial_s Z(s,\cdot)ds\Big\|_{L^2_{\mathbb{F}}(0,T;\mathbb{R}^m)}\\
&\equiv \mathbb{E}\int_0^T \Big|Z(t,r)-Z(0,r)-\Big(\int_0^t \partial_s Z(s,\cdot)ds\Big)(r)\Big|^2 dr = 0.
\end{aligned}$$

Therefore, it follows from (6.2.36) in Lemma 6.2.10, and notations in (6.2.43) that

$$\sup_{t\in[0,T]} \mathbb{E}\int_0^T \Big\{\big|Z(t,r)-\mathcal{Z}(t,r)\big|^2 + \big|Y(t,r)-\mathcal{Y}(t,r)\big|^2\Big\}dr = 0.$$

Recall that for given $t\in[0,T]$, there exists full measure Ω_t such that for any $\omega\in\Omega_t$, $r\mapsto Y(t,r)$ is continuous. Combining this with the continuity of $\mathcal{Y}$ in

(6.2.45), we then end up with

$$\sup_{t\in[0,T]} \mathbb{E} \sup_{r\in[0,T]} \left|Y(t,r) - \mathcal{Y}(t,r)\right|^2 = 0.$$

Consequently, we have

$$\mathbb{E}\int_r^T \left|g(t,s,Y(t,s),Z(t,s)) - g(t,s,\mathcal{Y}(t,s),\mathcal{Z}(t,s))\right|^2 ds = 0, \quad \forall r \in [0,T].$$

Therefore, $(\mathcal{Y}, \mathcal{Z})$ is a solution to BSDE (6.2.42).

Suppose (Y', Z') is another pair of solution. By the uniqueness of BSDEs, for any $t \in [0, T]$,

$$\mathbb{E} \sup_{s\in[0,T]} |Y(t,s) - Y'(t,s)|^2 + \mathbb{E}\int_0^T |Z(t,s) - Z'(t,s)|^2 ds = 0.$$

Let $\mathcal{Q}$ be the rational number in $[0, T]$ which is countable. For any $t \in \mathcal{Q}$, there exists a full measure $\Gamma_t \subset [0,T] \times \Omega$ such that for any (s,ω), $Z(t,s) = Z'(t,s)$. Let $\Gamma := \bigcap_{t\in\mathcal{Q}} \Gamma_t$ which still has full measure. Hence, for any $(s,\omega) \in \Gamma$, $Z(t,s) = Z'(t,s)$ for any $t \in \mathcal{Q}$. Given $(s,\omega) \in \Gamma$, for any $t_0 \in [0,T]$, by the continuity of $t \mapsto Z(t,s)$ and $t \mapsto Z'(t,s)$,

$$Z(t_0,s) = \lim_{t\to t_0,\, t\in\mathcal{Q}} Z(t,s) = \lim_{t\to t_0,\, t\in\mathcal{Q}} Z'(t,s) = Z'(t_0,s).$$

This implies that

$$\mathbb{E}\int_0^T \sup_{t\in[0,T]} |Z(t,s) - Z'(t,s)|^2 ds = 0.$$

Similarly, we have

$$\mathbb{E} \sup_{s\in[0,T]} \sup_{t\in[0,T]} |Y(t,s) - Y'(t,s)|^2 = 0.$$

The uniqueness is done. □

Remark 6.2.13 In the literature, it seems that under (H6.2.2), the regularity of (Y, Z) can be obtained as follows:

$$(Y(\cdot,\cdot), Z(\cdot,\cdot)) \in C^1([0,T]; L^2_{\mathbb{F}}(\Omega; C([0,T];\mathbb{R}^m))) \times C^1([0,T]; L^2_{\mathbb{F}}(0,T;\mathbb{R}^m)).$$

In contrast, here we obtain the following version:

$$(Y(\cdot,\cdot), Z(\cdot,\cdot)) \in L^2_{\mathbb{F}}(\Omega; C([0,T]^2; \mathbb{R}^m)) \times L^2_{\mathbb{F}}(0,T; C([0,T]; \mathbb{R}^m)).$$

Moreover, for almost (r,ω), $t \mapsto (Y(t,r), Z(t,r))$ is differentiable in the classical sense. As a result, our continuity and differentiability hold in the finite-dimensional Euclidean space, while the current analogue is shown in infinite-dimensional spaces $L^2_{\mathbb{F}}(\Omega; C([0,T]; \mathbb{R}))$ and $L^2_{\mathbb{F}}(0,T; \mathbb{R}^m)$.

After the previous preparations, it is time for us to give the comparison theorem of BSVIEs. Let us consider two BSVIEs as follows:

$$Y^i(t) = \psi^i(t) + \int_t^T g^i(t,s,Y^i(s),Z^i(t,s))ds - \int_t^T Z^i(t,s)dW(s), \quad t \in [0,T], \ i = 1,2. \tag{6.2.47}$$

Here, $\psi^i : [0,T] \times \Omega \to \mathbb{R}$, $g^i : [0,T]^2 \times \Omega \times \mathbb{R} \times \mathbb{R} \to \mathbb{R}$ are measurable mappings. We make the following assumptions.

(H6.2.3) Given $i = 1,2$, $\psi^i(0) \in L^2_{\mathcal{F}_T}(\Omega; \mathbb{R})$, for almost all $\omega \in \Omega$, $t \mapsto \psi^i(t,\omega)$ is differentiable (in the classical sense) with derivative $\psi^i_t(\cdot) \in C_{\mathcal{F}_T}([0,T]; L^2(\Omega; \mathbb{R}))$.

(H6.2.4) Given $i = 1,2$, suppose:

(i) (Differentiability w.r.t. t) $(s,y,z) \in [0,T] \times \mathbb{R} \times \mathbb{R}$, $t \mapsto g^i(t,s,y,z)$ is differentiable almost everywhere with derivative $g^i_t(t,s,y,z)$ satisfying $(t,z) \mapsto g^i_t(t,s,y,z)$ being continuous.

(ii) (Differentiability w.r.t. z) for $(t,s,y) \in [0,T]^2 \times \mathbb{R}$, $z \mapsto g^i(t,s,y,z)$ is differentiable almost everywhere with derivative $g^i_z(t,s,y,z)$ satisfying $(t,z) \mapsto g^i_z(t,s,y,z)$ being continuous.

(iii) (Integrability) for $(t,s) \in [0,T]^2$, $y, \bar{y}, z, \bar{z} \in \mathbb{R}$, and some $L > 0$,

$$\begin{aligned}
&\big|g^i(t,s,y,z) - g^i(t,s,\bar{y},\bar{z})\big| \leqslant L|y-\bar{y}| + L|z-\bar{z}|, \\
&\big|g^i_t(t,s,y,z)\big| \leqslant |g^i_t(t,s,0,0)| + L|y| + L|z|, \\
&\mathbb{E}\Big[\int_0^T |g^i(t,s,0,0)|ds\Big]^2 + \mathbb{E}\int_0^T \sup_{t\in[0,T]} |g^i_t(t,s,0,0)|^2 ds < \infty.
\end{aligned} \tag{6.2.48}$$

(H6.2.5) Suppose (H6.2.3) and (H6.2.4) hold, and $\psi^1(T) \geqslant \psi^2(T)$, and for $t, s \in [0,T]$, $y, z \in \mathbb{R}$,

$$g^1(s,s,y,z) \geqslant g^2(s,s,y,z), \quad \psi^1_t(t) \leqslant \psi^2_t(t), \quad g^1_t(t,s,y,z) \leqslant g^2_t(t,s,y,z).$$

In the above (H6.2.3)–(H6.2.4), we point out that the derivatives ψ^i_t, g^i_t hold in the classical Euclidean space but not in some abstract space, say, e.g., $L^2(\Omega; \mathbb{R})$. We state the following comparison theorem, the conclusion of which can fully cover the conventional nonlinear BSDE case.

Theorem 6.2.14 *Suppose* (H6.2.5) *hold. Then, for* $i = 1, 2$, (6.2.47) *admits a unique pair of* $(Y^i(\cdot), Z^i(\cdot))$ *such that* $Y^i(\cdot) \in L^2_{\mathbb{F}}(\Omega; C([0, T]; \mathbb{R}))$,

$$\mathbb{P}\big(\omega \in \Omega \mid Y^1(t, \omega) \geqslant Y^2(t, \omega), \quad \forall t \in [0, T]\big) = 1. \tag{6.2.49}$$

Proof Since the proof is length, we prefer to separate it into several steps.
Step 1. By the boundedness of g^i_y, g^i_z in (H6.2.3), we see that BSVIE (6.2.47) admits a unique pair of adapted solutions (Y^i, Z^i). To begin with, for $i = 1, 2$, $t \in [0, T]$, we introduce two classes of BSDEs:

$$\begin{cases} \lambda^i(t,r)=\psi^i(t)+\int_r^T g^i(t,s,Y^i(s),\theta^i(t,s))ds-\int_r^T \theta^i(t,s)dW(s), \\ \qquad\qquad\qquad\qquad\qquad\qquad\qquad\qquad r \in [0,T], \\ \partial_t\lambda^i(t,r) = \psi^i_t(t) + \int_r^T \big[g^i_z(t,s,Y^i(s),\theta^i(t,s))\partial_t\theta^i(t,s) \\ \qquad\qquad +g^i_t(t,s,Y^i(s),\theta^i(t,s))\big]ds - \int_r^T \partial_t\theta^i(t,s)dW(s), \\ \qquad\qquad\qquad\qquad\qquad\qquad\qquad\qquad r \in [t,T]. \end{cases} \tag{6.2.50}$$

For the first one, for given $t \in [0, T]$, by (H6.2.3)–(H6.2.4), it admits a unique

$$(\lambda^i(t,\cdot), \theta^i(t,\cdot)) \in L^2_{\mathbb{F}}(\Omega; C([0,T];\mathbb{R})) \times L^2_{\mathbb{F}}(0,T;\mathbb{R})$$

such that

$$\begin{aligned} &\mathbb{E}\sup_{r\in[0,T]}|\lambda^i(t,r)|^2 + \mathbb{E}\int_0^T |\theta^i(t,s)|^2 ds \\ &\leqslant K\mathbb{E}|\psi^i(t)| + K\mathbb{E}\Big[\int_0^T |g^i(t,s,Y^i(s),0)|ds\Big]^2 < \infty, \end{aligned}$$

which implies that

$$\begin{aligned} &\mathbb{E}\Big[\int_t^T |g^i_t(t,s,Y^i(s),\theta^i(t,s))|ds\Big]^2 \\ &\leqslant K\mathbb{E}\Big[\int_t^T |g^i_t(t,s,0,0)|ds\Big]^2 + K\mathbb{E}\int_t^T |Y^i(s)|^2 ds + K\mathbb{E}\int_t^T |\theta^i(t,s)|^2 ds < \infty. \end{aligned}$$

Therefore, for the second BSDE in (6.2.50), it admits a unique

$$(\partial_t\lambda^i(t,\cdot), \partial_t\theta^i(t,\cdot)) \in L^2_{\mathbb{F}}(\Omega; C([t,T];\mathbb{R})) \times L^2_{\mathbb{F}}(t,T;\mathbb{R}).$$

Step 2. In this step, we want to apply Lemma 6.2.12 to obtain a pair of (Λ^i, Θ^i) with higher regularity. To this end, we need to verify the requirements in (H6.2.2).

We first look at the case of $\psi^i(\cdot)$. For any $t \in [0, T]$, by (H6.2.3), for $i = 1, 2$,

$$\psi^i(t) - \psi^i(0) = \int_0^t \psi_r^i(r)dr, \quad \forall t \in [0, T]. \quad \text{a.s.}$$

This then implies that

$$\sup_{t\in[0,T]} \mathbb{E}|\psi^i(t)|^2 \leqslant \mathbb{E}|\psi^i(0)|^2 + T\mathbb{E}\int_0^T |\psi_r^i(r)|^2 dr < \infty.$$

Given $t_0 \in [0, T]$, we see that $r \mapsto f^i(r) \equiv \mathbb{E}|\psi_r^i(r) - \psi_r^i(t_0)|^2$ is continuous on $[0, T]$, and therefore

$$\begin{aligned}\lim_{t\downarrow t_0} \mathbb{E}\Big|\frac{\psi^i(t) - \psi^i(t_0)}{t - t_0} - \psi_r^i(t_0)\Big|^2 &= \lim_{t\downarrow t_0} \mathbb{E}\Big|\frac{1}{t - t_0}\int_{t_0}^t [\psi_r^i(r) - \psi_r^i(t_0)]dr\Big|^2 \\ &\leqslant \lim_{t\downarrow t_0} \frac{1}{t - t_0}\mathbb{E}\int_{t_0}^t [\psi_r^i(r) - \psi_r^i(t_0)]^2 dr = 0.\end{aligned}$$

The part of $t \uparrow t_0$ can be obtained similarly. Therefore, $t \mapsto \psi^i(t)$ is differentiable in $L^2_{\mathcal{F}_T}(\Omega; \mathbb{R})$ with derivative $\psi_t^i(\cdot) \in C_{\mathcal{F}_T}([0, T]; L^2(\Omega; \mathbb{R}))$.

Next, we turn to look at g^i. By definition, for any $0 \leqslant t_0 < t \leqslant T$,

$$\begin{aligned}&\mathbb{E}\int_0^T \Big|\frac{g^i(t, s, Y^i(s), \theta^i(t_0, s)) - g^i(t_0, s, Y^i(s), \theta^i(t_0, s))}{t - t_0} \\ &\qquad\qquad - g_r^i(t_0, s, Y^i(s), \theta^i(t_0, s))\Big|^2 ds \\ &= \mathbb{E}\int_0^T\Big|\int_{t_0}^t \frac{g_r^i(r, s, Y^i(s), \theta^i(t_0, s))}{t - t_0}dr - g_r^i(t_0, s, Y^i(s), \theta^i(t_0, s))\Big|^2 ds \\ &\leqslant \mathbb{E}\int_0^T\int_{t_0}^t \frac{|g_r^i(r, s, Y^i(s), \theta^i(t_0, s)) - g_r^i(t_0, s, Y^i(s), \theta^i(t_0, s))|^2}{|t - t_0|}drds \\ &= \mathbb{E}\int_{t_0}^t\int_0^T \frac{|g_r^i(r, s, Y^i(s), \theta^i(t_0, s)) - g_r^i(t_0, s, Y^i(s), \theta^i(t_0, s))|^2}{|t - t_0|}dsdr.\end{aligned} \tag{6.2.51}$$

By (H6.2.4) and dominated convergence theorem, we see the following continuity:

$$\begin{aligned}&\lim_{r\to r_0} \mathbb{E}\int_0^T |g_r^i(r, s, Y^i(s), \theta^i(t_0, s)) - g_r^i(t_0, s, Y^i(s), \theta^i(t_0, s))|^2 ds \\ &= \mathbb{E}\int_0^T |g_r^i(r_0, s, Y^i(s), \theta^i(t_0, s)) - g_r^i(t_0, s, Y^i(s), \theta^i(t_0, s))|^2 ds.\end{aligned}$$

Hence, the left hand of (6.2.51) converges to zero as $t \downarrow t_0$. The case of $t \uparrow t_0$ can be obtained similarly. Then, we have that $t \mapsto g^i(t,\cdot,Y^i(\cdot),\theta^i(t_0,\cdot))$ is differentiable in space $L^2_{\mathbb{F}}(0,T;\mathbb{R})$ at t_0 with derivative $g^i_t(t_0,\cdot,Y^i(\cdot),\theta^i(t_0,\cdot))$. Therefore, according to Proposition 3.2.2 of Chap. 3, we have

$$\lim_{t\to t_0}\mathbb{E}\int_0^T \left|\theta^i(t,s)-\theta^i(t_0,s)\right|^2 ds = 0.$$

Since $(t,z)\mapsto g^i_t(t,s,y,z)$ is continuous, and (6.2.48) holds, hence by dominated convergence theorem,

$$\mathbb{E}\int_0^T \left|g^i_t(t,s,Y^i(s),\theta^i(t,s)) - g^i_t(t_0,s,Y^i(s),\theta^i(t_0,s))\right|^2 ds \to 0, \quad t\to t_0.$$

To sum up the above arguments on ψ^i and g^i, the conditions in Lemma 6.2.12 are fulfilled. Then, the following pair of (Λ^i,Θ^i) defined by

$$\begin{cases} \Lambda^i(t,r)\equiv\lambda^i(0,r)+\displaystyle\int_0^t \bar\lambda^i(s,r)ds, & \forall t\in[0,T], \quad (r,\omega)\in[0,T]\times\Omega, \\ \Theta^i(t,r)\equiv\mu^i(0,r)+\displaystyle\int_0^t \bar\theta^i(s,r)ds, & \forall t\in[0,T], \quad (r,\omega)\in[0,T]\times\Omega, \end{cases} \tag{6.2.52}$$

is the unique solution of the first BSDE in (6.2.50) in

$$L^2_{\mathbb{F}}(\Omega;C([0,T]^2;\mathbb{R}))\times L^2_{\mathbb{F}}(0,T;C([0,T];\mathbb{R})).$$

Here, the process $\bar\lambda^i,\bar\theta^i\in L^2_{\mathbb{F},2}([0,T]\times[0,T]\times\Omega;\mathbb{R})$ is implied by Lemma 6.2.10 such that

$$\mathbb{E}\int_0^T\int_0^T \left|\partial_t\lambda^i(t,r)-\bar\lambda^i(t,r)\right|^2 drdt+\mathbb{E}\int_0^T\int_0^T \left|\partial_t\theta^i(t,r)-\bar\theta^i(t,r)\right|^2 drdt = 0.$$

Step 3. In this step, we prove that (6.2.47) admits a unique pair of adapted solution such that $Y^i(\cdot)\in L^2_{\mathbb{F}}(\Omega;C([0,T];\mathbb{R}))$.

We define $(\Lambda^i_d(\cdot),\Theta^i_\Delta(\cdot,\cdot))$ as follows:

$$\Lambda^i_d(t)\equiv\Lambda^i(t,t), \quad t\in[0,T], \quad \Theta^i_\Delta(t,s)\equiv\Theta^i(t,s), \quad t\leqslant s.$$

By the property of $\Lambda^i(\cdot,\cdot)$, one has $\Lambda^i_d(\cdot)\in L^2_{\mathbb{F}}(\Omega;C([0,T];\mathbb{R}))$, and

$$\Lambda^i_d(t) = \psi^i(t)+\int_t^T g^i(t,s,Y^i(s),\Theta^i_\Delta(t,s))ds - \int_t^T \Theta^i_\Delta(t,s)dW(s), \quad t\in[0,T], \quad i=1,2.$$

Thanks to the uniqueness of BSVIEs, we have

$$\mathbb{E}\int_0^T \big|Y^i(s)-\Lambda_d^i(s)\big|^2 ds = 0,$$

which implies that for any given $t\in[0,T]$,

$$\int_t^T g^i(t,s,\Lambda_d^i(s),\Theta_\Delta^i(t,s))ds = \int_t^T g^i(t,s,Y^i(s),\Theta_\Delta^i(t,s))ds, \quad a.s.$$

In other words, $(\Lambda_d^i(\cdot),\Theta_d^i(\cdot,\cdot))$ satisfies (6.2.47) with better regularity, i.e.,

$$\Lambda_d^i(t) = \psi^i(t) + \int_t^T g^i(t,s,\Lambda_d^i(s),\Theta_\Delta^i(t,s))ds - \int_t^T \Theta_\Delta^i(t,s)dW(s), \quad t\in[0,T], \quad i=1,2. \tag{6.2.53}$$

Next, let us look at the uniqueness. Suppose $(\mathcal{Y}^i,\mathcal{Z}^i)$ is another solution to (6.2.47) such that $\mathcal{Y}^i(\cdot)\in L_{\mathbb{F}}^2(\Omega;C([0,T];\mathbb{R}))$. On the one hand, by the continuity and integrability of $(\Lambda_d^i,\mathcal{Y}^i)$, we have

$$\begin{aligned}&\Big|\mathbb{E}|\mathcal{D}(s)|^2-\mathbb{E}|\mathcal{D}(s_0)|^2\Big|\\&\leqslant \Big[\mathbb{E}\big|\mathcal{D}(s)-\mathcal{D}(s_0)\big|^2\Big]^{\frac12}\Big[\mathbb{E}\big|\mathcal{D}(s)+\mathcal{D}(s_0)\big|^2\Big]^{\frac12}\to 0, \quad s\to s_0,\end{aligned}$$

where $\mathcal{D}(\cdot)\triangleq\Lambda_d^i(\cdot)-\mathcal{Y}^i(\cdot)$. Therefore, $s\mapsto\mathbb{E}|\mathcal{D}(s)|^2$ is continuous on $[0,T]$. Considering the fact that

$$\int_r^{r+\varepsilon}\mathbb{E}|\mathcal{D}(s)|^2 ds = 0, \quad \forall 0\leqslant r<r+\varepsilon\leqslant T,$$

we then have $\mathbb{E}|\mathcal{D}(s)|^2=0$ with any given $s\in[0,T]$. By the continuity of $s\mapsto\mathcal{D}(s)$, we finally have $\mathbb{P}\big(\mathcal{D}(s)=0, \ \ \forall s\in[0,T]\big)=1$, which implies the desired uniqueness.

Step 4. In this step, we will give an equivalent expression of (6.2.53) that is easier to obtain the comparison theorem. To this end, given $(\Lambda_d^i(\cdot),\Theta_\Delta^i(\cdot,\cdot))$, we introduce

$$\begin{aligned}\mathcal{M}^i(t,r) = \psi_t^i(t) + \int_r^T &\big[g_z^i(t,s,\Lambda_d^i(s),\Theta_\Delta^i(t,s))\mathcal{N}^i(t,s)\\&+g_t^i(t,s,\Lambda_d^i(s),\Theta_\Delta^i(t,s))\big]ds - \int_r^T \mathcal{N}^i(t,s)dW(s), \quad r\in[t,T].\end{aligned}$$

The following well-posedness is easy to see

$$(\mathcal{M}^i(t,\cdot),\mathcal{N}^i(t,\cdot))\in L_{\mathbb{F}}^2(\Omega;C([t,T];\mathbb{R}))\times L_{\mathbb{F}}^2(t,T;\mathbb{R}).$$

By the definition of Θ^i in (6.2.52), for almost (s,ω), we first observe that

$$t \mapsto \mathcal{G}^i(t,s) \equiv g^i(t,s,\Lambda_d^i(s),\Theta_\Delta^i(t,s))$$

is differentiable such that

$$\begin{aligned}
&g^i(t,s,\Lambda_d^i(s),\Theta_\Delta^i(t,s)) - g_i(s,s,\Lambda_d^i(s),\Theta_\Delta^i(s,s))\\
&= -\int_t^s \Big[g_t^i(r,s,\Lambda_d^i(s),\Theta_\Delta^i(r,s)) + g_z^i(r,s,\Lambda_d^i(s),\Theta_\Delta^i(r,s))\bar{\mu}^i(r,s)\Big]dr.
\end{aligned}$$

Hence, for any $t \in [0,T]$, by Fubini theorem and the relevant terms in (6.2.52)

$$\begin{aligned}
\Lambda_d^i(t) &= \psi^i(T) - \int_t^T \psi_r^i(r)dr + \int_t^T g^i(s,s,\Lambda_d^i(s),\Theta_\Delta^i(s,s))ds\\
&\quad -\int_t^T\int_r^T \Big[g_r^i(r,s,\Lambda_d^i(s),\Theta_\Delta^i(r,s)) + \partial_z g^i(r,s,\Lambda_d^i(s),\Theta_\Delta^i(r,s))\bar{\mu}^i(r,s)\Big]dsdr\\
&\quad -\int_t^T \theta^i(s,s)dW(s) + \int_t^T\int_r^T \bar{\theta}^i(r,s)dW(s)dr\\
&= \psi^i(T) + \int_t^T \big[g^i(s,s,\Lambda_d^i(s),\Theta_\Delta^i(s,s)) - \mathcal{M}^i(s,s)\big]ds - \int_t^T \Theta_\Delta^i(s,s)dW(s).
\end{aligned}$$

Given $\mathcal{M}^i(s,s)$, let us consider the following BSDE:

$$P^i(t) = \psi^i(T) + \int_t^T \big[g^i(s,s,P^i(s),Q^i(s)) - \mathcal{M}^i(s,s)\big]ds - \int_t^T Q^i(s)dW(s),$$

whose well-posedness leads by (H6.2.3)–(H6.2.4). By the uniqueness, we see that

$$\mathbb{P}\big(P^i(t) = \Lambda_d^i(t),\ \ \forall t \in [0,T]\big) = 1.$$

Thanks to (H6.2.5), by the comparison theorem of BSDEs, we see that $\mathcal{M}^1(t,t) \leqslant \mathcal{M}^2(t,t)$. a.s. Since $\psi^1(T) \geqslant \psi^2(T)$, using again the comparison theorem of BSDEs, one has $P^1(\cdot) \geqslant P^2(\cdot)$, and hence, $\Lambda_d^1(\cdot) \geqslant \Lambda_d^2(\cdot)$. The conclusion of (6.2.49) is obvious to see. □

3 Comparison Theorem for Type-II BSVIEs

In this subsection, we discuss the following Type-II BSVIEs:

$$Y(t) = \psi(t) + \int_t^T g(t,s,Y(s),Z(s,t))ds - \int_t^T Z(t,s)dW(s), \quad t \in [0,T]. \tag{6.3.1}$$

Note that since the generator $g(\cdot)$ depends on $Z(s,t)$, the notion of adapted solution in $\mathcal{H}^p_{\Delta^*}[0,T]$ will not be enough. Therefore, we adopt the notion of adapted M-solution to the above BSVIE. More precisely, an adapted M-solution is an adapted solution $(Y(\cdot), Z(\cdot\,,\cdot))$ which belongs to $\mathcal{M}^p[0,T]$. The following is a standard assumption for the BSVIE (6.3.1).

(H6.2.6) For $i=0,1$, the maps $g^i:\Delta^*\times\mathbb{R}^n\times\mathbb{R}^n\times\Omega\to\mathbb{R}^n$ are measurable, $s\mapsto g^i(t,s,y,\zeta)$ is $\mathbb{F}$-progressively measurable, $(y,\zeta)\mapsto g^i(t,s,y,\zeta)$ is uniformly Lipschitz, and $(t,s)\mapsto g^i(t,s,0,0)$ is uniformly bounded.

Under (H6.2.6), for any $\psi(\cdot)\in L^2_{\mathcal{F}_T}(0,T;\mathbb{R}^n)$, (6.3.1) admits a unique adapted M-solution $(Y(\cdot), Z(\cdot\,,\cdot))$. We will use a dual principle to prove the comparison theorem for adapted M-solution. Before going further, let us look at a simple example.

Example 6.3.1 Consider the following one-dimensional BSVIE:

$$Y(t)=\psi(t)+\int_t^T\frac{2T-t}{2T-s}Z(s,t)ds-\int_t^T Z(t,s)dW(s),\qquad t\in[0,T]. \tag{6.3.2}$$

We introduce the following FSVIE:

$$X(t)=1+\int_0^t\frac{2T-s}{2T-t}X(s)dW(s),\qquad t\in[0,T], \tag{6.3.3}$$

which is the equation in Example 6.1.10, and

$$\mathbb{P}\{\omega\in\Omega\bigm| X(t,\omega)<0\}>0,\qquad \forall t\in[0,T].$$

Hence, by taking

$$\psi(t)=\mathbf{1}_A(t,\omega),\qquad A=\{(t,\omega)\bigm| X(t,\omega)<0\},\ t\in[0,T],$$

and the duality principle (Theorem 5.4.1 of Chap. 4), we have

$$\mathbb{E}\int_0^T Y(t)dt=\mathbb{E}\int_0^T X(t)\mathbf{1}_A(t,\omega)dt<0,$$

which means that $Y(\cdot)\geqslant 0$ on $[0,T]$ could not be true, although $\psi(\cdot)\geqslant 0$.

The above example shows that comparison theorem may fail for linear Type-II BSVIEs if in the generator, the coefficient of $Z(s,t)$ depends both on t and s. The above example suggests us that if linear Type-II BSVIEs are considered for comparison of adapted M-solutions, the following should be a proper form:

$$
\begin{aligned}
Y(t) = \psi(t) + &\int_t^T \Big(A(t,s)Y(s) + C(t)Z(s,t)\Big)ds \\
&- \int_t^T Z(t,s)dW(s), \quad t \in [0,T].
\end{aligned} \tag{6.3.4}
$$

Note that $Z(t,s)$ does not appear in the drift term, and the coefficient $C(t)$ of $Z(s,t)$ is independent of s. For such an equation, we have the following result, which is comparable with Theorem 6.2.8.

Proposition 6.3.2 *Let* $A : [0,T] \times [0,T] \times \Omega \to \mathbb{R}^{n\times n}$ *and* $C : [0,T] \times \Omega \to \mathbb{R}^{n\times n}$ *be uniformly bounded and measurable such that:*

(i) (Continuity) for each $s \in [0,T]$*, there exists a continuous nondecreasing function* $\rho : [0,T] \to [0,\infty)$ *with* $\rho(0) = 0$*,*

$$
|A(s,t) - A(s,t')| \leqslant \rho(|t-t'|), \qquad t,t' \in [0,s], \text{ a.s., a.e. } s \in [0,T]. \tag{6.3.5}
$$

(ii) (Structural condition)

$$
A(t,s) \in \mathbb{R}^{n\times n}_{*+}, \quad t \in [0,s], \text{ a.s., a.e. } s \in [0,T],
$$

$$
A(s,\tau) - A(s,t) \in \mathbb{R}^{n\times n}_{+}, \quad \forall s \leqslant t \leqslant \tau \leqslant T, \text{ a.s., a.e. } s \in [0,T],
$$

$$
C(t) \in \mathbb{R}^{n\times n}_{d}, \quad \text{a.s., a.e. } t \in [0,T].
$$

Then, for any $\psi(\cdot) \in L^2_{\mathcal{F}_T}(0,T;\mathbb{R}^n)$ *with*

$$
\psi(t) \geqslant 0, \qquad \text{a.s, } t \in [0,T],
$$

the adapted M-solution $(Y(\cdot), Z(\cdot\,,\cdot))$ *of linear BSVIE* (6.3.4) *satisfies*

$$
\mathbb{E}_t \int_t^T Y(s)ds \geqslant 0, \qquad \forall t \in [0,T], \text{ a.s.} \tag{6.3.6}
$$

Proof Pick any $\eta(\cdot) \in L^2_{\mathbb{F}}(0,T;\mathbb{R}^n)$ with $\eta(\cdot) \geqslant 0$, consider the following linear FSVIE:

$$
X(t) = \varphi(t) + \int_0^t A(s,t)^\top X(s)ds + \int_0^t C(s)^\top X(s)dW(s), \qquad t \in [0,T],
$$

with

$$
\varphi(t) \triangleq \int_0^t \eta(s)ds, \qquad t \in [0,T].
$$

By our conditions on $A(\cdot\,,\cdot)$ and $C(\cdot)$, using Proposition 6.1.12, we have

$$X(t) \geqslant 0, \qquad \text{a.s. } t \in [0, T].$$

Then, by duality theorem (Theorem 5.4.1 of Chap. 4), one obtains

$$\begin{aligned}0 &\leqslant \mathbb{E}\int_0^T \langle \psi(t), X(t)\rangle dt = \mathbb{E}\int_0^T \langle \varphi(t), Y(t)\rangle dt \\ &= \mathbb{E}\int_0^T\int_0^t \langle \eta(s), Y(t)\rangle ds dt = \mathbb{E}\int_0^T \langle \eta(s), \int_s^T Y(t)dt\rangle ds.\end{aligned}$$

Thus, (6.3.6) follows since $\eta(\cdot)$ is arbitrary. □

Different from Theorem 6.2.8, in the above, we do not need the monotonicity of $t \mapsto \psi(t)$, and the conclusion (6.3.6) is weaker than (6.2.29).

Having the above result, we are able to get a comparison theorem for the following nonlinear BSVIEs ($i = 0, 1$):

$$\begin{aligned}Y^i(t) = \psi^i(t) &+ \int_t^T \Big(h^i(t, s, Y^i(s)) + C(t)Z^i(s, t)\Big)ds \\ &- \int_t^T Z^i(t, s)dW(s), \quad t \in [0, T].\end{aligned} \tag{6.3.7}$$

More precisely, the following theorem holds.

Theorem 6.3.3 *Let $g^i : \Delta^*[0, T] \times \mathbb{R}^n \times \mathbb{R}^n \times \Omega \to \mathbb{R}^n$ be the form of*

$$g^i(t, s, y, \zeta) \stackrel{\Delta}{=} h^i(t, s, y) + C(t)\zeta, \tag{6.3.8}$$

for some h^i and bounded $C(\cdot)$ such that:

(i) (Regular condition) $y \mapsto h^i(t, s, y)$ is Lipschitz and $h^i(\cdot, \cdot, 0)$ is bounded.
(ii) (Monotonicity) for any $y \in \mathbb{R}^n$, $0 \leqslant t \leqslant \tau \leqslant s \leqslant T$,

$$h^1(t, s, y) - h^0(t, s, y) \geqslant h^1(\tau, s, y) - h^0(\tau, s, y) \geqslant 0, \quad \text{a.s.}, \tag{6.3.9}$$

and for either $i = 0$ or $i = 1$,

$$h^i_y(t, s, y) \in \mathbb{R}^{n\times n}_{*+}, \quad h^i_y(t, s, y) - h^i_y(\tau, s, y) \in \mathbb{R}^{n\times n}_+, \quad t \leqslant \tau \leqslant s \leqslant T. \tag{6.3.10}$$

(iii) (Continuity) there exists a continuous nondecreasing function $\rho : [0, T] \to [0, \infty)$ with $\rho(0) = 0$

$$|h^0_y(t, s, y) - h^0_y(t', s, y)| \leqslant \rho(|t - t'|), \quad t, t' \in [0, s], \quad y \in \mathbb{R}^n, \text{ a.s.} \tag{6.3.11}$$

Then, for any $\psi^i(\cdot) \in L^2_{\mathcal{F}_T}(0,T;\mathbb{R}^n)$ *with*

$$\psi^1(t) - \psi^0(t) \geqslant \psi^1(\tau) - \psi^0(\tau) \geqslant 0, \qquad 0 \leqslant t \leqslant \tau \leqslant T, \text{ a.s.},$$

the adapted M-solutions $(Y^i(\cdot), Z^i(\cdot\,,\cdot))$ *of BSVIEs* (6.3.1) *satisfy*

$$\mathbb{E}_t \int_t^T Y^1(s)ds \geqslant \mathbb{E}_t \int_t^T Y^0(s)ds, \qquad t \in [0,T], \text{ a.s.} \tag{6.3.12}$$

Proof Observe the following:

$$\begin{aligned}
&Y^1(t) - Y^0(t)\\
&= \psi^1(t) - \psi^0(t) + \int_t^T \Big[h^1(t,s,Y^1(s)) - h^0(t,s,Y^1(s))\Big]ds\\
&\quad + \int_t^T \Big[A(t,s)\Big(Y^1(s) - Y^0(s)\Big) + C(t)\Big(Z^1(s,t) - Z^0(s,t)\Big)\Big]ds\\
&\quad - \int_t^T \Big(Z^1(t,s) - Z^0(t,s)\Big)dW(s),
\end{aligned}$$

where

$$A(t,s) = \int_0^1 h^0_y\big(t,s,Y^0(s) + \beta[Y^1(s) - y^0(s)]\big)d\beta.$$

Then, under our conditions, we have the comparison (6.3.12). □

4 Bibliographical Remarks

Comparison problems of solutions to SDEs have been discussed first by Skorokhod [179] in the 1960s. Under certain assumptions, he showed that the solution of diffusion equation is a monotonic function of the drift coefficient and then applied this fact to the uniqueness problem for solutions of SDEs. In 1973, Yamada [216] treated a comparison problem in a new method and improved Skorokhod's comparison theorem for solutions of SDEs. In 1976, Ikeda–Watanabe [109] further modified the result of Yamada so that it is more convenient for applications in discussing some stochastic optimal control problem. A little later, Mel'nikov [145] proved comparison theorems for SDEs governed by continuous martingales. In 1990, Ouknine–Rutkowski [155] discussed comparison for one-dimensional SDEs allowing the drift term to be discontinuous and the diffusion term to be degenerated. Note that in the abovementioned literature, any involved comparison SDEs have the same diffusions. Different from the above, in 1980, O'Brien [152] studied a comparison theorem for solutions of SDEs with different diffusion terms. On the other hand, in 2006, Peng–Zhu [166] treated comparison of solutions to two SDEs as the viability problem of the augmented SDE with constraint, obtained a necessary and sufficient condition for the comparison theorem. Other relevant study can be found in Gal'cuk–Davis [82], Yan [217], Huang [108], and so on.

The first comparison theorem for scalar BSDEs is due to Peng [163] in 1992, and then, it was generalized by Pardoux–Peng [159] and El Karoui–Peng–Quenez [68]. In 2006, Hu–Peng [104] gave a necessary and sufficient condition under which the comparison theorem holds for multidimensional BSDEs by means of "viability condition." Other relevant study can also be found in, e.g., Barles–Buckdahn–Pardoux [14], Pardoux [157], Cao–Yan [40], Liu–Ren [138], Cohen–Elliott–Pearce [50], and so on.

It seems to us that the earliest comparison theorems for FVIEs were due to Beesack [18] in 1969, which concerns with linear Volterra integral equations of the second kind. More general (allowing nonlinear dynamics) comparison theorems were contained in the monographs, e.g., Miller [148], Cochran [49], and Gripenberg–Londen–Staffans [88]. It is worthy of mentioning the work of Friedman [77, Section 4] in 1963, where solutions of two different FVIEs with weakly singular kernels were compared, which is useful in deriving upper and lower bounds on solutions of FVIEs. For the stochastic extension, a comparison theorem for solutions of stochastic dynamic equations with Volterra-type drifts was proved by Tudor [186] in 1989. Such a result was then extended by Ferreyra–Sundar [73] in 2000, where the diffusion coefficient is also of the Volterra type with the form $H(t)\sigma_j(s,x)$. It is important to note that a comparison of solutions to stochastic Volterra integral equations with general Volterra-type "diffusion" coefficients does not hold in general. A counterexample is given in the above context; see also Tudor [186].

The comparison theorem for BSVIEs is an interesting topic. Besides the interest from mathematical extension of that for BSDEs, there are some interesting possible applications. For example, it is used in Wang–Yong–Zhang [194] to prove the well-posedness of viscosity solution in the path-dependent framework. On the other hand, it is also widely used in mathematical finance problems. The first one lies in the application of dynamic risk measure for position processes. In the literature, the static risk measures for discrete-time or continuous-time processes have been studied in Cheridito–Delbaen–Kupper [46–48], Frittelli–Scandolo [81], and so on. It is natural that BSVIEs could be used to describe some continuous-time dynamic risk measures for not necessarily adapted stochastic processes. The second comes from the study of general yet realistic stochastic equilibrium utility problem. The stochastic differential utility was introduced and studied in Duffie–Epstein [63] and Lazrak–Quenez [134] where the inter-temporal consistency and Bellman principle of optimality are applicable. However, many real problems could be of time-inconsistent nature. In fact, quite a few experimental studies on time preference show that the standard assumption of time consistency is unrealistic. We believe that comparison theorems of BSVIEs play important roles in formulating general stochastic differential utility processes and investigating their properties such as comparative risk aversion, risk aversion, etc. Along this line, we refer to Chap. 11 for more details.

Most results presented in this chapter is based on the work of Wang–Yong [204]. However, in Sect. 2.3, we give some further relevant results which improve the analogue in [204] and enable us to fully cover the nonlinear BSDE case in, e.g., [163]. On the other hand, we point out that for comparison theorems of FSDEs and BSDEs, our proofs are different from those in Geiβ–Manthey [86], Hu–Peng [104], and Peng–Zhu [166], respectively.

Chapter 7
Representation of Adapted Solutions

In this chapter, we are looking for a representation of adapted solutions for BSVIEs (of Type-I and Type-II), in terms of solutions of FSDEs and PDEs. More precisely, suppose we are given the following Type-I BSVIE:

$$
\begin{aligned}
Y(t) = \psi(t, X(t), X(T)) &+ \int_t^T g(t, s, X(t), X(s), Y(s), Z(t, s))ds \\
&- \int_t^T Z(t, s)dW(s), \quad t \in [0, T],
\end{aligned}
\tag{7.1}
$$

where $X(\cdot)$ is the solution to an FSDE of the following form:

$$
\begin{cases} dX(s) = b(s, X(s))ds + \sigma(s, X(s))dW(s), & s \in [t, T], \\ X(t) = x. \end{cases}
\tag{7.2}
$$

In the above, b, σ, g, ψ are all proper deterministic functions. We try the following ansatz:

$$
Y(t) = \Theta(t, X(t)), \qquad Z(t, s) = \widetilde{\Theta}(t, s, X(t), X(s)), \quad 0 \leqslant t \leqslant s \leqslant T,
$$

with suitable maps Θ and $\widetilde{\Theta}$. If the above is successful with some nice maps Θ and $\widetilde{\Theta}$, then we could obtain some additional regularity of the adapted solution $(Y(\cdot), Z(\cdot\,,\cdot))$ to the BSVIE (7.1), for example, the continuity of $t \mapsto Y(t)$, and $(t, s) \mapsto Z(t, s)$. The same thing can also be tried for Type-II BSVIEs of the following form:

T. Wang, J. Yong, *Backward Stochastic Volterra Integral Equations*,
Probability Theory and Stochastic Modelling 109,
https://doi.org/10.1007/978-3-032-20848-4_7

$$
\begin{aligned}
Y(t) = \psi(t, X(t), X(T)) + &\int_t^T g(t, s, X(t), X(s), Y(s), Z(t, s), Z(s, t))ds \\
&- \int_t^T Z(t, s)dW(s), \quad t \in [0, T].
\end{aligned}
\tag{7.3}
$$

For notational simplicity, we assume that the Brownian motion is one dimensional.

1 Decoupling of FBSDEs and Representation of BSDEs

Let us begin with the following coupled linear ODE:

$$
\begin{cases}
\dot{X}(s) = A(s)X(s) + B(s)Y(s) + b(s), \\
\dot{Y}(s) = \widehat{A}(s)X(s) + \widehat{B}(s)Y(s) + g(s), \quad s \in [t, T], \\
X(t) = x, \qquad Y(T) = HX(T) + h.
\end{cases}
\tag{7.1.1}
$$

Suppose the above admits a unique solution $(X(\cdot), Y(\cdot))$ for any given $x, h, b(\cdot), g(\cdot)$. Then, we try the following ansatz:

$$
Y(s) = P(s)X(s) + \eta(s), \qquad s \in [t, T],
$$

for some $P(\cdot)$ and $\eta(\cdot)$. If this is true, then we should have

$$
\begin{aligned}
&\widehat{A}(s)X(s) + \widehat{B}(s)\big[P(s)X(s) + \eta(s)\big] + g(s) \\
&= \widehat{A}(s)X(s) + \widehat{B}(s)Y(s) + g(s) = \dot{Y}(s) \\
&= \dot{P}(s)X(s) + P(s)\big[A(s)X(s) + B(s)Y(s) + b(s)\big] + \dot{\eta}(s) \\
&= \dot{P}(s)X(s) + P(s)A(s)X(s) + P(s)B(s)\big[P(s)X(s) + \eta(s)\big] + P(s)b(s) + \dot{\eta}(s).
\end{aligned}
$$

Hence

$$
\begin{aligned}
0 = &\Big(\dot{P}(s) + P(s)A(s) - \widehat{B}(s)P(s) + P(s)B(s)P(s) - \widehat{A}(s)\Big)X(s) \\
&+ \dot{\eta}(s) + \big[P(s)B(s) - \widehat{B}(s)\big]\eta(s) + P(s)b(s) - g(s).
\end{aligned}
$$

Consequently, we should let $P(\cdot)$ and $\eta(\cdot)$ solve the following:

$$
\begin{cases}
\dot{P}(s) + P(s)A(s) - \widehat{B}(s)P(s) + P(s)B(s)P(s) - \widehat{A}(s) = 0, \quad s \in [t, T], \\
P(T) = H,
\end{cases}
\tag{7.1.2}
$$

and

$$
\begin{cases}
\dot{\eta}(s) + \big[P(s)B(s) - \widehat{B}(s)\big]\eta(s) + P(s)b(s) - g(s) = 0, \quad s \in [t, T], \\
\eta(T) = h.
\end{cases}
\tag{7.1.3}
$$

We call (7.1.2) *Riccati differential equation*, whose solution $P(\cdot)$ leads to the coupled system (7.1.1) decoupled. Such a method is referred to as the *invariant imbedding*. This method also applies to two-boundary point problems for nonlinear ODEs and much more beyond.

Let us now look at the case of FBSDEs. Consider the following system:

$$\begin{cases} dX(t)=b\big(t, X(t), Y(t), Z(t)\big)dt+\sigma\big(t, X(t), Y(t)\big)dW(t), \quad t\in[0, T], \\ dY(t) = -g\big(t, X(t), Y(t), Z(t)\big)dt + Z(t)dW(t), \quad t \in [0, T], \\ X(0) = x, \quad Y(T) = h(X(T)). \end{cases} \tag{7.1.4}$$

Suppose all the involved functions b, σ, g, h are deterministic. Inspired by the above, one could expect the following:

$$Y(t) = \Theta(t, X(t)), \qquad t \in [0, T],$$

for some differentiable function $\Theta : [0, T] \times \mathbb{R}^n \to \mathbb{R}^m$. If this is the case, then by Itô's formula, one must have

$$\begin{aligned} &-g(t, X(t), Y(t), Z(t))dt + Z(t)dW(t) = dY(t) \\ &\quad= \Big[\Theta_t(t, X(t)) + \frac{1}{2}\sigma(t, X(t), Y(t))^\top \Theta_{xx}(t, X(t))\sigma(t, X(t), Y(t)) \\ &\qquad+\Theta_x(t, X(t))b(t, X(t), Y(t), Z(t))\Big]dt + \Theta_x(t, X(t)\sigma(t, X(t), Y(t))dW(t), \end{aligned}$$

where

$$\sigma(t, x, y)^\top\Theta_{xx}(t, x)\sigma(t, x, y) \stackrel{\Delta}{=} \begin{pmatrix} \sigma(t, x, y)^\top \Theta^1_{xx}(t, x)\sigma(t, x, y) \\ \sigma(t, x, y)^\top \Theta^2_{xx}(t, x)\sigma(t, x, y) \\ \vdots \\ \sigma(t, x, y)^\top \Theta^m_{xx}(t, x)\sigma(t, x, y) \end{pmatrix}, \tag{7.1.5}$$

with

$$\Theta(t, x) \stackrel{\Delta}{=} \begin{pmatrix} \Theta^1(t, x) \\ \Theta^2(t, x) \\ \vdots \\ \Theta^m(t, x) \end{pmatrix}, \qquad (t, x) \in [0, T] \times \mathbb{R}^n.$$

Therefore, the following should hold:

$$Z(t) = \Theta_x(t, X(t))\sigma(t, X(t), Y(t)), \qquad t \in [0, T].$$

It is seen that $\Theta(\cdot\,,\cdot)$ should solve the following system of quasilinear parabolic *partial differential equations* (PDE, for short):

$$\begin{cases} \Theta_t(t,x)+\frac{1}{2}\sigma\big(t,x,\Theta(t,x)\big)^\top\Theta_{xx}(t,x)\sigma\big(t,x,\Theta(t,x)\big) \\ +\Theta_x(t,x)b\big(t,x,\Theta(t,x),\Theta_x(t,x)\sigma(t,x,\Theta(t,x))\big) \\ +g\big(t,x,\Theta(t,x),\Theta_x(t,x)\sigma(t,x,\Theta(t,x))\big)=0, \quad (t,x)\in[0,T]\times\mathbb{R}^n, \\ \Theta(T,x)=h(x), \qquad x\in\mathbb{R}^n. \end{cases} \tag{7.1.6}$$

Suppose the above PDE has a classical solution $\Theta(\cdot\,,\cdot)$. Then, we have the representation of the backward components $(Y(\cdot),Z(\cdot))$ in terms of the forward component $X(\cdot)$:

$$\begin{cases} Y(t)=\Theta\big(t,X(t)\big), \\ Z(t)=\Theta_x(t,X(t))\sigma\Big(t,X(t),\Theta\big(t,X(t)\big)\Big), \end{cases} \quad t\in[0,T]. \tag{7.1.7}$$

In the above, $\Theta(\cdot\,,\cdot)$ is called a *decoupling field* of the FBSDE (7.1.4), since we need only to solve the following FSDE:

$$\begin{cases} dX(t)=b\big(t,X(t),\Theta(t,X(t)),\Theta_x(t,X(t))\sigma(t,X(t),\Theta(t,X(t)))\big)dt \\ \qquad\qquad +\sigma\big(t,X(t),\Theta(t,X(t))\big)dW(t), \quad t\in[0,T], \\ X(0)=x. \end{cases} \tag{7.1.8}$$

We call (7.1.6) the *representation PDE* since the solution $\Theta(\cdot\,,\cdot)$ allows us to represent the backward component $(Y(\cdot),Z(\cdot))$ in terms of the forward component $X(\cdot)$. Thus, if the above representation (7.1.7) holds, then we get some more useful regularity information on $(Y(\cdot),Z(\cdot))$, especially for $Z(\cdot)$.

In the case that FBSDE (7.1.4) is decoupled, namely, b and σ are independent of $(Y(\cdot),Z(\cdot))$, then representation PDE (7.1.6) will read as follows:

$$\begin{cases} \Theta_t(t,x)+\frac{1}{2}\sigma(t,x)^\top\Theta_{xx}(t,x)\sigma(t,x)+\Theta_x(t,x)b(t,x) \\ \quad +g\big(t,x,\Theta(t,x),\Theta_x(t,x)\sigma(t,x)\big)=0, \quad (t,x)\in[0,T]\times\mathbb{R}^n, \\ \Theta(T,x)=h(x), \qquad x\in\mathbb{R}^n. \end{cases} \tag{7.1.9}$$

Since the leading term is linear, the solvability of this equation is much simpler. For such a case, the representation has a simpler form:

$$Y(t)=\Theta(t,X(t)), \qquad Z(t)=\Theta_x(t,X(t))\sigma(t,X(t)).$$

For this case, we do not call $\Theta(\cdot\,,\cdot)$ a decoupling field (since there is nothing to decouple).

It is natural to ask if we can establish a similar result for BSVIEs. However, we see that in the above, Itô's formula has played an essential role, whereas there is no such a direct formula for FSVIEs and BSVIEs. Hence, we have to take a different route.

2 Approximation of Type-I BSVIEs

This section is devoted to an approximation of Type-I BSVIEs by a sequence of BSDEs. On the one hand, such an approximation will be helpful for us to derive the representation of the adapted solutions to the Type-I BSVIEs. On the other hand, this will also be helpful for designing numerical scheme for such kind of BSVIEs. Before going further, let us first consider the following SDE:

$$X(s) = x + \int_t^s b(\tau, X(\tau))d\tau + \int_t^s \sigma(\tau, X(\tau))dW(\tau), \quad s \in [t, T]. \tag{7.2.1}$$

We need the following assumption concerning the FSDE (7.2.1).

(H7.2.1) The maps $b : [0, T] \times \mathbb{R}^n \to \mathbb{R}^n$, $\sigma : [0, T] \times \mathbb{R}^n \to \mathbb{R}^n$ are continuous, and there is a constant $L > 0$ such that

$$|b(t, x_1) - b(t, x_2)| + |\sigma(t, x_1) - \sigma(t, x_2)| \leqslant L|x_1 - x_2|, \\ \forall t \in [0, T], \ x_1, x_2 \in \mathbb{R}^n,$$

and

$$|b(t, 0)| + |\sigma(t, 0)| \leqslant L, \qquad \forall t \in [0, T].$$

From Proposition 2.3.11 in Chap. 2, we know that for any $(t, x) \in [0, T] \times \mathbb{R}^n$, FSDE (7.2.1) admits a unique solution $X(\cdot) \equiv X(\cdot\,; t, x) \in L^p_{\mathbb{F}}(\Omega; C([0, T]; \mathbb{R}^n))$ for any $p > 1$, and the following holds:

$$\mathbb{E}|X(s) - X(t)|^2 \leqslant K_0|s - t|, \qquad \forall s, t \in [0, T], \tag{7.2.2}$$

for some constant $K_0 > 0$ independent of s, t. Now, for such a given $X(\cdot)$, we consider the following Type-I BSVIE:

$$\begin{aligned} Y(t) = \psi(t, X(t), X(T)) &+ \int_t^T g(t, s, X(t), X(s), Y(s), Z(t, s))ds \\ &- \int_t^T Z(t, s)dW(s), \quad t \in [0, T]. \end{aligned} \tag{7.2.3}$$

We introduce the following assumptions:

(H7.2.2) Functions $\psi : [0, T] \times \mathbb{R}^n \times \mathbb{R}^n \to \mathbb{R}^m$ and $g : \Delta^*[0, T] \times \mathbb{R}^n \times \mathbb{R}^n \times \mathbb{R}^m \times \mathbb{R}^m \to \mathbb{R}^m$ are continuous. There exists a constant $L > 0$ such that

$$
\begin{aligned}
&|\psi(t, 0, 0)| + |g(t, s, 0, 0, 0, 0)| \leqslant L, \quad 0 \leqslant t \leqslant s \leqslant T,\\
&|\psi(t_1, \xi_1, x_1) - \psi(t_2, \xi_2, x_2)| \leqslant L\big(|t_1 - t_2| + |\xi_1 - \xi_2| + |x_1 - x_2|\big),\\
&\qquad\qquad\qquad\qquad \forall (t_1, \xi_1, x_1), (t_2, \xi_2, x_2) \in [0, T] \times \mathbb{R}^n \times \mathbb{R}^n,\\
&|g(t_1, s, \xi_1, x_1, y_1, z_1) - g(t_2, s, \xi_2, x_2, y_2, z_2)|\\
&\leqslant L\big(|t_1 - t_2| + |\xi_1 - \xi_2| + |x_1 - x_2| + |y_1 - y_2| + |z_1 - z_2|\big),\\
&\quad \forall (t_i, s, \xi_i, x_i, y_i, z_i) \in \Delta^*[0, T] \times \mathbb{R}^n \times \mathbb{R}^n \times \mathbb{R}^m \times \mathbb{R}^m, \quad i = 1, 2.
\end{aligned}
$$

Under (H7.2.2), for the unique solution $X(\cdot) \equiv X(\cdot\,; 0, x)$ to (7.2.1), it follows from Theorem 4.2.2 in Chap. 4 that BSVIE (7.2.3) admits a unique adapted solution $(Y(\cdot), Z(\cdot\,,\cdot))$ on $[0, T]$. Now, let $\mathcal{P}[0, T]$ be the set of all partitions Π of $[0, T]$ having the following form:

$$
\Pi : \quad 0 = t_0 < t_1 < t_2 < \cdot s < t_{N-1} < t_N = T, \tag{7.2.4}
$$

with some natural number $N > 1$. We define the *mesh size* $\|\Pi\|$ of Π by the following:

$$
\|\Pi\| = \max_{1 \leqslant i \leqslant N} (t_i - t_{i-1}).
$$

For a partition Π as above, let us make an observation. Keep in mind that when we discuss BSVIE (7.2.3), the process $X(\cdot)$ is given. Suppose $(Y(s), Z(t, s))$ has been determined for $t_{k+1} \leqslant t \leqslant s \leqslant T$ (see the region marked ① in the figure below). Then, for $t \in [t_k, t_{k+1})$, one has

$$
\begin{aligned}
Y(t) = \psi(t, X(t), X(T)) &+ \int_{t_{k+1}}^{T} g(t, s, X(t), X(s), Y(s), Z(t, s))ds\\
&+ \int_{t}^{t_{k+1}} g(t, s, X(t), X(s), Y(s), Z(t, s))ds - \int_{t}^{T} Z(t, s)dW(s).
\end{aligned}
$$

In the above, we see that $Y(s)$, $s \in [t_{k+1}, T]$, is known (Fig. 7.1). But, at the moment, $Z(t, s)$ has been determined only for $t_{k+1} \leqslant t \leqslant s \leqslant T$, and it is still unknown for $(t, s) \in [t_k, t_{k+1}) \times [t_{k+1}, T]$ (see the region marked ② in the Fig. 7.1). Also, $(Y(s), Z(t, s))$ is unknown for $t_k \leqslant t \leqslant s \leqslant t_{k+1}$ (the region marked ③ in the Fig. 7.1). Therefore, we are supposed to find $(Y(s), Z(t, s))$ for $t_k \leqslant t \leqslant t_{k+1}$ and $t \leqslant s \leqslant T$ (the regions marked ② and ③in the Fig. 7.1).

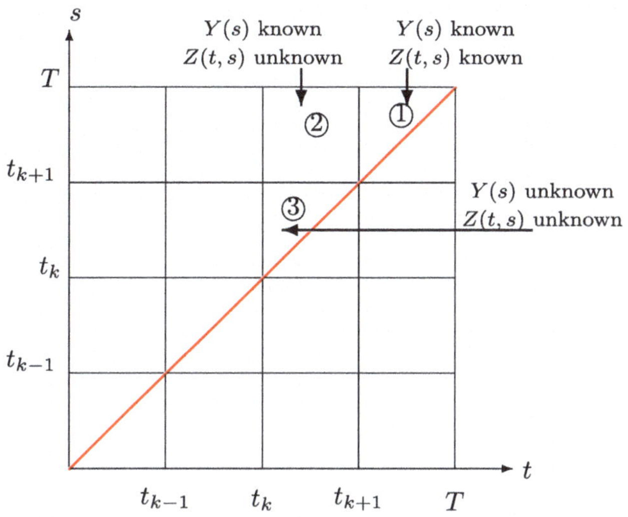

Fig. 7.1 Domains of (Y, Z)

We now construct an approximation of BSVIE (7.2.3). On $[t_{N-1}, T]$, we introduce the following BSDE:

$$\begin{cases} dY^{N-1}(s) = -g\big(t_{N-1}, s, X(t_{N-1}), X(s), Y^{N-1}(s), Z^{N-1}(s)\big)ds \\ \qquad\qquad +Z^{N-1}(s)dW(s), \qquad s \in [t_{N-1}, T], \\ Y^{N-1}(T) = \psi\big(t_{N-1}, X(t_{N-1}), X(T)\big). \end{cases} \tag{7.2.5}$$

Under (H7.2.2), it admits a unique adapted solution $(Y^{N-1}(\cdot), Z^{N-1}(\cdot))$. Next, on $[t_{N-2}, T]$, (not just on $[t_{N-2}, t_{N-1}]$), we introduce the following BSDE:

$$\begin{cases} dY^{N-2}(s) = -g\big(t_{N-2}, s, X(t_{N-2}), X(s), Y^{N-1}(s), Z^{N-2}(s)\big)ds \\ \qquad\qquad +Z^{N-2}(s)dW(s), \quad s \in [t_{N-1}, T), \\ dY^{N-2}(s) = -g\big(t_{N-2}, s, X(t_{N-2}), X(s), Y^{N-2}(s), Z^{N-2}(s)\big)ds \\ \qquad\qquad +Z^{N-2}(s)dW(s), \quad s \in [t_{N-2}, t_{N-1}), \\ Y^{N-2}(T) = \psi\big(t_{N-2}, X(t_{N-2}), X(T)\big), \\ Y^{N-2}(t_{N-1}) = Y^{N-2}(t_{N-1}+0). \end{cases} \tag{7.2.6}$$

In the above, $Y^{N-1}(s)$ is already determined by (7.2.5) on $[t_{N-1}, T]$, which has to stay unchanged. But, since t_{N-2} appears in g and ψ, $(Y^{N-2}(\cdot), Z^{N-2}(\cdot))$ and $(Y^{N-1}(\cdot), Z^{N-1}(\cdot))$ satisfy different BSDEs on $[t_{N-1}, T]$. Consequently, $Z^{N-2}(\cdot)$ is possibly different from $Z^{N-1}(\cdot)$ on $[t_{N-1}, T]$, and therefore, $Y^{N-2}(\cdot)$ could be different from $Y^{N-1}(\cdot)$ on $[t_{N-1}, T]$. Under (H7.2.2), the above BSDE admits a unique adapted solution on $[t_{N-2}, T]$.

In general, on $[t_k, T]$, by (H7.2.2), we have a unique adapted solution $(Y^k(\cdot), Z^k(\cdot))$ to the following BSDE:

$$\begin{cases} dY^k(s) = -g\big(t_k, s, X(t_k), X(s), Y^\ell(s), Z^k(s)\big)ds + Z^k(s)dW(s), \\ \qquad\qquad\qquad\qquad\qquad s \in [t_\ell, t_{\ell+1}), \quad k+1 \leqslant \ell \leqslant N-1, \\ dY^k(s) = -g\big(t_k, s, X(t_k), X(s), Y^k(s), Z^k(s)\big)ds + Z^k(s)dW(s), \\ \qquad\qquad\qquad\qquad\qquad s \in [t_k, t_{k+1}], \\ Y^k(T) = \psi\big(t_k, X(t_k), X(T)\big), \\ Y^k(t_\ell) = Y^k(t_\ell + 0), \quad k+1 \leqslant \ell \leqslant N-1. \end{cases} \tag{7.2.7}$$

Define

$$\begin{cases} Y^\Pi(s) \stackrel{\Delta}{=} \sum\limits_{k=0}^{N-2} Y^k(s) I_{[t_k, t_{k+1})}(s) + Y^{N-1}(s) I_{[t_{N-1}, T]}(s), \quad s \in [0, T], \\ Z^\Pi(t, s) \stackrel{\Delta}{=} \sum\limits_{k=0}^{N-2} Z^k(s) I_{[t_k, t_{k+1})}(t) + Z^{N-1}(s) I_{[t_{N-1}, T]}(t), \quad 0 \leqslant t \leqslant s \leqslant T. \end{cases}$$

The above defined $Y^\Pi(\cdot)$ possibly has discontinuity at $t_{N-1}, t_{N-2}, \cdot s, t_1$. With the above definition, we may rewrite (7.2.7) as

$$\begin{cases} dY^k(s) = -g\big(t_k, s, X(t_k), X(s), Y^\Pi(s), Z^k(s)\big)ds \\ \qquad\qquad\qquad\qquad +Z^k(s)dW(s), \qquad s \in [t_k, T], \\ Y^k(T) = \psi\big(t_k, X(t_k), X(T)\big), \\ Y^k(t_\ell) = Y^k(t_\ell + 0), \quad k+1 \leqslant \ell \leqslant N-1. \end{cases} \tag{7.2.8}$$

Here, we observe that for each $k = N, N-1, N-2, \cdot s, 2, 1$, $Y^k(\cdot)$ is continuous. Now, introduce

$$\tau^\Pi(t) \stackrel{\Delta}{=} \sum_{k=0}^{N-2} t_k I_{[t_k, t_{k+1})}(t) + t_{N-1} I_{[t_{N-1}, T]}(t), \qquad t \in [0, T]. \tag{7.2.9}$$

Then

$$0 \leqslant t - \tau^\Pi(t) \leqslant \|\Pi\|, \qquad t \in [0, T].$$

Hence, for any $t \in [0, T)$, let $t \in [t_k, t_{k+1})$, we have the following:

$$\begin{aligned} Y^\Pi(t) &= Y^k(t) \\ &= \psi\big(t_k, X(t_k), X(T)\big) + \int_t^{t_{k+1}} g\big(t_k, s, X(t_k), X(s), Y^k(s), Z^k(s)\big)ds \end{aligned}$$

$$+\sum_{\ell=k+1}^{N-1}\int_{t_\ell}^{t_{\ell+1}} g\big(t_k, s, X(t_k), X(s), Y^\ell(s), Z^k(s)\big)ds - \int_t^T Z^k(s)dW(s)$$

$$= \psi\big(\tau^\Pi(t), X(t), X(T)\big) + \int_t^T g\big(\tau^\Pi(t), s, X(t), X(s), Y^\Pi(s), Z^\Pi(t,s)\big)ds$$

$$-\int_t^T Z^\Pi(t,s)dW(s).$$

Therefore, $(Y^\Pi(\cdot), Z^\Pi(\cdot\,,\cdot))$ satisfies a BSDE on each (t_k, t_{k+1}), and satisfies a BSVIE on $[0, T]$. Note that since $t \mapsto g(\tau^\Pi(t), s, X(t), x, y, z)$ has possible jumps at t_k, the resulting $t \mapsto Y^\Pi(t)$ may also have jumps at t_k, regardless of its integral form. For the above constructed $(Y^\Pi(\cdot), Z^\Pi(\cdot\,,\cdot))$, we have the following proposition.

Proposition 7.2.1 *Suppose (H7.2.1)–(H7.2.2) hold, $X(\cdot)$ is the solution to the FSDE (7.2.1), $(Y(\cdot), Z(\cdot\,,\cdot))$ is the unique adapted solution of Type-I BSVIE* (7.1), *and $(Y^\Pi(\cdot), Z^\Pi(\cdot\,,\cdot))$ is constructed as above. Then*

$$\mathbb{E}\int_0^T |Y^\Pi(t) - Y(t)|^2 dt + \mathbb{E}\int_0^T\int_t^T |Z^\Pi(t,s) - Z(t,s)|^2 ds dt \leqslant K\|\Pi\|^2, \tag{7.2.10}$$

and for $0 \leqslant k \leqslant N-2$,

$$\begin{aligned}&\mathbb{E}\Big[\sup_{s\in[t_{k+1},T]} |Y^{k+1}(s) - Y^k(s)|^2\Big]\\ &\qquad +\mathbb{E}\Big(\int_{t_{k+1}}^T |Z^{k+1}(s) - Z^k(s)|^2 ds\Big) \leqslant K\|\Pi\|.\end{aligned} \tag{7.2.11}$$

In particular

$$\mathbb{E}|Y^\Pi(t_k) - Y^\Pi(t_k - 0)| \leqslant K\|\Pi\|, \qquad 1 \leqslant k \leqslant N-1.$$

Proof By the stability of adapted solution to BSVIEs (e.g., Theorem 4.2.2 in Chap. 4), we have

$$\begin{aligned}&\mathbb{E}\int_0^T |Y^\Pi(t) - Y(t)|^2 dt + \mathbb{E}\int_0^T\int_t^T |Z^\Pi(t,s) - Z(t,s)|^2 ds dt\\ &\leqslant K\mathbb{E}\int_0^T \Big|\psi\big(\tau^\Pi(t), X(t), X(T)\big) - \psi\big(t, X(t), X(T)\big)\Big|^2 dt\\ &\qquad +\mathbb{E}\int_0^T\int_t^T |g\big(\tau^\Pi(t), s, X(t), X(s), Y(s), Z(t,s)\big)\\ &\qquad\qquad -g\big(t, s, X(t), X(s), Y(s), Z(t,s)\big)|^2 ds dt\\ &\leqslant K\mathbb{E}\Big\{\int_0^T \|\Pi\|^2 dt + \int_0^T\int_t^T \|\Pi\|^2 ds dt\Big\} \leqslant K\|\Pi\|^2.\end{aligned}$$

This proves (7.2.10). Next, one has

$$\begin{cases} d\Big[Y^{k+1}(s) - Y^k(s)\Big] = -\Big[g(t_{k+1}, s, X(t_{k+1}), X(s), Y^{k+1}(s), Z^{k+1}(s)) \\ \qquad\qquad -g(t_k, s, X(t_k), X(s), Y^k(s), Z^k(s))\Big]ds \\ \qquad\qquad +\Big[Z^{k+1}(s) - Z^k(s)\Big]dW(s), \qquad s \in [t_{k+1}, T], \\ \Big[Y^{k+1}(T) - Y^k(T)\Big] = \psi(t_{k+1}, X(t_{k+1}), X(T)) - \psi(t_k, X(t_k), X(T)). \end{cases}$$

Hence

$$\begin{aligned} &\mathbb{E}\Big[\sup_{s\in[t_{k+1},T]} |Y^{k+1}(s) - Y^k(s)|^2\Big] + \mathbb{E}\Big(\int_{t_{k+1}}^T |Z^{k+1}(s) - Z^k(s)|^2 ds\Big) \\ &\leqslant K\mathbb{E}\Big\{|\psi(t_{k+1}, X(t_{k+1}), X(T)) - \psi(t_k, X(t_k), X(T))|^2 \\ &\quad +\Big(\int_{t_{k+1}}^T |g(t_{k+1}, s, X(t_{k+1}), X(s), Y^k(s), Z^k(s)) \\ &\qquad -g(t_k, s, X(t_k), X(s), Y^k(s), Z^k(s))|ds\Big)^2\Big\} \\ &\leqslant K\mathbb{E}\Big\{(t_{k+1} - t_k)^2 + |X(t_{k+1}) - X(t_k)|^2 \\ &\quad + \int_{t_{k+1}}^T \Big(|t_{k+1} - t_k| + |X(t_{k+1}) - X(t_k)|\Big)^2 ds\Big\} \leqslant K\|\Pi\|. \end{aligned}$$

This leads to our conclusion. □

3 Representation of Adapted Solutions for Type-I BSVIEs

In this section, we represent the adapted solution $(Y(\cdot), Z(\cdot\,,\cdot))$ of Type-I BSVIE (7.2.3) in terms of $X(\cdot)$, the solution to FSDE (7.2.1), and the solution to the corresponding representation PDE.

The first key point is to propose the appropriate form of representation PDE. To this end, we make some illustrations. Let $X(\cdot)$ be the solution of FSDE (7.2.1) and $\Pi \in \mathcal{P}[0, T]$ be of form (7.2.4). Let $(Y^k(\cdot), Z^k(\cdot))$ $(0 \leqslant k \leqslant N-1)$ and $(Y^\Pi(\cdot), Z^\Pi(\cdot\,,\cdot))$ be constructed as in Sect. 2. Under proper conditions, they can be represented by $X(\cdot)$ via certain Π-type PDE. By sending $\|\Pi\| \to 0$, the limit form should be the desired form of representation PDE. However, in the current setting, due to the dependence of g, ψ on ξ, there are some difficulties in doing that directly (see below for some explanations). Therefore, instead, we construct a sequence of auxiliary processes $(\bar{Y}^\Pi(\cdot), \bar{Z}^\Pi(\cdot\,,\cdot))$. On the one hand, it is close to $(Y^\Pi(\cdot), Z^\Pi(\cdot\,,\cdot))$. On the other hand, it can be represented by $X(\cdot)$, together with the solution of another new Π-PDE. Then, by sending $\|\Pi\| \to 0$, we shall derive the desired form of representation PDE and furthermore representation of $(Y(\cdot), Z(\cdot\,,\cdot))$ by $X(\cdot)$.

We split the above procedures into several steps.

Step 1. Construction and Representation of $(\bar{Y}^{N-1}(\cdot), \bar{Z}^{N-1}(\cdot))$ Consider the following BSDE:

$$\begin{cases} d\bar{Y}^{N-1}(s) = -g\big(t_{N-1}, s, X(t_{N-1}), X(s), \bar{Y}^{N-1}(s), \bar{Z}^{N-1}(s)\big)ds \\ \qquad\qquad +\bar{Z}^{N-1}(s)dW(s), \quad s \in [t_{N-1}, T], \\ \bar{Y}^{N-1}(T) = \psi\big(t_{N-1}, X(t_{N-1}), X(T)\big), \end{cases} \tag{7.3.1}$$

which coincides with BSDE (7.2.5). Thus, one has

$$(\bar{Y}^{N-1}(s), \bar{Z}^{N-1}(s)) = (Y^{N-1}(s), Z^{N-1}(s)), \qquad t_{N-1} \leqslant t \leqslant s \leqslant T. \tag{7.3.2}$$

Although in (7.3.1), the map

$$(s, x, y, z) \mapsto \big(g(t_{N-1}, s, X(t_{N-1}), x, y, z), \psi(t_{N-1}, X(t_{N-1}), x)\big)$$

is merely $\mathcal{F}_{t_{N-1}}$-measurable, not necessarily deterministic, since we are considering the BSDE on $[t_{N-1}, T]$, we have the following representation:

$$\begin{cases} \bar{Y}^{N-1}(s) = Y^{N-1}(s) = \Theta^{N-1}\big(s, X(t_{N-1}), X(s)\big), \\ \bar{Z}^{N-1}(s) = Z^{N-1}(s) = \Theta_x^{N-1}\big(s, X(t_{N-1}), X(s)\big)\sigma(s, X(s)), \\ \qquad\qquad\qquad\qquad\qquad\qquad s \in [t_{N-1}, T], \end{cases} \tag{7.3.3}$$

if $(s, x) \mapsto \Theta^{N-1}(s, \xi, x)$ is the classical solution to the following representation PDE:

$$\begin{cases} \Theta_s^{N-1}(s, \xi, x) + \frac{1}{2}\sigma(s, x)^\top \Theta_{xx}^{N-1}(s, \xi, x)\sigma(s, x) + \Theta_x^{N-1}(s, \xi, x)b(s, x) \\ \quad +g\big(t_{N-1}, s, \xi, x, \Theta^{N-1}(s, \xi, x), \Theta_x^{N-1}(s, \xi, x)\sigma(s, x)\big) = 0, \\ \qquad\qquad\qquad\qquad\qquad\qquad s \in [t_{N-1}, T], \\ \Theta^{N-1}(T, \xi, x) = \psi(t_{N-1}, \xi, x), \qquad x \in \mathbb{R}^n. \end{cases} \tag{7.3.4}$$

In the above, $\xi \in \mathbb{R}^n$ is treated as a parameter. With the representation (7.3.3), we can rewrite (7.3.1) as follows:

$$\begin{cases} d\bar{Y}^{N-1}(s) \\ = -g\big(t_{N-1}, s, X(t_{N-1}), X(s), \Theta^{N-1}(s, X(t_{N-1}), X(s)), \bar{Z}^{N-1}(s)\big)ds \\ \qquad\qquad +\bar{Z}^{N-1}(s)dW(s), \qquad s \in [t_{N-1}, T), \\ \bar{Y}^{N-1}(T) = \psi\big(t_{N-1}, X(t_{N-1}), X(T)\big). \end{cases} \tag{7.3.5}$$

Step 2. Construction and Representation of $(\bar Y^{N-2}(\cdot),\bar Z^{N-2}(\cdot))$ *(on* $[t_{N-2},T]$*)* On $[t_{N-1},T]$, we let $(\bar Y^{N-2}(\cdot),\bar Z^{N-2}(\cdot))$ be the adapted solution to the following BSDE:

$$\begin{cases} d\bar Y^{N-2}(s)\\ =-g\big(t_{N-2},s,X(t_{N-2}),X(s),\Theta^{N-1}(s,X(s),X(s)),\bar Z^{N-2}(s)\big)ds\\ \qquad\qquad +\bar Z^{N-2}(s)dW(s),\qquad s\in[t_{N-1},T],\\ \bar Y^{N-2}(T)=\psi\big(t_{N-2},X(t_{N-2}),X(T)\big).\end{cases}\tag{7.3.6}$$

Note that on $[t_{N-1},T]$, we have (7.3.2) and representation (7.3.3). Hence, by (7.2.5), we see that $(Y^{N-2}(\cdot),Z^{N-2}(\cdot))$ satisfies the following BSDE:

$$\begin{cases} dY^{N-2}(s)\\ =-g\big(t_{N-2},s,X(t_{N-2}),X(s),\Theta^{N-1}(s,X(t_{N-1}),X(s)),Z^{N-2}(s)\big)ds\\ \qquad\qquad +Z^{N-2}(s)dW(s),\qquad s\in[t_{N-1},T],\\ Y^{N-2}(T)=\psi\big(t_{N-2},X(t_{N-2}),X(T)\big).\end{cases}\tag{7.3.7}$$

We make two comparisons. First, (7.3.5) and (7.3.7) are different: $(t_{N-1},X(t_{N-1}))$ in the former is replaced by $(t_{N-2},X(t_{N-2}))$ in the latter. Second, (7.3.7) and (7.3.6) are different: $\Theta^{N-1}(s,X(t_{N-1}),\ X(s))$ in the former is replaced by $\Theta^{N-1}(s,X(s),X(s))$ in the latter. Note that in (7.3.7), both $X(t_{N-1})$ and $X(t_{N-2})$ appear. This will cause some difficulties in passing to the limit as $\|\Pi\|\to 0$ later on. This is exactly the difficulty that we will encounter if we use $(Y^{\Pi}(\cdot),Z^{\Pi}(\cdot))$ directly to get our representation. On the other hand, since $\|\Pi\|$ will be small, $X(t_{N-1})$ and $X(s)$ will be close (in some sense), for $s\in[t_{N-1},T]$, it should be harmless to replace $\Theta^{N-1}\big(s,X(t_{N-1}),X(s)\big)$ by $\Theta^{N-1}\big(s,X(s),X(s)\big)$ in the drift of the equation on $[t_{N-1},T]$. Let

$$\begin{aligned} g^{N-1}(s,y,z)&\triangleq g\big(t_{N-2},s,X(t_{N-1}),X(s),\Theta^{N-1}(s,X(t_{N-1}),X(s)),z\big),\\ \bar g^{N-1}(s,y,z)&\triangleq g\big(t_{N-2},s,X(t_{N-1}),X(s),y-\Theta^{N-1}(s,X(t_{N-1}),X(s))\\ &\qquad +\Theta^{N-1}(s,X(s),X(s)),z\big).\end{aligned}$$

Then, $(Y^{N-2}(\cdot),Z^{N-2}(\cdot))$ and $(\bar Y^{N-2}(\cdot),\bar Z^{N-2}(\cdot))$ satisfy

$$\begin{cases} dY^{N-2}(s)=-g^{N-1}(s,Y^{N-1}(s),Z^{N-2}(s))ds+Z^{N-2}(s)dW(s),\\ \qquad\qquad\qquad\qquad\qquad\qquad s\in[t_{N-1},T],\\ Y^{N-2}(T)=\psi(t_{N-2},X(t_{N-2}),X(T)),\end{cases}$$

and

$$\begin{cases} d\bar Y^{N-2}(s)=-\bar g^{N-1}(s,\bar Y^{N-1}(s),\bar Z^{N-2}(s))ds+\bar Z^{N-2}(s)dW(s),\\ \qquad\qquad\qquad\qquad\qquad\qquad s\in[t_{N-1},T],\\ \bar Y^{N-2}(T)=\psi(t_{N-2},X(t_{N-2}),X(T)).\end{cases}$$

Thus, if $\|\Theta_\xi^{N-1}\|_\infty^2 < \infty$, by the stability of adapted solutions to BSDEs in Chap. 3, we have

$$\begin{aligned}
&\mathbb{E}\Big[\sup_{s\in[t_{N-1},T]}|\bar{Y}^{N-2}(s)-Y^{N-2}(s)|^2\Big]+\mathbb{E}\int_{t_{N-1}}^T|\bar{Z}^{N-2}(s)-Z^{N-2}(s)|^2ds\\
&\leqslant K_1\mathbb{E}\Big(\int_{t_{N-1}}^T|g^{N-1}(s,\bar{Y}^{N-1}(s),\bar{Z}^{N-2}(s))-\bar{g}^{N-1}(s,\bar{Y}^{N-1}(s),\bar{Z}^{N-2}(s))|ds\Big)^2\\
&\leqslant K_1\mathbb{E}\Big(\int_{t_{N-1}}^T\big|g(t_{N-2},s,X(t_{N-2}),X(s),\Theta^{N-1}(s,X(t_{N-1}),X(s)),\bar{Z}^{N-2}(s))\\
&\qquad -g(t_{N-2},s,X(t_{N-2}),X(s),\Theta^{N-1}(s,X(s),X(s)),Z^{N-2}(s))\big|ds\Big)^2\\
&\leqslant K_1L^2\mathbb{E}\Big(\int_{t_{N-1}}^T|\Theta^{N-1}(s,X(s),X(s))-\Theta^{N-1}(s,X(t_{N-1}),X(s))|ds\Big)^2 \qquad (7.3.8)\\
&\leqslant K_1L^2\|\Theta_\xi^{N-1}\|_\infty^2\mathbb{E}\Big(\int_{t_{N-1}}^T|X(s)-X(t_{N-1})|ds\Big)^2\\
&\leqslant K_1L^2\|\Theta_\xi^{N-1}\|_\infty^2(T-t_{N-1})\int_{t_{N-1}}^T\mathbb{E}|X(s)-X(t_{N-1})|^2ds\\
&\leqslant K_1L^2\|\Theta_\xi^{N-1}\|_\infty^2K_0(T-t_{N-1})\int_{t_{N-1}}^T(s-t_{N-1})ds\\
&\leqslant \frac{K_0K_1L^2\|\Theta_\xi^{N-1}\|_\infty^2\|\Pi\|}{2}(T-t_{N-1})^2.
\end{aligned}$$

In the above, K_0 is the constant appearing in (7.2.2), and K_1 is a constant appearing in the stability estimate for the adapted solution of BSDEs, which can be made independent of the partition Π, under (H7.2.2).

Similar to the previous step, for (7.3.6), we have the following representation:

$$\begin{cases}\bar{Y}^{N-2}(s)=\Theta^{N-2}\big(s,X(t_{N-2}),X(s)\big),\\ \bar{Z}^{N-2}(s)=\Theta_x^{N-2}\big(s,X(t_{N-2}),X(s)\big)\sigma(s,X(s)),\end{cases}\quad s\in[t_{N-1},T], \qquad (7.3.9)$$

if $(s,x)\mapsto\Theta^{N-2}(s,\xi,x)$ is the classical solution to the following representation PDE:

$$\begin{cases}\Theta_s^{N-2}(s,\xi,x)+\frac{1}{2}\sigma(s,x)^\top\Theta_{xx}^{N-2}(s,\xi,x)\sigma(s,x)+\Theta_x^{N-2}(s,\xi,x)b(s,x)\\ \quad+g\big(t_{N-2},s,\xi,x,\Theta^{N-1}(s,x,x),\Theta_x^{N-2}(s,\xi,x)\sigma(s,x)\big)=0,\\ \qquad\qquad\qquad\qquad s\in[t_{N-1},T],\\ \Theta^{N-2}(T,\xi,x)=\psi(t_{N-2},\xi,x),\qquad x\in\mathbb{R}^n.\end{cases}$$

(7.3.10)

Note that equation (7.3.10) is different from (7.3.4), not just because t_{N-1} is replaced by t_{N-2} in g and ψ but also because $\Theta^{N-1}(s,\xi,x)$ is replaced by $\Theta^{N-1}(s,x,x)$ in g. We expect that $\Theta^{N-2}(s,\xi,x)\big|_{\xi=x}$ is close to $\Theta^{N-1}(s,\xi,x)\big|_{\xi=x}$, for $s\in[t_{N-1},T]$, when $t_{N-1}-t_{N-2}>0$ is small.

To complete the construction of $(\bar Y^{N-2}(\cdot),\bar Z^{N-2}(\cdot))$, we introduce the following BSDE on $[t_{N-2},t_{N-1})$:

$$\begin{cases} d\bar Y^{N-2}(s)=-g(t_{N-2},s,X(t_{N-2}),X(s),\bar Y^{N-2}(s),\bar Z^{N-2}(s))ds\\ \qquad\qquad\qquad +\bar Z^{N-2}(s)dW(s),\quad s\in[t_{N-2},t_{N-1}),\\ \bar Y^{N-2}(t_{N-1}-0)=\Theta^{N-2}(t_{N-1},X(t_{N-2}),X(t_{N-1})). \end{cases} \tag{7.3.11}$$

Now, on $[t_{N-2},t_{N-1})$, we have the following representation:

$$\begin{cases} \bar Y^{N-2}(s)=\Theta^{N-2}\big(s,X(t_{N-2}),X(s)\big),\\ \bar Z^{N-2}(s)=\Theta_x^{N-2}\big(s,X\big(t_{N-2}\big),X(s)\big)\sigma(s,X(s)),\end{cases}\quad s\in[t_{N-2},t_{N-1}), \tag{7.3.12}$$

if $(s,x)\mapsto\Theta^{N-2}(s,\xi,x)$ is the classical solution to the following representation PDE:

$$\begin{cases} \Theta_s^{N-2}(s,\xi,x)+\frac12\sigma(s,x)^\top\Theta_{xx}^{N-2}(s,\xi,x)\sigma(s,x)+\Theta_x^{N-2}(s,\xi,x)b(s,x)\\ \quad+g\big(t_{N-2},s,\xi,x,\Theta^{N-2}(s,\xi,x),\Theta_x^{N-2}(s,\xi,x)\sigma(s,x)\big)=0,\\ \qquad\qquad\qquad (s,x)\in[t_{N-2},t_{N-1})\times\mathbb{R}^n,\\ \Theta^{N-2}(t_{N-1},\xi,x)=\Theta^{N-2}(t_{N-1}+0,\xi,x),\qquad x\in\mathbb{R}^n. \end{cases} \tag{7.3.13}$$

Note that unlike (7.3.10), in the above, $\Theta^{N-2}(s,\xi,x)$ appears instead of $\Theta^{N-2}(s,x,x)$ in g. Next, since $(\bar Y^{N-2}(\cdot),\bar Z^{N-2}(\cdot))$ and $(Y^{N-2}(\cdot),Z^{N-2}(\cdot))$ satisfy the same equation on $[t_{N-2},t_{N-1})$ with different terminal conditions $\bar Y^{N-2}(t_{N-1})$ and $Y^{N-2}(t_{N-1})$, we must have, making use of (7.3.8),

$$\begin{aligned}&\mathbb{E}\Big[\sup_{s\in[t_{N-2},t_{N-1})}|\bar Y^{N-2}(s)-Y^{N-2}(s)|^2+\int_{t_{N-2}}^{t_{N-1}}|\bar Z^{N-2}(s)-Z^{N-2}(s)|^2ds\Big]\\ &\leqslant K_0\mathbb{E}\,|\bar Y^{N-2}(t_{N-1})-Y^{N-2}(t_{N-1})|^2\leqslant\frac{K_0^2K_1L^2\|\Theta_\xi^{N-1}\|_\infty^2\|\Pi\|}{2}(T-t_{N-1})^2.\end{aligned}$$

To summarize the above, we have

$$\begin{cases} \bar Y^{N-2}(s)=\Theta^{N-2}\big(s,X(t_{N-2}),X(s)\big),\\ \bar Z^{N-2}(s)=\Theta_x^{N-2}\big(s,X(t_{N-2}),X(s)\big)\sigma(s,X(s)),\end{cases}\quad s\in[t_{N-2},T], \tag{7.3.14}$$

if $(s, x) \mapsto \Theta^{N-2}(s, \xi, x)$ is the classical solution to the following system:

$$\begin{cases} \Theta_s^{N-2}(s,\xi,x)+\frac{1}{2}\sigma(s,x)^\top\Theta_{xx}^{N-2}(s,\xi,x)\sigma(s,x)+\Theta_x^{N-2}(s,\xi,x)b(s,x) \\ \quad +g\big(t_{N-2},s,\xi,x,\Theta^{N-1}(s,x,x),\Theta_x^{N-2}(s,\xi,x)\sigma(s,x)\big)=0, \\ \qquad\qquad\qquad\qquad (s,x)\in[t_{N-1},T]\times\mathbb{R}^n, \\ \Theta_s^{N-2}(s,\xi,x)+\frac{1}{2}\sigma(s,x)^\top\Theta_{xx}^{N-2}(s,\xi,x)\sigma(s,x)+\Theta_x^{N-2}(s,\xi,x)b(s,x) \\ \quad +g\big(t_{N-2},s,\xi,x,\Theta^{N-2}(s,\xi,x),\Theta_x^{N-2}(s,\xi,x)\sigma(s,x)\big)=0, \\ \qquad\qquad\qquad\qquad (s,x)\in[t_{N-2},t_{N-1})\times\mathbb{R}^n, \\ \Theta^{N-2}(T,\xi,x)=\psi(t_{N-2},\xi,x), \\ \Theta^{N-2}(t_{N-1},\xi,x)=\Theta^{N-2}(t_{N-1}+0,\xi,x). \end{cases} \tag{7.3.15}$$

Note that although $s \mapsto \bar{Y}^{N-2}(s)$ could be discontinuous at $s = t_{N-1}$, the function $s \mapsto \Theta^{N-2}(s, \xi, x)$ is continuous. Also, we point out that in the above system, the equations on $[t_{N-1}, T]$ and $[t_{N-2}, t_{N-1})$ are different: $\Theta^{N-1}(s, x, x)$ appears in g for the former and $\Theta^{N-2}(s, \xi, x)$ appears in g for the latter.

Step 3. Construction and Representation of $(\bar{Y}^{N-3}(\cdot), \bar{Z}^{N-3}(\cdot\,,\cdot))$ The above discussion seems not enough to obtain an inductive statement. In particular, we need to make sure that the estimate on the error between $(\bar{Y}^k(\cdot), \bar{Z}^k(\cdot))$ and $(Y^k(\cdot), Z^k(\cdot))$ will not be accumulated. Thus, let us construct $(\bar{Y}^{N-3}(\cdot), \bar{Z}^{N-3}(\cdot))$ on $[t_{N-3}, T]$. To this end, we consider the following BSDE on $[t_{N-2}, T]$:

$$\begin{cases} d\bar{Y}^{N-3}(s)=-g\big(t_{N-3},s,X(t_{N-3}),X(s),\Theta^{N-1}(s,X(s),X(s)), \\ \qquad\qquad \bar{Z}^{N-3}(s)\big)ds+\bar{Z}^{N-3}(s)dW(s), \quad s\in[t_{N-1},T], \\ d\bar{Y}^{N-3}(s)=-g\big(t_{N-3},s,X(t_{N-3}),X(s),\Theta^{N-2}(s,X(s),X(s)), \\ \qquad\qquad \bar{Z}^{N-3}(s)\big)ds+\bar{Z}^{N-3}(s)dW(s), \quad s\in[t_{N-2},t_{N-1}), \\ \bar{Y}^{N-3}(T)=\psi\big(t_{N-3},X(t_{N-3}),X(T)\big), \\ \bar{Y}^{N-3}(t_{N-1}-0)=\Theta^{N-3}\big(t_{N-1},X(t_{N-3}),X(t_{N-1})\big), \end{cases} \tag{7.3.16}$$

if $(s, x) \mapsto \Theta^{N-3}(s, \xi, x)$ is the classical solution to the following system:

$$\begin{cases} \Theta_s^{N-3}(s,\xi,x)+\frac{1}{2}\sigma(s,x)^\top\Theta_{xx}^{N-3}(s,\xi,x)\sigma(s,x)+\Theta_x^{N-3}(s,\xi,x)b(s,x) \\ \quad +g\big(t_{N-3},s,\xi,x,\Theta^{N-1}(s,x,x),\Theta_x^{N-3}(s,\xi,x)\sigma(s,x)\big)=0, \\ \qquad\qquad\qquad\qquad (s,x)\in[t_{N-1},T]\times\mathbb{R}^n, \\ \Theta_s^{N-3}(s,\xi,x)+\frac{1}{2}\sigma(s,x)^\top\Theta_{xx}^{N-3}(s,\xi,x)\sigma(s,x)+\Theta_x^{N-3}(s,\xi,x)b(s,x) \\ \quad +g\big(t_{N-3},s,\xi,x,\Theta^{N-2}(s,x,x),\Theta_x^{N-3}(s,\xi,x)\sigma(s,x)\big)=0, \\ \qquad\qquad\qquad\qquad (s,x)\in[t_{N-2},t_{N-1})\times\mathbb{R}^n, \\ \Theta^{N-3}(T,\xi,x)=\psi(t_{N-3},\xi,x), \\ \Theta^{N-3}(t_{N-1},\xi,x)=\Theta^{N-3}(t_{N-1}+0,\xi,x). \end{cases} \tag{7.3.17}$$

Then, we have the following representation:

$$\begin{cases} \bar{Y}^{N-3}(s) = \Theta^{N-3}(s, X(t_{N-3}), X(s)), \\ \bar{Z}^{N-3}(s) = \Theta_x^{N-3}(s, X(t_{N-3}), X(s))\sigma(s, X(s)), \end{cases} \quad s \in [t_{N-2}, T].$$

By (7.3.2) and (7.3.3), we see that

$$\begin{cases} dY^{N-3}(s) \\ = -g\big(t_{N-3}, s, X(t_{N-3}), X(s), \Theta^{N-1}(s, X(t_{N-1}), X(s)), Z^{N-3}(s)\big)ds \\ \qquad +Z^{N-3}(s)dW(s), \qquad s \in [t_{N-1}, T], \\ dY^{N-3}(s) \\ = -g\big(t_{N-3}, s, X(t_{N-3}), X(s), \Theta^{N-2}(s, X(t_{N-1}), X(s)), Z^{N-3}(s)\big)ds \\ \qquad +Z^{N-3}(s)dW(s), \qquad s \in [t_{N-2}, t_{N-1}), \\ Y^{N-3}(T) = \psi(t_{N-3}, X(t_{N-3}), X(T)). \end{cases}$$

Thus, by the stability of adapted solutions to BSDEs in Chap. 3, one has

$$\begin{aligned} &\mathbb{E}\Big[\sup_{s\in[t_{N-2},T]} |\bar{Y}^{N-3}(s) - Y^{N-3}(s)|^2 + \int_{t_{N-2}}^T |\bar{Z}^{N-3}(s) - Z^{N-3}(s)|^2 ds\Big] \\ &\leqslant K_1 \mathbb{E}\Big(\int_{t_{N-1}}^T \big|g(t_{N-3}, s, X(t_{N-3}), \Theta^{N-1}(s, X(s), X(s)), Z^{N-3}(s)) \\ &\qquad -g\big(t_{N-3}, s, X(t_{N-3}), \Theta^{N-1}(s, X(t_{N-1}), X(s)), Z^{N-3}(s)\big)\big|ds \\ &\qquad + \int_{t_{N-2}}^{t_{N-1}} \big|g(t_{N-3}, s, X(t_{N-3}), \Theta^{N-2}(s, X(s), X(s)), Z^{N-3}(s)) \\ &\qquad -g\big(t_{N-3}, s, X(t_{N-3}), \Theta^{N-2}(s, X(t_{N-2}), X(s)), Z^{N-3}(s)\big)\big|ds\Big)^2 \\ &\leqslant K_1 L^2 \mathbb{E}\Big(\int_{t_{N-1}}^T \big|\Theta^{N-1}(s, X(s), X(s)) - \Theta^{N-1}(s, X(t_{N-1}), X(s))\big|ds \\ &\qquad + \int_{t_{N-2}}^{t_{N-1}} \big|\Theta^{N-2}(s, X(s), X(s)) - \Theta^{N-2}(s, X(t_{N-1}), X(s))\big|ds\Big)^2 \\ &\leqslant K_1 L^2 \mathcal{M}_{1,2}^2 \mathbb{E}\Big(\int_{t_{N-1}}^T \big|X(s) - X(t_{N-1})\big|ds + \int_{t_{N-2}}^{t_{N-1}} \big|X(s) - X(t_{N-2})\big|ds\Big)^2 \\ &\leqslant K_1 L^2 \mathcal{M}_{1,2}^2 (T - t_{N-2}) K_0 \Big(\int_{t_{N-1}}^T (s - t_{N-1})ds + \int_{t_{N-2}}^{t_{N-1}} (s - t_{N-2})ds\Big) \\ &\leqslant \frac{K_0 K_1 L^2 \mathcal{M}_{1,2}^2}{2} (T - t_{N-2})\Big((T - t_{N-1})^2 + (t_{N-1} - t_{N-2})^2\Big) \\ &\leqslant \frac{K_0 K_1 L^2 \mathcal{M}_{1,2}^2}{2} (T - t_{N-2})^2 \|\varPi\|. \end{aligned}$$

Here, $\mathcal{M}_{1,2} \equiv \|\Theta_\xi^{N-1}\|_\infty \vee \|\Theta_\xi^{N-2}\|_\infty$ is supposed to be finite. Next, on $[t_{N-3}, t_{N-2})$, $(\bar{Y}^{N-3}(\cdot), \bar{Z}^{N-3}(\cdot))$ and $(Y^{N-3}(\cdot), Z^{N-3}(\cdot))$ satisfy the same equation with possibly different terminal conditions. Hence, we have

$$\begin{aligned}&\mathbb{E}\Big[\sup_{s\in[t_{N-3},t_{N-2})}|\bar{Y}^{N-3}(s)-Y^{N-3}(s)|^2+\int_{t_{N-3}}^{t_{N-2}}|\bar{Z}^{N-3}(s)-Z^{N-3}(s)|^2ds\Big]\\&\leqslant K_1\mathbb{E}|\bar{Y}^{N-3}(t_{N-2})-Y^{N-3}(t_{N-2})|^2\leqslant\frac{K_0K_1^2L^2\mathcal{M}_{1,2}^2}{2}(T-t_{N-2})^2\|\Pi\|.\end{aligned}$$

Also, we have

$$\begin{cases}\bar{Y}^{N-3}(s)=\Theta^{N-3}(s,X(t_{N-3}),X(s)),\\ \bar{Z}^{N-3}(s)=\Theta_x^{N-3}(s,X(t_{N-3}),X(s))\sigma(s,X(s)),\end{cases}\qquad s\in[t_{N-3},t_{N-2}),$$

if $\Theta^{N-3}(\cdot\,,\cdot)$ is the classical solution of the following system:

$$\begin{cases}\Theta_s^{N-3}(s,\xi,x)+\frac{1}{2}\sigma(s,x)^\top\Theta_{xx}^{N-3}(s,\xi,x)\sigma(s,x)+\Theta_x^{N-3}(s,\xi,x)b(s,x)\\ \quad+g\big(t_{N-3},s,\xi,x,\Theta^{N-3}(s,\xi,x),\Theta_x^{N-3}(s,\xi,x)\sigma(s,x)\big)=0,\\ \qquad\qquad\qquad\qquad\qquad\qquad (s,x)\in[t_{N-3},t_{N-2})\times\mathbb{R}^n,\\ \Theta^{N-3}(t_{N-2},\xi,x)=\Theta^{N-3}(t_{N-2}+0,\xi,x),\qquad x\in\mathbb{R}^n.\end{cases}\tag{7.3.18}$$

Step 4. Induction Now, we look at the general case. For each $k = 0, 1, \cdot s, N-1$, on $[t_k, T]$, we consider the following BSDE:

$$\begin{cases}d\bar{Y}^k(s)=-g\big(t_k,s,X(t_k),X(s),\Theta^\ell(s,X(s),X(s)),\bar{Z}^k(s)\big)ds\\ \qquad\qquad+\bar{Z}^k(s)dW(s),\quad s\in[t_\ell,t_{\ell+1}),\quad k+1\leqslant\ell\leqslant N-1,\\ d\bar{Y}^k(s)=-g\big(t_k,s,X(t_k),X(s),\bar{Y}^k(s),\bar{Z}^k(s)\big)ds\\ \qquad\qquad\qquad\qquad+\bar{Z}^k(s)dW(s),\qquad s\in[t_k,t_{k+1}),\\ \bar{Y}^k(T)=\psi\big(t_k,X(t_k),X(T)\big).\end{cases}\tag{7.3.19}$$

Then, the following representation holds:

$$\begin{cases}\bar{Y}^k(s)=\Theta^k(s,X(t_k),X(s)),\\ \bar{Z}^k(s)=\Theta_x^k\big(s,X(t_k),X(s)\big)\sigma(s,X(s)),\end{cases}\qquad s\in[t_k,T],\tag{7.3.20}$$

where $(s,x)\mapsto\Theta^k(s,\xi,x)$ is the solution to the following PDE:

$$\begin{cases}\Theta^k_s(s,\xi,x)+\frac{1}{2}\sigma(s,x)^\top\Theta^k_{xx}(s,\xi,x)\sigma(s,x)+\Theta^k_x(s,\xi,x)b(s,x)\\ \qquad+g\big(t_k,s,\xi,x,\Theta^\ell(s,x,x),\Theta^k_x(s,\xi,x)\sigma(s,x)\big)=0,\\ \qquad\qquad(s,x)\in[t_\ell,t_{\ell+1})\times\mathbb{R}^n,\quad k+1\leqslant\ell\leqslant N-1,\\ \Theta^k_s(s,\xi,x)+\frac{1}{2}\sigma(s,x)^\top\Theta^k_{xx}(s,\xi,x)\sigma(s,x)+\Theta^k_x(s,\xi,x)b(s,x)\\ \qquad+g\big(t_k,s,\xi,x,\Theta^k(s,\xi,x),\Theta^k_x(s,\xi,x)\sigma(s,x)\big)=0,\\ \qquad\qquad(s,x)\in[t_k,t_{k+1}]\times\mathbb{R}^n,\\ \Theta^k(T,\xi,x)=\psi(t_k,\xi,x),\quad\Theta^k(t_j,\xi,x)=\Theta^k(t_j+0,\xi,x),\\ \qquad\qquad j=N-1,\cdots,k+1,\quad x\in\mathbb{R}^n.\end{cases}\tag{7.3.21}$$

Step 5. Construction of the Auxiliary Sequence on $[0,T]$ We recall the definition of $\tau(t)$ (see (7.2.9)), and define

$$\bar\tau(t)\triangleq\sum_{k=0}^{N-2}t_{k+1}I_{[t_k,t_{k+1})}(t)+t_NI_{[t_{N-1},T]}(t),\qquad t\in[0,T].\tag{7.3.22}$$

Then

$$0\leqslant\bar\tau(t)-t\leqslant\|\Pi\|,\qquad\forall t\in[0,T],$$

and

$$[\tau(t),\bar\tau(t))=[t_k,t_{k+1}),\qquad\forall t\in[t_k,t_{k+1}),\ 0\leqslant k\leqslant N-1.$$

Let

$$\Theta^\Pi(t,s,\xi,x)\triangleq\sum_{k=0}^{N-1}\Theta^k(s,\xi,x)I_{[t_k,t_{k+1})}(t),\qquad 0\leqslant t\leqslant s\leqslant T,\quad x,\xi\in\mathbb{R}^n.$$

For $t\in[t_k,t_{k+1})$, with $k=0,1,\cdot s,N-1$, we have $s\in[\tau(t),T]=[t_k,T]$:

$$\begin{aligned}&\Theta^k(s,\xi,x)=\Theta^\Pi(t,s,\xi,x),\quad&&\Theta^k_s(s,\xi,x)=\Theta^\Pi_s(t,s,\xi,x),\\&\Theta^k_x(s,\xi,x)=\Theta^\Pi_x(t,s,\xi,x),\quad&&\Theta^k_{xx}(s,\xi,x)=\Theta^\Pi_{xx}(t,s,\xi,x),\end{aligned}$$

and for $s\in[\bar\tau(t),T]=[t_{k+1},T]$,

$$\sum_{\ell=k+1}^{N-1}\Theta^\ell(s,x,x)I_{[t_\ell,t_{\ell+1})}(s)=\Theta^\Pi(\tau(s),s,x,x).$$

Then, the above PDE system (7.3.21) can be written as

$$\begin{cases}\Theta^{\Pi}_s(t,s,\xi,x)+\frac{1}{2}\sigma(s,x)^\top\Theta^{\Pi}_{xx}(t,s,\xi,x)\sigma(s,x)+\Theta^{\Pi}_x(t,s,\xi,x)b(s,x)\\ \quad +g\big(\tau(t),s,\xi,x,\Theta^{\Pi}(\tau(s),s,x,x),\Theta^{\Pi}_x(t,s,\xi,x)\sigma(s,x)\big)=0,\\ \qquad\qquad (s,\xi,x)\in[\bar\tau(t),T]\times\mathbb{R}^n\times\mathbb{R}^n,\\ \Theta^{\Pi}_s(t,s,\xi,x)+\frac{1}{2}\sigma(s,x)^\top\Theta^{\Pi}_{xx}(t,s,\xi,x)\sigma(s,x)+\Theta^{\Pi}_x(t,s,\xi,x)b(s,x)\\ \quad +g\big(\tau(t),s,\xi,x,\Theta^{\Pi}(\tau(s),s,\xi,x),\Theta^{\Pi}_x(t,s,\xi,x)\sigma(s,x)\big)=0,\\ \qquad\qquad (s,\xi,x)\in[\tau(t),\bar\tau(t))\times\mathbb{R}^n\times\mathbb{R}^n,\\ \Theta^{\Pi}(\tau(t),T,\xi,x)=\psi(\tau(t),\xi,x),\qquad (t,\xi,x)\in[0,T]\times\mathbb{R}^n\times\mathbb{R}^n.\end{cases}\tag{7.3.23}$$

Also

$$\begin{aligned}&\mathbb{E}\Big[\sup_{s\in[t_{k+1},T]}|\bar Y^k(s)-Y^k(s)|^2+\int_{t_{k+1}}^T|\bar Z^k(s)-Z^k(s)|^2ds\Big]\\ &\leqslant K_1\mathbb{E}\Big(\sum_{\ell=k+1}^{N-1}\int_{t_\ell}^{t_{\ell+1}}\big|g(t_k,s,X(t_k),\Theta^\ell(s,X(s),X(s)),Z^k(s))\\ &\qquad -g\big(t_k,s,X(t_k),\Theta^\ell(s,X(t_\ell),X(s)),Z^k(s)\big)\big|ds\Big)^2\\ &\leqslant K_1L^2\mathbb{E}\Big(\sum_{\ell=k+1}^{N-1}\int_{t_\ell}^{t_{\ell+1}}\big|\Theta^\ell(s,X(s),X(s))-\Theta^\ell(s,X(t_\ell),X(s))\big|ds\Big)^2\\ &\leqslant K_1L^2\mathcal{M}_\Pi^2(T-t_{k+1})\Big(\textstyle\sum_{\ell=k+1}^{N-1}\int_{t_\ell}^{t_{\ell+1}}\mathbb{E}|X(s)-X(t_\ell)|^2ds\Big)\\ &\leqslant K_1L^2\mathcal{M}_\Pi^2(T-t_{k+1})K_1\Big(\textstyle\sum_{\ell=k+1}^{N-1}\int_{t_\ell}^{t_{\ell+1}}(s-t_\ell)ds\Big)\\ &\leqslant \frac{K_0K_1L^2\mathcal{M}_\Pi^2}{2}(T-t_{k+1})^2\|\Pi\|,\end{aligned}$$

and

$$\begin{aligned}&\mathbb{E}\Big[\sup_{s\in[t_k,t_{k+1})}|\bar Y^k(s)-Y^k(s)|^2+\int_{t_k}^{t_{k+1}}|\bar Z^k(s)-Z^k(s)|^2ds\Big]\\ &\leqslant K_1\mathbb{E}|\bar Y^k(t_{k+1})-Y^k(t_{k+1})|^2\leqslant\frac{K_0K_1^2L^2\mathcal{M}_\Pi^2}{2}(T-t_{k+1})^2\|\Pi\|.\end{aligned}$$

Here, $\mathcal{M}_\Pi\equiv\big\|\Theta_\xi^{N-1}\big\|_\infty\vee\big\|\Theta_\xi^{N-2}\big\|_\infty\vee\cdots\vee\big\|\Theta_\xi^{0}\big\|_\infty$ is supposed to be finite. Recall that

$$\begin{cases}Y^\Pi(s)\overset{\Delta}{=}\sum_{k=0}^{N-1}Y^k(s)I_{[t_k,t_{k+1})}(s), & s\in[0,T),\\ Z^\Pi(t,s)\overset{\Delta}{=}\sum_{k=0}^{N-1}Z^k(s)I_{[t_k,t_{k+1})}(t), & 0\leqslant t\leqslant s\leqslant T.\end{cases}$$

We also define

$$\begin{cases} \bar{Y}^{\Pi}(s) \triangleq \sum_{k=0}^{N-1} \Theta^k(s, X(t_k), X(s)) I_{[t_k, t_{k+1})}(s) \\ \qquad\quad = \Theta^{\Pi}\big(\bar{\tau}^{\Pi}(s), s, X(\bar{\tau}^{\Pi}(s)), X(s)\big), \\ \bar{Z}^{\Pi}(t, s) \triangleq \sum_{k=0}^{N-1} \Theta_x^k(s, X(t_k), X(s))\sigma(s, X(s)) I_{[t_k, t_{k+1})}(t) \\ \qquad\quad = \Theta^{\Pi}\big(t, s, X(\bar{\tau}^{\Pi}(t)), X(s)\big)\sigma(s, X(s)). \end{cases}$$

Consequently, for any $s \in [0, T)$, let $s \in [t_k, t_{k+1})$,

$$\begin{aligned} &\mathbb{E}|Y^{\Pi}(s) - \Theta^{\Pi}(s, s, X(s), X(s))|^2 \\ &\leqslant 2\mathbb{E}|Y^{\Pi}(s) - \bar{Y}^{\Pi}(s)|^2 + 2\mathbb{E}|\bar{Y}^{\Pi}(s) - \Theta^{\Pi}(s, s, X(s), X(s))|^2 \\ &= 2\mathbb{E}|Y^k(s) - \bar{Y}^k(s)|^2 + 2\mathbb{E}|\Theta^k(s, X(t_k), X(s)) - \Theta^k(s, X(s), X(s))|^2 \\ &\leqslant K_0 K_1^2 L^2 \mathcal{M}_{\Pi}^2 (T - t_{k+1})^2 \|\Pi\| + 2\mathcal{M}_{\Pi}^2 \mathbb{E}|X(t_k) - X(s)|^2 \\ &\leqslant K_0 K_1^2 L^2 \mathcal{M}_{\Pi}^2 T^2 \|\Pi\| + 2\mathcal{M}_{\Pi}^2 K_0 \|\Pi\| \leqslant K\|\Pi\|. \end{aligned} \tag{7.3.24}$$

Also

$$\begin{aligned} &\mathbb{E}\int_0^T \int_t^T |Z^{\Pi}(t, s) - \Theta_x^{\Pi}(t, s, X(\bar{\tau}^{\Pi}(t)), X(s))\sigma(s, X(s))|^2 ds \\ &\leqslant \sum_{k=0}^{N-1} \mathbb{E}\int_{t_k}^{t_{k+1}} \int_{t_k}^T \Big(|Z^k(s) - \bar{Z}^k(s)|^2 + |\Theta_x^k(s, X(t_k), X(s)) \\ &\qquad\qquad - \Theta_x^k(s, X(s), X(s))|^2\Big) ds dt \\ &\leqslant \sum_{k=0}^{N-1} 2\mathbb{E}\int_{t_k}^{t_{k+1}} K\|\Pi\| dt + \sum_{k=0}^{N-1} 2\mathbb{E}\int_{t_k}^{t_{k+1}} K\|\Pi\| dt \leqslant K\|\Pi\|. \end{aligned} \tag{7.3.25}$$

As $\|\Pi\| \to 0$, the limit form of (7.3.23) should be

$$\begin{cases} \Theta_s(t, s, \xi, x) + \frac{1}{2}\sigma(s, x)^{\top}\Theta_{xx}(t, s, \xi, x)\sigma(s, x) + \Theta_x(t, s, \xi, x)b(s, x) \\ \qquad + g\big(t, s, \xi, x, \Theta(s, s, x, x), \Theta_x(t, s, \xi, x)\sigma(s, x)\big) = 0, \\ \qquad\qquad (t, s, \xi, x) \in \Delta^*[0, T] \times \mathbb{R}^n \times \mathbb{R}^n, \\ \Theta(t, T, \xi, x) = \psi(t, \xi, x), \qquad (t, \xi, x) \in [0, T] \times \mathbb{R}^n \times \mathbb{R}^n. \end{cases} \tag{7.3.26}$$

Also by (7.3.24), (7.3.25), as well as Proposition 7.2.1, it is expected that

$$\begin{cases} Y(s) = \Theta(s, s, X(s), X(s)), & s \in [0, T], \\ Z(t, s) = \Theta_x(t, s, X(t), X(s))\sigma(s, X(s)), & (t, s) \in \Delta^*[0, T]. \end{cases} \tag{7.3.27}$$

This gives our desired representation. Next, we prove the result directly. More precisely, we have the following result.

Theorem 7.3.1 *Let* (H7.2.1)–(H7.2.2) *hold. Let* $\Theta : \Delta^*[0,T] \times \mathbb{R}^n \times \mathbb{R}^n \to \mathbb{R}^m$ *be the classical solution of representation PDE* (7.3.26)*. Let* $(Y(\cdot), Z(\cdot\,,\cdot))$ *be the adapted solution to the Type-I BSVIE (7.1) with* $X(\cdot)$ *being the solution to SDE (7.2.1). Then, we have* (7.3.27)*.*

Proof Fix $t \in [0,T)$, by Itô's formula to $s \mapsto \Theta(t,s,X(t),X(s))$ on $[t,T]$, we have

$$\begin{aligned} &d\Theta(t,s,X(t),X(s)) \\ &= \Theta_s(t,s,X(t),X(s))ds + \Theta_x(t,s,X(t),X(s))b(s,X(s))ds \\ &\quad +\Theta_x(t,s,X(t),X(s))\sigma(s,X(s))dW(s) \\ &\quad +\frac{1}{2}\sigma(s,X(s))^\top \Theta_{xx}(t,s,X(t),X(s))\sigma(s,X(s))ds. \end{aligned} \tag{7.3.28}$$

Since $\Theta(\cdot\,,\cdot\,,\cdot\,,\cdot)$ satisfies PDE (7.3.26), one has

$$\begin{aligned} &d\Theta(t,s,X(t),X(s)) \\ &= -g\Big(t,s,X(t),X(s),\Theta(s,s,X(s),X(s)),\Theta_x\big(t,s,X(t),X(s)\sigma(s,X(s))\big)\Big)ds \\ &\quad +\Theta_x(t,s,X(t),X(s))\sigma(s,X(s))dW(s), \end{aligned}$$

and $\Theta(t,T,X(t),X(T)) = \psi(t,X(t),X(T))$. Now, for $s \geqslant t$, we define

$$\begin{cases} \lambda(t,s) \triangleq \Theta(t,s,X(t),X(s)), \\ Z(t,s) \triangleq \Theta_x(t,s,X(t),X(s))\sigma(s,X(s)). \end{cases}$$

Then

$$\begin{aligned} \lambda(t,s) = \psi(t,X(t),X(T)) &+ \int_s^T g\big(t,r,X(t),X(r),\lambda(r,r),Z(t,r)\big)dr \\ &- \int_s^T Z(t,r)dW(r), \quad 0 \leqslant t \leqslant s \leqslant T. \end{aligned}$$

Let $t = s$ and $Y(s) \equiv \lambda(s,s)$, we then see that $(Y(\cdot), Z(\cdot\,,\cdot))$ satisfies BSVIE (7.1) and desired representation is obtained.

Here and after, for the classical solution of PDEs, both the differentiability and boundedness are required properly.

4 Representation of Adapted M-Solutions for Type-II BSVIEs

In this section, we are going to establish a representation of adapted M-solutions for Type-II BSVIE (7.3), where both $Z(t,s)$ and $Z(s,t)$ appear in the drift. We still have that $X(\cdot)$ is the solution to FSDE (7.2.1). We first present the following result which is interesting itself.

Proposition 7.4.1 *Let $\Lambda : [0,T]\times\mathbb{R}^n \to \mathbb{R}^m$ be continuous such that $\Lambda(\cdot, X(\cdot)) \in L^2_{\mathbb{F}}(0,T;\mathbb{R}^m)$ with $X(\cdot)$ satisfying (7.2.1). Let (H7.2.1) hold. Let the following PDE system admits a classical solution $\Gamma(\cdot\,,\cdot\,,\cdot)$:*

$$\begin{cases} \Gamma_s(t,s,x) + \Gamma_x(t,s,x)b(s,x) + \dfrac{1}{2}\sigma(s,x)^\top \Gamma_{xx}(t,s,x)\sigma(s,x) = 0, \\ \qquad\qquad\qquad\qquad\qquad\qquad 0 \leqslant s \leqslant t \leqslant T, \quad x \in \mathbb{R}^n, \\ \Gamma(t,t,x) = \Lambda(t,x), \qquad (t,x) \in [0,T]\times\mathbb{R}^n, \end{cases} \tag{7.4.1}$$

where the meaning of $\sigma(s,x)^\top \Gamma_{xx}(t,s,x)\sigma(s,x)$ is similar to (7.1.5), and Γ_x, Γ_{xx} are bounded. Then

$$\Lambda(t,X(t)) = \mathbb{E}\Lambda(t,X(t)) + \int_0^t \Gamma_x(t,s,X(s))\sigma(s,X(s))dW(s), \qquad t \in [0,T]. \tag{7.4.2}$$

Proof Fix any $t \in (0,T]$. By martingale representation theorem (Theorem 2.3.8, Chap. 2), one has

$$\Lambda(t,X(t)) = \mathbb{E}\Lambda(t,X(t)) + \int_0^t \zeta(t,s)dW(s), \qquad t \in [0,T], \tag{7.4.3}$$

for some $\zeta(t,\cdot) \in L^p_{\mathbb{F}}(\Omega; L^2(0,t;\mathbb{R}^m))$. We want to represent $\zeta(t,s)$ for $(t,s) \in \Delta_*[0,T]$ in terms of $X(\cdot)$. To this end, for any $t \in [0,T)$, we introduce the following (decoupled) FBSDE:

$$\begin{cases} dX(s) = b(s,X(s))ds + \sigma(s,X(s))dW(s), \qquad s \in [0,t], \\ d\eta(t,s) = \zeta(t,s)dW(s), \qquad s \in [0,t], \\ \eta(t,t) = \Lambda(t,X(t)), \end{cases} \tag{7.4.4}$$

where $t \in [0,T)$ is a parameter. We let

$$\eta(t,s) \stackrel{\Delta}{=} \Gamma(t,s,X(s)), \qquad s \in [0,t].$$

Then, at $s = t$, it is necessary that

$$\eta(t,t) = \Gamma(t,t,X(t)) = \Lambda(t,X(t)).$$

By Itô's formula, we have

$$
\begin{aligned}
\zeta(t,s)dW(s) = d\eta(t,s) = \Big[&\Gamma_s(t,s,X(s)) + \Gamma_x(t,s,X(s))b(s,X(s)) \\
&+\frac{1}{2}\sigma(s,X(s))^\top \Gamma_{xx}(t,s,X(s))\sigma(s,X(s))\Big]ds \\
&+\Gamma_x(t,s,X(s))\sigma(s,X(s))dW(s) \\
=\ &\Gamma_x(t,s,X(s))\sigma(s,X(s))dW(s).
\end{aligned}
$$

Hence, we have

$$
\zeta(t,s) = \Gamma_x(t,s,X(s))\sigma(s,X(s)), \qquad 0 \leqslant s \leqslant t,
$$

proving (7.4.2). □

From the above, we see that when $(t,s) \mapsto \Gamma_x(t,s,x)$ and $s \mapsto \sigma(s,x)$ are continuous, the map $(t,s) \mapsto \zeta(t,s)$ is continuous.

Now, we consider Type-II BSVIE. Let $(Y(\cdot), Z(\cdot\,,\cdot))$ be the adapted M-solution. Suppose

$$
Y(t) = \Lambda(t, X(t)), \qquad t \in [0,T],
$$

for some undetermined continuous function $\Lambda(\cdot\,,\cdot)$. By Proposition 7.4.1, we have

$$
Z(t,s) = \Gamma_x(t,s,X(s))\sigma(s,X(s)), \qquad 0 \leqslant s \leqslant t \leqslant T.
$$

Thus

$$
Z(s,t) = \Gamma_x(s,t,X(t))\sigma(t,X(t)), \qquad 0 \leqslant t \leqslant s \leqslant T.
$$

Then, we consider the following Type-I BSVIE:

$$
\begin{aligned}
Y(t) = \ &\psi(t,X(t),X(T)) \\
&+\int_t^T g\big(t,s,X(t),X(s),Y(s),Z(t,s),\Gamma_x(s,t,X(t))\sigma(s,X(s))\big)ds \\
&\quad -\int_t^T Z(t,s)dW(s), \qquad t \in [0,T].
\end{aligned}
\tag{7.4.5}
$$

If we let

$$
\widetilde{g}(t,s,\xi,x,y,z) \stackrel{\Delta}{=} g\big(t,s,\xi,x,y,z,\Gamma_x(s,t,\xi)\sigma(s,x)\big),
$$

then, (7.4.5) becomes

$$
\begin{aligned}
Y(t) = \psi(t,X(t),X(T)) + \int_t^T &\widetilde{g}\big(t,s,X(t),X(s),Y(s),Z(t,s)\big)ds \\
&-\int_t^T Z(t,s)dW(s), \qquad t \in [0,T].
\end{aligned}
\tag{7.4.6}
$$

Now, inspired by the study in the previous section, it is expected to have the following representation:

$$\begin{cases} Y(s) = \Theta(s, s, X(s), X(s)), & s \in [0, T], \\ Z(t, s) = \Theta_x(t, s, X(t), X(s))\sigma(s, X(s)), & 0 \leqslant t \leqslant s \leqslant T, \\ Z(t, s) = \Gamma_x(t, s, X(s))\sigma(s, X(s)), & 0 \leqslant s \leqslant t \leqslant T, \end{cases} \tag{7.4.7}$$

with

$$\begin{cases} \Gamma_s(t, s, x) + \Gamma_x(t, s, x)b(s, x) + \frac{1}{2}\sigma(s, x)^\top \Gamma_{xx}(t, s, x)\sigma(s, x) = 0, \\ \qquad\qquad 0 \leqslant s \leqslant t \leqslant T,\ x \in \mathbb{R}^n, \\ \Theta_s(t, s, \xi, x) + \frac{1}{2}\sigma(s, x)^\top \Theta_{xx}(t, s, \xi, x)\sigma(s, x) + \Theta_x(t, s, \xi, x)b(s, x) \\ +g(t, s, \xi, x, \Theta(t, s, x, x), \Theta_x(t, s, \xi, x)\sigma(s, x), \Gamma_x(s, t, \xi)\sigma(s, x)\big) = 0, \\ \qquad\qquad 0 \leqslant t \leqslant s \leqslant T, \quad x, \xi \in \mathbb{R}^n, \\ \Gamma(t, t, x) = \Theta(t, t, x, x), \qquad (t, x) \in [0, T] \times \mathbb{R}^n, \\ \Theta(t, T, \xi, x) = \psi(t, \xi, x), \qquad (t, \xi, x) \in [0, T] \times \mathbb{R}^n \times \mathbb{R}^n. \end{cases} \tag{7.4.8}$$

We now state the representation result as follows.

Theorem 7.4.2 *Let (H7.2.1)–(H7.2.2) hold, and Θ, Γ be the (bounded) classical solution of representation PDE (7.4.8). Let $(Y(\cdot), Z(\cdot\,, \cdot))$ be the adapted M-solution to Type-II BSVIE (7.3) with $X(\cdot)$ being the solution to SDE (7.2.1). Then, representation (7.4.7) holds.*

Proof For given $t \in [0, T)$, applying Itô's formula to $s \mapsto \Theta(t, s, X(t), X(s))$ on $[t, T]$, one has

$$\begin{aligned} &d\Theta(t, s, X(t), X(s)) \\ &= \Theta_s(t, s, X(t), X(s))ds + \Theta_x(t, s, X(t), X(s))b(s, X(s))ds \\ &\quad +\Theta_x(t, s, X(t), X(s))\sigma(s, X(s))dW(s) \\ &\quad +\frac{1}{2}\sigma(s, X(s))^\top \Theta_{xx}(t, s, X(t), X(s))\sigma(s, X(s))ds. \end{aligned}$$

Since Θ satisfies the second PDE of (7.4.8), one has

$$\begin{aligned} &d\Theta\big(t, s, X(t), X(s)\big) \\ &= \Theta_x\big(t, s, X(t), X(s)\big)\sigma\big(s, X(s)\big)dW(s) \\ &\quad -g\big(t, s, X(t), X(s), \Theta(s, s, X(s), X(s)), \Theta_x(t, s, X(t), X(s)\sigma(s, X(s)), \\ &\qquad \Gamma(s, t, X(t))\sigma(s, X(s))\big)ds, \end{aligned}$$

and

$$\Theta(t, T, X(t), X(T)) = \psi(t, X(t), X(T)).$$

For $s \geqslant t$, we set

$$\lambda(t,s) \triangleq \Theta(t,s,X(t),X(s)), \quad \bar{Z}(t,s) \triangleq \Theta_x(t,s,X(t),X(s))\sigma(s,X(s)).$$

Then

$$\begin{aligned}\lambda(t,s) = \psi(t,X(t),X(T)) + \int_s^T g\big(t,r,X(t),X(r),\lambda(r,r),\bar{Z}(t,r),\\ \Gamma(r,t,X(t))\sigma(r,X(r))\big)dr - \int_s^T \bar{Z}(t,r)dW(r).\end{aligned}$$

Let $t = s$ and $\bar{Y}(s) \triangleq \lambda(s,s)$, we then obtain

$$\begin{aligned}\bar{Y}(t) = \psi(t,X(t),X(T)) + \int_t^T g\big(t,r,X(t),X(r),\lambda(r,r),\bar{Z}(t,r),\\ \Gamma(r,t,X(t))\sigma(r,X(r))\big)dr - \int_t^T \bar{Z}(t,r)dW(r).\end{aligned} \tag{7.4.9}$$

Note that

$$\bar{Y}(t) = \Theta(t,t,X(t),X(t)) = \Gamma(t,t,X(t))$$

where Γ satisfies the first PDE in (7.4.8). By Proposition 7.4.1, we know that

$$\bar{Y}(t) = \mathbb{E}\bar{Y}(t) + \int_0^t \Gamma_x(t,s,X(s))\sigma(s,X(s))dW(s), \quad t \geqslant s.$$

Consequently, by defining $\bar{Z}(t,s) \triangleq \Gamma_x(t,s,X(s))\sigma(s,X(s))$ with $t \geqslant s$, we can rewrite above BSVIE as

$$\begin{aligned}\bar{Y}(t) = \psi(t,X(t),X(T)) + \int_t^T g\big(t,r,X(t),X(r),\lambda(r,r),\bar{Z}(t,r),\bar{Z}(r,t)\big)dr\\ - \int_t^T \bar{Z}(t,r)dW(r).\end{aligned}$$

The conclusion then follows easily from the uniqueness of BSVIEs. □

5 Well-Posedness of the Representation PDEs

In this section, we establish the well-posedness of the representation PDEs (7.3.26) and (7.4.8), in certain sense. Let us first look at the representation PDE (7.3.26) for Type-I BSVIEs, which is recalled here, for convenience:

$$\begin{cases} \Theta_s(t,s,\xi,x)+\frac{1}{2}\sigma(s,x)^\top\Theta_{xx}(t,s,\xi,x)\sigma(s,x)+\Theta_x(t,s,\xi,x)b(s,x) \\ \qquad +g(t,s,\xi,x,\Theta(s,s,x,x),\Theta_x(t,s,\xi,x)\sigma(s,x)\big)=0, \\ \qquad\qquad (t,s,\xi,x)\in\Delta^*[0,T]\times\mathbb{R}^n\times\mathbb{R}^n, \\ \Theta(t,T,\xi,x)=\psi(t,\xi,x), \qquad (t,\xi,x)\in[0,T]\times\mathbb{R}^n\times\mathbb{R}^n. \end{cases} \tag{7.5.1}$$

If we denote $\Theta \overset{\Delta}{=} (\Theta^1, \cdot s, \Theta^m)$ and

$$\frac{1}{2}\sigma(s,x)\sigma(s,x)^\top \overset{\Delta}{=} a(s,x)=\big(a_{ij}(s,x)\big),$$
$$b(s,x)\overset{\Delta}{=}(b_1(s,x),\cdot s,b_n(s,x))^\top,$$

then (7.5.1) can be rewritten as the following system (parameterized by $(t,\xi)\in[0,T)\times\mathbb{R}^n$):

$$\begin{cases} \Theta_s^k(t,s,\xi,x)+\sum_{i,j=1}^n a_{ij}(s,x)\Theta^k_{x_ix_j}(t,s,\xi,x)+\sum_{i=1}^n b_i(s,x)\Theta^k_{x_i}(t,s,\xi,x) \\ \quad +g^k(t,s,\xi,x,\Theta(s,s,x,x),\Theta_x(t,s,\xi,x)\sigma(s,x)\big)=0, \\ \qquad\qquad (s,x)\in[t,T]\times\mathbb{R}^n, \\ \Theta^k(t,T,\xi,x)=\psi^k(t,\xi,x), \qquad x\in\mathbb{R}^n, \qquad 1\leqslant k\leqslant m, \end{cases} \tag{7.5.2}$$

which is a quasilinear parabolic system for unknown functions $\Theta^1, \cdot s, \Theta^m$, with the same leading part for each equation. Eventually, we refer to Sect. 3 of Chap. 12 for more notational definitions and relevant conclusions.

5.1 The First Representation PDE

Now, let us return to the first representation PDE (7.5.1). We impose the following further assumption.

(H7.5.1) The maps $(s,x)\mapsto b(s,x)$, $\sigma(s,x)$, $\psi(t,\xi,x)$, and $g(t,s,\xi,x,y,z)$ are bounded and have all required differentiability with bounded derivatives. Moreover, there exists a constant $\bar\sigma>0$ such that

$$|\sigma(s,x)\xi|\geqslant\bar\sigma|\xi|^2, \qquad \forall(s,x,\xi)\in[0,T]\times\mathbb{R}^n\times\mathbb{R}^n. \tag{7.5.3}$$

The above assumption is much more than enough. However, in this paper, we prefer not to get into the most generality in this aspect, to reduce the complexity of presentation. Also, we will extend $(t,s)\mapsto g(t,s,\xi,x,y,z)$ from $\Delta^*[0,T]$ to $[0,T]^2$ by letting

$$g(t,s,\xi,x,y,z)\overset{\Delta}{=}g(s,t,\xi,x,y,z), \qquad \forall t,s\in[0,T].$$

We now state the following theorem.

Theorem 7.5.1 *Let (H7.5.1) hold. Then, for any* $(t,\xi)\in[0,T)\times\mathbb{R}^n$, *system* (7.5.1) *admits a unique classical solution* $\Theta(t,\cdot,\xi,\cdot)\in C^{1+\frac{\alpha}{2},2+\alpha}([t,T]\times\mathbb{R}^n)$, *and the following holds:*

$$\sup_{(t,\xi)\in[0,T]\times\mathbb{R}^n}|\Theta(t,\cdot,\xi,\cdot)|^{(2+\alpha)}_{[t,T]\times\mathbb{R}^n}\leqslant K\Big(1+\sup_{(t,\xi)\in[0,T]\times\mathbb{R}^n}|\psi(t,\xi,\cdot)|^{(2+\alpha)}_{\mathbb{R}^n}\Big). \tag{7.5.4}$$

Moreover, if $\widehat\Theta(t,\cdot,\xi,\cdot)$ *is the solution to the system* (7.5.1) *with the pair* (ψ,g) *replaced by* $(\widehat\psi,\widehat g)$ *that also satisfies (H7.5.1), then the following stability estimate holds:*

$$\begin{aligned}&\sup_{(t,\xi)\in[0,T]\times\mathbb{R}^n}|\Theta(t,\cdot,\xi,\cdot)-\widehat\Theta(t,\cdot,\xi,\cdot)|^{(2+\alpha)}_{[t,T]\times\mathbb{R}^n}\\&\leqslant K\sup_{(t,\xi)\in[0,T]\times\mathbb{R}^n}\Big(|\psi(t,\xi,\cdot)-\widehat\psi(t,\xi,\cdot)|^{(2+\alpha)}_{\mathbb{R}^n}\\&\qquad+|g(t,\cdot,\xi,\cdot)-\widehat g(t,\cdot,\xi,\cdot)|^{(\alpha)}_{[t,T]\times\mathbb{R}^n)}\Big),\quad\forall(t,\xi)\in[0,T)\times\mathbb{R}^n,\end{aligned} \tag{7.5.5}$$

where for $(t,s,\xi,x)\in\Delta^*[0,T]\times\mathbb{R}^n\times\mathbb{R}^n$,

$$\begin{aligned}g(t,s,\xi,x)&\stackrel{\Delta}{=}g\big(t,s,\xi,x,\Theta(s,s,x,x),\Theta_x(t,s,\xi,x)\sigma(s,x)\big),\\\widehat g(t,s,\xi,x)&\stackrel{\Delta}{=}\widehat g\big(t,x,\xi,x,\Theta(s,s,x,x),\Theta_x(t,s,\xi,x)\sigma(s,x)\big).\end{aligned} \tag{7.5.6}$$

Proof We split the proof into several steps.
Step 1. A Reduction. First, we let

$$\widetilde\Theta(t,s,\xi,x)\stackrel{\Delta}{=}\Theta(t,s,\xi,x)-\psi(t,\xi,x),\qquad(t,s)\in\Delta^*[0,T],\ x,\xi\in\mathbb{R}^n.$$

Then, $\Theta(\cdot,\cdot,\cdot,\cdot)$ is a solution of (7.5.1) if and only if $\widetilde\Theta(\cdot,\cdot,\cdot,\cdot)$ is a solution to the following:

$$\begin{cases}\widetilde\Theta_s(t,s,\xi,x)+\frac12\sigma(s,x)^\top\widetilde\Theta_{xx}(t,s,\xi,x)\sigma(s,x)+\widetilde\Theta_x(t,s,\xi,x)b(s,x)\\+\widetilde g(t,s,\xi,x,\widetilde\Theta(s,s,x,x),\widetilde\Theta_x(t,s,\xi,x)\sigma(s,x)\big)=0,\ (s,x)\in[t,T]\times\mathbb{R}^n,\\\widetilde\Theta(t,T,\xi,x)=0,\qquad x\in\mathbb{R}^n,\end{cases} \tag{7.5.7}$$

where

$$\begin{aligned}\widetilde g(t,s,\xi,x,y,z)&=g\big(t,s,\xi,x,y+\psi(t,\xi,x),\psi_x(t,\xi,x)\sigma(s,x)+z\sigma(s,x)\big)\\&\quad+\frac12\sigma(s,x)^\top\psi_{xx}(t,\xi,x)\sigma(s,x)+\psi_x(t,\xi,x)b(s,x),\qquad(s,x)\in[t,T]\times\mathbb{R}^n.\end{aligned}$$

Hence, without loss of generality, we may consider (7.5.1) with $\psi(t, \xi, x) \equiv 0$.
Step 2. The Solution Map of a Parabolic PDE. Let (see Fig. 7.2 below)

$$\begin{aligned}\widehat{\Delta}[S, T] &\stackrel{\Delta}{=} \Big([0, S] \times [S, T]\Big) \bigcup \Delta^*[S, T] \\ &\equiv \Big\{(t, s) \in [0, T]^2 \mid 0 \leqslant t \vee S \leqslant s \leqslant T\Big\}.\end{aligned} \tag{7.5.8}$$

Let $\mathscr{X}[S, T]$ be the set of all measurable functions $\theta : \widehat{\Delta}[S, T] \times \mathbb{R}^{2n} \to \mathbb{R}^m$ such that

$$\begin{aligned}&\|\theta\|_{\mathscr{X}[S,T]} \\ &\equiv \sup_{(t,\xi)\in[0,T]\times\mathbb{R}^n} \Big(|\theta(t,\cdot,\xi,\cdot)|^{(1+\alpha)}_{[t\vee S,T]\times\mathbb{R}^n} + |\theta_t(t,\cdot,\xi,\cdot)|^{(0)}_{[t\vee S,T]\times\mathbb{R}^n} \\ &\qquad + |\theta_\xi(t,\cdot,\xi,\cdot)|^{(0)}_{[t\vee S,T]\times\mathbb{R}^n} + |\theta_{xt}(t,\cdot,\xi,\cdot)|^{(0)}_{[t\vee S,T]\times\mathbb{R}^n} \\ &\qquad + |\theta_{x\xi}(t,\cdot,\xi,\cdot)|^{(0)}_{[t\vee S,T]\times\mathbb{R}^n}\Big) \\ &\equiv \sup_{(t,\xi)\in[0,T]\times\mathbb{R}^n} \Big(|\theta(t,\cdot,\xi,\cdot)|^{(0)}_{[t\vee S,T]\times\mathbb{R}^n} + |\theta_\eta(t,\cdot,\xi,\cdot)|^{(0)}_{[t\vee S,T]\times\mathbb{R}^n} \\ &\qquad + \langle\theta_\eta(t,\cdot,\xi,\cdot)\rangle^{(\alpha)}_{[t\vee S,T]\times\mathbb{R}^n} + \langle\theta(t,\cdot,\xi,\cdot)\rangle^{(\frac{1+\alpha}{2})}_{s,[t\vee S,T]\times\mathbb{R}^n} \\ &\qquad + |\theta_t(t,\cdot,\xi,\cdot)|^{(0)}_{[t\vee S,T]\times\mathbb{R}^n} + |\theta_\xi(t,\cdot,\xi,\cdot)|^{(0)}_{[t\vee S,T]\times\mathbb{R}^n} \\ &\qquad + |\theta_{xt}(t,\cdot,\xi,\cdot)|^{(0)}_{[t\vee S,T]\times\mathbb{R}^n} + |\theta_{x\xi}(t,\cdot,\xi,\cdot)|^{(0)}_{[t\vee S,T]\times\mathbb{R}^n}\Big) < \infty.\end{aligned} \tag{7.5.9}$$

Clearly, $\|\cdot\|_{\mathscr{X}[S,T]}$ is a norm under which $\mathscr{X}[S, T]$ is a Banach space (Fig. 7.2).

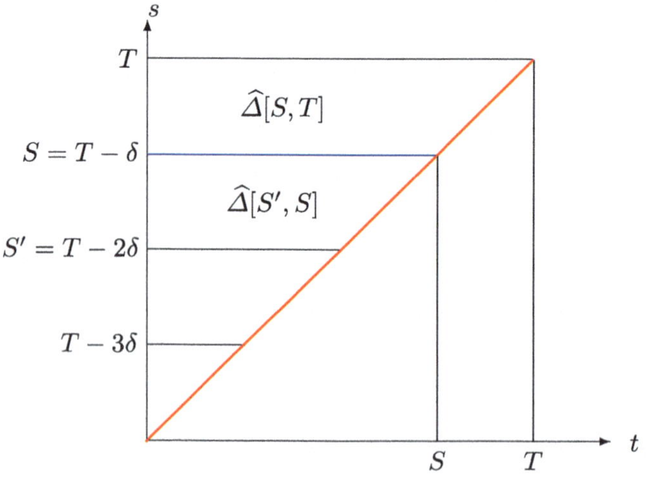

Fig. 7.2 Domain of double time variable

Let $S \in [0, T)$ be fixed. For any $\theta(\cdot\,,\cdot\,,\cdot\,,\cdot) \in \mathscr{X}[S, T]$, denote

$$g(t, \tau, \xi, \eta) \overset{\Delta}{=} g\big(t, \tau, \xi, \eta, \theta(\tau, \tau, \eta, \eta), \theta_\eta(t, \tau, \xi, \eta)\sigma(\tau, \eta)\big). \tag{7.5.10}$$

We claim that

$$|g(t, \cdot\,, \xi, \cdot)|^{(\alpha)}_{[t\vee S,T]\times\mathbb{R}^n} \leqslant K\big(1+\|\theta\|_{\mathscr{X}[S,T]}\big), \quad \forall (t, \xi) \in [S, T]\times\mathbb{R}^n. \tag{7.5.11}$$

Let us prove this. By boundedness of g, one has

$$\big|g\big(t, \tau, \xi, \eta, \theta(\tau, \tau, \eta, \eta), \theta_\eta(t, \tau, \xi, \eta)\sigma(\tau, \eta)\big)\big| \leqslant K. \tag{7.5.12}$$

Next, for $\tau, \tau' \in [t \vee S, T]$ and $\xi, \eta \in \mathbb{R}^n$,

$$\begin{aligned}
&\big|g(t, \tau, \xi, \eta, \theta(\tau, \tau, \eta, \eta), \theta_\eta(t, \tau, \xi, \eta)\sigma(\tau, \eta))\\
&\quad -g(t, \tau', \xi, \eta, \theta(\tau', \tau', \eta, \eta), \theta_\eta(t, \tau', \xi, \eta)\sigma(\tau', \eta))\big|\\
&\leqslant K\Big(|\tau-\tau'| + |\theta(\tau, \tau, \eta, \eta) - \theta(\tau', \tau', \eta, \eta)|\\
&\quad +|\theta_\eta(t, \tau, \xi, \eta)\sigma(\tau, \eta) - \theta_\eta(t, \tau', \xi, \eta)\sigma(\tau', \eta)|\Big)\\
&\leqslant K\Big(|\tau-\tau'| + \sup_{(\bar t,\bar\xi)\in[t\vee S,T]\times\mathbb{R}^n} |\theta_t(\bar t, \cdot\,, \bar\xi, \cdot)|^{(0)}_{[t\vee S,T]\times\mathbb{R}^n}|\tau-\tau'|\\
&\quad + \sup_{(\bar t,\bar\xi)\in[t\vee S,T]\times\mathbb{R}^n} \langle\theta(\bar t, \cdot\,, \bar\xi, \cdot)\rangle^{(\frac{\alpha}{2})}_{\tau,[t\vee S,T]\times\mathbb{R}^n}|\tau-\tau'|^{\frac{\alpha}{2}}\\
&\quad +|\theta_\eta(t, \cdot\,, \xi, \cdot)|^{(0)}_{[t\vee S,T]\times\mathbb{R}^n}|\tau-\tau'| + \langle\theta_\eta(t, \cdot\,, \xi, \cdot)\rangle^{(\frac{\alpha}{2})}_{\tau,[t\vee S,T]\times\mathbb{R}^n}|\tau-\tau'|^{\frac{\alpha}{2}}\Big)\\
&\leqslant K\Big(1+\|\theta\|_{\mathscr{X}[S,T]}\Big)|\tau-\tau'|^{\frac{\alpha}{2}}.
\end{aligned} \tag{7.5.13}$$

Likewise, for $\tau \in [t \vee S, T]$ and $\xi, \eta, \eta' \in \mathbb{R}^n$,

$$\begin{aligned}
&|g(t, \tau, \xi, \eta, \theta(\tau, \tau, \eta, \eta), \theta_\eta(t, \tau, \xi, \eta)\sigma(\tau, \eta))\\
&\quad -g(t, \tau, \xi, \eta', \theta(\tau, \tau, \eta', \eta'), \theta_\eta(t, \tau, \xi, \eta')\sigma(\tau, \eta'))|\\
&\leqslant K\Big(|\eta-\eta'| + |\theta(\tau, \tau, \eta, \eta) - \theta(\tau, \tau, \eta', \eta')|\\
&\quad +|\theta_\eta(t, \tau, \xi, \eta)\sigma(\tau, \eta) - \theta_\eta(t, \tau, \xi, \eta')\sigma(\tau, \eta')|\Big)\\
&\leqslant K\Big(|\eta-\eta'| + \sup_{(\bar t,\bar\xi)\in[t\vee S,T]\times\mathbb{R}^n} |\theta_\xi(\bar t, \cdot\,, \bar\xi, \cdot)|^{(0)}_{[t\vee S,T]\times\mathbb{R}^n}|\eta-\eta'|\\
&\quad + \sup_{(\bar t,\bar\xi)\in[t\vee S,T]\times\mathbb{R}^n} \langle\theta(\bar t, \cdot\,, \bar\xi, \cdot)\rangle^{(\alpha)}_{[t\vee S,T]\times\mathbb{R}^n}|\eta-\eta'|^{\alpha}\\
&\quad +|\theta_\eta(t, \cdot\,, \xi, \cdot)|^{(0)}_{[t\vee S,T]\times\mathbb{R}^n}|\eta-\eta'| + \langle\theta_\eta(t, \cdot\,, \xi, \cdot)\rangle^{(\alpha)}_{\eta,[t\vee S,T]\times\mathbb{R}^n}|\eta-\eta'|^{\alpha}\Big)\\
&\leqslant K\Big(1+\|\theta\|_{\mathscr{X}[S,T]}\Big)|\eta-\eta'|^{\alpha}.
\end{aligned} \tag{7.5.14}$$

Combining (7.5.12)–(7.5.14), we obtain (7.5.11). Now, for any $\theta(\cdot\,,\cdot\,,\cdot\,,\cdot) \in \mathscr{X}[S,T]$, consider the following linear parabolic system, parametrized by $(t,\xi) \in [0,T) \times \mathbb{R}^n$:

$$\begin{cases} \Theta_s(t,s,\xi,x)+\frac{1}{2}\sigma(s,x)^\top\Theta_{xx}(t,s,\xi,x)\sigma(s,x)+\Theta_x(t,s,\xi,x)b(s,x) \\ \quad +g(t,s,\xi,x,\theta(s,s,x,x),\theta_x(t,s,\xi,x)\sigma(s,x)\big)=0, \\ \qquad\qquad\qquad\qquad\qquad\qquad (s,x)\in[t\vee S,T]\times\mathbb{R}^n, \\ \Theta(t,T,\xi,x)=0, \qquad x\in\mathbb{R}^n. \end{cases} \tag{7.5.15}$$

Then, the corresponding solution $\Theta(t,\cdot\,,\xi,\cdot)$ uniquely exists and the following holds:

$$\begin{aligned} &\Theta(t,s,\xi,x) \\ &=\int_s^T\int_{\mathbb{R}^n}G(s,x;\tau,\eta)g\big(t,\tau,\xi,\eta,\theta(\tau,\tau,\eta,\eta),\theta_\eta(t,\tau,\xi,\eta)\sigma(\tau,\eta)\big)d\eta d\tau. \end{aligned} \tag{7.5.16}$$

Here, G is the fundamental solution of the parabolic system. We refer to Sect. 3 of Chap. 12 for more details. Due to (7.5.11), we have $\Theta(t,\cdot\,,\xi,\cdot) \in C^{2+\alpha}([t \vee S,T] \times \mathbb{R}^n)$. On the other hand, by (12.3.5) of Proposition 12.3.1 in Chap. 12, and the above (7.5.11), we have

$$\begin{aligned} &|\Theta(t,\cdot\,,\xi,\cdot)|^{(1+\alpha)}_{[t\vee S,T]\times\mathbb{R}^n} \\ &\leqslant K(T-S)^{\frac{\alpha}{2}}|g(t,\cdot\,,\xi,\cdot)|^{(\alpha)}_{[t\vee S,T]\times\mathbb{R}^n}\leqslant K(T-S)^{\frac{\alpha}{2}}(1+\|\theta\|_{\mathscr{X}[S,T]}). \end{aligned} \tag{7.5.17}$$

Next

$$\begin{aligned} &\Theta_t(t,s,\xi,x) \\ &=\int_s^T\int_{\mathbb{R}^n}G(s,x;\tau,\eta)g_t\big(t,\tau,\xi,\eta,\theta(\tau,\tau,\eta,\eta),\theta_\eta(t,\tau,\xi,\eta)\sigma(\tau,\eta)\big)d\eta d\tau \\ &\quad+\int_s^T\int_{\mathbb{R}^n}G(s,x;\tau,\eta)g_z\big(t,\tau,\xi,\eta,\theta(\tau,\tau,\eta,\eta),\theta_\eta(t,\tau,\xi,\eta)\sigma(\tau,\eta)\big) \\ &\qquad\qquad \theta_{\eta t}(t,\tau,\xi,\eta)\sigma(\tau,\eta)d\eta d\tau, \end{aligned}$$

and

$$\begin{aligned} &\Theta_{xt}(t,s,\xi,x) \\ &=\int_s^T\int_{\mathbb{R}^n}G_x(s,x;\tau,\eta)g_t\big(t,\tau,\xi,\eta,\theta(\tau,\tau,\eta,\eta),\theta_\eta(t,\tau,\xi,\eta)\sigma(\tau,\eta)\big)d\eta d\tau \\ &\quad+\int_s^T\int_{\mathbb{R}^n}G_x(s,x;\tau,\eta)g_z\big(t,\tau,\xi,\eta,\theta(\tau,\tau,\eta,\eta),\theta_\eta(t,\tau,\xi,\eta)\sigma(\tau,\eta)\big) \\ &\qquad\qquad \theta_{\eta t}(t,\tau,\xi,\eta)\sigma(\tau,\eta)d\eta d\tau. \end{aligned}$$

Using (12.3.3) of Proposition 12.3.1 in Chap. 12, one has

$$
\begin{aligned}
|\Theta_{xt}(t,s,\xi,x)| &\leqslant K\int_s^T\int_{\mathbb{R}^n}|G_x(s,x;\tau,\eta)|\big[1+|\theta_{\eta t}(t,\tau;\xi,\eta)|\big]d\eta d\tau\\
&\leqslant K\int_s^T\int_{\mathbb{R}^n}\frac{1}{(\tau-s)^{\frac{n+1}{2}}}e^{\frac{-\lambda|\eta-x|^2}{\tau-s}}d\eta d\tau\Big[1+|\theta_{xt}|^{(0)}_{\widehat{\Delta}[S,T]\times\mathbb{R}^{2n}}\Big]\\
&\leqslant K(T-S)^{\frac{1}{2}}+K(T-s)^{\frac{1}{2}}|\theta_{xt}|^{(0)}_{\widehat{\Delta}[S,T]\times\mathbb{R}^{2n}},
\end{aligned}
$$

where

$$
|\theta_{xt}|^{(0)}_{\widehat{\Delta}[S,T]\times\mathbb{R}^{2n}}\stackrel{\Delta}{=}\sup_{(t,s)\in\widehat{\Delta}[S,T],\xi,x\in\mathbb{R}^n}|\theta_{xt}(t,s,\xi,x)|.
$$

Similarly

$$
\begin{aligned}
|\Theta_t(t,s,\xi,x)| &\leqslant K\int_s^T\int_{\mathbb{R}^n}|G(s,x;\tau,\eta)|\big[1+|\theta_{\eta t}(t,\tau,\xi,\eta)|\big]d\eta d\tau\\
&\leqslant K\int_s^T\int_{\mathbb{R}^n}\frac{1}{(\tau-s)^{\frac{n}{2}}}e^{\frac{-\lambda|\eta-x|^2}{\tau-s}}d\eta d\tau\Big[1+|\theta_{xt}|^{(0)}_{\widehat{\Delta}[S,T]}\Big]\\
&\leqslant K(T-s)+K(T-s)|\theta_{xt}|^{(0)}_{\widehat{\Delta}[S,T]\times\mathbb{R}^{2n}}.
\end{aligned}
$$

Consequently, it holds that

$$
|\Theta_t|^{(0)}_{\widehat{\Delta}[S,T]\times\mathbb{R}^{2n}}+|\Theta_{xt}|^{(0)}_{\widehat{\Delta}[S,T]\times\mathbb{R}^{2n}}\leqslant K(T-S)^{\frac{1}{2}}\Big(1+|\theta_{xt}|^{(0)}_{\widehat{\Delta}[S,T]\times\mathbb{R}^{2n}}\Big). \tag{7.5.18}
$$

Likewise, we have

$$
|\Theta_\xi|^{(0)}_{\widehat{\Delta}[S,T]\times\mathbb{R}^{2n}}+|\Theta_{x\xi}|^{(0)}_{\widehat{\Delta}[S,T]\times\mathbb{R}^{2n}}\leqslant K(T-S)^{\frac{1}{2}}\Big(1+|\theta_{x\xi}|^{(0)}_{\widehat{\Delta}[S,T]\times\mathbb{R}^{2n}}\Big). \tag{7.5.19}
$$

Then, combining (7.5.17) with (7.5.18)–(7.5.19), one obtains

$$
\|\Theta\|_{\mathscr{X}[S,T]}\leqslant K(T-S)^{\frac{\alpha}{2}}\big(1+\|\theta\|_{\mathscr{X}[S,T]}\big), \tag{7.5.20}
$$

with $K>0$ being an absolute constant. Hence, we have defined a map $\mathcal{S}:\mathscr{X}[S,T]\to\mathscr{X}[S,T]$ by

$$
\mathcal{S}[\theta(\cdot,\cdot,\cdot,\cdot)]\stackrel{\Delta}{=}\Theta(\cdot,\cdot,\cdot,\cdot),\qquad\forall\theta(\cdot,\cdot,\cdot,\cdot)\in\mathscr{X}[S,T].
$$

Moreover, we shrink $T - S > 0$ (if necessary) so that $K(T-S)^{\frac{\alpha}{2}} \leqslant \frac{1}{2}$. Then, for any $M \geqslant 1$,

$$\|\Theta\|_{\mathscr{X}[S,T]} \leqslant \frac{1}{2}\big(1 + \|\theta\|_{\mathscr{X}[S,T]}\big) \leqslant M, \qquad \forall \|\theta\|_{\mathscr{X}[S,T]} \leqslant M.$$

Thus, $\mathcal{S}$ maps a ball in $\mathscr{X}[S,T]$, centered at 0 with radius M to itself.
Step 3. Contraction of the Solution Map. Let $\theta, \widehat{\theta} \in \mathscr{X}[S,T]$ such that

$$\|\theta\|_{\mathscr{X}[S,T]}, \ \|\widehat{\theta}\|_{\mathscr{X}[S,T]} \leqslant M,$$

with $S \in [0,T)$ and $M > 0$ obtained as above. Let

$$\begin{aligned} &\Theta(t,s,\xi,x) \stackrel{\Delta}{=} \mathcal{S}[\theta](t,s,\xi,x), \qquad \widehat{\Theta}(t,s,\xi,x) \stackrel{\Delta}{=} \mathcal{S}[\widehat{\theta}](t,s,\xi,x),\\ &v(t,s,\xi,x) \stackrel{\Delta}{=} \Theta(t,s,\xi,x) - \widehat{\Theta}(t,s,\xi,x). \end{aligned}$$

Then, $v(t,\cdot,\xi,\cdot)$ satisfies the following:

$$\begin{cases} v_s(t,s,\xi,x) + \frac{1}{2}\sigma(s,x)^\top v_{xx}(t,s,\xi,x)\sigma(s,x) + v_x(t,s,\xi,x)b(s,x) \\ \qquad + f(t,s,\xi,x) = 0, \quad (s,x) \in [t \vee S, T] \times \mathbb{R}^n, \\ v(t,T,\xi,x) = 0, \quad x \in \mathbb{R}^n, \end{cases} \tag{7.5.21}$$

with

$$\begin{aligned} f(t,s,\xi,x) \stackrel{\Delta}{=}\ & g(t,s,\xi,x,\theta(s,s,x,x),\theta_x(t,s,\xi,x)\sigma(s,x)) \\ & - g(t,s,\xi,x,\widehat{\theta}(s,s,x,x),\widehat{\theta}_x(t,s,\xi,x)\sigma(s,x)). \end{aligned}$$

Consequently

$$v(t,s,\xi,x) = \int_s^T \int_{\mathbb{R}^n} G(s,x;\tau,\eta) f(t,\tau,\xi,\eta)\, d\eta\, d\tau. \tag{7.5.22}$$

Similar to the proof of (7.5.11), we can show that

$$|f(t,\cdot,\xi,\cdot)|^{(\alpha)}_{[t\vee S,T]\times\mathbb{R}^n} \leqslant K(1+M)\|\theta - \widehat{\theta}\|_{\mathscr{X}[S,T]}, \quad \forall (t,\xi) \in [S,T] \times \mathbb{R}^n.$$

Then, one has

$$\|\Theta - \widehat{\Theta}\|_{\mathscr{X}[S,T]} \leqslant K(1+M)(T-S)^{\frac{\alpha}{2}} \|\theta - \widehat{\theta}\|_{\mathscr{X}[S,T]}, \tag{7.5.23}$$

for some absolute constant $K > 0$. Hence, by choosing $\delta = T - S > 0$ small, we obtain a contraction mapping $\mathcal{S}$ on $\mathscr{X}[T-\delta, T]$. Consequently, $\mathcal{S}$ has a unique fixed point which is the solution $\Theta(\cdot\,,\cdot\,,\cdot\,,\cdot)$ on $\widehat{\Delta}[T-\delta, T]\times\mathbb{R}^n\times\mathbb{R}^n$. In addition, it can be rewritten as follows:

$$\begin{cases}\Theta_s(t,s,\xi,x)+\frac{1}{2}\sigma(s,x)^\top\Theta_{xx}(t,s,\xi,x)\sigma(s,x)+\Theta_x(t,s,\xi,x)b(s,x)\\ \quad +g(t,s,\xi,x,\Theta(s,s,x,x),\Theta_x(t,s,\xi,x)\sigma(s,x)\big)=0,\\ \qquad\qquad\qquad\qquad\qquad\qquad\qquad\qquad (t,s)\in\widehat{\Delta}[S,T],\\ \Theta(t,T,\xi,x)=0, \qquad (t,\xi,x)\in[0,T]\times\mathbb{R}^n\times\mathbb{R}^n.\end{cases} \tag{7.5.24}$$

Note that $(t,\xi)\mapsto g(t,s,\xi,x,y,z)$ is assumed to be continuously differentiable with bounded derivatives. Therefore, by a standard argument, we know that

$$|\Theta_t(t,\cdot\,,\xi,\cdot)|^{(2+\alpha)}_{[t\vee(T-\delta),T]\times\mathbb{R}^n}+|\Theta_\xi(t,\cdot\,,\xi,\cdot)|^{(2+\alpha)}_{[t\vee(T-\delta),T]\times\mathbb{R}^n}<\infty.$$

Now, we denote

$$\bar{\psi}(t,\xi,x)\stackrel{\Delta}{=}\Theta(t,T-\delta,\xi,x), \qquad (t,\xi,x)\in[0,T-\delta]\times\mathbb{R}^n\times\mathbb{R}^n,$$

and consider the following equation:

$$\begin{cases}\Theta_s(t,s,\xi,x)+\frac{1}{2}\sigma(s,x)^\top\Theta_{xx}(t,s,\xi,x)\sigma(s,x)+\Theta_x(t,s,\xi,x)b(s,x)\\ \quad +g(t,s,\xi,x,\Theta(s,s,x,x),\Theta_x(t,s,\xi,x)\sigma(s,x)\big)=0,\\ \qquad\qquad\qquad\qquad\qquad\qquad (s,x)\in[t,T-\delta]\times\mathbb{R}^n,\ \xi\in\mathbb{R}^n,\\ \Theta(t,T-\delta,\xi,x)=\bar{\psi}(t,\xi,x), \qquad x\in\mathbb{R}^n.\end{cases} \tag{7.5.25}$$

Then, we may repeat the above procedure, to get a unique solution $\Theta(t,\cdot\,,\xi,\cdot)$ on $\widehat{\Delta}[T-2\delta, T-\delta]\times\mathbb{R}^n\times\mathbb{R}^n$. By continuing such a procedure, we obtain the existence and unique solution $\Theta(t,\cdot\,,\xi,\cdot)$ to the representation PDE (7.5.1), and (7.5.4) holds.

Step 4. Stability Estimates. Let $(\widehat{\psi},\widehat{g})$ be another pair of maps such that, together with b and σ, satisfy (H7.5.1) as well. Let $\widehat{\Theta}$ be the corresponding solution. Let

$$v(t,s,\xi,x)\stackrel{\Delta}{=}\Theta(t,s,\xi,x)-\widehat{\Theta}(t,s,\xi,x).$$

Then, $v(t,\cdot\,,\xi,\cdot)$ satisfies the following:

$$\begin{cases}v_s(t,s,\xi,x)+\frac{1}{2}\sigma(s,x)^\top v_{xx}(t,s,\xi,x)\sigma(s,x)+v_x(t,s,\xi,x)b(s,x)\\ \qquad +f(t,s,\xi,x)+\widehat{f}(t,s,\xi,x)=0, \quad (s,x)\in[t,T]\times\mathbb{R}^n,\\ v(t,T,\xi,x)=\varphi(t,\xi,x), \qquad x\in\mathbb{R}^n,\end{cases} \tag{7.5.26}$$

with

$$\begin{aligned} f(t,s,\xi,x) &\overset{\triangle}{=} \widehat{g}(t,s,\xi,x,\Theta(s,s,x,x),\Theta_x(t,s,\xi,x)\sigma(s,x)) \\ &\quad -\widehat{g}(t,s,\xi,x,\widehat{\Theta}(s,s,x,x),\widehat{\Theta}_x(t,s,\xi,x)\sigma(s,x)), \\ \bar{f}(t,s,\xi,x) &\overset{\triangle}{=} g(t,s,\xi,x,\Theta(s,s,x,x),\Theta_x(t,s,\xi,x)\sigma(s,x)) \\ &\quad -\widehat{g}(t,s,\xi,x,\Theta(s,s,x,x),\Theta_x(t,s,\xi,x)\sigma(s,x)), \\ \varphi(t,\xi,x) &\overset{\triangle}{=} \psi(t,\xi,x)-\widehat{\psi}(t,\xi,x). \end{aligned}$$

Then

$$\begin{aligned} v(t,s,\xi,x) = &\int_{\mathbb{R}^n} G(s,x;T,\eta)\varphi(t,\xi,\eta)d\eta + \int_s^T\int_{\mathbb{R}^n} G(s,x;\tau,\eta)\bar{f}(t,\tau,\xi,\eta)d\eta d\tau \\ &+ \int_s^T\int_{\mathbb{R}^n} G(s,x;\tau,\eta)f(t,\tau,\xi,\eta)d\eta d\tau. \end{aligned}$$

From (12.3.5) of Proposition 12.3.1 in Chap. 12, one has

$$\begin{aligned} &\Big|\int_{\mathbb{R}^n} G(\cdot\,,\cdot\,;T,\eta)\varphi(t,\xi,\eta)d\eta + \int_t^T\int_{\mathbb{R}^n} G(\cdot\,,\cdot\,;\tau,\eta)\bar{f}(t,\tau,\xi,\eta)d\eta d\tau\Big|^{(2+\alpha)}_{[t,T]\times\mathbb{R}^n} \\ &\leqslant K\Big(|\varphi(t,\xi,\cdot)|^{(2+\alpha)}_{\mathbb{R}^n} + |\bar{f}(t,\cdot\,,\xi,\cdot)|^{(\alpha)}_{[t,T]\times\mathbb{R}^n}\Big). \end{aligned}$$

Also, by Step 3 above, we have

$$\Big\|\int_\cdot^T\int_{\mathbb{R}^n} G(\cdot\,,\cdot;\tau,\eta)f(\cdot\,,\tau,\cdot\,,\eta)d\eta d\tau\Big\|_{\mathscr{X}[S,T]} \leqslant K(1+M)(T-S)^{\frac{\alpha}{2}}\|v\|_{\mathscr{X}[S,T]}.$$

Hence, for $T-S>0$ small, we obtain

$$\|v\|_{\mathscr{X}[S,T]} \leqslant K \sup_{(t,\xi)\in[S,T]\times\mathbb{R}^n}\Big(|\varphi(t,\xi,\cdot)|^{(2+\alpha)}_{\mathbb{R}^n} + |\bar{f}(t,\cdot\,,\xi,\cdot)|^{(\alpha)}_{[t,T]\times\mathbb{R}^n}\Big).$$

Repeating the same argument, we obtain

$$\|v\|_{\mathscr{X}[0,T]} \leqslant K \sup_{(t,\xi)\in[0,T]\times\mathbb{R}^n}\Big(|\varphi(t,\xi,\cdot)|^{(2+\alpha)}_{\mathbb{R}^n} + |\bar{f}(t,\cdot\,,\xi,\cdot)|^{(\alpha)}_{[0,T]\times\mathbb{R}^n}\Big).$$

Then, (7.5.23) follows. □

5.2 The Second Representation PDE

We now look at the second representation PDE. Similar to the previous subsection, without loss of generality, we again assume that $\psi(t,\xi,x)=0$. Thus, we consider the following family of parabolic systems (parameterized by $(t, \xi) \in (S, T) \times \mathbb{R}^n$):

$$
\begin{cases}
\Gamma_s(t,s,\xi) + \frac{1}{2}\sigma(s,\xi)^\top \Gamma_{\xi\xi}(t,s,\xi)\sigma(s,\xi) + \Gamma_\xi(t,s,\xi)b(s,\xi) = 0, \\
\qquad\qquad\qquad\qquad\qquad\qquad (s,x) \in [0,t] \times \mathbb{R}^n, \\
\Theta_s(t,s,\xi,x) + \frac{1}{2}\sigma(s,x)^\top \Theta_{xx}(t,s,\xi,x)\sigma(s,x) + \Theta_x(t,s,\xi,x)b(s,x) \\
+g(t,s,\xi,x,\Theta(s,s,x,x),\Theta_x(t,s,\xi,x)\sigma(s,x),\Gamma_\xi(s,t,\xi)\sigma(t,\xi)\big) = 0, \\
\qquad\qquad\qquad\qquad\qquad\qquad (s,x) \in [t,T] \times \mathbb{R}^n, \\
\Gamma(t,t,x) = \Theta(t,t,x,x), \qquad x \in \mathbb{R}^n, \\
\Theta(t,T,\xi,x) = 0, \qquad x \in \mathbb{R}^n.
\end{cases}
\tag{7.5.27}
$$

Note that for any given $(t, \xi) \in (0, T) \times \mathbb{R}^n$, the equation for $\Gamma(t, \cdot, \cdot)$ is to be solved on $[0, t]$, and the equation for $\Theta(t, \cdot, \xi, \cdot)$ is to be solved on $[t, T]$. The coupling appears at two places: $\Gamma_\xi(s, t, \xi)$ (with $0 \leqslant t \leqslant s$) appears in the equation for $\Theta(t, \cdot, \xi, \cdot)$, and $\Theta(t, t, x, x)$ appears as the terminal value for $\Gamma(t, s, x)$ at $s = t$.

See Fig. 7.3 for the domains in which $\Theta(t, x, \xi, x)$ and $\Gamma(t, s, \xi)$ are defined. Let us make some observations. Suppose $(\Gamma(\cdot, \cdot, \cdot), \Theta(\cdot, \cdot, \cdot, \cdot))$ is a solution to the above (7.5.27). Then, by Proposition 12.3.1 in Chap. 12,

$$
\Gamma(t,s,x) = \int_{\mathbb{R}^n} G(s,x;t,\eta)\Theta(t,t,\eta,\eta)d\eta, \quad 0 \leqslant s \leqslant t \leqslant T, \ x \in \mathbb{R}^n,
\tag{7.5.28}
$$

and thus

$$
\Gamma_\xi(\tau,t,\xi) = \int_{\mathbb{R}^n} G_\xi(t,\xi;\tau,\bar\eta)\Theta(\tau,\tau,\bar\eta,\bar\eta)d\bar\eta, \qquad 0 \leqslant t \leqslant \tau \leqslant T, \quad \xi \in \mathbb{R}^n.
$$

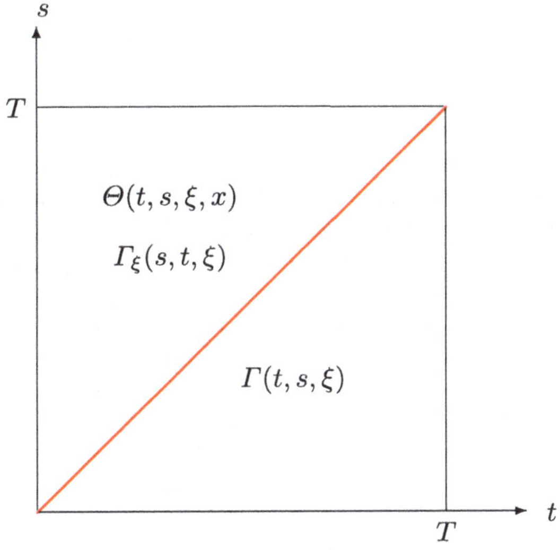

Fig. 7.3 Square area for time variables

On the other hand, for $(t, s, \xi, x) \in \Delta_*[0, T] \times \mathbb{R}^n \times \mathbb{R}^n$,

$$
\begin{aligned}
&\Theta(t, s, \xi, x) \\
&= \int_s^T \int_{\mathbb{R}^n} G(s, x; \tau, \eta) g\Big(t, \tau, \xi, \eta, \Theta(\tau, \tau, \eta, \eta), \Theta_\eta(t, \tau, \xi, \eta)\sigma(\tau, \eta), \\
&\qquad \Big[\int_{\mathbb{R}^n} G_\xi(t, \xi; \tau, \bar\eta)\Theta(\tau, \tau, \bar\eta, \bar\eta)d\bar\eta\Big]\sigma(\tau, \eta)\Big) d\eta d\tau.
\end{aligned} \tag{7.5.29}
$$

The above tells us that if (7.5.27) admits a classical solution (Γ, Θ), then Θ must be a solution to the above nonlinear integral equation (7.5.29). Conversely, if nonlinear integral equation (7.5.29) admits a smooth solution Θ, by defining Γ as (7.5.28), we have a solution (Γ, Θ) to the second representation PDE (7.5.27). Before studying the smooth solution of (7.5.29), let us introduce the following definition.

Definition 7.5.2 A pair of functions (Γ, Θ) is called a *mild solution* of (7.5.27) if Θ is a solution to the integral equation (7.5.29) and Γ is defined by (7.5.28).

Now, we introduce the following hypothesis for system (7.5.27).

(H7.5.2) The maps $(s, x) \mapsto b(s, x), \sigma(s, x)$, and $g(t, s, \xi, x, y, z, \zeta)$ are bounded and have all needed orders of bounded derivatives. Moreover, (7.5.3) holds for some $\bar\sigma > 0$.

The main result of this subsection is the following theorem.

Theorem 7.5.3 *Let (H7.5.2) hold. Then,* (7.5.27) *admits a unique mild solution* (Γ, Θ).

Proof Let $1 < p < 2$, $S \in [0, T)$ and recall $\widehat\Delta[S, T]$ defined by (7.5.8). Let $\mathscr{Y}[S, T]$ be the set of all functions $\theta : \widehat\Delta[S, T] \times \mathbb{R}^n \times \mathbb{R}^n \to \mathbb{R}^m$ such that

$$
\begin{aligned}
\|\theta\|_{\mathscr{Y}[S,T]} = \sup_{t\in[0,T]} \Big(\int_{t\vee S}^T \sup_{\xi,x\in\mathbb{R}^n} |\theta_x(t, s, \xi, x)|^p ds\Big)^{\frac{1}{p}} \\
+ \sup_{\substack{(t,s)\in\widehat\Delta[S,T]\\ \xi,x\in\mathbb{R}^n}} |\theta(t, s, \xi, x)| < \infty.
\end{aligned}
$$

Clearly, $\|\cdot\|_{\mathscr{Y}[S,T]}$ is a norm under which $\mathscr{Y}[S, T]$ is a Banach space.

For any $\theta \in \mathscr{Y}[S, T]$, and $(t, s, \xi, x) \in \widehat\Delta[S, T] \times \mathbb{R}^n \times \mathbb{R}^n$, we define

$$
\begin{aligned}
&\mathcal{S}[\theta](t, s, \xi, x) \\
&\triangleq \int_s^T \int_{\mathbb{R}^n} G(s, x; \tau, \eta) g\Big(t, \tau, \xi, \eta, \theta(\tau, \tau, \eta, \eta), \theta_\eta(t, \tau, \xi, \eta)\sigma(\tau, \eta), \\
&\qquad \Big[\int_{\mathbb{R}^n} G_\xi(t, \xi; \tau, \bar\eta)\theta(\tau, \tau, \bar\eta, \bar\eta)d\bar\eta\Big]\sigma(\tau, \eta)\Big) d\eta d\tau.
\end{aligned}
$$

Note that

$$\mathcal{S}[0](t,s,\xi,x)=\int_s^T\int_{\mathbb{R}^n}G(s,x;\tau,\eta)g(t,\tau,\xi,\eta,0,0,0)d\eta d\tau.$$

Thus

$$\mathcal{S}[0](s,s,x,x)=\int_s^T\int_{\mathbb{R}^n}G(s,x;\tau,\eta)g(s,\tau,x,\eta,0,0,0)d\eta d\tau,\qquad (s,x)\in[S,T]\times\mathbb{R}^n,$$

and for $(t,s,\xi,x)\in\widehat{\Delta}[S,T]\times\mathbb{R}^n\times\mathbb{R}^n$,

$$\mathcal{S}[0]_x(t,s,\xi,x)=\int_s^T\int_{\mathbb{R}^n}G_x(s,x;\tau,\eta)g(t,\tau,\xi,\eta,0,0,0)d\eta d\tau.$$

Then

$$\begin{aligned}|\mathcal{S}[0](t,s,\xi,x)|&\leqslant K\int_s^T\int_{\mathbb{R}^n}\frac{1}{(\tau-s)^{\frac{n}{2}}}e^{-\lambda\frac{|\eta-x|^2}{\tau-s}}d\eta d\tau\\&=K\int_s^T\int_{\mathbb{R}^n}e^{-\lambda z^2}dzd\tau\leqslant K.\end{aligned}$$

Also

$$\begin{aligned}|\mathcal{S}[0]_x(t,s,\xi,x)|&\leqslant K\int_s^T\int_{\mathbb{R}^n}\frac{1}{(\tau-s)^{\frac{n+1}{2}}}e^{-\lambda\frac{|\eta-x|^2}{\tau-s}}d\eta d\tau\\&=K\int_s^T\int_{\mathbb{R}^n}\frac{1}{(\tau-s)^{\frac{1}{2}}}e^{-\lambda z^2}dzd\tau\leqslant K.\end{aligned}$$

Hence, $\mathcal{S}[0]\in\mathscr{Y}[S,T]$.

Next, let $\theta,\widehat{\theta}\in\mathscr{Y}[S,T]$. We estimate the following:

$$\begin{aligned}&\big|\mathcal{S}[\theta](t,s,\xi,x)-\mathcal{S}[\widehat{\theta}](t,s,\xi,x)\big|\\&\leqslant K\int_s^T\int_{\mathbb{R}^n}|G(s,x;\tau,\eta)|\Big(|\theta(\tau,\tau,\eta,\eta)-\widehat{\theta}(\tau,\tau,\eta,\eta)|\\&\qquad+|\theta_\eta(t,\tau,\xi,\eta)-\theta_\eta(t,\tau,\xi,\eta)|\\&\qquad+\int_{\mathbb{R}^n}|G_\xi(t,\xi,\tau,\bar{\eta})|\,|\theta(\tau,\tau,\bar{\eta},\bar{\eta})-\widehat{\theta}(\tau,\tau,\bar{\eta},\bar{\eta})|d\bar{\eta}\Big)d\eta d\tau\end{aligned}$$

$$
\begin{aligned}
&\leqslant K\int_s^T\Big[\int_{\mathbb{R}^n}\frac{e^{-\lambda\frac{|\eta-x|^2}{\tau-s}}}{(\tau-s)^{\frac{n}{2}}}d\eta\Big]\sup_{x,\eta\in\mathbb{R}^n}|\theta_\eta(t,\tau,\xi,\eta)-\widehat{\theta}_\eta(t,\tau,\xi,\eta)|d\tau\\
&\quad+K\Big[\int_s^T\int_{\mathbb{R}^n}\Big(\frac{e^{-\lambda\frac{|\eta-x|^2}{\tau-s}}}{(\tau-s)^{\frac{n}{2}}}+\frac{e^{-\lambda\frac{|\eta-x|^2}{\tau-t}}}{(\tau-t)^{\frac{n+1}{2}}}\int_{\mathbb{R}^n}\frac{e^{-\lambda\frac{|\bar\eta-x|^2}{\tau-s}}}{(\tau-s)^{\frac{n}{2}}}d\bar\eta\Big)d\eta d\tau\Big]\\
&\qquad\cdot\sup_{(\tau,\eta)\in[s,T]\times\mathbb{R}^n}|\theta(\tau,\tau,\eta,\eta)-\widehat{\theta}(\tau,\tau,\eta,\eta)|\\
&\leqslant K\int_s^T\Big[\int_{\mathbb{R}^n}e^{-\lambda z^2}dz\Big]\sup_{\xi,x\in\mathbb{R}^n}|\theta_x(s,\tau,\xi,x)-\widehat{\theta}_x(s,\tau,\xi,x)|d\tau\\
&\quad+K\Big[\int_s^T\int_{\mathbb{R}^n}e^{-\lambda z^2}dz+\int_{\mathbb{R}^n}\frac{e^{-\lambda z^2}}{(\tau-t)^{\frac12}}\int_{\mathbb{R}^n}e^{-\lambda\bar z^2}d\bar z\Big)dzd\tau\Big]\\
&\qquad\cdot\sup_{(\tau,\eta)\in[s,T]\times\mathbb{R}^n}|\theta(\tau,\tau,\eta,\eta)-\widehat{\theta}(\tau,\tau,\eta,\eta)|\\
&\leqslant K\int_s^T\sup_{\xi,x\in\mathbb{R}^n}|\theta_x(s,\tau,\xi,x)-\widehat{\theta}_x(s,\tau,\xi,x)|d\tau\\
&\qquad+K\Big[(T-t)^{\frac12}-(s-t)^{\frac12}\Big]\sup_{(\tau,\eta)\in[s,T]\times\mathbb{R}^n}|\theta(\tau,\tau,\eta,\eta)-\widehat{\theta}(\tau,\tau,\eta,\eta)|\\
&\leqslant K(T-s)^{\frac{p-1}{p}}\Big(\int_s^T\sup_{\xi,x\in\mathbb{R}^n}|\theta_x(s,\tau,\xi,x)-\widehat{\theta}_x(s,\tau,\xi,x)|^pd\tau\Big)^{\frac1p}\\
&\quad+K(T-s)^{\frac12}\sup_{\substack{(t,\tau)\in\Delta[s,T]\\ \xi,x\in\mathbb{R}^n}}|\theta(t,\tau,\xi,x)-\widehat{\theta}(t,\tau,\xi,x)|.
\end{aligned}
$$

This leads to (note $\frac{p-1}{p}<\frac12$)

$$
\sup_{\substack{(t,s)\in\widehat{\Delta}[S,T]\\ \xi,x\in\mathbb{R}^n}}\big|\mathcal{S}[\theta](t,s,\xi,x)-\mathcal{S}[\widehat{\theta}](t,s,\xi,x)\big|\leqslant K(T-S)^{\frac{p-1}{p}}\|\theta-\widehat{\theta}\|_{\mathcal{Y}[S,T]}.
$$

Next

$$
\begin{aligned}
&\big|\mathcal{S}[\theta]_x(t,s,\xi,x)-\mathcal{S}[\widehat{\theta}]_x(t,s,\xi,x)\big|\\
&\leqslant K\int_s^T\int_{\mathbb{R}^n}|G_x(s,x;\tau,\eta)|\Big(|\theta(\tau,\tau,\eta,\eta)-\widehat{\theta}(\tau,\tau,\eta,\eta)|\\
&\quad+|\theta_\eta(t,\tau,\xi,\eta)-\widehat{\theta}_\eta(t,\tau,\xi,\eta)|\\
&\quad+\int_{\mathbb{R}^n}|G_\xi(t,\xi,\tau,\bar\eta)|\,|\theta(\tau,\tau,\bar\eta,\bar\eta)-\widehat{\theta}(\tau,\tau,\bar\eta,\bar\eta)|d\bar\eta\Big)d\eta d\tau\\
&=K\int_s^T\int_{\mathbb{R}^n}|G_x(s,x;\tau,\eta)|\,|\theta_\eta(t,\tau,\xi,\eta)-\widehat{\theta}_\eta(t,\tau,\xi,\eta)|d\eta d\tau\\
&\quad+K\int_s^T\int_{\mathbb{R}^n}\Big(|G_x(s,x;\tau,\eta)|+|G_\xi(t,\xi;\tau,\eta)|\int_{\mathbb{R}^n}|G_x(s,x;\tau,\bar\eta)|d\bar\eta\Big)\\
&\qquad\cdot|\theta(\tau,\tau,\eta,\eta)-\widehat{\theta}(\tau,\tau,\eta,\eta)|d\eta d\tau
\end{aligned}
$$

$$
\begin{aligned}
&\leqslant K\int_s^T\Big[\int_{\mathbb{R}^n}\frac{e^{-\lambda\frac{|\eta-x|^2}{\tau-s}}}{(\tau-s)^{\frac{n+1}{2}}}d\eta\Big]\sup_{\xi,\eta\in\mathbb{R}^n}|\theta_\eta(t,\tau,\xi,\eta)-\widehat{\theta}_\eta(t,\tau,\xi,\eta)|d\tau\\
&\quad+K\Big[\int_s^T\int_{\mathbb{R}^n}\Big(\frac{e^{-\lambda\frac{|\eta-x|^2}{\tau-s}}}{(\tau-s)^{\frac{n+}{2}}}+\frac{e^{-\lambda\frac{|\eta-x|^2}{\tau-t}}}{(\tau-t)^{\frac{n+1}{2}}}\int_{\mathbb{R}^n}\frac{e^{-\lambda\frac{|\bar\eta-x|^2}{\tau-s}}}{(\tau-s)^{\frac{n+1}{2}}}d\bar\eta\Big)d\eta d\tau\Big]\\
&\qquad\cdot\sup_{(\tau,\eta)\in[s,T]\times\mathbb{R}^n}|\theta(\tau,\tau,\eta,\eta)-\widehat{\theta}(\tau,\tau,\eta,\eta)|\\
&\leqslant K\int_s^T\frac{1}{(\tau-s)^{\frac12}}\sup_{\xi,x\in\mathbb{R}^n}|\theta_x(t,\tau,\xi,x)-\widehat{\theta}_x(t,\tau,\xi,x)|d\tau\\
&\quad+K\Big[\int_s^T\Big(\frac{1}{(\tau-s)^{\frac12}}+\frac{1}{(\tau-t)^{\frac12}(\tau-s)^{\frac12}}\Big)d\tau\Big]\\
&\qquad\cdot\sup_{\substack{(t,\tau)\in\widehat{\Delta}[s,T]\\ \xi,\eta\in\mathbb{R}^n}}|\theta(t,\tau,\xi,\eta)-\widehat{\theta}(t,\tau,\xi,\eta)|.
\end{aligned}
$$

Then, noting $1<p<2$, by Young's inequality, we have

$$
\begin{aligned}
&\Big(\int_{t\vee S}^T\sup_{\xi,x\in\mathbb{R}^n}|\mathcal{S}[\theta]_x(t,s,\xi,x)-\mathcal{S}[\widehat{\theta}]_x(t,s,\xi,x)|^p ds\Big)^{\frac1p}\\
&\leqslant K(T-S)^{\frac12}\Big[\Big(\int_{t\vee S}^T\sup_{\xi,x\in\mathbb{R}^n}|\theta_x(t,\tau,\xi,x)-\widehat{\theta}_x(t,\tau,\xi,x)|^p d\tau\Big)^{\frac1p}\\
&\qquad+\sup_{\substack{(t,s)\in\widehat{\Delta}[S,T]\\ \xi,x\in\mathbb{R}^n}}|\theta(t,s,\xi,x)-\widehat{\theta}(t,s,\xi,x)|\Big]\\
&\leqslant K(T-S)^{\frac12}\|\theta-\widehat{\theta}\|_{\mathscr{Y}[S,T]}.
\end{aligned}
$$

Combining the above, we obtain

$$
\|\mathcal{S}[\theta]-\mathcal{S}[\widehat{\theta}]\|_{\mathscr{Y}[S,T]}\leqslant K(T-S)^{\frac{p-1}{p}}\|\theta-\widehat{\theta}\|_{\mathscr{Y}[S,T]}.
$$

By taking $\widehat{\theta}=0$, we see that

$$
\|\mathcal{S}[\theta]\|_{\mathscr{Y}[S,T]}\leqslant\|\mathcal{S}[0]\|_{\mathscr{Y}[S,T]}+K(T-S)^{\frac{p-1}{p}}\|\theta\|_{\mathscr{Y}[S,T]},\qquad\forall\theta\in\mathscr{Y}[S,T].
$$

Consequently, $\mathcal{S}:\mathscr{Y}[S,T]\to\mathscr{Y}[S,T]$ is a contraction when $\delta=T-S>0$ is small. Hence, it admits a unique fixed point on $[T-\delta,T]$, which gives a solution to (7.5.16) on $\widehat{\Delta}[T-\delta,T]\times\mathbb{R}^{2n}$. By repeating the same argument, we will be able to get a unique solution of (7.5.16) on $\Delta_*[0,T]\times\mathbb{R}^{2n}$. Then, we define Γ by (7.5.28). This gives the existence of a mild solution (Γ,Θ) of (7.5.27).

The argument used to established the contractiveness of the solution map $\mathcal{S}$ also gives the uniqueness of the mild solution. □

6 Bibliographical Remarks

One way to solve coupled two-point boundary value problems of ODEs is to use the so-called *invariant embedding*. Such a method was first introduced in its primitive form by Ambarzumian in 1943 [10]. The basic idea of this method is to use the solution of an auxiliary equation so that the coupled two-point value problem is transferred into an initial value problem. The solution of such an auxiliary equation is called the decoupling field. Later, such a method was further developed by Bellman-Kalaba [24] and Chandrasekhar [42]. See also Scott [174], Bellman-Wing [25], and so on. In 1960, Kalman used the method to study linear quadratic (LQ, for short) optimal control problems [118] leading to the Riccati equation, the solution of which is exactly the decoupling field for the optimality system of the LQ problem.

As to the stochastic ODEs case, it is well-known that there is a very close relation between linear parabolic PDEs and FSDEs. More precisely, the solution of a linear parabolic equation can be represented by the expectation of an FSDE parameterized by the initial pair (t, x) of the time and state. This is called the Feynman-Kac formula discovered by Feynman [74] and Kac [117] in the later 1940s. In 1991, Peng discovered a nonlinear version of Feynman-Kac formula. More precisely, the solution of certain semilinear parabolic equation can be represented by the adapted solution to a proper BSDE [162]. In 1994, inspired by the invariant imbedding, Ma-Protter-Yong [141] established a decoupling field for coupled FBSDEs, namely, represent the backward components (Y, Z) by the forward component X via a solution of a parabolic PDE system (see also [142]).

Now, we turn to the adjustment of invariant embedding/decoupling idea in integral equation framework. First, we look at the backward case. In 2017, Yong modified the method of invariant imbedding to represent adapted solution of Type-I BSVIEs by the solution of the FSDE and a solution of representation PDE [222]. The results presented in the current chapter is based on Wang-Yong [205]. Later, this one idea was extended/adjusted into different scenarios; see Wang [191] for some extend BSVIE case, and Wang-Zheng [208] for the time inconsistent optimal control problem case. Second, we elaborate a little on the forward SVIE case. To adjust the ODEs [24] or SDEs [162] into this framework, Viens-Zhang [187] introduced the path-dependent PDEs rather than classical PDEs. This idea was then used in the linear quadratic control problem setting by Wang-Yong-Zhou [196] and forward-backward SVIE case by Wang-Yong-Zhang [194]. On the other hand, if we essentially decouple the linear forward SVIE and linear backward SDE type, Hamaguchi-Wang [94, 95] and Wang-Yong [206] developed some new techniques without involving path-dependent language and applied them into the linear quadratic problems, an extension of Kalman [118] to the VIEs case, and Pontryagin maximum principles (see Chap. 10), an extension of Peng [161] to SVIE framework. This point was further applied to the stochastic delay differential equations, to obtain second-order adjoint system for maximum principle in Meng-Shi-Wang-Zhang [147] and Riccati system for linear quadratic problem in Meng-Wang-Zhang [146].

Chapter 8
Type-I BSVIEs with Nonuniformly Lipschitz Generators

In this chapter, we will investigate Type-I BSVIEs with the generators being nonuniformly Lipschitz, which could be sublinear or superlinear in the argument(s) y and z. Some positive and negative results will be presented.

1 Osgood Type Generators

In this section, we discuss the following Type-I BSVIEs:

$$Y(t) = \psi(t) + \int_t^T g(t, s, Y(s), Z(t, s))ds - \int_t^T Z(t, s)dW(s), \quad t \in [0, T], \tag{8.1.1}$$

when the map $y \mapsto g(t, s, y, z)$ is not uniformly Lipschitz (to be made precisely below).

To begin with, let us look at a special one-dimensional Type-I BSVIE (or nonlinear BSDE):

$$Y(t) = \xi + \int_t^T f(Y(s))ds - \int_t^T Z(t, s)dW(s), \qquad t \in [0, T], \tag{8.1.2}$$

with continuously differentiable

$$f : \mathbb{R} \to [0, \infty), \quad f(0) = 0, \quad f(y) > 0, \quad \forall y > 0, \tag{8.1.3}$$

and $\xi \geqslant 0$. Note that the conditions on $f(\cdot)$ imply that 0 is the only point at which $f(\cdot)$ is zero. Unlike the BSVIEs we discussed in the previous section, here, we

T. Wang, J. Yong, *Backward Stochastic Volterra Integral Equations*,
Probability Theory and Stochastic Modelling 109,
https://doi.org/10.1007/978-3-032-20848-4_8

allow the map $y \mapsto f(y)$ to be nonuniformly Lipschitz. We have the following proposition which gives some kind of negative results.

Proposition 8.1.1 *Suppose $f(\cdot)$ satisfies the above conditions* (8.1.3).

(i) *Let the following hold for some $c > 0$:*

$$\int_c^\infty \frac{dy}{f(y)} < \infty. \tag{8.1.4}$$

Then, there exists constant ξ such that BSVIE (8.1.2) *cannot have a global (bounded) adapted solution on* $[0, T]$ *with deterministic $Y(\cdot)$.*

(ii) *Let the following hold for some $c_0 > 0$:*

$$\int_0^{c_0} \frac{dy}{f(y)} < \infty. \tag{8.1.5}$$

Then, BSVIE (8.1.2) *with $\xi = 0$ has uncountably many adapted solutions on* $[0, T]$.

Before presenting a proof of the above, we explain the meaning of the above results. Part (i) says that if $y \mapsto f(y)$ grows too fast (faster than a linear function, say) as $y \to \infty$, then the BSVIE (8.1.2) might not have a global adapted solution (for some free term ξ). In particular, it implies the failure of the existence of adapted solutions. Whereas part (ii) says that if $y \mapsto f(y)$ goes to zero too slow (slower than a linear function, say) as $y \to 0$, then the BSVIE (8.1.2) might have uncountably many adapted solutions. In particular, this implies the failure of the uniqueness of adapted solutions to the BSVIEs (8.1.2).

Proof (i) First, by (8.1.4), there exists a large constant $\bar{y} > 0$ such that

$$\int_{\bar{y}}^\infty \frac{dy}{f(y)} < T. \tag{8.1.6}$$

Next, with a deterministic constant $\xi \geqslant \bar{y}$, consider the following ODE terminal value problem on $[t_0, T]$ with $t_0 < T$:

$$\dot{Y}(t) = -f(Y(t)), \quad t \in [t_0, T], \qquad Y(T) = \xi. \tag{8.1.7}$$

Suppose $Y(\cdot)$ is a bounded solution of (8.1.7). We point out that by the continuous differentiability of $f(\cdot)$, $(Y(\cdot), 0)$ is the unique bounded adapted solution to the BSVIE (8.1.2). Since it is decreasing, $Y(t) > Y(T) = \xi$. Then, by separation of variables, we have (noting $f(y) > 0$, for $y > 0$)

$$\int_\xi^{Y(t)} \frac{dy}{f(y)} = T - t, \qquad t \in [t_0, T]. \tag{8.1.8}$$

Combining (8.1.6) and (8.1.8), one has

$$T > \int_{\bar{y}}^{\infty} \frac{dy}{f(y)} > \int_{\xi}^{Y(t_0)} \frac{dy}{f(y)} = T - t_0.$$

This gives $t_0 > 0$. In other words, (8.1.7) does not have a solution on $[0, T]$. This shows that global bounded adapted solution of BSVIE (8.1.2) does not exist.

(ii) For any $c \in (0, T]$, we define

$$t_c(y) = c - \int_0^y \frac{d\eta}{f(\eta)}, \qquad y \geqslant 0.$$

Since $f(y) > 0$, for any $y > 0$, under (8.1.5), the above is well-defined. Then

$$t_c'(y) = -\frac{1}{f(y)} < 0, \; y > 0.$$

Thus, the inverse function $Y_c(t)$ of $t_c(y)$ exists. This means

$$\int_0^{Y_c(t)} \frac{d\eta}{f(\eta)} = c - t.$$

Differentiating the above with respect to t, one gets

$$\dot{Y}_c(t) = -f(Y_c(t)), \quad t \in [0, c]; \qquad Y_c(c) = 0. \tag{8.1.9}$$

Defining $Y_c(t)$ to be 0 on $[c, T]$. Then, $Y_c(\cdot)$ is a solution to the following:

$$\begin{cases} \dot{Y}(t) = -f(Y(t)), & t \in [0, T], \\ Y(T) = 0. \end{cases}$$

Since $c \in [0, T)$ is arbitrary, BSVIE (8.1.2) has uncountably many adapted solutions $(Y_c(\cdot), 0)$. □

From the above, we see that the function $f(\cdot)$ plays important role for the well-posedness of Type-I BSVIE (8.1.2). This leads to the following definition.

Definition 8.1.2

(i) A function $\rho : [0, \infty) \to [0, \infty)$ is called an *Osgood function* if it is continuous, nondecreasing, $\rho(0) = 0$, and $\rho(r) > 0$ for all $r > 0$.

(ii) A function $g : \mathbb{R}^m \to \mathbb{R}^n$ is said to satisfy *Osgood condition* if

$$|g(y_1) - g(y_2)| \leqslant \rho\big(|y_1 - y_2|\big), \qquad \forall y_1, y_2 \in \mathbb{R}^m,$$

for some Osgood function $\rho(\cdot)$.

If $\rho(r) = Kr$, then it is an Osgood function, and it is uniform Lipschitz function. Hence, Osgood function is an extension of uniform Lipschitz function. Let us look at the following concrete examples of BSVIEs to get some more feeling.

Example 8.1.3 (i) Let $\alpha > 1$ and $a > 0$ such that

$$a^{\alpha-1}(\alpha - 1) > 1. \tag{8.1.10}$$

Consider the following BSVIE:

$$Y(t) = a + \int_t^1 |Y(s)|^\alpha ds - \int_t^1 Z(t,s)dW(s), \quad t \in [0,1]. \tag{8.1.11}$$

Then, the generator $g(t,s,y,z) = |y|^\alpha$ is growing faster than a linear function (as $y \to \infty$). Clearly, if $(Y(\cdot), Z(\cdot\,,\cdot))$ is an adapted solution such that $Y(\cdot)$ is deterministic, then by the uniqueness of BSVIEs, we have $Z(t,s) \equiv 0$. Consequently, the equation is equivalent to the ODE:

$$\begin{cases} \dot Y(t) = -|Y(t)|^\alpha, & t \in [0,1], \\ Y(1) = a. \end{cases} \tag{8.1.12}$$

We can check that

$$Y(t) = \frac{a}{[1 - a^{\alpha-1}(\alpha-1)(1-t)]^{\frac{1}{\alpha-1}}}, \quad t \in \Big(1 - \frac{1}{a^{\alpha-1}(\alpha-1)}, 1\Big]$$

is the unique solution of (8.1.12) on $(1 - \frac{1}{a^{\alpha-1}(\alpha-1)}, 1]$. From condition (8.1.10), we see that BSVIE (8.1.11) does not have a global adapted solution on the whole interval $[0,1]$ with deterministic $Y(\cdot)$.

(ii) Let $a > 1$ and $\beta > 1$ so that

$$(\log a)^{\beta-1}(\beta - 1) > 1. \tag{8.1.13}$$

Consider the following BSVIE:

$$Y(t) = a + \int_t^1 |Y(s)|\Big|\log|Y(s)|\Big|^\beta ds - \int_t^1 Z(t,s)dW(s), \quad t \in [0,1]. \tag{8.1.14}$$

Then, the generator is $g(t,s,y,z) = |y|(\log|y|)^\beta$ is growing faster than a linear function as $y \to \infty$. Similar to the above, we see that it suffices to consider the following ODE:

$$\begin{cases} \dot Y(t) = -|Y(t)|\Big|\log|Y(t)|\Big|^\beta, & t \in [0,1], \\ Y(1) = a. \end{cases} \tag{8.1.15}$$

We can check that

$$Y(t) = \exp\left\{\frac{\log a}{\left[1-(\log a)^{\beta-1}(\beta-1)(1-t)\right]^{\frac{1}{\beta-1}}}\right\},$$
$$t \in \left(1-\frac{1}{(\log a)^{\beta-1}(\beta-1)}, 1\right]$$

is the unique solution of (8.1.15) on $(1-\frac{1}{(\log a)^{\beta-1}(\beta-1)}, 1]$. From condition (8.1.13), we see that BSVIE (8.1.14) does not have a global adapted solution on the whole interval $[0, 1]$ with deterministic $Y(\cdot)$.

For this example, we have

$$f(y) = |y|^\alpha, \ (\alpha > 1), \quad \text{or} \quad f(y) = |y|\big|\log|y|\big|^\beta, \ (\beta > 1),$$

which is superlinear. Thus, (8.1.4) holds, leading to the nonexistence of global adapted solutions.

Example 8.1.4 (i) Let $\alpha \in (0, 1)$. Consider the following BSVIE:

$$Y(t) = \int_t^1 \frac{1}{1-\alpha} Y(s)^\alpha ds - \int_t^1 Z(t,s)dW(s), \qquad t \in [0,1]. \tag{8.1.16}$$

This is a case that $y \mapsto g(t,s,y,z)$ is approaching to 0 slower than a linear function. Then, for any $c \in [0, 1]$, the following is an adapted solution of the above:

$$Y(t) = (c-t)^{\frac{1}{1-\alpha}} \chi_{[0,c]}(t), \qquad Z(t,s) = 0. \tag{8.1.17}$$

In fact, for the above given $(Y(\cdot), Z(\cdot\,,\cdot))$, the BSVIE (8.1.16) reads

$$Y(t) = \int_t^1 \frac{1}{1-\alpha} Y(s)^\alpha ds, \qquad t \in [0,1],$$

which is equivalent to the following terminal value problem of an ODE:

$$\begin{cases} \dot Y(t) = -\dfrac{1}{1-\alpha} Y(t)^\alpha, & t \in [0,1], \\ Y(1) = 0. \end{cases}$$

We can check directly that for any $c \in [0, 1]$, $Y(\cdot)$ defined in (8.1.17) is a solution of the above. This means that BSVIE (8.1.16) has uncountably many adapted solutions.

(ii) Let $\beta > 1$ and consider the following BSVIE:

$$Y(t) = \int_t^1 \frac{1}{\beta-1} Y(s)\Big|\log|Y(s)|\Big|^\beta ds - \int_t^1 Z(t,s)dW(s), \quad t \in [0,1]. \tag{8.1.18}$$

In this case, $y \mapsto g(t,s,y,z) = \frac{1}{\beta-1} y\big|\log|y|\big|^\beta$ approaches to 0 slower than a linear function. For any $c \in (0,1]$, the following is an adapted solution of BSVIE (8.1.18):

$$Y(t) = \exp\Big(-\frac{1}{(c-t)^{\frac{1}{\beta-1}}}\Big)\chi_{[0,c)}(t), \quad Z(t,s) = 0. \tag{8.1.19}$$

In fact, similar to the above example, for the above given $(Y(\cdot), Z(\cdot\,,\cdot))$ with deterministic $Y(\cdot)$, the BSVIE (8.1.18) is equivalent to

$$\begin{cases} \dot Y(t) = \dfrac{1}{1-\beta} Y(t)\Big|\log|Y(t)|\Big|^\beta, & t \in [0,1], \\ Y(1) = 0. \end{cases}$$

Then, we can check that for any $c \in (0,1]$, $Y(\cdot)$ defined by (8.1.19) is a solution of the above terminal value problem of an ODE. Thus, BSVIE (8.1.18) admits uncountably many adapted solutions.

For this example, we have

$$f(y) = |y|^\alpha,\ (0<\alpha<1), \quad \text{or} \quad f(y) = |y|\big|\log|y|\big|^\beta,\ (\beta>1),$$

which is sublinear as $y \to 0$. Thus, (8.1.5) holds, leading to the nonuniqueness.

We now would like to look at the positive situation, namely, the adapted solution uniquely exists for the Type-I BSVIEs with non-Lipschitz conditions. To this end, let us first present the following elementary lemma.

Lemma 8.1.5 *Let $f : [0,b] \to \mathbb{R}$ be concave, i.e.,*

$$f(\lambda x + (1-\lambda)y) \geqslant \lambda f(x) + (1-\lambda) f(y), \quad \forall x, y \in [0,b],\ \lambda \in [0,1]. \tag{8.1.20}$$

Let $f(0) \geqslant 0$. Then, it is subadditive, i.e.,

$$f(x+y) \leqslant f(x) + f(y), \quad \forall x, y \in [0,b],\ \text{with } x+y \in [0,b]. \tag{8.1.21}$$

In the case that $f(\cdot)$ is nondecreasing, it is true that

$$0 \leqslant f(y) - f(x) \leqslant f(y-x), \quad \forall 0 \leqslant x \leqslant y \leqslant b. \tag{8.1.22}$$

Proof First of all, in (8.1.20), taking $y = 0$, we have

$$f(\lambda x) \geqslant \lambda f(x), \quad \forall x \in [0,b],\ \lambda \in [0,1].$$

Now, let $x, y \in [0, b]$ with $x + y \in (0, b]$ (the case $x + y = 0$ is trivial), we have

$$\begin{aligned} f(x) + f(y) &= f\Big((x+y)\frac{x}{x+y}\Big) + f\Big((x+y)\frac{y}{x+y}\Big) \\ &\geqslant \frac{x}{x+y} f(x+y) + \frac{y}{x+y} f(x+y) = f(x+y). \end{aligned}$$

This proves the first conclusion. Next, by the proved subadditivity, for any $0 \leqslant x \leqslant y \leqslant b$, we have

$$f(y) = f(x + y - x) \leqslant f(x) + f(y - x).$$

Hence, (8.1.22) follows. $\square$

Now, we look at some interesting special cases. For $\alpha \in (0, 1)$, $f(y) = y^\alpha$ is concave and increasing $[0, \infty)$. Hence, from the above Lemma 8.1.5,

$$|x^\alpha - y^\alpha| \leqslant |x - y|^\alpha, \qquad \forall x, y \in [0, \infty).$$

Next, for $\beta > 0$, let

$$f(y) = y\big|\log y\big|^\beta = \begin{cases} 0, & y = 0, \\ y\big(-\log y\big)^\beta, & y \in (0, 1], \\ y\big(\log y\big)^\beta, & y \in (1, \infty). \end{cases} \tag{8.1.23}$$

Then

$$f'(y) = \begin{cases} \big(-\log y\big)^\beta - \beta\big(-\log y\big)^{\beta-1}, & y \in (0, 1], \\ \big(\log y\big)^\beta + \beta\big(\log y\big)^{\beta-1}, & y \in (1, \infty), \end{cases} \tag{8.1.24}$$

and

$$f''(y) = \begin{cases} -\beta\dfrac{\big(-\log y\big)^{\beta-1}}{y} + \beta(\beta-1)\dfrac{\big(-\log y\big)^{\beta-2}}{y}, & y \in (0, 1), \\ \beta\dfrac{(\log y)^{\beta-1}}{y} + \beta(\beta-1)\dfrac{(\log y)^{\beta-2}}{y}, & y \in (1, \infty). \end{cases} \tag{8.1.25}$$

Clearly

$$f'(y) \begin{cases} > 0, & y \in (0, e^{-\beta}) \cup (1, \infty), \\ < 0, & y \in (e^{-\beta}, 1), \end{cases} \tag{8.1.26}$$

which means that $f(\cdot)$ is monotonically increasing on different intervals. For concavity/convexity, when $\beta > 1$,

$$f''(y) \begin{cases} > 0, & y \in (e^{1-\beta}, \infty), \\ < 0, & y \in (0, e^{1-\beta}), \end{cases} \tag{8.1.27}$$

and when $\beta \in (0, 1]$,

$$f''(y)\begin{cases} > 0, & y \in (e^{1-\beta}, \infty), \\ < 0, & y \in (0, 1) \cup (1, e^{1-\beta}). \end{cases} \tag{8.1.28}$$

Hence, $f(\cdot)$ is increasing on $[0, e^{-\beta}]$ and concave on $[0, e^{1-\beta} \wedge 1) \supseteq [0, e^{-\beta}]$, as long as $\beta > 0$ (regardless if $\beta > 1$ or $\beta \in (0, 1]$). Consequently

$$f(y) - f(x) \leqslant f(y - x), \qquad \forall 0 \leqslant x \leqslant y \leqslant e^{-\beta},$$

namely,

$$\Big|x\big|\log x\big|^\beta - y\big|\log y\big|^\beta\Big| \leqslant |x - y|\big|\log|x - y|\big|^\beta, \quad \forall x, y \in [0, e^{-\beta}]. \tag{8.1.29}$$

It seems to be nontrivial to prove (8.1.29) directly. On the other hand, for $c > 0$ large,

$$\int_c^\infty \frac{dy}{y^\alpha} = \begin{cases} \dfrac{1}{(\alpha - 1)c^{\alpha-1}} < \infty, & \alpha > 1, \\ \infty, & \alpha \in (0, 1], \end{cases} \tag{8.1.30}$$

$$\int_c^\infty \frac{dy}{y\big|\log y\big|^\beta} = \begin{cases} \dfrac{1}{(\beta - 1)|\log c|^{\beta-1}} < \infty, & \beta > 1, \\ \infty, & \beta \in (0, 1], \end{cases} \tag{8.1.31}$$

and for $c > 0$ small,

$$\int_0^c \frac{dy}{y^\alpha} = \begin{cases} \infty, & \alpha > 1, \\ \dfrac{c^{1-\alpha}}{1 - \alpha} < \infty, & \alpha \in (0, 1], \end{cases} \tag{8.1.32}$$

$$\int_0^c \frac{dy}{y\big|\log y\big|^\beta} = \begin{cases} \dfrac{1}{(\beta - 1)|\log c|^{\beta-1}} < \infty, & \beta > 1, \\ \infty, & \beta \in (0, 1]. \end{cases} \tag{8.1.33}$$

With the above observations, we now introduce the following hypothesis which is comparable with $(H4.2.1)_p$.

$(\mathbf{H8.1.1})_p$ Let $g : \Delta^*[0, T] \times \mathbb{R}^m \times \mathbb{R}^m \times \Omega \to \mathbb{R}^m$ be measurable such that $s \mapsto g(t, s, y, z)$ is $\mathbb{F}$-adapted and $g(t, s, 0, 0) = 0$. Let $L_y, L_z : \Delta^* \to \mathbb{R}$ satisfy

$$\sup_{t\in[0,T]} \int_0^T \Big(L_y(t, s)^{\frac{q(1+\varepsilon)}{q-1}} + L_z(t, s)^2\Big)ds < \infty, \tag{8.1.34}$$

for some $1 < q \leqslant p$, $\varepsilon > 0$, and a differentiable concave Osgood function $\rho : [0,\infty) \to [0,\infty)$ satisfy

$$\int_0^c \frac{dr}{\rho(r)} = +\infty, \tag{8.1.35}$$

such that

$$\begin{aligned}|g(t,s,y,z) - g(t,s,\bar y,\bar z)| \leqslant L_y(t,s)\rho\big(|y-\bar y|^q\big)^{\frac{1}{q}} + L_z(t,s)|z-\bar z|,\\ (t,s)\in\Delta^*[0,T],\ y,\bar y,z,\bar z\in\mathbb{R}^m, \text{ a.s.}\end{aligned} \tag{8.1.36}$$

For convenience, we call the above $g(\cdot)$ the *Osgood type generator*. Let us make some comments on the above hypothesis. First of all, since $r \mapsto \rho(r)$ is nondecreasing and concave, it must be sublinear. Hence, we may assume

$$0 < \rho(r) \leqslant K(1+r), \qquad \forall r > 0. \tag{8.1.37}$$

Then, for $c > 0$,

$$\int_c^\infty \frac{dr}{\rho(r)} \geqslant \int_c^\infty \frac{dr}{K(1+r)} = \infty. \tag{8.1.38}$$

This means the concavity of $\rho(\cdot)$ guarantees the conditions of nonexistence of global (adapted) solution to fail. Secondly, by $g(t,s,0,0)=0$, we have

$$\begin{aligned}|g(t,s,y,z)| &\leqslant L_y(t,s)\rho(|y|^q)^{\frac{1}{q}} + L_z(t,s)|z|\\ &\leqslant KL_y(t,s)(1+|y|) + L_z(t,s)|z|.\end{aligned} \tag{8.1.39}$$

Thus, the map $(y,z)\mapsto g(t,s,y,z)$ is sublinear. Let us analyze condition (8.1.36) further. To this end, we present the following lemma.

Lemma 8.1.6 *Let $\rho : [0,c] \to [0,\infty)$ be a concave Osgood function. Then, for any $1 \leqslant q < p$,*

$$q\Big(\frac{\rho(c)}{c}\Big)^{\frac{1}{q}} \int_0^{c^{\frac{1}{q}}} \frac{dr}{\rho(r^q)^{\frac{1}{q}}} \leqslant p\Big(\frac{\rho(c)}{c}\Big)^{\frac{1}{p}} \int_0^{c^{\frac{1}{p}}} \frac{dr}{\rho(r^p)^{\frac{1}{p}}}. \tag{8.1.40}$$

Consequently, if the left-hand side is infinite so is the right-hand side.

Proof Since $\rho(\cdot)$ is concave with $\rho(0)=0$, we have that

$$\rho(t) = \rho\Big(\frac{t}{c}c + \Big[1-\frac{t}{c}\Big]0\Big) \geqslant \frac{t}{c}\rho(c) + \Big[1-\frac{t}{c}\Big]\rho(0) = \frac{\rho(c)}{c}t, \quad \forall t\in[0,c].$$

Thus, applying the above with $t = \tau^q$, one has

$$\rho(\tau^q)^{\frac{1}{q}-\frac{1}{p}} \geqslant \Big(\frac{\rho(c)}{c}\Big)^{\frac{1}{q}-\frac{1}{p}}\tau^{1-\frac{q}{p}},$$

which leads to

$$\rho(\tau^q)^{\frac{1}{q}} \geqslant \Big(\frac{\rho(c)}{c}\Big)^{\frac{1}{q}-\frac{1}{p}}\rho(\tau^q)^{\frac{1}{p}}\tau^{1-\frac{q}{p}}.$$

Consequently, $r^p = \tau^q$:

$$\int_0^{c^{\frac{1}{p}}}\frac{dr}{\rho(r^p)^{\frac{1}{p}}} = \frac{q}{p}\int_0^{c^{\frac{1}{q}}}\frac{d\tau}{\rho(\tau^q)^{\frac{1}{p}}\tau^{1-\frac{q}{p}}} \geqslant \frac{q}{p}\Big(\frac{\rho(c)}{c}\Big)^{\frac{1}{q}-\frac{1}{p}}\int_0^{c^{\frac{1}{q}}}\frac{d\tau}{\rho(\tau^q)^{\frac{1}{q}}}.$$

Then, (8.1.40) follows. □

The above result roughly means that as $r \to 0$, $\rho(r^p)^{\frac{1}{p}}$ goes to zero faster that $\rho(r^q)^{\frac{1}{q}}$ if $1 \leqslant q < p$. Hence, for the map $y \mapsto g(t,s,y,z)$, the smaller the q, the weaker the condition (8.1.36).

Now, we are at the position of stating the following result.

Theorem 8.1.7 *Let* (H8.1.1)$_p$ *hold for some* $p > 1$. *Then, for any* $\psi(\cdot) \in L^p_{\mathcal{F}_T}(0,T;\mathbb{R}^m)$, *Type-I BSVIE* (8.1.1) *admits a unique adapted solution* $(Y(\cdot),\ Z(\cdot\,,\cdot)) \in \mathcal{H}^p_{\Delta*}[0,T]$ *such that the following estimate holds:*

$$\|(Y(\cdot),Z(\cdot\,,\cdot))\|^p_{\mathcal{H}^p_{\Delta*}[0,T]} \leqslant K\mathbb{E}\Big[1+\int_0^T|\psi(t)|^p dt\Big]. \tag{8.1.41}$$

Proposition 8.1.1, together with the previous examples, shows that (H8.1.1)$_p$ is very close to the necessary conditions for the well-posedness of BSVIE (8.1.1).

To prove the above theorem, we need some preparations. The following useful lemma is a version of the so-called *LaSalle-Bihari inequality*.

Lemma 8.1.8 *Let* $\rho : [0,\infty) \to [0,\infty)$ *be an Osgood function satisfying* (8.1.35) *for some* $c > 0$. *Let* $F : [0,\infty) \to [0,\infty)$ *and* $y : [0,T] \to [0,\infty)$ *be integrable such that*

$$y(t) \leqslant \int_t^T F(s)\rho\big(y(s)\big)ds < +\infty, \qquad a.e.\ t \in [0,T]. \tag{8.1.42}$$

Then, $y(t) = 0$, *a.e. on* $[0,T]$.

Proof Let

$$R(r) = \int_r^c \frac{ds}{\rho(s)}, \qquad r \in (0,c].$$

Then, for any $r > 0$, $R(r)$ is finite and by (8.1.35), we have

$$\lim_{r \downarrow 0} R(r) = \infty.$$

Define

$$V(t) = \int_t^T F(s)\rho\big(y(s)\big)ds, \qquad \forall t \in [0, T].$$

Since the integrand is nonnegative, $t \mapsto V(t)$ is continuous, nonincreasing, $V(T) = 0$, and

$$0 \leqslant y(t) \leqslant V(t), \qquad t \in [0, T].$$

Thus, it is suffice to prove that

$$V(t) = 0, \qquad \forall t \in [0, T].$$

We prove the above by contradiction. If there exists a $t' \in [0, T)$ such that $V(t') > 0$, then let $b \geqslant 0$ be the supremum of such $t' \in [0, T]$, i.e., for such a $b \in [0, T]$,

$$V(t) > 0, \quad t \in [0, b); \qquad V(t) = 0, \quad t \in [b, T].$$

Observe that for any $a \in [0, b)$ and almost all $t \in [0, a]$,

$$\frac{d}{dt}R\big(V(t)\big) = R'\big(V(t)\big)V'(t) = \frac{F(t)\rho\big(y(t)\big)}{\rho\big(V(t)\big)} \leqslant F(t).$$

Consequently

$$R\big(V(a)\big) \leqslant R\big(V(0)\big) + \int_0^a F(s)ds.$$

Sending $a \nearrow b$, we have $V(a) \searrow V(b) = 0$. Hence, the above leads to

$$\infty \leqslant R\big(V(0)\big) + \int_0^b F(s)ds,$$

which is a contradiction. This completes the proof. □

In the case that

$$\rho(r) = Kr, \qquad r \geqslant 0,$$

the LaSalle-Bihari inequality above is essentially the Gronwall's inequality. Thus, in some sense, the former is an extension of the latter.

Now, we are ready to prove Theorem 8.1.7.

Proof of Theorem 8.1.7 To begin with, we point out one useful fact. Since $q \leqslant p$, then $\frac{q}{q-1} \geqslant \frac{p}{p-1}$. Then, it follows from (8.1.34) that

$$\sup_{t\in[0,T]}\int_0^T L_y(t,s)^{\frac{p}{p-1}}ds \leqslant K \sup_{t\in[0,T]}\int_0^T L_y(t,s)^{\frac{q(1+\varepsilon)}{q-1}}ds < \infty. \tag{8.1.43}$$

We first prove the uniqueness. Let $(Y_i(\cdot), Z_i(\cdot\,,\cdot)) \in \mathcal{H}^p_{\Delta^*}[0,T]$ $(i = 1, 2)$ be two adapted solutions of BSVIE (8.1.1), i.e.,

$$Y_i(t) = \psi(t) + \int_t^T g(t,s,Y_i(s),Z_i(t,s))ds - \int_t^T Z_i(t,s)dW(s), \quad t \in [0,T]. \tag{8.1.44}$$

We introduce the following auxiliary Type-I BSVIE:

$$y_i(t) = \psi(t) + \int_t^T g(t,s,Y_i(s),z_i(t,s))ds - \int_t^T z_i(t,s)dW(s), \quad t \in [0,T]. \tag{8.1.45}$$

Note that the generator of the above is $(t,s,y,z) \mapsto g(t,s,Y_i(s),z)$ which is independent of y. Under (8.1.36), noting (8.1.37), one has that

$$|g(t,s,Y_i(s),0)| \leqslant L_y(t,s)\rho\big(|Y_i(s)|^q\big)^{\frac{1}{q}} \leqslant KL_y(t,s)\big(1+|Y_i(s)|\big). \tag{8.1.46}$$

This together with the above (8.1.43) indicates that

$$\begin{aligned}&\int_0^T \mathbb{E}\Big(\int_t^T |g(t,s,Y_i(s),0)|ds\Big)^p dt\\ &\leqslant K\mathbb{E}\Big[1+\sup_{t\in[0,T]}\Big(\int_0^T L_y(t,s)^{\frac{p}{p-1}}ds\Big)^{p-1}\Big(\int_0^T |Y_i(s)|^p ds\Big)\Big] < \infty.\end{aligned}$$

Further

$$|g(t,s,Y_i(s),z_1) - g(t,s,Y_i(s),z_2)| \leqslant L_z(t,s)|z_1 - z_2|. \tag{8.1.47}$$

Therefore, by Theorem 4.2.2 of Chap. 4, BSVIE (8.1.45) admits a unique adapted solution $(y_i(\cdot), z_i(\cdot\,,\cdot)) \in \mathcal{H}^p_{\Delta^*}[0,T]$. On the other hand, $(Y_i(\cdot), Z_i(\cdot\,,\cdot))$ is clearly an adapted solution of (8.1.45). Therefore, by uniqueness of auxiliary Type-I BSVIE (8.1.45), we must have

$$Y_i(t) = y_i(t), \qquad Z_i(t,s) = z_i(t,s), \qquad \text{a.e. } (t,s) \in \Delta^*[0,T].$$

Note that we do not claim, at this moment, that the adapted solution to BSVIE (4.0.1) (or (8.1.44)) is unique. Then by the stability estimate of adapted solutions to the auxiliary Type-I BSVIE (8.1.45) (see Chap. 4), we have (noting the concavity of $\rho(\cdot)$)

$$\begin{aligned}
&\mathbb{E}\Big[|Y_1(t)-Y_2(t)|^q+\Big(\int_t^T|Z_1(t,s)-Z_2(t,s)|^2ds\Big)^{\frac{q}{2}}\Big]\\
&\leqslant K\mathbb{E}\Big(\int_t^T|g(t,s,Y_1(s),Z_1(t,s))-g(t,s,Y_2(s),Z_1(t,s))|ds\Big)^q\\
&\leqslant K\mathbb{E}\Big(\int_t^T L_y(t,s)\rho(|Y_1(s)-Y_2(t)|^q)^{\frac{1}{q}}ds\Big)^q\\
&\leqslant K\mathbb{E}\Big(\int_t^T L_y(t,s)^{\frac{q}{q-1}}ds\Big)^{q-1}\int_t^T\rho(|Y_1(s)-Y_2(s)|^q)ds\\
&\leqslant K\int_t^T\rho\big(\mathbb{E}|Y_1(s)-Y_2(s)|^q\big)ds.
\end{aligned}$$

Then, uniqueness follows from LaSalle-Bihari inequality (Lemma 8.1.8).

Next, we look at the existence. Set $Y_0(\cdot)=0$, and define inductively $(Y_k(\cdot),Z_k(\cdot\,,\cdot))\in\mathcal{H}^p_{\Delta^*}[0,T]\subseteq\mathcal{H}^q_{\Delta^*}[0,T]$ the adapted solution to the following BSVIE, using Theorem 4.2.2 of Chap. 4:

$$Y_k(t)=\psi(t)+\int_t^T g(t,s,Y_{k-1}(s),Z_k(t,s))ds-\int_t^T Z_k(t,s)dW(s),\qquad t\in[0,T].$$

The following estimate holds: (noting the concavity of $\rho(\cdot)$)

$$\begin{aligned}
&\mathbb{E}\Big[|Y_k(t)|^p+\Big(\int_t^T|Z_k(t,s)|^2ds\Big)^{\frac{p}{2}}\Big]\leqslant K\mathbb{E}\Big[|\psi(t)|^p+\Big(\int_t^T|g(t,s,Y_{k-1}(s),0)|ds\Big)^p\Big]\\
&\leqslant K\mathbb{E}\Big[|\psi(t)|^p+\Big(\int_t^T L_y(t,s)K\big(1+|Y_{k-1}(s)|\big)ds\Big)^p\Big]\\
&\leqslant K\Big[1+\mathbb{E}|\psi(t)|^p+\Big(\int_t^T L_y(t,s)ds\Big)^p\\
&\qquad+\Big(\int_t^T L_y(t,s)^{\frac{p}{p-1}}ds\Big)^{p-1}\mathbb{E}\int_t^T|Y_{k-1}(s)|^pds\Big].
\end{aligned}$$

Then

$$\sup_{k\geqslant1}\mathbb{E}|Y_k(t)|^p\leqslant K\Big[1+\mathbb{E}|\psi(t)|^p+\int_t^T\sup_{k\geqslant1}\mathbb{E}|Y_k(s)|^pds\Big].$$

By Gronwall's inequality, one has

$$\sup_{k\geqslant1}\mathbb{E}|Y_k(t)|^p\leqslant K\Big\{1+\mathbb{E}|\psi(t)|^p+\int_t^T\mathbb{E}|\psi(\tau)|^pd\tau\Big\}.\tag{8.1.48}$$

Hence

$$\begin{aligned}&\int_0^T \sup_{k\geqslant 1}\mathbb{E}\Big[|Y_k(t)|^p+\Big(\int_t^T |Z_k(t,s)|^2ds\Big)^{\frac{p}{2}}\Big]dt\\&\leqslant K\mathbb{E}\Big[1+\int_0^T |\psi(t)|^p dt\Big]<\infty.\end{aligned}\tag{8.1.49}$$

Next, we have (since $q>1$)

$$\begin{aligned}&\mathbb{E}\Big[|Y_k(t)-Y_\ell(t)|^q+\Big(\int_t^T |Z_k(t,s)-Z_\ell(t,s)|^2ds\Big)^{\frac{q}{2}}\Big]\\&\leqslant K\mathbb{E}\Big(\int_t^T \big|g(t,s,Y_{k-1}(s),Z_k(t,s))-g(t,s,Y_{\ell-1}(s),Z_k(t,s))\big|ds\Big)^q\\&\leqslant K\mathbb{E}\Big(\int_t^T L_y(t,s)\rho\big(|Y_{k-1}(s)-Y_{\ell-1}(s)|^q\big)^{\frac{1}{q}}ds\Big)^q\\&\leqslant K\Big(\int_t^T L_y(t,s)^{\frac{q}{q-1}}ds\Big)^{q-1}\mathbb{E}\int_t^T \rho\big(|Y_{k-1}(s)-Y_{\ell-1}(s)|^q\big)ds\\&\leqslant K\int_t^T \rho\big(\mathbb{E}|Y_{k-1}(s)-Y_{\ell-1}(s)|^q\big)ds.\end{aligned}$$

Denote

$$\widehat{Y}_{k,\ell}(t)=Y_k(t)-Y_\ell(t),\quad \widehat{Z}_{k,\ell}(t,s)=Z_k(t,s)-Z_\ell(t,s).$$

By Fatou's lemma, the concavity, and the continuity of ρ, we obtain

$$\overline{\lim_{k,\ell}}\Big[\mathbb{E}|\widehat{Y}_{k,\ell}(t)|^q+\mathbb{E}\Big(\int_t^T |\widehat{Z}_{k,\ell}(t,s)|^2ds\Big)^{\frac{q}{2}}\Big]\leqslant K\int_\tau^T \rho\Big(\overline{\lim_{k,\ell}}\,\mathbb{E}|\widehat{Y}_{k,\ell}(s)|^q\Big)ds.$$

Using Lemma 8.1.8, for almost $s\in[0,T]$, we have $\overline{\lim_{k,\ell}}\,\mathbb{E}|\widehat{Y}_{k,\ell}(s)|^q=0$. Recall (8.1.48), it then follows from dominated convergence theorem that

$$\overline{\lim_{k,\ell}}\int_\tau^T \mathbb{E}|\widehat{Y}_{k,\ell}(t)|^q dt=0,\quad \forall \tau\in[0,T).$$

Thus, $(Y_k(\cdot),Z_k(\cdot\,,\cdot))$ is Cauchy in $\mathcal{H}^q_{\Delta^*}[0,T]$. Let

$$\lim_{k\to\infty}\mathbb{E}\Big[\int_0^T |Y_k(t)-Y(t)|^q dt+\int_0^T\Big(\int_t^T |Z_k(t,s)-Z(t,s)|^2ds\Big)^{\frac{q}{2}}dt\Big]=0,$$

for some $(Y(\cdot),Z(\cdot\,,\cdot))\in\mathcal{H}^q_{\Delta^*}[0,T]$ which is an adpated solution of (8.1.1) in $\mathcal{H}^q_{\Delta^*}[0,T]$. From (8.1.49), we see that actually $(Y(\cdot),Z(\cdot\,,\cdot))\in\mathcal{H}^p_{\Delta^*}[0,T]$. □

In the case that the generator $g(t, s, y, z) = g(t, s, y)$ is independent of z, we have the following result.

Theorem 8.1.9 *Let the generator of BSVIE be given by $g(t, s, y)$ which is independent of z. Let* (H8.1.1)$_p$ *hold for some $p > 1$, replacing* (8.1.36) *with the following:*

$$|g(t, s, y) - g(t, s, \bar{y})| \leqslant L_y(s)\rho\big(|y - \bar{y}|\big), \quad (t, s) \in \Delta^*[0, T], \; y, \bar{y} \in \mathbb{R}^m, \; a.s., \tag{8.1.50}$$

with

$$\int_0^T L_y(s)^{\frac{p(1+\varepsilon)}{p-1}} ds < \infty,$$

for some $\varepsilon > 0$. Then, the conclusions of Theorem 8.1.7 hold.

The proof is almost the same as that of Theorem 8.1.7, except that we directly have the following:

$$\begin{aligned} \mathbb{E}|Y_1(t) - Y_2(t)| &\leqslant \mathbb{E}\int_t^T |g(t, s, Y_1(s)) - g(t, s, Y_2(s))|ds \\ &\leqslant \int_t^T L_y(s)\rho\big(|Y_1(s) - Y_2(s)|\big)ds. \end{aligned}$$

Then, all other procedures can go through. The point here is that $q > 1$ in (8.1.36) is replaced by $q = 1$ in (8.1.50).

2 BMO Martingales and Relevant Results

2.1 A Motivation Example Related to BMO Martingales

In Chap. 4, the theory has been established for Type-I BSVIEs with the generator $g(t, s, y, z)$ having linear growth in (y, z). Moreover, from the last section, we see that if the map $y \mapsto g(t, s, y, z)$ has a superlinear growth (as $y \to \infty$), then one could not expect to have a global adapted solution in general.[1] It is naturally to ask can the map $z \mapsto g(t, s, y, z)$ be superlinear, but we still have global adapted solution?

The purpose of the rest of this chapter is to present some results of Type-I BSVIEs with the generator $g(t, s, y, z)$ having superlinear in z, including quadratic, sub-quadratic, and super-quadratic. Before our presentation of such theory, we need some preparations.

[1] This is the case even for ODEs.

We first consider the following one-dimensional BSDE:

$$Y(t) = \xi + \int_t^T g(s, Z(s))ds - \int_t^T Z(s)dW(s), \qquad t \in [0, T], \tag{8.2.1}$$

with $z \mapsto g(s, z)$ being up to quadratic growth. For convenience, we call such a BSDE a *quadratic backward stochastic differential equation* (QBSDE, for short). Note that one could consider the case that the generator g is also depending on y. But, as a bridge passing to Type-I BSVIEs, the above QBSDE is enough. Before going further, let us present the following example which motivates us to relate to BMO martingales theory.

Example 8.2.1 Consider the following one-dimensional BSDE:

$$\begin{cases} dY(t) = -\frac{1}{2}Z(t)^2 dt + Z(t)dW(t), & t \in [0, T], \\ Y(T) = \xi, \end{cases} \tag{8.2.2}$$

with ξ being $\mathcal{F}_T$-measurable and bounded. This is a BSDE with the generator of quadratic in z. To solve the above, we consider the following BSDE:

$$\eta(t) = e^{\xi} - \int_t^T \zeta(s)dW(s), \qquad t \in [0, T].$$

Obviously, by the boundedness of ξ (so is e^{ξ}), there exists a unique adapted solution (η, ζ) to the above. Actually, we have $\eta(t) = \mathbb{E}_t[e^{\xi}]$ with $t \in [0, T]$. Set

$$Y(t) = \log \eta(t) \equiv \log \mathbb{E}_t\big[e^{\xi}\big], \qquad Z(t) = \frac{\zeta(t)}{\eta(t)}, \quad t \in [0, T]. \tag{8.2.3}$$

Then

$$Y(T) = \log \mathbb{E}_T\big[e^{\xi}\big] = \log e^{\xi} = \xi.$$

Applying Itô's formula to $Y(\cdot)$, we have

$$dY(t) = -\frac{1}{2}\frac{\zeta(t)^2}{\eta(t)^2}dt + \frac{\zeta(t)}{\eta(t)}dW(t) = -\frac{1}{2}Z(t)^2 dt + Z(t)dW(t).$$

Hence, $(Y(\cdot), Z(\cdot))$ defined by (8.2.3) is an adapted solution to BSDE (8.2.2). Moreover, by noting $-\|\xi\|_\infty \leqslant \xi \leqslant \|\xi\|_\infty$, we have

$$|Y(t)| = |\log \eta(t)| \leqslant \|\xi\|_\infty.$$

Further, we apply Itô's formula to $t \mapsto e^{2Y(t)}$ to get the following:

$$d\big[e^{2Y(t)}\big] = 2e^{2Y(t)}Z(t)^2dt + 2e^{2Y(t)}Z(t)dW(t).$$

Thus, for any $\mathbb{F}$-stpping time $\tau \in \mathscr{T}[0,T]$, the set of all $\mathbb{F}$-stopping times valued in $[0,T]$,

$$2\mathbb{E}_\tau \int_\tau^T e^{2Y(s)}|Z(s)|^2ds = \mathbb{E}_\tau\Big[e^{2\xi} - e^{2Y(\tau)}\Big] \leqslant 2e^{2\|\xi\|_\infty}.$$

Hence

$$\begin{aligned}\mathbb{E}_\tau\Big|\int_\tau^T Z(s)dW(s)\Big|^2 &= \mathbb{E}_\tau \int_\tau^T |Z(s)|^2ds \\ &\leqslant \Big(e^{2\|\xi\|_\infty}\mathbb{E}_\tau \int_\tau^T e^{2Y(s)}|Z(s)|^2ds\Big) \leqslant e^{4\|\xi\|_\infty},\end{aligned} \tag{8.2.4}$$

which implies

$$\sup_{\tau\in\mathscr{T}[0,T]} \mathbb{E}_\tau\Big|\int_\tau^T Z(s)dW(s)\Big|^2 \leqslant e^{4\|\xi\|_\infty} < \infty. \tag{8.2.5}$$

On the other hand, let $(Y(\cdot), Z(\cdot))$ be an adapted solution of (8.2.2), meaning that $(Y(\cdot), Z(\cdot)) \in L^2_{\mathbb{F}}(\Omega; C([0,T];\mathbb{R})) \times L^2_{\mathbb{F}}(0,T;\mathbb{R})$ such that (8.2.2) holds. Then, applying Itô's formula to $t \mapsto e^{Y(t)}$, we get

$$e^{Y(t)} = e^{Y(0)} + \int_0^t e^{Y(s)}Z(s)dW(s), \qquad t \in [0,\tau_k],$$

where

$$\tau_k = \inf\{t \in [0,T] \bigm| Y(t) \geqslant k\}.$$

Thus

$$\mathbb{E}\big[e^{Y(\tau_k)}\big] = e^{Y(0)}, \qquad k \geqslant 1.$$

Since $\tau_k \nearrow T$ almost surely, as $k \to \infty$, by Fatou's lemma, we end up

$$\mathbb{E}\big[e^{\xi}\big] \leqslant e^{Y(0)}. \tag{8.2.6}$$

Remark 8.2.2 The above example reveals two facts: (i) QBSDEs could have an adapted solution $(Y(\cdot), Z(\cdot))$, provided the terminal state is bounded, with (8.2.5) holds, which has connection with BMO martingales, and (ii) if QBSDE has an adapted solution, then $\mathbb{E}[e^{\xi}] < \infty$ seems to be enough, which implies that the terminal term ξ is allowed to be unbounded, but not arbitrarily. Needless to say, such an example inspires us to present a general theory of QBSDEs (8.2.1) and further BSVIEs.

2.2 BMO Martingales

From Example 8.2.1, we see that if ξ is bounded, and $(Y(\cdot), Z(\cdot))$ is an adapted solution of (8.2.2), then $Z(\cdot)$ satisfies (8.2.5). This leads to the following consideration.

A local Lebesgue integrable function $f : \mathbb{R}^n \to \mathbb{R}$ is said to be of *bounded mean oscillation* (BMO, for short), if for every $x \in \mathbb{R}^n$,

$$\sup_Q \frac{1}{|Q|} \int_Q |f - f_Q| dx < \infty, \tag{8.2.7}$$

where the supremum is taken over all cubes in $\mathbb{R}^n$, $|Q|$ is the volume of Q, and f_Q is the average of f over Q:

$$f_Q \equiv \frac{1}{|Q|} \int_Q f(x) dx.$$

Therefore, $|f - f_Q|$ can be regarded as the oscillation of the function f within the cube Q around its average f_Q on Q. The term on the left-hand side of (8.2.7) inside of the supreme can be regarded as the mean oscillation of f.

We now introduce the following definition.

Definition 8.2.3 Let $p \geqslant 1$. A uniformly integrable (scalar-valued) martingale $M(\cdot)$ is called an *L^p-BMO martingale* (under probability $\mathbb{P}$) if (under $\mathbb{P}$)

$$\|M(\cdot)\|_{\mathrm{BMO}^p[0,T]} \equiv \sup_{\tau \in \mathscr{T}[0,T]} \big\| \big[\mathbb{E}_\tau |M(T) - M(\tau)|^p\big]^{\frac{1}{p}} \big\|_\infty < \infty. \tag{8.2.8}$$

Let us make some remarks. First of all, (8.2.8) defines a norm. We will see shortly that for any $p > 1$, the norm $\| \cdot \|_{\mathrm{BMO}^p[0,T]}$ is equivalent to $\| \cdot \|_{\mathrm{BMO}^1[0,T]}$. Thus, hereafter, we call any L^p-BMO martingale $M(\cdot)$ simply a BMO martingale and denote $\mathrm{BMO}^p[0, T] \equiv \mathrm{BMO}[0, T]$ to be the set of all BMO martingales on $[0, T]$ (under probability $\mathbb{P}$), without mentioning $p \geqslant 1$.

Next, if $M(\cdot)$ is a martingale, then $M(\tau) = \mathbb{E}_\tau[M(T)]$, for any $\tau \in \mathscr{T}[0, T]$. Therefore, (8.2.8) is equivalent to the following:

$$\sup_{\tau \in \mathscr{T}[0,T]} \operatorname*{esssup}_{\omega \in \Omega} \mathbb{E}_\tau \big| M(T) - \mathbb{E}_\tau[M(T)] \big|^p < \infty. \tag{8.2.9}$$

Comparing (8.2.9) and (8.2.7), one sees the reason of using the name BMO martingale. We now state the equivalence of the norms among different p.

Lemma 8.2.4 *Let $p \in (1, \infty)$. Then, there exists a constant $K > 0$ such that for any uniformly integrable martingale $M(\cdot)$, it holds*

$$\|M(\cdot)\|_{\mathrm{BMO}^1[0,T]} \leqslant \|M(\cdot)\|_{\mathrm{BMO}^p[0,T]} \leqslant K\|M(\cdot)\|_{\mathrm{BMO}^1[0,T]}. \tag{8.2.10}$$

Note that in the above, if one side is finite/infinite so is the other side. Also, a uniformly integrable martingale $M(\cdot)$ might not be a BMO martingale. Here is a simple example.

Example 8.2.5 Let

$$M(t) \equiv \int_0^t W(s)dW(s), \qquad t \in [0, T].$$

Then, $M(\cdot)$ is a martingale whose L^2-norm is bounded. Hence, we may regard it as an element in some set of uniformly integrable martingale. However, by Itô's formula,

$$\begin{aligned}
\mathbb{E}_\tau\Big[|M(T) - \mathbb{E}_\tau M(T)|^2\Big] &= \mathbb{E}_\tau \int_\tau^T W(s)^2 ds \\
&= \mathbb{E}_\tau \int_\tau^T \Big(W(\tau)^2 + 2\int_\tau^s W(t)dW(t) + s - \tau\Big)ds \\
&= (T-\tau)W(\tau)^2 + 2\mathbb{E}_\tau \int_\tau^T (T-t)W(t)dW(t) + \frac{1}{2}(T-\tau)^2 \\
&= (T-\tau)W(\tau)^2 + \frac{1}{2}(T-\tau)^2.
\end{aligned}$$

Clearly, the above is unbounded in $\tau \in \mathscr{T}[0, T]$. Thus, the above $M(\cdot)$ is not a BMO martingale.

Due to (8.2.10), for convenience, in what follows, when we are talking about BMO martingale, we refer to the case of $p = 2$, unless otherwise indicated. The following is a well-known property of BMO martingales.

Lemma 8.2.6 *If $M(\cdot)$ is a BMO martingale on $[0, T]$, then the corresponding Doléans-Dade exponential*

$$\mathcal{E}\{M\}(t) \equiv \exp\Big(M(t) - \frac{1}{2}\langle M\rangle(t)\Big), \quad 0 \leqslant t \leqslant T \tag{8.2.11}$$

is a uniformly integrable martingale, where $\langle M\rangle(\cdot)$ is the quadratic variation process of $M(\cdot)$ (see Chap. 2, Proposition 2.2.5 and (2.3.23))*.*

In the case

$$M(t) \equiv \int_0^t \sigma(s)dW(s), \qquad t \in [0, T],$$

for some $\sigma(\cdot) \in L^2_{\mathbb{F}}(0,T;\mathbb{R})$. Then, we have

$$\langle M\rangle(t) = \int_0^t \sigma(s)^2 ds, \qquad t \geqslant 0.$$

Also

$$X(t) \equiv \mathcal{E}\{M\}(t) = \exp\Big(\int_0^t \sigma(s)dW(s) - \frac{1}{2}\int_0^t \sigma(s)^2 ds\Big), \qquad t \geqslant 0.$$

By Itô's formula, $X(\cdot)$ satisfies the following:

$$\begin{cases} dX(t) = \sigma(t)X(t)dW(t), & t \geqslant 0, \\ X(0) = 1. \end{cases}$$

Clearly, $X(\cdot)$ is invertible, and $X(\cdot)^{-1}$ satisfies

$$\begin{cases} dX(t)^{-1} = \sigma(t)^2 X(t)^{-1}dt - \sigma(t)X(t)^{-1}dW(t), & t \geqslant 0, \\ X(0)^{-1} = 1. \end{cases}$$

The following result gives an equivalence between two relevant processes.

Lemma 8.2.7 *For any $K > 0$, there exist constants $c_1, c_2 > 0$ depending only on K such that for any BMO martingale $M(\cdot)$, and any one-dimensional BMO martingale $N(\cdot)$ satisfying $\|N(\cdot)\|_{\mathrm{BMO}_{\mathbb{P}}[0,T]} \leqslant K$, one has*

$$c_1\|M(\cdot)\|_{\mathrm{BMO}_{\mathbb{P}}[0,T]} \leqslant \|\overline{M}(\cdot)\|_{\mathrm{BMO}_{\widehat{\mathbb{P}}}[0,T]} \leqslant c_2\|M(\cdot)\|_{\mathrm{BMO}_{\mathbb{P}}[0,T]}, \tag{8.2.12}$$

where $d\widehat{\mathbb{P}} \equiv \mathcal{E}\{N\}(T)d\mathbb{P}$ and $\overline{M}(\cdot) \equiv M(\cdot) - \langle M, N\rangle(\cdot)$ with (see Proposition 2.2.5 of Chap. 2)

$$\langle M, N\rangle \equiv \frac{1}{4}\Big(\langle M+N, M+N\rangle - \langle M-N, M-N\rangle\Big).$$

Note that in the case

$$M(t) \equiv \int_0^t \sigma(s)dW(s), \qquad t \in [0,T], \qquad \sigma(\cdot) \in L^p_{\mathbb{F}}(\Omega; L^2(0,T;\mathbb{R}^n)),$$

we have

$$\langle M\rangle(t) = \int_0^t \sigma(s)^2 ds, \qquad t \geqslant 0.$$

In the similar spirit of Theorem 2.3.2 of Chap. 2, for $p > 1$, we have

$$\mathbb{E}_\tau\Big(\int_\tau^T |\sigma(s)|^2 ds\Big)^{\frac{p}{2}} \leqslant K_p \mathbb{E}_\tau\Big|\int_\tau^T \sigma(s)dW(s)\Big|^p \leqslant \overline{K}_p \mathbb{E}_\tau\Big(\int_\tau^T |\sigma(s)|^2 ds\Big)^{\frac{p}{2}}.$$

This, together with the definition of BMO martingales, motivates us to introduce some spaces of *BMO processes*. For $0 \leqslant t < T$, $\mathbb{H} \equiv \mathbb{R}^n, \mathbb{R}^m$, we define

$$\begin{aligned}
\overline{\text{BMO}}([t,T];\mathbb{H}) = \Big\{&\sigma : [t,T]\times\Omega \to \mathbb{H} \;\Big|\; \sigma(\cdot) \in L^2_{\mathbb{F}}(t,T;\mathbb{H}),\\
&\|\sigma(\cdot)\|^2_{\overline{\text{BMO}}([t,T];\mathbb{H})} \equiv \sup_{\tau\in\mathscr{T}[t,T]} \Big\|\mathbb{E}_\tau\Big(\int_\tau^T |\sigma(s)|^2_{\mathbb{H}} ds\Big)\Big\|_\infty < \infty\Big\},\\
\overline{\text{BMO}}(\Delta^*[t,T];\mathbb{H}) = \Big\{&\sigma : \Delta^*[t,T]\times\Omega \to \mathbb{H} \;\Big|\; \sigma(\cdot\,,\cdot) \in L^2_{\mathbb{F}}(\Delta^*[t,T];\mathbb{H}),\\
&\|\sigma(\cdot\,,\cdot)\|^2_{\overline{\text{BMO}}(\Delta^*[t,T];\mathbb{H})} \equiv \operatorname*{esssup}_{r\in[t,T]} \sup_{\tau\in\mathscr{T}[r,T]} \Big\|\mathbb{E}_\tau\Big(\int_\tau^T |\sigma(r,s)|^2_{\mathbb{H}} ds\Big)\Big\|_\infty < \infty\Big\},\\
\overline{\text{BMO}}([T_1,T_2]\times[T_3,T_4];\mathbb{H}) = \Big\{&\sigma : [T_1,T_2]\times[T_3,T_4]\times\Omega \to \mathbb{H} \;\Big|\\
&\sigma(\cdot\,,\cdot) \in L^2_{\mathbb{F}}([T_1,T_2]\times[T_3,T_4];\mathbb{H}),\quad \|\sigma(\cdot\,,\cdot)\|^2_{\overline{\text{BMO}}([T_1,T_2]\times[T_3,T_4];\mathbb{H})} \equiv\\
&\operatorname*{esssup}_{r\in[T_1,T_2]} \sup_{\tau\in\mathscr{T}[T_3,T_4]} \Big\|\mathbb{E}_\tau\Big(\int_\tau^T |\sigma(r,s)|^2_{\mathbb{H}} ds\Big)\Big\|_\infty < \infty\Big\},\quad T_i \in [0,T].
\end{aligned}$$

Obviously, for $\sigma \in \overline{\text{BMO}}([t,T];\mathbb{H})$, with $1 \leqslant i \leqslant n$, the i-th component $\sigma^i \in \overline{\text{BMO}}[t,T] \equiv \overline{\text{BMO}}([t,T];\mathbb{R})$.

We note that for $\sigma(\cdot) \in \overline{\text{BMO}}[t,T]$, if we let $\sigma(s) \equiv 0$, $s \in [0,t)$ extending $\sigma(\cdot)$, then the map $t \mapsto \int_0^t \sigma(s)dW(s)$ is a BMO martingale on $[0,T]$. In fact, for any $\tau \in \mathscr{T}[0,T]$, we have

$$\mathbb{E}_\tau \int_\tau^T |\sigma(s)|^2 ds = \mathbb{E}_\tau \int_\tau^{t\vee\tau} |\sigma(s)|^2 ds + \mathbb{E}_\tau \int_{t\vee\tau}^T |\sigma(s)|^2 ds.$$

For almost all $\omega \in \Omega$, if $\tau(\omega) > t$, then

$$\mathbb{E}_\tau \int_\tau^{t\vee\tau} |\sigma(s)|^2 ds = \mathbb{E}_\tau \int_\tau^{\tau} |\sigma(s)|^2 ds = 0.$$

Otherwise

$$\mathbb{E}_\tau \int_\tau^{t\vee\tau} |\sigma(s)|^2 ds = \mathbb{E}_\tau \int_\tau^{t} |\sigma(s)|^2 ds = 0,$$

where we use the fact that $\sigma(s) = 0$ for $s < t$. Therefore

$$\mathbb{E}_\tau \int_\tau^T |\sigma(s)|^2 ds = \mathbb{E}_\tau \Big\{\mathbb{E}_{t\vee\tau} \int_{t\vee\tau}^T |\sigma(s)|^2 ds\Big\} \leqslant \|\sigma(\cdot)\|^2_{\overline{\text{BMO}}[t,T]}.$$

By the arbitrariness of τ, we obtain that $\int_0^{\cdot}\sigma(s)dW(s)$ is a BMO martingale. Similarly, for $\sigma(\cdot\,,\cdot)\in\overline{\text{BMO}}(\Delta^*[t,T])$, if we let $\sigma(t,s)\equiv 0,\ s\in[0,t)$, extending $\sigma(\cdot\,,\cdot)$, then $s\mapsto\int_0^s\sigma(t,r)dW(r)$ is a BMO martingale on $[0,T]$ for almost all $t\in[0,T)$. We point out that any element in $\overline{\text{BMO}}[t,T]$ or in $\overline{\text{BMO}}(\Delta^*[t,T])$ itself (without Itô's integration) is not a BMO martingale. On the other hand, one sees easily that

$$L^\infty_{\mathbb{F}}(\Omega;L^2(0,T))\subsetneq\overline{\text{BMO}}[0,T]\subsetneq L^2_{\mathbb{F}}(0,T)\subsetneq L^1_{\mathbb{F}}(0,T),\tag{8.2.13}$$

with

$$\begin{aligned}&\Big|\mathbb{E}\int_0^T\sigma(s)ds\Big|\leqslant\mathbb{E}\int_0^T|\sigma(s)|ds\leqslant\sqrt{T}\Big(\mathbb{E}\int_0^T\sigma(s)^2ds\Big)^{\frac12}\\&\leqslant\sqrt{T}\Big(\sup_{\tau\in\mathscr{T}[0,T]}\Big\|\mathbb{E}_\tau\int_\tau^T\sigma(s)^2ds\Big\|_\infty\Big)^{\frac12}\equiv\sqrt{T}\|\sigma(\cdot)\|_{\overline{\text{BMO}}[0,T]}\\&\leqslant\sqrt{T}\Big\|\int_0^T\sigma(s)^2ds\Big\|_\infty^{\frac12}.\end{aligned}\tag{8.2.14}$$

Similar inclusions are true for the spaces on $\Delta^*[0,T]$. However

$$L^\infty_{\mathbb{F}}(\Omega;L^1(0,T))\not\subseteq\overline{\text{BMO}}[0,T];\quad\overline{\text{BMO}}[0,T]\not\subseteq L^\infty_{\mathbb{F}}(\Omega;L^1(0,T)).\tag{8.2.15}$$

Now, for any $\sigma_1(\cdot),\sigma_2(\cdot)\in\overline{\text{BMO}}[0,T]$, let

$$M_1(t)\equiv\int_0^t\sigma_1(s)dW(s),\qquad M_2(t)\equiv\int_0^t\sigma_2(s)dW(s),\qquad t\in[0,T].$$

Then, both $M_1(\cdot)$ and $M_2(\cdot)$ are BMO martingales. Further, it holds that

$$\begin{aligned}&N(t)\equiv M_1(t)-\langle M_1,M_2\rangle(t)\\&=\int_0^t\sigma_1(s)dW(s)-\int_0^t\sigma_1(s)\sigma_2(s)ds=\int_0^t\sigma_1(s)d\widehat{W}(s),\quad t\in[0,T],\end{aligned}$$

where

$$\widehat{W}(t)\equiv W(t)-\int_0^t\sigma_2(s)ds,\qquad t\in[0,T].$$

In this case, the exponential supermartingale $\widehat{\mathcal{E}}\{\sigma_2\}$ of $\sigma_2(\cdot)$ defined by

$$\widehat{\mathcal{E}}\{\sigma_2\}(t)\equiv\exp\Big(\int_0^t\sigma_2(s)dW(s)-\frac12\int_0^t\sigma_2(s)^2ds\Big),\qquad t\in[0,T]$$

is a uniformly integrable martingale (see Lemma 8.2.6). Hence, by Girsanov theorem (Theorem 2.3.9 of Chap. 2), we may define probability $\widehat{\mathbb{P}}$ by

$$d\widehat{\mathbb{P}} \equiv \mathcal{E}\{M_2\}(T)d\mathbb{P} = \widehat{\mathcal{E}}\{\sigma_2\}(T)d\mathbb{P},$$

and obtain that $\widehat{W}(\cdot)$ is a standard Brownian motion under $\widehat{\mathbb{P}}$. Further

$$\mathbb{E}_\tau^{\widehat{\mathbb{P}}}\big|N(T)-N(\tau)\big|^2 = \mathbb{E}_\tau^{\widehat{\mathbb{P}}}\int_\tau^T \sigma_1(s)^2ds = \mathbb{E}\Big[\widehat{\mathcal{E}}\{\sigma_2\}(T)\int_\tau^T \sigma_1(s)^2ds\Big]$$

where $\mathbb{E}^{\widehat{\mathbb{P}}}$ is the expectation under probability $\widehat{\mathbb{P}}$. Therefore, for any $\sigma_2 \in \overline{\mathrm{BMO}}[0,T]$ satisfying $\|\sigma_2\|_{\overline{\mathrm{BMO}}[0,T]} \leqslant \bar{K}$, for some constant $\bar{K}$, the above (8.2.12) reduces to

$$\begin{aligned}&c_1 \sup_{\tau\in\mathscr{T}[0,T]}\Big\|\mathbb{E}_\tau\int_\tau^T \sigma_1(s)^2ds\Big\|_\infty^{\frac{1}{2}} \leqslant \sup_{\tau\in\mathscr{T}[0,T]}\Big\|\mathbb{E}_\tau^{\widehat{\mathbb{P}}}\int_\tau^T \sigma_1(s)^2ds\Big\|_\infty^{\frac{1}{2}}\\&\leqslant c_2 \sup_{\tau\in\mathscr{T}[0,T]}\Big\|\mathbb{E}_\tau\int_\tau^T \sigma_1(s)^2ds\Big\|_\infty^{\frac{1}{2}},\end{aligned}\qquad(8.2.16)$$

where c_1, c_2 only depend on the constant $\bar{K}$.

To conclude this subsection, let us state the so-called *John-Nirenberg inequality*.

Lemma 8.2.8 *Suppose* $M(\cdot)\in \mathrm{BMO}[0,T]$ *with* $\|M(\cdot)\|_{\mathrm{BMO}[0,T]} < \frac{1}{4}$. *Then*

$$\mathbb{E}_\tau\Big[\exp\Big(|M(T)-M(\tau)|\Big)\Big] \leqslant \frac{1}{1-4\|M(\cdot)\|_{\mathrm{BMO}[0,T]}},\qquad \forall\tau\in\mathscr{T}[0,T].\qquad(8.2.17)$$

In applications, the following *extended John-Nirenberg inequality* is even more useful.

Lemma 8.2.9 *Let* $X(\cdot)\in L^2_{\mathbb{F}}(0,T;\mathbb{R})$ *satisfy*

$$\mathbb{E}_\tau\Big[\int_\tau^T |X(s)|^2ds\Big] \leqslant C_0,\qquad \tau\in\mathscr{T}[0,T],\qquad(8.2.18)$$

for some $C_0 > 0$. *Then*

$$\mathbb{E}_\tau\Big[\exp\Big(\varepsilon\int_\tau^T |X(s)|^2ds\Big)\Big] \leqslant \frac{1}{1-\varepsilon C_0},\quad \forall\tau\in\mathscr{T}[0,T],\ 0<\varepsilon<\frac{1}{C_0}.\qquad(8.2.19)$$

Note that if $X(\cdot)\in\overline{\mathrm{BMO}}[0,T]$, then by taking

$$C_0 \equiv \|X(\cdot)\|^2_{\overline{\mathrm{BMO}}[0,T]},\qquad \varepsilon\equiv\frac{1}{2\|X(\cdot)\|^2_{\overline{\mathrm{BMO}}[0,T]}},$$

we have for any $\tau \in \mathscr{T}[0,T]$,

$$\mathbb{E}_\tau\Big[\exp\Big(\frac{1}{2\|X(\cdot)\|^2_{\mathrm{BMO}[0.T]}}\int_\tau^T |X(s)|^2 ds\Big)\Big] \leqslant 2. \tag{8.2.20}$$

This inequality will be used several times later.

2.3 Some Further Results

Next, we present some further relevant results which will be useful for one-dimensional QBSDE (8.2.1).

The following lemma gives some useful approximation of general functions.

Lemma 8.2.10 *Let $g : \mathbb{R}^n \to \mathbb{R}$ be continuous and bounded by $K > 0$. Then*

$$g_j(x) \equiv \inf_{x' \in \mathbb{R}^n} \Big\{ g(x') + j|x - x'| \Big\} \tag{8.2.21}$$

is well-defined for $j \geqslant K$. The function $g_j(\cdot)$ is uniformly bounded by K, monotone increasing in j and

$$|g_j(x) - g_j(y)| \leqslant j|x - y|, \qquad \forall x, y \in \mathbb{R}^n, \tag{8.2.22}$$

$$\lim_{j\to\infty} g_j(x_j) = g(x), \qquad as \ \lim_{j\to\infty} x_j = x. \tag{8.2.23}$$

Proof Since $g(\cdot)$ is bounded by $K > 0$, for any $j \geqslant K$, one has

$$g(x') + j|x - x'| \geqslant -|g(x')| \geqslant -K, \qquad \forall x, x' \in \mathbb{R}^n.$$

Thus, $g_j(\cdot)$ is well-defined in $\mathbb{R}^n$ with $g_j(\cdot) \geqslant -K$. Also,

$$g_j(x) \leqslant \inf_{x' \in \mathbb{R}^n} \{|g(x')| + j|x - x'|\} \leqslant |g(x)| \leqslant K,$$

showing the boundedness of $g_j(\cdot)$ by K. Next, the monotonicity of $g_j(\cdot)$ in j is clear. Now, for $x, y \in \mathbb{R}^n$ and $x' \in \mathbb{R}^n$,

$$\begin{aligned} g(x') + j|x - x'| &= g(x') + j|y - x'| + j|x - x'| - j|y - x'| \\ &\geqslant g_j(y) - j|x - y|, \end{aligned}$$

which implies that

$$g_j(x) - g_j(y) \geqslant -j|x - y|.$$

By the symmetry of x and y, we obtain (8.2.22). Finally, let $x_j \to x$. For every j, by definition of $g_j(\cdot)$, there exists $y_j \in \mathbb{R}^n$ such that

$$g(x_j) \geqslant g_j(x_j) \geqslant g(y_j) + j|x_j - y_j| - \frac{1}{j}.$$

By the boundedness of $g(\cdot)$, it is easy to see

$$|y_j - x| \leqslant |y_j - x_j| + |x_j - x| \to 0, \qquad \text{as } j \to \infty.$$

Moreover, by (8.2.21), the definition of $g_j(\cdot)$,

$$g(x_j) \geqslant g_j(x_j) \geqslant g(y_j) - \frac{1}{j}.$$

This together with the continuity of $g(\cdot)$ and $x_j, y_j \to x$, leads to (8.2.23). □

The following is a standard result in functional analysis.

Lemma 8.2.11 *Let $X_k(\cdot) \in L^2_{\mathbb{F}}(0, T; \mathbb{R}^n)$ such that*

$$\sup_{k \geqslant 1} \mathbb{E} \int_0^T |X_k(t)|^2 dt < \infty.$$

Then, there exists a subsequence $\{X_{k_i}(\cdot)\}_{i \geqslant 1}$ converges weakly to some $X(\cdot) \in L^2_{\mathbb{F}}(0, T; \mathbb{R}^n)$, i.e.,

$$\lim_{i \to \infty} \mathbb{E} \int_0^T \langle X_{k_i}(t), \varphi(t) \rangle dt = \mathbb{E} \int_0^T \langle X(t), \varphi(t) \rangle dt, \quad \forall \varphi(\cdot) \in L^2_{\mathbb{F}}(0, T; \mathbb{R}^n).$$

Next, we introduce some more sets. For each $t \in [0, T]$, $\mu, \theta > 0$, $q \geqslant 1$, we introduce the following sets of processes:

$$\mathcal{E}^{\mu,\theta}_{\mathcal{F}_t}(\Omega; \mathbb{R}^n) = \Big\{\xi \in L^1_{\mathcal{F}_t}(\Omega; \mathbb{R}^n) \;\Big|\; \mathbb{E}\big[\exp\big(\mu|\xi|^\theta\big)\big] < \infty\Big\},$$

$$\mathcal{E}^{\mu,\theta}_{\mathcal{F}_T}(\Omega; L^q(t, T; \mathbb{R}^n)) = \Big\{\varphi(\cdot) \in L^q_{\mathcal{F}_T}(t, T; \mathbb{R}^n) \;\Big|\; \mathbb{E}\Big[e^{\mu\|\varphi(\cdot)\|^\theta_{L^q(t,T)}}\Big] < \infty\Big\},$$

$$\mathcal{E}^{\mu,\theta}_{\mathbb{F}}(\Omega; L^q(t, T; \mathbb{R}^n)) = \Big\{\varphi(\cdot) \in \mathcal{E}^{\mu,\theta}_{\mathcal{F}_T}(\Omega; L^q(t, T; \mathbb{R}^n)) \;\Big|\; \varphi(\cdot) \text{ is } \mathbb{F}\text{-progressively measurable}\Big\},$$

$$\mathcal{E}^{\mu,\theta}_{\mathcal{F}_T}(\Omega; C([t, T]; \mathbb{R}^n)) = \Big\{\varphi(\cdot) \in \mathcal{E}^{\mu,\theta}_{\mathcal{F}_T}(\Omega; L^\infty(t, T; \mathbb{R}^n)) \;\Big|\; \varphi(\cdot) \text{ is continuous on } [t, T]\Big\},$$

$$\mathcal{E}_{\mathbb{F}}^{\mu,\theta}(\Omega; C([t,T];\mathbb{R}^n)) = \Big\{\varphi(\cdot) \in \mathcal{E}_{\mathcal{F}_T}^{\mu,\theta}(\Omega; C([t,T];\mathbb{R}^n)) \mid \varphi(\cdot) \text{ is } \mathbb{F}\text{-progressively measurable}\Big\}.$$

Similarly, we define

$$\mathcal{E}_{\mathcal{F}_T}^{\mu,\theta}(\Omega; L^q(\Delta^*[t,T];\mathbb{R}^n)) = \Big\{\varphi : \Delta^*[t,T]\times\Omega \to \mathbb{R}^n \;\Big|\; \varphi(t,\cdot) \text{ is } \mathcal{F}_T\text{-measurable}, \operatorname*{esssup}_{\tau\in[t,T]} \mathbb{E}_\tau\Big[e^{\mu\|\varphi(\tau,\cdot)\|^\theta_{L^q(t,T)}}\Big] < \infty\Big\},$$

$$\mathcal{E}_{\mathbb{F}}^{\mu,\theta}(\Omega; L^q(\Delta^*[t,T];\mathbb{R}^n)) = \Big\{\varphi(\cdot) \in \mathcal{E}_{\mathcal{F}_T}^{\mu,\theta}(\Omega; L^q(\Delta^*[t,T];\mathbb{R}^n)) \mid \varphi(t,\cdot) \text{ is } \mathbb{F}\text{-progressively measurable on } [t,T]\Big\}.$$

Now, for any $\xi_1, \xi_2 \in \mathcal{E}_{\mathcal{F}_t}^{\mu,1}(\Omega;\mathbb{R}^n)$ and $\lambda \in (0,1)$,

$$\mathbb{E}\Big[e^{\mu|\lambda\xi_1+(1-\lambda)\xi_2|}\Big] \leqslant \mathbb{E}\Big[\Big(e^{\mu|\xi_1|}\Big)^\lambda\Big(e^{\mu|\xi_2|}\Big)^{1-\lambda}\Big] \leqslant \mathbb{E}\Big[\lambda e^{\mu|\xi_1|} + (1-\lambda)e^{\mu|\xi_2|}\Big].$$

Thus, $\mathcal{E}_{\mathcal{F}_t}^{\mu,1}(\Omega;\mathbb{R}^n)$ is a convex set but not a linear space. However, $\bigcap\limits_{\mu>0}\mathcal{E}_{\mathcal{F}_t}^{\mu,\theta}(\Omega;\mathbb{R}^n)$ is a linear space for any $\theta > 0$. As a matter of fact, for any

$$\xi_1,\xi_2 \in \bigcap_{\mu>0}\mathcal{E}_{\mathcal{F}_t}^{\mu,\theta}(\Omega;\mathbb{R}^n), \quad \alpha_1,\alpha_2\in\mathbb{R},$$

one has

$$\begin{aligned}
&\mathbb{E}\Big[e^{\mu|\alpha_1\xi_1+\alpha_2\xi_2|^\theta}\Big] \leqslant \mathbb{E}\Big[e^{\mu 2^{(\theta-1)^+}(|\alpha_1\xi_1|^\theta+|\alpha_2\xi_2|^\theta)}\Big]\\
&\leqslant \mathbb{E}\Big[\Big(e^{\mu 2^{(\theta-1)^+}(|\alpha_1|^\theta+|\alpha_2|^\theta)|\xi_1|^\theta}\Big)^{\frac{|\alpha_1|^\theta}{|\alpha_1|^\theta+|\alpha_2|^\theta}}\Big(e^{\mu 2^{(\theta-1)^+}(|\alpha_1|^\theta+|\alpha_2|^\theta)|\xi_2|^\theta}\Big)^{\frac{|\alpha_2|^\theta}{|\alpha_1|^\theta+|\alpha_2|^\theta}}\Big]\\
&\leqslant \frac{|\alpha_1|^\theta}{|\alpha_2|^\theta+|\alpha_2|^\theta}\mathbb{E}\Big[e^{\mu 2^{(\theta-1)^+}(|\alpha_1|^\theta+|\alpha_2|^\theta)|\xi_1|^\theta}\Big]\\
&\qquad + \frac{|\alpha_2|^\theta}{|\alpha_1|^\theta+|\alpha_2|^\theta}\mathbb{E}\Big[e^{\mu 2^{(\theta-1)^+}(|\alpha_1|^\theta+|\alpha_2|^\theta)|\xi_2|^\theta}\Big] < \infty.
\end{aligned}$$

The other sets defined above are similar. For instance, $\mathcal{E}_{\mathbb{F}}^{\mu,1}(\Omega; L^q(\Delta^*[t,T];\mathbb{R}^n))$ is a convex set but not a linear space. However, $\bigcap\limits_{\mu>0}\mathcal{E}_{\mathbb{F}}^{\mu,\theta}(\Omega; L^q(\Delta^*[t,T];\mathbb{R}^n))$ is a linear space, for any $\theta > 0$. Also, the following holds:

$$L^\infty_{\mathbb{F}}(\Omega; L^q(t,T;\mathbb{R}^n)) \subsetneq \mathcal{E}_{\mathbb{F}}^{\mu,\theta}(\Omega; L^q(t,T;\mathbb{R}^n)) \subsetneq L^q_{\mathbb{F}}(t,T;\mathbb{R}^n), \ \forall \mu,\theta>0. \tag{8.2.24}$$

Similar inclusions hold when $\mathbb{F}$ is replaced by $\mathcal{F}_T$. From the above, we see that elements in the above sets are not necessarily bounded.

Next, for fixed $t \in [0, T)$, we introduce $\mathcal{E}^{\mu,\theta}_{\mathbb{F},\infty}(\Omega; L^1(t, T; \mathbb{R}^n))$ which is the set of $\mathbb{F}$-adapted process $\varphi(\cdot)$ such that

$$\sup_{\tau\in\mathscr{T}[t,T]} \Big\|\mathbb{E}_\tau\Big[\exp\Big\{\mu\|\varphi(\cdot)\|^\theta_{L^q(\tau,T;\mathbb{R}^n)}\Big\}\Big]\Big\|_\infty < \infty.$$

By the definition, we see that

$$\mathcal{E}^{\mu,\theta}_{\mathbb{F},\infty}(\Omega; L^1(t, T; \mathbb{R}^n)) \subseteq \bigcap_{\mu>0} \mathcal{E}^{\mu,\theta}_{\mathbb{F}}(\Omega; L^{1+\delta}(t, T; \mathbb{R}^n)).$$

Now, we look at its connection with $\overline{\mathrm{BMO}}([t, T]; \mathbb{R}^n)$. For any $t \in [0, T)$, and $\varphi(\cdot) \in \overline{\mathrm{BMO}}([t, T]; \mathbb{R}^n)$, $\delta \in [0, 1)$, and $\mu > 0$, by Young's inequality, we have

$$\begin{aligned}
&\mu n^{\frac{3+\delta}{2}}|\varphi^i(s)|^{1+\delta} = \Big(\frac{|\varphi^i(s)|}{\sqrt{(1+\delta)\widetilde{M}}}\Big)^{1+\delta}\Big(\mu n^{\frac{3+\delta}{2}}\big[(1+\delta)\widetilde{M}\big]^{\frac{1+\delta}{2}}\Big)\\
&\leqslant \frac{1+\delta}{2}\frac{\varphi^i(s)^2}{(1+\delta)\widetilde{M}} + \frac{1-\delta}{2}\mu^{\frac{2}{1-\delta}}n^{\frac{3+\delta}{1-\delta}}(1+\delta)^{\frac{1+\delta}{1-\delta}}\widetilde{M}^{\frac{(1+\delta)}{1-\delta}}\\
&\equiv \frac{\varphi^i(s)^2}{2\widetilde{M}} + K_0(\delta, n, \mu)\widetilde{M}^{\frac{(1+\delta)}{1-\delta}} \equiv \frac{\varphi^i(s)^2}{2\widetilde{M}} + \widetilde{K}(\delta, n, \mu, \widetilde{M}), \quad s \in [t, T],
\end{aligned}$$

with $\varphi \equiv (\varphi^1, \cdot s\varphi^n)^\top$, and

$$\begin{aligned}
&\widetilde{M} \equiv \operatorname*{esssup}_{\tau\in\mathscr{T}[t,T]}\Big\|\sum_{1\leqslant i\leqslant n}\mathbb{E}_\tau\int_\tau^T|\varphi^i(s)|^2ds\Big\|_\infty = \big\|\varphi(\cdot)\big\|^2_{\overline{\mathrm{BMO}}([t,T];\mathbb{R}^n)},\\
&K_0(\delta, n, \mu) \equiv \frac{1-\delta}{2}(1+\delta)^{\frac{1+\delta}{1-\delta}}n^{\frac{3+\delta}{1-\delta}}\mu^{\frac{2}{1-\delta}},\\
&\widetilde{K}(\delta, n, \mu, \widetilde{M}) \equiv K_0(\delta, n, \mu)\widetilde{M}^{\frac{(1+\delta)}{1-\delta}}.
\end{aligned}\tag{8.2.25}$$

Define $\widetilde{\varphi}^i(\cdot) \equiv \frac{1}{\sqrt{2\widetilde{M}}}\varphi^i(\cdot)$. We have that $\|\widetilde{\varphi}^i(\cdot)\|_{\overline{\mathrm{BMO}}[t,T]} \leqslant \frac{1}{2}$. Thus, by John-Nirenberg inequality (8.2.20), we see that

$$\mathbb{E}_\tau\Big[\exp\Big(\frac{1}{2\widetilde{M}}\int_\tau^T|\varphi^i(s)|^2ds\Big)\Big] \leqslant 2, \quad \forall\tau \in \mathscr{T}[t, T].$$

Consequently, for any $\mu \geqslant 1$, and $\tau \in \mathscr{T}[t, T]$,

$$
\begin{aligned}
&\mathbb{E}_\tau\Big[\exp\Big(\mu\int_\tau^T|\varphi(s)|^{1+\delta}ds\Big)\Big]\leqslant\mathbb{E}_\tau\Big[\exp\Big(\mu n^{\frac{1+\delta}{2}}\sum_{i=1}^n\int_\tau^T|\varphi^i(s)|^{1+\delta}ds\Big)\Big]\\
&=\mathbb{E}_\tau\Big[\prod_{i=1}^n\exp\Big(\mu n^{\frac{1+\delta}{2}}\int_\tau^T|\varphi^i(s)|^{1+\delta}ds\Big)\Big]\\
&\leqslant\prod_{i=1}^n\Big\{\mathbb{E}_\tau\Big[\exp\Big(\mu n^{\frac{3+\delta}{2}}\int_\tau^T|\varphi^i(s)|^{1+\delta}ds\Big)\Big]\Big\}^{\frac{1}{n}}\\
&\leqslant\prod_{i=1}^n\Big\{\mathbb{E}_\tau\Big[\exp\Big(\int_\tau^T\frac{|\varphi^i(s)|^2}{2\widetilde{M}}ds\Big)\Big]\Big\}^{\frac{1}{n}}e^{(T-t)\widetilde{K}(\delta,n,\mu,\widetilde{M})}\\
&\leqslant 2e^{(T-t)\widetilde{K}(\delta,n,\mu,\widetilde{M})}<\infty.
\end{aligned}
$$

In the above first inequality, we use the fact:

$$
|x|^p=\Big(\sum_{i=1}^n|x_i|^2\Big)^{\frac{p}{2}}\leqslant n^{\frac{p}{2}}\sum_{i=1}^n|x_i|^p,\quad x=(x_1,\cdot s,x_n)^\top,\quad p\in(1,2),
$$

and in the second inequality, we use

$$
\mathbb{E}_\tau\Big[\prod_{i=1}^n|\xi_i|\Big]\leqslant\prod_{i=1}^n\Big[\mathbb{E}_\tau|\xi_i|^n\Big]^{\frac{1}{n}},
$$

with ξ_i being $\mathcal{F}_T$-measurable random variable. This implies that for any $\delta\in[0,1)$,

$$
\overline{\mathrm{BMO}}([t,T];\mathbb{R}^n)\subseteq\bigcap_{\mu>0}\mathcal{E}^{\mu,1}_{\mathbb{F},\infty}(\Omega;L^{1+\delta}(t,T;\mathbb{R}^n)),\tag{8.2.26}
$$

which shows that

$$
\overline{\mathrm{BMO}}([t,T];\mathbb{R}^n)\subseteq\bigcap_{\mu>0}\mathcal{E}^{\mu,1}_{\mathbb{F}}(\Omega;L^{1+\delta}(t,T;\mathbb{R}^n)).
$$

Similarly, for fixed $t\in[0,T)$, $\varphi(\cdot\,,\cdot)\in\overline{\mathrm{BMO}}(\Delta^*[t,T];\mathbb{R}^n)$, $r\in[t,T]$, $\tau\in\mathscr{T}[r,T]$,

$$
\begin{aligned}
&\mathbb{E}_\tau\Big[\exp\Big(\mu\int_\tau^T|\varphi(r,s)|^{1+\delta}ds\Big)\Big]\\
&\leqslant\prod_{i=1}^n\Big\{\mathbb{E}_\tau\Big[\exp\Big(\int_\tau^T\frac{|\varphi^i(r,s)|^2}{2\widetilde{M}}ds\Big)\Big]\Big\}^{\frac{1}{n}}e^{(T-r)\widetilde{K}(\delta,n,\mu,\widetilde{M})}\\
&\leqslant 2e^{(T-r)\widetilde{K}(\delta,n,\mu,\widetilde{M})}<\infty,
\end{aligned}\tag{8.2.27}
$$

with $\widetilde{K}$ in (8.2.25),

$$\widetilde{M} \equiv \operatorname*{esssup}_{r\in[t,T]} \operatorname*{esssup}_{\tau\in\mathscr{T}[r,T]} \Big\| \sum_{i=1}^{n} \mathbb{E}_\tau \int_\tau^T |\varphi^i(r,s)|^2 ds \Big\|_\infty .$$

Thus

$$\overline{\mathrm{BMO}}(\Delta^*[t,T];\mathbb{R}^n) \subseteq \bigcap_{\mu>0} \mathcal{E}_{\mathbb{F}}^{\mu,1}(\Omega; L^{1+\delta}(\Delta^*[t,T];\mathbb{R}^n)). \tag{8.2.28}$$

3 Well-Posedness of QBSDEs

In this section, we are going to discuss quadratic BSDEs.

3.1 One-Dimensional Cases with Bounded Terminal Conditions

We now discuss well-posedness of one-dimensional QBSDE (8.2.1). To begin with, we first introduce the following hypothesis on the generator.

(H8.3.1) Let $g : [0,T] \times \mathbb{R} \times \Omega \to \mathbb{R}$ be $\mathcal{B}([0,T]\times\mathbb{R})\otimes\mathcal{F}_T$-measurable such that $s \mapsto g(s,z)$ is $\mathbb{F}$-progressively measurable for all $z \in \mathbb{R}$. There exist $\alpha(\cdot) \in L_{\mathbb{F}}^\infty(\Omega; L^1(0,T;\mathbb{R}_+))$, $\beta(\cdot) \in \overline{\mathrm{BMO}}[0,T]$, and $\gamma \in [\frac{1}{4},\infty)$ such that

$$|g(t,0)| \leqslant \alpha(t), \qquad t \in [0,T], \quad \text{a.s.} \tag{8.3.1}$$

and

$$\begin{aligned} |g(t,z_1) - g(t,z_2)| \leqslant [\beta(t) + \gamma\big(|z_1| + |z_2|\big)]|z_1 - z_2|, \\ t \in [0,T],\ z_1.z_2 \in \mathbb{R}, \quad \text{a.s.} \end{aligned} \tag{8.3.2}$$

Combining the above (8.3.1) and (8.3.2), taking into account $\gamma \geqslant \frac{1}{4}$, one has

$$\begin{aligned} |g(t,z)| &\leqslant \alpha(t) + \beta(t)|z| + \gamma|z|^2 \leqslant \alpha(t) + \beta(t)^2 + \Big(\frac{1}{4} + \gamma\Big)|z|^2 \\ &\leqslant \alpha(t) + \beta(t)^2 + 2\gamma|z|^2, \qquad (t,z) \in [0,T]\times\mathbb{R}. \end{aligned} \tag{8.3.3}$$

We see that $\gamma \geqslant \frac{1}{4}$ is just for convenience, since $\gamma > 0$ will be enough. Note that in the above, the condition imposed on $\alpha(\cdot)$ means

$$\Big\| \int_0^T \alpha(s)ds \Big\|_{L^\infty_{\mathcal{F}_T}(\Omega)} < \infty, \tag{8.3.4}$$

which means that $t \mapsto \alpha(t)$ is merely L^1, almost surely, whereas condition (8.3.2) means that the "Lipschitz constant" of $z \mapsto g(t,z)$ is allowed to be random (and to linearly grow in z) and $t \mapsto \beta(t)$ is required to be L^2. Therefore, $\alpha(\cdot)$ and $\beta(\cdot)$ have different features (see (8.2.15)). The following gives an a priori estimate of adapted solutions to QBSDE (8.2.1).

Lemma 8.3.1 *Let* (H8.3.1) *hold, and* $\xi \in L^\infty_{\mathcal{F}_T}(\Omega)$. *Let* $(Y(\cdot), Z(\cdot)) \in L^\infty_{\mathbb{F}}(\Omega;\ C([0,T])) \times L^2_{\mathbb{F}}(0,T)$ *be an adapted solution to* (8.2.1). *Then*

$$\|Y(\cdot)\|_{L^\infty_{\mathbb{F}}(\Omega;C([0,T]))} \leqslant \|\xi\|_{L^\infty_{\mathcal{F}_T}(\Omega)} + \Big\| \int_0^T \alpha(s)ds \Big\|_{L^\infty_{\mathcal{F}_T}(\Omega)}, \tag{8.3.5}$$

and $Z(\cdot) \in \overline{\mathrm{BMO}}[0,T]$ *satisfying*

$$\begin{aligned} \|Z(\cdot)\|_{\overline{\mathrm{BMO}}[0,T]} \leqslant K e^{K(\|\xi\|_{L^\infty_{\mathcal{F}_T}(\Omega)} + \| \int_0^T \alpha(s)ds \|_{L^\infty_{\mathcal{F}_T}(\Omega)})} \\ \cdot\Big(1 + \Big\| \int_0^T [\alpha(s) + \beta(s)^2]ds \Big\|_{L^\infty_{\mathcal{F}_T}(\Omega)}\Big), \end{aligned} \tag{8.3.6}$$

for some constant $K > 0$, *depending on* γ.

Proof First, we pick a $\lambda > 0$, undetermined. Applying Itô's formula to $e^{\lambda Y(\cdot)}$, by (8.3.2), we have

$$\begin{aligned} de^{\lambda Y(s)} &= e^{\lambda Y(s)}\Big(-\lambda g(s, Z(s))ds + \lambda Z(s)dW(s) + \frac{1}{2}\lambda^2|Z(s)|^2 ds\Big) \\ &\geqslant -\lambda e^{\lambda Y(s)}\big[\alpha(s) + \beta(s)^2\big]ds + \lambda e^{\lambda Y(s)} Z(s)dW(s) \\ &\quad + \Big(\frac{1}{2}\lambda^2 - 2\gamma\lambda\Big)e^{\lambda Y(s)}|Z(s)|^2 ds. \end{aligned}$$

Consequently

$$\mathbb{E}_t \int_t^T \Big(\frac{1}{2}\lambda^2 - 2\gamma\lambda\Big)e^{\lambda Y(s)}|Z(s)|^2 ds \leqslant \mathbb{E}_t e^{\lambda\xi} + \mathbb{E}_t \int_t^T \lambda e^{\lambda Y(s)}\big[\alpha(s) + \beta(s)^2\big]ds.$$

We choose suitable λ such that $\frac{1}{2}\lambda^2 - 2\gamma\lambda = 1$, i.e.,

$$\lambda^2 - 4\gamma\lambda - 2 = 0 \qquad \Rightarrow \qquad \lambda = \frac{4\gamma + \sqrt{16\gamma^2 + 8}}{2} = 2\gamma + \sqrt{4\gamma^2 + 2},$$

and thus obtain

$$
\begin{aligned}
&e^{-\lambda\|Y(\cdot)\|_{L^\infty_{\mathbb{F}}(\Omega;C([t,T]))}}\mathbb{E}_t\int_t^T|Z(s)|^2ds\\
&\leqslant e^{\lambda\|\xi\|_{L^\infty_{\mathcal{F}_T}(\Omega)}}+\lambda e^{\lambda\|Y(\cdot)\|_{L^\infty_{\mathbb{F}}(\Omega;C([t,T]))}}\Big\|\int_t^T[\alpha(s)+\beta(s)^2]ds\Big\|_{L^\infty_{\mathcal{F}_T}(\Omega)}.
\end{aligned}
$$

Consequently

$$
\begin{aligned}
\mathbb{E}_t\int_t^T|Z(s)|^2ds\leqslant{}& e^{\lambda\big(\|\xi\|_{L^\infty_{\mathcal{F}_T}(\Omega)}+\|Y(\cdot)\|_{L^\infty_{\mathbb{F}}(\Omega;C([t,T]))}\big)}\\
&+\lambda e^{2\lambda\|Y(\cdot)\|_{L^\infty_{\mathbb{F}}(\Omega;C([t,T]))}}\Big\|\int_t^T\big[\alpha(s)+\beta(s)^2\big]ds\Big\|_{L^\infty_{\mathcal{F}_T}(\Omega)}.
\end{aligned}
$$

This implies $Z(\cdot)\in\overline{\mathrm{BMO}}[0,T]$. Next, we observe

$$
Y(t)=\xi+\int_t^T\Big[g(s,0)+g_z(s)Z(s)\Big]ds-\int_t^T Z(s)dW(s),
$$

where

$$
g_z(s)\equiv\frac{g(s,Z(s))-g(s,0)}{|Z(s)|^2}Z(s)\mathbf{1}_{\{|Z(s)|>0\}}.
$$

From (H8.3.1), $|g_z(\cdot)|\leqslant\beta(\cdot)+\gamma|Z(\cdot)|$ and hence $g_z(\cdot)\in\overline{\mathrm{BMO}}[0,T]$ (because $\beta(\cdot),Z(\cdot)\in\overline{\mathrm{BMO}}[0,T]$). Therefore, the process

$$
\widehat{W}(t)\equiv W(t)-\int_0^t g_z(s)ds,\qquad t\in[0,T],
$$

is a Brownian motion under probability $\widehat{\mathbb{P}}$ defined by $d\widehat{\mathbb{P}}\equiv\widehat{\mathcal{E}}\{g_z\}(T)d\mathbb{P}$ (see (2.3.23) of Chap. 2). Hence

$$
Y(t)=\xi+\int_t^T g(s,0)ds-\int_t^T Z(s)d\widehat{W}(s),
$$

which leads to

$$
Y(t)=\mathbb{E}_t^{\widehat{\mathbb{P}}}\Big[\xi+\int_t^T g(s,0)ds\Big].
$$

In the above, we have used the conclusion of Lemma 8.2.7. Then, (8.3.5) follows. Consequently, (8.3.6) also follows. □

The following gives a comparison for adapted solutions of QBSDE (8.2.1). In some sense, it is an extension of Theorem 6.1.6 of Chap. 6.

Lemma 8.3.2 *Suppose* $g_0(\cdot), g_1(\cdot)$ *satisfy* (H8.3.1), $\xi_0, \xi_1 \in L^\infty_{\mathcal{F}_T}(0,T;\mathbb{R})$. *Let* (Y_0, Z_0), $(Y_1, Z_1) \in L^\infty_{\mathbb{F}}(\Omega; C([0,T])) \times \overline{\mathrm{BMO}}[0,T]$ *be the corresponding adapted solutions to BSDE* (8.2.1). *Moreover*

$$\xi_0 \leqslant \xi_1, \quad g_0(s,z) \leqslant g_1(s,z), \quad (s,z) \in [0,T] \times \mathbb{R}, \quad a.s.$$

Then, for each $t \in [0,T]$, $Y_0(t) \leqslant Y_1(t)$, *almost surely.*

Proof Observe the following:

$$\begin{aligned} Y_1(t) - Y_0(t) &= \xi_1 - \xi_0 + \int_t^T [g_1(s, Z_1(s)) - g_0(s, Z_0(s))]ds \\ &\quad - \int_t^T [Z_1(s) - Z_0(s)]dW(s) \\ &= \xi_1 - \xi_0 + \int_t^T \Big(g_{1z}(s)[Z_1(s) - Z_0(s)] + g_1(s, Z_0(s)) - g_0(s, Z_0(s)) \Big) ds \\ &\quad - \int_t^T [Z_1(s) - Z_0(s)]dW(s), \end{aligned}$$

where

$$g_{1z}(s) \equiv \frac{g_1(s, Z_1(s)) - g_1(s, Z_0(s))}{Z_1(s) - Z_0(s)} \mathbf{1}_{\{Z_1(s) \neq Z_0(s)\}}.$$

By (H8.3.1), $g_{1z}(\cdot) \in \overline{\mathrm{BMO}}[0,T]$. Therefore, the process

$$\widehat{W}(t) \equiv W(t) - \int_0^t g_{1z}(s)ds, \quad t \in [0,T],$$

is a Brownian motion under probability $\widehat{\mathbb{P}}$ defined by $d\widehat{\mathbb{P}} \equiv \widehat{\mathcal{E}}\{g_{1z}\}(T)d\mathbb{P}$ (see (2.3.23) of Chap. 2, again). Hence

$$\begin{aligned} Y_1(t) - Y_0(t) &= \xi_1 - \xi_0 + \int_t^T [g_1(s, Z_0(s)) - g_0(s, Z_0(s))]ds \\ &\quad - \int_t^T [Z_1(s) - Z_0(s)]d\widehat{W}(s), \end{aligned}$$

which leads to

$$Y_1(t) - Y_0(t) = \mathbb{E}_t^{\widehat{\mathbb{P}}}\Big[\xi_1 - \xi_0 + \int_t^T [g_1(s, Z_0(s)) - g_0(s, Z_0(s))]ds\Big] \geqslant 0.$$

This proves the lemma. □

Next result gives the stability of the adapted solutions.

Lemma 8.3.3 *Suppose* $g_k(\cdot\,,\cdot)$ *satisfies* (H8.3.1) *uniformly in* $k \geqslant 1$*, i.e., there exist* $\alpha(\cdot) \in L^\infty_{\mathbb{F}}(\Omega; L^1(0,T))$*,* $\beta(\cdot) \in \overline{\text{BMO}}[0,T]$*, and* $\gamma \in [1,\infty)$ *such that*

$$|g_k(t,0)| \leqslant \alpha(t), \qquad t \in [0,T], \ \ a.s. \ \forall k \geqslant 1, \tag{8.3.7}$$

and

$$\begin{aligned} |g_k(t,z_1) - g_k(t,z_2)| \leqslant \big[\beta(t) + \gamma\big(|z_1| + |z_2|\big)\big]|z_1 - z_2|, \\ t \in [0,T], \ z_1, z_2 \in \mathbb{R}. \end{aligned} \tag{8.3.8}$$

In addition

$$\sup_{k\geqslant 1} \|\xi_k\|_\infty < \infty. \tag{8.3.9}$$

Let $(Y_k(\cdot), Z_k(\cdot)) \in L^\infty_{\mathbb{F}}(\Omega; C([0,T])) \times \overline{\text{BMO}}[0,T]$ *be the adapted solution of BSDE* (8.2.1) *associated with* $(\xi_k, g_k(\cdot))$*. Assume that*

$$\begin{aligned} &\xi_k \to \xi, \qquad in\ \mathbb{P}, \\ &g_k(t,z) \to g(t,z), \qquad a.s., \ \ a.e.\ t \in [0,T], \ \forall z \in \mathbb{R}. \end{aligned} \tag{8.3.10}$$

Moreover, $Y_k(\cdot) \uparrow Y(\cdot)$*. Then, there exists a process* $Z(\cdot) \in L^2_{\mathbb{F}}(0,T;\mathbb{R})$ *such that* $(Y(\cdot), Z(\cdot))$ *is the adapted solution of* (8.2.1) *associated with* $(\xi, g(\cdot))$*.*

Proof We split the proof into two steps.

Step 1 Convergence of $Z_k(\cdot)$*.*

By the conditions (8.3.7) and (8.3.9), applying Lemma 8.3.1, we then obtain

$$\sup_{k\geqslant 1} \|Y_k(\cdot)\|_\infty + \sup_{k\geqslant 1} \mathbb{E}\int_0^T |Z_k(s)|^2 ds < \infty. \tag{8.3.11}$$

Thus, it implies that $\|Y(\cdot)\|_\infty < \infty$. By Lemma 8.2.11, there exists a subsequence k_j such that $\{Z_{k_j}(\cdot)\}_{j\geqslant 1}$ converges weakly to some $Z(\cdot) \in L^2_{\mathbb{F}}(0,T;\mathbb{R})$. Let us denote

$$\begin{aligned} &\Delta Y_k(t) \equiv Y_k(t) - Y(t), \quad \Delta Y_{k,\ell}(t) \equiv Y_k(t) - Y_\ell(t), \\ &\Delta Z_k(t) \equiv Z_k(t) - Z(t), \quad \Delta Z_{k,\ell}(t) \equiv Z_k(t) - Z_\ell(t). \end{aligned}$$

We want to prove that the whole sequence is convergent:

$$\lim_{k\to\infty} \mathbb{E}\int_0^T |\Delta Z_k(s)|^2 ds = 0. \tag{8.3.12}$$

To this end, we first note that

$$\begin{aligned}
&\big|g_k(t,Z_k(t))-g_k(t,Z_\ell(t))\big|\leqslant\big[\beta(t)+\gamma\big(|Z_k(t)|+|Z_\ell(t)|\big)\big]|\Delta Z_{k,\ell}(t)|\\
&\leqslant\big[\beta(t)+\gamma\big(|\Delta Z_{k,\ell}(t)|+2|Z_\ell(t)|\big)\big]|\Delta Z_{k,\ell}(t)|\\
&\leqslant\big[\beta(t)+\gamma\big(|\Delta Z_{k,\ell}(t)|+2|\Delta Z_\ell(t)|+2|Z(t)|\big)\big]|\Delta Z_{k,\ell}(t)|\\
&\leqslant\beta(t)^2+4\gamma|\Delta Z_{k,\ell}(t)|^2+\gamma|\Delta Z_\ell(t)|^2+\gamma|Z(t)|^2,
\end{aligned}$$

where the last inequality uses the fact of $\gamma\geqslant 1$ and $2|xy|\leqslant x^2+y^2$. Then

$$\begin{aligned}
&\big|g_k(t,Z_k(t))-g_\ell(t,Z_\ell(t))\big|\\
&\leqslant\big|g_k(t,Z_k(t))-g_k(t,Z_\ell(t))\big|+\big|g_k(t,Z_\ell(t))-g_\ell(t,Z_\ell(t))\big|\\
&\leqslant\big|g_k(t,Z_\ell(t))-g_\ell(t,Z_\ell(t))\big|+\beta(t)^2+4\gamma|\Delta Z_{k,\ell}(t)|^2+\gamma|\Delta Z_\ell(t)|^2+\gamma|Z(t)|^2.
\end{aligned}$$

Now, we define

$$\varphi(y)\equiv\frac{1}{8K_0^2}\big(e^{4K_0y}-4K_0y-1\big),\qquad y\in\mathbb{R},\quad K_0\triangleq 4\gamma.$$

Then, one has

$$\begin{cases}\varphi(0)=0;\quad \varphi(y)>0, & y>0,\\ \varphi'(0)=0;\quad 0<\varphi'(y), & y>0;\\ \varphi''(y)=4K_0\varphi'(y)+2, & y\geqslant 0.\end{cases}$$

Applying Itô's formula to $\varphi(\Delta Y_{\ell,k}(\cdot))$, we obtain

$$\begin{aligned}
&d\varphi(\Delta Y_{\ell,k}(t))=-\varphi'(\Delta Y_{\ell,k}(t))\big[g_\ell(t,Z_\ell(t))-g_k(t,Z_k(t))\big]dt\\
&\quad+\varphi'(\Delta Y_{\ell,k}(t))\Delta Z_{\ell,k}(t)dW(t)+\frac{1}{2}\varphi''(\Delta Y_{\ell,k}(t))|\Delta Z_{\ell,k}(t)|^2dt\\
&\geqslant-\varphi'(\Delta Y_{\ell,k}(t))\Big(|g_k(t,Z_\ell(t))-g_\ell(t,Z_\ell(t))|\\
&\quad+\beta(t)^2+4\gamma|\Delta Z_{k,\ell}(t)|^2+\gamma|\Delta Z_\ell(t)|^2+\gamma|Z(t)|^2\Big)dt\\
&\quad+\varphi'(\Delta Y_{\ell,k}(t))\Delta Z_{\ell,k}(t)dW(t)+\big[2K_0\varphi'(\Delta Y_{\ell,k}(t))+1\big]|\Delta Z_{\ell,k}(t)|^2dt\\
&=-\varphi'(\Delta Y_{\ell,k}(t))\Big(|g_k(t,Z_\ell(t))-g_\ell(t,Z_\ell(t))|+\beta(t)^2+\gamma|\Delta Z_\ell(t)|^2+\gamma|Z(t)|^2\Big)dt\\
&\quad+\varphi'(\Delta Y_{\ell,k}(t))\Delta Z_{\ell,k}(t)dW(t)+\big[K_0\varphi'(\Delta Y_{\ell,k}(t))+1\big]|\Delta Z_{\ell,k}(t)|^2dt.
\end{aligned}$$

Consequently

$$\begin{aligned}
&\mathbb{E}\int_0^T\big[K_0\varphi'(\Delta Y_{\ell,k}(t))+1\big]|\Delta Z_{\ell,k}(t)|^2dt\\
&\leqslant\mathbb{E}\varphi(\Delta Y_{\ell,k}(T))-\varphi(\Delta Y_{\ell,k}(0))+\mathbb{E}\int_0^T\varphi'(\Delta Y_{\ell,k}(t))\\
&\qquad\Big(|g_k(t,Z_\ell(t))-g_\ell(t,Z_\ell(t))|+\beta(t)^2+\gamma|\Delta Z_\ell(t)|^2+\gamma|Z(t)|^2\Big)dt.
\end{aligned}$$

By the L^2 boundedness of $\{Z_k(\cdot)\}$, Mazur theorem, Fatou lemma, there exists a subsequence denoted by $\{Z_{k_r}\}_{r\geqslant 1}$, depending on ℓ, such that

$$\begin{aligned}
&\mathbb{E}\int_0^T \big[K_0\varphi'(\Delta Y_\ell(s))+1\big]\big|Z_\ell(s)-Z(s)\big|^2 ds\\
&\leqslant \varliminf_{r\to\infty}\Big\{\mathbb{E}\varphi(\Delta Y_{\ell,k_r}(T))-\varphi(\Delta Y_{\ell,k_r}(0))+\mathbb{E}\int_0^T \varphi'(\Delta Y_{\ell,k_r}(t))\\
&\qquad\Big(\big|g_{k_r}(t,Z_\ell(t))-g_\ell(t,Z_\ell(t))\big|+\beta(t)^2+\gamma|\Delta Z_\ell(t)|^2+\gamma|Z(t)|^2\Big)dt\Big\}.
\end{aligned}$$

Thanks to the boundedness of $\{Y_k(\cdot)\}$, $Y_k(\cdot)\uparrow Y(\cdot)$, and dominated convergence theorem, the above inequality implies that

$$\begin{aligned}
&\mathbb{E}\int_0^T \big[K_0\varphi'(\Delta Y_\ell(t))+1\big]|\Delta Z_\ell(t)|^2 dt\\
&\leqslant \mathbb{E}\varphi(\Delta Y_\ell(T))-\varphi(\Delta Y_\ell(0))+\mathbb{E}\int_0^T \varphi'(\Delta Y_\ell(t))\\
&\qquad\Big(\big|g(t,Z_\ell(t))-g_\ell(t,Z_\ell(t))\big|+\beta(t)^2+\gamma|\Delta Z_\ell(t)|^2+\gamma|Z(t)|^2\Big)dt.
\end{aligned}$$

Considering the fact of $K_0=4\gamma>1$, it further indicates that

$$\begin{aligned}
\mathbb{E}\int_0^T |\Delta Z_\ell(t)|^2 dt \leqslant\ & \mathbb{E}\varphi(\Delta Y_\ell(T))-\varphi(\Delta Y_\ell(0))+\mathbb{E}\int_0^T \varphi'(\Delta Y_\ell(t))\\
&\cdot\Big(\big|g(t,Z_\ell(t))-g_\ell(t,Z_\ell(t))\big|+\beta(t)^2+\gamma|Z(t)|^2\Big)dt.
\end{aligned}$$

Using again dominated convergence theorem, we obtain

$$\lim_{k\to\infty}\mathbb{E}\int_0^T |\Delta Z_k(s)|^2 ds=0. \tag{8.3.13}$$

The convergence of $Z_k(\cdot)$ follows.

Step 2 $(Y(\cdot),Z(\cdot))$ *is the adapted solution of (8.2.1).*

We claim that

$$\lim_{k\to\infty}\mathbb{E}\int_0^T |g_k(s,Z_k(s))-g(s,Z(s))|ds=0. \tag{8.3.14}$$

In fact, by the imposed condition (8.3.8), we have that

$$\begin{aligned}
&\big|g_k(t,Z_k(t))-g(t,Z(t))\big|\\
&\leqslant \big|g_k(t,Z(t))-g(t,Z(t))\big|+\big|g_k(t,Z_k(t))-g_k(t,Z(t))\big|\\
&\leqslant \big|g_k(t,Z(t))-g(t,Z(t))\big|+\big[\beta(t)+\gamma\big(|Z_k(t)|+|Z(t)|\big)\big]|\Delta Z_k(t)|.
\end{aligned}$$

For the second term on the right hand, from (8.3.11) and (8.3.12),

$$\begin{aligned}
&\mathbb{E}\int_0^T \big[\beta(t)+\gamma\big(|Z_k(t)|+|Z(t)|\big)\big]|\Delta Z_k(t)|dt \\
&\leqslant \Big[\mathbb{E}\int_0^T \big[\beta(t)+\gamma\big(|Z_k(t)|+|Z(t)|\big)\big]^2 dt\Big]^{\frac{1}{2}}\Big[\mathbb{E}\int_0^T |\Delta Z_k(t)|^2 dt\Big]^{\frac{1}{2}} \\
&\leqslant K\Big[\mathbb{E}\int_0^T |\Delta Z_k(t)|^2 dt\Big]^{\frac{1}{2}} \to 0.
\end{aligned}$$

For the first term on the right-hand side, we first observe that

$$|g_k(t,Z(t))| \leqslant |g_k(t,0)-g(t,0)|+\alpha(t)+\beta(t)+2\gamma|Z(t)|^2.$$

Then, using assumption (8.3.10), and dominated convergence theorem, we have that

$$\lim_{k\to\infty}\mathbb{E}\int_0^T |g_k(t,Z(t))-g(t,Z(t))|dt = 0.$$

Therefore, we can obtain (8.3.14). On the other hand, by (8.3.12), we have

$$\begin{aligned}
&\lim_{k\to\infty}\mathbb{E}\sup_{t\in[0,T]}\Big|\int_t^T \Delta Z_k(s)dW(s)\Big| \leqslant \lim_{k\to\infty}\Big(\mathbb{E}\sup_{t\in[0,T]}\Big|\int_t^T \Delta Z_k(s)dW(s)\Big|^2\Big)^{\frac{1}{2}} \\
&\leqslant K\lim_{k\to\infty}\Big(\mathbb{E}\int_0^T \big|\Delta Z_k(s)\big|^2 ds\Big)^{\frac{1}{2}} = 0.
\end{aligned}$$

Also, since the sequence $\{Y_k(\cdot)\}$ is bounded, by dominated convergence theorem,

$$\lim_{k\to\infty}\mathbb{E}\big|\Delta Y_k(t)\big| = 0, \quad \forall t\in[0,T].$$

Therefore, for any given $t\in[0,T]$, one has

$$\mathbb{E}\Big|Y(t)-\xi-\int_t^T g(s,Z(s))ds+\int_t^T Z(s)dW(s)\Big| = 0,$$

which implies that $(Y(\cdot),Z(\cdot))$ is the adapted solution of (8.2.1) associated with $(\xi,g(\cdot))$. □

We are now ready to establish the well-posedness of (one-dimensional) QBSDE (8.2.1).

Theorem 8.3.4 *Suppose* (H8.3.1) *holds and* $\xi\in L^\infty_{\mathcal{F}_T}(\Omega;\mathbb{R})$. *Then, QBSDE* (8.2.1) *admits a unique adapted solution:*

$$(Y(\cdot),Z(\cdot))\in L^\infty_{\mathbb{F}}(\Omega;C([0,T]))\times\overline{\mathrm{BMO}}[0,T].$$

Moreover, (8.3.5)–(8.3.6) *hold.*

Proof We first look at the existence. For any $k, \ell, j \geqslant 1$, we define

$$\begin{aligned} &g_k(t,z) \equiv g(t,z) \wedge k, \quad g_{k,\ell}(t,z) \equiv g_k(t,z) \vee (-\ell), \\ &g_{k,\ell,j}(t,z) \equiv \inf_{z' \in \mathbb{R}} \Big[g_{k,\ell}(t,z') + j|z-z'| \Big], \quad t \in [0,T], z \in \mathbb{R}. \end{aligned}$$

It is easy to see the following:

$$\begin{aligned} &g_k(t,z) \uparrow g(t,z), && \text{as } k \to \infty, \\ &g_{k,\ell}(t,z) \downarrow g_k(t,z), && \text{as } \ell \to \infty, \\ &-\ell \leqslant g_{k,\ell}(t,z) \leqslant k. \end{aligned}$$

It then follows from Lemma 8.2.10 that $g_{k,\ell,j}(t,z)$ is well-defined, uniformly bounded, increasing in j and k, and decreasing in ℓ. In addition, for any fixed k, ℓ, and $\lim_{j\to\infty} z_j = z$, we have

$$\begin{aligned} &\lim_{j\to\infty} g_{k,\ell,j}(t,z_j) = g_{k,\ell}(t,z), \quad \text{a.s.} \qquad -\ell \leqslant g_{k,\ell,j}(t,z) \leqslant k, \\ &|g_{k,\ell,j}(t,z) - g_{k,\ell,j}(t,z_1)| \leqslant j|z-z_1|, \quad \forall z, z_1 \in \mathbb{R}. \end{aligned}$$

Now, let us consider

$$Y_{k,\ell,j}(t) = \xi + \int_t^T g_{k,\ell,j}(s, Z_{k,\ell,j}(s))ds - \int_t^T Z_{k,\ell,j}(s)dW(s).$$

By the Lipschitz property of $g_{k,\ell,j}$, we obtain the existence and uniqueness of $(Y_{k,\ell,j}(\cdot), Z_{k,\ell,j}(\cdot))$ in the following space:

$$(Y_{k,\ell,j}(\cdot), Z_{k,\ell,j}(\cdot)) \in L^2_{\mathbb{F}}(\Omega; C([0,T])) \times L^2_{\mathbb{F}}(0,T).$$

In terms of the comparison theorem of BSDEs (see Theorem 6.1.6 in Chap. 6), $Y_{k,\ell,j}(\cdot)$ is increasing in j, decreasing in ℓ, and increasing in k. Hence, we can define

$$Y_{k,\ell}(\cdot) \equiv \lim_{j\to\infty} Y_{k,\ell,j}(\cdot), \quad Y_k(\cdot) \equiv \lim_{\ell\to\infty} Y_{k,\ell}(\cdot), \quad Y(\cdot) \equiv \lim_{k\to\infty} Y_k(\cdot).$$

Since $|g_{k,\ell}(s,z)| \leqslant k \vee \ell$, it follows that

$$|Y_{k,\ell,j}(\cdot)| \leqslant \|\xi\|_\infty + (k \vee \ell)T.$$

By Itô's formula on $|Y_{k,\ell,j_1}(\cdot) - Y_{k,\ell,j_2}(\cdot)|^2$,

$$\mathbb{E}\int_0^T |Z_{k,\ell,j_1}(s) - Z_{k,\ell,j_2}(s)|^2 ds \leqslant 2(k+\ell)\mathbb{E}\int_0^T |Y_{k,\ell,j_1}(s) - Y_{k,\ell,j_2}(s)| ds.$$

Note that as $j_1, j_2 \to \infty$, by Theorem 3.1.4 in Chap. 3, for any $s \in [0,T]$, $|Y_{k,\ell,j_1}(s) - Y_{k,\ell,j_2}(s)| \to 0$, almost surely. It then follows from dominated convergence theorem that

$$\mathbb{E}\int_0^T |Z_{k,\ell,j_1}(s) - Z_{k,\ell,j_2}(s)|^2 ds \to 0, \quad j_1, j_2 \to \infty.$$

As a result, there exists $Z_{k,\ell}(\cdot) \in L^2_{\mathbb{F}}(0,T)$ such that

$$\lim_{j\to\infty} \mathbb{E}\int_0^T |Z_{k,\ell,j}(s) - Z_{k,\ell}(s)|^2 ds = 0.$$

We claim that $(Y_{k,\ell}(\cdot), Z_{k,\ell}(\cdot))$ satisfies the following BSDE:

$$Y_{k,\ell}(t) = \xi + \int_t^T g_{k,\ell}(s, Z_{k,\ell}(s))ds - \int_t^T Z_{k,\ell}(s)dW(s).$$

Actually, from the above, there exists a subsequence $\{j_r\}$ such that

$$\lim_{r\to\infty} |Z_{k,\ell,j_r}(s) - Z_{k,\ell}(s)| = 0, \quad s \in [0,T], \text{ a.s. a.e.}$$

By the convergence of $g_{k,\ell,j}$ to $g_{k,\ell}$, we see that

$$\lim_{r\to\infty} g_{k,\ell,j_r}(s, Z_{k,\ell,j_r}(s)) = g_{k,\ell}(s, Z_{k,\ell}(s)), \quad \text{a.s.}$$

Using dominated convergence theorem, as $r \to \infty$,

$$\mathbb{E}\int_0^T \left|g_{k,\ell,j_r}(s, Z_{k,\ell,j_r}(s)) - g_{k,\ell}(s, Z_{k,\ell}(s))\right|^2 ds \to 0.$$

Note that

$$\lim_{r\to\infty} \mathbb{E}|Y_{k,\ell,j_r}(t) - Y_{k,\ell}(t)|^2 = 0, \quad \forall t \in [0,T],$$

and

$$\begin{aligned}
&\lim_{r\to\infty} \mathbb{E} \sup_{t\in[0,T]} \left|\int_t^T [Z_{k,\ell,j_r}(s) - Z_{k,\ell}(s)]dW(s)\right|^2 \\
&\leqslant K \lim_{r\to\infty} \mathbb{E}\int_0^T [Z_{k,\ell,j_r}(s) - Z_{k,\ell}(s)]^2 ds = 0.
\end{aligned}$$

Then, the desired conclusion follows from

$$\mathbb{E}\Big\{Y_{k,\ell,j}(t) - \xi - \int_t^T g_{k,\ell,j}(s, Z_{k,\ell,j}(s))ds + \int_t^T Z_{k,\ell,j}(s)dW(s)\Big\}^2 = 0.$$

For fixed $k \geqslant 1$, we claim that $g_{k,\ell}$ satisfies (H8.3.1), i.e., for $z_i \in \mathbb{R}$, $i = 1, 2$,

$$|g_{k,\ell}(t, z_1) - g_{k,\ell}(t, z_2)| \leqslant \big[\beta(t) + \gamma(|z_1| + |z_2|)\big]|z_1 - z_2|. \tag{8.3.15}$$

Given constant $b \in \mathbb{R}$, since $a \mapsto a \vee b = \frac{a+b+|a-b|}{2}$ and $a \mapsto a \wedge b = \frac{a+b-|a-b|}{2}$ are Lipschitz continuous with the Lipschitz constant 1, we have

$$|a_1 \wedge b - a_2 \wedge b| \leqslant |a_1 - a_2|, \quad |a_1 \vee c - a_2 \vee c| \leqslant |a_1 - a_2|.$$

Therefore, it follows that

$$|a_1 \vee c \wedge b - a_2 \vee c \wedge b| \leqslant |a_1 - a_2|.$$

Hence, we have the above (8.3.15). It is a direct calculation that

$$|g_{k,\ell}(\cdot\,, 0)| \leqslant |g_k(\cdot\,, 0)| \leqslant |g(\cdot\,, 0)|.$$

Recall that for fixed k, $g_{k,\ell}(\cdot\,, z) \to g_k(\cdot, z)$, as $\ell \to \infty$. By Lemma 8.3.3, we obtain the existence of $Z_k(\cdot) \in L^2_{\mathbb{F}}(0, T)$, as well as the result of $(Y_k(\cdot), Z_k(\cdot))$ being the adapted solution of

$$Y_k(t) = \xi + \int_t^T g_k(s, Z_k(s))ds - \int_t^T Z_k(s)dW(s).$$

Similarly as above, we can use Lemma 8.3.3, to obtain the existence of $Z(\cdot) \in L^2_{\mathbb{F}}(0, T)$. In addition, $(Y(\cdot), Z(\cdot))$ is the adapted solution of BSDE (8.2.1). From Lemma 8.3.1, we see actually $Z(\cdot) \in \overline{\text{BMO}}[0, T]$.

The uniqueness is implied by comparison theorem in Lemma 8.3.2. □

3.2 *Multidimensional Cases*

In this subsection, we mainly consider a multidimensional QBSDE. In other words, we allow the (8.2.1) to be multidimensional. We now present a lemma which gives an a priori estimate of adapted solution of (multidimensional) QBSDE (8.2.1).

Lemma 8.3.5 *Let $g : [0, T] \times \mathbb{R}^n \times \Omega \to \mathbb{R}^n$ be measurable such that for each $z \in \mathbb{R}^n$, $t \mapsto g(t, z)$ is $\mathbb{F}$-progressively measurable, satisfying*

$$\big|g^i(t, z)\big| \leqslant \alpha(t) + \gamma|z^i|^2, \quad (t, z) \in [0, T] \times \mathbb{R}^n,\ 1 \leqslant i \leqslant n, \tag{8.3.16}$$

for constant $\gamma > 0$, *and* $\mathbb{F}$*-adapted process* $\alpha(\cdot) \in \bigcap_{\mu \geqslant 1} \mathcal{E}_{\mathbb{F}}^{\mu,1}(\Omega; L^1(0,T;\mathbb{R}_+))$. *Let*

$$(Y(\cdot), Z(\cdot)) \in \bigcap_{\mu \geqslant 1} \mathcal{E}_{\mathbb{F}}^{\mu,1}(\Omega; C([0,T];\mathbb{R}^n)) \times L_{\mathbb{F}}^2(0,T;\mathbb{R}^n),$$

be an adapted solution to QBSDE (8.2.1), *and* $\xi \in \bigcap_{\mu \geqslant 1} \mathcal{E}_{\mathcal{F}_T}^{\mu,1}(\Omega;\mathbb{R}^n)$. *Then, for any* $\mu > 2\gamma$, *it is true that for some constant* $K > 0$ *depending on* γ, μ,

$$e^{\mu|Y^i(t)|} + \mathbb{E}_t \int_t^T |Z^i(s)|^2 ds \leqslant K\mathbb{E}_t\Big[e^{\mu|\xi^i|} + \int_t^T e^{\mu|Y^i(s)|}\alpha(s)ds\Big], \qquad 1 \leqslant i \leqslant n,\ t \in [0,T]. \tag{8.3.17}$$

Further, there exists constant $K > 0$ *depending on* γ, μ *such that for any* $\tau \in \mathscr{T}[0,T]$, *and* $1 \leqslant i \leqslant n$,

$$e^{\mu|Y^i(\tau)|} + \mathbb{E}_\tau \int_\tau^T |Z^i(s)|^2 ds \leqslant K\mathbb{E}_\tau e^{\mu\left(|\xi^i| + \int_\tau^T \alpha(s)ds\right)}. \tag{8.3.18}$$

Recall that $\alpha(\cdot) \in \bigcap_{\mu \geqslant 1} \mathcal{E}_{\mathbb{F}}^{\mu,1}(\Omega; L^1(0,T))$ means that for any $\mu \geqslant 1$,

$$\mathbb{E}\Big[\exp\Big(\mu\|\alpha(\cdot)\|_{L^1(0,T)}\Big)\Big] < \infty, \tag{8.3.19}$$

and $\|\alpha(\cdot)\|_{L^1(0,T)}$ is allowed to be unbounded (random variable). Thus, (8.3.19) is strictly weaker than (8.3.4). Also, ξ and $Y(\cdot)$ are actually allowed to be unbounded.

Proof For any $\mu > 2\gamma$, let

$$\varphi(\rho) \equiv e^{\mu\rho}, \qquad \rho \in [1,\infty).$$

We calculate

$$\varphi'(\rho) = \mu\varphi(\rho), \qquad \varphi''(\rho) = \mu^2\varphi(\rho) > 0.$$

Let $t \in [0,T)$ be fixed and $(Y(\cdot), Z(\cdot)) \in \bigcap_{\mu \geqslant 1} \mathcal{E}_{\mathbb{F}}^{\mu,1}(\Omega; C([0,T];\mathbb{R}^n)) \times L_{\mathbb{F}}^2(0,T;\mathbb{R}^n)$ be the adapted solution of QBSDE (8.2.1). For $i = 1, \cdots, n$, define

$$\mathcal{Y}^i(r) \equiv |Y^i(r)| + \int_t^r \bar{\alpha}(s)ds, \qquad r \in [t,T],$$

for some undetermined $\bar{\alpha}(\cdot) \in \bigcap_{\mu\geqslant 1} \mathcal{E}_{\mathbb{F}}^{\mu,1}(\Omega; L^1(0,T;\mathbb{R}_+))$. Then, by Itô-Tanaka's formula (Theorem 2.3.6 of Chap. 2),

$$
\begin{aligned}
d\mathcal{Y}^i(r) = \Big(-\operatorname{sgn}\big(Y^i(r)\big)g^i(r,Z(r)) + \bar{\alpha}(r)\Big)dr \\
+\operatorname{sgn}\big(Y^i(r)\big)Z^i(r)dW(r) + \frac{1}{2}dL^i(r),
\end{aligned}
$$

where $L^i(\cdot)$ is the local time of $|Y^i(\cdot)|$. Hence, applying Itô-Tanaka's formula to $s \mapsto \varphi(\mathcal{Y}^i(s))$ (see Corollary 2.3.7 of Chap. 2), one has

$$
\begin{aligned}
d\varphi(\mathcal{Y}^i(s)) &= \Big[\varphi'(\mathcal{Y}^i(s))\Big(-\operatorname{sgn}\big(Y^i(s)\big)g^i(s,Z(s)) + \bar{\alpha}(s)\Big) \\
&\quad +\frac{1}{2}\varphi''(\mathcal{Y}^i(s))|Z^i(s)|^2\Big]ds \\
&\quad +\varphi'(\mathcal{Y}^i(s))\operatorname{sgn}\big(Y^i(s)\big)Z^i(s)dW(s) + \frac{1}{2}\varphi'(\mathcal{Y}^i(s))dL^i(s) \\
&\geqslant e^{\mu\mathcal{Y}^i(s)}\Big[\Big(-\mu\gamma + \frac{\mu^2}{2}\Big)|Z^i(s)|^2 + \big(\bar{\alpha}(s) - \alpha(s)\big)\Big]ds \\
&\quad +\mu e^{\mu\mathcal{Y}^i(s)}\operatorname{sgn}\big(Y^i(s)\big)Z^i(s)dW(s), \quad s \in [t,T].
\end{aligned}
$$

For each $r \in [t,T]$ and $k \geqslant 1$, define the following stopping time:

$$
\sigma_k^r \equiv \inf\Big\{s \in [r,T] \;\Big|\; |Y(s)| + \int_r^s |Z(\tau)|^2 d\tau \geqslant k\Big\} \wedge T,
$$

with the convention $\inf\varnothing = +\infty$. Then, by integrating on $[r,\sigma_k^r]$ and taking expectation on both sides, for each $k \geqslant 1$ and $r \in [t,T]$, $\mu > 2\gamma$,

$$
\begin{aligned}
&e^{\mu\mathcal{Y}^i(r)} + \mathbb{E}_r\int_r^{\sigma_k^r} \mu e^{\mu\mathcal{Y}^i(s)}\Big(\frac{\mu}{2} - \gamma\Big)|Z^i(s)|^2 ds \\
&\leqslant \mathbb{E}_r e^{\mu\mathcal{Y}^i(\sigma_k^r)} + \mathbb{E}_r\int_r^{\sigma_k^r} e^{\mu\mathcal{Y}^i(s)}\big[\alpha(s) - \bar{\alpha}(s)\big]ds.
\end{aligned}
\tag{8.3.20}
$$

Hence, using dominated convergence theorem, sending $k \to \infty$ and letting $r = t$ yields

$$
\begin{aligned}
&e^{\mu|Y^i(t)|} + \mathbb{E}_t\int_t^T \mu e^{\mu\big(|Y^i(s)| + \int_t^s \bar{\alpha}(\tau)d\tau\big)}\Big(\frac{\mu}{2} - \gamma\Big)|Z^i(s)|^2 ds \\
&\leqslant \mathbb{E}_t e^{\mu\big(|\xi^i| + \int_t^T \bar{\alpha}(s)ds\big)} + \mathbb{E}_t\int_t^T e^{\mu\big(|Y^i(s)| + \int_t^s \bar{\alpha}(\tau)d\tau\big)}\big[\alpha(s) - \bar{\alpha}(s)\big]ds.
\end{aligned}
\tag{8.3.21}
$$

Now, taking $\bar{\alpha}(\cdot) = 0$, we have

$$
e^{\mu|Y^i(t)|} + \frac{\mu(\mu-2\gamma)}{2}\mathbb{E}_t\int_t^T e^{\mu|Y^i(s)|}|Z^i(s)|^2 ds \leqslant \mathbb{E}_t e^{\mu|\xi^i|} + \mathbb{E}_t\int_t^T e^{\mu|Y^i(s)|}\alpha(s)ds.
$$

Then, (8.3.17) follows. Next, we take $\bar\alpha(\cdot)=\alpha(\cdot)$ to get

$$e^{\mu|Y^i(t)|}+\frac{\mu(\mu-2\gamma)}{2}\mathbb{E}_t\int_t^T e^{\mu\left(|Y^i(s)|+\int_t^s\alpha(\tau)d\tau\right)}|Z^i(s)|^2ds\leqslant\mathbb{E}_te^{\mu\left(|\xi^i|+\int_t^T\alpha(s)ds\right)}.$$

Then, (8.3.18) follows with $\tau=t$. The general stopping time case can be obtained similarly. □

We point out that both (8.3.17) and (8.3.18) will play important yet different roles in our later discussions. In (8.3.18), both $Y^i(\cdot)$ and $Z^i(\cdot)$ can be dominated by ξ^i and $\alpha(\cdot)$ simultaneously, but in the exponential order. In (8.3.17), such an exponential order requirement for $\alpha(\cdot)$ is not needed, but $Y^i(\cdot)$ does.

To conclude this section, as an application of the previous multi-dimensional result, we would like to look at the well-posedness of one-dimensional QBSDE (8.2.1), under a slightly different hypothesis. So, we come back to one-dimensional QBSDE (8.2.1). We introduce the following.

(H8.3.2) Let (H8.3.1) hold except (8.3.1), which is replaced by the following:

$$|g(s,z)|\leqslant\alpha(s)+\gamma|z|^2,\qquad s\in[0,T],\ z\in\mathbb{R},\tag{8.3.22}$$

for some $\gamma>0$ and

$$\alpha(\cdot)\in\bigcap_{\mu\geqslant1}\mathcal{E}^{\mu,1}_{\mathbb{F},\infty}(\Omega;L^1(0,T;\mathbb{R}_+)),\qquad\text{a.s.}\tag{8.3.23}$$

Note that condition (8.3.23) is strictly weaker than (8.3.4) (as $\int_0^T\alpha(s)ds$ could be unbounded). From (8.2.26), we know that when $\alpha(\cdot)\in\overline{\text{BMO}}[0,T]$, (8.3.23) holds for all $\mu>0$.

Theorem 8.3.6 *Suppose* (H8.3.2) *holds and* ξ *is bounded. Then, QBSDE* (8.2.1) *has a unique adapted solution* $(Y(\cdot),Z(\cdot))\in L^\infty_{\mathbb{F}}(\Omega;C([0,T]))\times\overline{\text{BMO}}[0,T]$. *In addition, for any* $\mu>2\gamma$, $r\in[0,T]$, *there exists a constant* $K>0$ *depending only on* γ, μ,

$$e^{\mu|Y(t)|}+\mathbb{E}_t\int_t^T|Z(s)|^2ds\leqslant K\mathbb{E}_t\Big[e^{\mu|\xi|}+\int_t^Te^{\mu|Y(s)|}\alpha(s)ds\Big],\quad t\in[r,T],\tag{8.3.24}$$

and for any $\tau\in\mathscr{T}[r,T]$,

$$e^{\mu|Y(\tau)|}+\mathbb{E}_\tau\int_\tau^T|Z(s)|^2ds\leqslant K\mathbb{E}_te^{\mu\left(|\xi|+\int_\tau^T\alpha(s)ds\right)}.\tag{8.3.25}$$

Proof For integers $k, \ell \geqslant 1$, and $r \in [0, T)$, we define

$$\begin{aligned} g_{k,\ell}(t, z) &\equiv \mathbf{1}_{[0,\sigma_k]}(t)g^+(t, z) - \mathbf{1}_{[0,\sigma_\ell]}(t)g^-(t, z), \qquad t \in [r, T], \\ \sigma_k &\equiv \inf\Big\{t \in [r, T] \;\Big|\; \int_r^t |g(s, 0)|ds \geqslant k\Big\}. \end{aligned}$$

Therefore, for $G_{k,\ell}(t) \equiv \big[\mathbf{1}_{[0,\sigma_k]}(t) + \mathbf{1}_{[0,\sigma_\ell]}(t)\big]|g(t, 0)|$, by the above (8.3.22),

$$|g_{k,\ell}(t, z)| \leqslant G_{k,\ell}(t) + 2\gamma|z|^2, \qquad \int_r^T G_{k,\ell}(t)dt \leqslant k + \ell.$$

According to Theorem 8.3.4, for any $r \in [0, T)$, there exists a unique pair $(Y_{k,\ell}(\cdot), Z_{k,\ell}(\cdot))$ of BSDE (8.2.1) on $[r, T]$ corresponding with $(\xi, g_{k,\ell}(\cdot))$ such that

$$(Y_{k,\ell}(\cdot), Z_{k,\ell}(\cdot)) \in L^\infty_{\mathbb{F}}(\Omega; C([r, T])) \times \overline{\text{BMO}}[r, T].$$

Now, by Lemma 8.3.2, we see that $Y_{k,\ell}(\cdot)$ is nondecreasing in k and nonincreasing in ℓ on $[r, T]$. Further, by Lemma 8.3.5, for any $\mu > 2\gamma$, there exists a $K > 0$ depending only on γ, μ such that

$$e^{\mu|Y_{k,\ell}(t)|} + \mathbb{E}_t\int_t^T |Z_{k,\ell}(s)|^2 ds \leqslant K\mathbb{E}_t\Big[e^{\mu\big(|\xi| + \int_t^T \alpha(r)dr\big)}\Big], \quad t \in [r, T]. \tag{8.3.26}$$

Finally, thanks to Lemma 8.3.3, we obtain an adapted solution $(Y(\cdot), Z(\cdot))$ to the QBSDE (8.2.1) such that $Y(\cdot)$ is bounded and $Z(\cdot) \in L^2_{\mathbb{F}}(r, T)$.

Once the existence of $(Y(\cdot), Z(\cdot))$ is obtained, by means of Lemma 8.3.5, we immediately obtain (8.3.17)–(8.3.18), and $Z(\cdot) \in \overline{\text{BMO}}[r, T]$. The above estimates (8.3.24) and (8.3.25) follow from Lemma 8.3.5.

For the uniqueness, suppose $(Y'(\cdot), Z'(\cdot))$ is another adapted solution such that $Y'(\cdot)$ is bounded and $Z'(\cdot) \in \overline{\text{BMO}}[0, T]$. If we define

$$\widehat{Y}(s) \equiv Y(s) - Y'(s), \qquad \widehat{Z}(s) \equiv Z(s) - Z'(s),$$

then it is easy to see

$$\widehat{Y}(t) = \int_t^T g_z(s)\widehat{Z}(s)ds - \int_t^T \widehat{Z}(s)dW(s), \quad t \in [0, T],$$

where

$$g_z(s) \equiv \big[g(s, Z(s)) - g(s, Z'(s))\big]\widehat{Z}(s)\mathbf{1}_{\{\widehat{Z}(s)\neq 0\}}, \quad s \in [0, T].$$

By (H8.3.1), we see that $g_z(\cdot) \in \overline{\text{BMO}}[0, T]$ since

$$\big|g_z(s)\big| \leqslant K\big[\beta(t) + |Z(s)| + |Z'(s)|\big].$$

As a result,

$$\widehat{W}(t) \equiv W(t) - \int_0^t g_z(s)ds, \quad t \in [0, T],$$

is a Brownian motion under probability $\widehat{\mathbb{P}}$ defined by $\widehat{\mathbb{P}}(d\omega) \equiv \widehat{\mathcal{E}}\{g_z\}(T)\mathbb{P}(d\omega)$. Hence, $\widehat{Y}(\cdot) = 0$, $\widehat{Z}(\cdot) = 0$, which leads to the uniqueness. □

Example 8.3.7 Suppose $g(t, z) = \beta(t)^\kappa z + |z|^2$ where $\kappa \in (0, 1)$, and $\beta(\cdot) \in \overline{\text{BMO}}[0, T]$. Then, it is easy to see that

$$\big|g(t, z)\big| \leqslant \beta(t)^{2\kappa} + 2|z|^2.$$

From Young's inequality, for any $\mu > 0$, we have

$$\mu\beta(t)^{2\kappa} \leqslant \frac{\beta(t)^2}{2\|\beta\|^2_{\overline{\text{BMO}}[0,T]}} + K(\mu, \kappa)\|\beta\|^{\frac{2\kappa}{1-\kappa}}_{\overline{\text{BMO}}[0,T]},$$

where $K(\mu, \kappa)$ is a constant depends on μ, κ. By John-Nirenberg inequality for BMO martingale, for each $t \in [0, T]$, and $\tau \in [t, T]$, (see (8.2.20))

$$\mathbb{E}_\tau\Big[\exp\Big(\frac{1}{2\|\beta\|^2_{\overline{\text{BMO}}[0,T]}}\int_\tau^T |\beta(s)|^2 ds\Big)\Big] \leqslant 2.$$

Therefore, for any $\mu > 0$, and $t \in [0, T]$,

$$\mathbb{E}\Big[\exp\Big(\mu\int_t^T |\beta(s)|^{2\kappa} ds\Big)\Big] \leqslant 2\exp\Big[K(\mu, \kappa)\|\beta\|^{\frac{2\kappa}{1-\kappa}}_{\overline{\text{BMO}}[0,T]}\Big] < \infty.$$

This shows $\beta(\cdot)^\kappa \in \bigcap_{\mu \geqslant 1} \mathcal{E}_{\mathbb{F}}^{\mu,1}(\Omega; L^1(0, T; \mathbb{R}_+))$, and thus, (H8.3.2) holds.

4 Quadratic Type-I BSVIEs with Bounded Free Term

Our purpose is to look at Type-I BSVIE which is rewritten here for convenience:

$$Y(t) = \psi(t) + \int_t^T g(t, s, Y(s), Z(t, s))ds - \int_t^T Z(t, s)dW(s), \quad t \in [0, T]. \tag{8.4.1}$$

In this section, we first look at the case that the free term $\psi(\cdot)$ is bounded. The case that $\psi(\cdot)$ being possibly unbounded will be much more complicated, which will be discussed in Sect. 5 later.

4.1 A Simple Case

In this subsection, we look at the following simple Type-I QBSVIE in one dimension:

$$Y(t) = \psi(t) + \int_t^T g(t,s,Z(t,s))ds - \int_t^T Z(t,s)dW(s), \quad t \in [0,T]. \tag{8.4.2}$$

We introduce the following hypothesis, which is comparable with (H8.3.1).

(H8.4.1) Map $g : \Delta^*[0,T] \times \mathbb{R} \times \Omega \to \mathbb{R}$ is measurable so that $s \mapsto g(t,s,z)$ is $\mathbb{F}$-progressively measurable on $[t,T]$ for each $(t,z) \in [0,T) \times \mathbb{R}$. There exists an $\alpha : \Delta^*[0,T] \times \Omega \to [0,\infty)$ such that $s \mapsto \alpha(t,s)$ is $\mathbb{F}$-adapted on $[t,T]$, satisfying

$$\operatorname*{esssup}_{t\in[0,T]} \sup_{\tau\in\mathscr{T}[t,T]} \Big\| \mathbb{E}_\tau \Big[e^{\mu\|\alpha(t,\cdot)\|_{L^1(\tau,T)}} \Big] \Big\|_\infty < \infty, \quad \forall \mu > 0, \tag{8.4.3}$$

a $\beta(\cdot\,,\cdot) \in \overline{\mathrm{BMO}}(\Delta^*[0,T])$, and a constant $\gamma > 0$ such that

$$|g(t,s,z)| \leqslant \alpha(t,s) + \gamma|z|^2, \quad (t,s,z) \in \Delta^*[0,T] \times \mathbb{R}, \tag{8.4.4}$$

$$\begin{aligned} |g(t,s,z) - g(t,s,\bar z)| \leqslant \Big(\beta(t,s) + \gamma\big(|z| + |\bar z|\big)\Big)|z - \bar z|, \\ (t,s) \in \Delta^*[0,T],\ z, \bar z \in \mathbb{R}. \end{aligned} \tag{8.4.5}$$

The main feature of the above is that the generator g is independent of variable y. For this case, we have the following simple result, which gives a priori estimates of the adapted solutions.

Lemma 8.4.1 *Suppose* (H8.4.1) *holds, and* $\psi(\cdot) \in L^\infty_{\mathcal{F}_T}(0,T;\mathbb{R})$. *Then, Type-I QBSVIE* (8.4.2) *admits a unique adapted solution* $(Y(\cdot), Z(\cdot\,,\cdot)) \in L^\infty_{\mathbb{F}}(0,T) \times \overline{\mathrm{BMO}}(\Delta^*[0,T])$. *Moreover, for any* $\mu > 2\gamma$, *there exists a constant* $K > 0$ *depending on* γ, μ *such that for almost* $t \in [0,T]$, *and any* $T_1 \in [t,T]$,

$$\begin{aligned} & e^{\mu|Y(t)|} \leqslant K\mathbb{E}_t e^{\mu\left(|\psi(t)| + \int_t^T \alpha(t,s)ds\right)}, \\ & \big\|Z(t,\cdot)\big\|_{\overline{\mathrm{BMO}}[T_1,T]} \leqslant K \sup_{\tau\in\mathscr{T}[T_1,T]} \Big\| \mathbb{E}_\tau e^{\mu\left(|\psi(t)| + \|\alpha(t,\cdot)\|_{L^1(\tau,T)}\right)} \Big\|_\infty. \end{aligned} \tag{8.4.6}$$

Proof For fixed $t \in [0,T]$, and any $T_1 \in [t,T]$, we consider the following QBSDE on $[T_1,T]$:

$$\lambda(t,r) = \psi(t) + \int_r^T g(t,s,\nu(t,s))ds - \int_r^T \nu(t,s)dW(s), \quad \forall r \in [T_1,T]. \tag{8.4.7}$$

Thanks to (H8.4.1) and $\psi(\cdot) \in L^\infty_{\mathcal{F}_T}(0,T)$, by applying Theorem 8.3.6, we see that QBSDE (8.4.7) admits a unique adapted solution $(\lambda(t,\cdot), \nu(t,\cdot))$ on $[T_1,T]$. In addition, the following is satisfied with $\mu > 2\gamma$:

$$\begin{aligned} &e^{\mu|\lambda(t,r)|} \leqslant K\mathbb{E}_r e^{\mu\left(|\psi(t)|+\int_r^T \alpha(t,s)ds\right)}, \quad r \in [T_1,T], \\ &\mathbb{E}_\tau \int_\tau^T |\nu(t,s)|^2 ds \leqslant K\mathbb{E}_\tau e^{\mu\left(|\psi(t)|+\int_\tau^T \alpha(t,s)ds\right)}, \quad \tau \in \mathscr{T}[T_1,T]. \end{aligned} \tag{8.4.8}$$

Here, K only depends on γ, μ. Define

$$Y(t) \equiv \lambda(t,t), \quad t \in [0,T], \quad Z(t,s) \equiv \nu(t,s), \quad (t,s) \in \Delta^*[0,T].$$

We obtain an adapted solution for QBSVIE (8.4.2) on $[0,T]$ and (8.4.6) holds.

Next, we look at the uniqueness. Suppose $(\widetilde{Y}(\cdot), \widetilde{Z}(\cdot)) \in L^\infty_{\mathbb{F}}(0,T;\mathbb{R}) \times \overline{\mathrm{BMO}}(\Delta^*[0,T];\mathbb{R})$ is another adapted solution. Define

$$\widehat{Y}(s) \equiv Y(s) - \widetilde{Y}(s), \quad \widehat{Z}(t,s) \equiv Z(t,s) - \widetilde{Z}(t,s),$$

then it is easy to see

$$\widehat{Y}(t) = \int_t^T g_z(t,s)\widehat{Z}(t,s)ds - \int_t^T \widehat{Z}(t,s)dW(s), \quad t \in [0,T],$$

where

$$g_z(t,s) \equiv \frac{g(t,s,Z(t,s)) - g(t,s,\widetilde{Z}(t,s))}{\widehat{Z}(t,s)} \mathbf{1}_{\{\widehat{Z}(t,s)\neq 0\}}, \quad s \in [t,T].$$

By (8.4.5), we see that $g_z(t,\cdot) \in \overline{\mathrm{BMO}}[t,T]$ since

$$\left|g_z(t,s)\right| \leqslant \beta(t,s) + \gamma\left(|Z(t,s)| + |\widetilde{Z}(t,s)|\right), \quad s \geqslant t.$$

As a result, by extending $g_z(t,s)$ to be zero for $s \in [0,t)$, and defining

$$\widehat{W}(t,s) \equiv W(s) - \int_0^s g_z(t,r)dr, \quad s \in [0,T],$$

we see that $\widehat{W}(t,\cdot)$ is a Brownian motion under probability $\widehat{\mathbb{P}}^t$ defined by $\widehat{\mathbb{P}}^t(d\omega) \equiv \mathcal{E}\{g_z(t,\cdot)\}(T)\mathbb{P}(d\omega)$. Hence

$$\widehat{Y}(t) = -\int_t^T \widehat{Z}(t,s)d\widehat{W}(t,s), \quad t \in [0,T],$$

from which $\widehat{Y}(\cdot) = 0$, and hence, $\widehat{Z}(\cdot,\cdot) = 0$, which leads to the uniqueness. □

4.2 Local Existence of Adapted Solutions

In this subsection, we consider the well-posedness of general (that the generator g is allowed to depend on y) Type-I (multidimensional) QBSVIE (8.4.1) in small intervals. Let us begin with the following hypotheses on the generator $g(\cdot)$.

(H8.4.2) Map $g : \Delta^*[0,T] \times \mathbb{R}^n \times \mathbb{R}^n \times \Omega \to \mathbb{R}^n$ is measurable so that $s \mapsto g(t,s,y,z)$ is $\mathbb{F}$-progressively measurable on $[t,T]$ for each $(t,y,z) \in [0,T)\times\mathbb{R}^n\times\mathbb{R}^n$. There exists an $\alpha(\cdot\,,\cdot) \in L^\infty(0,T;L^\infty_{\mathbb{F}}(\Omega;L^1(\cdot\,,T)))$, a modulus of continuity $\phi(\cdot)$, and constants $\gamma > 0$, $\delta \in [0,1)$ such that

$$\big|g^i(t,s,y,z)\big| \leqslant \alpha(t,s) + \phi(|y|) + \gamma\Big(\sum_{j\neq i}|z^j|^{1+\delta} + |z^i|^2\Big),$$
$$\forall (t,s,y,z) \in \Delta^*[0,T]\times\mathbb{R}^n\times\mathbb{R}^n,\ i = 1,2,\cdot s,n. \tag{8.4.9}$$

Further, it holds that

$$\begin{aligned}&|g^i(t,s,y,z) - g^i(t,s,\bar y,\bar z)|\\ &\leqslant \phi(|y|\vee|\bar y|)\Big[\big(1+|z|+|\bar z|\big)\big(|y-\bar y| + |z^i - \bar z^i|\big)\\ &\qquad + \big(1+|z|^\delta+|\bar z|^\delta\big)\sum_{j\neq i}|z^j - \bar z^j|\Big],\\ &\forall (t,s)\in\Delta^*[0,T],\ y,\bar y,z,\bar z\in\mathbb{R}^n,\ i=1,2,\cdot s,n.\end{aligned} \tag{8.4.10}$$

Note that for the generator $g(\cdot)$, only the i-th component $g^i(\cdot)$ can quadratically depends on z^i, and it relies on z^j (with $j \neq i$) merely sub-quadractically. The following gives the local well-posedness of Type-I QBSVIE.

Theorem 8.4.2 *Let* (H8.4.2) *hold and* $\psi(\cdot) \in L^\infty_{\mathcal{F}_T}(0,T;\mathbb{R}^n)$. *Then, there exists a* $\widetilde T \in [0,T)$ *such that Type-I QBSVIE* (8.4.1) *admits a unique adapted solution* $(Y(\cdot), Z(\cdot\,,\cdot)) \in L^\infty_{\mathbb{F}}(\widetilde T,T;\mathbb{R}^n)\times\overline{\mathrm{BMO}}(\Delta^*[\widetilde T,T];\mathbb{R}^n)$.

Proof We split the lengthy proof into several steps.

Step 1 Construction of the solution map satisfying certain estimates.

Suppose $(y(\cdot), z(\cdot\,,\cdot)) \in L^\infty_{\mathbb{F}}(0,T;\mathbb{R}^n)\times\overline{\mathrm{BMO}}(\Delta^*[0,T];\mathbb{R}^n)$. Given constant $T_\varepsilon \in [0,T]$, for $i = 1,2,\cdots,n$, consider one-dimensional QBSVIE:

$$Y^i(t) = \psi^i(t) + \int_t^T g^i(t,s,y(s),z[Z^i](t,s))ds - \int_t^T Z^i(t,s)dW(s),$$
$$t\in[T_\varepsilon,T], \tag{8.4.11}$$

where for $(t,s)\in\Delta^*[T_\varepsilon,T]$,

$$z[Z^i](t,s) \equiv (z^1(t,s),\cdot s, z^{i-1}(t,s), Z^i(t,s), z^{i+1}(t,s),\cdot s, z^n(t,s))^\top.$$

In other words, the i-th component of z is replaced by Z^i. From (H8.4.2), for any $\zeta, \bar{\zeta} \in \mathbb{R}$, it is easy to see

$$
\begin{aligned}
&\big|g^i(t,s,y(s),z[\zeta](t,s))\big| \leqslant \alpha(t,s)+\phi(|y(s)|)+\gamma\big(|z(t,s)|^{1+\delta}+|\zeta|^2\big),\\
&\big|g^i(t,s,y(s),z[\zeta](t,s))-g^i(t,s,y(s),z[\bar{\zeta}](t,s))\big|\\
&\leqslant \phi(|y(s)|)\big[1+2|z(t,s)|+|\zeta|+|\bar{\zeta}|\big]|\zeta-\bar{\zeta}|.
\end{aligned}
$$

Since $z(\cdot,\cdot) \in \overline{\mathrm{BMO}}(\Delta^*[0,T];\mathbb{R}^n)$, by (8.2.27), for almost all $t \in [T_\varepsilon, T]$, $\tau \in \mathscr{T}[t,T]$, $\mu > 1$,

$$
\mathbb{E}_\tau\Big[\exp\Big(\mu\int_\tau^T |z(t,s)|^{1+\delta}ds\Big)\Big] \leqslant 2\exp\Big\{(T-t)K_0(\delta,n,\mu)\widetilde{M}^{\frac{(1+\delta)}{1-\delta}}\Big\}<\infty,
\tag{8.4.12}
$$

with K in (8.2.25),

$$
\widetilde{M} \equiv \operatorname*{esssup}_{t\in[T_\varepsilon,T]}\ \operatorname*{esssup}_{\tau\in\mathscr{T}[t,T]} \Big\|\sum_{i=1}^n \mathbb{E}_\tau\int_\tau^T |z^i(t,s)|^2ds\Big\|_\infty = \big\|z(\cdot,\cdot)\big\|^2_{\overline{\mathrm{BMO}}(\Delta^*[T_\varepsilon,T];\mathbb{R}^n)}.
$$

Together with $y(\cdot)$ being bounded and

$$
\sup_{t\in[0,T]} \Big\|\int_t^T \alpha(t,s)ds\Big\|_{L^2_{\mathcal{F}_T}(\Omega;\mathbb{R})} < \infty,
$$

we see that for each $i = 1, 2, \cdot s, n$, $g^i(\cdot)$ satisfies (H8.4.1) with $\alpha(t,s)$, $\beta(t,s)$, and γ being

$$
\alpha(t,s)+\phi(|y(s)|)+\gamma|z(t,s)|^{1+\delta}, \qquad 1+2|z(t,s)|, \qquad 1,
$$

respectively. Thus, from Lemma 8.4.1, the above QBSVIE (8.4.11) admits a unique solution $(Y^i(\cdot), Z^i(\cdot,\cdot)) \in L^\infty_{\mathbb{F}}(T_\varepsilon,T;\mathbb{R}) \times \overline{\mathrm{BMO}}(\Delta^*[T_\varepsilon,T];\mathbb{R})$, with $Y_i(t) = \lambda^i(t,t)$ where

$$
\lambda^i(t,r) = \psi^i(t)+\int_r^T g^i(t,s,y(s),z[Z^i](t,s))ds-\int_r^T Z^i(t,s)dW(s), \qquad T_\varepsilon \leqslant t \leqslant r \leqslant T.
\tag{8.4.13}
$$

More precisely, with

$$
Y(t) \equiv (Y^1(t), \cdot s, Y^n(t)), \qquad Z(t,s) \equiv (Z^1(t,s), \cdot s, Z^n(t,s)),
$$

we have the following equation is satisfied:

$$
\begin{aligned}
Y(t) = \psi(t) + \int_t^T g(t, s, y(s), z[Z](t, s))ds - \int_t^T Z(t, s)dW(s), \\
t \in [T_\varepsilon, T],
\end{aligned}
\tag{8.4.14}
$$

where for $(t, s) \in \Delta^*[T_\varepsilon, T]$ and $(y(\cdot), z(\cdot\,, \cdot)) \in L^\infty_{\mathbb{F}}(T_\varepsilon, T; \mathbb{R}^n) \times \overline{\text{BMO}}(\Delta^*[T_\varepsilon, T]; \mathbb{R}^n)$,

$$
\begin{aligned}
&g(t, s, y(s), z[Z](t, s)) \\
&\equiv \big(g^1(t, s, y(s), z[Z^1](t, s)), \cdot s, g^n(t, s, y(s), z[Z^n](t, s))\big).
\end{aligned}
$$

Therefore, we have defined a solution map $\Gamma : (y(\cdot), z(\cdot\,, \cdot)) \mapsto (Y(\cdot), Z(\cdot\,, \cdot))$, from $L^\infty_{\mathbb{F}}(T_\varepsilon, T; \mathbb{R}^n) \times \overline{\text{BMO}}(\Delta^*[T_\varepsilon, T]; \mathbb{R}^n)$ to itself. We now establish some estimates. First, we have that for $\mu \equiv \max\{2, 3\gamma\}$,

$$
\begin{aligned}
&\exp\Big(\|Y^i(\cdot)\|_{L^\infty_{\mathbb{F}}(T_\varepsilon, T; \mathbb{R})}\Big) + \|Z^i(\cdot\,, \cdot)\|^2_{\overline{\text{BMO}}(\Delta^*[T_\varepsilon, T]; \mathbb{R})} \\
&\leqslant \bar K \exp\Big\{\mu\big\|\psi^i(\cdot)\big\|_{L^\infty_{\mathcal{F}_T}(0,T)} + \mu\Big\|\int_\cdot^T \alpha(\cdot\,, s)ds\Big\|_{L^\infty_{\mathcal{F}_T}(0,T)} \\
&\quad + \mu\big\|\phi(|y(\cdot)|)\big\|_{L^\infty_{\mathbb{F}}(T_\varepsilon, T; \mathbb{R})}(T - T_\varepsilon) \\
&\quad + (T - T_\varepsilon)K_0(\delta, n, \gamma\mu)\Big\|z(\cdot, \cdot)\Big\|^{\frac{2(1+\delta)}{1-\delta}}_{\overline{\text{BMO}}(\Delta^*[T_\varepsilon, T]; \mathbb{R}^n)}\Big\}.
\end{aligned}
\tag{8.4.15}
$$

Here, K_0 is defined in (8.2.25) with μ by $\gamma\mu$, while constant $\bar K$ only depends on γ, μ. Then, we have

$$
\begin{aligned}
&\exp\Big(\|Y(\cdot)\|_{L^\infty_{\mathbb{F}}(T_\varepsilon, T; \mathbb{R}^n)}\Big) + \|Z(\cdot\,, \cdot)\|^2_{\overline{\text{BMO}}(\Delta^*[T_\varepsilon, T]; \mathbb{R}^n)} \\
&\leqslant \bar K \exp\Big[\bar K\Big\{\big\|\psi(\cdot)\big\|_{L^\infty_{\mathcal{F}_T}(0,T; \mathbb{R}^n)} + \Big\|\int_\cdot^T \alpha(\cdot\,, s)ds\Big\|_{L^\infty_{\mathcal{F}_T}(0,T)} \\
&\quad + \big\|\phi(|y(\cdot)|)\big\|_{L^\infty_{\mathbb{F}}(T_\varepsilon, T; \mathbb{R}^n)}(T - T_\varepsilon) \\
&\quad + \|z(\cdot\,, \cdot)\|^{2\frac{1+\delta}{1-\delta}}_{\overline{\text{BMO}}(\Delta^*[T_\varepsilon, T]; \mathbb{R}^n)}(T - T_\varepsilon)\Big\}\Big].
\end{aligned}
\tag{8.4.16}
$$

Here, $\bar K$ only depends on n, δ, γ.

Step 2 Construction of the convex (and closed) set $\mathscr{B}[T_\varepsilon, T]$ so that

$$
\Gamma(\mathscr{B}[T_\varepsilon, T]) \subseteq \mathscr{B}[T_\varepsilon, T]. \tag{8.4.17}
$$

To construct $\mathscr{B}[T_\varepsilon, T]$, we let

$$
\|y(\cdot)\|_{L^\infty_{\mathbb{F}}(T_\varepsilon, T; \mathbb{R}^n)},\ \|z(\cdot\,, \cdot)\|_{\overline{\text{BMO}}(\Delta^*[T_\varepsilon, T]; \mathbb{R}^n)} \leqslant L.
$$

Then, (8.4.16) implies

$$\begin{aligned}
&\exp\Big(\|Y(\cdot)\|_{L^\infty_{\mathbb{F}}(T_\varepsilon,T;\mathbb{R}^n)}\Big)+\|Z(\cdot\,,\cdot)\|^2_{\overline{\mathrm{BMO}}(\Delta^*[T_\varepsilon,T];\mathbb{R}^n)}\\
&\leqslant \bar K\exp\Big[\bar K\Big\{\|\psi(\cdot)\|_{L^\infty_{\mathcal{F}_T}(0,T;\mathbb{R}^n)}+\Big\|\int_\cdot^T\alpha(\cdot\,,s)ds\Big\|_{L^\infty_{\mathcal{F}_T}(0,T)}\\
&\qquad+\phi(L)(T-T_\varepsilon)+L^{2\frac{1+\delta}{1-\delta}}(T-T_\varepsilon)\Big\}\Big]\\
&=\widetilde K\exp\Big\{\bar K\Big(\phi(L)+L^{2\frac{1+\delta}{1-\delta}}\Big)(T-T_\varepsilon)\Big\},
\end{aligned}\tag{8.4.18}$$

where the last $\widetilde K$ depends on $\|\psi(\cdot)\|_{L^\infty_{\mathcal{F}_T}(0,T;\mathbb{R}^n)}$, $\Big\|\int_\cdot^T\alpha(\cdot\,,s)ds\Big\|_{L^\infty_{\mathcal{F}_T}(0,T)}$, $\bar K$. To have (8.4.17), we require

$$\widetilde K e^{\bar K(\phi(L)+L^{2\frac{1+\delta}{1-\delta}})(T-T_\varepsilon)}\leqslant L.\tag{8.4.19}$$

This will be true if $T-T_\varepsilon>0$ is small enough and $L\geqslant 2\widetilde K$. Then, with such $L>0$ and $T_\varepsilon<T$, we may define convex and closed set:

$$\begin{aligned}
\mathscr{B}[T_\varepsilon,T]\equiv\Big\{&(y(\cdot),z(\cdot\,,\cdot))\in L^\infty_{\mathbb{F}}(T_\varepsilon,T;\mathbb{R}^n)\times\overline{\mathrm{BMO}}(\Delta^*[T_\varepsilon,T];\mathbb{R}^n)\;\Big|\\
&\|y(\cdot)\|_{L^\infty_{\mathbb{F}}(T_\varepsilon,T;\mathbb{R}^n)},\ \|z(\cdot\,,\cdot)\|_{\overline{\mathrm{BMO}}(\Delta^*[T_\varepsilon,T];\mathbb{R}^n)}\leqslant L\Big\}.
\end{aligned}\tag{8.4.20}$$

Given such a $\mathscr{B}[T_\varepsilon,T]$, we naturally have (8.4.17). Note that if we replace T_ε by a $T'_\varepsilon\in(T_\varepsilon,T)$, then the $\mathscr{B}[T'_\varepsilon,T]$ satisfies corresponding (8.4.17). Thus, shrinking $T-T_\varepsilon$ will keep (8.4.17).

Step 3. Contraction property of solution map Γ on $\mathscr{B}[T_\varepsilon,T]$.

Let $(y(\cdot),z(\cdot\,,\cdot)),(\widetilde y(\cdot),\widetilde z(\cdot\,,\cdot))\in\mathscr{B}[T_\varepsilon,T]$, and let

$$(Y(\cdot),Z(\cdot\,,\cdot))\equiv\Gamma(y(\cdot),z(\cdot\,,\cdot)),\qquad(\widetilde Y(\cdot),\widetilde Z(\cdot\,,\cdot))\equiv\Gamma(\widetilde y(\cdot),\widetilde z(\cdot\,,\cdot)).$$

For $t\in[T_\varepsilon,T]$, $s\in[t,T]$, let

$$\begin{aligned}
&\Delta Y(t)=Y(t)-\widetilde Y(t)\equiv(Y^1(t)-\widetilde Y^1(t),\cdot s,Y^n(t)-\widetilde Y^n(t))^\top,\\
&\Delta Z(t,s)=Z(t,s)-\widetilde Z(t,s)\equiv(Z^1(t,s)-\widetilde Z^1(t,s),\cdot s,Z^n(t,s)-\widetilde Z^n(t,s))^\top.
\end{aligned}$$

We may similarly define $\Delta y(\cdot)$, $\Delta z(\cdot,\cdot)$, $\Delta Z^i(\cdot,\cdot)$, $\Delta\lambda^i(\cdot,\cdot)$, etc. Observe that

$$\begin{aligned}
&g^i(t,s,y(s),z[Z^i](t,s))-g^i(t,s,\widetilde y(s),\widetilde z[\widetilde Z^i](t,s))\\
&=g^i(t,s,y(s),z[Z^i](t,s))-g^i(t,s,y(s),z[\widetilde Z^i](t,s))\\
&\qquad+g^i(t,s,y(s),z[\widetilde Z^i](t,s))-g^i(t,s,\widetilde y(s),\widetilde z[\widetilde Z^i](t,s))\\
&=g^i_{z^i}(t,s)\Delta Z^i(t,s)+\mathcal{G}^i(t,s)
\end{aligned}$$

where

$$g_{z^i}^i(t,s) \equiv \frac{g^i(t,s,y(s),z[Z^i](t,s)) - g^i(t,s,y(s),z[\widetilde{Z}^i](t,s))}{\Delta Z^i(t,s)} \mathbf{1}_{\{\Delta Z^i(t,s)\neq 0\}},$$
$$\mathcal{G}^i(t,s) \equiv g^i(t,s,y(s),z[\widetilde{Z}^i](t,s)) - g^i(t,s,\widetilde{y}(s),\widetilde{z}[\widetilde{Z}^i](t,s)).$$

In view of (8.4.10) and boundedness of $y(\cdot)$, we have

$$\begin{aligned} \big|g_{z^i}^i(t,s)\big| &\leqslant \phi(|y(s)|)\big(1+2|z(t,s)|+|Z^i(t,s)|+|\widetilde{Z}^i(t,s)|\big) \\ &\leqslant K\big(1+|z(t,s)|+|Z(t,s)|+|\widetilde{Z}(t,s)|\big), \qquad s\in(t,T], \end{aligned} \tag{8.4.21}$$

and

$$\begin{aligned} &\big|\mathcal{G}^i(t,s)\big| \\ &\leqslant \phi(|y(s)|\vee|\widetilde{y}(s)|)\Big[\big(1+|z(t,s)|+|\widetilde{z}(t,s)|+2|\widetilde{Z}(t,s)|\big)|y(s)-\widetilde{y}(s)| \\ &\quad +\big(1+|z(t,s)|^\delta+|\widetilde{z}(t,s)|^\delta+2|\widetilde{Z}(t,s)|^\delta\big)\sum_{j\neq i}|z^j(t,s)-\widetilde{z}^j(t,s)|\Big] \\ &\leqslant K\Big[\big(1+|z(t,s)|+|\widetilde{z}(t,s)|+|\widetilde{Z}(t,s)|\big)|\Delta y(s)| \\ &\quad +\big(1+|z(t,s)|^\delta+|\widetilde{z}(t,s)|^\delta+|\widetilde{Z}(t,s)|^\delta\big)|\Delta z(t,s)|\Big], \qquad s\in(t,T], \end{aligned} \tag{8.4.22}$$

where K only depends on $\|y(\cdot)\|_{L^\infty_{\mathbb{F}}(0,T;\mathbb{R}^n)}$, $\phi(\cdot)$. Since $z(\cdot\,,\cdot), Z(\cdot\,,\cdot), \widetilde{Z}(\cdot\,,\cdot) \in \overline{\text{BMO}}(\Delta^*[T_\varepsilon,T];\mathbb{R}^n)$, we see from (8.4.21) that

$$\operatorname*{esssup}_{t\in[0,T]}\ \operatorname*{esssup}_{\tau\in\mathscr{T}[t,T]} \Big\|\mathbb{E}_\tau\int_\tau^T |g_{z^i}^i(t,s)|^2 ds\Big\|_\infty < \infty. \tag{8.4.23}$$

For each $i=1,\cdots,n$ and almost every $t\in[T_\varepsilon,T]$, by (8.4.13),

$$\Delta\lambda^i(t,r) = \int_r^T\big[g_{z^i}^i(t,s)\Delta Z^i(t,s)+\mathcal{G}^i(t,s)\big]ds - \int_r^T \Delta Z^i(t,s)dW(s), \qquad r\in[t,T]. \tag{8.4.24}$$

Extend $g_{z^i}^i(t,s)$ to be zero for $s\in[0,t)$ and define

$$N_i(t,r) \equiv \int_0^r g_{z^i}^i(t,s)dW(s), \qquad r\in[0,T].$$

In view of (8.4.23), we know that $N_i(t,\cdot)$ is a BMO martingale under $\mathbb{P}$ on $[0,T]$ and that

$$\begin{aligned}\|N_i(t,\cdot)\|_{\overline{\mathrm{BMO}}[0,T]} &\leqslant K\Big(1+\|Z(\cdot,\cdot)\|_{\overline{\mathrm{BMO}}(\Delta[T_\varepsilon,T])}\\ &\quad+\|\widetilde{Z}(\cdot,\cdot)\|_{\overline{\mathrm{BMO}}(\Delta[T_\varepsilon,T])}+\|z(\cdot,\cdot)\|_{\overline{\mathrm{BMO}}(\Delta[T_\varepsilon,T])}\Big)\leqslant K.\end{aligned} \tag{8.4.25}$$

Note that the first K only depends on $\|y(\cdot)\|_{L^\infty_{\mathbb{F}}(0,T;\mathbb{R}^n)}$, $\phi(\cdot)$, while the second K have further dependence on BMO norm of Z, $\widetilde{Z}$, and z. Recall that (y,z), (Y,Z), $(\widetilde{Y},\widetilde{Z})$ belong to $\mathscr{B}[T_\varepsilon,T]$ with their norms being bounded by L in (8.4.19). Therefore, the constant K in (8.4.25) depends only on the norm of given coefficients (ψ, α, n, γ, and δ). Then, by Lemma 8.2.6, the stochastic exponential $\widehat{\mathcal{E}}\{g_z^i(t,\cdot)\}(\cdot)$ is a uniformly integrable martingale on $[0,T]$ under $\mathbb{P}$.

Now, we define the probability measure $\mathbb{P}^{t,i}$ on $(\Omega,\mathcal{F}_T)$ by the following:

$$d\mathbb{P}^{t,i}\equiv\widehat{\mathcal{E}}[g_z^i(t,\cdot)](T)d\mathbb{P}.$$

We extend $\Delta Z^i(t,s)$ to be zero for $s\in[0,t)$ and obtain that

$$M_i(t,r)\equiv\int_0^r\Delta Z^i(t,s)dW(s),\qquad r\in[0,T],$$

is a BMO martingale under $\mathbb{P}$ on $[0,T]$. Then, by Girsanov theorem, the process

$$\widehat{W}_i(t,r)\equiv W(r)-\int_0^r g^i_{z^i}(t,s)ds,\qquad r\in[0,T], \tag{8.4.26}$$

is a standard Brownian motion under the probability measure $\mathbb{P}^{t,i}$. As a result, by Lemma 8.2.7, the process

$$\widehat{M}_i(t,r)\equiv\int_0^r\Delta Z^i(t,s)d\widehat{W}_i(t,s),\quad r\in[0,T],$$

is a BMO martingale under the probability measure $\mathbb{P}^{t,i}$. Moreover, by (8.4.26), we know that (8.4.24) can be written as

$$\Delta\lambda^i(t,r)+\int_r^T\Delta Z^i(t,s)d\widehat{W}_i(t,s)=\int_r^T\mathcal{G}^i(t,s)ds,\quad T_\varepsilon\leqslant t\leqslant r\leqslant T. \tag{8.4.27}$$

Let $\mathbb{E}_r^{t,i}[\,\cdot\,]$ denote the conditional expectation operator with respect to $\mathcal{F}_r$ under the probability measure $\mathbb{P}^{t,i}$. Taking square and applying the operator $\mathbb{E}_t^{t,i}[\,\cdot\,]$ in both sides of (8.4.27) yields that

$$|\Delta\lambda^i(t,r)|^2+\mathbb{E}_r^{t,i}\Big[\int_r^T|\Delta Z^i(t,s)|^2ds\Big]=\mathbb{E}_r^{t,i}\Big[\Big|\int_r^T\mathcal{G}^i(t,s)ds\Big|^2\Big]. \tag{8.4.28}$$

For $i = 1, \cdots, n$, it follows from (8.4.22) that for almost every $t \in [T_\varepsilon, T]$ and each $r \in [t, T]$,

$$\Big(\int_r^T \big|\mathcal{G}^i(t,s)\big|ds\Big)^2 \leqslant K\Big(\int_r^T \big(1 + |z(t,s)| + |\widetilde{z}(t,s)| + |\widetilde{Z}(t,s)|\big)|\Delta y(s)|ds\Big)^2$$
$$+K\Big(\int_r^T \big(1 + |z(t,s)|^\delta + |\widetilde{z}(t,s)|^\delta + |\widetilde{Z}(t,s)|^\delta\big)|\Delta z(t,s)|ds\Big)^2.$$

By Hölder's inequality, we deduce that for each $r \in [t, T]$,

$$\begin{aligned}
&\Big(\int_r^T \big(1 + |z(t,s)| + |\widetilde{z}(t,s)| + |\widetilde{Z}(t,s)|\big)|\Delta y(s)|ds\Big)^2 \\
&\leqslant K\Big(T - T_\varepsilon\Big)\int_r^T \big(1 + |z(t,s)|^2 + |\widetilde{z}(t,s)|^2 + |\widetilde{Z}(t,s)|^2\big)|\Delta y(s)|^2 ds \\
&= K(T - T_\varepsilon)\int_r^T \vartheta(t,s)|\Delta y(s)|^2 ds,
\end{aligned} \tag{8.4.29}$$

where

$$\vartheta(t,s) \equiv 1 + |z(t,s)|^2 + |\widetilde{z}(t,s)|^2 + |\widetilde{Z}(t,s)|^2, \qquad s \in [t, T].$$

Similarly

$$\begin{aligned}
&\Big(\int_r^T \big(1 + |z(t,s)|^\delta + |\widetilde{z}(t,s)|^\delta + |\widetilde{Z}(t,s)|^\delta\big)|\Delta z(t,s)|ds\Big)^2 \\
&\leqslant \Big(\int_r^T \big(1 + |z(t,s)|^\delta + |\widetilde{z}(t,s)|^\delta + |\widetilde{Z}(t,s)|^\delta\big)^2 ds\Big)\Big(\int_r^T |\Delta z(t,s)|^2 ds\Big) \\
&\leqslant \Big(T - T_\varepsilon\Big)^{1-\delta}\Big(\int_r^T |\Delta z(t,s)|^2 ds\Big) \\
&\qquad \Big(\int_r^T \big(1 + |z(t,s)|^\delta + |\widetilde{z}(t,s)|^\delta + |\widetilde{Z}(t,s)|^\delta\big)^{\frac{2}{\delta}} ds\Big)^\delta \\
&\leqslant K(T - T_\varepsilon)^{1-\delta}\Big(\int_r^T \vartheta(t,s)ds\Big)^\delta\Big(\int_r^T |\Delta z(t,s)|^2 ds\Big) \\
&\leqslant K(T - T_\varepsilon)^{1-\delta}\Big(1 + \int_r^T \vartheta(t,s)ds\Big)\Big(\int_r^T |\Delta z(t,s)|^2 ds\Big).v
\end{aligned} \tag{8.4.30}$$

Thus, for each $i = 1, \cdots, n$, we substitute (8.4.29) and (8.4.30) into (8.4.28) to obtain that for $T_\varepsilon \leqslant t \leqslant r \leqslant T$,

$$\begin{aligned}
&|\Delta\lambda^i(t,r)|^2 + \mathbb{E}_r^{t,i}\Big[\int_r^T |\Delta Z^i(t,s)|^2 ds\Big] \\
&\leqslant K(T - T_\varepsilon)\mathbb{E}_r^{t,i}\Big[\int_r^T \vartheta(t,s)|\Delta y(s)|^2 ds\Big] \\
&+K(T - T_\varepsilon)^{1-\delta}\mathbb{E}_r^{t,i}\Big[\Big(1 + \int_r^T \vartheta(t,s)ds\Big)\int_r^T |\Delta z(t,s)|^2 ds\Big].
\end{aligned} \tag{8.4.31}$$

For each $i = 1, \cdot s, n$, $t \in [T_\varepsilon, T]$, in view of the definitions of $\vartheta(\cdot\,,\cdot)$, $\mathbb{E}^{t,i}[\,\cdot\,]$ and $\mathbb{P}^{t,i}$ together with (8.2.16) and (8.4.26), there exists constant K only depending on given coefficients (ψ, α, n, γ, and δ) such that for each $\tau \in \mathscr{T}[t, T]$,

$$\left\|\Delta Z^i(t,\cdot)\right\|^2_{\overline{\mathrm{BMO}}[t,T]} \leqslant K \sup_{\tau\in\mathscr{T}[t,T]} \left\|\mathbb{E}^{t,i}_\tau\Big[\int_\tau^T |\Delta Z^i(t,s)|^2 ds\Big]\right\|_\infty, \tag{8.4.32}$$

and

$$\begin{aligned}
&\mathbb{E}^{t,i}_\tau\Big[\int_\tau^T \vartheta(t,s)ds\Big]\\
&\leqslant K\Big(1+\|z(t,\cdot)\|^2_{\overline{\mathrm{BMO}}[t,T]}+\|\widetilde{z}(t,\cdot)\|^2_{\overline{\mathrm{BMO}}[t,T]}+\|\widetilde{Z}(t,\cdot)\|^2_{\overline{\mathrm{BMO}}[t,T]}\Big)\leqslant K,
\end{aligned}$$

$$\begin{aligned}
&\mathbb{E}^{t,i}_\tau\Big[\Big(1+\int_\tau^T \vartheta(t,s)ds\Big)^2\Big]\\
&\leqslant K\Big(1+\|z(t,\cdot)\|^4_{\overline{\mathrm{BMO}}[t,T]}+\|\widetilde{z}(t,\cdot)\|^4_{\overline{\mathrm{BMO}}[t,T]}+\|\widetilde{Z}(t,\cdot\|^4_{\overline{\mathrm{BMO}}[t,T]}\Big)\leqslant K,\\
&\mathbb{E}^{t,i}_\tau\Big[\Big(\int_\tau^T |\Delta z(t,s)|^2 ds\Big)^2\Big]\leqslant K\left\|\Delta z(t,\cdot)\right\|^4_{\overline{\mathrm{BMO}}[t,T]}.
\end{aligned}$$

Then, by virtue of Hölder's inequality, we have

$$\mathbb{E}^{t,i}_\tau\Big[\int_\tau^T \vartheta(t,s)|\Delta y(s)|^2 ds\Big]\leqslant K\left\|\Delta y(\cdot)\right\|^2_{L^\infty_{\mathbb{F}}(T_\varepsilon,T;\mathbb{R}^n)}, \tag{8.4.33}$$

and

$$\mathbb{E}^{t,i}_\tau\Big[\Big(1+\int_\tau^T \vartheta(t,s)ds\Big)\int_\tau^T |\Delta z(t,s)|^2 ds\Big]\leqslant K\left\|\Delta z(t,\cdot)\right\|^2_{\overline{\mathrm{BMO}}[t,T]}. \tag{8.4.34}$$

Note that (8.4.31) holds still when r is replaced with any $\tau \in \mathscr{T}[t, T]$. It follows from (8.4.32)–(8.4.34) that (recall that $\lambda^i(t,t) = Y^i(t)$)

$$\begin{aligned}
&\left\|\Delta Y(\cdot)\right\|^2_{L^\infty_{\mathbb{F}}(T_\varepsilon,T;\mathbb{R}^n)}+\left\|\Delta Z(\cdot,\cdot)\right\|^2_{\overline{\mathrm{BMO}}(\Delta^*[T_\varepsilon,T])}\\
&\leqslant K(T-T_\varepsilon)\left\|\Delta y(\cdot)\right\|^2_{L^\infty_{\mathbb{F}}(T_\varepsilon,T;\mathbb{R}^n)}+K(T-T_\varepsilon)^{1-\delta}\left\|\Delta z(\cdot,\cdot)\right\|^2_{\overline{\mathrm{BMO}}(\Delta^*[T_\varepsilon,T])}.
\end{aligned}$$

At this moment, we point out that K only depends on given coefficients (ψ, α, n, γ, and δ). We choose T_ε which is small enough such that

$$\|(Y(\cdot)-\widetilde{Y}(\cdot), Z(\cdot\,,\cdot)-\widetilde{Z}(\cdot\,,\cdot))\|_{\mathscr{B}[T_\varepsilon,T]}\leqslant\frac{1}{2}\|(y(\cdot)-\widetilde{y}(\cdot), z(\cdot\,,\cdot)-\widetilde{z}(\cdot\,,\cdot))\|_{\mathscr{B}[T_\varepsilon,T]}.$$

This means that Γ is a contraction mapping on the convex closed $\mathscr{B}[T_\varepsilon, T]$, which is just the desired result. The proof is then complete. □

4.3 Global Existence of Adapted Solutions

In this subsetion, we are going to establish the global existence of adapted solutions. The basic idea was explained in Chap. 4, Sect. 1. The previous subsection gives us the existence of adapted solution $(Y(\cdot), Z(\cdot\,,\cdot))$ in the region ①. To get a BSVIE in the region ③, we need to determine $Z(\cdot\,,\cdot)$ in the region ② or solve a stochastic Fredholm integral equation (SFDE). We thus have the following result.

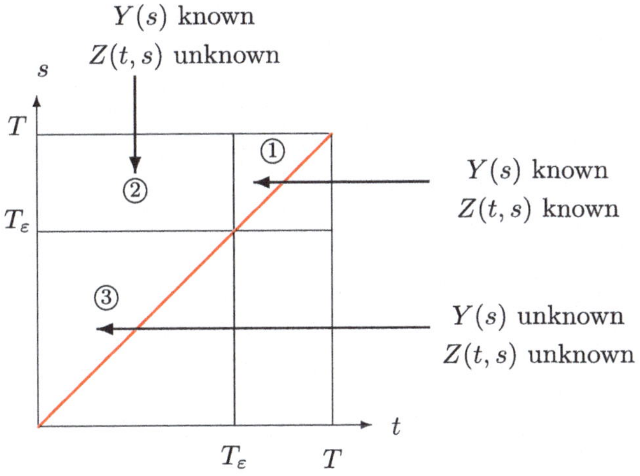

Theorem 8.4.3 *Let* (H8.4.2) *hold,* $\psi(\cdot) \in L^\infty_{\mathcal{F}_T}(0, T; \mathbb{R}^n)$*, and for some* $T_\varepsilon \in [0, T)$*,* $Y(\cdot) \in L^\infty_{\mathbb{F}}(T_\varepsilon, T; \mathbb{R}^n)$ *satisfies BSVIE* (8.4.1) *on* $[T_\varepsilon, T]$*. Then, there exists* $T'_\varepsilon \geqslant T_\varepsilon$ *such that the following SFDE*

$$\psi^{T'_\varepsilon}(t)=\psi(t)+\int_{T'_\varepsilon}^{T} g(t, s, Y(s), Z(t, s))ds-\int_{T'_\varepsilon}^{T} Z(t, s)dW(s), \qquad t\in[0,T'_\varepsilon], \tag{8.4.35}$$

admits a unique adapted solution

$$(\psi^{T'_\varepsilon}(\cdot), Z(\cdot\,,\cdot)) \in L^\infty_{\mathcal{F}_{T'_\varepsilon}}(0, T'_\varepsilon; \mathbb{R}^n) \times \overline{\text{BMO}}([0, T'_\varepsilon] \times [T'_\varepsilon, T]; \mathbb{R}^n).$$

Note that combining the result of Theorems 8.4.2 and 8.4.3, we see that for $t \in [0, T'_\varepsilon]$,

$$\begin{aligned} Y(t) &= \psi(t) + \int_t^T g(t, s, Y(s), Z(t, s))ds - \int_t^T Z(t, s)dW(s) \\ &= \psi(t) + \int_{T'_\varepsilon}^T g(t, s, Y(s), Z(t, s))ds - \int_{T'_\varepsilon}^T Z(t, s)dW(s) \\ &\quad + \int_t^{T'_\varepsilon} g(t, s, Y(s), Z(t, s))ds - \int_t^{T'_\varepsilon} Z(t, s)dW(s) \\ &= \psi^{T'_\varepsilon}(t) + \int_t^{T'_\varepsilon} g(t, s, Y(s), Z(t, s))ds - \int_t^{T'_\varepsilon} Z(t, s)dW(s), \end{aligned}$$

with the free term $\psi^{T'_\varepsilon}(\cdot)$ being $\mathcal{F}_{T'_\varepsilon}$-measurable. Hence, this is a BSVIE on $[0, T'_\varepsilon]$. Therefore, if we can archive that the step size from T to T'_ε fixed positively, by backward induction, we will have the global existence. We now sketch the proof of the above theorem, as it is very similar to that of Theorem 8.4.2.

Sketch of the Proof Given undetermined $T'_\varepsilon \geqslant T_\varepsilon$, and $z \in \overline{\text{BMO}}([0, T'_\varepsilon] \times [T'_\varepsilon, T]; \mathbb{R}^n)$, consider the following system:

$$\begin{aligned} \lambda^i(t, r) = \psi^i(t) + \int_r^T g^i(t, s, Y(s), z[Z^i](t, s))ds - \int_r^T Z^i(t, s)dW(s), \\ r \in [T'_\varepsilon, T], \quad t \in [0, T'_\varepsilon], \quad i = 1, 2, \cdots, n, \end{aligned} \tag{8.4.36}$$

where for $(t, s) \in [0, T'_\varepsilon] \times [T'_\varepsilon, T]$,

$$z[Z^i](t, s) \equiv (z^1(t, s), \cdot s, z^{i-1}(t, s), Z^i(t, s), z^{i+1}(t, s), \cdot s, z^n(t, s))^\top.$$

With the help of Lemma 8.4.1, for each fixed $1 \leqslant i \leqslant n$, we obtain the existence and uniqueness of

$$(\lambda^i(t, \cdot), Z^i(t, \cdot)) \in L^\infty_{\mathbb{F}}\big(\Omega; C([T'_\varepsilon, T])\big) \times \overline{\text{BMO}}[T'_\varepsilon, T].$$

Moreover, similar to (8.4.15), for any $r \in [T'_\varepsilon, T]$, and almost all $t \in [0, T'_\varepsilon]$, there exists a $\bar{K}$ depending on γ:

$$\begin{aligned} &e^{|\lambda^i(t,r)|} + \|Z^i(t, \cdot)\|_{\overline{\text{BMO}}[T'_\varepsilon, T]} \\ &\leqslant \bar{K} \exp\Big[\bar{K}\Big\{\|\psi^i(\cdot)\|_{L^\infty_{\mathcal{F}_T}(0,T)} + \Big\|\int_\cdot^T \alpha(\cdot\,, s)ds\Big\|_{L^\infty_{\mathcal{F}_T}(0,T)} \\ &\quad + \|\phi(|Y(\cdot)|)\|_{L^\infty_{\mathbb{F}}(T'_\varepsilon, T; \mathbb{R})}(T - T'_\varepsilon) \\ &\quad + (T - T'_\varepsilon)K_0(\delta, n, \gamma\mu)\|z(\cdot, \cdot)\|^{\frac{2(1+\delta)}{1-\delta}}_{\overline{\text{BMO}}([0, T'_\varepsilon] \times [T'_\varepsilon, T]; \mathbb{R}^n)}\Big\}\Big]. \end{aligned}$$

As a result, by taking $r = T'_\varepsilon$ in (8.4.36), we obtain

$$(\psi^{T'_\varepsilon}(\cdot), Z(\cdot\,,\cdot)) \in L^\infty_{\mathcal{F}_{T'_\varepsilon}}(0, T'_\varepsilon; \mathbb{R}) \times \overline{\text{BMO}}([0, T'_\varepsilon] \times [T'_\varepsilon, T]; \mathbb{R})$$

with $\psi^{T'_\varepsilon}(\cdot) \equiv \lambda^i(\cdot, T_\varepsilon)$. Consequently, the following system

$$\psi^{T'_\varepsilon}(t) = \psi(t) + \int_{T'_\varepsilon}^T g(t, s, Y(s), z[Z](t, s))ds - \int_{T'_\varepsilon}^T Z(t, s)dW(s), \qquad t \in [0, T_\varepsilon],$$

admits a unique solution

$$(\psi^{T'_\varepsilon}(\cdot), Z(t, \cdot)) \in L^\infty_{\mathcal{F}_{T'_\varepsilon}}(T'_\varepsilon, T; \mathbb{R}^n) \times \overline{\text{BMO}}([T'_\varepsilon, T]; \mathbb{R}^n)$$

with

$$\begin{aligned}
&\lambda \equiv (\lambda^1, \cdots, \lambda^n), \quad Z \equiv (Z^1, \cdots, Z^n),\\
&g(t, s, Y(s), z[Z](t, s))\\
&\equiv \big(g^1(t, s, Y(s), z[Z^1](t, s)), \cdots, g^n(t, s, Y(s), z[Z^n](t, s))\big).
\end{aligned}$$

In addition

$$\begin{aligned}
&\|Z\|^2_{\overline{\text{BMO}}([0,T'_\varepsilon]\times[T'_\varepsilon,T];\mathbb{R}^n)}\\
&\equiv \operatorname*{esssup}_{t\in[0,T'_\varepsilon]} \operatorname*{esssup}_{\tau\in\mathscr{T}[T'_\varepsilon,T]} \Big\| \mathbb{E}_\tau\Big[\sum_{1\leqslant i\leqslant n} \int_\tau^T |Z^i(t, s)|^2 ds \Big] \Big\|_\infty\\
&\leqslant \bar{K} \exp\Big[\bar{K} \Big\{ \big\|\psi(\cdot)\big\|_{L^\infty_{\mathcal{F}_T}(0,T;\mathbb{R}^n)} + \Big\| \int_\cdot^T \alpha(\cdot\,, s)ds \Big\|_{L^\infty_{\mathcal{F}_T}(0,T)}\\
&\quad + \big\|\phi(|Y(\cdot)|)\big\|_{L^\infty_{\mathbb{F}}(T'_\varepsilon,T;\mathbb{R}^n)}(T - T_\varepsilon)\\
&\quad + \|z(\cdot\,,\cdot)\|^{2\frac{1+\delta}{1-\delta}}_{\overline{\text{BMO}}([0,T'_\varepsilon]\times[T'_\varepsilon,T];\mathbb{R}^n)}(T - T'_\varepsilon) \Big\} \Big].
\end{aligned} \tag{8.4.37}$$

Here, $\bar{K}$ only depends on n, δ, γ. As a result, we define solution map $\varGamma$ as $\varGamma(z) = Z$.

Next, we construct a convex (and closed) set $\mathscr{B}([0, T'_\varepsilon] \times [T'_\varepsilon, T])$ so that

$$\varGamma\big(\mathscr{B}([0, T'_\varepsilon] \times [T'_\varepsilon, T])\big) \subseteq \mathscr{B}\big([0, T'_\varepsilon] \times [T'_\varepsilon, T]\big). \tag{8.4.38}$$

To construct $\mathscr{B}([0, T'_\varepsilon] \times [T'_\varepsilon, T])$, we let $\|z(\cdot\,,\cdot)\|_{\overline{\text{BMO}}([0,T'_\varepsilon]\times[T'_\varepsilon,T];\mathbb{R}^n)} \leqslant L$ with L to be determined. Then, (8.4.37) implies

$$\|Z\|^2_{\overline{\text{BMO}}([0,T'_\varepsilon]\times[T'_\varepsilon,T];\mathbb{R}^n)} \leqslant \widetilde{K} \exp\Big\{ \bar{K}\Big(L^{2\frac{1+\delta}{1-\delta}}\Big)(T - T'_\varepsilon) \Big\},$$

where the last $\widetilde{K}$ depends on $\|\psi(\cdot)\|_{L^\infty_{\mathcal{F}_T}(0,T;\mathbb{R}^n)}$, $\left\|\int_\cdot^T \alpha(\cdot\,,s)ds\right\|_{L^\infty_{\mathcal{F}_T}(0,T)}$, $\phi(\cdot)$, $\|Y(\cdot)\|_{L^\infty_{\mathbb{F}}(0,T;\mathbb{R}^n)}$ and $\bar{K}$. To have (8.4.38), we require

$$\widetilde{K}e^{\bar{K}L^{2\frac{1+\delta}{1-\delta}}(T-T'_\varepsilon)} \leqslant L.$$

This will be true if $T-T'_\varepsilon > 0$ is small enough and $L \geqslant 2\widetilde{K}$. Then, with such $L > 0$ and $T'_\varepsilon < T$, we may define convex and closed set:

$$\mathscr{B}[0,T'_\varepsilon]\times[T'_\varepsilon,T] \equiv \Big\{z(\cdot\,,\cdot)\in\overline{\mathrm{BMO}}([0,T'_\varepsilon]\times[T'_\varepsilon,T];\mathbb{R}^n)\ \Big|\ \|z(\cdot\,,\cdot)\|_{\overline{\mathrm{BMO}}([0,T'_\varepsilon]\times[T'_\varepsilon,T];\mathbb{R}^n)} \leqslant L\Big\}.$$

Given such a $\mathscr{B}([0,T'_\varepsilon]\times[T'_\varepsilon,T])$, we naturally have (8.4.38). Note that if we replace T_ε by a $T'_\varepsilon\in(T_\varepsilon,T)$, then the $\mathscr{B}([0,T'_\varepsilon]\times[T'_\varepsilon,T])$ satisfies corresponding (8.4.38). Thus, shrinking $T-T'_\varepsilon$ will keep (8.4.38).

We continue to prove that the map Γ is a contraction in $\mathscr{B}([0,T'_\varepsilon]\times[T'_\varepsilon,T])$ with proper T'_ε. Now, for any $z(\cdot\,,\cdot),\widetilde{z}(\cdot\,,\cdot)\in\mathscr{B}([0,T'_\varepsilon]\times[T'_\varepsilon,T])$, set

$$Z(\cdot\,,\cdot)\equiv\big[\Gamma(z)\big](\cdot\,,\cdot),\qquad \widetilde{Z}(\cdot\,,\cdot)\equiv\big[\Gamma(\widetilde{z})\big](\cdot\,,\cdot).$$

For almost every $t\in[0,T'_\varepsilon]$, we have

$$\|Z(t,\cdot)\|^2_{\overline{\mathrm{BMO}}[T'_\varepsilon,T]},\quad \|\widetilde{Z}(t,\cdot)\|^2_{\overline{\mathrm{BMO}}[T'_\varepsilon,T]} \leqslant L. \tag{8.4.39}$$

Next, for $i=1,\cdots,n$, and $t\in[0,T'_\varepsilon]$, $r\in[T'_\varepsilon,T]$, let (with $\rho\equiv\lambda,z,Z$)

$$\Delta\rho^i(t,r)\equiv\rho^i(t,r)-\widetilde{\rho}^i(t,r),\quad \Delta\rho(t,r)\equiv(\Delta\rho^1(t,r),\cdots,\Delta\rho^n(t,r))^\top.$$

Define (with $0\leqslant t\leqslant T'_\varepsilon\leqslant s\leqslant T$)

$$\begin{aligned} g^i_z(t,s)&\equiv\frac{g^i(t,s,Y(s),z[Z^i](t,s))-g^i(t,s,Y(s),z[\widetilde{Z}^i](t,s))}{\Delta Z^i(t,s)}\mathbf{1}_{\{\Delta Z^i(t,s)\neq0\}},\\ \mathcal{G}^i(t,s)&\equiv g^i(t,s,Y(s),z[\widetilde{Z}^i](t,s))-g^i(t,s,Y(s),\widetilde{z}[\widetilde{Z}^i](t,s)).\end{aligned}$$

Define the probability measure $\mathbb{P}^{t,i}$ on $(\Omega,\mathcal{F}_T)$ by $d\mathbb{P}^{t,i}\equiv\widehat{\mathcal{E}}[g^i_z(t,\cdot)](T)d\mathbb{P}$ and the process (with extension of $g^i_z(t,s)$ to be zero for $s\in[0,T'_\varepsilon]$)

$$\widetilde{W}_i(t,r)\equiv W(r)-\int_0^r g^i_z(t,s)ds,\ \ r\in[0,T]. \tag{8.4.40}$$

Similar to Theorem 8.4.2, we have

$$\begin{aligned}&\Big\|\Delta Z(\cdot,\cdot)\Big\|^2_{\overline{\mathrm{BMO}}([0,T'_\varepsilon]\times[T'_\varepsilon,T];\mathbb{R}^n)}\\&\leqslant K(T-T'_\varepsilon)^{1-\delta}\Big\|\Delta z(\cdot,\cdot)\Big\|^2_{\overline{\mathrm{BMO}}([0,T'_\varepsilon]\times[T'_\varepsilon,T];\mathbb{R}^n)}.\end{aligned}$$

Here, K only depends on given coefficients (ψ, α, n, γ, and δ). Picking proper T_ε, we obtain that Γ is a contraction mapping, which is just the desired result. □

As mentioned before, we now want to keep the step size uniformly positive (not going to zero). To this end, we introduce the following further assumption.

(H8.4.3) There exist $\alpha(\cdot\,,\cdot)\in L^\infty(0,T;L^\infty_{\mathbb{F}}(\Omega;L^1(\cdot\,,T;\mathbb{R})))$, and constants $\beta\geqslant\gamma>0$, $\delta\in[0,1)$ such that

$$\begin{aligned}&-\alpha(t,s)-\beta\Big(|y|+\sum_{j\neq i}|z^j|^{1+\delta}\Big)+\gamma|z^i|^2\\&\leqslant g^i(t,s,y,z)\leqslant\alpha(t,s)+\beta\Big(|y|+\sum_{j\neq i}|z^j|^{1+\delta}+|z^i|^2\Big),\\&\qquad\qquad\forall(t,s,y,z)\in\Delta^*[0,T]\times\mathbb{R}^n\times\mathbb{R}^n,\ i=1,2,\cdot s,n.\end{aligned}\tag{8.4.41}$$

In (8.4.9), if $\phi(y)=\beta y$, then (8.4.41) holds. Thus, the latter is obviously stronger than the former. Condition (8.4.41) (and also (8.4.9)) means that the i-th component g^i of the generator g has exactly quadratic growth in the i-th component z^i of z and has sub-quadratical growth in the other components of z. Such a condition is referred to as the *diagonally strictly quadratic* condition of $z\mapsto g(t,s,y,z)$. Roughly speaking, (8.4.41) amounts to saying that

$$0<\gamma\leqslant\varliminf_{|z^i|\to\infty}\frac{g^i(t,s,y,z)}{|z^i|^2}\leqslant\varlimsup_{|z^i|\to\infty}\frac{g^i(t,s,y,z)}{|z^i|^2}\leqslant\beta,\quad 1\leqslant i\leqslant n.\tag{8.4.42}$$

Alternatively, we might need the following hypothesis.

(H8.4.4) There exist $\alpha(\cdot\,,\cdot)\in L^\infty(0,T;L^\infty_{\mathbb{F}}(\Omega;L^2(\cdot\,,T;\mathbb{R})))$, constant $\beta>0$, such that for some $k=0,1,2,\cdot s,n$ and each $(t,s,y,z)\in\Delta^*[0,T]\times\mathbb{R}^n\times\mathbb{R}^n$, it holds that

$$\begin{cases}|g^i(t,s,y,z)|\leqslant\alpha(t,s)+\beta\big(|y|+|z^i|^2\big), & 1\leqslant i\leqslant k,\\ |g^i(t,s,y,z)|\leqslant\alpha(t,s)+\beta\Big(|y|+\displaystyle\sum_{j=k+1}^n|z^j|\Big), & k+1\leqslant i\leqslant n.\end{cases}\tag{8.4.43}$$

Here, $k = 0$ means the first condition in (8.4.43) is absent and $k = n$ means the second condition of (8.4.43) is absent.

We point out that between (H8.4.3) and (H8.4.4), none implies the other. As a matter fact, under (H8.4.4), instead of (8.4.42), we have

$$\overline{\lim_{|z^i|\to\infty}} \frac{|g^i(t,s,y,z)|}{|z^i|^2} \leqslant \beta, \qquad 1 \leqslant i \leqslant n.$$

Thus, under (H8,4,4), the map $z^i \mapsto g^i(t,s,y,z)$ might not be coercive. Thus, (H8.4.3) seems to be stronger than (H8.4.4). On the other hand, the second relation in (8.4.43) says that the map $z^k \mapsto g^i(t,s,y,z)$ is at most of linear growth, for $i > k$. This seems that (H8.4.4) is stronger. Summarizing the above, we see that none of (H8.4.3) and (H8.4.4) is stronger.

Theorem 8.4.4 *Let* (H8.4.3) *or* (H8.4.4) *hold and* $\psi(\cdot) \in L^\infty_{\mathcal{F}_T}(0,T;\mathbb{R}^n)$. *Suppose there exists some* $\delta \in (0,T]$ *such that BSVIE* (8.4.1) *admits a unique adapted solution:*

$$(Y(\cdot), Z(\cdot\,,\cdot)) \in L^\infty_{\mathbb{F}}(T-\delta,T;\mathbb{R}^n) \times \overline{\mathrm{BMO}}(\Delta^*[T-\delta,T];\mathbb{R}^n).$$

Then, there exists a constant $K_0 > 0$ *depending on* $(\psi,\alpha,\beta,\gamma,\delta,n,T)$, *but not on* δ, *such that*

$$\|Y(\cdot)\|_{L^\infty_{\mathbb{F}}(T-\delta,T;\mathbb{R}^n)} + \|Z(\cdot\,,\cdot)\|_{\overline{\mathrm{BMO}}(\Delta^*[T-\delta,T];\mathbb{R}^n)} \leqslant K_0. \tag{8.4.44}$$

Proof Given the above $(Y(\cdot), Z(\cdot\,,\cdot))$, let us consider the following parameterized BSDE on $[T-\delta,T]$, i.e., for $t \in [T-\delta,T]$,

$$\eta(t,r) = \psi(t) + \int_r^T g(t,s,Y(s),Z(t,s))ds - \int_r^T \zeta(t,s)dW(s), \quad r \in [t,T]. \tag{8.4.45}$$

We observe that both (H8.4.3) and (H8.4.4) imply that, for each $i = 1,2,\cdots,n$, and some constant K depending only on β,γ,δ,

$$\big|g^i(t,s,y,z)\big| \leqslant \alpha(t,s) + K\big(|y| + \sum_{j\neq i}|z^j|^{1+\delta} + |z^i|^2\big). \tag{8.4.46}$$

Since $(Y(\cdot), Z(\cdot\,,\cdot)) \in L^\infty_{\mathbb{F}}(T-\delta,T;\mathbb{R}^n) \times \overline{\mathrm{BMO}}([T-\delta,T];\mathbb{R}^n)$, then by (8.4.46), for each $i = 1,2,\cdots,n$, and $r \in [t,T]$,

$$
\begin{aligned}
&\mathbb{E}_r\int_r^T |g^i(t,s,Y(s),Z(t,s))|ds\\
&\leqslant K\Big[\mathbb{E}_r\int_r^T \alpha(t,s)ds+\mathbb{E}_r\int_r^T |Y(s)|ds\\
&\qquad+\mathbb{E}_r\int_r^T |Z(t,s)|^{1+\delta}ds+\mathbb{E}_r\int_r^T |Z^i(t,s)|^2ds\Big]\\
&\leqslant K\Big[1+\mathbb{E}_r\int_r^T |Y(s)|ds+\mathbb{E}_r\int_r^T |Z(t,s)|^2ds\Big]\leqslant K.
\end{aligned}
\tag{8.4.47}
$$

Therefore, the above (8.4.45) admits a unique solution $(\eta(t,\cdot),\zeta(t,\cdot))$ such that

$$
(\eta(t,\cdot),\zeta(t,\cdot))\in L^\infty_{\mathbb{F}}(t,T;\mathbb{R}^n)\times\overline{\mathrm{BMO}}([t,T];\mathbb{R}^n).
$$

By the imposed uniqueness assumption of BSVIE (8.4.1) in $[T-\delta,T]$, we see that

$$
\eta(t,t)=Y(t),\quad \zeta(t,s)=Z(t,s),\quad T-\delta\leqslant s\leqslant t\leqslant T. \tag{8.4.48}
$$

Step 1 In this step, we prove the conclusion under condition (H8.4.3).

First of all, for any given $k\geqslant 1$, by Young's inequality we have that

$$
k|\zeta(t,s)|^{1+\delta}\leqslant\frac{1}{\widetilde M}|\zeta(t,s)|^2+K,\quad s\in[t,T], \tag{8.4.49}
$$

where K is a constant depending on $\widetilde M$, δ, and k, and

$$
\widetilde M\equiv 2\sup_{1\leqslant i\leqslant n}\ \operatorname*{esssup}_{t\in[0,T]}\ \operatorname*{esssup}_{\tau\in\mathscr{T}[t,T]}\Big\|\mathbb{E}_\tau\int_\tau^T|\zeta^i(t,s)|^2ds\Big\|_\infty<\infty.
$$

Define process $\widetilde\zeta^i(t,\cdot)$ as $\frac{1}{\sqrt{\widetilde M}}\zeta^i(t,\cdot)$ on $[t,T]$, we see that $\|\widetilde\zeta^i(t,\cdot)\|_{\overline{\mathrm{BMO}}[t,T]}\leqslant\frac12$. By the extended John-Nirenberg inequality (Lemma 8.2.9), we see that

$$
\mathbb{E}_r\exp\Big\{\frac{1}{\widetilde M}\int_r^T|\zeta^i(t,s)|^2ds\Big\}\leqslant\frac{1}{1-\frac{1}{2\widetilde M}\widetilde M}=2,\qquad\forall r\in[t,T].
$$

Therefore, for any $k\geqslant 1$,

$$
\begin{aligned}
&\mathbb{E}_r\exp\Big\{k\int_r^T|\zeta(t,s)|^{1+\delta}ds\Big\}=\mathbb{E}_r\exp\Big\{\frac{k}{n}\sum_{i=1}^n\int_r^T|\zeta^i(t,s)|^{1+\delta}ds\Big\}\\
&=\mathbb{E}_r\Big[\prod_{i=1}^n\exp\Big\{k\int_r^T|\zeta^i(t,s)|^{1+\delta}ds\Big\}\Big]\leqslant\prod_{i=1}^n\Big[\mathbb{E}_r\exp\Big\{k\int_r^T|\zeta^i(t,s)|^{1+\delta}ds\Big\}\Big]^{\frac1n}\\
&\leqslant\prod_{i=1}^n\Big[\mathbb{E}_r\exp\Big\{\int_r^T\frac{|\zeta^i(t,s)|^2}{\widetilde M}ds\Big\}\Big]^{\frac1n}e^K\leqslant 2e^K<\infty.
\end{aligned}
$$

This together with the integrability of $\alpha(t,\cdot)$, $y(\cdot)$ indicates that $\alpha_0(t,\cdot) \in \bigcap\limits_{\mu\geqslant 1} \mathcal{E}_{\mathbb{F}}^{\mu,\frac{2\delta}{1+\delta}}(\Omega; L^1(t,T;\mathbb{R}_+))$, where

$$\alpha_0(t,s) \equiv \alpha(t,s) + \beta\big(|Y(s)| + |Z(t,s)|^{1+\delta}\big).$$

By virtue of Lemma 8.3.5, we have for each $i = 1, 2, \cdot s, n$, a.e. $t \in [0,T]$ and each $r \in [t,T]$, noting that $\psi(\cdot)$ is bounded:

$$\begin{aligned}
&\exp\Big(\gamma|\eta^i(t,r)|\Big) + \mathbb{E}_r\Big[\int_r^T |\zeta^i(t,s)|^2 ds\Big]\\
&\leqslant K\mathbb{E}_r\Big[\exp\Big(K|\psi^i(t)| + K\int_r^T \big[\alpha(t,s) + \beta\big(|Y(s)| + |Z(t,s)|^{1+\delta}\big)\big]ds\Big)\Big]\\
&\leqslant K\exp\Big(K\int_r^T \|Y(s)\|_\infty ds\Big)\mathbb{E}_r\Big[\exp\Big(K\int_r^T |Z(t,s)|^{1+\delta}ds\Big)\Big].
\end{aligned}\tag{8.4.50}$$

Here, K depends on the given coefficients, say, $\alpha(\cdot)$, β, $\psi(\cdot)$, γ, and δ. Note that on the right-hand side, the term $\int_r^T |Z(t,s)|^{1+\delta}ds$ appears in the exponent, which cannot be directly controlled by the term $\int_r^T |Z(t,s)|^2 ds$. We now make use of the diagonally strictly quadratic condition (8.4.41). Let $i = 1,\cdots,n$, and $k \geqslant 1$. For almost every $t \in [0,T]$ and each $r \in [t,T]$, we define the following: $\mathbb{F}$-stopping time

$$\sigma_k^{t,r,i} \equiv \inf\Big\{\tau \in [r,T] \mid \int_r^\tau |Z^i(t,s)|^2 ds \geqslant k\Big\} \wedge T,$$

with the convention $\inf \varnothing = +\infty$, and define

$$X_k^{t,r,i} \equiv \eta^i(t,r) - \eta^i(t,\sigma_k^{t,r,i}) + \int_r^{\sigma_k^{t,r,i}} \big[\alpha(t,s) + \beta\big(|Y(s)| + |Z(t,s)|^{1+\delta}\big)\big]ds.$$

Then, it follows from the first inequality in (8.4.41), the result of (8.4.48), and BSDE (8.4.45) that

$$\begin{aligned}
&\gamma\int_r^{\sigma_k^{t,r,i}} |\zeta^i(t,s)|^2 ds = \gamma\int_r^{\sigma_k^{t,r,i}} |Z^i(t,s)|^2 ds\\
&\leqslant \int_r^{\sigma_k^{t,r,i}} \big[\alpha(t,s) + \beta\big(|Y(s)| + |Z(t,s)|^{1+\delta}\big) + g^i(t,s,Y(s),Z(t,s))\big]ds\\
&= X_k^{t,r,i} + \int_r^{\sigma_k^{t,r,i}} \zeta^i(t,s)dW(s), \quad r \in [t,T].
\end{aligned}\tag{8.4.51}$$

This yields that for each $\varepsilon > 0$,

$$
\begin{aligned}
&\exp\Big(\gamma\varepsilon\int_r^{\sigma_k^{t,r,i}} |\zeta^i(t,s)|^2 ds\Big)\\
&\leqslant \exp\Big(\varepsilon X_k^{t,r,i}\Big)\cdot\exp\Big(\frac{3}{2}\varepsilon^2\int_r^{\sigma_k^{t,r,i}} |\zeta^i(t,s)|^2 ds\Big)\\
&\quad\cdot\exp\Big(\varepsilon\int_r^{\sigma_k^{t,r,i}}\zeta^i(t,s)dW(s)-\frac{3}{2}\varepsilon^2\int_r^{\sigma_k^{t,r,i}}|\zeta^i(t,s)|^2ds\Big),\quad r\in[t,T].
\end{aligned}
\tag{8.4.52}
$$

Observe that the process

$$
H_k^{t,r,i}(\tau)\equiv\exp\Big(3\varepsilon\int_r^{\tau\wedge\sigma_k^{t,r,i}}\zeta^i(t,s)dW(s)-\frac{9}{2}\varepsilon^2\int_r^{\tau\wedge\sigma_k^{t,r,i}}|\zeta^i(t,s)|^2ds\Big)
$$

is a positive martingale on $[r, T]$ with $H_k^{t,r,i}(r) = 1$. Taking conditional expectation on both sides of (8.4.52) and applying Hölder's inequality, we obtain that for each $r \in [t, T]$,

$$
\begin{aligned}
&\mathbb{E}_r\Big[\exp\Big(\gamma\varepsilon\int_r^{\sigma_k^{t,r,i}}|\zeta^i(t,s)|^2ds\Big)\Big]\\
&\leqslant\Big(\mathbb{E}_r\Big[\exp\Big(3\varepsilon X_k^{t,r,i}\Big)\Big]\Big)^{\frac{1}{3}}\Big(\mathbb{E}_r\Big[\exp\Big(\frac{9}{2}\varepsilon^2\int_r^{\sigma_k^{t,r,i}}|\zeta^i(t,s)|^2ds\Big)\Big]\Big)^{\frac{1}{3}}.
\end{aligned}
$$

Consequently, for $\varepsilon \leqslant \frac{2\gamma}{9}$, we have

$$
\Big\{\mathbb{E}_r\Big[\exp\Big(\gamma\varepsilon\int_r^{\sigma_k^{t,r,i}}|\zeta^i(t,s)|^2ds\Big)\Big]\Big\}^{\frac{2}{3}}\leqslant\Big\{\mathbb{E}_r\Big[\exp\Big(3\varepsilon X_k^{t,r,i}\Big)\Big]\Big\}^{\frac{1}{3}},\qquad r\in[t,T].
$$

In view of the definitions of $\sigma_k^{t,r,i}$ and $X_k^{t,r,i}$, sending $k \to \infty$ in the above inequality yields that for each $\varepsilon \in (0, \frac{2\gamma}{9}]$ and almost every $t \in [0, T]$,

$$
\begin{aligned}
&\Big\{\mathbb{E}_r\Big[\exp\Big(\gamma\varepsilon\int_r^{T}|\zeta^i(t,s)|^2ds\Big)\Big]\Big\}^{2}\leqslant\lim_{k\to\infty}\mathbb{E}_r\Big[\exp\Big(3\varepsilon X_k^{t,r,i}\Big)\Big]\\
&=\mathbb{E}_r\Big\{\exp\Big[3\varepsilon\Big(\eta^i(t,r)-\psi^i(t)+\int_r^T\big[\alpha(t,s)+\beta\big(|Y(s)|+|Z(t,s)|^{1+\delta}\big)\big]ds\Big)\Big]\Big\}\\
&\leqslant K\exp\Big\{3\varepsilon\|\eta(t,r)\|_\infty+K\int_r^T\|Y(s)\|_\infty ds\Big\}\mathbb{E}_r\Big[\exp\Big(3\varepsilon\beta\int_r^T|Z(t,s)|^{1+\delta}ds\Big)\Big],\\
&\hspace{25em} r\in[t,T].
\end{aligned}
$$

Thus, it follows from Hölder's inequality that for almost every $t \in [0, T]$, with $\varepsilon_0 \equiv \frac{2\gamma}{9}$,

$$\begin{aligned}
&\Big\{\mathbb{E}_r\Big[\exp\Big(\frac{\gamma\varepsilon_0}{n}\int_r^T |\zeta(t,s)|^2 ds\Big)\Big]\Big\}^2 \\
&= \Big\{\mathbb{E}_r\Big[\exp\Big(\frac{\gamma\varepsilon_0}{n}\sum_{i=1}^n\int_r^T |\zeta^i(t,s)|^2 ds\Big)\Big]\Big\}^2 \\
&= \Big\{\mathbb{E}_r\Big[\prod_{i=1}^n\exp\Big(\frac{\gamma\varepsilon_0}{n}\int_r^T |\zeta^i(t,s)|^2 ds\Big)\Big]\Big\}^2 \\
&\leqslant \Big\{\prod_{i=1}^n\Big\{\mathbb{E}_r\Big[\exp\Big(\frac{\gamma\varepsilon_0}{n}\int_r^T |\zeta^i(t,s)|^2 ds\Big)\Big]^n\Big\}^{\frac{1}{n}}\Big\}^2 \qquad (8.4.53)\\
&= \Big\{\prod_{i=1}^n\Big\{\mathbb{E}_r\Big[\exp\Big(\gamma\varepsilon_0\int_r^T |\zeta^i(t,s)|^2 ds\Big)\Big]\Big\}^{\frac{1}{n}}\Big\}^2 \\
&\leqslant K\exp\Big[3\varepsilon_0\|\eta(t,r)\|_\infty + K\int_r^T \|Y(s)\|_\infty ds\Big] \\
&\qquad\cdot\Big\{\mathbb{E}_r\Big[\exp\Big(3\varepsilon_0\beta\int_r^T |Z(t,s)|^{1+\delta} ds\Big)\Big]\Big\}, \qquad r \in [t,T].
\end{aligned}$$

By Young's inequality we have that

$$3\varepsilon_0\beta|Z(t,s)|^{1+\delta} \leqslant \frac{\gamma\varepsilon_0}{n}|Z(t,s)|^2 + K, \quad s \in [t,T]. \tag{8.4.54}$$

Here, K depends on n, ε_0, γ, and β. Coming back to (8.4.53), by (8.4.54) and (8.4.48), for a.e. $t \in [0,T]$, $r \in [t,T]$,

$$\begin{aligned}
&\Big\{\mathbb{E}_r\Big[\exp\Big(\frac{\gamma\varepsilon_0}{n}\int_r^T |\zeta(t,s)|^2 ds\Big)\Big]\Big\}^2 \\
&\leqslant K\exp\Big(3\varepsilon_0\|\eta(t,r)\|_\infty + K\int_r^T \|Y(s)\|_\infty ds\Big)\mathbb{E}_r\Big[\exp\Big(\frac{\gamma\varepsilon_0}{n}\int_r^T |Z(t,s)|^2 ds\Big)\Big] \\
&= K\exp\Big(3\varepsilon_0\|\eta(t,r)\|_\infty + K\int_r^T \|Y(s)\|_\infty ds\Big)\mathbb{E}_r\Big[\exp\Big(\frac{\gamma\varepsilon_0}{n}\int_r^T |\zeta(t,s)|^2 ds\Big)\Big],
\end{aligned}$$

which leads to

$$\begin{aligned}
&\mathbb{E}_r\Big[\exp\Big(\frac{\gamma\varepsilon_0}{n}\int_r^T |\zeta(t,s)|^2 ds\Big)\Big] \\
&\leqslant K\exp\Big(3\varepsilon_0\|\eta(t,r)\|_\infty + K\int_r^T \|Y(s)\|_\infty ds\Big),
\end{aligned} \tag{8.4.55}$$

with K depending on n, ε_0, γ, β, $\psi(\cdot)$, and $\alpha(\cdot,\cdot)$. Therefore, for any $\nu > 0$, noting (by Young's inequality again)

$$\nu|\zeta(t,s)|^{1+\delta} \leqslant \frac{\gamma\varepsilon_0}{n^2}|\zeta(t,s)|^2 + K,$$

by Hölder's inequality and (8.4.55), we have

$$\begin{aligned}&\mathbb{E}_r\Big[\exp\Big(\nu\int_r^T|\zeta(t,s)|^{1+\delta}ds\Big)\Big]\leqslant K\mathbb{E}_r\Big[\exp\Big(\frac{\gamma\varepsilon_0}{n^2}\int_r^T|\zeta(t,s)|^2ds\Big)\Big]\\&\leqslant K\Big\{\mathbb{E}_r\Big[\exp\Big(\frac{\gamma\varepsilon_0}{n}\int_r^T|\zeta(t,s)|^2ds\Big)\Big]\Big\}^{\frac{1}{n}}\\&\leqslant K\exp\Big[\frac{3\varepsilon_0}{n}\|\eta(t,r)\|_\infty+K\int_r^T\|Y(s)\|_\infty ds\Big],\quad r\in[t,T].\end{aligned}\tag{8.4.56}$$

Hence, it follows from (8.4.50) and (8.4.56) that for $i=1,\cdots,n$ and a.e. $t\in[0,T]$,

$$\begin{aligned}&\exp\Big(\gamma|\eta^i(t,r)|\Big)+\mathbb{E}_r\Big[\int_r^T|\zeta^i(t,s)|^2ds\Big]\\&\leqslant K\exp\Big(K\int_r^T\|Y(s)\|_\infty ds\Big)\mathbb{E}_r\Big[\exp\Big(K\int_r^T|\zeta(t,s)|^{1+\delta}ds\Big)\Big]\\&\leqslant K\exp\Big(K\int_r^T\|Y(s)\|_\infty ds\Big)\exp\Big[\frac{3\varepsilon_0}{n}\|\eta(t,r)\|_\infty+K\int_r^T\|Y(s)\|_\infty ds\Big]\\&\leqslant K\exp\Big(\frac{3\varepsilon_0}{n}\|\eta(t,r)\|_\infty+K\int_r^T\|Y(s)\|_\infty ds\Big),\quad r\in[t,T].\end{aligned}\tag{8.4.57}$$

Note that $3\varepsilon_0=\frac{2\gamma}{3}<\gamma$. The above leads to

$$\|\eta(t,r)\|_\infty\leqslant K_0\Big(1+\int_r^T\|Y(s)\|_\infty ds\Big),\quad r\in[t,T],\tag{8.4.58}$$

for some absolute constant $K_0>0$, independent of $Y(\cdot)$, in particular. Let $t=r$, we have

$$\|Y(t)\|_\infty\leqslant K_0\Big(1+\int_r^T\|Y(s)\|_\infty ds\Big),\qquad r\in[t,T].$$

Then, by Gronwall's inequality, we see the uniform bound of $Y(\cdot)$. Also, by enlarging $K_0>0$ if necessary, we have that (see (8.4.57))

$$\mathbb{E}_r\Big(\int_r^T|\zeta^i(t,s)|^2ds\Big)\leqslant K_0\exp\Big(2K_0\int_r^T\|Y(s)\|_\infty ds\Big),\quad r\in[t,T].\tag{8.4.59}$$

Hence, the conclusion holds since

$$\big\|Z(\cdot\,,\cdot)\big\|_{\overline{\mathrm{BMO}}(\Delta^*[0,T])}=\operatorname*{esssup}_{t\in[0,T]}\big\|\zeta(t,\cdot)\big\|_{\overline{\mathrm{BMO}}[t,T]}\leqslant K.$$

Step 2 We prove the conclusion under condition (H8.4.4). For each $i = 1, 2, \cdot s, k$, by virtue of Lemma 8.3.5, taking

$$\alpha_0(t, s) \equiv \alpha(t, s) + \beta|Y(s)|,$$

we have a.e. $t \in [0, T]$ and each $r \in [t, T]$:

$$\begin{aligned}
&\exp\Big(|\eta^i(t, r)|\Big) + \mathbb{E}_r\Big[\int_r^T |\zeta^i(t, s)|^2 ds\Big] \\
&\leqslant K\mathbb{E}_r\Big[\exp\Big(K|\psi^i(t)| + K\int_r^T \big[\alpha(t, s) + \beta|Y(s)|\big]ds\Big)\Big] \\
&\leqslant K\exp\Big(K\int_r^T \|Y(s)\|_\infty ds\Big).
\end{aligned} \tag{8.4.60}$$

Unlike the above (8.4.50), there is no ζ term on the right hand. Taking $t = r$, for $i = 1, 2, \cdots, k$, we have

$$\begin{aligned}
&\big\|Y^i(t)\big\|_\infty \leqslant K + K\int_t^T \big\|Y(s)\big\|_\infty ds, \quad t \in [0, T], \\
&\mathbb{E}_t\Big[\int_t^T |\zeta^i(t, s)|^2 ds\Big] \leqslant K\exp\Big(K\int_r^T \|Y(s)\|_\infty ds\Big).
\end{aligned} \tag{8.4.61}$$

Next, let us treat the part when $i = k + 1, \cdot s, n$. Recall that

$$\eta^i(t, r) = \psi^i(t) + \int_r^T g^i(t, s, Y(s), Z(t, s))ds - \int_r^T \zeta^i(t, s)dW(s).$$

By Itô's formula to $e^{\rho s}|\eta^i(t, s)|^2$ on $[r, T]$, we have

$$\begin{aligned}
d\Big[e^{\rho s}|\eta^i(t, s)|^2\Big] &= \rho e^{\rho s}|\eta^i(t, s)|^2 ds - 2e^{\rho s}\eta^i(t, s)g^i(t, s, Y(s), Z(t, s))ds \\
&\quad +2e^{\rho s}\eta^i(t, s)\zeta^i(t, s)dW(s) + e^{\rho s}|\zeta^i(t, s)|^2 ds.
\end{aligned}$$

Therefore

$$\begin{aligned}
&e^{\rho r}|\eta^i(t, r)|^2 + \int_r^T e^{\rho s}|\zeta^i(t, s)|^2 ds + \int_r^T \rho e^{\rho s}|\eta^i(t, s)|^2 \\
&= e^{\rho T}|\psi^i(t)|^2 + 2\int_r^T e^{\rho s}\eta^i(t, s)g^i(t, s, Y(s), Z(t, s))ds \\
&\quad -2\int_r^T e^{\rho s}\eta^i(t, s)\zeta^i(t, s)dW(s).
\end{aligned}$$

Notice that for any undetermined constant K,

$$
\begin{aligned}
&2\int_r^T e^{\rho s}\eta^i(t,s)g^i(t,s,Y(s),Z(t,s))ds\\
&\leqslant \frac{1}{K}\int_r^T e^{\rho s}|\eta^i(t,s)|^2ds + K\int_r^T e^{\rho s}|g^i(t,s,Y(s),Z(t,s))|^2ds\\
&\leqslant \frac{1}{K}\int_r^T e^{\rho s}|\eta^i(t,s)|^2ds + K\int_r^T e^{\rho s}\Big[|\alpha(t,s)|+\beta|Y(s)|+\beta\sum_{j=k+1}^n|Z^j(t,s)|\Big]^2ds.
\end{aligned}
$$

As a result, for $i = k+1,\cdots,n$,

$$
\begin{aligned}
&\Big[e^{\rho r}|\eta^i(t,r)|^2\Big] + \mathbb{E}_r\int_r^T e^{\rho s}|\zeta^i(t,s)|^2ds + \mathbb{E}_r\int_r^T \rho e^{\rho s}|\eta^i(t,s)|^2\\
&\leqslant \mathbb{E}_r\Big[e^{\rho T}|\psi^i(t)|^2\Big] + \frac{1}{K}\mathbb{E}_r\int_r^T e^{\rho s}|\eta^i(t,s)|^2ds\\
&\quad + K\mathbb{E}_r\int_r^T e^{\rho s}\Big[|\alpha(t,s)|+\beta|Y(s)|+\beta\sum_{j=k+1}^n|Z^j(t,s)|\Big]^2ds,
\end{aligned}
$$

which then implies that

$$
\begin{aligned}
&\sum_{i=k+1}^n\Big[e^{\rho r}|\eta^i(t,r)|^2 + \mathbb{E}_r\int_r^T e^{\rho s}|\zeta^i(t,s)|^2ds + \mathbb{E}_r\int_r^T \rho e^{\rho s}|\eta^i(t,s)|^2\Big]\\
&\leqslant \sum_{i=k+1}^n \mathbb{E}_r\Big[e^{\rho T}|\psi^i(t)|^2\Big] + \sum_{i=k+1}^n\frac{1}{K}\mathbb{E}_r\int_r^T e^{\rho s}|\eta^i(t,s)|^2ds\\
&\quad + nK\mathbb{E}_r\int_r^T e^{\rho s}\Big[|\alpha(t,s)|+\beta|Y(s)|+\beta\sum_{j=k+1}^n|Z^j(t,s)|\Big]^2ds.
\end{aligned}
$$

Then, we choose $K\equiv\frac{1}{2n\beta}$, and $\rho\equiv 4n\beta$, and obtain that

$$
\begin{aligned}
&\sum_{i=k+1}^n\Big[e^{\rho r}|\eta^i(t,r)|^2\Big] + \frac{1}{2}\sum_{i=k+1}^n\mathbb{E}_r\int_r^T e^{\rho s}|\zeta^i(t,s)|^2ds\\
&\leqslant \mathbb{E}_r\Big[e^{\rho T}|\psi(t)|^2\Big] + nK\mathbb{E}_r\int_r^T e^{\rho s}\Big[|\alpha(t,s)|+\beta|Y(s)|\Big]^2ds.
\end{aligned}
$$

By taking $t = r$, we have

$$
\begin{aligned}
&\sum_{i=k+1}^n\Big[e^{\rho t}|Y^i(t)|^2\Big] + \frac{1}{2}\sum_{i=k+1}^n\mathbb{E}_t\int_t^T e^{\rho s}|\zeta^i(t,s)|^2ds\\
&\leqslant \mathbb{E}_t\Big[e^{\rho T}|\psi(t)|^2\Big] + nK\mathbb{E}_t\int_t^T e^{\rho s}\Big[|\alpha(t,s)|+\beta|Y(s)|\Big]^2ds.
\end{aligned}
$$

Recall that the above (8.4.61), we have

$$\|Y(t)\|_\infty^2 \leqslant K + K\int_t^T \|Y(s)\|_\infty^2 ds, \quad t \in [0, T]. \tag{8.4.62}$$

Then, by Gronwall's inequality, we have the result of $Y(\cdot)$. The case of $Z(\cdot\,,\cdot)$ is easy to see. □

Now, we arrive at the main conclusion of this section.

Theorem 8.4.5 *Suppose* (H8.4.2) *holds and either* (H8.4.3) *or* (H8.4.4) *holds,* $\psi(\cdot) \in L^\infty(0, T; \mathbb{R}^n)$. *Then, BSVIE* (8.4.1) *admits a unique adapted solution:*

$$(Y(\cdot), Z(\cdot\,,\cdot)) \in L^\infty_{\mathbb{F}}(0, T; \mathbb{R}^n) \times \overline{\mathrm{BMO}}(\Delta^*[0, T]; \mathbb{R}^n).$$

Proof Given T_ε in Theorem 8.4.2 and T'_ε in Theorem 8.4.3, we know that $\bar{T} \equiv T_\varepsilon \vee T'_\varepsilon = T'_\varepsilon$. By Theorem 8.4.2, we obtain the well-posedness for BSVIE (8.4.1) in $[\bar{T}, T]$. Next, we consider

$$\begin{aligned}\psi^{\bar{T}}(t) = \psi(t) &+ \int_{\bar{T}}^T g(t, r, Y(r), Z(t, r))dr \\ &- \int_{\bar{T}}^T Z(t, r)dW(r), \quad t \in [0, \bar{T}].\end{aligned} \tag{8.4.63}$$

According to Theorems 8.4.3 and 8.4.4, the above (8.4.63) admits a unique pair of solution:

$$(\psi^{\bar{T}}(\cdot), Z(\cdot\,,\cdot)) \in L^\infty_{\mathcal{F}_{\bar{T}}}(0, \bar{T}; \mathbb{R}^n) \times \overline{\mathrm{BMO}}([0, \bar{T}] \times [\bar{T}, T]; \mathbb{R}^n),$$

with a uniform estimate (8.4.44). Thus, we can keep the step size $T - \bar{T}$ positively away from zero. Then, by induction, one can obtain the solvability in $[0, T]$. □

5 Unbounded Free Terms and Sub-quadratic Type-I BSVIEs

It is natural to ask if the free term is unbounded, can we still have well-posedness of Type-I BSVIEs with quadratic growth in z? Further, in Chap. 4, we have established a theory for Type-I BSVIEs with the generator having at most linear growth in z and allowing the free term to be (unbounded and) merely L^p integrable ($p > 1$). Thus, we may ask what if the generator g is sub-quadratic, up to quadratic (still superlinear) in z and the free term is unbounded? The purpose of this section is to investigate such kind of questions to some extent.

5.1 BSDEs with Sub-quadratic Growth

In this subsection, we look at some relevant results for BSDEs with sub-quadratic generators. For convenience, we rewrite the above BSDE (8.2.1) in the following:

$$Y(t) = \xi + \int_t^T g(s, Z(s))ds - \int_t^T Z(s)dW(s), \quad t \in [0, T]. \tag{8.5.1}$$

Here, the BSDE could be one dimensional or multidimensional. Let us first introduce the following hypothesis.

(H8.5.1) Let $n = 1$ and the map $g : [0, T] \times \mathbb{R} \times \Omega \to \mathbb{R}$ be measurable such that $t \mapsto g(t, z)$ is $\mathbb{F}$-progressively measurable, for all $z \in \mathbb{R}$. There exist some constants $\gamma > 0$, $\delta \in (0, 1]$, and a map

$$\alpha(\cdot) \in \bigcap_{\mu \geqslant 1} \mathcal{E}_{\mathbb{F}}^{\mu, \frac{2\delta}{1+\delta}}(\Omega; L^1(0, T; \mathbb{R}_+)), \tag{8.5.2}$$

such that

$$|g(t, z)| \leqslant \alpha(t) + \gamma |z|^{1+\delta}, \quad \forall (t, z) \in [0, T] \times \mathbb{R}, \tag{8.5.3}$$

and

$$\begin{aligned} g(t, \theta z + (1-\theta)\bar{z}) \leqslant \theta g(t, z) + (1-\theta)\big(\alpha(t) + \gamma |\bar{z}|^{1+\delta}\big), \\ \forall (t, z, \bar{z}) \in [0, T] \times \mathbb{R} \times \mathbb{R}, \ \theta \in (0, 1). \end{aligned} \tag{8.5.4}$$

In particular, the above is true if (8.5.3) holds and $z \mapsto g(t, z)$ is convex for all $t \in [0, T]$.

From the last statement of (H8.5.1), we see that (8.5.4), together with (8.5.3), can be regarded as an extension of the convexity of the map $z \mapsto g(t, z)$. Now, if the generator $g(\cdot)$ satisfies assumption *(H8.5.1)* with (8.5.4) replaced by

$$\begin{aligned} g(t, \theta z + (1-\theta)\bar{z}) \geqslant \theta g(t, z) - (1-\theta)\big(\alpha(t) + \gamma |\bar{z}|^{1+\delta}\big), \\ \forall (t, z, \bar{z}) \in [0, T] \times \mathbb{R} \times \mathbb{R}, \ \theta \in (0, 1), \end{aligned} \tag{8.5.5}$$

then by defining

$$\bar{g}(t, z) \equiv -g(t, z), \quad \forall (t, z) \in [0, T] \times \mathbb{R},$$

(H8.5.1) holds for $\bar{g}(\cdot)$. The main result of this subsection is the following.

Theorem 8.5.1 *Let* (H8.5.1) *hold. Then, for any* $\xi \in \bigcap_{\mu\geqslant 1} \mathcal{E}_{\mathcal{F}_T}^{\mu,\frac{2\delta}{1+\delta}}(\Omega;\mathbb{R})$, *BSDE* (8.5.1) *admits a unique adapted solution:*

$$(Y(\cdot),Z(\cdot))\in\Big[\bigcap_{\mu\geqslant 1}\mathcal{E}_{\mathbb{F}}^{\mu,\frac{2\delta}{1+\delta}}(\Omega;C([0,T];\mathbb{R}))\Big]\times\Big[\bigcap_{p\geqslant 1}L_{\mathbb{F}}^{p}(\Omega;L^2(0,T;\mathbb{R}))\Big].$$

Since $\alpha(\cdot)$ could be unbounded, for given $z\in\mathbb{R}$, the map $t\mapsto g(t,z)$ could be unbounded. Also, $z\mapsto g(t,z)$ could be quadratic or sub-quadratically superlinear. Hence, the above result gives the existence and uniqueness of the adapted solution of BSDE (8.5.1) for such a generator with possibly unbounded terminal term $\xi\in\bigcap_{\mu\geqslant 1}\mathcal{E}_{\mathcal{F}_T}^{\mu,\frac{2\delta}{1+\delta}}(\Omega;\mathbb{R})$.

To prove Theorem 8.5.1, we need a few lemmas. The following lemma gives a priori estimate of the adapted solutions. It is an extension of Lemma 8.3.5, which will play crucial roles below. We point out condition (8.5.4) is not necessary in these lemmas.

Lemma 8.5.2 *Suppose that* ξ *is an* $\mathcal{F}_T$*-measurable random variable and random field* $g:[0,T]\times\mathbb{R}^n\times\Omega\to\mathbb{R}^n$ *is measurable such that for each* $z\in\mathbb{R}^n$, $t\mapsto g(t,z)$ *is* $\mathbb{F}$*-progressively measurable. Let* $(Y(\cdot),Z(\cdot))$ *be an adapted solution of BSDE* (8.5.1) *such that* $Y(\cdot)\in\bigcap_{\mu\geqslant 1}\mathcal{E}_{\mathbb{F}}^{\mu,\frac{2\delta}{1+\delta}}(\Omega;C([0,T];\mathbb{R}^n))$ *and*

$$\begin{aligned}&sgn\big(Y^i(s)\big)g^i(s,Z(s))\leqslant\alpha(s)+\beta|Z^i(s)|^{1+\delta},\\&\qquad\qquad\qquad s\in[0,T],\ 1\leqslant i\leqslant n,\end{aligned}\tag{8.5.6}$$

for some constants $\beta>0$, $\delta\in(0,1]$, *and some map* $\alpha(\cdot)$ *satisfying* (8.5.2). *Then, there exists a constant* $K>0$ *depending only on* (δ,β,T) *such that for each* $i=1,\cdot s,n$,

$$e^{|Y^i(t)|^{\frac{2\delta}{1+\delta}}}+\mathbb{E}_t\Big[\int_t^T|Z^i(s)|^2ds\Big]\leqslant K\mathbb{E}_t\Big[e^{K\left(|\xi^i|+\int_t^T\alpha(\tau)d\tau\right)^{\frac{2\delta}{1+\delta}}}\Big],\ t\in[0,T],\tag{8.5.7}$$

and for each $p>0$, *there exists a constant* $K>0$ *depending only on* (δ,β,T) *and* p *such that for each* $i=1,\cdot s,n$,

$$\begin{aligned}&\mathbb{E}\Big[\sup_{s\in[t,T]}e^{p|Y^i(s)|^{\frac{2\delta}{1+\delta}}}\Big]+\mathbb{E}\Big[\Big(\int_t^T|Z^i(s)|^2ds\Big)^{\frac{p}{2}}\Big]\\&\leqslant K\mathbb{E}\Big[e^{K\left(|\xi^i|+\int_t^T\alpha(\tau)d\tau\right)^{\frac{2\delta}{1+\delta}}}\Big],\qquad t\in[0,T].\end{aligned}\tag{8.5.8}$$

Proof Let $t \in [0,T)$ be fixed, $(Y(\cdot), Z(\cdot))$ be the adapted solution of QBSDE (8.2.1). Define

$$\varphi(r,\rho) \equiv e^{\mu(r)\rho^\theta}, \qquad (r,\rho) \in [t,T] \times [\rho_0, \infty),$$

with $\mu(\cdot) \geqslant 1$, $\mu'(r) > 0$, $\rho_0 > 1$, and $\theta \in (0,1]$, all being undetermined. Note that the smaller the θ, the smaller the $\varphi(r,\rho)$. We calculate

$$\begin{aligned}
&\varphi_r(r,\rho) = \varphi(r,\rho)\mu'(r)\rho^\theta, \qquad \varphi_\rho(r,\rho) = \varphi(r,\rho)\theta\mu(r)\rho^{\theta-1},\\
&\varphi_{\rho\rho}(r,\rho) = \varphi(r,\rho)\Big(\theta^2\mu(r)^2\rho^{2\theta-2} + \theta(\theta-1)\mu(r)\rho^{\theta-2}\Big)\\
&\qquad = \varphi(r,\rho)\theta\mu(r)\rho^{\theta-2}\Big(\theta\mu(r)\rho^\theta - (1-\theta)\Big)\\
&\qquad \geqslant \varphi(r,\rho)\theta\mu(r)\rho^{\theta-2}(\theta\rho_0^\theta - 1) > 0,
\end{aligned}$$

provided $\rho_0 > \theta^{-\frac{1}{\theta}}$. For such a case, $\rho \mapsto \varphi(r,\rho)$ is strictly convex on $[\rho_0, \infty)$. Note that $\lim\limits_{\theta\downarrow 0}\theta^{-\frac{1}{\theta}} = +\infty$. Thus, in this approach, when θ is getting small, ρ_0 is getting large. For $i = 1, \cdots, n$, define

$$\mathcal{Y}^i(r) \equiv |Y^i(r)| + \kappa + \int_t^r \alpha(s)ds, \quad r \in [t,T],$$

with $\kappa \geqslant \rho_0$, and denote by $L^i(\cdot)$ the local time of $Y^i(\cdot)$ at 0. In view of (8.5.6), applying Itô-Tanaka's formula (Chap. 2, Theorem 2.3.6) to $s \mapsto \varphi(s, \mathcal{Y}^i(s))$, one has

$$\begin{aligned}
d\varphi\big(s,\mathcal{Y}^i(s)\big) &= \Big(\varphi_s\big(s,\mathcal{Y}^i(s)\big) + \varphi_\rho\big(s,\mathcal{Y}^i(s)\big)\big(-\operatorname{sgn}\big(Y^i(s)\big)g^i(s,Z(s)) + \alpha_0(s)\big)\\
&\quad + \frac{1}{2}\varphi_{\rho\rho}\big(s,\mathcal{Y}^i(s)\big)|Z^i(s)|^2\Big)ds\\
&\quad + \varphi_\rho\big(s,\mathcal{Y}^i(s)\big)dL^i(s) + \varphi_\rho\big(s,\mathcal{Y}^i(s)\big)\operatorname{sgn}\big(Y^i(s)\big)Z^i(s)dW(s)\\
&\geqslant \varphi\big(s,\mathcal{Y}^i(s)\big)\Big[\mu'(s)\mathcal{Y}^i(s)^\theta - \theta\mu(s)\mathcal{Y}^i(s)^{\theta-1}\beta|Z^i(s)|^{1+\delta}\\
&\quad + \frac{\theta\mu(s)}{2}\mathcal{Y}^i(s)^{\theta-2}\Big(\theta\mu(s)\mathcal{Y}^i(s)^\theta - (1-\theta)\Big)|Z^i(s)|^2\Big]ds\\
&\quad + \varphi\big(s,\mathcal{Y}^i(s)\big)\theta\mu(s)\mathcal{Y}^i(s)^{\theta-1}\operatorname{sgn}\big(Y^i(s)\big)Z^i(s)dW(s), \qquad s \in [t,T].
\end{aligned}$$

For each $r \in [t,T]$ and $\ell \geqslant 1$, define the following stopping time:

$$\sigma_\ell^r \equiv \inf\Big\{s \in [r,T]\,\Big|\,|Y(s)| + \int_r^s |Z(\tau)|^2 d\tau \geqslant \ell\Big\} \wedge T$$

with the convention $\inf\varnothing = +\infty$. Then, by integrating on $[r, \sigma_\ell^r]$, we get that for each $\ell \geqslant 1$ and $r \in [t, T]$,

$$
\begin{aligned}
\varphi(r, \mathcal{Y}^i(r)) + X_r^\ell \leqslant \varphi(\sigma_\ell^r, \mathcal{Y}^i(\sigma_\ell^r)) + \int_r^{\sigma_\ell^r} \varphi(s, \mathcal{Y}^i(s))\Big[& -\mu'(s)\mathcal{Y}^i(s)^\theta \\
& + \frac{\theta\mu(s)}{2}\mathcal{Y}^i(s)^{\theta-2}\Big(2\beta\mathcal{Y}^i(s)|Z^i(s)|^{1+\delta} \\
& - \theta\mu(s)\mathcal{Y}^i(s)^\theta|Z^i(s)|^2 + (1-\theta)|Z^i(s)|^2\Big)\Big]ds,
\end{aligned}
\tag{8.5.9}
$$

where

$$
X_r^\ell \equiv \int_r^{\sigma_\ell^r} \varphi(s, \mathcal{Y}^i(s))\theta\mu(s)\mathcal{Y}^i(s)^{\theta-1}\operatorname{sgn}\big(Y^i(s)\big)Z^i(s)dW(s).
$$

We split the rest of the proof into two cases for $\delta \in (0, 1)$ and $\delta = 1$.

Case 1 For $\delta \in (0, 1)$, one has

$$
\begin{aligned}
2\beta\mathcal{Y}^i(s)|Z^i(s)|^{1+\delta} &= 2\beta\Big(\mathcal{Y}^i(s)^\theta|Z^i(s)|^2\Big)^{\frac{1+\delta}{2}}\mathcal{Y}^i(s)^{1-\theta\frac{1+\delta}{2}} \\
&\leqslant \frac{\theta}{2}\mathcal{Y}^i(s)^\theta|Z^i(s)|^2 + K_0\mathcal{Y}^i(s)^{(1-\theta\frac{1+\delta}{2})\frac{2}{1-\delta}} \\
&= \frac{\theta}{2}\mathcal{Y}^i(s)^\theta|Z^i(s)|^2 + K_0\mathcal{Y}^i(s)^{\frac{2-\theta(1+\delta)}{1-\delta}}, \quad s \in [t, T].
\end{aligned}
$$

In the above, $K_0 > 0$ is an absolute constant, only depending on β, δ, θ. We now require that

$$
\frac{2-\theta(1+\delta)}{1-\delta} \leqslant 2,
$$

which is equivalent to $\theta \geqslant \frac{2\delta}{1+\delta}$. Hence, the best possibility is to choose $\theta = \frac{2\delta}{1+\delta}$. Then, with such a θ, we have (noting $\mu(s) \geqslant 1$ and $\mathcal{Y}^i(s) \geqslant k$) that for each $\ell \geqslant 1$ and $r \in [t, T]$,

$$
\begin{aligned}
& \varphi(r, \mathcal{Y}^i(r)) + X_r^\ell \\
& \leqslant \varphi(\sigma_\ell^r, \mathcal{Y}^i(\sigma_\ell^r)) + \int_r^{\sigma_\ell^r} \varphi(s, \mathcal{Y}^i(s))\mathcal{Y}^i(s)^\theta\Big[-\Big(\mu'(s) - \frac{\theta K_0}{2}\mu(s)\Big) \\
& \qquad - \frac{\theta\mu(s)}{2}\Big(\frac{\theta}{2}\mathcal{Y}^i(s)^\theta - (1-\theta)\Big)|Z^i(s)|^2\Big)\Big]ds \\
& \leqslant \varphi(\sigma_\ell^r, \mathcal{Y}^i(\sigma_\ell^r)) + \int_r^{\sigma_\ell^r} \varphi(s, \mathcal{Y}^i(s))\mathcal{Y}^i(s)^\theta\Big[-\Big(\mu'(s) - \frac{\theta K_0}{2}\mu(s)\Big)\Big) \\
& \qquad - \frac{\theta\mu(s)}{2}\Big(\frac{\theta}{2}k^\theta - 1\Big)|Z^i(s)|^2\Big)\Big]ds.
\end{aligned}
\tag{8.5.10}
$$

Now, we choose $k \geqslant 1 + \theta^{-\frac{1}{\theta}}$ and $\mu(\cdot) \geqslant 1$ so that

$$\frac{\theta}{2}\Big(\frac{\theta}{2}k^{\theta} - 1\Big) \geqslant 1, \qquad \mu'(s) \geqslant \frac{\theta K_0}{2}\mu(s), \quad s \in [t, T]. \tag{8.5.11}$$

Then, for each $\ell \geqslant 1$ and $r \in [t, T]$, in view of $\mathbb{E}_r[X_r^\ell] = 0$, by taking conditional expectation with respect to $\mathcal{F}_r$ in (8.5.10), we can obtain that

$$\varphi(r, \mathcal{Y}^i(r)) + \mathbb{E}_r\Big[\int_r^{\sigma_\ell^r} |Z^i(s)|^2 ds\Big] \leqslant \mathbb{E}_r\big[\varphi\big(\sigma_\ell^r, \mathcal{Y}^i(\sigma_\ell^r)\big)\big], \tag{8.5.12}$$

with κ depending on δ. By Fatou's lemma and Lebesgue's dominated convergence theorem, as well as the integrability of $Y(\cdot)$ and $\alpha(\cdot)$, sending ℓ to infinity in (8.5.11) yields the following:

$$\begin{aligned} &e^{\mu(r)\big(|Y^i(r)|+\kappa+\int_t^r \alpha(\tau)d\tau\big)^{\frac{2\delta}{1+\delta}}} + \mathbb{E}_r\Big[\int_r^T |Z^i(s)|^2 ds\Big] \\ &\leqslant \mathbb{E}_r\Big[e^{\mu(T)\big(|\xi^i|+\kappa+\int_t^T \alpha(\tau)d\tau\big)^{\frac{2\delta}{1+\delta}}}\Big], \quad r \in [t, T], \end{aligned}$$

which is just the following result

$$e^{\mu(t)\big(|Y^i(t)|+\kappa\big)^{\frac{2\delta}{1+\delta}}} + \mathbb{E}_t\Big[\int_t^T |Z^i(s)|^2 ds\Big] \leqslant \mathbb{E}_t\Big[e^{\mu(T)\big(|\xi^i|+\kappa+\int_t^T \alpha(\tau)d\tau\big)^{\frac{2\delta}{1+\delta}}}\Big] \tag{8.5.13}$$

for the case $\delta \in (0, 1)$ when $r = t$. Furthermore, it follows from Doob's martingale inequality that for each $p > 1$,

$$\begin{aligned} &\mathbb{E}\Big[\sup_{r\in[t,T]} e^{p\mu(r)\big(|Y^i(r)|+\kappa+\int_t^r \alpha(\tau)d\tau\big)^{\frac{2\delta}{1+\delta}}}\Big] \\ &\leqslant \mathbb{E}\Big[\sup_{r\in[t,T]} \Big\{\mathbb{E}_r\Big[e^{\mu(T)\big(|\xi^i|+\kappa+\int_t^T \alpha(\tau)d\tau\big)^{\frac{2\delta}{1+\delta}}}\Big]\Big\}^p\Big] \\ &\leqslant \Big(\frac{p}{p-1}\Big)^p \mathbb{E}\Big[e^{p\mu(T)\big(|\xi^i|+\kappa+\int_t^T \alpha(\tau)d\tau\big)^{\frac{2\delta}{1+\delta}}}\Big]. \end{aligned} \tag{8.5.14}$$

On the other hand, by (8.5.10) and (8.5.11), we can deduce that for each $\ell \geqslant 1$ and $p > 1$,

$$\mathbb{E}\Big[\Big(\int_r^{\sigma_\ell^r} |Z^i(s)|^2 ds\Big)^{\frac{p}{2}}\Big] \leqslant 2^{p-1}\Big\{\mathbb{E}\Big[\Big(\varphi(\sigma_\ell^r, \mathcal{Y}^i(\sigma_\ell^r))\Big)^{\frac{p}{2}}\Big] + \mathbb{E}\big[|X_r^\ell|^{\frac{p}{2}}\big]\Big\}, \quad r \in [t, T].$$

By virtue of BDG's inequality, the definition of X_r^ℓ and Hölder's inequality and in view of $\mathcal{Y}^i(s) \geqslant \kappa \geqslant 1$ and $\theta \in (0,1)$, we observe that

$$
\begin{aligned}
&2^{p-1}\mathbb{E}\big[|X_r^\ell|^{\frac{p}{2}}\big] \leqslant 2^{p-1}\mu(T)^{\frac{p}{2}}\mathbb{E}\Big[\Big(\int_r^{\sigma_\ell^r}\big[\varphi\big(s,\mathcal{Y}^i(s)\big)\big]^2|Z^i(s)|^2ds\Big)^{\frac{p}{4}}\Big]\\
&\leqslant 2^{p-1}\mu(T)^{\frac{p}{2}}\mathbb{E}\Big[\sup_{s\in[r,\sigma_\ell^r]}\big[\big(\varphi\big(s,\mathcal{Y}^i(s)\big)\big)^{\frac{p}{2}}\big]\Big(\int_r^{\sigma_\ell^r}|Z^i(s)|^2ds\Big)^{\frac{p}{4}}\Big]\\
&\leqslant 2^{2p-3}\mu(T)^p\mathbb{E}\Big[\sup_{s\in[r,\sigma_\ell^r]}\big(\varphi\big(s,\mathcal{Y}^i(s)\big)\big)^p\Big]+\frac{1}{2}\mathbb{E}\Big[\Big(\int_r^{\sigma_\ell^r}|Z^i(s)|^2ds\Big)^{\frac{p}{2}}\Big],\quad r\in[t,T].
\end{aligned}
$$

We have

$$
\begin{aligned}
&\mathbb{E}\Big[\Big(\int_r^{\sigma_\ell^r}|Z^i(s)|^2ds\Big)^{\frac{p}{2}}\Big]\\
&\leqslant 2^p\mathbb{E}\Big[\big(\varphi(\sigma_\ell^r,\mathcal{Y}^i(\sigma_\ell^r))\big)^{\frac{p}{2}}\Big]+2^{2p-2}(\mu(T))^p\mathbb{E}\Big[\sup_{s\in[r,\sigma_\ell^r]}\big(\varphi\big(s,\mathcal{Y}^i(s)\big)\big)^p\Big]\\
&\leqslant (8\mu(T))^p\mathbb{E}\Big[\sup_{s\in[r,\sigma_\ell^r]}\big(\varphi\big(s,\mathcal{Y}^i(s)\big)\big)^p\Big],\qquad r\in[t,T],
\end{aligned}
$$

and then (letting $\ell \to \infty$ and $r = t$)

$$
\begin{aligned}
&\mathbb{E}\Big[\Big(\int_t^{T}|Z^i(s)|^2ds\Big)^{\frac{p}{2}}\Big]\leqslant (8\mu(T))^p\mathbb{E}\Big[\sup_{s\in[t,T]}\big(\varphi\big(s,\mathcal{Y}^i(s)\big)\big)^p\Big]\\
&=(8\mu(T))^p\mathbb{E}\Big[\sup_{r\in[t,T]}e^{p\mu(r)\left(|Y^i(r)|+\kappa+\int_t^r\alpha(\tau)d\tau\right)^{\frac{2\delta}{1+\delta}}}\Big].
\end{aligned}
\tag{8.5.15}
$$

Combining (8.5.14) and (8.5.15) yields

$$
\begin{aligned}
&\mathbb{E}\Big[\sup_{s\in[t,T]}e^{p\mu(s)\left(|Y^i(s)|+\kappa+\int_t^s\alpha(\tau)d\tau\right)^{\frac{2\delta}{1+\delta}}}\Big]+\mathbb{E}\Big[\Big(\int_t^{T}|Z^i(s)|^2ds\Big)^{\frac{p}{2}}\Big]\\
&\leqslant\Big(\frac{p}{p-1}\Big)^p\big((8\mu(T))^p+1\big)\mathbb{E}\Big[e^{p\mu(T)\left(|\xi^i|+\kappa+\int_t^T\alpha(\tau)d\tau\right)^{\frac{2\delta}{1+\delta}}}\Big],\quad\forall p>1,
\end{aligned}
\tag{8.5.16}
$$

for the case $\delta \in (0,1)$. Finally, (8.5.7) and (8.5.8) follow from (8.5.13) and (8.5.16) as well for the case of $\delta \in (0,1)$.

Case 2 For $\delta = 1$, we have that $\theta = \frac{2\delta}{1+\delta} = 1$, and it follows from (8.5.9) that for each $\ell \geqslant 1$,

$$
\begin{aligned}
&\varphi(r, \mathcal{Y}^i(r)) + X_r^\ell \\
&\leqslant \varphi(\sigma_\ell^r, \mathcal{Y}^i(\sigma_\ell^r)) + \int_r^{\sigma_\ell^r} \varphi(s, \mathcal{Y}^i(s))\Big[- \mathcal{Y}^i(s)\mu'(s) \\
&\qquad + \frac{\mu(s)}{2}\mathcal{Y}^i(s)^{-1}\Big(2\beta\mathcal{Y}^i(s)|Z^i(s)|^2 - \mu(s)\mathcal{Y}^i(s)|Z^i(s)|^2\Big)\Big]ds \\
&= \varphi(\sigma_\ell^r, \mathcal{Y}^i(\sigma_\ell^r)) + \int_r^{\sigma_\ell^r} \varphi(s, \mathcal{Y}^i(s))\Big[- \mathcal{Y}^i(s)\mu'(s) \\
&\qquad + \frac{\mu(s)}{2}(2\beta - \mu(s))|Z^i(s)|^2\Big]ds, \qquad r \in [t, T].
\end{aligned}
\tag{8.5.17}
$$

Hence, by choosing

$$
\mu'(s) > 0, \qquad \frac{\mu(s)}{2}\big(\mu(s) - 2\beta\big) \geqslant 1, \quad s \in [0, T],
$$

and taking conditional expectation with respect to $\mathcal{F}_r$ in (8.5.17), we get (8.5.12) for the case $\delta = 1$, and sending m to infinity and letting $r = t$ yields (8.5.13) for the case $\delta = 1$. Then, a similar argument to that from (8.5.14) to (8.5.15) yields (8.5.16) for the case $\delta = 1$. Finally, (8.5.7) and (8.5.8) follow from (8.5.13) and (8.5.16) as well for the case of $\delta = 1$. □

With a similar proof, we can show that if the assumptions in Lemma 8.5.2 hold with (8.5.6) replaced by

$$
\mathbf{1}_{\{Y^i(s)>0\}}g^i(s, Z(s)) \leqslant \alpha(s) + \beta|Z^i(s)|^{1+\delta}, \quad \text{a.s.},
$$

then there exist a $\kappa \geqslant 1$ depending on δ and a differentiable function $\mu : [0, T] \to [1, \infty)$ depending on δ, β, T, such that for each $i = 1, \cdots, n$,

$$
\begin{aligned}
&e^{\mu(t)\left((Y^i(t))^+ + \kappa\right)^{\frac{2\delta}{1+\delta}}} + \mathbb{E}_t\Big[\int_t^T \mathbf{1}_{\{Y^i(s)>0\}}|Z^i(s)|^2 ds\Big] \\
&\leqslant \mathbb{E}_t\Big[e^{\mu(T)\left((\xi^i)^+ + \kappa + \int_t^T \alpha(\tau)d\tau\right)^{\frac{2\delta}{1+\delta}}}\Big], \qquad t \in [0, T].
\end{aligned}
$$

Consequently, there exists a constant $K > 0$ depending only on (δ, β, T) such that for each $i = 1, \cdots, n$,

$$
\begin{aligned}
&e^{[(Y^i(t))^+]^{\frac{2\delta}{1+\delta}}} + \mathbb{E}_t\Big[\int_t^T \mathbf{1}_{\{Y^i(s)>0\}}|Z^i(s)|^2 ds\Big] \\
&\leqslant K\mathbb{E}_t\Big[e^{K\left((\xi^i)^+ + \int_t^T \alpha(\tau)d\tau\right)^{\frac{2\delta}{1+\delta}}}\Big], \qquad t \in [0, T].
\end{aligned}
$$

The following result is comparable with Lemma 8.3.3.

Lemma 8.5.3 *Suppose there exist* $\bar{\alpha}(\cdot) \in L^\infty_{\mathbb{F}}(\Omega; L^1(0,T))$ *and* $\gamma \in [1,\infty)$, $\delta \in (0,1]$ *such that for* $k \geqslant 1$,

$$|g_k(t,z)| \leqslant \bar{\alpha}(t) + \gamma|z|^{1+\delta}, \qquad t \in [0,T],\ z \in \mathbb{R}. \tag{8.5.18}$$

In addition

$$\sup_{k\geqslant 1} \|\xi_k\|_\infty < \infty. \tag{8.5.19}$$

Let $(Y_k(\cdot), Z_k(\cdot)) \in L^\infty_{\mathbb{F}}(\Omega; C([0,T])) \times \overline{\text{BMO}}[0,T]$ *be the adapted solution of BSDE* (8.5.1) *associated with* $(\xi_k, g_k(\cdot))$. *Assume that*

$$\begin{aligned} &\xi_k \to \xi, && \text{in } \mathbb{P}, \\ &g_k(t,z_k) \to g(t,z), && \text{a.s., a.e. } t \in [0,T],\ \forall z_k \to z \in \mathbb{R}. \end{aligned} \tag{8.5.20}$$

Moreover, $Y_k(\cdot) \uparrow Y(\cdot)$. *Then, there exists a process* $Z(\cdot) \in L^2_{\mathbb{F}}(0,T;\mathbb{R})$ *such that* $(Y(\cdot), Z(\cdot))$ *is the adapted solution of* (8.5.1) *associated with* $(\xi, g(\cdot))$.

We have replaced the Lipschitz type condition (8.3.8) in Lemma 8.3.3 by the above superlinear growth condition (8.5.18). In addition, the convergence of $g_k(\cdot)$ in (8.5.20) is slightly different from the above (8.3.10).

Proof The proof is similar as the above Lemma 8.3.3. For reader's convenience, we give the analogue details.

First, by conditions (8.5.18) and (8.5.19), as well as Lemma 8.5.2, we then obtain

$$\sup_{k\geqslant 1} \|Y_k(\cdot)\|_\infty + \sup_{k\geqslant 1} \mathbb{E}\int_0^T |Z_k(s)|^2 ds < \infty, \tag{8.5.21}$$

which implies that $\|Y(\cdot)\|_\infty < \infty$. Using Lemma 8.2.11, there exists a subsequence k_i such that $Z_{k_i}(\cdot)$ converges weakly to some $Z(\cdot) \in L^2_{\mathbb{F}}(0,T;\mathbb{R})$. Let us denote

$$\begin{aligned} &\Delta Y_k(t) \equiv Y_k(t) - Y(t), \quad \Delta Y_{\ell,k}(t) \equiv Y_\ell(t) - Y_k(t), \\ &\Delta Z_k(t) \equiv Z_k(t) - Z(t), \quad \Delta Z_{\ell,k}(t) \equiv Z_\ell(t) - Z_k(t). \end{aligned}$$

We would like to prove that

$$\lim_{k\to\infty} \mathbb{E}\int_0^T |\Delta Z_k(s)|^2 ds = 0. \tag{8.5.22}$$

To this end, by (8.5.18), we first note that

$$\begin{aligned}&\big|g_\ell(t,Z_\ell(t))-g_k(t,Z_k(t))\big|\\&\qquad\leqslant 2\bar\alpha(t)+\gamma\big(|Z_\ell(t)|^2+|Z_k(t)|^2\big)\\&\qquad\leqslant 2\bar\alpha(t)+\gamma\big(2|\Delta Z_{\ell,k}(t)|^2+3|Z_k(t)|^2\big)\\&\qquad\leqslant 2\bar\alpha(t)+2\gamma|\Delta Z_{\ell,k}(t)|^2+6\gamma|\Delta Z_k(t)|^2+6\gamma|Z(t)|^2,\end{aligned}$$

with $\delta=1$. For the case of $0<\delta<1$, by Young inequality,

$$\begin{aligned}&\big|g_\ell(t,Z_\ell(t))-g_k(t,Z_k(t))\big|\\&\qquad\leqslant 2\bar\alpha(t)+\gamma\big(|Z_\ell(t)|^{1+\delta}+|Z_k(t)|^{1+\delta}\big)\\&\qquad\leqslant 2\alpha(t)+\gamma\big(|Z_\ell(t)|^2+|Z_k(t)|^2\big)\\&\qquad\leqslant 2\alpha(t)+2\gamma|\Delta Z_{\ell,k}(t)|^2+6\gamma|\Delta Z_k(t)|^2+6\gamma|Z(t)|^2,\end{aligned}$$

where $\alpha(\cdot)\equiv\bar\alpha(\cdot)+K$, with constant K depending on γ,δ. Now, we define

$$\varphi(y)\equiv\frac{1}{8K_0^2}\big(e^{4K_0y}-4K_0y-1\big),\qquad y\in\mathbb{R},\quad K_0\equiv 6\gamma.$$

Then, one has

$$\begin{cases}\varphi(0)=0; & \varphi(y)>0,\quad y>0,\\ \varphi'(0)=0; & 0<\varphi'(y),\quad y>0;\\ \varphi''(y)=4K_0\varphi'(y)+2, & \qquad y\geqslant 0.\end{cases}$$

Applying Itô's formula to $\varphi(\Delta Y_{\ell,k}(\cdot))$, we obtain

$$\begin{aligned}d\varphi(\Delta Y_{\ell,k}(t))&=-\varphi'(\Delta Y_{\ell,k}(t))\big[g_\ell(t,Z_\ell(t))-g_k(t,Z_k(t))\big]dt\\&\quad+\varphi'(\Delta Y_{\ell,k}(t))\Delta Z_{\ell,k}(t)dW(t)+\frac{1}{2}\varphi''(\Delta Y_{\ell,k}(t))|\Delta Z_{\ell,k}(t)|^2dt\\&\geqslant-|\varphi'(\Delta Y_{\ell,k}(t))|\Big(\alpha(t)+K_0\big(|\Delta Z_{\ell,k}(t)|^2+|\Delta Z_k(t)|^2+|Z(t)|^2\big)\Big)dt\\&\quad+\varphi'(\Delta Y_{\ell,k}(t))\Delta Z_{\ell,k}(t)dW(t)+\big[2K_0\varphi'(\Delta Y_{\ell,k}(t))+1\big]|\Delta Z_{\ell,k}(t)|^2dt\\&=-|\varphi'(\Delta Y_{\ell,k}(t))|\Big(\alpha(t)+K_0\big(|\Delta Z_k(t)|^2+|Z(t)|^2\big)\Big)dt\\&\quad+\varphi'(\Delta Y_{\ell,k}(t))\Delta Z_{\ell,k}(t)dW(t)+\big[K_0\varphi'(\Delta Y_{\ell,k}(t))+1\big]|\Delta Z_{\ell,k}(t)|^2dt.\end{aligned}$$

Consequently

$$\begin{aligned}\mathbb{E}\int_0^T\big[K_0\varphi'(\Delta Y_{\ell,k}(t))+1\big]|\Delta Z_{\ell,k}(t)|^2dt&\leqslant\mathbb{E}\varphi(\Delta Y_{\ell,k}(T))-\varphi(\Delta Y_{\ell,k}(0))\\&\quad+\mathbb{E}\int_0^T K_0\varphi'(\Delta Y_{\ell,k}(t))\big(|\Delta Z_k(t)|^2+|Z(t)|^2+\alpha(t)\big)dt.\end{aligned}$$

By the L^2-boundedness of $\{Z_k(\cdot)\}$, Mazur theorem, and Fatou lemma, there exists a subsequence $\{(Y_{\ell_r,k}(\cdot), Z_{\ell_r,k}(\cdot))\}$ such that

$$\begin{aligned}
&\mathbb{E}\int_0^T \big[K_0\varphi'(\Delta Y_k(t))+1\big]|\Delta Z_k(t)|^2 dt\\
&\leqslant \underline{\lim}_{r\to\infty}\mathbb{E}\int_0^T \big[K_0\varphi'(\Delta Y_{\ell_r,k}(t))+1\big]|\Delta Z_{\ell_r,k}(t)|^2 dt\\
&\leqslant \underline{\lim}_{r\to\infty}\Big[\mathbb{E}\varphi(\Delta Y_{\ell_r,k}(T))-\varphi(\Delta Y_{\ell_r,k}(0))\\
&\qquad+\mathbb{E}\int_0^T K_0\varphi'(\Delta Y_{\ell_r,k}(t))\big(|\Delta Z_k(t)|^2+|Z(t)|^2+\alpha(t)\big)dt\Big].
\end{aligned}$$

Thanks to the boundedness of $\{Y_k(\cdot)\}$, $Y_k(\cdot)\uparrow Y(\cdot)$, and dominated convergence theorem, the above inequality implies that

$$\begin{aligned}
&\mathbb{E}\int_0^T \big[K_0\varphi'(\Delta Y_k(t))+1\big]|\Delta Z_k(t)|^2 dt\\
&\leqslant \mathbb{E}\varphi(\Delta Y_k(T))-\varphi(\Delta Y_k(0))\\
&\qquad+\mathbb{E}\int_0^T K_0\varphi'(\Delta Y_k(t))\big(|\Delta Z_k(t)|^2+|Z(t)|^2+\alpha(t)\big)dt,
\end{aligned}$$

which then leads to

$$\begin{aligned}
\mathbb{E}\int_0^T |\Delta Z_k(t)|^2 dt &\leqslant \mathbb{E}\varphi(\Delta Y_k(T))-\varphi(\Delta Y_k(0))\\
&\qquad+\mathbb{E}\int_0^T K_0\varphi'(\Delta Y_k(t))\big(|Z(t)|^2+\alpha(t)\big)dt.
\end{aligned}$$

Using again dominated convergence theorem, we obtain the above (8.5.22).

Next, we prove that $(Y(\cdot), Z(\cdot))$ is the adapted solution of (8.5.1). First, from the above (8.5.22), there exists a subsequence $\{Z_{k_r}\}_{r\geqslant 1}$ such that $Z_{k_r}(s)\to Z(s)$, $r\to\infty$, a.s. a.e. and

$$\Big[\mathbb{E}\int_0^T |Z_{k_{r+1}}(t)-Z_{k_r}(t)|^2 dt\Big]^{\frac{1}{2}}\leqslant\frac{1}{2^r},\qquad r\in\mathbb{N}.$$

We define

$$\bar{Z}(t)\equiv|Z_{k_1}(t)|+\sum_{r=1}^{\infty}\big|Z_{k_{r+1}}(t)-Z_{k_r}(t)\big|,\quad t\in[0,T].$$

Hence

$$\begin{aligned}&\Big[\mathbb{E}\int_0^T|\bar Z(t)|^2dt\Big]^{\frac12}\leqslant\Big[\mathbb{E}\int_0^T|Z_{k_1}(t)|^2dt\Big]^{\frac12}+\sum_{r=1}^{\infty}\Big[\mathbb{E}\int_0^T|Z_{k_{r+1}}(t)-Z_{k_r}(t)|^2dt\Big]^{\frac12}\\&\leqslant\Big[\mathbb{E}\int_0^T|Z_{k_1}(t)|^2dt\Big]^{\frac12}+\sum_{r=1}^{\infty}\frac{1}{2^r}<\infty.\end{aligned}$$

Therefore, we have

$$\sup_{r\geqslant 1}|Z_{k_r}(t)|\leqslant|Z_{k_1}(t)|+\sup_{k\geqslant 1}\sum_{r=1}^{k}\big|Z_{k_{r+1}}(t)-Z_{k_r}(t)\big|\leqslant\bar Z(t).$$

By the imposed condition (8.5.20), we have that

$$\big|g_{k_r}(t,Z_{k_r}(t))-g(t,Z(t))\big|\to 0,\quad r\to\infty.$$

In addition

$$|g_{k_r}(t,Z_{k_r}(t))|\leqslant\alpha(t)+\gamma|Z_{k_r}(t)|^2\leqslant\alpha(t)+\gamma|\bar Z(t)|^2.$$

Then, by dominated convergence theorem, we arrive at

$$\lim_{r\to\infty}\mathbb{E}\int_0^T\big|g_{k_r}(s,Z_{k_r}(s))-g(s,Z(s))\big|ds=0. \tag{8.5.23}$$

By (8.5.22), we have

$$\begin{aligned}&\lim_{k\to\infty}\mathbb{E}\sup_{t\in[0,T]}\Big|\int_t^T\Delta Z_k(s)dW(s)\Big|\leqslant\lim_{k\to\infty}\Big\{\mathbb{E}\sup_{t\in[0,T]}\Big|\int_t^T\Delta Z_k(s)dW(s)\Big|^2\Big\}^{\frac12}\\&\leqslant K\lim_{k\to\infty}\Big\{\mathbb{E}\int_0^T\big|\Delta Z_k(s)\big|^2ds\Big\}^{\frac12}=0.\end{aligned}$$

Since the sequence $\{Y_k(\cdot)\}$ is bounded, by dominated convergence theorem,

$$\lim_{k\to\infty}\mathbb{E}\big|\Delta Y_k(t)\big|=0.$$

Therefore, for any given $t\in[0,T]$, one has

$$\mathbb{E}\Big|Y(t)-\xi-\int_t^T g(s,Z(s))ds+\int_t^T Z(s)dW(s)\Big|=0,$$

which implies that $(Y(\cdot),Z(\cdot))$ is the adapted solution of BSDE (8.5.1) associated with $(\xi,g(\cdot))$. □

Lemma 8.5.4 *Suppose $g(\cdot\,,\cdot)$ is bounded, $z \mapsto g(t,z)$ is continuous, and $\xi \in L^\infty_{\mathcal{F}_T}(\Omega;\mathbb{R})$. Then, BSDE* (8.5.1) *admits a unique minimal adapted solution:*

$$(Y,Z) \in L^\infty_{\mathbb{F}}(\Omega; C([0,T])) \times L^2_{\mathbb{F}}(0,T)$$

in the sense that for any other solution (Y',Z') we have $Y \leqslant Y'$.

Suppose (Y_0,Z_0), (Y_1,Z_1) are the corresponding adapted minimal solutions to BSDE (8.5.1) *with $(\xi_0, g_0(\cdot))$, $(\xi_1, g_1(\cdot))$. Moreover*

$$\xi_0 \leqslant \xi_1, \quad g_0(s,z) \leqslant g_1(s,z), \quad (s,z) \in [0,T]\times\mathbb{R}, \quad a.s.$$

Then, for each $t \in [0,T]$, $Y_0(t) \leqslant Y_1(t)$, almost surely.

Proof We first look at the existence of minimal solution. For any $k \geqslant 1$, we define

$$g_k(t,z) \equiv \inf_{z'\in\mathbb{R}} \Big[g(t,z') + k|z-z'|\Big]. \tag{8.5.24}$$

It then follows from Lemma 8.2.10 that $g_k(t,z)$ is well-defined, increasing in k. In addition, for any $\lim\limits_{k\to\infty} z_k = z$,

$$\lim_{k\to\infty} g_k(t,z_k) = g(t,z), \quad \text{a.s.}, \quad |g_k(t,z) - g_k(t,z_1)| \leqslant k|z-z_1|.$$

Now, let us consider

$$Y_k(t) = \xi + \int_t^T g_k(s, Z_k(s))ds - \int_t^T Z_k(s)dW(s), \quad t \in [0,T].$$

By the Lipschitz property of g_k and the boundedness of ξ, we obtain the existence and uniqueness of

$$(Y_k, Z_k) \in L^\infty_{\mathbb{F}}(\Omega; C([0,T])) \times L^2_{\mathbb{F}}(0,T).$$

In terms of the comparison theorem of BSDEs, Y_k is increasing in k. Hence, we can define $Y \equiv \lim\limits_{k\to\infty} Y_k$. Since $|g(\cdot,\cdot)| \leqslant M$ for some constant M, it follows that

$$|Y_k| \leqslant \|\xi\|_\infty + MT.$$

By Itô's formula to $|Y_{k_1} - Y_{k_2}|^2$ on $[0,T]$,

$$\mathbb{E}\int_0^T |Z_{k_1}(s) - Z_{k_2}(s)|^2 ds \leqslant 2M\mathbb{E}\int_0^T |Y_{k_1}(s) - Y_{k_2}(s)|ds.$$

Note that as $k_1, k_2 \to \infty$, $|Y_{k_1}(s)-Y_{k_2}(s)| \to 0$. a.s. It then follows from dominated convergence theorem that

$$\mathbb{E}\int_0^T |Z_{k_1}(s) - Z_{k_2}(s)|^2 ds \to 0, \quad k_1, k_2 \to \infty.$$

As a result, there exists $Z \in L^2_{\mathbb{F}}(0, T)$ such that

$$\lim_{k\to\infty} \mathbb{E}\int_0^T |Z_k(s) - Z(s)|^2 ds = 0.$$

We claim that (Y, Z) satisfies BSDE (8.5.1). Actually, from the above, there exists a subsequence $\{k_j\}$ such that

$$\lim_{j\to\infty} |Z_{k_j}(s) - Z(s)| = 0, \quad s \in [0, T], \text{ a.s. a.e.}$$

By the convergence of $g_k(\cdot)$ to $g(\cdot)$, we see that

$$\lim_{j\to\infty} g_{k_j}(s, Z_{k_j}(s)) = g(s, Z(s)), \quad \text{a.s.}$$

Using dominated convergence theorem, as $j \to \infty$,

$$\mathbb{E}\int_0^T \Big|g_{k_j}(s, Z_{k_j}(s)) - g(s, Z(s))\Big|^2 ds \to 0.$$

Note that

$$\lim_{j\to\infty} \mathbb{E}|Y_{k_j}(t) - Y(t)|^2 = 0, \quad \forall t \in [0, T],$$

and

$$\begin{aligned}&\lim_{j\to\infty} \mathbb{E} \sup_{t\in[0,T]} \Big|\int_t^T [Z_{k_j}(s) - Z(s)]dW(s)\Big|^2 \\ &\leqslant K \lim_{j\to\infty} \mathbb{E}\int_0^T [Z_{k_j}(s) - Z(s)]^2 ds = 0.\end{aligned}$$

Then, the desired conclusion follows from

$$\mathbb{E}\Big\{Y_k(t) - \xi - \int_t^T g_k(s, Z_k(s))ds + \int_t^T Z_k(s)dW(s)\Big\}^2 = 0.$$

Suppose $(\bar{Y}(\cdot), \bar{Z}(\cdot))$ is another solution, and we show that the above $(Y(\cdot), Z(\cdot))$ is minimal in the sense of $Y(t) \leqslant \bar{Y}(t)$, for all $t \in [0, T)$, almost surely. In fact

$$
\begin{aligned}
Y_k(t) - \bar{Y}(t) &= \int_t^T \big[g_k(s, Z_k(s)) - g(s, \bar{Z}(s))\big]ds - \int_t^T \big[Z_k(s) - \bar{Z}(s)\big]dW(s) \\
&= \varphi(t) + \int_t^T \frac{\big[g_k(s, Z_k(s)) - g_k(s, \bar{Z}(s))\big]}{[Z_k(s) - \bar{Z}(s)]\mathbf{1}_{\{Z_k(s) - \bar{Z}(s) \neq 0\}}}\big[Z_k(s) - \bar{Z}(s)\big]ds \\
&\quad - \int_t^T \big[Z_k(s) - \bar{Z}(s)\big]dW(s)
\end{aligned}
$$

where

$$
\varphi(t) \equiv \int_t^T \big[g_k(s, \bar{Z}(s)) - g(s, \bar{Z}(s))\big]ds \leqslant 0, \quad t \in [0, T].
$$

By the comparison theorem of BSDEs (see Theorem 6.1.6 in Chap. 6), we have $Y_k(\cdot) \leqslant \bar{Y}(\cdot)$ which implies that $Y(\cdot) \leqslant \bar{Y}(\cdot)$. The uniqueness of minimal solution is easy to see.

Finally, let us look at the comparison theorem of minimal solution $(Y_0(\cdot), Z_0(\cdot))$, $(Y_1(\cdot), Z_1(\cdot))$ associated with $(\xi_0, g_0(\cdot))$, $(\xi_1, g_1(\cdot))$. For $g_0(\cdot)$, we define $g_{0,k}(t, z)$ in the similar spirit as in (8.5.24), and consider the corresponding solution $(Y_{0,k}, Z_{0,k})$ satisfying

$$
Y_{0,k}(t) = \xi_0 + \int_t^T g_{0,k}(s, Z_{0,k}(s))ds - \int_t^T Z_{0,k}(s)dW(s), \quad t \in [0, T].
$$

Since (Y_0, Z_0) is minimal solution, we have $Y_0 = \lim_{k\to\infty} Y_{0,k}$. Notice that

$$
\begin{aligned}
Y_{0,k}(t) - Y_1(t) &= \xi_0 - \xi_1 + \int_t^T \big[g_{0,k}(s, Z_{0,k}(s)) - g_1(s, Z_1(s))\big]ds \\
&\quad - \int_t^T \big[Z_{0,k}(s) - Z_1(s)\big]dW(s) \\
&= \bar{\varphi}(t) + \int_t^T \big[g_{0,k}(s, Z_{0,k}(s)) - g_{0,k}(s, Z_1(s))\big]ds \\
&\quad - \int_t^T \big[Z_{0,k}(s) - Z_1(s)\big]dW(s)
\end{aligned}
$$

where

$$
\bar{\varphi}(t) \equiv \xi_0 - \xi_1 + \int_t^T \big[g_{0,k}(s, Z_1(s)) - g_1(s, Z_1(s))\big]ds \leqslant 0, \quad t \in [0, T].
$$

Then, by comparison theorem of BSDEs, we have $Y_{0,k} \leqslant Y_1$, which leads to $Y_0(\cdot) \leqslant Y_1(\cdot)$. □

Now, let us give the proof of Theorem 8.5.1.

Proof First, let us consider the case when ξ and g are nonnegative. For integers $\ell \geqslant 1$, and $r \in [0, T)$, we define

$$g_\ell(t, z) \equiv g(t, z) \wedge \ell, \quad t \in [r, T], \quad \xi_\ell \equiv \xi \wedge \ell.$$

We consider the solution of the following BSDE:

$$Y_\ell(t) = \xi_\ell + \int_t^T g_\ell(s, Z_\ell(s))ds - \int_t^T Z_\ell(s)dW(s), \quad t \in [0, T].$$

According to Lemma 8.5.4, for each $\ell \in \mathbb{N}$, it admits a minimal solution $(Y_\ell, Z_\ell) \in L^\infty_{\mathbb{F}}(\Omega; C([0, T]; \mathbb{R})) \times L^2_{\mathbb{F}}(0, T; \mathbb{R})$ such that $Y_\ell(\cdot)$ is increasing in ℓ.

Notice that

$$|g_\ell(s, z)| \leqslant |g(s, z)| \leqslant \alpha(s) + \gamma |z|^{1+\delta}.$$

Then, according to Lemma 8.5.2, for each $t \in [0, T]$,

$$e^{|Y_\ell(t)|^{\frac{2\delta}{1+\delta}}} + \mathbb{E}_t\Big[\int_t^T |Z_\ell(s)|^2 ds\Big] \leqslant K\mathbb{E}_t\Big[e^{K\left(|\xi| + \int_t^T \alpha(\tau)d\tau\right)^{\frac{2\delta}{1+\delta}}}\Big] \equiv \Psi(t). \tag{8.5.25}$$

Since the right hand $\Psi(\cdot)$ is not bounded, we introduce the following stopping time:

$$\tau_k \equiv \inf\{t \in [0, T], \quad \Psi(t) \geqslant k\} \wedge T.$$

Let us take a closer look at the above stopping time. For any $k \in \mathbb{N}$, there exists a $\mathbb{P}$-null set $\mathcal{N}_k$ such that for any $\omega \in \Omega \setminus \mathcal{N}_k$, $\tau_k(\omega)$ is well-defined. Let $\mathcal{N} \equiv \bigcup_{k \in \mathbb{N}} \mathcal{N}_k$ be an $\mathbb{P}$-null set. Given $\omega \in \Omega \setminus \mathcal{N}$, $\tau_k(\omega)$ is well-defined for any $k \in \mathbb{N}$. In addition, for such choice of ω, by the continuity of $t \mapsto \Psi(t)$, there exists $\widetilde{k}(\omega)$ such that $\tau_{\widetilde{k}(\omega)}(\omega) = T$. In other words

$$\mathbb{P}\Big(\omega \in \Omega;\, \tau_{\widetilde{k}(\omega)}(\omega) = T\Big) = 1. \tag{8.5.26}$$

For given $k \in \mathbb{N}$, by defining

$$Y_{\ell,k}(t) \equiv Y_\ell(t \wedge \tau_k), \quad Z_{\ell,k}(t) \equiv Z_\ell(t)\mathbf{1}_{\{t \leqslant \tau_k\}},$$

we see that

$$Y_{\ell,k}(t) = Y_\ell(\tau_k) + \int_t^T \mathbf{1}_{\{s \leqslant \tau_k\}} g_\ell(s, Z_{\ell,k}(s))ds - \int_t^T Z_{\ell,k}(s)dW(s), \quad t \in [0, T].$$

At this moment, let us use Lemma 8.5.3 to obtain a solution to the following BSDE:

$$Y_k(t) = Y_k(\tau_k) + \int_t^T \mathbf{1}_{\{s\leqslant\tau_k\}} g(s, Z_k(s))ds - \int_t^T Z_k(s)dW(s), \quad t \in [0, T],$$

such that

$$Y_k(t) \equiv \sup_{\ell\geqslant 1} Y_\ell(t \wedge \tau_k), \qquad \lim_{\ell\to\infty} \mathbb{E}\int_0^T |Z_{\ell,k}(t) - Z_k(t)|^2 dt = 0.$$

Notice that for any given $k \in \mathbb{N}$, we have

$$\begin{aligned}
&\mathbb{P}\big(\omega \in \Omega; t \mapsto Y_k(t,\omega) \text{ is continuous for any } t \in [0,T]\big) = 1,\\
&(\lambda \otimes \mathbb{P})\big((t,\omega) \in [0,T] \times \Omega; Z_k(t,\omega) \text{ is well-defined}\big) = T,
\end{aligned}$$

where λ is Lebesgue measure on $[0, T]$. Using similar arguments as the above (8.5.26), we have

$$\begin{aligned}
&\mathbb{P}\big(\omega \in \Omega; t \mapsto Y_k(t,\omega) \text{ is continuous for any } t \in [0,T],\ k \in \mathbb{N}\big) = 1,\\
&(\lambda \otimes \mathbb{P})\big((t,\omega) \in [0,T] \times \Omega; Z_k(t,\omega) \text{ is well-defined for any } k \in \mathbb{N}\big) = T.
\end{aligned}$$

Taking (8.5.26) into consideration as well, the following processes (Y, Z) on $[0, T]$ are well-defined:

$$\begin{aligned}
Y(t) &\equiv Y_1(t)\mathbf{1}_{\{t\leqslant\tau_1\}} + \sum_{\ell\geqslant 2}^{\widetilde{k}} Y_\ell(t)\mathbf{1}_{[\tau_{\ell-1},\tau_\ell]}(t),\\
Z(t) &\equiv Z_1(t)\mathbf{1}_{\{t\leqslant\tau_1\}} + \sum_{\ell\geqslant 2}^{\widetilde{k}} Z_\ell(t)\mathbf{1}_{[\tau_{\ell-1},\tau_\ell]}(t).
\end{aligned}$$

In addition, $\int_0^T |Z(t)|^2 dt < \infty$, a.s., and (Y, Z) is a solution to BSDE (8.5.1) in the sense that $t \mapsto Y(t)$ is continuous on $[0, T]$, $Z(\cdot)$ belongs to $L^2(0, T)$, a.s., $t \mapsto g(t, Z(t))$ belongs to $L^1(0, T)$ and (8.5.1) holds $\mathbb{P}$-a.s.

For the general case, we can use approximation $\xi^{\ell,p} \equiv \xi^+ \wedge \ell - \xi^- \wedge p$ and $g^{\ell,p}(t, z) \equiv g^{\ell,p}(t, z)^+ \wedge \ell - g^{\ell,p}(t, z)^- \wedge p$.

Next, let us prove that

$$(Y(\cdot), Z(\cdot)) \in \Big[\bigcap_{\mu\geqslant 1} \mathcal{E}_{\mathbb{F}}^{\mu,\frac{2\delta}{1+\delta}}(\Omega; C([0,T];\mathbb{R}))\Big] \times \Big[\bigcap_{p\geqslant 1} L_{\mathbb{F}}^p(\Omega; L^2(0,T;\mathbb{R}))\Big].$$

By the construction of $Y(\cdot)$ and (8.5.25), we see that

$$e^{|Y(t)|^{\frac{2\delta}{1+\delta}}} \leqslant K\mathbb{E}_t\Big[e^{K\big(|\xi|+\int_t^T\alpha(\tau)d\tau\big)^{\frac{2\delta}{1+\delta}}}\Big],$$

then the conclusion of $Y(\cdot)$ is followed by the assumptions of ξ and $\alpha(\cdot)$.

Following the same arguments as that in obtaining (8.5.12), for $\theta \equiv \frac{2\delta}{1+\delta}$, by choosing suitable $\kappa \geqslant 1+\theta^{-\frac{1}{\theta}}$, $\mu(\cdot) \geqslant 1$, we have

$$\int_0^{\sigma_j} |Z(s)|^2 ds \leqslant \varphi(\sigma_j, \mathcal{Y}(\sigma_j)) - X^j,$$

where $\varphi(r,\rho) \equiv e^{\mu(r)\rho^\theta}$

$$\begin{aligned}
&\sigma_j \equiv \inf\Big\{s \in [0,T] \;\Big|\; |Y(s)| + \int_0^s |Z(\tau)|^2 d\tau \geqslant j\Big\} \wedge T,\\
&\mathcal{Y}(r) \equiv |Y(r)| + \kappa + \int_0^r \alpha(s)ds, \quad r \in [0,T],\\
&X^j \equiv \int_0^{\sigma_j} \varphi\big(s,\mathcal{Y}(s)\big)\theta\mu(s)\mathcal{Y}(s)^{\theta-1}\mathrm{sgn}\,\big(Y(s)\big)Z(s)dW(s).
\end{aligned}$$

Therefore, for any $p \geqslant 1$,

$$\Big[\int_0^{\sigma_j} |Z(s)|^2 ds\Big]^{\frac{p}{2}} \leqslant K\Big[\varphi(\sigma_j,\mathcal{Y}(\sigma_j))\Big]^{\frac{p}{2}} + [X^j]^{\frac{p}{2}}.$$

Notice that

$$\begin{aligned}
\mathbb{E}[X^j]^{\frac{p}{2}} &\leqslant K\mathbb{E}\Big[\sup_{s\in[0,T]} |\varphi(s,\mathcal{Y}(s))|^{\frac{p}{2}}\Big\{\int_0^{\sigma_j} |Z(s)|^2 ds\Big\}^{\frac{p}{4}}\Big]\\
&\leqslant \frac{p}{2}\mathbb{E}\Big[\int_0^{\sigma_j} |Z(s)|^2 ds\Big]^{\frac{p}{2}} + K\mathbb{E}\sup_{s\in[0,T]} |\varphi(s,\mathcal{Y}(s))|^p,
\end{aligned}$$

we thus have

$$\mathbb{E}\Big[\int_0^{\sigma_j} |Z(s)|^2 ds\Big]^{\frac{p}{2}} \leqslant K\mathbb{E}\sup_{s\in[0,T]} |\varphi(s,\mathcal{Y}(s))|^p.$$

Sending j to infinity, we obtain the desired conclusion of $Z(\cdot)$ by the assumption of $\alpha(\cdot)$ and ξ.

Eventually, we look at the uniqueness. Suppose $(\bar{Y},\bar{Z})$ is another pair of solution. Let $\nu \in (0,1)$, and define

$$\widetilde{Y} \equiv \frac{Y-\nu\bar{Y}}{1-\nu}, \quad \widetilde{Z} \equiv \frac{Z-\nu\bar{Z}}{1-\nu}.$$

We have

$$\widetilde{Y}(t) = \widetilde{Y}(T) - \int_t^T \delta g(s)ds - \int_t^T \widetilde{Z}(s)dW(s),$$

where by (8.5.4), which is the only place that we use this condition,

$$\begin{aligned}\delta g(s) &\equiv \frac{g(s, Z(s)) - \nu g(s, \bar{Z}(s))}{1-\nu}\\ &= \frac{1}{1-\nu}\Big[g\Big(s, \nu\bar{Z}(s) + (1-\nu)\frac{Z(s)-\nu\bar{Z}(s)}{1-\nu}\Big) - \nu g(s, \bar{Z}(s))\Big]\\ &\leqslant K|\widetilde{Z}(s)|^{1+\delta}.\end{aligned}$$

According to Lemma 8.5.2, we have

$$e^{|\widetilde{Y}(t)|^{\frac{2\delta}{1+\delta}}} + \mathbb{E}_t\Big[\int_t^T |\widetilde{Z}(s)|^2 ds\Big] \leqslant K\mathbb{E}_t\Big[e^{K|\widetilde{\xi}|^{\frac{2\delta}{1+\delta}}}\Big], \qquad t \in [0, T].$$

Notice that $\widetilde{\xi} = \xi$, by the assumption of ξ, for any $t \in [0, T]$, we see that

$$\big|Y(t) - \nu\bar{Y}(t)\big|^{\frac{2\delta}{1+\delta}} \leqslant K\mathbb{E}_t\Big[e^{K|\widetilde{\xi}|^{\frac{2\delta}{1+\delta}}}\Big](1-\nu)^{\frac{2\delta}{1+\delta}}.$$

Sending $\nu \to 1$, we see that $Y(\cdot) = \bar{Y}(\cdot)$. Similarly, we have

$$\mathbb{E}\int_0^T \big|Z(s) - \nu\bar{Z}(s)\big|^2 ds \leqslant K\mathbb{E}\Big[e^{K|\widetilde{\xi}|^{\frac{2\delta}{1+\delta}}}\Big](1-\nu)^2.$$

By dominated convergence theorem, we see that $Z(\cdot) = \bar{Z}(\cdot)$. □

5.2 *Results for BSVIEs*

In this subsection, we are going to look at Type-I BSVIE (8.4.1) (with free term being possibly unbounded). We first introduce the following assumption on the generator $g(\cdot)$.

(H8.5.2) The map $g : \Delta^*[0, T] \times \mathbb{R}^n \times \mathbb{R}^n \to \mathbb{R}^n$ is measurable so that $s \mapsto g(t, s, y, z)$ is $\mathbb{F}$-progressively measurable on $[t, T]$ for each $(t, y, z) \in [0, T) \times \mathbb{R}^n \times \mathbb{R}^n$. There exist constants $\beta > 0$, $\delta \in (0, 1]$ and a nonnegative process $\alpha : \Delta^*[0, T] \times \Omega \to \mathbb{R}_+$ with $s \mapsto \alpha(t, s)$ being $\mathbb{F}$-progressively measurable on $[t, T]$, and with

$$\mathbb{E}\Big\{\operatorname*{esssup}_{t\in[0,T]}\Big[\exp\Big(p\Big|\int_t^T \alpha(t, s)ds\Big|^{\frac{2\delta}{1+\delta}}\Big)\Big]\Big\} < \infty, \quad \forall p \geqslant 1,$$

such that $g(t,s,y,z) \equiv (g^1(t,s,y,z^1),\cdots,g^n(t,s,y,z^n))^\top$ with

$$\begin{aligned}&|g^i(t,s,y,z^i)|\leqslant\alpha(t,s)+\beta\big(|y|+|z^i|^{1+\delta}\big),\\&\qquad\qquad\forall(t,s,y,z)\in\Delta^*[0,T]\times\mathbb{R}^n\times\mathbb{R}^n,\ 1\leqslant i\leqslant n,\\&|g(t,s,y,z)-g(t,s,\bar y,z)|\leqslant\beta|y-\bar y|,\\&\qquad\qquad\forall(t,s)\in\Delta^*[0,T],\ y,\bar y,z\in\mathbb{R}^n,\end{aligned}\tag{8.5.27}$$

and

$$\begin{aligned}&g^i(t,s,y,\theta z^i+(1-\theta)\bar z^i)\\&\leqslant\theta g^i(t,s,y,z^i)+(1-\theta)\big[\alpha(t,s)+\beta\big(|y|+|\bar z^i|^{1+\delta}\big)\big],\\&\qquad\forall(t,s)\in\Delta^*[0,T],\ y,z,\bar z\in\mathbb{R}^n,\ \theta\in(0,1),\ 1\leqslant i\leqslant n.\end{aligned}\tag{8.5.28}$$

Relevant to the above, let us introduce the following hypothesis.

(H8.5.2)′ Let (H8.5.2) hold with (8.5.28) replaced by the following: For each $i = 1,2,\cdot s,n$, either

$$\begin{aligned}&g^i(t,s,y,\theta z^i+(1-\theta)\bar z^i)\\&\leqslant\theta g^i(t,s,y,z^i)+(1-\theta)\big[\alpha(t,s)+\beta\big(|y|+|\bar z^i|^{1+\delta}\big)\big],\\&\qquad\qquad\forall(t,s)\in\Delta^*[0,T],\ y,z,\bar z\in\mathbb{R}^n,\ \theta\in(0,1),\end{aligned}\tag{8.5.29}$$

or

$$\begin{aligned}&g^i(t,s,y,\theta z^i+(1-\theta)\bar z^i)\\&\geqslant\theta g^i(t,s,y,z^i)-(1-\theta)\big[\alpha(t,s)+\beta\big(|y|+|\bar z^i|^{1+\delta}\big)\big],\\&\qquad\qquad\forall(t,s)\in\Delta^*[0,T],\ y,z,\bar z\in\mathbb{R}^n,\ \theta\in(0,1).\end{aligned}\tag{8.5.30}$$

Although (H8.5.2)′ seems to be a little general than (H8.5.2), we now show that actually they are equivalent after some minor modifications. In fact, by relabeling, we may assume that for some $k = 0,1,2,\cdot s,n$, one has (8.5.29) holds for $1\leqslant i\leqslant k$, and (8.5.30) holds for $k+1\leqslant i\leqslant n$. If $k=0$, it means that for all $i=1,2,\cdot s,n$, (8.5.30) holds, and if $k=n$, it means that for all $i=1,2,\cdot s,n$, (8.5.29) holds. Now, we define

$$\begin{aligned}&\bar z=(\bar z^1,\cdot s,\bar z^k,\bar z^{k+1},\cdot s,\bar z^n)=(z^1,\cdot s,z^k,-z^{k+1},\cdot s,-z^n),\\&\bar y=(\bar y^1,\cdot s,\bar y^k,\bar y^{k+1},\cdot s,\bar y^n)=(y^1,\cdot s,y^k,-y^{k+1},\cdot s,-y^n),\\&\bar g^i(t,s,y,z^i)=g^i(t,s,\bar y,\bar z^i),\quad 1\leqslant i\leqslant k,\\&\bar g^i(t,s,y,z^i)=-g^i(t,s,\bar y,\bar z^i),\quad k+1\leqslant i\leqslant n.\end{aligned}$$

Then, (H8.5.2)′ holds for $g(t,s,y,z)$ if and only if (H8.5.2) holds for $\bar g(t,s,\bar y,\bar z)$.

Next, we introduce the following hypothesis.

(H8.5.3) Let (H8.5.2) hold with (8.5.28) replaced by the following: for each $i = 1, 2, \cdot s, n$, and each $(t, s, y) \in \Delta^*[0, T] \times \mathbb{R}^n$, $z^i \mapsto g^i(t, s, y, z^i)$ is convex, or concave.

It is clear that when (H8.5.3) holds, then (H8.5.2)$'$ must hold for g. Hence, (H8.5.2) (or (H8.5.2)$'$) is not very restrictive.

For notational convenience, we will frequently use the following function:

$$\Phi_\lambda(x; \mu) \equiv \exp(\mu x^\lambda), \quad \forall x, \mu \geqslant 0, \ \lambda > 0. \tag{8.5.31}$$

Note that for each $x_1, x_2, \mu \geqslant 0$ and $\lambda \in (0, 1]$,

$$\Phi_\lambda(x_1 + x_2; \mu) \leqslant \Phi_\lambda(x_1; \mu)\Phi_\lambda(x_2; \mu). \tag{8.5.32}$$

Lemma 8.5.5 *Let* (H8.5.2) *hold and* $\lambda \equiv \frac{2\delta}{1+\delta}$. *Then, for any*

$$\psi(\cdot) \in \bigcap_{p\geqslant 1} \mathcal{E}^{p,\lambda}_{\mathcal{F}_T}(\Omega; L^\infty(0, T; \mathbb{R}^n)), \quad y(\cdot) \in \bigcap_{p\geqslant 1} \mathcal{E}^{p,\lambda}_{\mathbb{F}}(\Omega; L^\infty(0, T; \mathbb{R}^n)),$$

the following BSVIE

$$Y(t) = \psi(t) + \int_t^T g(t, s, y(s), Z(t, s))ds - \int_t^T Z(t, s)dW(s), \quad t \in [0, T], \tag{8.5.33}$$

admits a unique adapted solution $(Y(\cdot), Z(\cdot\,, \cdot))$ *with*

$$\begin{aligned} Y(\cdot) &\in \bigcap_{p\geqslant 1} \mathcal{E}^{p,\lambda}_{\mathbb{F}}(\Omega; L^\infty(0, T; \mathbb{R}^n)), \\ Z(\cdot\,, \cdot) &\in \bigcap_{p\geqslant 1} L^\infty(0, T; L^p_{\mathbb{F}}(\Omega; L^2(\cdot\,, T; \mathbb{R}^n))), \end{aligned}$$

such that $\eta(t, \cdot) \in \bigcap_{p\geqslant 1} \mathcal{E}^{p,\lambda}_{\mathbb{F}}(\Omega; C([t, T]; \mathbb{R}^n))$ *for almost every* $t \in [0, T]$, *where*

$$\eta(t, r) \equiv Y(t) - \int_t^r g(t, s, y(s), Z(t, s))ds + \int_t^r Z(t, s)dW(s), \quad r \in [t, T]. \tag{8.5.34}$$

Proof Fix $y(\cdot) \in \bigcap_{p\geqslant 1} \mathcal{E}^{p,\lambda}_{\mathbb{F}}(\Omega; L^\infty(0, T; \mathbb{R}^n))$. For each $i = 1, \cdot s, n$, and almost every $t \in [0, T]$, it follows from (8.5.27) and (8.5.28) that

$$|g^i(t, s, y(s), z)| \leqslant \alpha(t, s) + \beta|y(s)| + \beta|z|^{1+\delta}, \quad \forall (s, z) \in [t, T] \times \mathbb{R}, \tag{8.5.35}$$

and

$$\begin{aligned} & g^i(t,s,y(s),\theta z^i+(1-\theta)\bar z^i) \\ & \leqslant \theta g^i(t,s,y(s),z^i)+(1-\theta)\big[\alpha(t,s)+\beta|y(s)|+\beta|\bar z^i|^{1+\delta}\big], \\ & \qquad\qquad \forall (t,s)\in\Delta^*[0,T],\ z,\bar z\in\mathbb{R}^n,\ \theta\in(0,1),\ 1\leqslant i\leqslant n. \end{aligned} \tag{8.5.36}$$

Furthermore, due to integrability of $\alpha(\cdot\,,\cdot)$, $\psi(\cdot)$ and $y(\cdot)$, by Hölder's inequality, we have

$$\mathbb{E}\Big[\Phi_\lambda\Big(|\psi^i(t)|+\int_t^T\big(\alpha(t,s)+\beta|y(s)|\big)ds;\,p\Big)\Big]<+\infty,\quad \forall p>1. \tag{8.5.37}$$

Then, by Theorem 8.5.1, we know that the following one-dimensional parameterized BSDE

$$\eta^i(t,r)=\psi^i(t)+\int_r^T g^i(t,s,y(s),\zeta^i(t,s))ds-\int_r^T\zeta^i(t,s)dW(s),\qquad r\in[t,T], \tag{8.5.38}$$

admits a unique adapted solution $(\eta^i(t,\cdot),\zeta^i(t,\cdot))$, taking value in $\mathbb{R}^2$ such that

$$\mathbb{E}\Big[\Phi_\lambda\Big(\sup_{s\in[t,T]}|\eta^i(t,s)|;\,p\Big)+\Big(\int_t^T|\zeta^i(t,s)|^2dt\Big)^{\frac{p}{2}}\Big]<+\infty,\quad \forall p>1.$$

Let

$$\eta(t,\cdot)\equiv(\eta^1(t,\cdot),\cdot s,\eta^n(t,\cdot))^\top \ \text{ and }\ \zeta(t,\cdot)\equiv(\zeta^1(t,\cdot),\cdot s,\zeta^n(t,\cdot)^\top. \tag{8.5.39}$$

Then, by Hölder's inequality, we know that for almost every $t\in[0,T]$, $(\eta(t,\cdot),\zeta(t,\cdot))$ is the unique adapted solution of the following BSDE:

$$\eta(t,r)=\psi(t)+\int_r^T g(t,s,y(s),\zeta(t,s))ds-\int_r^T\zeta(t,s)dW(s),\qquad r\in[t,T], \tag{8.5.40}$$

in the space $\big[\bigcap\limits_{p\geqslant 1}\mathcal{E}_{\mathbb{F}}^{p,\lambda}(\Omega;C([t,T];\mathbb{R}^n))\big]\times\big[\bigcap\limits_{p\geqslant 1}L_{\mathbb{F}}^{p}(\Omega;L^2(t,T;\mathbb{R}^n))\big]$. Let

$$Y(t)\equiv\eta(t,t),\quad Z(t,s)\equiv\zeta(t,s),\quad (t,s)\in\Delta^*[0,T]. \tag{8.5.41}$$

Then, it follows from (8.5.40) and (8.5.41) that $(Y(\cdot),Z(\cdot,\cdot))$ solves BSVIE (8.5.33) and that the $\eta(t,\cdot)$ defined in (8.5.34) equals to that in (8.5.40) and then it belongs to $\bigcap\limits_{p\geqslant 1}\mathcal{E}_{\mathbb{F}}^{p,\lambda}(\Omega;C([t,T];\mathbb{R}^n))$.

Furthermore, define

$$\Psi_\lambda(\nu) \equiv \operatorname*{esssup}_{t\in[0,T]} \Phi_\lambda\Big(|\psi(t)| + \int_t^T \alpha(t,s)ds;\, \nu\Big), \quad \nu \in [0,+\infty). \tag{8.5.42}$$

In view of (8.5.35)–(8.5.38), by using Lemma 8.5.2, we can deduce that for each $i = 1, \cdots, n$, almost every $t \in [0,T]$ and each $p > 1$,

$$\begin{aligned}
\Phi_\lambda\Big(|Y^i(t)|;1\Big) &= \Phi_\lambda\Big(|\eta^i(t,t)|;1\Big)\\
&\leqslant K\mathbb{E}_t\Big[\Phi_\lambda\Big(|\psi^i(t)| + \int_t^T (\alpha(t,s)+\beta|y(s)|)ds;\, K\Big)\Big]\\
&\leqslant K\mathbb{E}_t\Big[\Psi_\lambda(K)\Phi_\lambda\Big(\int_t^T |y(s)|ds;\, K\Big)\Big]
\end{aligned}$$

and

$$\begin{aligned}
&\mathbb{E}\Big[\Phi_\lambda\Big(\sup_{s\in[t,T]} |\eta^i(t,s)|;\, p\Big)\Big] + \mathbb{E}\Big[\Big(\int_t^T |Z^i(t,s)|^2 ds\Big)^{\frac{p}{2}}\Big]\\
&\leqslant K\mathbb{E}\Big[\Psi_\lambda(K_p)\Phi_\lambda\Big(\int_t^T |y(s)|ds;\, K_p\Big)\Big],
\end{aligned}$$

and then, by virtue of Jensen's inequality, Doob's martingale inequality, and Hölder's inequality together with (8.5.39), (8.5.32), and (8.5.41), we have

$$\begin{aligned}
&\Phi_\lambda\Big(|Y(t)|;1\Big) \leqslant \Phi_\lambda\Big(\sum_{i=1}^n |Y^i(t)|;1\Big) = \prod_{i=1}^n \Phi_\lambda\Big(|Y^i(t)|;1\Big)\\
&\leqslant \Big\{K\mathbb{E}_t\Big[\Psi_\lambda(K)\Phi_\lambda\Big(\int_t^T |y(s)|ds;\, K\Big)\Big]\Big\}^n\\
&\leqslant K\mathbb{E}_t\Big[\Psi_\lambda(K)\Phi_\lambda\Big(\int_t^T |y(s)|ds;\, K\Big)\Big],
\end{aligned} \tag{8.5.43}$$

and

$$\begin{aligned}
&\mathbb{E}\Big[\Phi_\lambda\Big(\operatorname*{esssup}_{t\in[0,T]} |Y(t)|;\, p\Big)\Big] = \mathbb{E}\Big[\operatorname*{esssup}_{t\in[0,T]} \Phi_\lambda\Big(|Y(t)|;\, p\Big)\Big]\\
&\leqslant K_p\mathbb{E}\Big\{\operatorname*{esssup}_{t\in[0,T]} \Big(\mathbb{E}_t\Big[\Psi_\lambda(K)\Phi_\lambda\Big(\int_0^T |y(s)|ds;\, K\Big)\Big]\Big)^p\Big\}\\
&\leqslant K_p\mathbb{E}\Big[\Psi_\lambda(K_p)\Phi_\lambda\Big(\int_0^T |y(s)|ds;\, K_p\Big)\Big].
\end{aligned} \tag{8.5.44}$$

Further

$$
\begin{aligned}
&\mathbb{E}\Big[\Phi_\lambda\Big(\sup_{s\in[t,T]}|\eta(t,s)|;\,p\Big)\Big]+\mathbb{E}\Big[\Big(\int_t^T|Z(t,s)|^2ds\Big)^{\frac{p}{2}}\Big]\\
&\leqslant \mathbb{E}\Big[\prod_{i=1}^n\Phi_\lambda\Big(\sup_{s\in[t,T]}|\eta^i(t,s)|;\,p\Big)\Big]+n^{p-1}\sum_{i=1}^n\mathbb{E}\Big[\Big(\int_t^T|Z^i(t,s)|^2ds\Big)^{\frac{p}{2}}\Big]\\
&\leqslant \prod_{i=1}^n\Big\{\mathbb{E}\Big[\Phi_\lambda\Big(\sup_{s\in[t,T]}|\eta^i(t,s)|;\,np\Big)\Big]\Big\}^{\frac{1}{n}}\\
&\qquad\qquad\qquad +K_p\mathbb{E}\Big[\Psi_\lambda(K_p)\Phi_\lambda\Big(\int_t^T|y(s)|ds;\,K_p\Big)\Big]\\
&\leqslant K_p\mathbb{E}\Big[\Psi_\lambda(K_p)\Phi_\lambda\Big(\int_t^T|y(s)|ds;\,K_p\Big)\Big].
\end{aligned}
\tag{8.5.45}
$$

Therefore, in view of the assumptions on $\psi(\cdot)$, $\alpha(\cdot\,,\cdot)$, and $y(\cdot)$, it follows from (8.5.44), (8.5.45), and Hölder's inequality that

$$
\begin{aligned}
&(Y(\cdot), Z(\cdot\,,\cdot))\\
&\in\Big[\bigcap_{p\geqslant 1}\mathcal{E}_{\mathbb{F}}^{p,\lambda}(\Omega;L^\infty(0,T;\mathbb{R}^n))\Big]\times\Big[\bigcap_{p\geqslant 1}L^\infty(0,T;L_{\mathbb{F}}^p(\Omega;L^2(\cdot\,,T;\mathbb{R}^n)))\Big].
\end{aligned}
$$

Finally, we look at the uniqueness. Suppose

$$
\begin{aligned}
&(\widetilde{Y}(\cdot), \widetilde{Z}(\cdot\,,\cdot))\\
&\in\Big[\bigcap_{p\geqslant 1}\mathcal{E}_{\mathbb{F}}^{p,\lambda}(\Omega;L^\infty(0,T;\mathbb{R}^n))\Big]\times\Big[\bigcap_{p\geqslant 1}L^\infty(0,T;L_{\mathbb{F}}^p(\Omega;L^2(\cdot\,,T;\mathbb{R}^n)))\Big]
\end{aligned}
$$

is another pair of adapted solution of BSVIE (8.5.33) such that $\widetilde{\eta}(t,\cdot)\in\bigcap_{p\geqslant 1}\mathcal{E}_{\mathbb{F}}^{p,\lambda}(\Omega;C([t,T];\mathbb{R}^n))$ for almost every $t\in[0,T]$, where

$$
\widetilde{\eta}(t,r)\equiv\widetilde{Y}(t)-\int_t^r g(t,s,U(s),\widetilde{Z}(t,s))ds+2n\int_t^r\widetilde{Z}(t,s)dW(s),\qquad r\in[t,T].
$$

Then, for every almost $t\in[0,T]$,

$$
(\widetilde{\eta}(t,\cdot),\widetilde{Z}(t,\cdot))\in\Big[\bigcap_{p\geqslant 1}\mathcal{E}_{\mathbb{F}}^{p,\lambda}(\Omega;C([t,T];\mathbb{R}^n))\Big]\times\Big[\bigcap_{p\geqslant 1}L_{\mathbb{F}}^p(\Omega;L^2(t,T;\mathbb{R}^n))\Big]
$$

satisfies (8.5.40). By the uniqueness of BSDEs, we see that $\widetilde{\eta}(t,\cdot)=\eta(t,\cdot)$, $\widetilde{Z}(t,\cdot)=Z(t,\cdot)$ for almost every $t\in[0,T]$, which implies that $Y(\cdot)=\widetilde{Y}(\cdot)$. This completes the proof. □

The following is the main result of this part.

Theorem 8.5.6 *Let* (H8.5.2) *hold and* $\lambda \equiv \frac{2\delta}{1+\delta}$. *Then, for any*

$$\psi(\cdot) \in \bigcap_{p\geqslant 1} \mathcal{E}^{p,\lambda}(\Omega; L^\infty(0,T;\mathbb{R}^n)),$$

the Type-I BSVIE (8.4.1) *admits a unique pair of adapted solution:*

$$\begin{aligned}&(Y(\cdot), Z(\cdot\,,\cdot))\\&\in\Big[\bigcap_{p\geqslant 1}\mathcal{E}_{\mathbb{F}}^{p,\lambda}(\Omega; L^\infty(0,T;\mathbb{R}^n))\Big]\times\Big[\bigcap_{p\geqslant 1}L^\infty(0,T;L_{\mathbb{F}}^p(\Omega;L^2(\cdot\,,T;\mathbb{R}^n)))\Big],\end{aligned} \tag{8.5.46}$$

such that $\eta(t,\cdot) \in \bigcap_{p\geqslant 1} \mathcal{E}_{\mathbb{F}}^{p,\lambda}(\Omega; C([t,T];\mathbb{R}^n))$ *for every almost* $t \in [0,T]$, *where*

$$\begin{aligned}\eta(t,r) \equiv Y(t) - \int_t^r g(t,s,Y(s),Z(t,s))ds\\ + \int_t^r Z(t,s)dW(s), \quad r\in[t,T].\end{aligned} \tag{8.5.47}$$

Proof Set $(Y_0(\cdot), Z_0(\cdot\,,\cdot)) = (0,0)$ and define recursively the sequence $\{(Y_k(\cdot), Z_k(\cdot\,,\cdot))\}_{k=1}^\infty$ by the adapted solutions of the following BSVIEs given through the way in Lemma 8.5.5, with $t\in[0,T]$:

$$\begin{aligned}Y_{k+1}(t) = \psi(t) + \int_t^T g(t,s,Y_k(s),Z_{k+1}(t,s))ds\\ - \int_t^T Z_{k+1}(t,s)dW(s).\end{aligned} \tag{8.5.48}$$

And, for almost every $t \in [0,T]$ and $\ell \geqslant 0$, by virtue of Theorem 8.5.1, we let $(\eta_{k+1}(t,\cdot), \zeta_{k+1}(t,\cdot))$ be the unique adapted solution in the space $\Big[\bigcap_{p\geqslant 1}\mathcal{E}_{\mathbb{F}}^{p,\lambda}(\Omega;C([t,T];\mathbb{R}^n))\Big] \times \Big[\bigcap_{p\geqslant 1} L_{\mathbb{F}}^p(\Omega;L^2(t,T;\mathbb{R}^n))\Big]$ of the following BSDE:

$$\begin{aligned}\eta_{k+1}(t,r) = \psi(t) + \int_r^T g(t,s,Y_k(s),\zeta_{k+1}(t,s))ds\\ - \int_r^T \zeta_{k+1}(t,s)dW(s), \quad r\in[t,T].\end{aligned} \tag{8.5.49}$$

It then follows from Lemma 8.5.5 and (8.5.48) that

$$\begin{aligned}Y_{k+1}(t) &= \eta_{k+1}(t,t),\\ Z_{k+1}(t,s) &= \zeta_{k+1}(t,s),\end{aligned} \qquad \text{a.e. } (t,s)\in\Delta^*[0,T]. \tag{8.5.50}$$

In addition, for almost every $t \in [0, T]$ and each $r \in [t, T]$, we have

$$\eta_{k+1}(t, r) = Y_{k+1}(t) - \int_t^r g(t, s, Y_k(s), \zeta_{k+1}(t, s))ds + \int_t^r \zeta_{k+1}(t, s)dW(s).$$

Then, in view of (8.5.49) and (8.5.50), by using Lemma 8.5.2 together with a similar argument to (8.5.43) and (8.5.45), we can deduce that for almost every $t \in [0, T]$ and for each $k \geqslant 0$ and $p > 1$,

$$\Phi_\lambda\Big(|Y_{k+1}(t)|, 1\Big) \leqslant K\mathbb{E}_t\Big[\Psi_\lambda(K)\Phi_\lambda\Big(\int_t^T |Y_k(s)|ds;\, K\Big)\Big], \tag{8.5.51}$$

and

$$\begin{aligned}&\mathbb{E}\Big[\Phi_\lambda\Big(\sup_{s\in[t,T]}|\eta_{k+1}(t, s)|;\, p\Big)\Big] + \mathbb{E}\Big[\Big(\int_t^T |Z_{k+1}(t, s)|^2 ds\Big)^{\frac{p}{2}}\Big]\\ &\leqslant K_p\mathbb{E}\Big[\Psi_\lambda(K_p)\Phi_\lambda\Big(\int_0^T |Y_k(s)|ds;\, K_p\Big)\Big],\end{aligned} \tag{8.5.52}$$

where and hereafter, $\Psi_\lambda(\cdot)$ and $\Phi_\lambda(\cdot\,;\cdot)$ is defined respectively in (8.5.42) and (8.5.31).

In the sequel, we will show that $\{(Y_k(\cdot), Z_k(\cdot, \cdot))\}_{k=1}^\infty$ is a Cauchy sequence and converges to a process $(Y(\cdot), Z(\cdot\,, \cdot))$ with

$$\begin{aligned}Y(\cdot) &\in \bigcap_{p\geqslant 1} \mathcal{E}_{\mathbb{F}}^{p,\lambda}(\Omega; L^\infty(0, T; \mathbb{R}^n)),\\ Z(\cdot\,, \cdot) &\in \bigcap_{p\geqslant 1} L^\infty(0, T; L_{\mathbb{F}}^p(\Omega; L^2(\cdot\,, T; \mathbb{R}^n))),\end{aligned}$$

which is a desired adapted solution of BSVIE (8.4.1). First, we prove that for each $p > 1$,

$$\sup_{k\geqslant 1}\mathbb{E}\Big[\Phi_\lambda\Big(\operatorname*{esssup}_{t\in[0,T]}|Y_k(t)|;\, p\Big)\Big] < +\infty, \tag{8.5.53}$$

and

$$\begin{aligned}\sup_{k\geqslant 1}\operatorname*{esssup}_{t\in[0,T]}\Big\{&\mathbb{E}\Big[\Phi_\lambda\Big(\sup_{s\in[t,T]}|\eta_k(t, s)|;\, p\Big)\Big]\\ &+\mathbb{E}\Big[\Big(\int_t^T |Z_k(t, s)|^2 ds\Big)^{\frac{p}{2}}\Big]\Big\} < +\infty.\end{aligned} \tag{8.5.54}$$

In fact, in view of (8.5.51), by Jensen's inequality, we deduce that for almost every $t \in [0, T]$ and for each $k \geqslant 0$ and $\kappa > 0$,

$$\Phi_\lambda\Big(e^{\kappa t}|Y_{k+1}(t)|;1\Big) \leqslant \bar{K}\mathbb{E}_t\Big[\Psi_\lambda(\bar{K})\Phi_\lambda\Big(e^{\kappa t}\int_t^T |Y_k(s)|ds;K\Big)\Big] \tag{8.5.55}$$

with $\bar{K}$ being a constant depending on κ. Observe that

$$e^{\kappa t}\int_t^T |Y_k(s)|ds \leqslant e^{\kappa t}\int_t^T e^{-\kappa s}ds \sup_{s\in[t,T]}\big(e^{\kappa s}|Y_k(s)|\big) \leqslant \frac{1}{\kappa}\sup_{s\in[0,T]}\big(e^{\kappa s}|Y_k(s)|\big).$$

We have

$$\Phi_\lambda\Big(e^{\kappa t}|Y_{k+1}(t)|;1\Big) \leqslant \bar{K}\mathbb{E}_t\Big[\Psi_\lambda(\bar{K})\Phi_\lambda\Big(\sup_{s\in[0,T]}\big(e^{\kappa s}|Y_k(s)|\big);\frac{K}{\kappa^\lambda}\Big)\Big],$$

and by Doob's martingale inequality and Hölder's inequality,

$$\begin{aligned}
&\mathbb{E}\Big[\operatorname*{esssup}_{t\in[0,T]}\Phi_\lambda\Big(e^{\kappa t}|Y_{k+1}(t)|;p\Big)\Big]\\
&\leqslant \bar{K}\Big(\frac{p}{p-1}\Big)^p\mathbb{E}\Big[\Psi_\lambda(p\bar{K})\Phi_\lambda\Big(\sup_{s\in[0,T]}\big(e^{\kappa s}|Y_k(s)|\big);\frac{pK}{\kappa^\lambda}\Big)\Big]\\
&\leqslant \bar{K}_p\sqrt{\mathbb{E}[\Psi_\lambda(\bar{K}_p)]}\Big\{\mathbb{E}\Big[\Phi_\lambda\Big(\operatorname*{esssup}_{s\in[0,T]}\big(e^{\kappa s}|Y_k(s)|\big);\frac{2pK}{\kappa^\lambda}\Big)\Big]\Big\}^{\frac{1}{2}}, \quad \forall p>1,
\end{aligned}$$

where $\bar{K}_p$ being a variable constant further depending on κ and p. Taking $\kappa_0 \equiv (2K)^{\frac{1}{\lambda}}$ in the last inequality yields that for each $p > 1$,

$$\begin{aligned}
&\mathbb{E}\Big[\Phi_\lambda\Big(\operatorname*{esssup}_{t\in[0,T]}\big(e^{\kappa_0 t}|Y_{k+1}(t)|\big);p\Big)\Big]\\
&\leqslant \widehat{K}_p\sqrt{\mathbb{E}[\Psi_\lambda(\widehat{K}_p)]}\Big\{\mathbb{E}\Big[\Phi_\lambda\Big(\operatorname*{esssup}_{t\in[0,T]}\big(e^{\kappa_0 t}|Y_k(t)|\big);p\Big)\Big]\Big\}^{\frac{1}{2}},
\end{aligned}$$

with $\widehat{K}_p$ being a constant depending on k_0 and p. Then, by induction,

$$\begin{aligned}
&\mathbb{E}\Big[\Phi_\lambda\Big(\operatorname*{esssup}_{t\in[0,T]}\big(e^{\kappa_0 t}|Y_{k+1}(t)|\big);p\Big)\Big]\leqslant \widehat{K}_p\Big\{\sqrt{\mathbb{E}[\Psi_\lambda(\widehat{K}_p)]}\Big\}^{1+\frac{1}{2}+\cdot s+\frac{1}{2^k}}\\
&\cdot\Big\{\mathbb{E}\Big[\Phi_\lambda\Big(\operatorname*{esssup}_{t\in[0,T]}\big(e^{\kappa_0 t}|Y_0(t)|\big);p\gamma\Big)\Big]\Big\}^{\frac{1}{2^{k+1}}} \leqslant \widehat{K}_p\mathbb{E}\Big[\Psi_\lambda(\widehat{K}_p)\Big].
\end{aligned} \tag{8.5.56}$$

Therefore, (8.5.53) follows from (8.5.56). Furthermore, in view of (8.5.52), (8.5.53), and the assumptions on $\psi(\cdot)$ and $\alpha(\cdot\,,\cdot)$, it follows from Hölder's inequality that (8.5.54) is also true.

Next, for each fixed $k, \ell \geqslant 1$, $\theta \in (0,1)$ and $(t,s) \in \Delta^*[0,T]$, define

$$\begin{aligned}\delta_\theta \eta_{k,\ell}(t,s) &\equiv \frac{\eta_{k+\ell}(t,s) - \theta\eta_k(t,s)}{1-\theta},\\ \delta_\theta Y_{k,\ell}(t) &\equiv \frac{Y_{k+\ell}(t) - \theta Y_k(t)}{1-\theta},\\ \delta_\theta Z_{k,\ell}(t,s) &\equiv \frac{Z_{k+\ell}(t,s) - \theta Z_k(t,s)}{1-\theta}.\end{aligned}$$

Then, for almost every $t \in [0,T]$, $\delta_\theta Y_{k,\ell}(t) = \delta_\theta \eta_{k,\ell}(t,t)$, and

$$\begin{aligned}&(\delta_\theta \eta_{k,\ell}(t,\cdot), \delta_\theta Z_{k,\ell}(t,\cdot))\\ &\in \Big[\bigcap_{p\geqslant 1} \mathcal{E}_{\mathbb{F}}^{p,\lambda}(\Omega; C([t,T];\mathbb{R}^n))\Big] \times \Big[\bigcap_{p\geqslant 1} L_{\mathbb{F}}^p(\Omega; L^2(t,T;\mathbb{R}^n))\Big]\end{aligned}$$

solves the following system of BSDEs: for $i = 1, \cdots, n$ and $r \in [t,T]$,

$$\begin{aligned}\delta_\theta \eta^i_{k,\ell}(t,r) = \psi^i(t) + \int_r^T &\delta_\theta g^i_{k,\ell}\big(t,s,\delta_\theta Z^i_{k,\ell}(t,s)\big)ds\\ &- \int_r^T \delta_\theta Z^i_{k,\ell}(t,s)dW(s),\end{aligned}$$

where for each $(t,s,z) \in \Delta^*[0,T] \times \mathbb{R}$,

$$\begin{aligned}\delta_\theta g^i_{k,\ell}(t,s,z) \equiv &\frac{1}{1-\theta}\Big(g^i\big(t,s,Y_{k+\ell-1)}(s),(1-\theta)z+\theta Z^i_k(t,s)\big)\\ &-\theta g^i\big(t,s,Y_{k-1}(s),Z^i_k(t,s)\big)\Big).\end{aligned} \tag{8.5.57}$$

It follows from (8.5.57), (8.5.27), and (8.5.28) that for each $(t,s,z) \in \Delta^*[0,T] \times \mathbb{R}$,

$$\begin{aligned}\delta_\theta g^i_{k,\ell}(t,s,z) &\leqslant \frac{1}{1-\theta}\Big(\beta|Y_{k+\ell-1}(s) - Y_{k-1}(s)|\\ &\qquad +(1-\theta)\big(\alpha(t,s) + \beta|Y_{k-1}(s)| + \beta|z|^{1+\delta}\big)\Big)\\ &\leqslant \beta|\delta_\theta Y_{k-1,\ell}(s)| + \beta|Y_{k-1}(s)| + \alpha(t,s) + \beta|Y_{k-1}(s)| + \beta|z|^{1+\delta}\\ &= \alpha(t,s) + 2\beta|Y_{k-1}(s)| + \beta|\delta_\theta Y_{k-1,\ell}(s)| + \beta|z|^{1+\delta}.\end{aligned}$$

Then, we deduce that for almost every $t \in [0,T]$ and each $i = 1, \cdot s, n$,

$$\begin{aligned}\Phi_\lambda\Big(\Big[\delta_\theta \eta^i_{k,\ell}(t,r)\Big]^+;1\Big) \leqslant K\mathbb{E}_r\Big[\Phi_\lambda\Big((\psi^i(t))^+ + \int_r^T \big(\alpha(t,s)\\ +|Y_{k-1}(s)| + |\delta_\theta Y_{k-1,\ell}(s)|\big)ds; K\Big)\Big], \quad r \in [t,T],\end{aligned} \tag{8.5.58}$$

and then

$$\begin{aligned}\Phi_\lambda\Big(\Big[\delta_\theta Y^i_{k,\ell}(t)\Big]^+;1\Big)\leqslant K\mathbb{E}_t\Big[\Phi_\lambda\Big((\psi^i(t))^+ + \int_t^T\big(\alpha(t,s)\\ +|Y_{k-1}(s)|+|\delta_\theta Y_{k-1,\ell}(s)|\big)ds;K\Big)\Big],\quad t\in[0,T].\end{aligned}\tag{8.5.59}$$

On the other hand, define

$$\begin{aligned}\delta_\theta\widetilde{\eta}_{k,\ell}(t,s)&\equiv\frac{\eta_k(t,s)-\theta\eta_{(k+\ell}(t,s)}{1-\theta},\\ \delta_\theta\widetilde{Y}_{k,\ell}(t)&\equiv\frac{Y_k(t)-\theta Y_{k+\ell}(t)}{1-\theta},\\ \delta_\theta\widetilde{Z}_{k,\ell}(t,s)&\equiv\frac{Z_k(t,s)-\theta Z_{k+\ell}(t,s)}{1-\theta}.\end{aligned}$$

The same computation as above yields that for almost every $t\in[0,T]$ and each $i=1,\cdots,n$,

$$\begin{aligned}&\Phi_\lambda\Big(\Big[\delta_\theta\widetilde{\eta}^i_{k,\ell}(t,r)\Big]^+;1\Big)\\ &\leqslant K\mathbb{E}_r\Big[\Phi_\lambda\Big((\psi^i(t))^+ + \int_r^T\big(\alpha(t,s)+|Y_{k+\ell-1}(s)|\\ &\qquad\qquad +|\delta_\theta\widetilde{Y}_{k-1,\ell}(s)|\big)ds;K\Big)\Big],\quad r\in[t,T],\end{aligned}\tag{8.5.60}$$

and then

$$\begin{aligned}&\Phi_\lambda\Big((\delta_\theta\widetilde{Y}^i_{k,\ell}(t))^+;1\Big)\\ &\leqslant K\mathbb{E}_t\Big[\Phi_\lambda\Big((\psi^i(t))^+ + \int_t^T\big(\alpha(t,s)+|Y_{k+\ell-1}(s)|\\ &\qquad\qquad +|\delta_\theta\widetilde{Y}_{k-1,\ell}(s)|\big)ds;K\Big)\Big].\end{aligned}\tag{8.5.61}$$

Furthermore, observe that for almost every $t\in[0,T]$ and for each $i=1,\cdots,n$ and $r\in[t,T]$, we have

$$\begin{aligned}&\Big(\delta_\theta Y^i_{k,\ell}(t)\Big)^-=\frac{(Y^i_{k+\ell}(t)-\theta Y^i_k(t))^-}{1-\theta}=\frac{(\theta Y^i_k(t)-Y^i_{k+\ell}(t))^+}{1-\theta}\\ &\leqslant\frac{\theta\big(Y^i_k(t)-\theta Y^i_{k+\ell}(t)\big)^+ +(1-\theta^2)|Y^i_{k+\ell}(t)|}{1-\theta}\\ &\leqslant\Big(\delta_\theta\widetilde{Y}^i_{k,\ell}(t)\Big)^+ +2|Y_{k+\ell}(t)|,\end{aligned}\tag{8.5.62}$$

and similarly

$$\begin{cases} \Big(\delta_\theta \widetilde{Y}^i_{k,\ell}(t)\Big)^- \leqslant \Big(\delta_\theta Y^i_{k.\ell}(t)\Big)^+ + 2|Y_k(t)|, \\ \Big(\delta_\theta \eta^i_{k,\ell}(t,r)\Big)^- \leqslant \Big(\delta_\theta \widetilde{\eta}^i_{k,\ell}(t,r)\Big)^+ + 2|\eta_{k+\ell}(t,r)|, \\ \Big(\delta_\theta \widetilde{\eta}^i_{k,\ell}(t,r)\Big)^- \leqslant \Big(\delta_\theta \eta^i_{k,\ell}(t,r)\Big)^+ + 2|\eta_k(t,r)|. \end{cases} \tag{8.5.63}$$

It then follows from (8.5.59), (8.5.61), (8.5.62), (8.5.63), (8.5.32), and Jensen's inequality that

$$\begin{aligned} &\Phi_\lambda\Big(|\delta_\theta Y^i_{k,\ell}(t)|;1\Big) \leqslant \Phi_\lambda\Big(\Big(\delta_\theta Y^i_{k,\ell}(t)\Big)^+;1\Big)\Phi_\lambda\Big(\Big(\delta_\theta Y^i_{k,\ell}(t)\Big)^-;1\Big) \\ &\leqslant K\mathbb{E}_t\Big[\Phi_\lambda\Big(|\psi(t)|+|Y_{k=\ell}(t)|+\int_t^T\big(\alpha(t,s)+|Y_{k-1}(s)|+|Y_{k+\ell-1}(s)| \\ &\quad +\delta_\theta Y_{k-1,\ell}(s)|+|\delta_\theta \widetilde{Y}_{k-1,\ell)}(s)|\big)ds;K\Big)\Big] \end{aligned}$$

and

$$\begin{aligned} &\Phi_\lambda\Big(|\delta_\theta \widetilde{Y}^i_{k.\ell}(t)|;1\Big) \leqslant \Phi_\lambda\Big(\big(\delta_\theta \widetilde{Y}^i_{k,\ell}(t)\big)^+;1\Big)\Phi_\lambda\Big(\big(\delta_\theta \widetilde{Y}^i_{k,\ell}(t)\big)^-;1\Big) \\ &\leqslant K\mathbb{E}_t\Big[\Phi_\lambda\Big(|\psi(t)|+|Y_k(t)|+\int_t^T\big(\alpha(t,s)+|Y_{k-1}(s)|+|Y_{k+\ell-1}(s)| \\ &\qquad +|\delta_\theta Y_{k-1,\ell)}(s)|+|\delta_\theta \widetilde{Y}_{k-1,\ell}(s)|\big)ds;K\Big)\Big]. \end{aligned}$$

Consequently, by (8.5.32) and Jensen's inequality again, we get that for almost every $t \in [0,T]$,

$$\begin{aligned} &\Phi_\lambda\Big(|\delta_\theta Y^i_{k,\ell}(t)|+|\delta_\theta \widetilde{Y}^i_{k,\ell}(t)|;1\Big) \\ &\leqslant K\mathbb{E}_t\Big[\Phi_\lambda\Big(|\psi(t)|+|Y_k(t)|+|Y_{k+\ell}(t)| \\ &\qquad +\int_t^T\big(\alpha(t,s)+|Y_{k-1}(s)|+|Y_{k+\ell-1}(s)|\big)ds \\ &\qquad +\int_t^T\big(|\delta_\theta Y_{k-1,\ell}(s)|+|\delta_\theta \widetilde{Y}_{k-1,\ell}(s)|\big)ds;K\Big)\Big], \quad i=1,\cdot s,n, \end{aligned}$$

and then

$$\begin{aligned} &\Phi_\lambda\Big(|\delta_\theta Y_{k,\ell}(t)|+|\delta_\theta \widetilde{Y}_{k,\ell}(t)|;1\Big) \\ &\leqslant K\mathbb{E}_t\Big[\Phi_\lambda\Big(\operatorname*{esssup}_{s\in[0,T]}\big(|Y_k(s)|+|Y_{k+\ell}(s)|\big) \\ &\qquad +\int_0^T\big(|Y_{k-1}(s)|+|Y_{k+\ell-1}(s)|\big)ds \\ &\qquad +\int_t^T\big(|\delta_\theta Y_{k-1,\ell}(s)|+|\delta_\theta \widetilde{Y}_{k-1,\ell}(s)|\big)ds;K\Big)\Psi_\lambda(K)\Big]. \end{aligned} \tag{8.5.64}$$

On the other hand, in view of (8.5.58), (8.5.60), and (8.5.63), the same computation as above yields that for almost every $t \in [0, T]$ and each $r \in [t, T]$,

$$
\begin{aligned}
&\Phi_\lambda\Big(|\delta_\theta \eta_{k,\ell}(t,r)| + |\delta_\theta \widetilde{\eta}_{k,\ell}(t,r)|;\, 1\Big) \\
&\leqslant K\mathbb{E}_r\Big[\Phi_\lambda\Big(\operatorname*{esssup}_{s\in[t,T]}\big(|\eta_k(t,s)| + |\eta_{k+\ell}(t,s)|\big) \\
&\qquad + \int_0^T \big(|Y_{k-1}(s)| + |Y_{k+\ell-1}(s)|\big)ds \\
&\qquad + \int_0^T \big(|\delta_\theta Y_{k-1,\ell}(s)| + |\delta_\theta \widetilde{Y}_{k-1,\ell}(s)|\big)ds;\, K\Big)\Psi_\lambda(K)\Big].
\end{aligned}
\tag{8.5.65}
$$

Observing the similarity of (8.5.64) and (8.5.51), we can induce with respect to k and use a similar argument as that obtaining (8.5.56) to derive that there exists a positive constant $\kappa_0 > 0$ depending only on (β, δ, n, T) such that for each $q > 1$ and $\theta \in (0, 1)$,

$$
\begin{aligned}
&\mathbb{E}\Big[\Phi_\lambda\Big(\operatorname*{esssup}_{t\in[0,T]}\big[e^{\kappa_0 t}\big(|\delta_\theta Y_{k,\ell}(t)| + |\delta_\theta \widetilde{Y}_{k,\ell}(t)|\big)\big];\, q\Big)\Big] \\
&\leqslant \bar{K}(q)\Big\{\mathbb{E}\Big[\Phi_\lambda\Big(\operatorname*{esssup}_{t\in[0,T]}\big[e^{\kappa_0 t}\big(|\delta_\theta Y_{1,\ell}(t)| + |\delta_\theta \widetilde{Y}_{1,\ell}(t)|\big)\big];\, q\Big)\Big]\Big\}^{\frac{1}{2^k}},
\end{aligned}
\tag{8.5.66}
$$

where, in view of (8.5.53), (8.5.32), and assumptions on $\psi(\cdot)$ and $\alpha(\cdot, , \cdot)$ together with Hölder's inequality,

$$
\begin{aligned}
M(q) \equiv \widehat{K}_q \sup_{k,\ell\geqslant 1} \mathbb{E}\Big[\Psi_\lambda(\widehat{K}_q)\Phi_\lambda\Big(&\operatorname*{esssup}_{s\in[0,T]}\big(|Y_k(s) + |Y_{k+\ell}(s)|\big) \\
&+\int_0^T \big(|Y_{k-1}(s)| + |Y_{k+\ell-1}(s)|\big)ds;\, \widehat{K}_q\Big)\Big] < +\infty,
\end{aligned}
$$

with $\widehat{K}_q$ being a variable constant further depending on k_0 and q. In view of (8.5.53), it follows from (8.5.66) that for each $q > 1$ and $\theta \in (0, 1)$,

$$
\varlimsup_{k\to\infty} \sup_{\ell\geqslant 1} \mathbb{E}\Big[\Phi_\lambda\Big(\operatorname*{esssup}_{t\in[0,T]}\big(|\delta_\theta Y_{k,\ell}(t)| + |\delta_\theta \widetilde{Y}_{k,\ell}(t)|\big);\, q\Big)\Big] \leqslant M(q).
\tag{8.5.67}
$$

Thus, for each $\theta \in (0, 1)$,

$$
\varlimsup_{k\to\infty} \sup_{\ell\geqslant 1} \mathbb{E}\Big[\operatorname*{esssup}_{t\in[0,T]} |Y_{k+\ell}(t) - \theta Y_k(t)|^\lambda\Big] \leqslant \frac{(1-\theta)^\lambda}{2} M(2),
$$

and then, in view of (8.5.53) again,

$$\begin{aligned}&\overline{\lim_{k\to\infty}}\sup_{\ell\geqslant 1}\mathbb{E}\Big[\operatorname*{esssup}_{t\in[0,T]}|Y_{k+\ell}(t)-Y_k(t)|^{\lambda}\Big]\\&\leqslant(1-\theta)^{\lambda}\Big\{\frac{M(2)}{2}+\sup_{k\geqslant 1}\mathbb{E}\Big[\operatorname*{esssup}_{t\in[0,T]}|Y_k(t)|\Big]\Big\}<\infty.\end{aligned}$$

Sending θ to 1 in the last inequality, we get that

$$\overline{\lim_{k\to\infty}}\sup_{\ell\geqslant 1}\mathbb{E}\Big[\operatorname*{esssup}_{t\in[0,T]}|Y_{k+\ell}(t)-Y_k(t)|^{\lambda}\Big]=0.$$

Using again (8.5.53) and Hölder's inequality, we have for each $q>1$,

$$\begin{aligned}&\lim_{k\to\infty}\sup_{\ell\geqslant 1}\mathbb{E}\Big[\operatorname*{esssup}_{t\in[0,T]}|Y_{k+\ell}(t)-Y_k(t)|^{q}\Big]\\&\leqslant\lim_{k\to\infty}\sup_{\ell\geqslant 1}\Big\{\mathbb{E}\Big[\operatorname*{esssup}_{t\in[0,T]}|Y_{k+\ell}(t)-Y_k(t)|^{2q-\lambda}\Big]\Big\}^{\frac{1}{2}}\\&\qquad\cdot\Big\{\mathbb{E}\Big[\operatorname*{esssup}_{t\in[0,T]}|Y_{k+\ell}(t)-Y_k(t)|^{\lambda}\Big]\Big\}^{\frac{1}{2}}\\&\leqslant\sup_{k\geqslant 1}\Big\{\mathbb{E}\Big[\operatorname*{esssup}_{t\in[0,T]}\Phi_{\lambda}\Big(|Y_k(t)|;K_q\Big)\Big]\Big\}^{\frac{1}{2}}\\&\qquad\cdot\lim_{k\to\infty}\sup_{\ell\geqslant 1}\Big\{\mathbb{E}\Big[\operatorname*{esssup}_{t\in[0,T]}|Y_{k+\ell}(t)-Y_k(t)|^{\lambda}\Big]\Big\}^{\frac{1}{2}}=0,\end{aligned}$$

and then

$$\begin{aligned}&\lim_{k\to\infty}\sup_{\ell\geqslant 1}\mathbb{E}\Big[\operatorname*{esssup}_{t\in[0,T]}\Phi_{\lambda}\Big(|Y_{k+\ell}(t)-Y_k(t)|;q\Big)-1\Big]\\&\leqslant\lim_{k\to\infty}\sup_{\ell\geqslant 1}\mathbb{E}\Big[\operatorname*{esssup}_{t\in[0,T]}\{\Phi_{\lambda}\Big(|Y_{k+\ell}(t)-Y_k(t)|;q\Big)q|Y_{k+\ell}(t)-Y_k(t)|^{\lambda}\}\Big]\\&\leqslant\sup_{k\geqslant 1}\mathbb{E}\Big\{\Big[\operatorname*{esssup}_{t\in[0,T]}\Phi_{\lambda}\Big(|Y_k(t)|;2q\Big)\Big]\Big\}^{\frac{1}{2}}\\&\qquad\cdot\lim_{k\to\infty}\sup_{\ell\geqslant 1}\Big\{\mathbb{E}\Big[q^2\operatorname*{esssup}_{t\in[0,T]}|Y_{k+\ell}(t)-Y_k(t)|^{2\lambda}\Big]\Big\}^{\frac{1}{2}}=0.\end{aligned}$$

Consequently, there is an adapted process $Y(\cdot)\in\bigcap_{p\geqslant 1}\mathcal{E}_{\mathbb{F}}^{p,\lambda}(\Omega;L^{\infty}(0,T;\mathbb{R}^n))$ such that for each $q>1$,

$$\lim_{k\to\infty}\mathbb{E}\Big[\operatorname*{esssup}_{t\in[0,T]}|Y_k(t)-Y(t)|^{q}\Big]=0, \tag{8.5.68}$$

and

$$\lim_{k\to\infty} \mathbb{E}\Big[\operatorname*{esssup}_{t\in[0,T]} \Phi_\lambda\Big(|Y_k(t)-Y(t)|;q\Big)\Big]=1. \tag{8.5.69}$$

Finally, coming back to (8.5.65), in view of (8.5.53), (8.5.54), and (8.5.67), by virtue of Doob's martingale inequality and Hölder's inequality, we can deduce that for each $q>1$ and $\theta\in(0,1)$,

$$\overline{\lim_{k\to\infty}}\ \sup_{\ell\geqslant 1}\operatorname*{esssup}_{t\in[0,T]} \mathbb{E}\Big[\Phi_\lambda\Big(\sup_{r\in[t,T]}\big(|\delta_\theta\eta_{k,\ell}(t,r)|+|\delta_\theta\widetilde{\eta}_{k,\ell}(t,r)|\big);q\Big)\Big]\leqslant \widehat{M}(q),$$

where $\widehat{M}(q)$ is a positive constant depending on q and being independent of θ. Thus, a similar proof to (8.5.68) and (8.5.69) yields that for almost every $t\in[0,T]$, there is an adapted process $\eta(t,\cdot)\in\bigcap\limits_{p\geqslant 1}\mathcal{E}_{\mathbb{F}}^{p,\lambda}(\Omega;C([t,T];\mathbb{R}^n))$ such that for each $q>1$,

$$\lim_{k\to\infty}\operatorname*{esssup}_{t\in[0,T]}\mathbb{E}\Big(\sup_{s\in[t,T]}|\eta_k(t,s)-\eta(t,s)|^q\Big)=0, \tag{8.5.70}$$

and

$$\lim_{k\to\infty}\operatorname*{esssup}_{t\in[0,T]}\mathbb{E}\Big[\Phi_\lambda\Big(\sup_{s\in[t,T]}|\eta_k(t,s)-\eta(t,s)|;q\Big)\Big]=1. \tag{8.5.71}$$

Furthermore, it follows from Itô's formula that for almost every $t\in[0,T]$ and for each $k,\ell\geqslant 1$,

$$\begin{aligned}&\mathbb{E}\Big[\int_t^T|Z_{k+\ell}(t,s)-Z_k(t,s)|^2ds\Big]\\&\leqslant 2\mathbb{E}\Big[\sup_{s\in[t,T]}|\eta_{k+\ell}(t,s)-\eta_k(t,s)|\cdot\Delta^{k,\ell}(t)\Big],\end{aligned} \tag{8.5.72}$$

with

$$\Delta^{k,\ell}(t)\equiv\int_t^T\sum_{i=1}^n\big|g^i(t,s,Y_{k+\ell-1}(s),Z^i_{k+\ell}(t,s))-g^i(t,s,Y_{k-1}(s),Z^i_k(t,s))\big|ds.$$

And, by virtue of the first inequality in (8.5.27), the assumption on $\alpha(\cdot,\cdot)$, (8.5.53) and (8.5.54), we get that

$$\sup_{k,\ell\geqslant 1}\operatorname*{esssup}_{t\in[0,T]}\mathbb{E}\Big[|\Delta^{k,\ell}(t)|^2\Big]<\infty. \tag{8.5.73}$$

Then, applying Hölder's inequality to (8.5.72) and using (8.5.70) and (8.5.73) lead to that

$$\lim_{k\to\infty}\sup_{\ell\geqslant 1}\operatorname*{esssup}_{t\in[0,T]}\mathbb{E}\Big[\int_t^T|Z_{k+\ell}(t,s)-Z_k(t,s)|^2ds\Big]=0,$$

from which together with (8.5.54) it follows that

$$\lim_{k\to\infty}\sup_{\ell\geqslant 1}\operatorname*{esssup}_{t\in[0,T]}\mathbb{E}\Big[\Big(\int_t^T|Z_{k+\ell}(t,s)-Z_k(t,s)|^2ds\Big)^{\frac{p}{2}}\Big]=0,\quad \forall p>1.$$

Consequently, there exists a process $Z(\cdot\,,\cdot)\in\bigcap_{p\geqslant 1}L^\infty(0,T;L^p_{\mathbb{F}}(\Omega;L^2(\cdot\,,T;\mathbb{R}^n)))$ such that

$$\lim_{k\to\infty}\operatorname*{esssup}_{t\in[0,T]}\mathbb{E}\Big[\Big(\int_t^T|Z_k(t,s)-Z(t,s)|^2ds\Big)^{\frac{q}{2}}\Big]=0,\qquad \forall q>1.\qquad (8.5.74)$$

Thus, in view of (8.5.68), (8.5.69), and (8.5.74), by sending ℓ to infinity in (8.5.48), we deduce that $(Y(\cdot),Z(\cdot,\cdot))$ is a desired solution. In addition, for the above $\eta(t,\cdot)$ in (8.5.71), it is easy to see that

$$\eta(t,r)=\psi(t)+\int_r^T g(t,s,Y(s),Z(t,s))ds-\int_r^T Z(t,s)dW(s),\quad r\in[t,T],$$

from which we see that (8.5.47) holds.

Next, we look at the uniqueness. Suppose $(\widetilde Y(\cdot),\widetilde Z(\cdot,\cdot))$ is another adapted solution such that $\widetilde\eta(t,\cdot)\in\bigcap_{p\geqslant 1}\mathcal{E}^{p,\lambda}_{\mathbb{F}}(\Omega;C([t,T];\mathbb{R}^n))$ for almost every $t\in[0,T]$, where

$$\widetilde\eta(t,r)\equiv\widetilde Y(t)-\int_t^r g(t,s,\widetilde Y(s),\widetilde Z(t,s))ds+\int_t^r\widetilde Z(t,s)dW(s),\qquad r\in[t,T].$$

For each fixed $\theta\in(0,1)$ and $(t,s)\in\Delta^*[0,T]$, define

$$\begin{cases}\delta_\theta\eta(t,s)\equiv\dfrac{\eta(t,s)-\theta\widetilde\eta(t,s)}{1-\theta},\quad \delta_\theta Y(t)\equiv\dfrac{Y(t)-\theta\widetilde Y(t)}{1-\theta},\\ \delta_\theta\widetilde\eta(t,s)\equiv\dfrac{\widetilde\eta(t,s)-\theta\eta(t,s)}{1-\theta},\quad \delta_\theta\widetilde Y(t)\equiv\dfrac{\widetilde Y(t)-\theta Y(t)}{1-\theta},\\ \delta_\theta Z(t,s)\equiv\dfrac{Z(t,s)-\theta\widetilde Z(t,s)}{1-\theta},\quad \delta_\theta\widetilde Z(t,s)\equiv\dfrac{\widetilde Z(t,s)-\theta Z(t,s)}{1-\theta}.\end{cases}$$

Following the similar arguments from (8.5.57) to (8.5.67), for each $q > 1$ and $\theta \in (0, 1)$, we have

$$\mathbb{E}\Big[\Phi_\lambda\Big(\operatorname*{esssup}_{t\in[0,T]}\big(|\delta_\theta Y(t)| + |\delta_\theta \widetilde{Y}(t)|\big); q\Big)\Big] \leqslant \widetilde{M}(q),$$

where $\widetilde{M}(q)$ is a positive constant depending on q and being independent of θ. Thus, for each $\theta \in (0, 1)$, it follows that

$$\mathbb{E}\Big[\operatorname*{esssup}_{t\in[0,T]} |Y(t) - \theta\widetilde{Y}(t)|^\lambda\Big] \leqslant \frac{(1-\theta)^\lambda}{2}\widetilde{M}(2),$$

and then

$$\mathbb{E}\Big[\operatorname*{esssup}_{t\in[0,T]} |Y(t) - \widetilde{Y}(t)|^\lambda\Big] \leqslant (1-\theta)^\lambda \lambda\Big\{\frac{\widetilde{M}(2)}{2} + \mathbb{E}\Big[\operatorname*{esssup}_{t\in[0,T]} |\widetilde{Y}(t)|\Big]\Big\} < +\infty.$$

Sending θ to 1, we see that $Y(\cdot) = \widetilde{Y}(\cdot)$. Finally, the conclusions of

$$Z(\cdot\,,\cdot) = \widetilde{Z}(\cdot\,,\cdot), \qquad \eta(\cdot\,,\cdot) = \widetilde{\eta}(\cdot\,,\cdot)$$

follow from Lemma 8.5.5. The proof is complete. □

Remark 8.5.7 Lemma 8.5.5 and Theorem 8.5.6 are obtained under the condition that $g^i(t, s, y, z)$ varies with (t, s, y) and the ith component z^i of $z \in \mathbb{R}^n$ for each $i = 1, \cdots, n$. When g^i also depends on the z^j for some $j \neq i$, the complexity of the problem increases significantly.

6 Super-Quadratic BSVIEs

In the previous sections, we discuss the case when the generator $g(t, s, y, z)$ of the BSVIEs is at most of quadratic growth in z. In this part, we want to look at the case that $z \mapsto g(t, s, y, z)$ is super-quadratic.

As before, let us first look at the situation of BSDE. More precisely, we consider the following BSDE:

$$Y(t) = \xi + \int_t^T f(Z(s))ds - \int_t^T Z(s)dW(s), \quad t \in [0, T], \tag{8.6.1}$$

where the free term ξ and the generator f satisfy the following assumption:

(H8.6.1) Random variable ξ is bounded $\mathcal{F}_T$-measurable and $\mathbb{R}$-valued, and $f : \mathbb{R} \to \mathbb{R}^+$ is convex and differentiable with

$$f(0) = 0 \quad \text{and} \quad \overline{\lim_{|z|\to+\infty}} \frac{f(z)}{|z|^2} = \infty.$$

In the case that (H8.6.1) holds, a pair of $\mathbb{F}$-adapted processes $(Y(\cdot), Z(\cdot))$ taking value in $\mathbb{R} \times \mathbb{R}$ is called a bounded adapted solution of (8.6.1) if they satisfy BSDE (8.6.1) for each $t \in [0, T]$ such that $Y(\cdot)$ is bounded and

$$\mathbb{E}\left[\int_0^T f(Z(t))dt\right] < +\infty.$$

We have the following nonexistence result.

Theorem 8.6.1 *Suppose* (H8.6.1) *holds. Then, there exists a* $\xi \in L^{\infty}_{\mathcal{F}_T}(\Omega; \mathbb{R})$ *such that BSDE* (8.6.1) *has no bounded adapted solution.*

Proof We split the proof into several steps.

Step 1 Construction of measure $\mathbb{Q} \ll \mathbb{P}$, but $\mathbb{Q} \not\sim \mathbb{P}$. In other words, $\mathbb{Q}$ is absolutely continuous with respect to $\mathbb{P}$, but it is not equivalent to $\mathbb{P}$.

By (H8.6.1), there exists a sequence z_k such that

$$\lim_{k\to\infty} |z_k| = \infty, \qquad f(z_k) \geqslant k|z_k|^2, \quad k \geqslant 1.$$

Without loss of generality, let us assume that $z_k \geqslant 1$ for all $k \geqslant 1$. Then

$$kz_k^2 \leqslant f(z_k) - f(0) = f'(\zeta_k)z_k \leqslant f'(z_k)z_k,$$

since $f(\cdot)$ is convex and thus $f'(\cdot)$ is increasing. Also

$$f(z_k) = f(z_k) - f(0) = f'(\zeta_k)z_k \leqslant f'(z_k)z_k.$$

Clearly

$$\alpha \equiv \sum_{k=1}^{\infty} \frac{1}{z_k f'(z_k)k^2} \leqslant \sum_{k=1}^{\infty} \frac{1}{z_k^2 k^3} < \infty.$$

Denote

$$\delta_k \equiv \frac{T}{\alpha z_k f'(z_k)k^2}, \qquad k \geqslant 1.$$

Then

$$\sum_{i=1}^{\infty} \delta_k = T.$$

Next, we let

$$Z(s) \equiv \sum_{k=1}^{\infty} z_k \mathbf{1}_{(s_{k-1},s_k]}(s), \qquad s_k \equiv \sum_{i=1}^{k} \delta_i, \quad s_0 = 0.$$

Then, it follows that

$$\begin{aligned}\int_0^T f(Z(s))ds &= \sum_{k=1}^{\infty} \int_{s_{k-1}}^{s_k} f(Z(s))ds = \sum_{k=1}^{\infty} f(z_k)\delta_k \\ &\leqslant \sum_{k=1}^{\infty} z_k f'(z_k)\delta_k = \frac{T}{\alpha} \sum_{k=1}^{\infty} \frac{1}{k^2} < \infty,\end{aligned}$$

and

$$\int_0^T |Z(s)|^2 ds = \sum_{k=1}^{\infty} |z_k|^2 \delta_k \leqslant \frac{T}{\alpha} \sum_{k=1}^{\infty} \frac{1}{k^3} < \infty.$$

Next, by letting $q(\cdot) \equiv f'(Z(\cdot))$, we have

$$\int_0^T |q(s)|^2 ds = \sum_{k=1}^{\infty} |f'(z_k)|^2 \delta_k \geqslant \sum_{k=1}^{\infty} k f'(z_k) z_k \delta_k = \frac{T}{\alpha} \sum_{k=1}^{\infty} \frac{1}{k} = \infty.$$

Recall that, for $t \in [0, T]$,

$$\mathcal{E}\{q\}(t) \equiv \exp\Big(\int_0^t q(s)dW(s) - \frac{1}{2} \int_0^t |q(s)|^2 ds \Big).$$

Thus

$$\mathcal{E}\{q\}(t) \begin{cases} > 0 & \text{well-defined}, \quad t \in [0, T), \\ = 0, & \qquad\qquad\quad\;\; t = T. \end{cases}$$

Next, let

$$Y(t) \equiv \int_0^t f(Z(s))ds - \int_0^t Z(s)dW(s), \qquad t \geqslant 0,$$

and

$$\tau_k \equiv \inf\big\{t \in [0, T] \bigm| |Y(t)| \geqslant k,\ \mathcal{E}\{q\}(t) \geqslant k\big\}.$$

Then, $\tau_k \in \mathscr{T}[0, T]$. We may let $k \geqslant 1$ large so that $\mathbb{P}(A_k) > 0$, where $A_k \equiv (\tau_k = T)$. This is possible since otherwise, one would have $\mathcal{E}\{q\}(T) < \infty$, a contradiction. Now, we set

$$q_k(t) \equiv f'(Z(t))\mathbf{1}_{[0,\tau_k]}(t), \qquad t \in [0, T],$$

and set

$$d\mathbb{Q} \equiv \mathcal{E}\{q_k\}(T)d\mathbb{P}.$$

Hence, we have $\mathbb{Q} \ll \mathbb{P}$. However, note that

$$\mathbb{Q}(A_k) = \int_{A_k} \mathcal{E}\{q_k\}(\tau_k)d\mathbb{P} = \int_{A_k} \mathcal{E}\{q_k\}(T)d\mathbb{P} = 0. \tag{8.6.2}$$

This means $\mathbb{Q} \not\sim \mathbb{P}$. On the other hand, we have $\mathbb{Q} \sim \mathbb{P}$ on $\mathcal{F}_t$ with $t \in [0, T)$.

By setting

$$Y^k(t) \equiv Y(\tau_k \wedge t),\quad Z^k(t) \equiv Z(t)\mathbf{1}_{[0,\tau_k]}(t),\quad \xi \equiv Y(\tau_k) \in L^\infty_{\mathcal{F}_T}(\Omega; \mathbb{R}),$$

we see that $(Y^k(\cdot), Z^k(\cdot))$ is a bounded adapted solution of (8.6.1).

Step 2 Given the above ξ and A_k, let us define $\eta \equiv \xi + \mathbf{1}_{A_k}$. Notice that $\mathbf{1}_{A_k} \in L^\infty_{\mathcal{F}_T}(\Omega; \mathbb{R})$ such that

$$\mathbf{1}_{A_k} \geqslant 0,\ \ a.s.,\ \ \mathbb{P}(\{\omega \in \Omega \bigm| \mathbf{1}_{A_k}(\omega) > 0\}) > 0,\ \ \mathbf{1}_{A_k}\mathcal{E}\{q_k\}(T) = 0.$$

In addition, under (8.6.2), we have $\xi = \eta$, $\mathbb{Q}$ a.s. Next, we prove that if $(\bar{Y}, \bar{Z})$ is a solution associated with (g, η), then $\bar{Y}$ is unbounded.

To begin with, let us make some preparations as follows. Define

$$U(\xi, \tau) \equiv \operatorname{essinf}\Big\{\mathbb{E}^{\widetilde{\mathbb{P}}}\Big[\xi + \int_\tau^T f_L(q(s))ds \bigm| \mathcal{F}_\tau\Big] \Bigm| \widetilde{\mathbb{P}} \sim \mathbb{P}\Big\}, \qquad \tau \in \mathscr{T}[0, T],$$

where $f_L(\cdot)$ is the Legendre transform of $f(\cdot)$ defined as

$$\begin{aligned} &f_L(z) \equiv \sup_{x\in\mathbb{R}}\big[zx - f(x)\big],\ \ z \in \mathbb{R},\\ &f_L(f'(x)) = f'(x)x - f(x),\quad x \in \mathbb{R}, \end{aligned} \tag{8.6.3}$$

$q(\cdot)$ is the density process associated with $\widetilde{\mathbb{P}}$ such that

$$\mathbb{E}_{\mathbb{P}}\Big[\frac{d\widetilde{\mathbb{P}}}{d\mathbb{P}}\big|\mathcal{F}_{\cdot}\Big]=\mathcal{E}\{q\}_{\cdot},\quad \mathbb{E}^{\widetilde{\mathbb{P}}}\int_0^T f_L(q(s))ds<\infty. \tag{8.6.4}$$

Obviously, $f_L(0)=0$, $U(\xi,\cdot)$ is bounded since $|U(\xi,\cdot)|\leqslant\|\xi\|_\infty$. We claim that $U(\xi,\cdot)=Y^k(\cdot)$, and

$$U(\xi,t)=\mathbb{E}^{\mathbb{Q}}\Big[\xi+\int_t^T f_L(q_k(s))ds\Big|\mathcal{F}_t\Big],\qquad t<T. \tag{8.6.5}$$

First, by (8.6.3), the definitions of q_k, $Z(\cdot)$, together with the nonnegativeness of $f(\cdot)$, we see that

$$\begin{aligned}
&\mathbb{E}^{\mathbb{Q}}\int_0^T f_L(q_k(s))ds=\mathbb{E}^{\mathbb{Q}}\int_0^{\tau_k} f_L(Z(s))ds\\
&=\mathbb{E}^{\mathbb{Q}}\int_0^{\tau_k} f'(Z(s))Z(s)ds-\mathbb{E}^{\mathbb{Q}}\int_0^{\tau_k} f(Z(s))ds\\
&\leqslant\mathbb{E}^{\mathbb{Q}}\int_0^T f'(Z(s))Z(s)ds=\sum_{k=1}^\infty f'(z_k)z_k\delta_k=\sum_{k=1}^\infty\frac{T}{\alpha k^2}<\infty.
\end{aligned}$$

Therefore, the right hand of (8.6.5) is well-defined. Since $(Y^k(\cdot),Z^k(\cdot))$ is the bounded solution of (8.6.1), the following holds true for any $t<T$ almost surely under $\mathbb{Q}$,

$$\begin{aligned}
U(\xi,t)&\leqslant\mathbb{E}^{\mathbb{Q}}\Big[\xi+\int_t^T f_L(q_k(s))ds\Big|\mathcal{F}_t\Big]=\mathbb{E}^{\mathbb{Q}}\Big[\xi+\int_{t\wedge\tau_k}^{\tau_k} f_L(q_k(s))ds\Big|\mathcal{F}_t\Big]\\
&=Y^k(t)+\mathbb{E}^{\mathbb{Q}}\Big[\int_{t\wedge\tau_k}^{\tau_k}[f_L(q_k(s))+f(Z(s))]ds-\int_{t\wedge\tau_k}^{\tau_k}Z(s)dW(s)\Big|\mathcal{F}_t\Big]\\
&=Y^k(t)+\mathbb{E}^{\mathbb{Q}}\Big[\int_{t\wedge\tau_k}^{\tau_k}[f_L(q_k(s))+f(Z(s))-Z(s)q_k(s)]ds\Big|\mathcal{F}_t\Big]=Y^k(t).
\end{aligned} \tag{8.6.6}$$

In the last equality, we use the definition of $q_k(\cdot)$ as well as the second result in (8.6.3). Since $\mathbb{Q}\sim\mathbb{P}$ on $\mathcal{F}_t$ for $t<T$, we see that (8.6.6) holds true for any $t<T$. On the other hand, for any $\widetilde{\mathbb{P}}\sim\mathbb{P}$ such that the corresponding density process $q(\cdot)$ satisfies (8.6.4), since $(Y^k(\cdot),Z^k(\cdot))$ is the bounded solution of (8.6.1), we have

$$\begin{aligned}
&\mathbb{E}^{\widetilde{\mathbb{P}}}\Big[\xi+\int_t^T f_L(q(s))ds\Big|\mathcal{F}_t\Big]\\
&=Y^k(t)+\mathbb{E}^{\widetilde{\mathbb{P}}}\Big[\int_t^T[f(Z(s))+f_L(q(s))]ds-\int_t^T Z(s)dW(s)\Big|\mathcal{F}_t\Big]\\
&=Y^k(t)+\mathbb{E}^{\widetilde{\mathbb{P}}}\Big[\int_t^T[f(Z(s))+f_L(q(s))-Z(s)q(s)]ds\Big|\mathcal{F}_t\Big]\geqslant Y^k(t).
\end{aligned}$$

By the arbitrariness of $q(\cdot)$, we see that $U(\xi, t) \geqslant Y^k(t)$ with $t \in [0, T]$. This together with (8.6.6) shows that $U(\xi, \cdot) = Y^k(\cdot)$ as well as (8.6.5).

Recall the above η which equals to ξ under $\mathbb{Q}$, we see that

$$\begin{aligned} U(\eta, t) &\leqslant \mathbb{E}^{\mathbb{Q}}\Big[\eta + \int_t^T f(q_k(s))ds\Big|\mathcal{F}_t\Big] \\ &= \mathbb{E}^{\mathbb{Q}}\Big[\xi + \int_t^T f(q_k(s))ds\Big|\mathcal{F}_t\Big] = U(\xi, t), \quad t < T. \end{aligned}$$

On the other hand, $U(\xi, t) \leqslant U(\eta, t)$ for any $t \in [0, T)$ by the monotonicity of $U(\cdot)$. Hence, $U(\xi, t) = U(\eta, t)$. a.s. for any $t < T$.

Recall the above $(\bar{Y}, \bar{Z})$, if it is bounded, then for any $t < T$,

$$\bar{Y}(t) \leqslant U(\xi, t) = U(\eta, t) = Y^k(t),$$

and therefore

$$\eta = Y(T) = \lim_{t \to T} Y(t) \leqslant \lim_{t \to T} Y^k(t) = Y^k(T) = \xi, \quad \mathbb{P} - a.s.$$

This is a contradiction to the fact of $\mathbb{P}(\{\omega \in \Omega, \big|\eta(\omega) > \xi(\omega)\}) > 0$. Therefore, $\bar{Y}(\cdot)$ is not bounded. □

The above shows the negative result of boundedness. At this moment, one may ask: *once the existence of solution is done, how about the uniqueness?* We have the following result.

Theorem 8.6.2 *Suppose* (H8.6.1) *holds and* (8.6.1) *admits a bounded solution* (Y, Z). *Then, for each* $y < Y(0)$, *BSDE* (8.6.1) *admits another bounded solution* $(\bar{Y}(\cdot), \bar{Z}(\cdot))$ *satisfying* $\bar{Y}(0) = y$.

Proof Suppose $\delta_n \equiv T(1 - 2^{-n})$ with $n \geqslant 0$, then we have $[0, T) = \bigcup_{n \geqslant 0} [\delta_n, \delta_{n+1})$. We use the induction method. In other words, suppose we have another bounded solution constructed on $[0, \delta_n)$ with $\bar{Y}(0) = y < Y(0)$ such that $\bar{Y}(\delta_n) \leqslant Y(\delta_n)$. a.s. Next, we separate the proof into two steps.

Step 1 We define a process $M(\cdot)$ on $[\delta_n, \delta_{n+1})$ such that $\lim_{t \to \delta_{n+1}} M(t) = +\infty$, a.s.

To begin with, let us make some preparations. Given the above n, the super-quadratic assumption of $z \mapsto f(z)$ implies the existence of sequence $\{\alpha_k\}_{k \geqslant 0}$ such that for constant $\theta \in (0, 1)$,

$$f(\alpha_k) \geqslant 4^n \alpha_k^2, \quad \alpha_k^2 \geqslant \frac{1}{(\theta^k - \theta^{k+1})\theta^k x_n}, \quad x_n \equiv \delta_{n+1} - \delta_n = 2^{-(n+1)}T.$$

Next, let us define

$$\sigma(t) \equiv \sum_{k=0}^{\infty} \alpha_k \mathbf{1}_{[\delta_{n+1}-\theta^k x_n, \delta_{n+1}-\theta^{k+1} x_n)}(t), \quad t \in [0, T),$$
$$M(t) \equiv \int_{\delta_n}^{t} f(\sigma(s))ds - \int_{\delta_n}^{t} \sigma(s)dW(s), \quad t \in [\delta_n, \delta_{n+1}).$$

For this $\sigma(\cdot)$ and the corresponding $f(\sigma(\cdot))$, we have

$$\lim_{t\to\delta_{n+1}} \Phi^n(t) \equiv \lim_{t\to\delta_{n+1}} \int_{\delta_n}^{t} \sigma(s)^2 ds = \infty, \qquad \lim_{t\to\delta_{n+1}} \int_{\delta_n}^{t} f(\sigma(s))ds = \infty,$$

since for any $N \in \mathbb{N}$, and $t \in [\delta_{n+1} - \theta^{N+1}x_n, \delta_{n+1} - \theta^{N+2}x_n)$,

$$\begin{aligned} &\Phi(t) \equiv \int_{\delta_n}^{t} \sigma(s)^2 ds \geqslant \sum_{k=0}^{N} \alpha_k^2(\theta^k - \theta^{k+1})x_n \geqslant \sum_{k=0}^{N} \frac{1}{\theta^k}, \\ &\int_{\delta_n}^{t} f(\sigma(s))ds \geqslant 4^n \int_{\delta_n}^{t} \sigma(s)^2 ds \geqslant 4^n \sum_{k=0}^{N} \frac{1}{\theta^k}. \end{aligned} \tag{8.6.7}$$

We observe that $\Phi(\cdot)$ is strictly increasing, and the following process

$$\widetilde{W}(t) \equiv \int_{\delta_n}^{\Phi^{-1}(t)} \sigma(s)dW(s), \quad t \in [\delta_n, \delta_{n+1}),$$

is a time-changed Brownian motion w.r.t. filtration $\{\widetilde{\mathcal{F}}\}$ generated by $\widetilde{W}(\cdot)$ such that

$$\lim_{t\to\delta_{n+1}} \frac{\widetilde{W}(\Phi(t))}{\Phi(t)} = 0, \quad \mathbb{P} - \text{a.s.}$$

It then follows from (8.6.7) that

$$M(t) \geqslant 4^n\Phi(t) - \widetilde{W}(\Phi(t)) = \Phi(t)\Big[4^n - \frac{\widetilde{W}(\Phi(t))}{\Phi(t)}\Big] \to +\infty, \qquad t \to \delta_{n+1},$$

which leads to our desired conclusion of $M(\cdot)$.

Step 2 Since we have obtained the solution $(\bar{Y}, \bar{Z})$ on $[0, \delta_n)$, we will construct the remaining part of $(\bar{Y}, \bar{Z})$ on $[\delta_n, T)$.

Let us first find a pair of $(\widehat{Y}_n, \widehat{Z}_n)$ on $[\delta_n, \delta_{n+1}]$. The case of $[\delta_k, \delta_{k+1})$ with $k \geqslant n+1$ can be obtained similarly with $\widehat{Y}_k(\delta_{k+1}) = \widehat{Y}_{k+1}(\delta_{k+1})$. In this manner, we define the pair of processes on $[0, T)$ as follows:

$$
\begin{aligned}
\bar{Y}(t) &\equiv \bar{Y}(t)\mathbf{1}_{[0,\delta_n)}(t) + \sum_{k\geqslant n}^{\infty} \widehat{Y}_k(t)\mathbf{1}_{[\delta_k,\delta_{k+1}]}(t),\\
\bar{Z}(t) &\equiv \bar{Z}(t)\mathbf{1}_{[0,\delta_n)}(t) + \sum_{k\geqslant n}^{\infty} \widehat{Z}_k(t)\mathbf{1}_{[\delta_k,\delta_{k+1}]}(t).
\end{aligned}
\tag{8.6.8}
$$

As preparations, for any $\varepsilon > 0$, we introduce three stopping times $\tau_{1,n}$, $\tau_{2,n}$, $\tau_{3,n}$ as follows:

$$
\begin{aligned}
\tau_{1,n} &\equiv \inf\Big\{t \geqslant \delta_n \Big| M(t) = -2^{-(n+1)}\varepsilon\Big\} \wedge \delta_{n+1},\\
\tau_{2,n} &\equiv \inf\Big\{t \geqslant \delta_n \Big| \bar{Y}(\delta_n) + M(t) \geqslant Y(t)\Big\} \wedge \delta_{n+1},\\
\tau_{3,n} &\equiv \inf\Big\{t \geqslant \tau_{1,n} \Big| \bar{Y}(\delta_n) - 2^{-(n+1)}\varepsilon = Y(t)\Big\} \wedge \delta_{n+1}.
\end{aligned}
$$

Here, $\tau_{1,n}$, $\tau_{2,n}$ are stopping times when $\bar{Y}(\delta_n) + M(\cdot)$ touches $\bar{Y}(\delta_n) - 2^{-(n+1)}\varepsilon$ and $Y(\cdot)$ on $[\delta_n, \delta_{n+1})$, respectively. As to $\tau_{2,n}$, we also have $\mathbb{P}(\tau_2 < \delta_{n+1}) = 1$ by the result of $M(\cdot)$ in Step 1.

Given the bounded solution (Y, Z), we define $\widehat{Z}_n(\cdot)$ on $[\delta_n, \delta_{n+1})$ as follows:

$$
\begin{aligned}
\widehat{Z}_n(t) &\equiv \mathbf{1}_{\{\tau_{1,n}<\tau_{2,n},\ \tau_{3,n}<\delta_{n+1}\}}\big(\sigma(t)\mathbf{1}_{\{t\leqslant\tau_{1,n}\}} + Z(t)\mathbf{1}_{\{t>\tau_{3,n}\}}\big)\\
&\quad + \mathbf{1}_{\{\tau_{1,n}<\tau_{2,n},\ \tau_{3,n}=\delta_{n+1}\}}\sigma(t)\mathbf{1}_{\{t\leqslant\tau_{1,n}\}}\\
&\quad + \mathbf{1}_{\{\tau_{1,n}\geqslant\tau_{2,n}\}}\big(\sigma(t)\mathbf{1}_{\{t\leqslant\tau_{2,n}\}} + Z(t)\mathbf{1}_{\{t>\tau_{2,n}\}}\big)\\
&= Z(t)\big[\mathbf{1}_{\{\tau_{1,n}<\tau_{2,n},\ \tau_{3,n}<\delta_{n+1},\ t>\tau_{3,n}\}} + \mathbf{1}_{\{\tau_{1,n}\geqslant\tau_{2,n},\ t>\tau_{2,n}\}}\big]\\
&\quad + \mathbf{1}_{\{t\leqslant\tau_{1,n}\wedge\tau_{2,n}\}}\sigma(t).
\end{aligned}
$$

One can check that $\widehat{Z}_n(\cdot)$ is predictable process on $[\delta_n, \delta_{n+1})$. Now, we define

$$
\widehat{Y}_n(t) \equiv \bar{Y}(\delta_n) + \int_{\delta_n}^{t} f(\bar{Z}(s))ds - \int_{\delta_n}^{t} \bar{Z}(s)dW(s), \quad t \in [\delta_n, \delta_{n+1}).
$$

By the definition of $(\widehat{Y}_n, \widehat{Z}_n)$ on $[\delta_n, \delta_{n+1})$, we have

$$
\begin{aligned}
&\widehat{Y}_n(t)\mathbf{1}_{\{\tau_{2,n}\leqslant\tau_{1,n},\ t\geqslant\tau_{2,n}\}} = Y(t)\mathbf{1}_{\{\tau_{2,n}\leqslant\tau_{1,n},\ t\geqslant\tau_{2,n}\}},\\
&\widehat{Y}_n(t)\mathbf{1}_{\{\tau_{2,n}>\tau_{1,n},\ \tau_{3,n}\geqslant t\geqslant\tau_{1,n}\}} = \big(\widehat{Y}_n(\delta_n) - 2^{-(n+1)}\varepsilon\big)\mathbf{1}_{\{\tau_{2,n}>\tau_{1,n},\ \tau_{3,n}\geqslant t\geqslant\tau_{1,n}\}},\\
&\widehat{Y}_n(t)\mathbf{1}_{\{\tau_{1,n}<\tau_{2,n},\ t>\tau_{3,n}\}} = Y(t)\mathbf{1}_{\{\tau_{1,n}<\tau_{2,n},\ t>\tau_{3,n}\}},
\end{aligned}
$$

from which we see that

$$
\begin{aligned}
\widehat{Y}_n(\delta_{n+1}) &= Y(\delta_{n+1})\big(\mathbf{1}_{\{\tau_{2,n}\leqslant\tau_{1,n}\}} + \mathbf{1}_{\{\tau_{1,n}<\tau_{2,n},\ \tau_{3,n}<\delta_{n+1}\}}\big)\\
&\quad + \big(\widehat{Y}_n(\delta_n) - 2^{-(n+1)}\varepsilon\big)\mathbf{1}_{\{\tau_{1,n}<\tau_{2,n},\ \tau_{3,n}=\delta_{n+1}\}}.
\end{aligned}
\tag{8.6.9}
$$

Since $\widehat{Y}_n(\delta_n) = \bar{Y}(\delta_n) \leqslant Y(\delta_n)$, hence we have $\widehat{Y}_n(\delta_{n+1}) \leqslant Y(\delta_{n+1})$. At this moment, for arbitrary $\gamma > 0$, let us define D_n, $\widehat{D}_n$ and $D_{\gamma,n}$ as follows:

$$\begin{aligned} D_n &\equiv \{\omega \in \Omega \big| \tau_{1,n} < \tau_{2,n},\ \tau_{3,n} = \delta_{n+1}\}, \\ \widehat{D}_n &\equiv \{\omega \in \Omega \big| \widehat{Y}_n(\delta_{n+1}, \omega) \neq Y(\delta_{n+1}, \omega)\}, \\ D_{\gamma,n} &\equiv \{\omega \in \Omega \big| \big|\widehat{Y}_n(\delta_{n+1}, \omega) - Y(\delta_{n+1}, \omega)\big| \geqslant \gamma\}. \end{aligned}$$

By (8.6.9) it is easy to see that if $D_{\gamma,n} \subset \widehat{D}_n \subset D_n$ which implies that $\mathbb{P}(D_{\gamma,n}) \leqslant \mathbb{P}(\widehat{D}_n) \leqslant \mathbb{P}(D_n)$, by the definition of $\tau_{i,n}$, $i = 1, 2, 3$, as well as submartingale inequality, we arrive at

$$\begin{aligned} &\mathbb{P}(D_{\gamma,n}) \leqslant \mathbb{P}(\widehat{D}_n) \leqslant \mathbb{P}(D_n) \\ &\leqslant \mathbb{P}\big(\{\omega \in \Omega \big| \exists t \in [\delta_n, \delta_{n+1}) \text{ such that } M(t, \omega) < -2^{-(n+1)}\varepsilon\}\big) \\ &= \mathbb{P}\big(\{\omega \in \Omega \big| \exists t \in [\delta_n, \delta_{n+1}) \text{ such that } 4^n \Phi(t) - \widetilde{W}(\Phi(t)) < -2^{-(n+1)}\varepsilon\}\big) \\ &= \mathbb{P}\big(\{\omega \in \Omega \big| \exists r \in [\delta_n, \delta_{n+1}) \text{ such that } 4^n r - \widetilde{W}(r) < -2^{-(n+1)}\varepsilon\}\big) \leqslant e^{-2^n\varepsilon}. \end{aligned}$$

Therefore, by Borel-Cantelli lemma and the fact of $\sum\limits_{n=0}^{\infty} \exp\{-2^n\varepsilon\} < \infty$, we have $\mathbb{P}(\bigcap\limits_{n=0}^{\infty} \bigcup\limits_{k\geqslant n} D_{\gamma,k}) = 0$ with any $\gamma > 0$, which implies that

$$\lim_{n\to\infty} \big|\widehat{Y}_n(\delta_{n+1}) - Y(\delta_{n+1})\big| = 0, \quad \mathbb{P} - a.s.$$

Since $\bar{Y}(\delta_{n+1}) = \widehat{Y}_n(\delta_{n+1})$, and $\delta_{n+1} \to T$, as $n \to \infty$, we then obtain that $\bar{Y}(T) = Y(T) = \xi$. Consequently, $(\bar{Y}, \bar{Z})$ is another solution. By (8.6.8) and the construction of $\widehat{Y}_k(\cdot)$, we see that $\bar{Y}(\cdot)$ is also bounded. □

Now, let us consider the following BSVIE:

$$Y(t) = \psi(t) + \int_t^T g(t, Z(t,s))ds - \int_t^T Z(t,s)dW(s), \quad t \in [0, T]. \tag{8.6.10}$$

We discuss its bounded solution $(Y(\cdot), Z(\cdot\,,\cdot))$ taking value in $\mathbb{R} \times \mathbb{R}$ by which we mean that $(Y(\cdot), Z(\cdot, \cdot))$ is a solution of (8.6.10) such that $Y(\cdot)$ is bounded and for each $t \in [0, T]$,

$$\mathbb{E}\left[\int_t^T g(t, Z(t,s))ds\right] < +\infty.$$

Similar as (H8.6.1), for the coefficients in (8.6.10), we use the following assumptions.

(H8.6.2) The free term $\psi(\cdot)$ is a bounded, $\mathcal{F}_T$-measurable, and $\mathbb{R}$-valued process, and the generator $g : [0, T] \times \mathbb{R} \to \mathbb{R}^+$ is convex and differentiable in the second variable with, for each $t \in [0, T]$,

$$g(t,0)=0, \quad \text{and} \quad \overline{\lim_{|z|\to\infty}} \frac{g(t,z)}{|z|^2}=\infty.$$

Similar as Theorem 8.6.1, for BSVIE (8.6.10), we have some negative results.

Theorem 8.6.3 *Suppose* (H8.6.2) *holds with* g *independent of* t. *Then, there exists a bounded and continuous* $\mathcal{F}_T$*-measurable process* $\psi(\cdot)$ *such that BSVIE* (8.6.10) *has no bounded and continuous solution satisfying* $Y(T)=\psi(T)$.

Proof With $f(\cdot)$ replaced by $g(\cdot)$, suppose η is the bounded random variable in the step 2 proof of Theorem 8.6.1 such that the following BSDE

$$Y(t)=\eta+\int_t^T g(Z(s))ds-\int_t^T Z(s)dW(s), \quad t\in[0,T], \tag{8.6.11}$$

has no bounded solution $Y(\cdot)$. For convenience, we keep the meanings of all the notations in the proof of Theorem 8.6.1. For example, $\mathbb{Q}$ is the original probability satisfying (8.6.2).

Next, let $h(\cdot)$ be a bounded and continuous function on $[0, T]$, $\psi(\cdot)\equiv\eta+h(\cdot)$. We consider BSVIE

$$P(t)=\psi(t)+\int_t^T g(Q(t,s))ds-\int_t^T Q(t,s)dW(s), \quad t\in[0,T]. \tag{8.6.12}$$

Suppose (8.6.12) admits a bounded and continuous solution satisfying $P(T)=\psi(T)$. We define $\widehat{P}(\cdot)\equiv P(\cdot)-h(\cdot)$. Then, we see that $(\widehat{P}, Q)$ satisfies

$$\widehat{P}(t)=\eta+\int_t^T g(Q(t,s))ds-\int_t^T Q(t,s)dW(s), \quad t\in[0,T]. \tag{8.6.13}$$

Like the above (8.6.3), we define the Legendre transform of g as follows:

$$g_L(z)\equiv\sup_{x\in\mathbb{R}}\big[zx-g(x)\big], \quad z\in\mathbb{R}. \tag{8.6.14}$$

With $\mathbb{Q}$, q_k having the same interpretations as in the proof of Theorem 8.6.1, we have

$$\begin{aligned}
&\mathbb{E}^{\mathbb{Q}}\Big[\eta+\int_t^T g_L(q_k(s))ds\Big|\mathcal{F}_t\Big]\\
&=\widehat{P}(t)+\mathbb{E}^{\mathbb{Q}}\Big[\int_t^T\big[g(Q(t,s))+g_L(q_k(s))\big]ds-\int_t^T Q(t,s)dW(s)\Big|\mathcal{F}_t\Big]\\
&=\widehat{P}(t)+\mathbb{E}^{\mathbb{Q}}\Big[\int_t^T\big[g(Q(t,s))+g_L(q_k(s))-Q(t,s)q_k(s)\big]ds\Big|\mathcal{F}_t\Big]\geqslant\widehat{P}(t).
\end{aligned}$$

Recall the proof of Theorem 8.6.1, for any $t < T$, we then have

$$\widehat{P}(t) \leqslant \mathbb{E}^{\mathbb{Q}}\Big[\eta + \int_t^T g_L(q_k(s))ds \Big| \mathcal{F}_t\Big] = \mathbb{E}^{\mathbb{Q}}\Big[\xi + \int_t^T g_L(q_k(s))ds \Big| \mathcal{F}_t\Big].$$

Following the idea in the step 2 proof of the above theorem, by sending $t \to T$, we obtain $\eta \leqslant \xi$, a contradiction to the fact of $\mathbb{P}(\{\omega \in \Omega | \eta(\omega) > \xi(\omega)\}) > 0$. □

As to the case when g depends on t, let us look at the following negative example.

Example 8.6.4 Recall $\eta \in L^\infty_{\mathcal{F}_T}(\Omega; \mathbb{R})$ is the random variable in (8.6.11) associated with g. We consider the following BSVIE on $[0, 2T]$:

$$\begin{aligned} &Y(t) = \psi(t) + \int_t^{2T} \widetilde{g}(t, Z(t,s))ds - \int_t^{2T} Z(t,s)dW(s), \\ &\psi(t) \equiv \eta I_{[0,T]}(t), \quad \widetilde{g}(t,z) \equiv I_{[0,T]}(t) g(z). \end{aligned} \tag{8.6.15}$$

If (8.6.15) admits a bounded solution on $[0, 2T]$, then for $t \in (T, 2T]$, it is easy to see

$$(Y(t), Z(t,s)) = (0, 0), \quad T \leqslant t \leqslant s \leqslant 2T,$$

and for $t \in [0, T]$,

$$Y(t) = \eta + \int_t^T g(Z(t,s))ds - \int_t^{2T} Z(t,s)dW(s).$$

Taking conditional expectation $\mathbb{E}_T$ on both sides, we have

$$Y(t) = \eta + \int_t^T g(Z(t,s))ds - \int_t^T Z(t,s)dW(s).$$

Using the same arguments as in Theorem 8.6.3, we obtain that $Y(\cdot)$ cannot be bounded and continuous.

As to the general case of g depending on t, let us look at the following result.

Theorem 8.6.5 *Suppose* (H8.6.2) *holds. Then, there exists a bounded* $\mathcal{F}_T$-*measurable process* $\psi(\cdot)$ *such that BSVIE* (8.6.10) *has no solution* (Y, Z) *satisfying for any* $t \in [0, T]$, $\lambda(t, \cdot) \in L^\infty_{\mathbb{F}}(\Omega; C([t, T]; \mathbb{R}))$,

$$\lambda(t,r) \equiv Y(t) - \int_t^r g(t, Z(t,s))ds + \int_t^r Z(t,s)dW(s), \quad r \in [t, T]. \tag{8.6.16}$$

Proof Let us take $\psi(\cdot) \equiv \bar{\xi}$, where the bounded random variable $\bar{\xi}$ is determined by $g(0, \cdot)$ such that the following BSDE

$$\bar{Y}(t) = \bar{\xi} + \int_t^T g(0, \bar{Z}(s))ds - \int_t^T \bar{Z}(s)dW(s), \quad t \in [0, T],$$

has no bounded solution according to Theorem 8.6.1. We prove the conclusion by contradiction. Suppose (8.6.10) admits a solution $(Y(\cdot), Z(\cdot, \cdot))$ such that $\lambda(t, \cdot)$ is bounded. It is easy to see that for each $t \in [0, T]$, $(\lambda(t, \cdot), Z(t, \cdot))$ satisfies the following BSDE:

$$\lambda(t, r) = \bar{\xi} + \int_r^T g(t, Z(t, s))ds - \int_r^T Z(t, s)dW(s), \quad r \in [t, T].$$

Let $t = 0$, we see that $(\lambda(0, \cdot), Z(0, \cdot))$ solves

$$\lambda(0, r) = \bar{\xi} + \int_r^T g(0, Z(0, s))ds - \int_r^T Z(0, s)dW(s), \quad r \in [0, T],$$

and $\lambda(0, \cdot)$ is bounded. Obviously, it is contradicted with the choice of $\bar{\xi}$. □

Remark 8.6.6 In the BSDE case when g, Z do not depend on t, we see that $\lambda(t, \cdot)$ is also t-independent and $\lambda(t, r) = Y(r)$. In this case, the above result consists with Theorem 8.6.1.

Theorem 8.6.5 indicates that one should carefully impose delicate conditions on $\psi(\cdot)$ to guarantee the boundedness of λ. At this moment, one may ask the following question: if the existence is given, how about the uniqueness? Similar as Theorem 8.6.2, we have the following result,

Theorem 8.6.7 *Suppose* (H8.6.2) *holds and BSVIE* (8.6.10) *admits a solution* (Y, Z) *such that for* $t \in [0, T]$, *the above* $\lambda(t, \cdot)$ *in* (8.6.16) *belongs to* $L^\infty_{\mathbb{F}}(\Omega; C([t, T]; \mathbb{R}))$. *Then, there exists another bounded solution* $(\bar{Y}(\cdot), \bar{Z}(\cdot))$ *such that the corresponding* $\bar{\lambda}(t, \cdot) \in L^\infty_{\mathbb{F}}(\Omega; C([t, T]; \mathbb{R}))$.

Proof By the given assumptions, we have

$$\lambda(t, r) = \psi(t) + \int_r^T g(t, Z(t, s))ds - \int_r^T Z(t, s)dW(s), \quad r \geqslant t. \tag{8.6.17}$$

In other words, for given $t \in [0, T)$, $(\lambda(t, \cdot), Z(t, \cdot))$ satisfies a BSDE associated with $(\psi(t), g(t, \cdot))$. By Theorem 8.6.2, there exists pair of processes $(\bar{\lambda}(t, \cdot), \bar{Z}(t, \cdot))$ satisfying (8.6.17), i.e.,

$$\bar{\lambda}(t, r) = \psi(t) + \int_r^T g(t, \bar{Z}(t, s))ds - \int_r^T \bar{Z}(t, s)dW(s), \quad r \geqslant t,$$

and

$$\mathbb{E}\int_0^T |\bar{Z}(t,s) - Z(t,s)|^2 ds > 0.$$

Let $t = r$, and define $\bar{Y}(t) \equiv \bar{\lambda}(t,t)$, we then have

$$\bar{Y}(t) = \psi(t) + \int_t^T g(t, \bar{Z}(t,s))ds - \int_t^T \bar{Z}(t,s)dW(s), \quad t \geqslant 0.$$

In other words, $(\bar{Y}, \bar{Z})$ is another different solution of BSVIE (8.6.10). The conclusion of $\bar{\lambda}$ is easy to see. □

7 Bibliographical Remarks

In the current chapter, to study Type-I BSVIEs, some non-Lipschitz requirements of $(y, z) \mapsto g(t, s, y, z)$ are imposed in two different ways. Section 1 is devoted to the case of $y \mapsto g(t, s, y, z)$, where the Osgood function and LaSalle–Bihari inequality are involved. Theorem 8.1.7 is an adjustment to the relevant study in Shi–Wang [175] and Wang–Zhang [211]. The remaining section is devoted to the non-Lips condition of $z \mapsto g(t, s, y, z)$ which is much more complex than the above one.

A martingale in BMO is a probabilistic version of a function of bounded mean oscillation introduced by F. John and L. Nirenberg. Our preparation on BMO is borrowed from the monograph Kazamaki [122]. Lemma 8.2.4 comes from [122, Corollary 2.1], Lemma 8.2.6 comes from [122, Theorem 2.3], Lemma 8.2.7 comes from [122, Theorem 3.1], and Lemma 8.2.8 is [122, Theorem 2.1].

Quadratic BSDEs (i.e., $z \mapsto g(t, s, y, z)$ being quadratic growth) were firstly studied in [124] where the terminal condition is assumed to be bounded. There have been many further studies on this subject. We refer to, e.g., Briand–Hu [36], [37], Delbaen–Hu–Richou [56], and Barrieu–El Karoui [15]. Note that all these works require the solution to be scalar. There have been some recent studies on multidimensional quadratic BSDEs, where some special structures of the generator are required; see, e.g., Tevzadze [182], Hu–Tang [105], Jamneshan–Kupper–Luo [114], and Frei and dos Reis [80] (a counterexample in multidimensional case). BSDEs with super-quadratic growth generator was studied in Delbaen-Hu-Bao [55] where some negative results were given.

Starting from Sect. 2, we consider the Type-I BSVIEs with $z \mapsto g(t, s, y, z)$ being superlinear growth. As preparation, we provide some conclusions for a particular case of BSDEs. When the terminal condition is bounded, the proof of well-posedness for BSDEs is adopted from that in [37] and Zhang [226]. Partial ideas of proving Lemma 8.3.1 come from Theorem 7.2.1 in [226]. The method of proving Lemma 8.3.3 and Theorem 8.3.4 is adopted from Lemma 7.3.2 in [226], while the assumptions are weaker than theirs. The result of Theorem 8.3.6 seems to be new, while partial ideas of the proof are inspired by [37]. To prove the well-posedness of BSDE with unbounded terminal conditions, we rely on Lemma 8.5.2 which was given in Fan-Wang-Yong [71]. Theorem 8.5.1, the proof of which is adapted from [37], shows the existence and uniqueness of adapted solution for BSDE when the terminal condition is allowed to be unbounded, and the generator has somewhat convex or concave type condition on the solution. For the sake of completeness, we also present some results on super-quadratic BSDEs, the proofs of which are adapted from [55].

For the well-posedness of Type-I BSVIEs, the basic induction ideas are inspired by the four step arguments in [220] and are different from the corresponding version in [71]. By means of this method, some subtle and delicate estimates are needed, the proof of which are adapted from that in Fan-Hu-Tang [70]. Eventually, we point out some other works on quadratic BSVIEs. In Wang–Sun–Yong [192], the one-dimensional well-posedness was established with bounded free terms. As to the multidimensional case, it was studied in Hernández [99] with essentially small bounded free terms. In the current chapter, these limitations on dimension and free terms can be relaxed in proper manners.

Chapter 9
Mean-Field BSVIEs

In this chapter, we discuss the following BSVIE:

$$\begin{aligned} Y(t) = \psi(t) + \int_t^T g\Big(t,s,Y(s),Z(t,s),Z(s,t),\Gamma\big(t,s,Y(s),Z(t,s),Z(s,t)\big)\Big)ds \\ - \int_t^T Z(t,s)dW(s), \quad t \in [0,T]. \end{aligned} \tag{9.1}$$

Here, for some map $\theta(\cdot)$ (see the following sections for precise meaning of the above),

$$\Gamma(t,s,Y,Z,\widehat{Z}) \stackrel{\Delta}{=} \mathbb{E}\Big[\theta(t,s,y,z,\hat{z},Y,Z,\widehat{Z})\Big]_{(y,z,\hat{z})=(Y,Z,\widehat{Z})} \tag{9.2}$$

with $(Y, Z, \widehat{Z})$ being any suitable random variables. We call (9.1) a *mean-field backward stochastic Volterra integral equation* of Type-II (Type-II MF-BSVIE, for short). Similar to the previous chapters, we call g the generator and $\psi(\cdot)$ the free term. In the case that $Z(s,t)$ is absent, the equation is called a Type-I MF-BSVIE. In this chapter, we will extend some results of Chaps. 4 and 5 to the above equations.

For $H \stackrel{\Delta}{=} \mathbb{R}^n$, $p > 1$, and any σ-field $\mathcal{G} \subseteq \mathcal{F}$, let

$$\begin{aligned} L^p_{\mathcal{G}}(\Omega; H) &\stackrel{\Delta}{=} \Big\{\xi : \Omega \to H \;\Big|\; \xi \text{ is} \mathcal{G}\text{-measurable}, \mathbb{E}|\xi|^p < \infty\Big\}, \\ L^p_{\mathcal{G}}(0,T; H) &\stackrel{\Delta}{=} \Big\{X : [0,T]\times\Omega \to H \;\Big|\; \forall s \in [0,T], X(s) \text{ is} \mathcal{G}\text{-measurable}, \\ &\qquad \mathbb{E}\int_0^T |X(s)|^p ds < \infty\Big\}. \end{aligned}$$

T. Wang, J. Yong, *Backward Stochastic Volterra Integral Equations*,
Probability Theory and Stochastic Modelling 109,
https://doi.org/10.1007/978-3-032-20848-4_9

Next, let $(\Omega^2, \mathcal{F}^2, \mathbb{P}^2) \triangleq (\Omega \times \Omega, \mathcal{F} \otimes \mathcal{F}, \mathbb{P} \otimes \mathbb{P})$ be the completion of the product probability space of the original $(\Omega, \mathcal{F}, \mathbb{P})$ with itself, where we define the filtration as $\mathbb{F}^2 \triangleq \{\mathcal{F}_t \otimes \mathcal{F}_t,\ t \in [0, T]\}$ with $\mathcal{F}_t \otimes \mathcal{F}_t$ being the completion of $\mathcal{F}_t \times \mathcal{F}_t$. It is worthy of noting that any random variable $\xi = \xi(\omega)$ defined on Ω can be extended naturally to Ω^2 as $\xi'(\omega, \omega') \equiv \xi(\omega)$, for $(\omega, \omega') \in \Omega^2$. Throughout this chapter, all the equalities and inequalities, involving random variables, are in the sense of $\mathbb{P}(d\omega) \times \mathbb{P}(d\omega')$ almost surely on Ω^2 or in the sense of $\mathbb{P}(d\omega)$ almost surely on Ω. Similar to the above, for any σ-field $\mathcal{G} \subseteq \mathcal{F}^2$, we define

$$L^1_{\mathcal{G}}(\Omega^2; H) \triangleq \Big\{\xi : \Omega^2 \to H \;\Big|\; \xi \text{ is} \mathcal{G}\text{-measurable}, \\ \mathbb{E}^2|\xi| \equiv \int_{\Omega^2} |\xi(\omega, \omega')|\mathbb{P}(d\omega')\mathbb{P}(d\omega) < \infty\Big\}.$$

For any $\eta \in L^1_{\mathcal{F}^2}(\Omega^2; H)$, we denote

$$\mathbb{E}'\eta(\omega, \cdot) \triangleq \int_{\Omega} \eta(\omega, \omega')\mathbb{P}(d\omega') \in L^1_{\mathcal{F}}(\Omega; H).$$

Note that if $\eta(\omega, \omega') \equiv \eta(\omega')$, then

$$\mathbb{E}'\eta = \int_{\Omega} \eta(\omega')\mathbb{P}(d\omega') = \int_{\Omega} \eta(\omega)\mathbb{P}(d\omega) = \mathbb{E}\eta.$$

In what follows, $\mathbb{E}'$ will be used when we need to distinguish ω' from ω, which is the case when both ω and ω' appear at the same time. Also for notational simplicity, in this chapter we suppose the Brownian motion is one-dimensional.

1 Mean-Field FSVIEs

Let us begin with the following SDE in $\mathbb{R}$

$$\begin{cases} dX(t) = \mathbb{E}\big[b(\xi, X(t))\big]_{\xi = X(t)} dt + dW(t), & t \in [0, T], \\ X(0) = x, \end{cases} \tag{9.1.1}$$

where $b : \mathbb{R} \times \mathbb{R} \to \mathbb{R}$ is a suitable function. In the above,

$$\mathbb{E}\big[b(\xi, X(t))\big]_{\xi = X(t)} = \int_{\Omega} b(X(t, \omega), X(t, \omega'))\mathbb{P}(d\omega'). \tag{9.1.2}$$

Equation (9.1.1) is called a *McKean-Vlasov* SDE.

Inspired by (9.1.1), one can consider the following more general SDE:

$$\begin{cases} dX(t)=b\big(t, X(t), \mathbb{E}[\theta^b(t, \xi, X(t))]_{\xi=X(t)}\big)dt \\ \qquad\qquad +\sigma\big(t, X(t), \mathbb{E}[\theta^\sigma(t, \xi, X(t))]_{\xi=X(t)}\big)dW(t), \quad t \in [0, T], \\ X(0) = x, \end{cases} \tag{9.1.3}$$

where θ^b and θ^σ are some suitable maps. We call the above a *mean-field* (forward) SDE (MF-FSDE, for short). From (9.1.3), we see that (9.1.1) is a special case of (9.1.3). Note also that (9.1.3) is an extension of classical Itô type SDEs. Due to the dependence of b and σ on $\mathbb{E}[\theta^b(t, \xi, X(t))]_{\xi=X(t)}$ and $\mathbb{E}[\theta^\sigma(t, \xi, X(t))]_{\xi=X(t)}$, respectively, MF-FSDE (9.1.3) is *nonlocal* with respect to the event $\omega \in \Omega$. The equivalent integral form of (9.1.3) is as follows:

$$\begin{aligned} X(t) = x + &\int_0^t b\big(s, X(s), \mathbb{E}[\theta^b(s, \xi, X(s))]_{\xi=X(s)}\big)ds \\ &+ \int_0^t \sigma\big(s, X(s), \mathbb{E}[\theta^\sigma(s, \xi, X(s))]_{\xi=X(s)}\big)dW(s), \quad t \in [0, T]. \end{aligned} \tag{9.1.4}$$

Inspired by the relation between SDEs and SVIEs, a natural extension of the above is the following:

$$\begin{aligned} X(t) = \varphi(t) + &\int_0^t b\big(t, s, X(s), \mathbb{E}[\theta^b(t, s, \xi, X(s))]_{\xi=X(s)}\big)ds \\ &+ \int_0^t \sigma\big(t, s, X(s), \mathbb{E}[\theta^\sigma(t, s, \xi, X(s))]_{\xi=X(s)}\big)dW(s), \quad t \in [0, T]. \end{aligned} \tag{9.1.5}$$

We call the above a mean-field (forward) stochastic Volterra integral equation (MF-FSVIE, for short). It is worthy of pointing out that when the drift b and diffusion σ in (9.1.5) are independent of the nonlocal terms $\mathbb{E}[\theta^b(t, s, \xi, X(s))]_{\xi=X(s)}$ and $\mathbb{E}[\theta^\sigma(t, s, \xi, X(s))]_{\xi=X(s)}$, respectively, (9.1.5) is reduced to an FSVIE.

We now further introduce the following equation:

$$\begin{aligned} X(t) = \varphi(t) + &\int_0^t b\big(t, s, X(s), \Gamma^b(t, s, X(s))\big)ds \\ &+ \int_0^t \sigma\big(t, s, X(s), \Gamma^\sigma(t, s, X(s))\big)dW(s), \quad t \in [0, T], \end{aligned} \tag{9.1.6}$$

where

$$\begin{cases} \Gamma^b(t, s, X) = \mathbb{E}'\big[\theta^b(t, s, \xi, X)\big]_{\xi=X} \\ \qquad\qquad \equiv \int_\Omega \theta^b(t, s, \omega, \omega', X(\omega), X(\omega'))\mathbb{P}(d\omega'), \\ \Gamma^\sigma(t, s, X) = \mathbb{E}'\big[\theta^\sigma(t, s, \xi, X)\big]_{\xi=X} \\ \qquad\qquad \equiv \int_\Omega \theta^\sigma(t, s, \omega, \omega', X(\omega), X(\omega'))\mathbb{P}(d\omega'). \end{cases} \tag{9.1.7}$$

Note that θ^b and θ^σ are allowed to depend on both ω and ω' and to have different dimensions, which makes (9.1.6) a slightly more general than (9.1.5).

An $\mathbb{F}$-adapted process $X(\cdot)$ is called a solution to (9.1.6) if (9.1.6) is satisfied in the usual Itô sense. To guarantee the well-posedness of (9.1.6), let us introduce the following hypotheses on the maps (b, σ) and $(\theta^b, \theta^\sigma)$, respectively.

(H9.1.1) Given $m_1,\ m_2 \in \mathbb{N}$, the maps $b : \Delta_*[0,T] \times \Omega \times \mathbb{R}^n \times \mathbb{R}^{m_1} \to \mathbb{R}^n$ and $\sigma : \Delta_*[0,T] \times \Omega \times \mathbb{R}^n \times \mathbb{R}^{m_2} \to \mathbb{R}^n$ are measurable, and for all $(t, x, \gamma, \gamma') \in [0,T] \times \mathbb{R}^n \times \mathbb{R}^{m_1} \times \mathbb{R}^{m_2}$, the map

$$(s, \omega) \mapsto (b(t, s, \omega, x, \gamma), \sigma(t, s, \omega, x, \gamma'))$$

is $\mathbb{F}$-progressively measurable on $[0,t]$. Moreover, there exists some constant $L > 0$ such that

$$\begin{aligned}&|b(t,s,\omega,x_1,\gamma_1) - b(t,s,\omega,x_2,\gamma_2)|\\&\quad +|\sigma(t,s,\omega,x_1,\gamma_1') - \sigma(t,s,\omega,x_2,\gamma_2')|\\&\leqslant L\big(|x_1-x_2| + |\gamma_1-\gamma_2| + |\gamma_1'-\gamma_2'|\big),\\&\qquad (t,s,\omega) \in \Delta_*[0,T]\times\Omega,\ (x_i,\gamma_i,\gamma_i') \in \mathbb{R}^n\times\mathbb{R}^{m_1}\times\mathbb{R}^{m_2},\ i=1,2.\end{aligned} \tag{9.1.8}$$

Moreover,

$$\begin{aligned}&|b(t,s,\omega,x,\gamma)| + |\sigma(t,s,\omega,x,\gamma')| \leqslant L\big(1+|x|+|\gamma|+|\gamma'|\big),\\&\qquad (t,s,\omega,x,\gamma,\gamma') \in \Delta_*[0,T]\times\Omega\times\mathbb{R}^n\times\mathbb{R}^{m_1}\times\mathbb{R}^{m_2}.\end{aligned} \tag{9.1.9}$$

(H9.1.2) The maps $\theta^b : \Delta_*[0,T] \times \Omega^2 \times \mathbb{R}^{2n} \to \mathbb{R}^{m_1}$ and $\theta^\sigma : \Delta_*[0,T] \times \Omega^2 \times \mathbb{R}^{2n} \to \mathbb{R}^{m_2}$ are measurable, and for all $(t, x, x') \in [0,T] \times \mathbb{R}^{2n}$, the map

$$(s, \omega, \omega') \mapsto (\theta^b(t, s, \omega, \omega', x, x'), \theta^\sigma(t, s, \omega, \omega', x, x'))$$

is $\mathbb{F}^2$-progressively measurable on $[0,t]$. Moreover, there exists some constant $L > 0$ such that

$$\begin{aligned}&|\theta^b(t,s,\omega,\omega',x_1,x_1') - \theta^b(t,s,\omega,\omega',x_2,x_2')|\\&\quad +|\theta^\sigma(t,s,\omega,\omega',x_1,x_1') - \theta^\sigma(t,s,\omega,\omega',x_2,x_2')|\\&\leqslant L\big(|x_1-x_2| + |x_1'-x_2'|\big),\\&\qquad (t,s,\omega,\omega') \in \Delta_*[0,T]\times\Omega^2,\ \ (x_i,x_i') \in \mathbb{R}^{2n},\quad i=1,2,\end{aligned} \tag{9.1.10}$$

and

$$\begin{aligned}&|\theta^b(t,s,\omega,\omega',x,x')| + |\theta^\sigma(t,s,\omega,\omega',x,x')| \leqslant L\big(1+|x|+|x'|\big),\\&\qquad (t,s,\omega,\omega') \in \Delta_*[0,T]\times\Omega^2,\quad x,x' \in \mathbb{R}^n.\end{aligned} \tag{9.1.11}$$

Notice that in (H9.1.1) and (H9.1.2), we did not assume, e.g., $b(t,s,0,0)=0$. This is different from the analogue case in Chap. 4. Of course, this does not cause essential change. We will also need the following assumptions.

(H9.1.1)′ In addition to (H9.1.1), there exists some modulus of continuity $\rho:[0,\infty)\to[0,\infty)$ such that

$$\begin{aligned}&\big|b(t_1,s,x,\gamma)-b(t_2,s,x,\gamma)\big|+\big|\sigma(t_1,s,x,\gamma')-\sigma(t_2,s,x,\gamma')\big|\\&\leqslant\rho(|t_1-t_2|)L\big(1+|x|+|\gamma|+|\gamma'|\big),\\&\qquad\forall\,(t_1,t_2)\in[s,T],\ x\in\mathbb{R}^n,\ \gamma\in\mathbb{R}^{m_1},\ \gamma'\in\mathbb{R}^{m_2}.\ \text{a.s.}\end{aligned}$$

(H9.1.2)′ In addition to (H9.1.2), the map $t\mapsto(\theta^b(t,s,\omega,\omega',x,x'),\theta^\sigma(t,s,\omega,\omega',x,x'))$ is continuous on $[s,T]$, and there exists some modulus of continuity $\rho:[0,\infty)\to[0,\infty)$ such that

$$\begin{aligned}&\big|\theta^b(t_1,s,x,x')-\theta^b(t_2,s,x,x')\big|+\big|\theta^\sigma(t_1,s,x,x')-\theta^\sigma(t_2,s,x,x')\big|\\&\leqslant\rho(|t_1-t_2|)L\big(1+|x|+|x'|\big),\\&\qquad\qquad\forall\,(t_1,t_2)\in[s,T],\ x,\ x'\in\mathbb{R}^n.\ \text{a.s.}\end{aligned}$$

The following result is concerned on MF-FSVIE (9.1.6).

Theorem 9.1.1 *Let* (H9.1.1)–(H9.1.2) *hold. Then for any* $p\geqslant2$, *and* $\varphi(\cdot)\in L^p_{\mathbb{F}}(0,T;\mathbb{R}^n)$, *MF-FSVIE (9.1.6) admits a unique solution* $X(\cdot)\in L^p_{\mathbb{F}}(0,T;\mathbb{R}^n)$, *and the following estimate holds:*

$$\mathbb{E}\int_0^T|X(t)|^pdt\leqslant K\Big(1+\mathbb{E}\int_0^T|\varphi(t)|^pdt\Big).\tag{9.1.12}$$

Further, for $i=1,2$, *let* $X_i(\cdot)\in L^p_{\mathbb{F}}(0,T;\mathbb{R}^n)$ *be the solutions of (9.1.6) corresponding to* $\varphi_i(\cdot)\in L^p_{\mathbb{F}}(0,T;\mathbb{R}^n)$ *and* $b_i(\cdot),\sigma_i(\cdot),\theta_i^b(\cdot),\theta_i^\sigma(\cdot)$ *satisfying* (H9.1.1)–(H9.1.2). *For* $i=1,2$, *let*

$$\begin{cases}\Gamma_i^b(t,s,X)=\mathbb{E}'\Big[\theta_i^b(t,s,\xi,X)\Big]_{\xi=X}\equiv\displaystyle\int_\Omega\theta_i^b\big(t,s,\omega,\omega',X(\omega),X(\omega')\big)\mathbb{P}(d\omega'),\\ \Gamma_i^\sigma(t,s,X)=\mathbb{E}'\Big[\theta_i^\sigma(t,s,\xi,X)\Big]_{\xi=X}\equiv\displaystyle\int_\Omega\theta_i^\sigma\big(t,s,\omega,\omega',X(\omega),X(\omega')\big)\mathbb{P}(d\omega').\end{cases}$$

Then the following stability estimate holds:

$$\begin{aligned}\mathbb{E}\int_0^T|X_1(t)-X_2(t)|^pdt&\leqslant K\Big\{\mathbb{E}\int_0^T|\varphi_1(t)-\varphi_2(t)|^pdt\\&+\mathbb{E}\int_0^T\Big(\int_0^t\Big|b_1\big(t,s,X_1(s),\Gamma_1^b(t,s,X_1(s))\big)\\&\qquad-b_2\big(t,s,X_1(s),\Gamma_2^b(t,s,X_1(s))\big)\Big|ds\Big)^pdt\\&+\mathbb{E}\int_0^T\Big(\int_0^t\Big|\sigma_1\big(t,s,X_1(s),\Gamma_1^\sigma(t,s,X_1(s))\big)\\&\qquad-\sigma_2\big(t,s,X_1(s),\Gamma_2^\sigma(t,s,X_1(s))\big)\Big|^2ds\Big)^{\frac p2}dt\Big\}.\end{aligned}\tag{9.1.13}$$

Moreover, let (H9.1.1)$'$–(H9.1.2)$'$ *hold. Then for any* $p \geqslant 2$, *and any* $\varphi(\cdot) \in C_{\mathbb{F}}([0,T]; L^p(\Omega;\mathbb{R}^n))$, *the unique solution* $X(\cdot) \in C_{\mathbb{F}}([0,T]; L^p(\Omega;\mathbb{R}^n))$, *and estimate (9.1.12) can be improved to the following:*

$$\sup_{t\in[0,T]} \mathbb{E}|X(t)|^p \leqslant K\Big\{1 + \sup_{t\in[0,T]} \mathbb{E}|\varphi(t)|^p\Big\}. \tag{9.1.14}$$

Also, for $i = 1,2$, *let* $X_i(\cdot) \in C_{\mathbb{F}}(0,T; L^p(\Omega;\mathbb{R}^n))$ *be the solutions of (9.1.6) corresponding to* $\varphi_i(\cdot) \in C_{\mathbb{F}}(0,T; L^p(\Omega;\mathbb{R}^n))$ *and* $b_i(\cdot), \sigma_i(\cdot), \theta_i^b(\cdot), \theta_i^\sigma(\cdot)$ *satisfying* (H9.1.1)$'$–(H9.1.2)$'$. *Then the stability estimate (9.1.13) can be improved to the following:*

$$\begin{aligned}\sup_{t\in[0,T]} \mathbb{E}|X_1(t) - X_2(t)|^p &\leqslant K\Big\{ \sup_{t\in[0,T]} \mathbb{E}|\varphi_1(t) - \varphi_2(t)|^p \\ &+ \sup_{t\in[0,T]} \mathbb{E}\Big(\int_0^t \Big|b_1\big(t,s,X_1(s),\Gamma_1^b(t,s,X_1(s))\big) \\ &\qquad - b_2\big(t,s,X_1(s),\Gamma_2^b(t,s,X_1(s))\big)\Big|ds\Big)^p \\ &+ \sup_{t\in[0,T]} \mathbb{E}\Big(\int_0^t \Big|\sigma_1\big(t,s,X_1(s),\Gamma_1^\sigma(t,s,X_1(s))\big) \\ &\qquad - \sigma_2\big(t,s,X_1(s),\Gamma_2^\sigma(t,s,X_1(s))\big)\Big|^2 ds\Big)^{\frac{p}{2}}\Big\}.\end{aligned} \tag{9.1.15}$$

Proof By (H9.1.2), making use of (9.1.11), for any $X(\cdot) \in L^p_{\mathbb{F}}(0,T;\mathbb{R}^n)$, we have

$$\big|\Gamma^b(t,s,X(s))\big| + \big|\Gamma^\sigma(t,s,X(s))\big| \leqslant L\big(1 + \mathbb{E}|X(s)| + |X(s)|\big).$$

Thus, if $X(\cdot) \in L^p_{\mathbb{F}}(0,T;\mathbb{R}^n)$ is a solution to (9.1.6) with $\varphi(\cdot) \in L^p_{\mathbb{F}}(0,T;\mathbb{R}^n)$, then by (9.1.9),

$$\begin{aligned}\mathbb{E}|X(t)|^p &\leqslant 3^{p-1}\mathbb{E}\Big\{|\varphi(t)|^p + \Big|\int_0^t b\big(t,s,X(s),\Gamma^b(t,s,X(s))\big)ds\Big|^p \\ &\qquad + \Big|\int_0^t \sigma\big(t,s,X(s),\Gamma^\sigma(t,s,X(s))\big)dW(s)\Big|^p\Big\} \\ &\leqslant 3^{p-1}\Big\{\mathbb{E}|\varphi(t)|^p + \mathbb{E}\Big(\int_0^t L[1 + |X(s)| + |\Gamma^b(t,s,X(s))|]ds\Big)^p \\ &\qquad + \mathbb{E}\Big(\int_0^t L^2[1 + |X(s)| + |\Gamma^\sigma(t,s,X(s))|]^2 ds\Big)^{\frac{p}{2}}\Big\} \\ &\leqslant K\Big\{1 + \mathbb{E}|\varphi(t)|^p + \mathbb{E}\int_0^t |X(s)|^p ds\Big\}.\end{aligned} \tag{9.1.16}$$

Consequently,

$$\int_0^t \mathbb{E}|X(r)|^p dr \leqslant K\Big\{1+\int_0^t \mathbb{E}|\varphi(r)|^p dr + \mathbb{E}\int_0^t \Big[\int_0^r |X(s)|^p ds\Big]dr\Big\},\qquad 0\leqslant t\leqslant T.$$

Using Gronwall's inequality, we obtain (9.1.12).

Now, let $\delta>0$ be undetermined. For any $x(\cdot)\in L^p_{\mathbb{F}}(0,\delta;\mathbb{R}^n)$, define

$$\begin{aligned}\mathcal{G}(x(\cdot))(t) \overset{\Delta}{=} \varphi(t) &+ \int_0^t b\big(t,s,x(s),\Gamma^b(t,s,x(s))\big)ds\\ &+\int_0^t \sigma\big(t,x,x(s),\Gamma^\sigma(t,s,x(s))\big)dW(s),\qquad t\in[0,\delta].\end{aligned}$$

Then we have

$$\begin{aligned}&\mathbb{E}\int_0^\delta \big|\mathcal{G}(x(\cdot))(t)\big|^p dt\\ &\leqslant K\mathbb{E}\Big\{\int_0^\delta |\varphi(t)|^p dt + \int_0^\delta\Big[\int_0^t \Big(1+|x(s)|+|\Gamma^b(t,s,x(s))|\Big)ds\Big]^p dt\\ &\qquad\qquad + \int_0^\delta\Big|\int_0^t \sigma\big(t,s,x(s),\Gamma^\sigma(t,s,x(s))\big)dW(s)\Big|^p dt\Big\}\\ &\leqslant K\Big\{\mathbb{E}\int_0^\delta |\varphi(t)|^p dt + \mathbb{E}\int_0^\delta |x(t)|^p dt\Big\}.\end{aligned}$$

Thus, $\mathcal{G}\;:\;L^p_{\mathbb{F}}(0,\delta;\mathbb{R}^n)\;\to\;L^p_{\mathbb{F}}(0,\delta;\mathbb{R}^n)$. Next, for any $x_1(\cdot),x_2(\cdot)\in L^p_{\mathbb{F}}(0,\delta;\mathbb{R}^n)$, we have (making use of (9.1.8) and (9.1.10))

$$\begin{aligned}&\mathbb{E}\int_0^\delta \big|\mathcal{G}(x_1(\cdot))(t)-\mathcal{G}(x_2(\cdot))(t)\big|^p dt\\ &\leqslant 2^{p-1}\Big\{\mathbb{E}\int_0^\delta\Big[\int_0^t L\Big(|x_1(s)-x_2(s)|\\ &\qquad\qquad +\big|\Gamma^b(t,s,x_1(s))-\Gamma^b(t,s,x_2(s))\big|\Big)ds\Big]^p dt\\ &\quad+\int_0^\delta \mathbb{E}\Big[\int_0^t L^2\Big(|x_1(s)-x_2(s)|^2\\ &\qquad\qquad +\big|\Gamma^\sigma(t,s,x_1(s))-\Gamma^\sigma(t,s,x_2(s))\big|^2\Big)ds\Big]^{\frac{p}{2}}dt\Big\}\\ &\leqslant K_0\delta\mathbb{E}\int_0^\delta |x_1(t)-x_2(t)|^p dt,\end{aligned}$$

with $K_0 > 0$ being an absolute constant (only depending on L and p). Then letting $\delta \overset{\Delta}{=} \frac{1}{2K_0}$, we see that $\mathcal{G} : L^p_{\mathbb{F}}(0, \delta; \mathbb{R}^n) \to L^p_{\mathbb{F}}(0, \delta; \mathbb{R}^n)$ is a contraction. Hence, MF-FSVIE (9.1.6) admits a unique solution $X(\cdot) \in L^p_{\mathbb{F}}(0, \delta; \mathbb{R}^n)$.

Next, for $t \in [\delta, 2\delta]$, we write (9.1.5) as

$$\begin{aligned} X(t) = \widehat{\varphi}(t) &+ \int_\delta^t b\big(t, s, X(s), \Gamma^b(t, s, X(s))\big)ds \\ &+ \int_\delta^t \sigma\big(t, s, X(s), \Gamma^\sigma(t, s, X(s))\big)dW(s), \end{aligned} \tag{9.1.17}$$

with

$$\begin{aligned} \widehat{\varphi}(t) \overset{\Delta}{=} \varphi(t) &+ \int_0^\delta b\big(t, s, X(s), \Gamma^b(t, s, X(s))\big)ds \\ &+ \int_0^\delta \sigma\big(t, s, X(s), \Gamma^\sigma(t, s, X(s))\big)dW(s). \end{aligned}$$

Then a similar argument as above applies to obtain a unique solution of (9.1.17) on $[\delta, 2\delta]$. It is important to note that the step-length $\delta > 0$ is uniform. Hence, by induction, we obtain the unique solvability of (9.1.6) on $[0, T]$.

Now, for $i = 1, 2$, let $X_i(\cdot) \in L^p_{\mathbb{F}}(0, T; \mathbb{R}^n)$ be the solutions of (9.1.6) corresponding to $\varphi_i(\cdot) \in L^p_{\mathbb{F}}(0, T; \mathbb{R}^n)$ and $b_i(\cdot), \sigma_i(\cdot), \theta_i^b(\cdot), \theta_i^\sigma(\cdot)$ (satisfying (H9.1.1)–(H9.1.2)). Then

$$\begin{aligned} \mathbb{E}|X_1(t) - X_2(t)|^p \leqslant 3^{p-1}\mathbb{E}\Big\{&|\varphi_1(t) - \varphi_2(t)|^p \\ &+\Big(\int_0^t \Big|b_1\big(t, s, X_1(s), \Gamma_1^b(t, s, X_1(s))\big) \\ &\qquad -b_2\big(t, s, X_2(s), \Gamma_2^b(t, s, X_2(s))\big)\Big|ds\Big)^p \\ &+\Big|\int_0^t \Big(\sigma_1\big(t, s, X_1(s), \Gamma_1^\sigma(t, s, X_1(s))\big) \\ &\qquad -\sigma_2\big(t, s, X_2(s), \Gamma_2^\sigma(t, s, X_2(s))\big)\Big)dW(s)\Big|^p\Big\} \\ \leqslant K\Big\{&\mathbb{E}|\varphi_1(t) - \varphi_2(t)|^p + \mathbb{E}\int_0^t |X_1(s) - X_2(s)|^p ds \\ &+\mathbb{E}\Big(\int_0^t \Big|b_1\big(t, s, X_1(s), \Gamma_1^b(t, s, X_1(s))\big) \\ &\qquad -b_2\big(t, s, X_1(s), \Gamma_2^b(t, s, X_1(s))\big)\Big|ds\Big)^p \\ &+\mathbb{E}\Big|\int_0^t \Big(\sigma_1\big(t, s, X_1(s), \Gamma_1^\sigma(t, s, X_1(s))\big) \\ &\qquad -\sigma_2\big(t, s, X_1(s), \Gamma_2^\sigma(t, s, X_1(s))\big)\Big)dW(s)\Big|^p\Big\}. \end{aligned}$$

Then we can obtain estimate (9.1.13).

Similar to Proposition 3.4.4 in Chap. 3, the conclusions under (H9.1.1)′–(H9.1.2)′ are easy to obtain. □

To conclude this section, let us look at linear MF-FSVIEs, by which we mean the following:

$$\begin{aligned} X(t) = \varphi(t) &+ \int_0^t \Big(A_0(t,s)X(s) + \mathbb{E}'\big[C_0(t,s)X(s)\big]\Big)ds \\ &+ \int_0^t \Big(A_1(t,s)X(s) + \mathbb{E}'\big[C_1(t,s)X(s)\big]\Big)dW(s), \quad t \in [0,T]. \end{aligned} \tag{9.1.18}$$

For such an equation, we introduce the following hypotheses.

(H9.1.3) The following maps are measurable and uniformly bounded

$$A_0, A_1 : \Delta_*[0,T] \times \Omega \to \mathbb{R}^{n\times n}, \qquad C_0, C_1 : \Delta_*[0,T] \times \Omega^2 \to \mathbb{R}^{n\times n}.$$

For any $t \in [0,T]$, $s \mapsto (A_0(t,s), A_1(t,s))$ is $\mathbb{F}$-progressively measurable on $[0,t]$, and $s \mapsto (C_0(t,s), C_1(t,s))$ is $\mathbb{F}^2$-progressively measurable on $[0,t]$.

(H9.1.3)′ In addition to (H9.1.3), the maps

$$t \mapsto (A_0(t,s,\omega), A_1(t,s,\omega), C_0(t,s,\omega,\omega'), C_1(t,s,\omega,\omega'))$$

are continuous with proper modulus.

Clearly, by defining

$$\begin{cases} b(t,s,\omega,x,\gamma) \overset{\Delta}{=} A_0(t,s,\omega)x + \gamma, & \theta^b(t,s,\omega,\omega',x,x') \overset{\Delta}{=} C_0(t,s,\omega,\omega')x', \\ \sigma(t,s,\omega,x,\gamma') \overset{\Delta}{=} A_1(t,s,\omega)x + \gamma', & \theta^\sigma(t,s,\omega,\omega',x,x') \overset{\Delta}{=} C_1(t,s,\omega,\omega')x', \end{cases}$$

we see that (9.1.18) is a special case of (9.1.6). Moreover, (H9.1.3) implies (H9.1.1)–(H9.1.2), and (H9.1.3)′ implies (H9.1.1)′–(H9.1.2)′.

We point out that linear MF-FSVIE (9.1.18) is general enough in some sense. To see this, let us formally look at the variational equation of (9.1.6) which is quite important in discussing the optimal control theories. More precisely, let $X^\delta(\cdot)$ be the unique solution of (9.1.6) with $\varphi(\cdot)$ replaced by $\varphi(\cdot) + \delta\bar{\varphi}(\cdot)$. We formally let

$$\bar{X}(t) \overset{\Delta}{=} \lim_{\delta\to 0} \frac{X^\delta(t) - X(t)}{\delta}, \quad t \in [0,T].$$

Then $\bar{X}(\cdot)$ should satisfy the following linear MF-FSVIE:

$$
\begin{aligned}
\bar{X}(t) = \bar{\varphi}(t) + \int_0^t \Big(& b_x(t,s)\bar{X}(s) \\
& + b_y(t,s)\mathbb{E}'\big[\theta_x^b(t,s)\bar{X}(s,\omega) + \theta_{x'}^b(t,s)\bar{X}(s,\omega')\big]\Big) ds \\
+ \int_0^t \Big(& \sigma_x(t,s)\bar{X}(s) \\
& + \sigma_{y'}(t,s)\mathbb{E}'\big[\theta_x^\sigma(t,s)\bar{X}(s,\omega) + \theta_{x'}^\sigma(t,s)\bar{X}(s,\omega')\big]\Big) dW(s),
\end{aligned}
\tag{9.1.19}
$$

where

$$
\begin{cases}
b_x(t,s) \overset{\Delta}{=} b_x\big(t,s,\omega,\Gamma^b(t,s,X(s,\omega))\big), \\
b_y(t,s) \overset{\Delta}{=} b_y\big(t,s,\omega,\Gamma^b(t,s,X(s,\omega))\big), \\
\theta_x^b(t,s) \overset{\Delta}{=} \theta_x^b\big(t,s,\omega,\omega',X(s,\omega),X(s,\omega')\big), \\
\theta_{x'}^b(t,s) \overset{\Delta}{=} \theta_{x'}^b\big(t,s,\omega,\omega',X(s,\omega),X(s,\omega')\big),
\end{cases}
\tag{9.1.20}
$$

and similar rule also holds for $\sigma_x(t,s)$, $\sigma_{y'}(t,s)$, $\theta_x^\sigma(t,s)$, and $\theta_{x'}^\sigma(t,s)$. It is interesting to note that (9.1.19) can be written as follows:

$$
\begin{aligned}
\bar{X}(t) = \bar{\varphi}(t) + \int_0^t \Big\{ & \Big(b_x(t,s) + b_y(t,s)\mathbb{E}'\big[\theta_x^b(t,s)\big]\Big)\bar{X}(s) \\
& + \mathbb{E}'\big[b_y(t,s)\theta_{x'}^b(t,s)\bar{X}(s)\big]\Big\} ds \\
+ \int_0^t \Big\{ & \Big(\sigma_x(t,s) + \sigma_y(t,s)\mathbb{E}'\big[\theta_x^\sigma(t,s)\big]\Big)\bar{X}(s,\omega) \\
& + \mathbb{E}'\big[\sigma_y(t,s)\theta_{x'}^\sigma(t,s)\bar{X}(s,\omega')\big]\Big\} dW(s),
\end{aligned}
\tag{9.1.21}
$$

which is a special case of (9.1.18).

2 Preliminary Results on MF-BSVIEs

In this section, we will make some preliminaries for the later study on mean-field BSVIEs (9.1).

Let $\theta : \Delta^*[0,T] \times \Omega^2 \times \mathbb{R}^{6n} \to \mathbb{R}^m$ be some suitable map (see below for precise conditions) and define

$$
\begin{aligned}
& \Gamma(t,s,Y,Z,\widehat{Z}) \equiv \mathbb{E}'\big[\theta(t,s,y,z,\widehat{z},Y,Z,\widehat{Z})\big]_{(y,z,\hat{z})=(Y,Z,\widehat{Z})} \\
& = \int_\Omega \theta\big(t,s,\omega,\omega',Y(\omega),Z(\omega),\widehat{Z}(\omega),Y(\omega'),Z(\omega'),\widehat{Z}(\omega')\big)\mathbb{P}(d\omega'),
\end{aligned}
\tag{9.2.1}
$$

for all reasonable random variables $(Y, Z, \widehat{Z})$. Hereafter, when we talk about MF-BSVIE (9.1), the mapping Γ is defined by (9.2.1). With such a mapping, we have

$$
\begin{aligned}
&\Gamma(t,s,Y(s),Z(t,s),Z(s,t)) \equiv \Gamma(t,s,\omega,Y(s,\omega),Z(t,s,\omega),Z(s,t,\omega)) \\
&= \int_\Omega \theta\big(t,s,\omega,\omega',Y(s,\omega),Z(t,s,\omega),Z(s,t,\omega), \\
&\qquad\qquad Y(s,\omega'),Z(t,s,\omega'),Z(s,t,\omega')\big)\mathbb{P}(d\omega').
\end{aligned}
$$

Clearly, the operator Γ is *nonlocal* in the sense that the value $\Gamma\big(t,s,\omega,Y(s,\omega),Z(t,s,\omega),Z(s,t,\omega)\big)$ of $\Gamma\big(t,s,Y(s),Z(t,s),Z(s,t)\big)$ at ω depends on the whole set

$$
\big\{(Y(s,\omega'),Z(t,s,\omega'),Z(s,t,\omega'))\ \big|\ \omega'\in\Omega\big\},
$$

not just on $(Y(s,\omega),Z(t,s,\omega),Z(s,t,\omega))$. To get some feeling about such an operator, let us look at a simple but nontrivial special case.

Example 9.2.1 Let

$$
\begin{aligned}
&\theta(t,s,\omega,\omega',y,z,\hat z,y',z',\hat z') \\
&\stackrel{\Delta}{=} \theta_0(t,s,\omega)+A_0(t,s,\omega)y+B_0(t,s,\omega)z+C_0(t,s,\omega)\hat z \\
&\qquad +A_1(t,s,\omega,\omega')y'+B_1(t,s,\omega,\omega')z'+C_1(t,s,\omega,\omega')\hat z'.
\end{aligned}
$$

We should carefully distinguish ω' and ω in the above. Then (suppressing ω)

$$
\begin{aligned}
&\Gamma\big(t,s,Y(s),Z(t,s),Z(s,t)\big) \\
&=\theta_0(t,s)+A_0(t,s)Y(s)+B_0(t,s)Z(t,s)+C_0(t,s)Z(s,t) \\
&\qquad +\mathbb{E}'\big[A_1(t,s)Y(s)\big]+\mathbb{E}'\big[B_1(t,s)Z(t,s)\big]+\mathbb{E}'\big[C_1(t,s)Z(s,t)\big],
\end{aligned}
$$

where, for example,

$$
\mathbb{E}'\big[B_1(t,s)Z(t,s)\big] \stackrel{\Delta}{=} \int_\Omega B_1(t,s,\omega,\omega')Z(t,s,\omega')\mathbb{P}(d\omega').
$$

For such a case, $(Y(\cdot),Z(\cdot\,,\cdot))\mapsto\Gamma\big(\cdot\,,\cdot\,,Y(\cdot),Z(\cdot\,,\cdot),Z(\cdot\,,\cdot)\big)$ is affine.

Having some feeling about the operator Γ from the above, let us look at some useful properties of the operator Γ in general. To this end, we make the following assumption.

(H9.2.1)$_q$ The map $\theta : \Delta^*[0,T] \times \Omega^2 \times \mathbb{R}^{6n} \to \mathbb{R}^m$ is measurable and for all $(t, y, z, \hat{z}, y', z', \hat{z}') \in [0,T] \times \mathbb{R}^{6n}$, the map $(s, \omega, \omega') \mapsto \theta(t, s, \omega, \omega', y, z, \hat{z}, y', z', \hat{z}')$ is $\mathbb{F}^2$-progressively measurable on $[t, T]$. Moreover, there exist constants $L > 0$ and $q \geqslant 2$ such that

$$\begin{aligned}&\big|\theta(t, s, \omega, \omega', y_1, z_1, \hat{z}_1, y_1', z_1', \hat{z}_1') - \theta(t, s, \omega, \omega', y_2, z_2, \hat{z}_2, y_2', z_2', \hat{z}_2')\big|\\&\leqslant L\Big(|y_1-y_2|+|z_1-z_2|+|\hat{z}_1-\hat{z}_2|+|y_1'-y_2'|+|z_1'-z_2'|+|\hat{z}_1'-\hat{z}_2'|\Big),\\&\qquad (t, s, \omega, \omega') \in \Delta^*[0, T] \times \Omega^2,\ (y_i, z_i, \hat{z}_i, y_i', z_i', \hat{z}_i') \in \mathbb{R}^{6n}, i = 1, 2,\end{aligned} \tag{9.2.2}$$

and

$$\begin{aligned}&\big|\theta(t, s, \omega, \omega', y, z, \hat{z}, y', z', \hat{z}')\big|\\&\leqslant L\Big(1 + |y| + |z| + |\hat{z}|^{\frac{2}{q}} + |y'| + |z'| + |\hat{z}'|^{\frac{2}{q}}\Big),\\&\qquad (t, s, \omega, \omega') \in \Delta^*[0, T] \times \Omega^2,\ (y, z, \hat{z}, y', z', \hat{z}') \in \mathbb{R}^{6n}.\end{aligned} \tag{9.2.3}$$

Remark 9.2.2 In the above, we may replace constant L by some function $L(\cdot,\cdot)$ with certain integrability in Chaps. 4 and 1. However, for the simplicity of presentation, we prefer to take a constant L. Also, we note that $(\hat{z}, \hat{z}') \mapsto \theta(t, s, \omega, \omega', y, z, \hat{z}, y', z', \hat{z}')$ is assumed to grow no more than $|\hat{z}|^{\frac{2}{q}} + |\hat{z}'|^{\frac{2}{q}}$. If $q = 2$, then the growth is linear, and if $q > 2$, the growth is sublinear. This condition is very subtle in showing that the solution $(Y(\cdot), Z(\cdot\,,\cdot))$ of MF-BSVIE belongs to $\mathcal{H}^q[0, T]$ (see Chap. 5, Sect. 1). We would like to mention that (H9.2.1)$_\infty$ is understood as that (9.2.3) is replaced by the following

$$\begin{aligned}&\big|\theta(t, s, \omega, \omega', y, z, \hat{z}, y', z', \hat{z}')\big| \leqslant L\big(1 + |y| + |z| + |y'| + |z'|\big),\\&\qquad (t, s, \omega, \omega') \in \Delta^*[0, T] \times \Omega^2,\ (y, z, \hat{z}, y', z', \hat{z}') \in \mathbb{R}^{6n}.\end{aligned} \tag{9.2.4}$$

Under (H9.2.1)$_q$, for any $(Y(\cdot), Z(\cdot\,,\cdot)) \in \mathcal{H}^q[0, T]$, we see that for each $t \in [0, T]$, the map

$$\begin{aligned}(s, \omega) &\mapsto \Gamma\big(t, s, \omega, Y(s), Z(t, s), Z(s, t)\big)\\&\equiv \int_\Omega \theta\big(t, s, \omega, \omega', Y(s, \omega), Z(t, s, \omega), Z(s, t, \omega),\\&\qquad\qquad Y(s, \omega'), Z(t, s, \omega'), Z(s, t, \omega')\big)\mathbb{P}(d\omega')\end{aligned}$$

is $\mathbb{F}$-progressively measurable on $[t, T]$. Also,

$$\begin{aligned}&\big|\theta(t, s, \omega, \omega', Y(s, \omega), Z(t, s, \omega), Z(s, t, \omega), y, z, \hat{z})\big|\\&\leqslant L\big(1 + |Y(s, \omega)| + |Z(t, s, \omega)| + |Z(s, t, \omega)|^{\frac{2}{q}} + |y| + |z| + |\hat{z}|^{\frac{2}{q}}\big).\end{aligned} \tag{9.2.5}$$

Consequently,

$$
\begin{aligned}
&\big|\Gamma\big(t,s,Y(s),Z(t,s),Z(s,t)\big)\big| \\
&\leqslant L\big(1+|Y(s)|+|Z(t,s)|+|Z(s,t)|^{\frac{2}{q}} \\
&\qquad +\mathbb{E}|Y(s)|+\mathbb{E}|Z(t,s)|+\mathbb{E}|Z(s,t)|^{\frac{2}{q}}\big).
\end{aligned}
\tag{9.2.6}
$$

Likewise, for any $(Y_1(\cdot), Z_1(\cdot\,,\cdot)), (Y_2(\cdot), Z_2(\cdot\,,\cdot)) \in \mathcal{H}^q[0,T]$, we have

$$
\begin{aligned}
&\big|\Gamma\big(t,s,Y_1(s),Z_1(t,s),Z_1(s,t)\big)-\Gamma\big(t,s,Y_2(s),Z_2(t,s),Z_2(s,t)\big)\big| \\
&\leqslant L\big(|Y_1(s)-Y_2(s)|+|Z_1(t,s)-Z_2(t,s)| \\
&\quad +|Z_1(s,t)-Z_2(s,t)|+\mathbb{E}|Y_1(s)-Y_2(s)| \\
&\quad +\mathbb{E}|Z_1(t,s)-Z_2(t,s)|+\mathbb{E}|Z_1(s,t)-Z_2(s,t)|\big).
\end{aligned}
\tag{9.2.7}
$$

The above two estimates will play an interesting role later.

Next let us give the main assumptions of $\theta(\cdot)$ and $g(\cdot)$ in MF-BSVIE (9.1).

(H9.2.2)$_q$ The map $\theta : \Delta^*[0,T]\times\Omega^2\times\mathbb{R}^{6n} \to \mathbb{R}^m$ satisfies (H9.2.1)$_q$. The map $g : \Delta^*[0,T]\times\Omega\times\mathbb{R}^{3n}\times\mathbb{R}^m \to \mathbb{R}^n$ is measurable, and for all $(t,y,z,\widehat{z},\gamma) \in [0,T]\times\mathbb{R}^{3n}\times\mathbb{R}^m$, the map $(s,\omega)\mapsto g(t,s,\omega,y,z,\widehat{z},\gamma)$ is $\mathbb{F}$-progressively measurable. Moreover, there exist constants $L>0$ and $q\geqslant 2$ such that

$$
\begin{aligned}
&\big|g(t,s,\omega,y_1,z_1,\widehat{z}_1,\gamma_1)-g(t,s,\omega,y_2,z_2,\widehat{z}_2,\gamma_2)\big| \\
&\leqslant L\big(|y_1-y_2|+|z_1-z_2|+|\widehat{z}_1-\widehat{z}_2|+|\gamma_1-\gamma_2|\big), \\
&\qquad (t,s,\omega)\in\Delta^*[0,T]\times\Omega,\ (y_i,z_i,\widehat{z}_i,\gamma_i)\in\mathbb{R}^{3n}\times\mathbb{R}^m,\ i=1,2,
\end{aligned}
\tag{9.2.8}
$$

and

$$
\begin{aligned}
&\big|g(t,s,\omega,y,z,\widehat{z},\gamma)\big| \leqslant L\big(1+|y|+|z|+|\widehat{z}|^{\frac{2}{q}}+|\gamma|\big), \\
&\qquad (t,s,\omega)\in\Delta^*[0,T]\times\Omega,\ (y,z,\widehat{z},\gamma)\in\mathbb{R}^{3n}\times\mathbb{R}^m.
\end{aligned}
\tag{9.2.9}
$$

Similar to (H9.2.1)$_\infty$ in the previous section, (H9.2.2)$_\infty$ is understood that (9.2.4) holds and (9.2.9) is replaced by the following:

$$
\begin{aligned}
&\big|g(t,s,\omega,y,z,\widehat{z},\gamma)\big| \leqslant L\big(1+|y|+|z|+|\gamma|\big), \\
&\qquad (t,s,\omega)\in\Delta^*[0,T]\times\Omega,\ (y,z,\widehat{z},\gamma)\in\mathbb{R}^{3n}\times\mathbb{R}^m.
\end{aligned}
\tag{9.2.10}
$$

We now introduce the following definition (which is comparable with Chap. 5, Definition 5.1.2).

Definition 9.2.3 A pair of $(Y(\cdot), Z(\cdot\,,\cdot)) \in \mathcal{H}^p[0,T]$ is called an *adapted M-solution* of MF-BSVIE (9.1) if (9.1) is satisfied in the Itô sense and the following holds:

$$
Y(t)=\mathbb{E}Y(t)+\int_0^t Z(t,s)dW(s), \qquad t\in[0,T].
\tag{9.2.11}
$$

It is clear that (9.2.11) implies

$$Y(t)=\mathbb{E}\big[Y(t)\mid\mathcal{F}_S\big]+\int_S^t Z(t,s)dW(s),\qquad 0\leqslant S\leqslant t\leqslant T. \tag{9.2.12}$$

Recall that in Chap. 5, we define $\mathcal{M}^p[0,T]$ as the set of all elements $(y(\cdot),z(\cdot\,,\cdot))\in\mathcal{H}^p[0,T]$ satisfying

$$y(t)=\mathbb{E}\big[y(t)\mid\mathcal{F}_S\big]+\int_S^t z(t,s)dW(s),\qquad t\in[S,T],\quad S\in[0,T). \tag{9.2.13}$$

Obviously, $\mathcal{M}^p[0,T]$ is a closed subspace of $\mathcal{H}^p[0,T]$.

In addition, for $p>1$, let us recall (5.1.7) in Chap. 5 as follows:

$$\begin{aligned}&\mathbb{E}\Big(\int_S^t|z(t,s)|^2ds\Big)^{\frac{p}{2}}\leqslant\mathbb{E}\Big[\big|\mathbb{E}_S[y(t)]\big|^p+\Big(\int_S^t|z(t,s)|^2ds\Big)^{\frac{p}{2}}\Big]\\&\leqslant K\mathbb{E}\Big|\mathbb{E}_S[y(t)]+\int_S^t z(t,s)dW(s)\Big|^p=K\mathbb{E}|y(t)|^p.\end{aligned} \tag{9.2.14}$$

As a result, like that in Chap. 5, we can use the following as an equivalent norm in $\mathcal{M}^p[0,T]$:

$$\|(y(\cdot),z(\cdot\,,\cdot))\|_{\mathcal{M}^p[0,T]}\equiv\mathbb{E}\Big\{\int_0^T e^{\beta t}|y(t)|^pdt+\int_0^T e^{\beta t}\Big(\int_t^T|z(t,s)|^2ds\Big)^{\frac{p}{2}}dt\Big\}^{\frac{1}{p}}.$$

Sometimes we use $\mathcal{M}^{p,\beta}[0,T]$ for $\mathcal{M}^p[0,T]$ to emphasize the involved parameter β.

3 Well-Posedness of MF-BSVIEs

In this section, we are going to establish the well-posedness of Type-II MF-BSVIEs.

3.1 A Special Case of Type-I MF-BSVIEs

In this subsection, we first consider the following special MF-BSVIE:

$$\begin{aligned}Y(t)=\psi(t)&+\int_t^T\widetilde{g}\big(t,s,Z(t,s),\widetilde{\Gamma}(t,s,Z(t,s))\big)ds\\&-\int_t^T Z(t,s)dW(s),\quad t\in[0,T],\end{aligned} \tag{9.3.1}$$

where for $(t,s)\in\Delta^*[0,T]$,

$$\begin{cases}\widetilde g(t,s,z,\gamma)\stackrel{\Delta}{=}g(t,s,y(s),z,z(s,t),\gamma),\quad z\in\mathbb{R}^n,\quad \gamma\in\mathbb{R}^m,\\ \widetilde\Gamma(t,s,Z(t,s))\stackrel{\Delta}{=}\Gamma\big(t,s,y(s),Z(t,s),z(s,t)\big)\\ \qquad\equiv\mathbb{E}'\Big[\theta\big(t,s,y',z',\widehat z',y(s),Z(t,s),z(s,t)\big)\Big]_{(y',z',\widehat z')=(y(s),Z(t,s),z(s,t))},\end{cases}$$

and some given $(y(\cdot),z(\cdot\,,\cdot))\in\mathcal{M}^p[0,T]$. Therefore,

$$\begin{aligned}&\widetilde g\big(t,s,Z(t,s),\widetilde\Gamma(t,s,Z(t,s))\big)\\&\quad=g\big(t,s,y(s),Z(t,s),z(s,t),\Gamma(t,s,y(s),Z(t,s),z(s,t))\big).\end{aligned}$$

Note that we may take much more general $\widetilde g(\cdot)$ and $\widetilde\Gamma(\cdot)$. However, the above setting is already sufficient for our purpose. In addition, by restricting such a case, we can avoid stating a lengthy assumption similar to (H9.2.2)$_q$. We now state and prove the following result concerning MF-BSVIE (9.3.1).

Proposition 9.3.1 *Let* (H9.2.2)$_q$ *hold. Then for any* $p>1$, $\psi(\cdot)\in L^p_{\mathcal{F}_T}(0,T;\mathbb{R}^n)$, $(y(\cdot),z(\cdot\,,\cdot))\in\mathcal{M}^p[0,T]$, *MF-BSVIE (9.3.1) admits a unique M-solution* $(Y(\cdot),Z(\cdot\,,\cdot))\in\mathcal{M}^p[0,T]$. *Moreover, the following estimate holds:*

$$\begin{aligned}&\mathbb{E}\Big[|Y(t)|^p+\Big(\int_t^T|Z(t,s)|^2ds\Big)^{\frac p2}\Big]\\&\leqslant K\mathbb{E}\Big[|\psi(t)|^p+\Big(\int_t^T|\widetilde g(t,s,0,\widetilde\Gamma(t,s,0))|ds\Big)^p\Big].\end{aligned}\tag{9.3.2}$$

Further, for $i=1,2$, *let* $\psi_i(\cdot)\in L^p_{\mathcal{F}_T}(0,T;\mathbb{R}^n)$, $(y_i(\cdot),z_i(\cdot\,,\cdot))\in\mathcal{M}^p[0,T]$, *and*

$$\begin{cases}\widetilde g_i\Big(t,s,Z(t,s),\widetilde\Gamma_i\big(t,s,Z(t,s)\big)\Big)\\ =g_i\Big(t,s,y_i(s),Z(t,s),z_i(s,t),\Gamma_i\big(t,s,y_i(s),Z(t,s),z_i(s,t)\big)\Big),\\ \Gamma_i\big(t,s,y_i(s),Z(t,s),z_i(s,t)\big)\\ =\mathbb{E}'\Big[\theta_i\big(t,s,y',z',\widehat z',y_i(s),Z(t,s),z_i(s,t)\big)\Big]_{(y',z',\hat z')=\big(y_i(s),Z(t,s),z_i(s,t)\big)},\end{cases}$$

with $g_i(\cdot)$ *and* $\theta_i(\cdot)$ *satisfying* (H9.2.2)$_q$. *Then the corresponding M-solutions* $(Y_i(\cdot),Z_i(\cdot))$ *satisfy the following stability estimate:*

$$\begin{aligned}&\mathbb{E}\Big[|Y_1(t)-Y_2(t)|^p+\Big(\int_t^T|Z_1(t,s)-Z_2(t,s)|^2ds\Big)^{\frac p2}\Big]\\&\leqslant K\mathbb{E}\Big[|\psi_1(t)-\psi_2(t)|^p+\Big(\int_t^T\big|\widetilde g_1\big(t,s,Z_1(t,s),\widetilde\Gamma_1(t,s,Z_1(t,s))\big)\\&\qquad\qquad-\widetilde g_2\big(t,s,Z_1(t,s),\widetilde\Gamma_2(t,s,Z_1(t,s))\big)\big|ds\Big)^p\Big].\end{aligned}\tag{9.3.3}$$

Proof Consider the following MF-BSDE (parameterized by $t \in [0, T)$):

$$\eta(r) = \psi(t)+\int_r^T \widetilde{g}\big(t, s, \zeta(s), \widetilde{\Gamma}(t, s, \zeta(s))\big)ds-\int_r^T \zeta(s)dW(s), \quad r\in[t,T]. \tag{9.3.4}$$

If $p \in (1, 2]$, it follows from $(\text{H9.2.2})_q$ that □

$$\begin{aligned}&\mathbb{E}\int_0^T \Big(\int_t^T \big|\widetilde{g}(t, s, 0, \widetilde{\Gamma}(t, s, 0))\big|ds\Big)^p dt\\&\leqslant K\mathbb{E}\int_0^T \Big(\int_t^T \big(|y(s)| + |z(s, t)| + \mathbb{E}|y(s)| + \mathbb{E}|z(s, t)|\big)ds\Big)^p dt + K\\&\leqslant K\mathbb{E}\int_0^T \int_t^T |y(s)|^p dsdt + K\mathbb{E}\int_0^T \Big(\int_0^s |z(s, t)|^2 dt\Big)^{\frac{p}{2}} ds + K < \infty.\end{aligned}$$

As to the case of $p = q > 2$, similarly we have

$$\begin{aligned}&\mathbb{E}\int_0^T \Big(\int_t^T \big|\widetilde{g}(t, s, 0, \widetilde{\Gamma}(t, s, 0))\big|ds\Big)^q dt\\&\leqslant K\mathbb{E}\int_0^T \Big(\int_t^T \big(|y(s)| + |z(s, t)|^{\frac{2}{q}} + \mathbb{E}|y(s)| + \mathbb{E}|z(s, t)|^{\frac{2}{q}}\big)ds\Big)^q dt + K\\&\leqslant K\mathbb{E}\int_0^T \int_t^T |y(s)|^q dsdt + K\mathbb{E}\int_0^T \int_t^T |z(s, t)|^2 dsdt + K < \infty.\end{aligned}$$

Similar to a standard argument for BSDEs (e.g., Theorem 3.1.4 in Chap. 3), making use of contraction mapping theorem, we can show that the above MF-BSDE admits a unique adapted solution

$$(\eta(\cdot), \zeta(\cdot)) \equiv \Big(\eta(\cdot\,; t, \psi(\cdot)), \zeta(\cdot\,; t, \psi(\cdot))\Big).$$

Moreover, the following estimate holds:

$$\begin{aligned}&\mathbb{E}\Big[\sup_{r\in[t,T]} |\eta(r; t, \psi(\cdot))|^p + \Big(\int_t^T |\zeta(s; t, \psi(\cdot))|^2 ds\Big)^{\frac{p}{2}}\Big]\\&\leqslant K\mathbb{E}\Big[|\psi(t)|^p + \Big(\int_t^T |\widetilde{g}(t, s, 0, \widetilde{\Gamma}(t, s, 0))|ds\Big)^p\Big].\end{aligned} \tag{9.3.5}$$

Further, for $i = 1, 2$, let $\psi_i(\cdot) \in L^p_{\mathcal{F}_T}(0, T; \mathbb{R}^n)$, $(y_i(\cdot), z_i(\cdot\,, \cdot)) \in \mathcal{M}^p[0, T]$, and

$$
\begin{cases}
\widetilde{g}_i\big(t,s,Z(t,s),\widetilde{\Gamma}_i(t,s,Z(t,s))\big) \\
\qquad = g_i\Big(t,s,y_i(s),Z(t,s),z_i(s,t),\Gamma_i\big(t,s,y_i(s),Z(t,s),z_i(s,t)\big)\Big), \\
\Gamma_i\big(t,s,y_i(s),Z(t,s),z_i(s,t)\big) \\
= \mathbb{E}'\Big[\theta_i\big(t,s,y',z',\widehat{z}',y_i(s),Z(t,s),z_i(s,t)\big)\Big]_{(y',z',\widehat{z}')=\big(y_i(s),Z(t,s),z_i(s,t)\big)}
\end{cases}
$$

with $g_i(\cdot)$ and $\theta_i(\cdot)$ satisfying (H9.2.2)$_q$ and $(\eta_i(\cdot),\zeta_i(\cdot))$ be the adapted solutions of the corresponding BSDE. It follows that

$$
\begin{aligned}
&\mathbb{E}\Big[\sup_{r\in[t,T]}|\eta_1(r)-\eta_2(r)|^p\Big]+\mathbb{E}\Big(\int_t^T|\zeta_1(s)-\zeta_2(s)|^2ds\Big)^{\frac{p}{2}} \\
&\leqslant K\mathbb{E}\Big[|\psi_1(t)-\psi_2(t)|^p+\Big(\int_t^T\big|\widetilde{g}_1\big(t,s,\zeta_1(s),\widetilde{\Gamma}_1(t,s,\zeta_1(s))\big) \\
&\qquad\qquad -\widetilde{g}_2\big(t,s,\zeta_1(s),\widetilde{\Gamma}_2(t,s,\zeta_1(s))\big)\big|ds\Big)^p\Big].
\end{aligned}
\tag{9.3.6}
$$

Now, we define

$$
Y(t)\stackrel{\Delta}{=}\eta(t;t,\psi(\cdot)),\quad Z(t,s)\stackrel{\Delta}{=}\zeta(s;t,\psi(\cdot)),\qquad (t,s)\in\Delta^*[0,T],
$$

and $Z(t,s)$ on $\Delta_*[0,T]$ through the martingale representation:

$$
Y(t)=\mathbb{E}Y(t)+\int_0^t Z(t,s)dW(s),\qquad t\in[0,T].
$$

Then $(Y(\cdot),Z(\cdot\,,\cdot))\in\mathcal{M}^p[0,T]$ is the unique M-solution to (9.3.1). Estimates (9.3.2) and (9.3.3) follow easily from (9.3.5) and (9.3.6), respectively. □

Note that the cases that we are interested in are $p=2,q$. We will use them below.

3.2 A General Form of Type-II MF-BSVIEs

Now, we consider our Type-II MF-BSVIEs. For convenience, let us rewrite (9.1) here:

$$
\begin{aligned}
Y(t)=\psi(t)+&\int_t^T g\big(t,s,Y(s),Z(t,s),Z(s,t),\Gamma(t,s,Y(s),Z(t,s),Z(s,t))\big)ds \\
&-\int_t^T Z(t,s)dW(s),\qquad t\in[0,T],
\end{aligned}
\tag{9.3.7}
$$

with

$$\begin{aligned}&\Gamma(t,s,Y(s),Z(t,s),Z(s,t))=\mathbb{E}'\big[\theta(t,s,Y(s),Z(t,s),Z(s,t))\big]\\&=\int_\Omega\theta\big(t,s,\omega,\omega',Y(s,\omega),Z(t,s,\omega),Z(s,t,\omega),\\&\qquad\qquad Y(s,\omega'),Z(t,s,\omega'),Z(s,t,\omega')\big)\mathbb{P}(d\omega').\end{aligned}\tag{9.3.8}$$

Our main result of this section is the following.

Theorem 9.3.2 *Let* (H9.2.2)$_q$ *hold with* $2\leqslant q<\infty$. *Then for any* $\psi(\cdot)\in L^q_{\mathcal{F}_T}(0,T;\mathbb{R}^n)$, *Type-II MF-BSVIE (9.3.7) admits a unique adapted M-solution* $(Y(\cdot),Z(\cdot\,,\cdot))\in\mathcal{M}^q[0,T]$, *and the following estimate holds:*

$$\|(Y(\cdot),Z(\cdot\,,\cdot))\|_{\mathcal{M}^q[0,T]}\leqslant K\Big(1+\|\psi(\cdot)\|_{L^q_{\mathcal{F}_T}(0,T;\mathbb{R}^n)}\Big).\tag{9.3.9}$$

Moreover, for $i=1,2$, *let* $g_i(\cdot)$ *and* $\theta_i(\cdot)$ *satisfy* (H9.2.2)$_q$, *and* $\psi_i(\cdot)\in L^q_{\mathcal{F}_T}(0,T;\mathbb{R}^n)$. *Let* $(Y_i(\cdot),Z_i(\cdot\,,\cdot))\in\mathcal{M}^q[0,T]$ *be the corresponding adapted M-solutions. Then*

$$\begin{aligned}&\|(Y_1(\cdot),Z_1(\cdot\,,\cdot))-(Y_2(\cdot),Z_2(\cdot\,,\cdot))\|^2_{\mathcal{M}^2[0,T]}\\&\leqslant K\mathbb{E}\Big\{\int_0^T|\psi_1(t)-\psi_2(t)|^2dt+\int_0^T\Big(\int_t^T|(g_1-g_2)(t,s)|ds\Big)^2dt\Big\},\end{aligned}\tag{9.3.10}$$

where

$$\begin{aligned}&(g_1-g_2)(t,s)\\&\overset{\Delta}{=}g_1\big(t,s,Y_1(s),Z_1(t,s),Z_1(s,t),\Gamma_1(t,s,Y_1(s),Z_1(t,s),Z_1(s,t))\big)\\&\quad-g_2\big(t,s,Y_1(s),Z_1(t,s),Z_1(s,t),\Gamma_2(t,s,Y_1(s),Z_1(t,s),Z_1(s,t))\big),\end{aligned}$$

with

$$\begin{aligned}&\Gamma_i(t,s,Y_i(s),Z_i(t,s),Z_i(s,t))\\&\overset{\Delta}{=}\mathbb{E}'\big[\theta_i(t,s,Y_i(s),Z_i(t,s),Z_i(s,t),y,z,\widehat{z})\big]_{(y,z,\hat{z})=(Y_i(s),Z_i(t,s),Z_i(s,t))}\\&\equiv\int_\Omega\theta_i\big(t,s,\omega,\omega',Y_i(s,\omega),Z_i(t,s,\omega),Z_i(s,t,\omega),\\&\qquad\qquad Y_i(s,\omega'),Z_i(t,s,\omega'),Z_i(s,t,\omega')\big)\mathbb{P}(d\omega').\end{aligned}$$

Proof We split the proof into several steps.
Step 1. Existence and uniqueness of M-solutions of (9.3.7) in $\mathcal{M}^p[0,T]$ with $p\in(1,2]$.

Let $\psi(\cdot)\in L^p_{\mathcal{F}_T}(0,T;\mathbb{R}^n)$ be given. For any $(y(\cdot),z(\cdot\,,\cdot))\in\mathcal{M}^p[0,T]$, we consider the following MF-BSVIE

$$
\begin{aligned}
Y(t) = \psi(t) + \int_t^T & g\big(t, s, y(s), Z(t, s), z(s, t), \Gamma(t, s, y(s), Z(t, s), z(s, t))\big)ds \\
& - \int_t^T Z(t, s)dW(s), \qquad t \in [0, T].
\end{aligned}
$$

According to Proposition 9.3.1, there exists a unique adapted M-solution $(Y(\cdot), Z(\cdot\,,\cdot)) \in \mathcal{M}^p[0, T]$. Moreover, the following estimate holds: (making use of (9.2.6) and (9.2.7))

$$
\begin{aligned}
& \mathbb{E}\Big\{|Y(t)|^p + \Big(\int_t^T |Z(t, s)|^2 ds\Big)^{\frac{p}{2}}\Big\} \\
& \leqslant K\mathbb{E}\Big\{|\psi(t)|^p + \Big(\int_t^T \big|g\big(t, s, y(s), 0, z(s, t), \Gamma(t, s, y(s), 0, z(s, t))\big)\big|ds\Big)^p\Big\} \\
& \leqslant K\mathbb{E}\Big\{1 + |\psi(t)|^p + \Big[\int_t^T \big(|y(s)| + |z(s, t)| + |\Gamma(t, s, y(s), 0, z(s, t))|\big)ds\Big]^p\Big\} \\
& \leqslant K\mathbb{E}\Big\{1 + |\psi(t)|^p + \Big[\int_t^T \big(|y(s)| + |z(s, t)| + \mathbb{E}|y(s)| + \mathbb{E}|z(s, t)| \\
& \qquad\qquad + |y(s)| + |z(s, t)|\big)ds\Big]^p\Big\} \\
& \leqslant K\mathbb{E}\Big\{1 + |\psi(t)|^p + \Big[\int_t^T \big(|y(s)| + |z(s, t)|\big)ds\Big]^p\Big\} \\
& \leqslant K\mathbb{E}\Big\{1 + |\psi(t)|^p + \int_t^T |y(s)|^p ds + \int_t^T |z(s, t)|^p ds\Big\}.
\end{aligned}
$$

Consequently, (making use of (9.2.14) for $(y(\cdot), z(\cdot\,,\cdot)) \in \mathcal{M}^p[0, T]$)

$$
\begin{aligned}
& \|(Y(\cdot), Z(\cdot\,,\cdot))\|^p_{\mathcal{H}^p[0,T]} \equiv \mathbb{E}\Big\{\int_0^T |Y(t)|^p dt + \int_0^T \Big(\int_0^T |Z(t, s)|^2 ds\Big)^{\frac{p}{2}} dt\Big\} \\
& \leqslant K\mathbb{E}\Big\{1 + \int_0^T |\psi(t)|^p dt + \int_0^T \int_t^T |y(s)|^p ds dt + \int_0^T \int_t^T |z(s, t)|^p ds dt\Big\} \\
& \leqslant K\mathbb{E}\Big\{1 + \int_0^T |\psi(t)|^p dt + \int_0^T |y(t)|^p dt + \int_0^T \int_0^t |z(t, s)|^p ds dt\Big\} \\
& \leqslant K\mathbb{E}\Big\{1 + \int_0^T |\psi(t)|^p dt + \int_0^T |y(t)|^p dt + \int_0^T \Big(\int_0^t |z(t, s)|^2 ds\Big)^{\frac{p}{2}} dt\Big\} \\
& \leqslant K\mathbb{E}\Big\{1 + \int_0^T |\psi(t)|^p dt + \int_0^T |y(t)|^p dt\Big\} \\
& \leqslant K\Big\{1 + \|\psi(\cdot)\|^p_{L^p_{\mathcal{F}_T}(0,T;\mathbb{R}^n)} + \|(y(\cdot), z(\cdot\,,\cdot))\|_{\mathcal{M}^p[0,T]}\Big\}.
\end{aligned}
$$

Hence, if we define $\Theta\big(y(\cdot), z(\cdot\,,\cdot)\big) \stackrel{\Delta}{=} \big(Y(\cdot), Z(\cdot\,,\cdot)\big)$, then Θ maps from $\mathcal{M}^p[0, T]$ to itself. We now show that the mapping Θ is contractive. To this end, take any $(y_i(\cdot), z_i(\cdot\,,\cdot)) \in \mathcal{M}^p[0, T]$ $(i = 1, 2)$, and let

$$
\big(Y_i(\cdot), Z_i(\cdot\,,\cdot)\big) = \Theta\big(y_i(\cdot), z_i(\cdot\,,\cdot)\big).
$$

Then by Proposition 9.3.1, we have (note (9.2.7))

$$
\begin{aligned}
&\mathbb{E}\Big[|Y_1(t)-Y_2(t)|^p+\Big(\int_t^T|Z_1(t,s)-Z_2(t,s)|^2ds\Big)^{\frac{p}{2}}\Big]\\
&\leqslant K\mathbb{E}\Big(\int_t^T\Big|g\big(t,s,y_1(s),Z_1(t,s),z_1(s,t),\Gamma(t,s,y_1(s),Z_1(t,s),z_1(s,t))\big)\\
&\qquad-g\big(t,s,y_2(s),Z_1(t,s),z_2(s,t),\Gamma(t,s,y_2(s),Z_1(t,s),z_2(s,t))\big)\Big|ds\Big)^p\\
&\leqslant K\mathbb{E}\Big[\int_t^T\Big(|y_1(s)-y_2(s)|+|z_1(s,t)-z_2(s,t)|\\
&\qquad+\big|\Gamma\big(t,s,y_1(s),Z_1(t,s),z_1(s,t)\big)-\Gamma\big(t,s,y_2(s),Z_1(t,s),z_2(s,t)\big)\big|\Big)ds\Big]^p\\
&\leqslant K\mathbb{E}\Big[\int_t^T\Big(|y_1(s)-y_2(s)|+|z_1(s,t)-z_2(s,t)|+|y_1(s)-y_2(s)|\\
&\qquad+|z_1(s,t)-z_2(s,t)|+\mathbb{E}|y_1(s)-y_2(s)|+\mathbb{E}|z_1(s,t)-z_2(s,t)|\Big)ds\Big]^p\\
&\leqslant K\mathbb{E}\Big[\int_t^T\Big(|y_1(s)-y_2(s)|+|z_1(s,t)-z_2(s,t)|\Big)ds\Big]^p\\
&\leqslant K\mathbb{E}\Big[\int_t^T|y_1(s)-y_2(s)|^pds+\Big(\int_t^T|z_1(s,t)-z_2(s,t)|ds\Big)^p\Big].
\end{aligned}
$$

Hence,

$$
\begin{aligned}
&\|\Theta(y_1(\cdot),z_1(\cdot\,,\cdot))-\Theta(y_2(\cdot),z_2(\cdot\,,\cdot))\|^p_{\mathcal{M}^{p,\beta}[0,T]}\\
&\equiv\int_0^T e^{\beta t}\mathbb{E}\Big[|Y_1(t)-Y_2(t)|^p+\Big(\int_t^T|Z_1(t,s)-Z_2(t,s)|^2ds\Big)^{\frac{p}{2}}\Big]dt\\
&\leqslant K\int_0^T e^{\beta t}\mathbb{E}\Big[\int_t^T|y_1(s)-y_2(s)|^pds+\Big(\int_t^T|z_1(s,t)-z_2(s,t)|ds\Big)^p\Big]dt\\
&\leqslant\frac{K}{\beta}\mathbb{E}\int_0^T e^{\beta t}|y_1(t)-y_2(t)|^pdt\\
&\qquad+K\Big(\frac{p}{\beta q}\Big)^{\frac{p}{q}}\mathbb{E}\int_0^T\int_0^s e^{\beta s}|z_1(s,t)-z_2(s,t)|^pdtds\\
&\leqslant\Big(\frac{K}{\beta}+K\Big(\frac{1}{\beta}\Big)^{\frac{p}{q}}\Big)\mathbb{E}\int_0^T e^{\beta t}|y_1(t)-y_2(t)|^pdt\\
&\leqslant\Big(\frac{K}{\beta}+K\Big(\frac{1}{\beta}\Big)^{\frac{p}{q}}\Big)\|(y_1(\cdot),z_1(\cdot\,,\cdot))-(y_2(\cdot),z_2(\cdot\,,\cdot))\|^p_{\mathcal{M}^{p,\beta}[0,T]}.
\end{aligned}
$$

Since the constant $K>0$ that appears in the right-hand side of the above is independent of β, by choosing $\beta>0$ large enough, we obtain that Θ is a contraction. Hence, there exists a unique fixed point $(Y(\cdot),Z(\cdot\,,\cdot))\in\mathcal{M}^p[0,T]$, which is the unique adapted M-solution of (9.1).

Step 2. The adapted M-solution $(Y(\cdot),Z(\cdot\,,\cdot))\in\mathcal{M}^q[0,T]$ if $\psi(\cdot)\in L^q_{\mathcal{F}_T}(0,T;\mathbb{R}^n)$ with $q\in[2,\infty)$.

Let $\psi(\cdot)\in L^q_{\mathcal{F}_T}(0,T;\mathbb{R}^n)\subseteq L^2_{\mathcal{F}_T}(0,T;\mathbb{R}^n)$. According to Step 1, there exists a unique adapted M-solution $(Y(\cdot),Z(\cdot\,,\cdot))\in\mathcal{M}^2[0,T]$. We want to show that in the current case, $(Y(\cdot), Z(\cdot\,,\cdot))$ is actually in $\mathcal{M}^q[0,T]$. To show this, for the obtained adapted M-solution $(Y(\cdot), Z(\cdot\,,\cdot))$, let us consider the following MF-BSVIE:

$$\begin{aligned}\widetilde{Y}(t)=\psi(t)+&\int_t^T g\big(t,s,\widetilde{Y}(s),\widetilde{Z}(t,s),Z(s,t),\Gamma(t,s,\widetilde{Y}(s),\widetilde{Z}(t,s),Z(s,t))\big)ds\\ &-\int_t^T \widetilde{Z}(t,s)dW(s),\qquad\qquad t\in[0,T].\end{aligned}$$

For any $(y(\cdot), z(\cdot\,,\cdot)) \in \mathcal{M}^p[0,T]$, by Proposition 9.3.1 (with $p=q$), the following MF-BSVIE admits a unique adapted M-solution $(\widetilde{Y}(\cdot), \widetilde{Z}(\cdot\,,\cdot)) \in \mathcal{M}^q[0,T]$:

$$\begin{aligned}\widetilde{Y}(t)=\psi(t)+&\int_t^T g\Big(t,s,y(s),\widetilde{Z}(t,s),Z(s,t),\\ &\qquad\Gamma(t,s,y(s),\widetilde{Z}(t,s),Z(s,t))\Big)ds\\ &-\int_t^T \widetilde{Z}(t,s)dW(s),\qquad\qquad t\in[0,T].\end{aligned}\tag{9.3.11}$$

Thus, if we define $\widetilde{\Theta}(y(\cdot), z(\cdot\,,\cdot)) \stackrel{\Delta}{=} (\widetilde{Y}(\cdot), \widetilde{Z}(\cdot\,,\cdot))$, then $\widetilde{\Theta}:\mathcal{M}^q[0,T]\to\mathcal{M}^q[0,T]$. We now show that $\widetilde{\Theta}$ is a contraction on $\mathcal{M}^q[0,T]$ (compare that Θ in Step 1 is a contraction on $\mathcal{M}^2[0,T]$). To this end, let $(y_i(\cdot), z_i(\cdot\,,\cdot)) \in \mathcal{M}^q[0,T]$ and let

$$(\widetilde{Y}_i(\cdot), \widetilde{Z}_i(\cdot\,,\cdot)) = \widetilde{\Theta}(y_i(\cdot), z_i(\cdot\,,\cdot)),\qquad i=1,2.$$

Then by Proposition 9.3.1 (with $p=q$), we have

$$\begin{aligned}&\mathbb{E}\Big[|\widetilde{Y}_1(t)-\widetilde{Y}_2(t)|^q+\Big(\int_t^T|\widetilde{Z}_1(t,s)-\widetilde{Z}_2(t,s)|^2ds\Big)^{\frac{q}{2}}\Big]\\ &\leqslant K\mathbb{E}\Big(\int_t^T\big|g\big(t,s,y_1(s),\widetilde{Z}_1(t,s),Z(s,t),\Gamma(t,s,y_1(s),\widetilde{Z}_1(t,s),Z(s,t))\big)\\ &\qquad -g\big(t,s,y_2(s),\widetilde{Z}_1(t,s),Z(s,t),\Gamma(t,s,y_2(s),\widetilde{Z}_1(t,s),Z(s,t))\big)\big|ds\Big)^q\\ &\leqslant K\mathbb{E}\Big[\int_t^T\Big(|y_1(s)-y_2(s)|+|\Gamma(t,s,y_1(s),\widetilde{Z}_1(t,s),Z(s,t))\\ &\qquad\qquad -\Gamma(t,s,y_2(s),\widetilde{Z}_1(t,s),Z(s,t))|\Big)ds\Big]^q\\ &\qquad\leqslant K\mathbb{E}\Big[\int_t^T\Big(|y_1(s)-y_2(s)|+\mathbb{E}|y_1(s)-y_2(s)|\Big)ds\Big]^q\\ &\qquad\leqslant K\mathbb{E}\int_t^T|y_1(s)-y_2(s)|^q ds.\end{aligned}$$

Then

$$\begin{aligned}&\mathbb{E}\Big[\int_0^T e^{\beta t}|\widetilde{Y}_1(t)-\widetilde{Y}_2(t)|^q dt+\int_0^T e^{\beta t}\Big(\int_t^T|\widetilde{Z}_1(t,s)-\widetilde{Z}_2(t,s)|^2 ds\Big)^{\frac{q}{2}}dt\Big]\\&\leqslant K\mathbb{E}\int_0^T e^{\beta t}\int_t^T|y_1(s)-y_2(s)|^q dsdt=K\mathbb{E}\int_0^T\int_0^s e^{\beta t}|y_1(s)-y_2(s)|^q dtds\\&\leqslant\frac{K}{\beta}\int_0^T e^{\beta t}|y_1(t)-y_2(t)|^q dt.\end{aligned}$$

Hence, $\widetilde{\Theta}$ is a contraction on $\mathcal{M}^q[0,T]$ (with the equivalent norm), which implies that (9.3.11) admits a unique adapted M-solution $(\widetilde{Y}(\cdot),\widetilde{Z}(\cdot\,,\cdot))\in\mathcal{M}^q[0,T]\subseteq\mathcal{M}^2[0,T]$. Then by the uniqueness of adapted solutions in $\mathcal{M}^2[0,T]$ of (9.3.11), it is necessary that

$$(Y(\cdot),Z(\cdot\,,\cdot))=(\widetilde{Y}(\cdot),\widetilde{Z}(\cdot\,,\cdot))\in\mathcal{M}^q[0,T].$$

Step 3. Some estimates.

According to Proposition 9.3.1, we have

$$\begin{aligned}&\mathbb{E}\Big\{|Y(t)|^q+\Big(\int_t^T|Z(t,s)|^2 ds\Big)^{\frac{q}{2}}\Big\}\\&\leqslant K\mathbb{E}\Big\{|\psi(t)|^q+\Big(\int_t^T\big|g\big(t,s,Y(s),0,Z(s,t),\Gamma(t,s,Y(s),0,Z(s,t))\big)\big|ds\Big)^q\Big\}\\&\leqslant K\mathbb{E}\Big\{1+|\psi(t)|^q+\Big[\int_t^T\Big(|Y(s)|+|Z(s,t)|^{\frac{2}{q}}\Big)ds\Big]^q\Big\}\\&\leqslant K\mathbb{E}\Big\{1+|\psi(t)|^q+\int_t^T|Y(s)|^q ds+\Big(\int_t^T|Z(s,t)|^{\frac{2}{q}}ds\Big)^q\Big\}.\end{aligned}$$

Then

$$\begin{aligned}&\mathbb{E}\Big\{\int_0^T e^{\beta t}|Y(t)|^q dt+\int_0^T e^{\beta t}\Big(\int_t^T|Z(t,s)|^2 ds\Big)^{\frac{q}{2}}dt\Big\}\\&\leqslant K\mathbb{E}\Big\{\int_0^T e^{\beta t}\Big(1+|\psi(t)|^q\Big)dt+\int_0^T e^{\beta t}\int_t^T|Y(s)|^q dsdt\\&\qquad+\int_0^T e^{\beta t}\Big(\int_t^T|Z(s,t)|^{\frac{2}{q}}ds\Big)^q dt\Big\}.\end{aligned}$$

Note that

$$\begin{aligned}&\mathbb{E}\int_0^T e^{\beta t}\Big(\int_t^T|Z(s,t)|^{\frac{2}{q}}ds\Big)^q dt=\mathbb{E}\int_0^T e^{\beta t}\Big(\int_t^T e^{-\frac{\beta}{q}s}e^{\frac{\beta}{q}s}|Z(s,t)|^{\frac{2}{q}}ds\Big)^q dt\\&\leqslant\mathbb{E}\int_0^T e^{\beta t}\Big(\int_t^T e^{-\frac{\beta}{q-1}s}ds\Big)^{q-1}\Big(\int_t^T e^{\beta s}|Z(s,t)|^2 ds\Big)dt\end{aligned}$$

$$
\begin{aligned}
&= \mathbb{E}\int_0^T e^{\beta t}\Big[\frac{q-1}{\beta}\Big(e^{\frac{-\beta t}{q-1}} - e^{-\frac{\beta T}{q-1}}\Big)\Big]^{q-1}\Big(\int_t^T e^{\beta s}|Z(s,t)|^2 ds\Big)dt \\
&\leqslant \frac{(q-1)^{q-1}}{\beta^{q-1}}\mathbb{E}\int_0^T\int_0^s e^{\beta s}|Z(s,t)|^2 dtds \leqslant \frac{(q-1)^{q-1}}{\beta^{q-1}}\mathbb{E}\int_0^T e^{\beta t}|Y(t)|^2 dt.
\end{aligned}
$$

Hence,

$$
\begin{aligned}
&\mathbb{E}\Big\{\int_0^T e^{\beta t}|Y(t)|^q dt + \int_0^T e^{\beta t}\Big(\int_t^T |Z(t,s)|^2 ds\Big)^{\frac{q}{2}} dt\Big\} \\
&\leqslant K\mathbb{E}\Big\{\int_0^T e^{\beta t}\Big(1+|\psi(t)|^q\Big)dt + \frac{1}{\beta}\int_0^T e^{\beta t}|Y(t)|^q dt + \frac{1}{\beta^{q-1}}\int_0^T e^{\beta t}|Y(t)|^2 dt\Big\}.
\end{aligned}
$$

By choosing $\beta > 0$ large, we obtain

$$
\mathbb{E}\Big\{\int_0^T e^{\beta t}|Y(t)|^q dt + \int_0^T e^{\beta t}\Big(\int_t^T |Z(t,s)|^2 ds\Big)^{\frac{q}{2}} dt\Big\} \leqslant K\mathbb{E}\Big(1+\int_0^T e^{\beta t}|\psi(t)|^q dt\Big).
$$

Then (9.3.9) follows.

Finally, let $\psi_i(\cdot) \in L^q_{\mathcal{F}_T}(0,T;\mathbb{R}^n)$, $g_i(\cdot)$ and $\theta_i(\cdot)$ satisfy (H9.2.2)$_q$. Observe that

$$
\begin{aligned}
&Y_1(t) \\
&= \psi_1(t) + \int_t^T g_1\big(t,s,Y_1(s),Z_1(t,s),Z_1(s,t),\Gamma_1(t,s,Y_1(s),Z_1(t,s),Z_1(s,t))\big)ds \\
&\quad - \int_t^T g_2\big(t,s,Y_1(s),Z_1(t,s),Z_1(s,t),\Gamma_2(t,s,Y_1(s),Z_1(t,s),Z_1(s,t))\big)ds \\
&\quad + \int_t^T g_2\big(t,s,Y_1(s),Z_1(t,s),Z_1(s,t),\Gamma_2(t,s,Y_1(s),Z_1(t,s),Z_1(s,t))\big)ds \\
&\quad - \int_t^T g_2\big(t,s,Y_2(s),Z_1(t,s),Z_2(s,t),\Gamma_2(t,s,Y_2(s),Z_1(t,s),Z_2(s,t))\big)ds \\
&\quad + \int_t^T g_2\big(t,s,Y_2(s),Z_1(t,s),Z_2(s,t),\Gamma_2(t,s,Y_2(s),Z_1(t,s),Z_2(s,t))\big)ds \\
&\quad - \int_t^T Z_1(t,s)dW(s) \\
&\equiv \widehat{\psi}_1(t) + \int_t^T g_2\big(t,s,Y_2(s),Z_1(t,s),Z_2(s,t),\Gamma_2(t,s,Y_2(s),Z_1(t,s),Z_2(s,t))\big)ds \\
&\quad - \int_t^T Z_1(t,s)dW(s),
\end{aligned}
$$

with $\widehat{\psi}_1(\cdot)$ defined in an obvious way. Then by Proposition 9.3.1 (with $p = 2$), we obtain

$$
\begin{aligned}
&\mathbb{E}\Big[|Y_1(t)-Y_2(t)|^2+\int_t^T|Z_1(t,s)-Z_2(t,s)|^2ds\Big]\leqslant K\mathbb{E}|\widehat{\psi}_1(t)-\psi_2(t)|^2\\
&\leqslant K\mathbb{E}\Big\{|\psi_1(t)-\psi_2(t)|^2\\
&\qquad+\Big(\int_t^T\big|g_1\big(t,s,Y_1(s),Z_1(t,s),Z_1(s,t),\Gamma_1(t,s,Y_1(s),Z_1(t,s),Z_1(s,t))\big)\\
&\qquad\quad-g_2\big(t,s,Y_1(s),Z_1(t,s),Z_1(s,t),\Gamma_2(t,s,Y_1(s),Z_1(t,s),Z_1(s,t))\big)\big|ds\Big)^2\\
&\qquad+\Big(\int_t^T\big|g_2\big(t,s,Y_1(s),Z_1(t,s),Z_1(s,t),\Gamma_2(t,s,Y_1(s),Z_1(t,s),Z_1(s,t))\big)\\
&\qquad\quad-g_2\big(t,s,Y_2(s),Z_1(t,s),Z_2(s,t),\Gamma_2(t,s,Y_2(s),Z_1(t,s),Z_2(s,t))\big)\big|ds\Big)^2\Big\}\\
&\leqslant K\mathbb{E}\Big\{|\psi_1(t)-\psi_2(t)|^2+\Big(\int_t^T|(g_1-g_2)(t,s)|ds\Big)^2\\
&\qquad+\Big[\int_t^T\Big(|Y_1(s)-Y_2(s)|+|Z_1(s,t)-Z_2(s,t)|\Big)ds\Big]^2\Big\}.
\end{aligned}
$$

Then similar to the proof of the contraction for Θ, we can obtain our stability estimate (9.3.10). □

Let us make some remarks on the above result, together with its proof.

First of all, we have seen that the growth of the maps

$$
\hat{z}\mapsto g(t,s,y,z,\hat{z}),\qquad(\hat{z},\hat{z}')\mapsto\theta(t,s,y,z,\hat{z},y',z',\hat{z}')\tag{9.3.12}
$$

plays an important role in proving the well-posedness of MF-BSVIEs, especially for the case of $p>2$. If (H9.2.2)$_\infty$ holds, then for any $p>1$, as long as $\psi(\cdot)\in L^p_{\mathcal{F}_T}(0,T;\mathbb{R}^n)$, (9.3.7) admits a unique adapted M-solution $(Y(\cdot),Z(\cdot\,,\cdot))\in\mathcal{M}^p[0,T]$. On the other hand, if the maps in (9.3.12) grow linearly, the adapted M-solution $(Y(\cdot),Z(\cdot\,,\cdot))$ of (9.3.7) may not be in $\mathcal{M}^p[0,T]$ for $p>2$, even if $\psi(\cdot)\in L^p_{\mathcal{F}_T}(0,T;\mathbb{R}^n)$. This can be seen from Example 5.2.3 in Chap. 5.

To conclude this section, let us look at a class of linear MF-BSVIE which takes the following form:

$$
\begin{aligned}
Y(t)=\psi(t)+\int_t^T&\Big(\bar{A}_0(t,s)Y(s)+\bar{B}_0(t,s)Z(t,s)+\bar{C}_0(t,s)Z(s,t)\\
&+\mathbb{E}'\big[\bar{A}_1(t,s)Y(s)+\bar{B}_1(t,s)Z(t,s)+\bar{C}_1(t,s)Z(s,t)\big]\Big)ds\\
&-\int_t^T Z(t,s)dW(s),\quad t\in[0,T].
\end{aligned}\tag{9.3.13}
$$

For the coefficients, we adopt the following hypothesis.

(H9.3.1) The maps

$$
\bar{A}_0,\bar{B}_0,\bar{C}_0:\Delta^*[0,T]\times\Omega\to\mathbb{R}^{n\times n},\qquad\bar{A}_1,\bar{B}_1,\bar{C}_1:\Delta^*[0,T]\times\Omega^2\to\mathbb{R}^{n\times n}
$$

are measurable and uniformly bounded. Moreover, for any $t \in [0, T]$, $s \mapsto (\bar{A}_0(t,s), \bar{B}_0(t,s), \bar{C}_0(t,s))$ is $\mathbb{F}$-progressively measurable on $[t, T]$, and $s \mapsto (\bar{A}_1(t,s), \bar{B}_1(t,s), \bar{C}_1(t,s))$ is $\mathbb{F}^2$-progressively measurable on $[t, T]$.

It is not hard to see that under (H9.3.1), we have $(\text{H9.2.2})_q$ with $q = 2$. Hence, we have the following corollary.

Corollary 9.3.3 *Let* (H9.3.1) *hold. Then for any* $\psi(\cdot) \in L^2_{\mathcal{F}_T}(0, T; \mathbb{R}^n)$, (9.3.13) *admits a unique adapted M-solution* $(Y(\cdot), Z(\cdot\,,\cdot)) \in \mathcal{M}^2[0, T]$.

4 Two Interesting Topics

In this section, inspired by the study in Chaps. 6 and 7, we study two interesting topics of MF-BSVIEs, i.e., the comparison theorems and representation theorems.

4.1 Comparison Theorem of MF-BSVIEs

In this subsection, let us discuss the comparison theorem for MF-BSVIE. For simplicity, we consider the following form of MF-BSVIE

$$Y(t) = \psi(t) + \int_t^T \Gamma^g\big(t, s, Y(s), Z(t,s)\big) - \int_t^T Z(t,s)dW(s), \quad t \in [0, T],$$

where for $(t, s) \in \Delta^*[0, T]$,

$$\Gamma^g(t, s, Y(s), Z(t,s)) \triangleq \mathbb{E}'\big[g(t, s, y, z, Y(s), Z(t,s))\big]\Big|_{(y,z)=\big(Y(s),Z(t,s)\big)}.$$

The general study of (9.3.7) (with the absence of $Z(s, t)$) can be obtained in the essentially same manner. For $i = 0, 1$, suppose there exists $(Y_i, Z_i) \in \mathcal{H}^2_{\Delta^*}[0, T]$ satisfying the following MF-BSVIE

$$Y_i(t) = \psi_i(t) + \int_t^T \Gamma_i^g(t, s, Y_i(s), Z_i(t,s)) - \int_t^T Z_i(t,s)dW(s), \quad t \in [0, T],$$

where for $(t, s) \in \Delta^*[0, T]$,

$$\Gamma_i^g(t, s, Y_i(s), Z_i(t,s)) \triangleq \mathbb{E}'\big[g_i(t, s, y, z, Y_i(s), Z_i(t,s))\big]\Big|_{(y,z)=\big(Y_i(s),Z_i(t,s)\big)}.$$

Theorem 9.4.1 *Let* $(\text{H9.2.1})_2$ *hold with* $g(\cdot)$ *replaced by* $g_i(\cdot)$. *In addition, for any* (t,s,y,z,y',z'), *one of the* $g_i(\cdot)$ *is independent of* z', *and the other is nondecreasing in* y' *and* y.

Let $\psi_i(\cdot) \in L^2_{\mathcal{F}_T}(0,T;\mathbb{R}^n)$ *and* $(Y_i(\cdot), Z_i(\cdot\,,\cdot)) \in \mathcal{H}^2_{\Delta^*}[0,T]$ *be the unique adapted solution associated with* $(\psi_i(\cdot), g_i(\cdot))$. *If* $\psi_0(\cdot) \leqslant \psi_1(\cdot)$, *and*

$$g_0(t,s,y,z,y',z') \leqslant g_1(t,s,y,z,y',z'),$$

then, it holds that

$$Y_0(t) \leqslant Y_1(t), \qquad a.s.\ t \in [0,T]. \tag{9.4.1}$$

Proof In the following, without loss of generality, let us assume $g_0(\cdot)$ is nondecreasing in y' and y, $g_1(\cdot)$ is independent of z'. We define

$$\begin{aligned}&\widehat{Y}(\cdot) \stackrel{\Delta}{=} Y_0(\cdot) - Y_1(\cdot),\ \widehat{Z}(\cdot\,,\cdot) \stackrel{\Delta}{=} Z_0(\cdot\,,\cdot) - Z_1(\cdot\,,\cdot),\\ &\widehat{\Gamma}^g(t,s,y_0,z_0,y_1,z_1) \stackrel{\Delta}{=} \Gamma_0^g(t,s,y_0,z_0) - \Gamma_1^g(t,s,y_1,z_1).\end{aligned}$$

□

As to $\widehat{\Gamma}^g$, we have

$$\begin{aligned}&\widehat{\Gamma}^g(t,s,y_0,z_0,y_1,z_1)\\ &= \mathbb{E}'\big[g_0\big(t,s,y_0,z_0,Y_0(s),Z_0(t,s)\big)\big] - \mathbb{E}'\big[g_1\big(t,s,y_1,z_1,Y_1(s)\big)\big]\\ &= \mathbb{E}'\big[g_0\big(t,s,y_0,z_0,Y_0(s),Z_0(t,s)\big) - g_0\big(t,s,y_1,z_0,Y_0(s),Z_0(t,s)\big)\big]\\ &\quad + \mathbb{E}'\big[g_0\big(t,s,y_1,z_0,Y_0(s),Z_0(t,s)\big) - g_0\big(t,s,y_1,z_1,Y_0(s),Z_0(t,s)\big)\big]\\ &\quad + \mathbb{E}'\big[g_0\big(t,s,y_1,z_1,Y_0(s),Z_0(t,s)\big) - g_0\big(t,s,y_1,z_1,Y_1(s),Z_0(t,s)\big)\big]\\ &\quad + \mathbb{E}'\big[g_0\big(t,s,y_1,z_1,Y_1(s),Z_0(t,s)\big) - g_1\big(t,s,y_1,z_1,Y_1(s)\big)\big]\\ &= \mathbb{E}'\big[g_{0,y}(t,s)\big][y_0 - y_1] + \Delta g_1(t,s) + \mathbb{E}'\big[g_{0,Y}(t,s)[Y_0(s) - Y_1(s)]\big] + \Delta g_{0,1}(t,s),\end{aligned}$$

where

$$\begin{aligned}&g_{0,y}(t,s)\\ &\stackrel{\Delta}{=} \frac{g_0\big(t,s,y_0,z_0,Y_0(s),Z_0(t,s)\big) - g_0\big(t,s,y_1,z_0,Y_0(s),Z_0(t,s)\big)}{y_0 - y_1}\mathbf{1}_{\{y_0 \neq y_1\}},\end{aligned}$$

$$\begin{aligned}&g_{0,Y}(t,s)\\ &\stackrel{\Delta}{=} \frac{g_0\big(t,s,y_1,z_1,Y_0(s),Z_0(t,s)\big) - g_0\big(t,s,y_1,z_1,Y_1(s),Z_0(t,s)\big)}{Y_0(s) - Y_1(s)}\mathbf{1}_{\{Y_0(s) \neq Y_1(s)\}}.\end{aligned}$$

By the imposed assumptions, we have

$$g_{0,Y}(t,s) \geqslant 0, \qquad g_{0,y}(t,s) \geqslant 0, \qquad \Delta g_{0,1}(t,s) \leqslant 0.$$

For $\Delta g_1(t,s)$, we have

$$|\Delta g_1(t,s)| \leqslant L|z_0 - z_1|.$$

Therefore, with $\widehat{y} \stackrel{\Delta}{=} y_0 - y_1, \widehat{z} \stackrel{\Delta}{=} z_0 - z_1$,

$$\begin{aligned}\widehat{\Gamma}^g(t,s,y_0,z_0,y_1,z_1) &\leqslant \mathbb{E}'\big[g_{0,Y}(t,s)\widehat{Y}(s)^+\big] + \mathbb{E}'\big[g_{0,y}(t,s)\big]\widehat{y}^+ + L|\widehat{z}| \\ &\leqslant L\Big\{\mathbb{E}'\big[\widehat{Y}(s)^+\big] + \widehat{y}^+ + |\widehat{z}|\Big\}.\end{aligned} \tag{9.4.2}$$

Given the above $(Y_i(\cdot), Z_i(\cdot,\cdot))$, let us consider the following BSDE parameterized by t,

$$\eta_i(t,r) = \psi_i(t) + \int_r^T \Gamma_i^g(t,s,Y_i(s),\zeta_i(t,s))ds - \int_r^T \zeta_i(t,s)dW(s), \qquad r \in [t,T].$$

By the uniqueness of BSVIEs, it is easy to see

$$\eta_i(t,t) = Y_i(t), \quad t \in [0,T], \quad \zeta_i(t,s) = Z_i(t,s), \quad (t,s) \in \Delta^*[0,T], \quad i = 0,1.$$

Similar as the above, we define

$$\begin{aligned}&\widehat{\eta}(\cdot,\cdot) \stackrel{\Delta}{=} \eta_0(\cdot,\cdot) - \eta_1(\cdot,\cdot), \quad \widehat{\zeta}(\cdot,\cdot) \stackrel{\Delta}{=} \zeta_0(\cdot,\cdot) - \zeta_1(\cdot,\cdot), \\ &\widehat{\psi}(\cdot) \stackrel{\Delta}{=} \psi_0(\cdot) - \psi_1(\cdot).\end{aligned}$$

Therefore, we have

$$\begin{aligned}\widehat{\eta}(t,r) = \widehat{\psi}(t) + \int_r^T &\widehat{\Gamma}^g\big(t,s,Y_0(s),\zeta_0(t,s),Y_1(s),\zeta_1(t,s)\big)ds \\ &- \int_r^T \widehat{\zeta}(t,s)dW(s), \quad r \in [t,T].\end{aligned}$$

Notice that from Tanaka's formula, we have

$$\begin{aligned}d\big[\widehat{\eta}(t,s)\big]^+ = \mathbf{1}_{\{\widehat{\eta}(t,s)>0\}}\Big[&-\widehat{\Gamma}^g(t,s,Y_0(s),z_0(t,s),Y_1(s),z_1(t,s))ds \\ &+\widehat{\zeta}(t,s)dW(s)\Big] + \frac{1}{2}dL(t,s).\end{aligned}$$

Here $L(t,\cdot)$ is the local time of $\widehat{\eta}(t,\cdot)$ at 0 and is an increasing continuous process (see Chap. 2, Sect. 3), satisfying

$$\int_0^r |\widehat{\eta}(t,s)|dL(t,s) = 0, \qquad \text{a.s.} \quad r \in [0,T].$$

Therefore, by Itô's formula, for undetermined constant $\beta > 0$, we have

$$
\begin{aligned}
&d\Big[e^{\beta s}\big[(\widehat{\eta}(t,s))^+\big]^2\Big]\\
&= \beta e^{\beta s}\big[(\widehat{\eta}(t,s))^+\big]^2 ds + 2e^{\beta s}\big(\widehat{\eta}(t,s)\big)^+ d\big(\widehat{\eta}(t,s)\big)^+ + e^{\beta s}\mathbf{1}_{\{\widehat{\eta}(t,s)>0\}}|\widehat{\zeta}(t,s)|^2 ds\\
&= \beta e^{\beta s}\big[(\widehat{\eta}(t,s))^+\big]^2 ds - 2e^{\beta s}\big(\widehat{\eta}(t,s)\big)^+\widehat{\Gamma}^g(t,s,Y_0(s),\zeta_0(t,s),Y_1(s),\zeta_1(t,s))ds\\
&\quad +2e^{\beta s}\big(\widehat{\eta}(t,s)\big)^+\widehat{\zeta}(t,s)dW(s) + e^{\beta s}\mathbf{1}_{\{\widehat{\eta}(t,s)>0\}}\big|\widehat{\zeta}(t,s)\big|^2 ds,
\end{aligned}
$$

from which we have

$$
\begin{aligned}
&e^{\beta r}\big[(\widehat{\eta}(t,r))^+\big]^2 + \int_r^T \beta e^{\beta s}\big[(\widehat{\eta}(t,s))^+\big]^2 ds + \int_r^T e^{\beta s}\mathbf{1}_{\{\widehat{\eta}(t,s)>0\}}|\widehat{\zeta}(t,s)|^2 ds\\
&= e^{\beta T}\big[(\widehat{\eta}(t,T))^+\big]^2 + \int_r^T 2e^{\beta s}\big(\widehat{\eta}(t,s)\big)^+\widehat{\Gamma}^g(t,s,Y_0(s),\zeta_0(t,s),Y_1(s),\zeta_1(t,s))ds\\
&\quad -\int_r^T 2e^{\beta s}\big(\widehat{\eta}(t,s)\big)^+\widehat{\zeta}(t,s)dW(s).
\end{aligned}
$$

Recall the above (9.4.2); we have that

$$
\begin{aligned}
&2\big(\widehat{\eta}(t,s)\big)^+\widehat{\Gamma}^g(t,s,Y_0(s),\zeta_0(t,s),Y_1(s),\zeta_1(t,s))\\
&\leqslant 2L\big(\widehat{\eta}(t,s)\big)^+\Big\{\mathbb{E}'\big[\widehat{Y}(s)^+\big] + \widehat{Y}(s)^+ + \big|\widehat{Z}(t,s)\big|\Big\}\\
&\leqslant L\big(\mathbb{E}'\widehat{Y}(s)^+\big)^2 + L\big(\widehat{\eta}(t,s)^+\big)^2 + L\big[\widehat{Y}(s)^+\big]^2 + L\big(\widehat{\eta}(t,s)^+\big)^2\\
&\quad +\mathbf{1}_{\{\widehat{\eta}(t,s)>0\}}\big|\widehat{Z}(t,s)\big|^2 + L^2\big(\widehat{\eta}(t,s)^+\big)^2,
\end{aligned}
$$

from which we have

$$
\begin{aligned}
&\mathbb{E}\int_r^T 2e^{\beta s}\big(\widehat{\eta}(t,s)\big)^+\widehat{\Gamma}^g(t,s,Y_0(s),\zeta_0(t,s),Y_1(s),\zeta_1(t,s))ds\\
&\leqslant [2L+2L^2]\mathbb{E}\int_r^T e^{\beta s}\big(\widehat{\eta}(t,s)^+\big)^2 ds + 2L\mathbb{E}\int_r^T e^{\beta s}\big[\widehat{Y}(s)^+\big]^2 ds\\
&\quad +\mathbb{E}\int_r^T e^{\beta s}\mathbf{1}_{\{\widehat{\eta}(t,s)>0\}}\big[\widehat{Z}(t,s)\big]^2 ds.
\end{aligned}
$$

Therefore, we have

$$
\begin{aligned}
&e^{\beta r}\mathbb{E}\big[(\widehat{\eta}(t,r))^+\big]^2 + \mathbb{E}\int_r^T \beta e^{\beta s}\big[(\widehat{\eta}(t,s))^+\big]^2 ds + \mathbb{E}\int_r^T e^{\beta s}\mathbf{1}_{\{\widehat{\eta}(t,s)>0\}}|\widehat{\zeta}(t,s)|^2 ds\\
&\leqslant e^{\beta T}\mathbb{E}\big[(\widehat{\eta}(t,T))^+\big]^2 + [2L+L^2]\mathbb{E}\int_r^T e^{\beta s}\big(\widehat{\eta}(t,s)^+\big)^2 ds\\
&\quad +2L\mathbb{E}\int_r^T e^{\beta s}\big[\widehat{Y}(s)^+\big]^2 ds + \mathbb{E}\int_r^T e^{\beta s}\mathbf{1}_{\{\widehat{\eta}(t,s)>0\}}\big[\widehat{Z}(t,s)\big]^2 ds.
\end{aligned}
$$

Now we choose $\beta = 2L + L^2 + 1$; we have

$$e^{\beta r}\mathbb{E}\big[\big(\eta(t,r)\big)^+\big]^2 \leqslant e^{\beta T}\mathbb{E}\big[\big(\widehat{\eta}(t,T)\big)^+\big]^2 + 2L\mathbb{E}\int_r^T e^{\beta s}\big[\widehat{Y}(s)^+\big]^2 ds.$$

By taking $t = r$; we have

$$e^{\beta t}\mathbb{E}\big[\big(\widehat{Y}(t)\big)^+\big]^2 \leqslant e^{\beta T}\mathbb{E}\big[\big(\widehat{\eta}(t,T)\big)^+\big]^2 + 2L\mathbb{E}\int_t^T e^{\beta s}\big[\widehat{Y}(s)^+\big]^2 ds.$$

By Gronwall inequality,

$$e^{\beta t}\mathbb{E}\big[\big(\widehat{Y}(t)\big)^+\big]^2 \leqslant K\mathbb{E}\big[\big(\widehat{\eta}(t,T)\big)^+\big]^2 = 0,$$

which indicates that $Y_0(\cdot) \leqslant Y_1(\cdot)$. □

To conclude this part, let us make some remarks. First, the above proof relies on Tanaka's formula and relevant arguments, which is different from Theorem 6.2.3 in Chap. 6. Second, one also discusses the comparison theorem under certain conditions when $Y(\cdot)$ has some proper continuity as in Theorem 6.2.3 in Chap. 6. Also for the analogue of Propositions 6.2.8 and 6.3.2 in Chap. 6, the proofs are similar. We prefer to leave these details for the interested readers.

4.2 Representation Theorem of MF-BSVIEs

In this section, we give a representation of the solutions to MF-BSVIEs by means of a class of mean-field SDEs and a class of nonlocal PDEs.

Consider a mean-field SDE of the form

$$\begin{cases} dX^{x_0}(s) = \Gamma^b(s, X^{x_0}(s))ds + \Gamma^\sigma(s, X^{x_0}(s))dW(s), & s \in [0,T], \\ X^{x_0}(0) = x_0, \end{cases} \tag{9.4.3}$$

where for $(s,x) \in [0,T] \times \mathbb{R}^n$,

$$\Gamma^b(s,x) \triangleq \mathbb{E}'\big[b\big(s,x,X^{x_0}(s)\big)\big], \quad \Gamma^\sigma(s,x) \triangleq \mathbb{E}'\big[\sigma\big(s,x,X^{x_0}(s)\big)\big].$$

For the above coefficients, we introduce the following assumptions.

(H9.4.1) The maps $b, \sigma : [0,T] \times \Omega \times \mathbb{R}^n \times \mathbb{R}^n \to \mathbb{R}^n$ are measurable, and for all $(x,y) \in \mathbb{R}^{2n}$, $(s,\omega') \mapsto (b(s,\omega',x,y), \sigma(s,\omega',x,y))$ is $\mathbb{F}$-progressively measurable on $[0,T]$. In addition, there exists constant L such that

$$\begin{aligned} &|b(s,x_1,y_1) - b(s,x_2,y_2)| + |\sigma(s,x_1,y_1) - \sigma(s,x_2,y_2)| \\ &\leqslant L(|x_1 - x_2| + |y_1 - y_2|), \quad (s,\omega') \in [0,T] \times \Omega, \quad x_i, y_i \in \mathbb{R}^n. \end{aligned}$$

In addition,

$$|b(s,x,\gamma)|+|\sigma(s,x,\gamma)|\leqslant L(1+|x|+|\gamma|),\quad s\in[0,T],\quad x,\ \gamma\in\mathbb{R}^n.$$

By following the similar ideas as in Theorem 9.1.1, we deduce the existence and uniqueness of the process $X^{x_0}(\cdot)\in L^2_{\mathbb{F}}(\Omega;C([0,T];\mathbb{R}^n))$ for the mean-field SDE (9.4.3). Here the solution can have continuous path due to the differential system character. In the following, when there is no confusion we simplify $X^{x_0}(\cdot)$ as $X(\cdot)$. Next let us look at the BSVIE,

$$\begin{aligned}Y^{x_0}(r)=&\Gamma^{\psi}\big(r,X^{x_0}(T)\big)+\int_r^T\Gamma^g\big(r,s,X^{x_0}(s),Y^{x_0}(s),Z^{x_0}(r,s)\big)ds\\&-\int_r^T Z^{x_0}(r,s)dW(s),\quad r\in[0,T].\end{aligned}\tag{9.4.4}$$

where for $(r,s)\in\Delta^*[0,T]$, $x,y,z\in\mathbb{R}^n$,

$$\begin{aligned}&\Gamma^{\psi}(r,x)\triangleq\mathbb{E}'\big[\psi\big(r,x,X^{x_0}(T)\big)\big],\\&\Gamma^g(r,s,x,y,z)\triangleq\mathbb{E}'\big[g\big(r,s,x,X^{x_0}(s),y,z,Y^{x_0}(s),Z^{x_0}(r,s)\big)\big].\end{aligned}$$

In the following, we define $(Y(\cdot),Z(\cdot))\triangleq(Y^{x_0}(\cdot),Z^{x_0}(\cdot))$. We introduce the following assumptions on $\psi(\cdot)$ and $g(\cdot)$.

(H9.4.2) $\psi:[0,T]\times\Omega\times\mathbb{R}^n\times\mathbb{R}^n\to\mathbb{R}^n$ is an $\mathcal{F}_T$-measurable stochastic process and $g:\Omega\times\Delta[0,T]\times\mathbb{R}^{2n}\times\mathbb{R}^{4m}\to\mathbb{R}^n$ is a measurable mapping such that $(s,\omega')\mapsto g(t,s,\omega',x,x',y,z,y',z')$ is $\mathcal{F}$-progressively measurable on $[t,T]$. In addition, there exists a constant L such that

$$\begin{aligned}&\big|g(t,s,x_1,\widehat{x}_1,y_1,z_1,\widehat{y}_1,\widehat{z}_1)-g(t,s,x_2,\widehat{x}_2,y_2,z_2,\widehat{y}_2,\widehat{z}_2)\big|\\&\leqslant L\big(|x_1-x_2|+|\widehat{x}_1-\widehat{x}_2|+|y_1-y_2|+|\widehat{y}_1-\widehat{y}_2|+|z_1-z_2|+|\widehat{z}_1-\widehat{z}_2|\big),\\&\qquad\qquad(s,\omega')\in[0,T]\times\Omega,\quad x_i,\widehat{x}_i,y_i,\widehat{y}_i,z_i,\widehat{z}_i\in\mathbb{R}^n,\end{aligned}$$

and

$$\begin{aligned}&\big|g(t,s,x,\widehat{x},y,z,\widehat{y},\widehat{z})\big|\leqslant L\big(1+|x|+|\widehat{x}|+|y|+|\widehat{y}|+|z|+|\widehat{z}|\big),\\&\qquad\qquad(s,\omega')\in[0,T]\times\Omega,\quad x,\widehat{x},y,\widehat{y},z,\widehat{z}\in\mathbb{R}^n.\end{aligned}$$

Under (H9.4.2), by Theorem 9.3.2, it is easy to see that MF-BSVIE (9.4.4) admits a unique adapted solution $(Y(\cdot),Z(\cdot))$. We introduce the following PDE

$$\begin{cases}\Theta_s(t,s,x)+\dfrac{1}{2}\Gamma^{\sigma}(s,x)^{\top}\Theta_{xx}(t,s,x)\Gamma^{\sigma}(s,x)+\Theta_x(t,s,x)\Gamma^b(s,x)\\\qquad+\widehat{\Gamma}^g\big(t,s,x,\Theta(s,s,x),\Theta_x(t,s,x)\sigma(s,x)\big)=0,\\\qquad\qquad(t,s,x)\in\Delta^*[0,T]\times\mathbb{R}^n,\\\Theta(t,T,x)=\Gamma^{\psi}(t,x),\qquad(t,x)\in[0,T]\times\mathbb{R}^n,\end{cases}\tag{9.4.5}$$

where for $(t, s, x, y, z) \in \Delta^*[0, T] \times \mathbb{R}^n \times \mathbb{R}^{2m}$,

$$\begin{aligned}&\widehat{\Gamma}^g(t, s, x, y, z)\\&\quad\stackrel{\triangle}{=} \mathbb{E}'\big[g\big(t, s, x, X(s), y, z, \Theta(s, s, X(s)), \Theta_x(t, s, X(s))\Gamma^\sigma(s, X(s))\big)\big].\end{aligned}$$

Theorem 9.4.2 *Let* (H9.4.1)–(H9.4.2) *hold. Let* $\Theta : \Delta^*[0, T] \times \mathbb{R}^n \times \mathbb{R}^n \to \mathbb{R}$ *be the classical solution of the representation PDE (9.4.5). Let* $(Y(\cdot), Z(\cdot\,,\cdot))$ *be the adapted solution to the Type-I BSVIE (9.4.4) with* $X(\cdot)$ *being the solution to SDE (9.4.3). Then the following representation holds*

$$Y(s) \stackrel{\triangle}{=} \Theta(s, s, X(s)), \quad Z(r, s) \stackrel{\triangle}{=} \Theta_x(r, s, X(s))\Gamma^\sigma(s, X(s)). \tag{9.4.6}$$

Proof For fixed $r \in [0, T]$, applying Itô's formula to $s \mapsto \Theta(r, s, X(s))$ on $[r, T]$, we have

$$\begin{aligned}d\Theta(r, s, X(s)) &= \Theta_s(r, s, X(s))ds + \Theta_x(r, s, X(s))\Gamma^b(s, X(s))ds\\&\quad+\Theta_x(r, s, X(s))\Gamma^\sigma(s, X(s))dW(s)\\&\quad+\frac{1}{2}\big[\Gamma^\sigma(s, X(s))\big]^\top\Theta_{xx}(r, s, X(s))\Gamma^\sigma(s, X(s))ds.\end{aligned}$$

Since $\Theta(\cdot\,,\cdot\,,\cdot\,,\cdot)$ satisfies PDE (9.4.5), one has

$$\begin{aligned}d\Theta(r, s, X(s)) &= -\widehat{\Gamma}^g\big(r, s, X(s), \Theta(s, s, X(s)), \Theta_x(r, s, X(s))\Gamma^\sigma(s, X(s))\big)ds\\&\quad+\Theta_x(r, s, X(s))\Gamma^\sigma(s, X(s))dW(s),\end{aligned}$$

and $\Theta(r, T, X(T)) = \Gamma^\psi(r, X(T))$. Now, for $s \geqslant r \geqslant 0$, we define

$$\begin{cases}\eta(r, s) \stackrel{\triangle}{=} \Theta(r, s, X(s)),\\ Z(r, s) \stackrel{\triangle}{=} \Theta_x(r, s, X(s))\Gamma^\sigma(s, X(s)).\end{cases}$$

In particular, let $r = s$; we have $\eta(s, s) = \Theta(s, s, X(s))$. Recall that for $0 \leqslant r \leqslant s \leqslant T$,

$$\begin{aligned}&\widehat{\Gamma}^g\big(r, s, X(s), \eta(s, s), Z(r, s)\big)\\&= \mathbb{E}'\big[g\big(r, s, x, X(s), y, z, \Theta(s, s, X(s)), \Theta_x(r, s, X(s))\\&\qquad\cdot\Gamma^\sigma(s, X(s))\big)\big]_{x=X(s),y=\eta(s,s),z=Z(r,s)}\\&= \mathbb{E}'\big[g\big(r, s, x, X(s), y, z, \eta(s, s), Z(r, s)\big)\big]_{x=X(s),y=\eta(s,s),z=Z(r,s)}.\end{aligned}$$

For $0 \leqslant r \leqslant \tau \leqslant T$, we have

$$\begin{aligned}\Theta(r, \tau, X(\tau)) &= \Gamma^\psi(r, X(T)) - \int_\tau^T Z(r, s)dW(s)\\&\quad+\int_\tau^T \mathbb{E}'\big[g\big(r, s, x, X(s), y, z, \eta(s, s), Z(r, s)\big)\big]_{x=X(s),y=\eta(s,s),z=Z(r,s)}ds.\end{aligned}$$

Let $r = \tau$; we have

$$\eta(r,r) = \Gamma^{\psi}(r, X(T)) - \int_r^T Z(r,s)dW(s)$$
$$+ \int_r^T \mathbb{E}'\big[g\big(r,s,x,X(s),y,z,\eta(s,s),Z(r,s)\big)\big]_{x=X(s),y=\eta(s,s),z=Z(r,s)} ds.$$

By the uniqueness of mean-field BSVIE (9.4.4), we have the above (9.4.6). □

□

Similar to Theorem 7.4.2 in Chap. 7, we can establish a result for Type-II MF-BSVIEs. We prefer to leave the details to the interested readers.

5 Bibliographical Remarks

McKean-Vlasov SDE was suggested by Kac in 1956 as a stochastic toy model for the Vlasov kinetic equation of plasma, and the study of which was initiated by McKean in 1966. Since then, many authors made contributions on McKean-Vlasov type SDEs and applications; see, for examples, Dawson [54], Sznitman [181], Chan [41], and Ahmed-Ding [8]. In recent years, related topics and problems have attracted more and more attentions; see, for examples, Huang-Li-Wang [106], Huang-Malhamé-Caines [107], Lasry-Lions [132], Borkar-Kumar [35], Kotelenez-Kurtz [126], and so on.

On the other hand, mean-field backward stochastic differential equations (MF-BSDEs, for short) were introduced in Buckdahn-Li-Peng [39] where a class of nonlocal partial differential equations are studied with the help of an MF-BSDE and a McKean-Vlasov forward equation. Buckdahn-Djehiche-Li-Peng [38] studied the solution of a mean-field backward stochastic differential equation driven by a forward stochastic differential of McKean-Vlasov type that can be approximated by the solution of some decoupled forward-backward equation with coefficients governed by N independent copies. They also proved a central limit theorem for MF-BSDEs where the limit equation is a forward-backward stochastic differential equation of mean-field type which is not only governed by a Brownian motion but also by an independent Gaussian field.

The contents from Sects. 1 to 3 are adapted from Shi-Wang-Yong [176], while the study of Sect. 4 is slight (partial) generalization of the results in previous chapters. In fact, we see that MF-BSVIE not only includes MF-BSDEs (which, of course, also includes standard BSDEs) introduced in [39], [38] but also generalizes BSVIEs studied in the previous chapters in a natural way. Besides, investigating MF-BSVIEs allows us to meet the need in the study of optimal control problem for MF-FSVIEs. In fact, when studying Pontryagin maximum principle, the adjoint equation is naturally be an MF-BSVIE which well-posedness is not avoidable. We refer to Shi-Wang-Yong [176] for the maximum principle of MF-SVIE with convex control domain. Also for mean-field games of FSVIEs, MF-BSVIEs should also be useful. We refer to Huang-Li-Wang [106] for a special class of mean-field linear quadratic games of FSVIEs. On the other hand, the works of Meng-Wang-Zhang [146] and Meng-Shi-Wang-Zhang [147] indicate that BSVIEs are also quite useful in studying optimal control problem of stochastic delay differential equations (SDDEs). Besides, FSVIEs are also closely connected with stochastic (time) fractional differential system and stochastic evolution equations; see Wang-Zheng [209]. In the same spirits, when studying the mean-field games problem or mean-field control problem of these systems, the MF-BSVIEs should play important roles.

Chapter 10
Optimal Control for FSVIEs

1 General Considerations

Let us start with a classical stochastic optimal control problem (for an FSDE). To this end, we consider the following controlled FSDE:

$$\begin{cases} dX(s) = b(s, X(s), u(s))ds + \sigma(s, X(s), u(s))dW(s), & s \in [0, T], \\ X(0) = x, \end{cases} \tag{10.1.1}$$

with *Bolza cost functional*:

$$J^0(u(\cdot)) = \mathbb{E}\Big[h(X(T)) + \int_0^T g(t, X(t), u(t))dt\Big], \tag{10.1.2}$$

where b, σ, h, g are some suitable maps, the *control* $u(\cdot)$ is taken from some suitable set $\mathscr{U}[0, T]$ (to be clearly specified later), and the *state process* $X(\cdot)$ is valued in $\mathbb{R}^n$. In the above, we omit the dependence of cost functional J^0 on x for notational simplicity. Also the components of $X(\cdot)$ could be wealth, owned commodities/assets, inventory of products, and some economic factors (interest rates, unemployment rate) and so on. Also, $u(\cdot)$ can be regarded as some kind of investment, production effort, labor force, transaction of assets, etc. Then our classical optimal control problem can be stated as follows.

Problem (C)^0 For given $x \in \mathbb{R}^n$, find a $\bar{u}(\cdot) \in \mathscr{U}[0, T]$, called an *optimal control*, such that

$$J^0(\bar{u}(\cdot)) = \inf_{u(\cdot)\in\mathscr{U}[0,T]} J^0(u(\cdot)). \tag{10.1.3}$$

T. Wang, J. Yong, *Backward Stochastic Volterra Integral Equations*,
Probability Theory and Stochastic Modelling 109,
https://doi.org/10.1007/978-3-032-20848-4_10

Now let us make some further analysis on equation (10.1.1) which can be written as follows:

$$X(t) = x + \int_0^t b(s, X(s), u(s))ds + \int_0^t \sigma(s, X(s), u(s))dW(s), \qquad t \in [0, T]. \tag{10.1.4}$$

We observe the following:

$$X(t + \Delta t) - X(t) = \int_t^{t+\Delta t} b(s, X(s), u(s))ds + \int_t^{t+\Delta t} \sigma(s, X(s), u(s)dW(s).$$

This means the increment $X(t+\Delta t) - X(t)$ of the state on $[t, t+\Delta t]$ is completely determined by the corresponding "driving force" in this time interval. Thus, the "infinitesimal" rate of the change for the state is *memoryless*. However, in reality, memory cannot be avoided. For example, the current production level and rate usually depends on some renovation of production equipment some time ago, the profit of investment usually depends on the transactions some time before, the air pollution is caused by some bad production strategies some years ago, etc. To model various possible situations with memory, instead of (10.1.4), we may consider the following controlled FSVIE:

$$X(t) = \varphi(t) + \int_0^t b(t, s, X(s), u(s))ds + \int_0^t \sigma(t, s, X(s), u(s))dW(s), \qquad t \in [0, T]. \tag{10.1.5}$$

Unlike (10.1.4), due to the dependence of $\big(b(t, s, x, u), \sigma(t, s, x, u)\big)$ on t, even for the case $\varphi(t) \equiv x$, we have

$$\begin{aligned} &X(t + \Delta t) - X(t) \\ &= \int_t^{t+\Delta t} b(t + \Delta t, s, X(s), u(s))ds + \int_t^{t+\Delta t} \sigma(t + \Delta t, s, X(s), u(s)dW(s) \\ &\quad + \int_0^t \big[b(t + \Delta t, s, X(s), u(s)) - b(t, s, X(s), u(s))\big]ds \\ &\quad + \int_0^t \big[\sigma(t + \Delta t, s, X(s), u(s)) - \sigma(t, s, X(s), u(s))\big]dW(s), \end{aligned}$$

which depends not only on the values of the "driving force" in $[t, t+\Delta t]$ but also on those in the whole interval $[0, t]$ up to the current time t. Therefore, with suitable choices of b and σ, it is possible to model certain memory effects through (10.1.5).

Based on these arguments, we now discuss optimal control problems for FSVIE (10.1.5), where $b, \sigma : \Delta_*[0, T] \times \mathbb{R}^n \times U \times \Omega \to \mathbb{R}^n$ are given maps and

$\varphi(\cdot) \in C_{\mathbb{F}}([0, T]; L^2(\Omega; \mathbb{R}^n))$; U is a subset in $\mathbb{R}^m$. The cost functional is defined to be the following Bolza form:

$$J(u(\cdot)) = \mathbb{E}\Big[h(X(T)) + \int_0^T g(t, X(t), u(t))dt\Big], \tag{10.1.6}$$

with $g : [0, T] \times \mathbb{R}^n \times U \times \Omega \to \mathbb{R}$ and $h : \mathbb{R}^n \times \Omega \to \mathbb{R}$ being given maps. Under proper conditions (see the later sections in this chapter), (10.1.5) admits a unique solution $X(\cdot) \in C_{\mathbb{F}}([0, T]; L^p(\Omega; \mathbb{R}^n))$ (see Chap. 3, Proposition 3.4.4). Thus, the cost functional $J(u(\cdot))$ is well-defined. We point out that $J(u(\cdot))$ should depend on $\varphi(\cdot)$. For notational simplicity, we omit such an obvious reliance. Based on the above, our optimal control problem can be stated as follows.

Problem (C) Find a $\bar{u}(\cdot) \in \mathscr{U}[0, T]$ such that

$$J(\bar{u}(\cdot)) = \inf_{u(\cdot)\in\mathscr{U}[0,T]} J(u(\cdot)). \tag{10.1.7}$$

Any $\bar{u}(\cdot)$ satisfying (10.1.7) is called an optimal control of Problem (C), the corresponding state process $\bar{X}(\cdot)$ is called an optimal state process, and $(\bar{X}(\cdot), \bar{u}(\cdot))$ is called an optimal pair. The difference between this problem and Problem (C)^0 is that the state equations are different.

For the above optimal control problem, one of the central problems is to obtain the necessary conditions for optimal controls, which might lead to a construction of optimal controls. In this chapter, we are going to achieve this for two cases that the control domain U is convex or is not assumed so. It turns out that these two situations are significantly different: The latter is much more difficult. For simplicity, we assume that $W(\cdot)$ is a one-dimensional standard Brownian motion.

2 Maximum Principle for Convex Control Domains

In this section, we establish necessary conditions for optimal controls, called *maximum principle*, for the case that the control domain U is convex in $\mathbb{R}^m$. To begin with, in the current subsection, we define

$$\mathscr{U}[0, T] \equiv L^2_{\mathbb{F}}(0, T; U) \equiv \Big\{u : [0, T] \times \Omega \to U \;\Big|\; u(\cdot) \text{ is} \mathbb{F}\text{-progressively measurable}, \;\; \mathbb{E}\int_0^T |u(t)|^2 dt < \infty\Big\}. \tag{10.2.1}$$

We now introduce the following assumption. The conditions assumed are more than enough. One can relax many of them. But we prefer these strong conditions to make the presentation simpler.

(H10.2.1) Let $U \subseteq \mathbb{R}^m$ be convex, $b, \sigma : \Delta_*[0,T] \times \mathbb{R}^n \times U \times \Omega \to \mathbb{R}^n$, $g : \Delta_*[0,T] \times \mathbb{R}^n \times U \times \Omega \to \mathbb{R}$, and $h : \mathbb{R}^n \times \Omega \to \mathbb{R}$ be measurable such that $s \mapsto \big(b(t,s,x,u), \sigma(t,s,x,u), g(s,x,u)\big)$ is $\mathbb{F}$-progressively measurable on $[0,t]$, for each $t \in [0,T]$, and differentiable in the variables (x,u), with the derivatives being uniformly bounded and continuous in all the arguments. Moreover, there exists an $L > 0$ and modulus ρ such that

$$
\begin{aligned}
&|b(t,s,0,u)| + |\sigma(t,s,0,u)| \leqslant L, \qquad (t,s,u) \in \Delta_*[0,T] \times U,\\
&|f(t,s,x,u) - f(\bar t,s,x,u)| \leqslant \rho(|t-\bar t|)\big(1+|x|\big),\\
&\qquad\qquad f \equiv b,\sigma,\quad t,\bar t \in [s,T],\quad s \in [0,T),\quad (x,u) \in \mathbb{R}^n \times U.
\end{aligned}
\tag{10.2.2}
$$

By Proposition 3.4.4 in Chap. 3, we see that under (H10.2.1), for any $\varphi(\cdot) \in C_{\mathbb{F}}([0,T]; L^2(\Omega;\mathbb{R}^n))$ and $u(\cdot) \in \mathscr{U}[0,T]$, (10.1.5) admits a unique solution $X(\cdot) \in C_{\mathbb{F}}([0,T]; L^2(\Omega;\mathbb{R}^n))$ and $J(u(\cdot))$ is well-defined.

We recall Convention 1 from Chap. 1 Sect. 2 concerning differentiable maps. We now prove the following theorem called *Pontryagin's maximum principle* with convex control domains.

Theorem 10.2.1 *Let* (H10.2.1) *hold, and* $(\bar X(\cdot), \bar u(\cdot))$ *be an optimal pair of Problem* (C). *Then there exists a unique* $\zeta(\cdot)$ *satisfying*

$$
h_x(\bar X(T))^\top = \mathbb{E}\big[h_x(\bar X(T))^\top\big] + \int_0^T \zeta(s)dW(s), \tag{10.2.3}
$$

and a unique adapted M-solution $\big(\bar Y(\cdot), \bar Y_0(\cdot), \bar Z(\cdot\,,\cdot), \bar Z_0(\cdot\,,\cdot)\big)$ *of the following BSVIEs: for* $t \in [0,T]$,

$$
\left\{
\begin{aligned}
\bar Y(t) &= g_x(t,\bar X(t),\bar u(t))^\top + b_x(T,t,\bar X(t),\bar u(t))^\top h_x(\bar X(T))^\top\\
&\qquad + \sigma_x(T,t,\bar X(t),\bar u(t))^\top \zeta(t) + \int_t^T \big[b_x(s,t,\bar X(t),\bar u(t))^\top \bar Y(s)\\
&\qquad + \sigma_x(s,t,\bar X(t),\bar u(t))^\top \bar Z(s,t)\big]ds - \int_t^T \bar Z(t,s)dW(s),\\
\bar Y_0(t) &= g_u(t,\bar X(t),\bar u(t))^\top + b_u(T,t,\bar X(t),\bar u(t))^\top h_x(\bar X(T))^\top\\
&\qquad + \sigma_u(T,t,\bar X(t),\bar u(t))^\top \zeta(t) + \int_t^T \big[b_u(s,t,\bar X(t),\bar u(t))^\top \bar Y(s)\\
&\qquad + \sigma_u(s,t,\bar X(t),\bar u(t))^\top \bar Z(s,t)\big]ds - \int_t^T \bar Z_0(t,s)dW(s),
\end{aligned}
\right.
\tag{10.2.4}
$$

such that

$$
\big\langle \bar Y_0(t), u - \bar u(t)\big\rangle \geqslant 0, \qquad \forall u \in U,\ t \in [0,T],\ \text{a.s.} \tag{10.2.5}
$$

Note that the second equation in (10.2.4) is a BSVIE in which the drift and the free term do not depend on $(\bar{Y}_0(\cdot), \bar{Z}_0(\cdot\,,\cdot))$.

Proof Let $(\bar{X}(\cdot), \bar{u}(\cdot))$ be an optimal pair of Problem (C). Take any $u(\cdot) \in \mathscr{U}[0, T]$. Since U is convex, so is $\mathscr{U}[0, T]$. Thus, for any $\varepsilon \in (0, 1)$,

$$u_\varepsilon(\cdot) \triangleq \bar{u}(\cdot) + \varepsilon[u(\cdot) - \bar{u}(\cdot)] \in \mathscr{U}[0, T]. \tag{10.2.6}$$

This means that the above *convex perturbation* is admissible. Let $X_\varepsilon(\cdot)$ be the solution of (10.1.5) corresponding to $u_\varepsilon(\cdot)$. Define

$$\xi_\varepsilon(t) = \frac{X_\varepsilon(t) - \bar{X}(t)}{\varepsilon}, \qquad t \in [0, T]. \tag{10.2.7}$$

Then $\xi_\varepsilon(\cdot) \to \xi(\cdot)$ in $L^2_{\mathbb{F}}(0, T; \mathbb{R}^n)$ with $\xi(\cdot)$ satisfying the following:

$$\begin{aligned}\xi(t) =& \int_0^t \Big\{b_x(t, s, \bar{X}(s), \bar{u}(s))\xi(s) + b_u(t, s, \bar{X}(s), \bar{u}(s))\big[u(s) - \bar{u}(s)\big]\Big\}ds\\ &+ \int_0^t \Big\{\sigma_x(t, s, \bar{X}(s), \bar{u}(s))\xi(s)\\ &\qquad + \sigma_u(t, s, \bar{X}(s), \bar{u}(s))\big[u(s) - \bar{u}(s)\big]\Big\}dW(s)\\ \equiv& \,\bar{\varphi}(t) + \int_0^t b_x(t, s, \bar{X}(s), \bar{u}(s))\xi(s)ds\\ &+ \int_0^t \sigma_x(t, s, \bar{X}(s), \bar{u}(s))\xi(s)dW(s),\end{aligned} \tag{10.2.8}$$

where

$$\begin{aligned}\bar{\varphi}(t) =& \int_0^t b_u(t, s, \bar{X}(s), \bar{u}(s))\big[u(s) - \bar{u}(s)\big]ds\\ &+ \int_0^t \sigma_u(t, s, \bar{X}(s), \bar{u}(s))\big[u(s) - \bar{u}(s)\big]dW(s), \qquad t \in [0, T].\end{aligned} \tag{10.2.9}$$

Now, let $(\bar{Y}(\cdot), \bar{Y}_0(\cdot), \bar{Z}(\cdot\,,\cdot), \bar{Z}_0(\cdot\,,\cdot))$ be the unique adapted M-solution to BSVIE (10.2.4). By the optimality of $(\bar{X}(\cdot), \bar{u}(\cdot))$, and Theorem 5.4.1 (duality principle) of Chap. 5, we have

$$\begin{aligned}0 \leqslant& \frac{J(u_\varepsilon(\cdot)) - J(\bar{u}(\cdot))}{\varepsilon}\\ \to& \,\mathbb{E}\Big\{h_x(\bar{X}(T))\xi(T) + \int_0^T \{g_x(t, \bar{X}(t), \bar{u}(t))\xi(t)\\ &\qquad + g_u(t, \bar{X}(t), \bar{u}(t))[u(t) - \bar{u}(t)]\}dt\Big\}\\ =& \,\mathbb{E}\Big\{h_x(\bar{X}(T))\bar{\varphi}(T) + \int_0^T \Big[\big\langle \bar{Y}(t), \bar{\varphi}(t)\big\rangle + g_u(t, \bar{X}(t), \bar{u}(t))\big[u(t) - \bar{u}(t)\big]\Big]dt\Big\}\end{aligned}$$

$$
\begin{aligned}
&= \mathbb{E}\int_0^T \Big\langle b_u(T,t,\bar{X}(t),\bar{u}(t))^\top h_x(\bar{X}(T))^\top + \sigma_u(T,t,\bar{X}(t),\bar{u}(t))^\top \bar{\zeta}(t) \\
&\qquad + \int_t^T \big[b_u(s,t,\bar{X}(t),\bar{u}(t))^\top \bar{Y}(s) + \sigma_u(s,t,\bar{X}(t),\bar{u}(t))^\top \bar{Z}(s,t)\big]ds \\
&\qquad\qquad + g_u(t,\bar{X}(t),\bar{u}(t))^\top, u(t)-\bar{u}(t)\Big\rangle dt \\
&= \mathbb{E}\int_0^T \big\langle \bar{Y}_0(t), u(t)-\bar{u}(t)\big\rangle dt.
\end{aligned}
$$

Since the above holds for all $u(\cdot) \in \mathscr{U}[0,T]$, we obtain (10.2.5). □

Now, we define

$$
\begin{aligned}
H(t,x,u,\eta,\zeta,\bar{Y}(\cdot),\bar{Z}(\cdot\,,t)) &\stackrel{\Delta}{=} g(t,x,u)^\top + b(T,t,x,u)^\top\eta + \sigma(T,t,x,u)^\top\zeta \\
&\quad + \int_t^T b(s,t,x,u)^\top \bar{Y}(s)ds + \int_t^T \sigma(s,t,x,u)^\top \bar{Z}(s,t)ds.
\end{aligned}
\tag{10.2.10}
$$

Then (10.2.5) can be written as

$$
\Big\langle \mathbb{E}_t H_u\big(t,\bar{X}(t),\bar{u}(t),h_x(X(T))^\top,\zeta(t),\bar{Y}(\cdot),\bar{Z}(\cdot\,,t)\big), u-\bar{u}(t)\Big\rangle \geqslant 0. \tag{10.2.11}
$$

We call $H(\cdot)$ defined by (10.2.10) the *Hamiltonian* of our optimal control problem, call (10.2.5) (or (10.2.11)) the *maximum condition*, and call the first BSVIE in (10.2.4) the *first-order adjoint equation*, along the optimal pair $(\bar{X}(\cdot),\bar{u}(\cdot))$.

By putting (10.1.5) and (10.2.4)–(10.2.5) together (dropping bars), we obtain the following *optimality system*:

$$
\begin{cases}
X(t) = \varphi(t) + \displaystyle\int_0^t b(t,s,X(s),u(s))ds + \int_0^t \sigma(t,s,X(s),u(s))dW(s), \\
Y(t) = g_x(t,X(t),u(t))^\top + b_x(T,t,X(t),u(t))^\top h_x(X(T))^\top \\
\qquad + \sigma_x(T,t,X(t),u(t))^\top \zeta(t) + \displaystyle\int_t^T \Big[b_x(s,t,X(t),u(t))^\top Y(s) \\
\qquad + \sigma_x(s,t,X(t),u(t))^\top Z(s,t)\Big]ds - \displaystyle\int_t^T Z(t,s)dW(s), \\
Y_0(t) = g_u(t,X(t),u(t))^\top + b_u(T,t,X(t),u(t))^\top h_x(X(T))^\top \\
\qquad + \sigma_u(T,t,X(t),u(t))^\top \zeta(t) + \displaystyle\int_t^T \Big[b_u(s,t,X(t),u(t))^\top Y(s) \\
\qquad + \sigma_u(s,t,X(t),u(t))^\top Z(s,t)\Big]ds - \displaystyle\int_t^T Z_0(t,s)dW(s), \\
h_x(X(T))^\top = \mathbb{E}_t[h_x(X(T))^\top] + \displaystyle\int_t^T \zeta(s)dW(s), \\
\big\langle Y_0(t), u-u(t)\big\rangle \geqslant 0, \qquad \forall u \in U,\ t \in [0,T],\ \text{a.s.}
\end{cases}
\tag{10.2.12}
$$

This is a coupled systems of FSVIE and BSVIE. The coupling is through the maximum condition (via $u(\cdot)$). We call (10.2.12) a *forward-backward stochastic Volterra integral equation* (FBSVIE, for short).

Remark 10.2.2 Note that in case $U = [-1, 1]$, the maximum condition (10.2.11) is equivalent to the following:

$$\begin{aligned}\bar u(t)=\operatorname{sgn}\Big\{ -\mathbb{E}_t\Big[g_u(t,\bar X(t),\bar u(t))^\top+b_u(T,t,\bar X(t),\bar u(t))^\top h_x(\bar X(T))^\top\\ +\sigma_u(T,t,\bar X(t),\bar u(t))^\top\bar\zeta(t)+\int_t^T\Big(b_u(s,t,\bar X(t),\bar u(t))^\top\bar Y(s)\\ +\sigma_u(s,t,\bar X(t),\bar u(t))^\top\bar Z(s,t)\Big)ds\Big]\Big\},\quad t\in[0,T],\ \text{a.s.}\end{aligned}\tag{10.2.13}$$

We may call (10.2.13) a *stochastic bang-bang principle* for Problem (C).

To conclude this subsection, we look at a special case by returning back to Problem $(C)^0$, where the state equation (10.1.5) reduces to an FSDE. Then, the equation for $(\bar Y(\cdot), \bar Z(\cdot\,,\cdot))$ reads

$$\begin{aligned}\bar Y(t)&=g_x(t,\bar X(t),\bar u(t))^\top+b_x(t,\bar X(t),\bar u(t))^\top h_x(\bar X(T))^\top\\ &\quad+\sigma_x(t,\bar X(t),\bar u(t))^\top\zeta(t)+\int_t^T\Big[b_x(t,\bar X(t),\bar u(t))^\top\bar Y(s)\\ &\quad+\sigma_x(t,\bar X(t),\bar u(t))^\top\bar Z(s,t)\Big]ds-\int_t^T\bar Z(t,s)dW(s)\\ &=g_x(t,\bar X(t),\bar u(t))^\top+b_x(t,\bar X(t),\bar u(t))^\top\Big[h_x(\bar X(T))^\top+\int_t^T\bar Y(s)ds\Big]\\ &\quad+\sigma_x(t,\bar X(t),\bar u(t))^\top\Big[\zeta(t)+\int_t^T\bar Z(s,t)ds\Big]-\int_t^T\bar Z(t,s)dW(s)\\ &\equiv g_x(t,\bar X(t),\bar u(t))^\top+b_x(t,\bar X(t),\bar u(t))^\top\mathcal{Y}(t)\\ &\quad+\sigma_x(t,\bar X(t),\bar u(t))^\top\mathcal{Z}(t)-\int_t^T\bar Z(t,s)dW(s),\end{aligned}\tag{10.2.14}$$

where for $t \in [0, T]$,

$$\mathcal{Y}(t)=h_x(\bar X(T))^\top+\int_t^T\bar Y(s)ds,\qquad \mathcal{Z}(t)=\zeta(t)+\int_t^T\bar Z(s,t)ds.$$

Applying $\mathbb{E}_t$ to the both sides of (10.2.14), we have (note $\bar Y(\cdot)$ is $\mathbb{F}$-adapted)

$$\begin{aligned}\bar Y(t)&=\mathbb{E}_t[\bar Y(t)]\\ &=g_x(t,\bar X(t),\bar u(t))^\top+b_x(t,\bar X(t),\bar u(t))^\top\mathbb{E}_t\Big[h_x(\bar X(T))^\top+\int_t^T\bar Y(s)ds\Big]\\ &\qquad+\sigma_x(t,\bar X(t),\bar u(t))^\top\mathbb{E}_t\Big[\zeta(t)+\int_t^T\bar Z(s,t)ds\Big]\\ &\equiv g_x(t,\bar X(t),\bar u(t))^\top+b_x(t,\bar X(t),\bar u(t))^\top p(t)+\sigma_x(t,\bar X(t),\bar u(t))^\top q(t),\end{aligned}$$

where for $t \in [0, T]$,

$$p(t) = \mathbb{E}_t[\mathcal{Y}(t)], \qquad q(t) = \mathcal{Z}(t).$$

Then given $t \in [0, T]$,

$$\begin{aligned}
p(t) &= \mathbb{E}_t[\mathcal{Y}(t)] = \mathbb{E}_t\Big[h_x(\bar{X}(T))^\top + \int_t^T \bar{Y}(s)ds\Big] \\
&= h_x(\bar{X}(T))^\top - \int_t^T \zeta(s)dW(s) + \int_t^T \mathbb{E}_t[\bar{Y}(s)]ds \\
&= h_x(\bar{X}(T))^\top - \int_t^T \zeta(s)dW(s) + \int_t^T \Big(\bar{Y}(s) - \int_t^T \bar{Z}(s, r)dW(r)\Big)ds \\
&= h_x(\bar{X}(T))^\top + \int_t^T \bar{Y}(s)ds - \int_t^T \Big(\zeta(s) + \int_t^T \bar{Z}(r, s)dr\Big)dW(s) \\
&= h_x(\bar{X}(T))^\top + \int_t^T \bar{Y}(s)ds - \int_t^T q(s)dW(s).
\end{aligned}$$

Hence, $(p(\cdot), q(\cdot))$ is the adapted solution of the following BSDE:

$$\begin{aligned}
p(t) = h_x(\bar{X}(T))^\top + \int_t^T &\big[g_x(s, \bar{X}(s), \bar{u}(s))^\top + b_x(s, \bar{X}(s), \bar{u}(s))^\top p(s) \\
&+ \sigma_x(s, \bar{X}(s), \bar{u}(s))^\top q(s)\big]ds - \int_t^T q(s)dW(s).
\end{aligned} \tag{10.2.15}$$

Next, for $t \in [0, T]$, we look at $Y_0(t)$ as follows:

$$\begin{aligned}
Y_0(t) &= g_u(t, X(t), u(t))^\top + b_u(t, X(t), u(t))^\top h_x(X(T))^\top \\
&\quad + \sigma_u(t, X(t), u(t))^\top \zeta(t) + \int_t^T \Big[b_u(t, X(t), u(t))^\top Y(s) \\
&\quad + \sigma_u(t, X(t), u(t))^\top Z(s, t)\Big]ds - \int_t^T Z_0(t, s)dW(s) \\
&= g_u(t, X(t), u(t))^\top + b_u(t, X(t), u(t))^\top \mathcal{Y}(t) \\
&\quad + \sigma_u(t, X(t), u(t))^\top \mathcal{Z}(t) - \int_t^T Z_0(t, s)dW(s).
\end{aligned}$$

Thus, by the fact of $Y_0(\cdot)$ being $\mathbb{F}$-adapted, taking $\mathbb{E}_t$ in the above, we obtain

$$\begin{aligned}
Y_0(t) = g_u(t, X(t), u(t))^\top + b_u(t, X(t), u(t))^\top p(t) \\
+ \sigma_u(t, X(t), u(t))^\top q(t).
\end{aligned} \tag{10.2.16}$$

Hence, we may state the following result.

Proposition 10.2.3 *Let $(\bar{X}(\cdot), \bar{u}(\cdot))$ be an optimal pair of optimal control problem for FSDE (10.1.1), with a convex control domain. Then, under all required* *conditions, the BSDE* (10.2.15) *admits a unique adapted solution* $(p(\cdot), q(\cdot))$ *such that by defining* $Y_0(\cdot)$ *as* (10.2.16), *it holds*

$$\langle Y_0(t), u - u(t)\rangle \geqslant 0, \qquad \forall u \in U,\ t \in [0,T],\ \text{a.s.}$$

3 Maximum Principle for General Control Domains

We now consider the case that $U \in \mathbb{R}^m$ is not assumed to be convex. In the current subsection, we define

$$\mathscr{U}[0,T] \equiv \Big\{u : [0,T] \times \Omega \to U \;\big|\; u(\cdot) \text{ is } \mathbb{F}\text{-progressively measurable}\Big\}. \tag{10.3.1}$$

In this case, if $\bar{u}(\cdot) \in \mathscr{U}[0,T]$ is an optimal control, and $u(\cdot) \in \mathscr{U}[0,T]$ is another admissible control, then (10.2.6) is not true in general, i.e., convex perturbation might not be admissible. Therefore, we need to use a different method to derive necessary optimality condition.

For reader's convenience, we rewrite the controlled FSVIE (10.1.5) and cost functional (10.1.6) as follows:

$$X(t) = \varphi(t) + \int_0^t b(t,s,X(s),u(s))ds + \int_0^t \sigma(t,s,X(s),u(s))dW(s), \quad t \in [0,T], \tag{10.3.2}$$

and

$$J(u(\cdot)) = \mathbb{E}\Big[h(X(T)) + \int_0^T g(t,X(t),u(t))dt\Big]. \tag{10.3.3}$$

To begin with, let us introduce some notations and spaces used in the sequel. We recall the upper and lower triangle domains by the following:

$$\begin{aligned} \Delta^*[\tau,T] &\stackrel{\Delta}{=} \big\{(r,s) \in [\tau,T]^2 \;\big|\; \tau \leqslant r < s \leqslant T\big\}, \\ \Delta_*[\tau,T] &\stackrel{\Delta}{=} \big\{(r,s) \in [\tau,T]^2 \;\big|\; \tau \leqslant s < r \leqslant T\big\}, \end{aligned} \qquad \tau \in [0,T).$$

For $\mathbb{H} \equiv \mathbb{R}^m$, $\mathbb{R}^n$, etc., inspired by the definition of $L^2_{\mathbb{F}}\big(\Omega; L^2(\Delta^*[\tau,T];\mathbb{H})\big)$ in Sect. 4 of Chap. 2, we introduce the following space which is quite useful in the sequel.

$$\begin{aligned} L^2_{\mathbb{F}}\big(\Omega; L^2(\Delta_*[\tau,T];\mathbb{H})\big) \stackrel{\Delta}{=} \Big\{&\zeta : \Delta_*[\tau,T] \times \Omega \to \mathbb{H} \;\big|\; \text{a.e. } r \in [\tau,T], \\ &\zeta(r,\cdot) \in L^2_{\mathbb{F}}(\Omega; L^2(\tau,r;\mathbb{H})),\ \mathbb{E}\int_\tau^T\!\!\int_\tau^r |\zeta(r,s)|^2 ds dr < \infty\Big\}. \end{aligned}$$

For FSVIE (10.3.2), we introduce the following assumptions.

(H10.3.1) Let $T > 0$ and $\varnothing \neq U \subset \mathbb{R}^m$ be measurable. Let $b, \sigma : \Delta_*[0, T] \times \mathbb{R}^n \times U \times \Omega \to \mathbb{R}^n$ be measurable such that the map $s \mapsto \big(b(t, s, x, u), \sigma(t, s, x, u)\big)$ is $\mathbb{F}$-progressively measurable on $[0, t]$, and the map $x \mapsto (b(t, s, x, u), \sigma(t, s, x, u))$ is twice continuously differentiable with the first and second order derivatives being uniformly bounded and continuous uniformly in all its arguments. Moreover, for some $L > 0$, (10.2.2) holds.

Under (H10.3.1), for $\varphi(\cdot) \in C_{\mathbb{F}}([0, T]; L^p(\Omega; \mathbb{R}^n))$ with $p > 1$, and any $u(\cdot) \in \mathscr{U}[0, T]$, it follows from Proposition 3.4.4 of Chap. 3 that (10.3.2) admits a unique solution $X(\cdot) \equiv X(\cdot\,; u(\cdot)) \in C_{\mathbb{F}}([0, T]; L^p(\Omega; \mathbb{R}^n))$. For the involved functions h and g in (10.3.3), we make the following assumptions.

(H10.3.2) Let $h : \mathbb{R}^n \times \Omega \to \mathbb{R}$, $g : [0, T] \times \mathbb{R}^n \times U \times \Omega \to \mathbb{R}$ be measurable such that $x \mapsto (h(x), g(t, x, u))$ is twice continuously differentiable with

$$|g_x(t, 0, u)| + |h_{xx}(x)| + |g_{xx}(t, x, u)| \leqslant L, \qquad (t, x, u) \in [0, T] \times \mathbb{R}^n \times U,$$

for some constant $L > 0$ and $(t, x, u) \mapsto g_{xx}(t, x, u)$ being continuous uniformly in all its arguments.

It is seen that under (H10.3.1)–(H10.3.2), Problem (C) is well-formulated. We point out that (H10.3.1)–(H10.3.2) can be relaxed somehow, for example, one may allow b, σ, as well as g to have some growth in u. We prefer not to get into the most generality to keep the presentation simple.

3.1 Variational Equations of the State Process

In this subsection, we are going to present the variations of the state process under the so-called spike variation of the control process. For Problem (C), such a result will lead to the first order optimality condition in the form of variational inequality.

To state the main result of this section, let $(\bar{X}(\cdot), \bar{u}(\cdot))$ be any fixed state-control pair of the state equation, which could be optimal, in particular. In the following, let $u \in U$, $\tau \in [0, T)$, and $\varepsilon > 0$ be sufficiently small such that $\tau + \varepsilon \leqslant T$. Then we define

$$u^\varepsilon(\cdot) \stackrel{\Delta}{=} u\mathbf{1}_{[\tau,\tau+\varepsilon]}(\cdot) + \bar{u}(\cdot)\mathbf{1}_{[0,T]\setminus[\tau,\tau+\varepsilon]}(\cdot). \tag{10.3.4}$$

This is called a *spike variation* of $\bar{u}(\cdot)$ in the direction of u on $[\tau, \tau + \varepsilon]$. Let $X^\varepsilon(\cdot)$ be the state process corresponding to $u^\varepsilon(\cdot)$. Our goal in this subsection is to obtain expansions of $X^\varepsilon(\cdot) - \bar{X}(\cdot)$ in ε.

To begin with, we introduce the following abbreviations: $(1 \leqslant i \leqslant n)$

$$
\begin{aligned}
& b_x(t,s) \triangleq b_x(t,s,\bar{X}(s),\bar{u}(s)), \quad b^i_{xx}(t,s) \triangleq b^i_{xx}(t,s,\bar{X}(s),\bar{u}(s)), \\
& \delta b(t,s) \triangleq b(t,s,\bar{X}(s),u) - b(t,s,\bar{X}(s),\bar{u}(s)), \\
& \delta b_x(t,s) \triangleq b_x(t,s,\bar{X}(s),u) - b_x(t,s,\bar{X}(s),\bar{u}(s)), \\
& \delta b^i_{xx}(t,s) \triangleq b^i_{xx}(t,s,\bar{X}(s),u) - b^i_{xx}(t,s,\bar{X}(s),\bar{u}(s)), \quad 1 \leqslant i \leqslant n, \\
& \sigma_x(t,s) \triangleq \sigma_x(t,s,\bar{X}(s),\bar{u}(s)), \quad \sigma^i_{xx}(t,s) \triangleq \sigma^i_{xx}(t,s,\bar{X}(s),\bar{u}(s)), \\
& \delta\sigma(t,s) \triangleq \sigma(t,s,\bar{X}(s),u) - \sigma(t,s,\bar{X}(s),\bar{u}(s)), \\
& \delta\sigma_x(t,s) \triangleq \sigma_x(t,s,\bar{X}(s),u) - \sigma_x(t,s,\bar{X}(s),\bar{u}(s)), \\
& \delta\sigma^i_{xx}(t,s) \triangleq \sigma^i_{xx}(t,s,\bar{X}(s),u) - \sigma^i_{xx}(t,s,\bar{X}(s),\bar{u}(s)), \quad 1 \leqslant i \leqslant n.
\end{aligned}
\tag{10.3.5}
$$

Then, under (H10.3.1), the stochastic processes $b_x(\cdot)$, $b^i_{xx}(\cdot)$, $\sigma_x(\cdot)$, $\sigma^i_{xx}(\cdot)$, $\delta b_x(\cdot)$, $\delta b^i_{xx}(\cdot)$, $\delta\sigma_x(\cdot)$, $\delta\sigma^i_{xx}(\cdot)$ are bounded, and by (10.2.2), the boundedness of $b_x(\cdot)$, $\sigma_x(\cdot)$ and $\sigma^i_{xx}(\cdot)$, one has

$$
\big|\delta b(t,s)\big|,\ \big|\delta\sigma(t,s)\big| \leqslant K\big(1+|\bar{X}(s)|\big), \qquad (t,s)\in\Delta_*[0,T].
$$

We now introduce the following variational equations:

$$
\begin{aligned}
X^\varepsilon_1(t) = & \int_0^t b_x(t,s)X^\varepsilon_1(s)ds \\
& + \int_0^t \big[\sigma_x(t,s)X^\varepsilon_1(s) + \delta\sigma(t,s)\mathbf{1}_{[\tau,\tau+\varepsilon]}(s)\big]dW(s), \quad t\in[0,T],
\end{aligned}
\tag{10.3.6}
$$

and

$$
\begin{aligned}
X^\varepsilon_2(t) = & \int_0^t \Big(b_x(t,s)X^\varepsilon_2(s) + \frac{1}{2}b_{xx}(t,s)X^\varepsilon_1(s)^2 + \delta b(t,s)\mathbf{1}_{[\tau,\tau+\varepsilon]}(s)\Big)ds \\
& + \int_0^t \Big(\sigma_x(t,s)X^\varepsilon_2(s) + \frac{1}{2}\sigma_{xx}(t,s)X^\varepsilon_1(s)^2 \\
& \qquad + \delta\sigma_x(t,s)X^\varepsilon_1(s)\mathbf{1}_{[\tau,\tau+\varepsilon]}(s)\Big)dW(s), \quad t\in[0,T],
\end{aligned}
\tag{10.3.7}
$$

where

$$
b_{xx}(t,s)X^\varepsilon_1(s)^2 \triangleq \begin{pmatrix} X^\varepsilon_1(s)^\top b^1_{xx}(t,s)X^\varepsilon_1(s) \\ \vdots \\ X^\varepsilon_1(s)^\top b^n_{xx}(t,s)X^\varepsilon_1(s) \end{pmatrix},
\tag{10.3.8}
$$

and $\sigma_{xx}(t,s)X^\varepsilon_1(s)^2$ is similar. The following proposition is concerned with the above equations.

Proposition 10.3.1 *Let* (H10.3.1) *hold. Then the variational equations* (10.3.6) *and* (10.3.7) *admit unique strong solutions* $X_1^\varepsilon(\cdot)$ *and* $X_2^\varepsilon(\cdot)$. *Further, the following are satisfied:*

$$\sup_{s\in[0,T]}\mathbb{E}\big|X_1^\varepsilon(s)\big|^k = O\big(\varepsilon^{\frac{k}{2}}\big), \qquad k\geqslant 1, \tag{10.3.9}$$

and

$$\sup_{s\in[0,T]}\mathbb{E}\big|X_2^\varepsilon(s)\big|^k = O\big(\varepsilon^{k}\big), \qquad k\geqslant 2. \tag{10.3.10}$$

Proof First of all, (10.3.6) and (10.3.7) are linear FSVIEs with uniform Lipschitz conditions. Hence, they admit unique strong solutions $X_1^\varepsilon(\cdot)$ and $X_2^\varepsilon(\cdot)$. Next, for any $k\geqslant 1$,

$$\begin{aligned}\sup_{t\in[0,T]}\mathbb{E}\big|X_1^\varepsilon(t)\big|^k &\leqslant K\sup_{t\in[0,T]}\mathbb{E}\Big(\int_0^t\big|\delta\sigma(t,s)\big|^2\mathbf{1}_{[\tau,\tau+\varepsilon]}(s)ds\Big)^{\frac{k}{2}}\\ &\leqslant K\mathbb{E}\Big(\int_\tau^{\tau+\varepsilon}\big(1+\big|\bar X(s)\big|^2\big)ds\Big)^{\frac{k}{2}}\leqslant K\varepsilon^{\frac{k}{2}},\end{aligned} \tag{10.3.11}$$

and for any $k\geqslant 2$,

$$\begin{aligned}\sup_{t\in[0,T]}\mathbb{E}\big|X_2^\varepsilon(t)\big|^k &\leqslant K\Big\{\sup_{t\in[0,T]}\mathbb{E}\Big|\int_0^t\Big(\frac{1}{2}b_{xx}(t,s)X_1^\varepsilon(s)^2+\delta b(t,s)\mathbf{1}_{[\tau,\tau+\varepsilon]}(s)\Big)ds\Big|^k\\ &\quad+\sup_{t\in[0,T]}\mathbb{E}\Big[\int_0^t\Big(\frac{1}{2}\sigma_{xx}(t,s)X_1^\varepsilon(s)^2+\delta\sigma_x(t,s)X_1^\varepsilon(s)\mathbf{1}_{[\tau,\tau+\varepsilon]}(s)\Big)^2ds\Big]^{\frac{k}{2}}\Big\}\\ &\leqslant K\mathbb{E}\Big[\Big(\int_0^T\big|X_1^\varepsilon(s)\big|^2ds\Big)^k+\Big(\int_\tau^{\tau+\varepsilon}\big[1+\big|\bar X(s)\big|\big]ds\Big)^k\\ &\quad+\Big(\int_0^T\big|X_1^\varepsilon(s)\big|^4ds\Big)^{\frac{k}{2}}+\Big(\int_\tau^{\tau+\varepsilon}\big|X_1^\varepsilon(s)\big|^2ds\Big)^{\frac{k}{2}}\Big]\\ &\leqslant K\mathbb{E}\Big[\int_0^T\big|X_1^\varepsilon(s)\big|^{2k}ds+\varepsilon^k+\varepsilon^{k-1}\int_\tau^{\tau+\varepsilon}\big|X_1^\varepsilon(s)\big|^kds\\ &\quad+\int_0^T\big|X_1^\varepsilon(s)\big|^{2k}ds+\varepsilon^{\frac{k-2}{2}}\int_\tau^{\tau+\varepsilon}\big|X_1^\varepsilon(s)\big|^kds\Big]\leqslant K\varepsilon^k.\end{aligned}$$

This completes the proof. □

The main result of this subsection can be stated as follows.

Theorem 10.3.2 *Let* (H10.3.1) *hold. Then for any* $k \geqslant 2$,

$$\begin{cases} \sup\limits_{s\in[0,T]} \mathbb{E}\big|X^\varepsilon(s)-\bar X(s)\big|^k = O\big(\varepsilon^{\frac{k}{2}}\big),\\ \sup\limits_{s\in[0,T]} \mathbb{E}\big|X^\varepsilon(s)-\bar X(s)-X_1^\varepsilon(s)\big|^k = O\big(\varepsilon^{k}\big),\\ \sup\limits_{s\in[0,T]} \mathbb{E}\big|X^\varepsilon(s)-\bar X(s)-X_1^\varepsilon(s)-X_2^\varepsilon(s)\big|^k = o\big(\varepsilon^{k}\big).\end{cases} \tag{10.3.12}$$

Proof The proof is lengthy and technical. We split the proof into several steps.

Step 1. Estimation of $X^\varepsilon(\cdot)-\bar X(\cdot)$.

Let us look at the following

$$\begin{aligned} X^\varepsilon(t)-\bar X(t) &= \int_0^t \Big(b\big(t,s,X^\varepsilon(s),u^\varepsilon(s)\big)-b\big(t,s,\bar X(s),\bar u(s)\big)\Big)ds\\ &\quad+\int_0^t \Big(\sigma\big(t,s,X^\varepsilon(s),u^\varepsilon(s)\big)-\sigma\big(t,s,\bar X(s),\bar u(s)\big)\Big)dW(s)\\ &=\int_0^t \Big(\widetilde b_x^\varepsilon(t,s)\big[X^\varepsilon(s)-\bar X(s)\big]+\delta b(t,s)\mathbf{1}_{[\tau,\tau+\varepsilon]}(s)\Big)ds\\ &\quad+\int_0^t \Big(\widetilde\sigma_x^\varepsilon(t,s)\big[X^\varepsilon(s)-\bar X(s)\big]+\delta \sigma(t,s)\mathbf{1}_{[\tau,\tau+\varepsilon]}(s)\Big)dW(s), \end{aligned} \tag{10.3.13}$$

where $\delta b(t,s)$ and $\delta\sigma(t,s)$ are given by (10.3.5), and

$$\begin{cases} \widetilde b_x^\varepsilon(t,s) \triangleq \displaystyle\int_0^1 b_x\big(t,s,\bar X(s)+\alpha[X^\varepsilon(s)-\bar X(s)],u^\varepsilon(s)\big)d\alpha,\\ \widetilde \sigma_x^\varepsilon(t,s) \triangleq \displaystyle\int_0^1 \sigma_x\big(t,s,\bar X(s)+\alpha[X^\varepsilon(s)-\bar X(s)],u^\varepsilon(s)\big)d\alpha.\end{cases} \tag{10.3.14}$$

We see that (10.3.13) is a linear FSVIE for $X^\varepsilon(\cdot)-\bar X(\cdot)$, with zero initial condition. By (H10.3.1), $\widetilde b_x^\varepsilon(\cdot\,,\cdot)$ and $\widetilde\sigma_x^\varepsilon(\cdot\,,\cdot)$ are bounded. Making use the stability estimate of FSVIE (Proposition 3.4.4 of Chap. 3), for $k \geqslant 1$, we have

$$\begin{aligned} &\sup_{t\in[0,T]} \mathbb{E}\big|X^\varepsilon(t)-\bar X(t)\big|^k\\ &\leqslant K\Big\{\sup_{t\in[0,T]}\mathbb{E}\Big[\Big(\int_\tau^{\tau+\varepsilon}|\delta b(t,s)|ds\Big)^k\Big]+\sup_{t\in[0,T]}\mathbb{E}\Big[\Big(\int_\tau^{\tau+\varepsilon}|\delta\sigma(t,s)|^2ds\Big)^{\frac{k}{2}}\Big]\Big\}\\ &\leqslant K\Big[\mathbb{E}\Big(\int_\tau^{\tau+\varepsilon}(1+|\bar X(s)|)ds\Big)^k+\mathbb{E}\Big(\int_\tau^{\tau+\varepsilon}(1+|\bar X(s)|^2)ds\Big)^{\frac{k}{2}}\Big]\\ &\leqslant K\Big(\varepsilon^{k-1}\int_\tau^{\tau+\varepsilon}(1+\mathbb{E}|\bar X(s)|^k)ds+\varepsilon^{\frac{k-2}{2}}\int_\tau^{\tau+\varepsilon}(1+\mathbb{E}|\bar X(s)|^k)ds\Big)\\ &\leqslant K\big[\varepsilon^k+\varepsilon^{\frac{k}{2}}\big]\Big(1+\sup_{s\in[0,T]}\mathbb{E}|\bar X(s)|^k\Big)\leqslant K\varepsilon^{\frac{k}{2}}. \end{aligned}$$

This gives the first relation in (10.3.12).

Step 2. Estimation of $X^\varepsilon(\cdot)-\bar X(\cdot)-X_1^\varepsilon(\cdot)$.

We observe that (see (10.3.14))

$$\begin{aligned}
\widetilde b_x^{\varepsilon,i}(t,s)&\equiv\int_0^1 b_x^i\big(t,s,\bar X(s)+\alpha[X^\varepsilon(s)-\bar X(s)],u^\varepsilon(s)\big)d\alpha\\
&=\int_0^1\Big(b_x^i\big(t,s,\bar X(s)+\alpha[X^\varepsilon(s)-\bar X(s)],u^\varepsilon(s)\big)\\
&\qquad -b_x^i(t,s,\bar X(s),u^\varepsilon(s))\Big)d\alpha+b_x^i(t,s,\bar X(s),\bar u(s))\\
&\qquad +\big[b_x^i(t,s,\bar X(s),u)-b_x^i(t,s,\bar X(s),\bar u(s))\big]\mathbf{1}_{[\tau,\tau+\varepsilon]}(s)\\
&=\Big[\int_0^1\!\!\int_0^1\alpha\Big(b_{xx}^i\big(t,s,\bar X(s)+\alpha\beta[X^\varepsilon(s)-\bar X(s)],u^\varepsilon(s)\big)\\
&\qquad -b_{xx}^i(t,s,\bar X(s),u^\varepsilon(s))\Big)d\beta d\alpha\Big]\big[X^\varepsilon(s)-\bar X(s)\big]\\
&\qquad +\frac12\Big(b_{xx}^i(t,s)+\delta b_{xx}^i(t,s)\mathbf{1}_{[\tau,\tau+\varepsilon]}(s)\Big)\big[X^\varepsilon(s)-\bar X(s)\big]\\
&\qquad +b_x^i(t,s)+\delta b_x^i(t,s)\mathbf{1}_{[\tau,\tau+\varepsilon]}(s)\\
&\equiv b_x^i(t,s)+\delta b_x^i(t,s)\mathbf{1}_{[\tau,\tau+\varepsilon]}(s)+\Big[\frac12\Big(b_{xx}^i(t,s)+\delta b_{xx}^i(t,s)\mathbf{1}_{[\tau,\tau+\varepsilon]}(s)\Big)\\
&\qquad +\Delta_\varepsilon b_{xx}^i(t,s)\Big]\big[X^\varepsilon(s)-\bar X(s)\big],
\end{aligned}\tag{10.3.15}$$

where $b_x^i(t,s)$, $b_{xx}^i(t,s)$, $\delta b_x^i(t,s)$, $\delta b_{xx}^i(t,s)$ are given in (10.3.5), and

$$\begin{aligned}
\Delta_\varepsilon b_{xx}^i(t,s)\overset{\Delta}{=}\int_0^1\int_0^1\alpha\Big[&b_{xx}^i\big(t,s,\bar X(s)+\alpha\beta[X^\varepsilon(s)-\bar X(s)],u^\varepsilon(s)\big)\\
&-b_{xx}^i(t,s,\bar X(s),u^\varepsilon(s))\Big]d\beta d\alpha.
\end{aligned}$$

Similar to (10.3.15),

$$\begin{aligned}
\widetilde\sigma_x^{i,\varepsilon}(t,s)&=\sigma_x^i(t,s)+\delta\sigma_x^i(t,s)\mathbf{1}_{[\tau,\tau+\varepsilon]}(s)\\
&\quad+\Big[\frac12\Big(\sigma_{xx}^i(t,s)+\delta\sigma_{xx}^i(t,s)\mathbf{1}_{[\tau,\tau+\varepsilon]}(s)\Big)+\Delta_\varepsilon\sigma_{xx}^i(t,s)\Big]\big[X^\varepsilon(s)-\bar X(s)\big],
\end{aligned}$$

where $\sigma_x^i(t,s)$, $\sigma_{xx}^i(t,s)$, $\delta\sigma_x^i(t,s)$, $\delta\sigma_{xx}^i(t,s)$ are as in (10.3.5), and

$$\begin{aligned}
\Delta_\varepsilon\sigma_{xx}^i(t,s)\overset{\Delta}{=}\int_0^1\int_0^1\alpha\Big[&\sigma_{xx}^i\big(t,s,\bar X(s)+\alpha\beta[X^\varepsilon(s)\bar X(s)],u^\varepsilon(s)\big)\\
&-\sigma_{xx}^i(t,s,\bar X(s),u^\varepsilon(s))\Big]d\beta d\alpha.
\end{aligned}$$

Thus, one may also rewrite (10.3.13) as follows

$$
\begin{aligned}
&X^\varepsilon(t)-\bar X(t)\\
&=\int_0^t\Big\{b_x(t,s)\big[X^\varepsilon(s)-\bar X(s)\big]\\
&+\Big[\frac{1}{2}\Big(b_{xx}(t,s)+\delta b_{xx}(t,s)\mathbf{1}_{[\tau,\tau+\varepsilon]}(s)\Big)+\Delta_\varepsilon b_{xx}(t,s)\Big]\big[X^\varepsilon(s)-\bar X(s)\big]^2\\
&+\delta b_x(t,s)\mathbf{1}_{[\tau,\tau+\varepsilon]}(s)\big[X^\varepsilon(s)-\bar X(s)\big]+\delta b(t,s)\mathbf{1}_{[\tau,\tau+\varepsilon]}(s)\Big\}ds\\
&+\int_0^t\Big\{\sigma_x(t,s)\big[X^\varepsilon(s)-\bar X(s)\big]\\
&+\Big[\frac{1}{2}\Big(\sigma_{xx}(t,s)+\delta\sigma_{xx}(t,s)\mathbf{1}_{[\tau,\tau+\varepsilon]}(s)\Big)+\Delta_\varepsilon\sigma_{xx}(t,s)\Big]\big[X^\varepsilon(s)-\bar X(s)\big]^2\\
&+\delta\sigma_x(t,s)\mathbf{1}_{[\tau,\tau+\varepsilon]}(s)\big[X^\varepsilon(s)-\bar X(s)\big]+\delta\sigma(t,s)\mathbf{1}_{[\tau,\tau+\varepsilon]}(s)\Big\}dW(s),
\end{aligned}
\tag{10.3.16}
$$

where the terms involving $[X^\varepsilon(\cdot)-\bar X(\cdot)]^2$ are defined similar to (10.3.8). Consequently, by subtracting (10.3.6) from the above, we have

$$
\begin{aligned}
&X^\varepsilon(t)-\bar X(t)-X_1^\varepsilon(t)\\
&=\int_0^t\Big(b_x(t,s)\big[X^\varepsilon(s)-\bar X(s)-X_1^\varepsilon(s)\big]+R_1^{\varepsilon,b}(t,s)\Big)ds\\
&+\int_0^t\Big(\sigma_x(t,s)\big[X^\varepsilon(s)-\bar X(s)-X_1^\varepsilon(s)\big]+R_1^{\varepsilon,\sigma}(t,s)\Big)dW(s),
\end{aligned}
\tag{10.3.17}
$$

where

$$
\begin{aligned}
R_1^{\varepsilon,b}(t,s)=&\Big[\frac{1}{2}\Big(b_{xx}(t,s)+\delta b_{xx}(t,s)\mathbf{1}_{[\tau,\tau+\varepsilon]}(s)\Big)+\Delta_\varepsilon b_{xx}(t,s)\Big]\big[X^\varepsilon(s)-\bar X(s)\big]^2\\
&+\delta b_x(t,s)\mathbf{1}_{[\tau,\tau+\varepsilon]}(s)\big[X^\varepsilon(s)-\bar X(s)\big]+\delta b(t,s)\mathbf{1}_{[\tau,\tau+\varepsilon]}(s),
\end{aligned}
$$

and

$$
\begin{aligned}
R_1^{\varepsilon,\sigma}(t,s)=&\frac{1}{2}\Big(\sigma_{xx}(t,s)+\delta\sigma_{xx}(t,s)\mathbf{1}_{[\tau,\tau+\varepsilon]}(s)\Big)+\Delta_\varepsilon\sigma_{xx}(t,s)\Big]\big[X^\varepsilon(s)-\bar X(s)\big]^2\\
&+\delta\sigma_x(t,s)\mathbf{1}_{[\tau,\tau+\varepsilon]}(s)\big[X^\varepsilon(s)-\bar X(s)\big].
\end{aligned}
$$

We see that (10.3.17) is a linear FSVIE for $X^\varepsilon(\cdot)-\bar X(\cdot)-X_1^\varepsilon(\cdot)$, with zero initial condition. Hence, for $k\geqslant 2$,

$$
\begin{aligned}
&\sup_{t\in[0,T]}\mathbb{E}\big|X^\varepsilon(t)-\bar X(t)-X_1^\varepsilon(t)\big|^k\\
&\leqslant K\sup_{t\in[0,T]}\mathbb{E}\Big[\Big(\int_0^t|R_1^{\varepsilon,b}(t,s)|ds\Big)^k+\Big(\int_0^t|R_1^{\varepsilon,\sigma}(t,s)|^2ds\Big)^{\frac{k}{2}}\Big]\\
&\leqslant K\sup_{t\in[0,T]}\mathbb{E}\Big\{\int_0^t\Big[\Big(1+\mathbf{1}_{[\tau,\tau+\varepsilon]}(s)+\big|\Delta_\varepsilon b_{xx}(t,s)\big|\Big)\big|X^\varepsilon(s)-\bar X(s)\big|^2\\
&\qquad\qquad+\Big(1+|\bar X(s)|+\big|X^\varepsilon(s)-\bar X(s)\big|\Big)\mathbf{1}_{[\tau,\tau+\varepsilon]}(s)\Big]ds\Big\}^k\\
&\quad+K\sup_{t\in[0,T]}\mathbb{E}\Big\{\int_0^t\Big[\Big(1+\mathbf{1}_{[\tau,\tau+\varepsilon]}(s)+\big|\Delta_\varepsilon\sigma_{xx}(t,s)\big|\Big)^2|X^\varepsilon(s)-\bar X(s)|^4\\
&\qquad\qquad+\big|X^\varepsilon(s)-\bar X(s)\big|^2\mathbf{1}_{[\tau,\tau+\varepsilon]}(s)\Big]ds\Big\}^{\frac{k}{2}}\\
&\leqslant K\mathbb{E}\Big[\int_0^T\Big(\big|X^\varepsilon(s)-\bar X(s)\big|^2+\mathbf{1}_{[\tau,\tau+\varepsilon]}(s)\Big)ds\Big]^k\\
&\quad+K\mathbb{E}\Big[\int_0^T\Big(\big|X^\varepsilon(s)-\bar X(s)\big|^4+\big|X^\varepsilon(s)-\bar X(s)\big|^2\mathbf{1}_{[\tau,\tau+\varepsilon]}(s)\Big)ds\Big]^{\frac{k}{2}}\leqslant K\varepsilon^k.
\end{aligned}
$$

This proves the second relation in (10.3.12).

Step 3. Estimation of $X^\varepsilon(\cdot)-\bar X(\cdot)-X_1^\varepsilon(\cdot)-X_2^\varepsilon(\cdot)$.

Subtracting (10.3.7) from (10.3.17), we get

$$
\begin{aligned}
&X^\varepsilon(t)-\bar X(t)-X_1^\varepsilon(t)-X_2^\varepsilon(t)\\
&=\int_0^t\Big(b_x(t,s)\big[X^\varepsilon(s)-\bar X(s)-X_1^\varepsilon(s)-X_2^\varepsilon(s)\big]+R_2^{\varepsilon,b}(t,s)\Big)ds\\
&\quad+\int_0^t\Big(\sigma_x(t,s)\big[X^\varepsilon(s)-\bar X(s)-X_1^\varepsilon(s)-X_2^\varepsilon(s)\big]+R_2^{\varepsilon,\sigma}(t,s)\Big)dW(s),
\end{aligned}
\tag{10.3.18}
$$

where

$$
\begin{aligned}
R_2^{\varepsilon,b}(t,s)&=\Big[\frac12\Big(b_{xx}(t,s)+\delta b_{xx}(t,s)\mathbf{1}_{[\tau,\tau+\varepsilon]}(s)\Big)+\Delta_\varepsilon b_{xx}(t,s)\Big]\big[X^\varepsilon(s)-\bar X(s)\big]^2\\
&\qquad+\delta b_x(t,s)\mathbf{1}_{[\tau,\tau+\varepsilon]}(s)[X^\varepsilon(s)-\bar X(s)]-\frac12 b_{xx}(t,s)X_1^\varepsilon(s)^2\\
&=\frac12 b_{xx}(t,s)\big[X^\varepsilon(s)-\bar X(s)-X_1^\varepsilon(s)\big]\big[X^\varepsilon(s)-\bar X(s)+X_1^\varepsilon(s)\big]\\
&\quad+\Big(\delta b_x(t,s)\big[X^\varepsilon(s)-\bar X(s)\big]+\frac12\delta b_{xx}(t,s)\big[X^\varepsilon(s)-\bar X(s)\big]^2\Big)\mathbf{1}_{[\tau,\tau+\varepsilon]}(s)\\
&\quad+\Delta_\varepsilon b_{xx}(t,s)\big[X^\varepsilon(s)-\bar X(s)\big]^2,
\end{aligned}
$$

and

$$
\begin{aligned}
R_2^{\varepsilon,\sigma}(t,s) =& \Big[\frac{1}{2}\Big(\sigma_{xx}(t,s)+\delta\sigma_{xx}(t,s)\mathbf{1}_{[\tau,\tau+\varepsilon]}s)\Big) + \Delta_\varepsilon\sigma_{xx}(t,s)\Big]\big[X^\varepsilon(s)-\bar{X}(s)\big]^2 \\
&+\delta\sigma_x(t,s)\mathbf{1}_{[\tau,\tau+\varepsilon]}(s)\big[X^\varepsilon(s)-\bar{X}(s)-X_1^\varepsilon(s)\big]-\frac{1}{2}\sigma_{xx}(t,s)X_1^\varepsilon(s)^2 \\
=& \frac{1}{2}\sigma_{xx}(t,s)\big[X^\varepsilon(s)-\bar{X}(s)-X_1^\varepsilon(s)\big]\big[X^\varepsilon(s)-\bar{X}(s)+X_1^\varepsilon(s)\big] \\
&+\Big(\delta\sigma_x(t,s)\big[X^\varepsilon(s)-\bar{X}(s)-X_1^\varepsilon(s)\big] \\
&+\frac{1}{2}\delta\sigma_{xx}(t,s)\big[X^\varepsilon(s)-\bar{X}(s)\big]^2\Big)\mathbf{1}_{[\tau,\tau+\varepsilon]}(s)+\frac{1}{2}\Delta_\varepsilon\sigma_{xx}(t,s)\big[X^\varepsilon(s)-\bar{X}(s)\big]^2.
\end{aligned}
$$

Consequently, by the stability estimate of FSVIEs (e.g., Proposition 3.4.4 of Chap. 3), we have (for $k\geqslant 2$)

$$
\begin{aligned}
&\sup_{t\in[0,T]}\mathbb{E}\big|X(t)-\bar{X}(t)-X_1^\varepsilon(t)-X_2^\varepsilon(t)\big|^k \\
&\leqslant K\sup_{t\in[0,T]}\mathbb{E}\Big[\Big(\int_0^t\big|R_2^{\varepsilon,b}(t,s)\big|ds\Big)^k+\Big(\int_0^t\big|R_2^{\varepsilon,\sigma}(t,s)\big|^2ds\Big)^{\frac{k}{2}}\Big] \\
&\leqslant K\Big\{\mathbb{E}\Big[\int_0^T\Big(\big|X^\varepsilon(s)-\bar{X}(s)-X_1^\varepsilon(s)\big|\big|X^\varepsilon(s)-\bar{X}(s)+X_1^\varepsilon(s)\big|\Big)ds\Big]^k \\
&\quad+\mathbb{E}\Big[\int_\tau^{\tau+\varepsilon}\Big(\big|X^\varepsilon(s)-\bar{X}(s)\big|+\big|X^\varepsilon(s)-\bar{X}(s)\big|^2\Big)ds\Big]^k \\
&\quad+\mathbb{E}\sup_{t\in[0,T]}\Big[\int_0^T\big|\Delta_\varepsilon b_{xx}(t,s)\big|\big|X^\varepsilon(s)-\bar{X}(s)\big|^2ds\Big]^k \\
&\quad+\mathbb{E}\Big[\int_0^T\Big(\big|X^\varepsilon(s)-\bar{X}(s)-X_1^\varepsilon(s)\big|^2\big|X^\varepsilon(s)-\bar{X}(s)+X_1^\varepsilon(s)\big|^2\Big)ds\Big]^{\frac{k}{2}} \\
&\quad+\mathbb{E}\Big[\int_\tau^{\tau+\varepsilon}\Big(\big|X^\varepsilon(s)-\bar{X}(s)-X_1^\varepsilon(s)\big|^2+\big|X^\varepsilon(s)-\bar{X}(s)\big|^4\Big)ds\Big]^{\frac{k}{2}} \\
&\quad+\mathbb{E}\sup_{t\in[0,T]}\Big[\int_0^T\big|\Delta_\varepsilon\sigma_{xx}(t,s)\big|^2\big|X^\varepsilon(s)-\bar{X}(s)\big|^4ds\Big]^{\frac{k}{2}}\Big\} \\
&\leqslant K\Big[\varepsilon^{\frac{3k}{2}}+\sup_{t\in[0,T]}\mathbb{E}\int_0^T\Big(|\Delta_\varepsilon b_{xx}(t,s)|^k+|\Delta_\varepsilon\sigma_{xx}(t,s)|^k\Big)|X^\varepsilon(s)-\bar{X}(s)|^kds\Big]=o(\varepsilon^k).
\end{aligned}
$$

In the above, we need the continuity of the map $x\mapsto(b_{xx}(t,s,x,u),\sigma_{xx}(t,s,x,u))$ uniform in all its arguments. This proves the third relation in (10.3.12). □

Essentially, (10.3.12) are some sort of Taylor expansion of the state process with respect to the spike variation of the controls.

3.2 Variation of the Cost Functional and the First Order Adjoint Equation

Our next goal is to derive the variation of the cost functional under spike variation of the controls. To this end, we introduce the following notations:

$$\begin{aligned} &h_x \stackrel{\Delta}{=} h_x(\bar{X}(T)), \qquad h_{xx} \stackrel{\Delta}{=} h_{xx}(\bar{X}(T)),\\ &g_x(s) \stackrel{\Delta}{=} g_x(s, \bar{X}(s), \bar{u}(s)), \qquad g_{xx}(s) \stackrel{\Delta}{=} g_{xx}(s, \bar{X}(s), \bar{u}(s)). \end{aligned} \tag{10.3.19}$$

Also, we introduce the following equations:

$$\begin{cases} \mathbb{E}_t[h_x^\top] = h_x^\top - \displaystyle\int_t^T \zeta(s)dW(s), \qquad t \in [0, T],\\ Y(t) = g_x(t)^\top + b_x(T,t)^\top h_x^\top + \sigma_x(T,t)^\top \zeta(t)\\ \qquad +\displaystyle\int_t^T \Big(b_x(s,t)^\top Y(s) + \sigma_x(s,t)^\top Z(s,t)\Big)ds - \int_t^T Z(t,s)dW(s),\\ \qquad\qquad t \in [0, T]. \end{cases} \tag{10.3.20}$$

The second equation in the above is called the *first order adjoint equation*. The first equation of the above is just a martingale representation. The second is a Type-II BSVIE which has a unique adapted M-solution. It is clear that the M-adapted solution $(Y(\cdot), Z(\cdot\,,\cdot))$ of Type-II BSVIE in (10.3.20) is only depending on the pair $(\bar{X}(\cdot), \bar{u}(\cdot))$ (and independent of the spike variation of the control). Finally, let us recall the Hamiltonian defined by (10.2.10) which is rewritten here: (with t being replaced by s for the convenience below)

$$\begin{aligned} &H(s, x, u, \eta, \zeta, \bar{Y}(\cdot), \bar{Z}(\cdot\,, s))\\ &\stackrel{\Delta}{=} g(s, x, u) + \big\langle \eta, b(T, s, x, u)\big\rangle + \big\langle \zeta, \sigma(T, s, x, u)\big\rangle\\ &\quad + \int_s^T \big\langle \bar{Y}(r), b(r, s, x, u)\big\rangle dr + \int_s^T \big\langle \bar{Z}(r, s), \sigma(r, s, x, u)\big\rangle dr. \end{aligned} \tag{10.3.21}$$

Now, we are ready to state and prove the main result of this subsection.

Proposition 10.3.3 *Let* (H10.3.1)–(H10.3.2) *hold. Let $u^\varepsilon(\cdot)$ be defined by* (10.3.4) *and $X^\varepsilon(\cdot)$ be the corresponding state. Then*

$$J(u^\varepsilon(\cdot)) - J(\bar{u}(\cdot)) = \frac{1}{2}\mathscr{E}(\varepsilon) + \mathbb{E}\int_\tau^{\tau+\varepsilon} \Delta H(t)dt + o(\varepsilon), \tag{10.3.22}$$

where

$$\begin{aligned} \Delta H(t) \stackrel{\Delta}{=} \mathbb{E}_t\Big[&H\big(t, \bar{X}(t), u, \mathbb{E}_t[h_x^\top], \zeta(t), Y(\cdot), Z(\cdot\,, t)\big)\\ &-H\big(t, \bar{X}(t), \bar{u}(t), \mathbb{E}_t[h_x^\top], \zeta(t), Y(\cdot), Z(\cdot\,, t)\big)\Big], \end{aligned} \tag{10.3.23}$$

and

$$
\begin{aligned}
\mathscr{E}(\varepsilon) &\overset{\triangle}{=} \mathbb{E}\Big[X_1^\varepsilon(T)^\top h_{xx}\big(\bar{X}(T)\big)X_1^\varepsilon(T) \\
&\quad+\int_0^T X_1^\varepsilon(t)^\top H_{xx}\big(t,\bar{X}(t),\bar{u}(t),\mathbb{E}_t[h_x^\top],\zeta(t),Y(\cdot),Z(\cdot\,,t)\big)X_1^\varepsilon(t)dt\Big] \\
&\equiv \mathbb{E}\Big[X_1^\varepsilon(T)^\top h_{xx}X_1^\varepsilon(T)+\int_0^T X_1^\varepsilon(t)^\top H_{xx}(t)X_1^\varepsilon(t)dt\Big].
\end{aligned}
\tag{10.3.24}
$$

Proof Denote

$$
\begin{aligned}
&\delta g(t) \overset{\triangle}{=} g\big(t,\bar{X}(t),u\big)-g\big(t,\bar{X}(t),\bar{u}(t)\big),\\
&\delta g_x(t) \overset{\triangle}{=} g_x\big(t,\bar{X}(t),u\big)-g_x\big(t,\bar{X}(t),\bar{u}(t)\big),\\
&\delta g_{xx}(t) \overset{\triangle}{=} g_{xx}\big(t,\bar{X}(t),u\big)-g_{xx}\big(t,\bar{X}(t),\bar{u}(t)\big).
\end{aligned}
\tag{10.3.25}
$$

We observe the following:

$$
\begin{aligned}
J\big(u^\varepsilon(\cdot)\big)-J\big(\bar{u}(\cdot)\big) &= \mathbb{E}\Big[h\big(X^\varepsilon(T)\big)-h\big(\bar{X}(T)\big) \\
&\qquad+\int_0^T\Big(g(t,X^\varepsilon(t),u^\varepsilon(t))-g(t,\bar{X}(t),\bar{u}(t))\Big)dt\Big] \\
&= \mathbb{E}\Big[\widetilde{h}_x^\varepsilon[X^\varepsilon(T)-\bar{X}(T)] \\
&\quad+\int_0^T\Big(\widetilde{g}_x^\varepsilon(t)\big[X^\varepsilon(t)-\bar{X}(t)\big]+\delta g(t)\mathbf{1}_{[\tau,\tau+\varepsilon]}(t)\Big)dt\Big],
\end{aligned}
\tag{10.3.26}
$$

where

$$
\begin{aligned}
&\widetilde{h}_x^\varepsilon = \int_0^1 h_x\big(\bar{X}(T)+\alpha[X^\varepsilon(T)-\bar{X}(T)]\big)d\alpha,\\
&\delta g(t) = g\big(t,\bar{X}(t),u\big)-g\big(t,\bar{X}(t),\bar{u}(t)\big),\\
&\widetilde{g}_x^\varepsilon(t) = \int_0^1 g_x\big(s,\bar{X}(s)+\alpha[X^\varepsilon(s)-\bar{X}(s)],u^\varepsilon(s)\big)d\alpha.
\end{aligned}
$$

Now, similar to (10.3.15), we have

$$
\widetilde{h}_x^\varepsilon = h_x+\Big(\frac{1}{2}h_{xx}+\Delta_\varepsilon h_{xx}\Big)\big[X^\varepsilon(T)-\bar{X}(T)\big],
$$

with h_x, h_{xx} given in (10.3.19), and

$$
\Delta_\varepsilon h_{xx} \overset{\triangle}{=} \int_0^1\!\!\int_0^1 \alpha\Big[h_{xx}\big(\bar{X}(T)+\alpha\beta[X^\varepsilon(T)-\bar{X}(T)]\big)-h_{xx}\big(\bar{X}(T)\big)\Big]d\beta d\alpha.
$$

Also, similar to (10.3.15),

$$\begin{aligned}\widetilde{g}^{\varepsilon}_x(t) &= g_x(t) + \delta g_x(t)\mathbf{1}_{[\tau,\tau+\varepsilon]}(t)\\ &+\Big[\frac{1}{2}\Big(g_{xx}(t) + \delta g_{xx}(t)\mathbf{1}_{[\tau,\tau+\varepsilon]}(t)\Big) + \Delta_\varepsilon g_{xx}(t)\Big]\big[X^\varepsilon(t) - \bar{X}(t)\big],\end{aligned}$$

with $g_x(t)$, $g_{xx}(t)$, $\delta g_x(t)$, $\delta g_{xx}(t)$ given (10.3.19), and

$$\begin{aligned}\Delta_\varepsilon g_{xx}(t) \triangleq \int_0^1\int_0^1 \alpha\Big[&g_{xx}\big(t, \bar{X}(t) + \alpha\beta[X^\varepsilon(t) - \bar{X}(t)], u^\varepsilon(t)\big)\\ &-g_{xx}\big(t, \bar{X}(t), u^\varepsilon(t)\big)\Big]d\beta d\alpha.\end{aligned}$$

Hence,

$$\begin{aligned}&J\big(u^\varepsilon(\cdot)\big) - J\big(\bar{u}(\cdot)\big)\\ &=\mathbb{E}\Big\{h_x\big[X^\varepsilon(T)-\bar{X}(T)\big]+\Big(\frac{1}{2}h_{xx} + \Delta_\varepsilon h_{xx}\Big)\big[X^\varepsilon(T)-\bar{X}(T)\big]^2\\ &\quad+\int_0^T\Big[g_x(t)\big[X^\varepsilon(t)-\bar{X}(t)\big]+\Big(\frac{1}{2}g_{xx}(t)+\Delta_\varepsilon g_{xx}(t)\Big)\big[X^\varepsilon(t)-\bar{X}(t)\big]^2\\ &\quad+\Big(\delta g(t) + \delta g_x(t)\big[X^\varepsilon(t) - \bar{X}(t)\big]\\ &\quad+\frac{1}{2}\delta g_{xx}(t)\big[X^\varepsilon(t) - \bar{X}(t)\big]^2\Big)\mathbf{1}_{[\tau,\tau+\varepsilon]}(t)\Big]dt\Big\}\\ &= \mathbb{E}\Big[h_x\big[X_1^\varepsilon(T)+X_2^\varepsilon(T)\big]+\frac{1}{2}h_{xx}X_1^\varepsilon(T)^2+\int_0^T\Big(g_x(t)\big[X_1^\varepsilon(t)+X_2^\varepsilon(t)\big]\\ &\quad+\frac{1}{2}g_{xx}(t)X_1^\varepsilon(t)^2 + \delta g(t)\mathbf{1}_{[\tau,\tau+\varepsilon]}(t)\Big)dt\Big] +o(\varepsilon),\end{aligned}\tag{10.3.27}$$

with

$$h_{xx}\big[X^\varepsilon(T) - \bar{X}(T)\big]^2 = \big[X^\varepsilon(T) - \bar{X}(T)\big]^\top h_{xx}\big[X^\varepsilon(T) - \bar{X}(T)\big].$$

Other terms involving quadratic terms are similar. In the above, for terms involving $h(\cdot)$, we look at term-by-term. For the first term, we have

$$\begin{aligned}&\mathbb{E}\Big[h_x\big[X^\varepsilon(T) - \bar{X}(T)\big]\Big]\\ &= \mathbb{E}\Big[h_x\big[X_1^\varepsilon(T)+ X_2^\varepsilon(T)\big]\Big] + \mathbb{E}\Big[h_x\big[X^\varepsilon(T)-\bar{X}(T)-X_1^\varepsilon(T)-X_2^\varepsilon(T)\big]\Big]\\ &= \mathbb{E}\Big[h_x\big[X_1^\varepsilon(T) + X_2^\varepsilon(T)\big]\Big] + o(\varepsilon).\end{aligned}$$

For the second term, we have

$$
\begin{aligned}
&\mathbb{E}\Big[\frac{1}{2}h_{xx}\big[X^{\varepsilon}(T)-\bar{X}(T)\big]^2\Big]\\
&=\mathbb{E}\Big[\frac{1}{2}h_{xx}\big[X_1^{\varepsilon}(T)+X_2^{\varepsilon}(T)\big]^2+\frac{1}{2}h_{xx}\big[X^{\varepsilon}(T)-\bar{X}(T)-X_1^{\varepsilon}(T)-X_2^{\varepsilon}(T)\big]^2\\
&\qquad+h_{xx}\big[X_1^{\varepsilon}(T)+X_2^{\varepsilon}(T)\big]\big[X^{\varepsilon}(T)-\bar{X}(T)-X_1^{\varepsilon}(T)-X_2^{\varepsilon}(T)\big]\Big]\\
&=\frac{1}{2}\mathbb{E}\Big[h_x\big[X_1^{\varepsilon}(T)+X_2^{\varepsilon}(T)\big]^2+o(\varepsilon)\Big]=\frac{1}{2}\mathbb{E}\Big[h_xX_1^{\varepsilon}(T)^2+o(\varepsilon)\Big].
\end{aligned}
$$

For the third term, we have

$$
\begin{aligned}
&\Big|\mathbb{E}\Big[\Delta_{\varepsilon}h_{xx}\big[X_1^{\varepsilon}(T)-\bar{X}(T)\big]^2\Big]\Big|\\
&\leqslant K\big|\Delta_{\varepsilon}h_{xx}\big|\Big(\big|X^{\varepsilon}(T)-\bar{X}(T)-X_1^{\varepsilon}(T)-X_2^{\varepsilon}(T)|^2+|X_1^{\varepsilon}(T)\big|^2+\big|X_2^{\varepsilon}(T)\big|^2\Big)\\
&\leqslant K\big|\Delta_{\varepsilon}h_{xx}\big|\varepsilon=o(\varepsilon).
\end{aligned}
$$

Here, we have used the uniform continuity of $x\mapsto h_{xx}(x)$. Similarly, for the terms involving $g(\cdot)$, we also look at each term. For the first term, we have

$$
\mathbb{E}\Big[g_x(t)\big[X^{\varepsilon}(t)-\bar{X}(t)\big]\Big]=\mathbb{E}\Big[g_x(t)\big[X_1^{\varepsilon}(t)+X_2^{\varepsilon}(t)\big]\Big]+o(\varepsilon).
$$

For the second term, we have

$$
\begin{aligned}
&\mathbb{E}\Big(\frac{1}{2}g_{xx}(t)\big[X^{\varepsilon}(t)-\bar{X}(t)\big]^2\Big)\\
&=\mathbb{E}\Big[\frac{1}{2}g_{xx}(t)\Big(\big[X^{\varepsilon}(t)-\bar{X}(t)-X_1^{\varepsilon}(t)\big]^2+X_1^{\varepsilon}(t)^2\\
&\qquad+2X^{\varepsilon}(t)\big[X^{\varepsilon}(t)-\bar{X}(t)-X_1^{\varepsilon}(t)\big]\Big)\Big]=\mathbb{E}\Big[\frac{1}{2}g_{xx}(t)X^{\varepsilon}(t)^2+o(\varepsilon)\Big].
\end{aligned}
$$

For the third term, we have

$$
\Big|\mathbb{E}\int_0^T\Delta_{\varepsilon}g_{xx}(t)\big[X^{\varepsilon}(t)-\bar{X}(t)\big]^2dt\Big|\leqslant\varepsilon\mathbb{E}\int_0^T\big|\Delta_{\varepsilon}g_{xx}(t)\big|dt=o(\varepsilon).
$$

Here, we have used the assumption that $x\to g_{xx}(t,x,u)$ is continuous uniformly in all its arguments. For the fifth and sixth terms, we have

$$
\begin{aligned}
&\Big|\mathbb{E}\int_{\tau}^{\tau+\varepsilon}\Big(\delta g_x(t)\big[X^{\varepsilon}(t)-\bar{X}(t)\big]+\frac{1}{2}\delta g_{xx}(t)\big[X^{\varepsilon}(t)-\bar{X}(t)\big]^2\Big)dt\Big|\\
&\leqslant K\mathbb{E}\int_{\tau}^{\tau+\varepsilon}\Big(\big|X^{\varepsilon}(t)-\bar{X}(t)\big|+\big|X^{\varepsilon}(t)-\bar{X}(t)\big|^2\Big)dt\leqslant K\varepsilon^{\frac{3k}{2}}.
\end{aligned}
$$

We see from (10.3.22) that the expansion of the cost functional involves $X_1^{\varepsilon}(\cdot)+X_2^{\varepsilon}(\cdot)$, which depends on the spike variation of the controls. Our next goal is to use duality to remove $X_2^\varepsilon(\cdot)$. By (10.3.6) and (10.3.7), we have

$$
\begin{aligned}
&X_1^\varepsilon(t)+X_2^\varepsilon(t)\\
&=\int_0^t b_x(t,s)\big[X_1^\varepsilon(s)+X_2^\varepsilon(s)\big]ds+\int_0^t \sigma_x(t,s)\big[X_1^\varepsilon(s)+X_2^\varepsilon(s)\big]dW(s)\\
&\quad+\int_0^t \Big(\frac{1}{2}b_{xx}(t,s)X_1^\varepsilon(s)^2+\delta b(t,s)\mathbf{1}_{[\tau,\tau+\varepsilon]}(s)\Big)ds\\
&\quad+\int_0^t\Big(\frac{1}{2}\sigma_{xx}(t,s)X_1^\varepsilon(s)^2+\big[\delta\sigma(t,s)+\delta\sigma_x(t,s)X_1^\varepsilon(s)\big]\mathbf{1}_{[\tau,\tau+\varepsilon]}(s)\Big)dW(s),\\
&\hspace{22em} t\in[0,T].
\end{aligned}
\tag{10.3.28}
$$

Next we let $\zeta(\cdot)$ satisfy the first equation in (10.3.20) and $(Y(\cdot),Z(\cdot\,,\cdot))$ be the adapted M-solution of the second equation in (10.3.20), for which

$$
Y(t)=\mathbb{E}Y(t)+\int_0^t Z(t,s)dW(s),\qquad (t,s)\in\Delta_*[0,T].
$$

Then by the duality between (10.3.28) and the second equation in (10.3.20), as in Theorem 5.4.3 of Chap. 5, one has

$$
\begin{aligned}
&\mathbb{E}\int_0^T \big\langle X_1^\varepsilon(t)+X_2^\varepsilon(t),g_x(t)^\top+b_x(T,t)^\top h_x^\top+\sigma_x(T,t)^\top\zeta(t)\big\rangle dt\\
&=\mathbb{E}\int_0^T\Big\langle Y(t),\int_0^t\Big(\frac{1}{2}b_{xx}(t,s)X_1^\varepsilon(s)^2+\delta b(t,s)\mathbf{1}_{[\tau,\tau+\varepsilon]}(s)\Big)ds\\
&\quad+\int_0^t\Big(\frac{1}{2}\sigma_{xx}(t,s)X_1^\varepsilon(s)^2+\big[\delta\sigma(t,s)+\delta\sigma_x(t,s)X_1^\varepsilon(s)\big]\mathbf{1}_{[\tau,\tau+\varepsilon]}(s)\Big)dW(s)\Big\rangle dt\\
&=\mathbb{E}\int_0^T\!\!\int_s^T\Big\langle Y(t),\Big(\frac{1}{2}b_{xx}(t,s)X_1^\varepsilon(s)^2+\delta b(t,s)\mathbf{1}_{[\tau,\tau+\varepsilon]}(s)\Big)\Big\rangle dtds\\
&\quad+\mathbb{E}\int_0^T\!\!\int_s^T\Big\langle Z(t,s),\Big(\frac{1}{2}\sigma_{xx}(t,s)X_1^\varepsilon(s)^2+\big[\delta\sigma(t,s)+\delta\sigma_x(t,s)X_1^\varepsilon(s)\big]\mathbf{1}_{[\tau,\tau+\varepsilon]}(s)\Big)\Big\rangle dtds.
\end{aligned}
$$

Also, by (10.3.28) and the first equation in (10.3.20),

$$
\begin{aligned}
&\mathbb{E}\Big(h_x\big[X_1^\varepsilon(T)+X_2^\varepsilon(T)\big]\Big)\\
&=\mathbb{E}\Big[\int_0^T h_x\Big(b_x(T,s)\big[X_1^\varepsilon(s)+X_2^\varepsilon(s)\big]+\frac{1}{2}b_{xx}(T,s)X_1^\varepsilon(s)^2+\delta b(T,s)\mathbf{1}_{[\tau,\tau+\varepsilon]}(s)\Big)ds\\
&\qquad+\int_0^T\zeta(s)^\top\Big(\sigma_x(T,s)\big[X_1^\varepsilon(s)+X_2^\varepsilon(s)\big]+\frac{1}{2}\sigma_{xx}(T,s)X_1^\varepsilon(s)^2\\
&\qquad\qquad+\big(\delta\sigma(T,s)+\delta\sigma_x(T,s)X_1^\varepsilon(s)\big)\mathbf{1}_{[\tau,\tau+\varepsilon]}(s)\Big)ds\Big]\\
&=\mathbb{E}\int_0^T\Big[\big\langle b_x(T,s)^\top h_x^\top+\sigma_x(T,s)^\top\zeta(s),X_1^\varepsilon(s)+X_2^\varepsilon(s)\big\rangle\\
&\qquad+\frac{1}{2}h_xb_{xx}(T,s)X_1^\varepsilon(s)^2+\frac{1}{2}\zeta(s)^\top\sigma_{xx}(T,s)X_1^\varepsilon(s)^2\\
&\qquad+\Big(h_x\delta b(T,s)+\zeta(s)^\top\big(\delta\sigma(T,s)+\delta\sigma_x(T,s)X_1^\varepsilon(s)\big)\Big)\mathbf{1}_{[\tau,\tau+\varepsilon]}(s)\Big]ds.
\end{aligned}
$$

Thus,

$$
\begin{aligned}
&\mathbb{E}\Big(h_x\big[X_1^\varepsilon(T)+X_2^\varepsilon(T)\big]+\int_0^T g_x(t)\big[X_1^\varepsilon(t)+X_2^\varepsilon(t)\big]dt\Big)\\
&=\mathbb{E}\int_0^T\int_s^T\Big\langle Y(t),\Big(\frac{1}{2}b_{xx}(t,s)X_1^\varepsilon(s)^2+\delta b(t,s)\mathbf{1}_{[\tau,\tau+\varepsilon]}(s)\Big)\Big\rangle dtds\\
&\quad+\mathbb{E}\int_0^T\int_s^T\Big\langle Z(t,s),\Big(\frac{1}{2}\sigma_{xx}(t,s)X_1^\varepsilon(s)^2\\
&\qquad\qquad+\big[\delta\sigma(t,s)+\delta\sigma_x(t,s)X_1^\varepsilon(s)\big]\mathbf{1}_{[\tau,\tau+\varepsilon]}(s)\Big)\Big\rangle dtds\\
&\quad+\int_0^T\Big[\frac{1}{2}h_xb_{xx}(T,s)X_1^\varepsilon(s)^2+\frac{1}{2}\zeta(s)^\top\sigma_{xx}(T,s)X_1^\varepsilon(s)^2\\
&\quad+\big[h_x\delta b(T,s)+\zeta(s)^\top\big(\delta\sigma(T,s)+\delta\sigma_x(T,s)X_1^\varepsilon(s)\big)\big]\mathbf{1}_{[\tau,\tau+\varepsilon]}(s)\Big]ds\\
&=\mathbb{E}\Big\{\int_0^T\int_s^T\Big[\frac{1}{2}\Big(\big\langle Y(t),b_{xx}(t,s)X_1^\varepsilon(s)^2\big\rangle+\big\langle Z(t,s),\sigma_{xx}(t,s)X_1^\varepsilon(s)^2\big\rangle\Big)\\
&\quad+\big(\big\langle Y(t),\delta b(t,s)\big\rangle+\langle Z(t,s),\delta\sigma(t,s)+\delta\sigma_x(t,s)X_1^\varepsilon(s)\rangle\big)\mathbf{1}_{[\tau,\tau+\varepsilon]}(s)dtds\Big]\\
&\quad+\int_0^T\Big[\frac{1}{2}\Big(\big\langle h_x^\top,b_{xx}(T,s)X_1^\varepsilon(s)^2\big\rangle+\big\langle\zeta(s),\sigma_{xx}(T,s)X_1^\varepsilon(s)^2\big\rangle\Big)\\
&\quad+\Big(\big\langle h_x^\top,\delta b(T,s)\big\rangle+\big\langle\zeta(s),\delta\sigma(T,s)+\delta\sigma_x(T,s)X_1^\varepsilon(s)\big\rangle\Big)\mathbf{1}_{[\tau,\tau+\varepsilon]}(s)\Big]ds\Big\}\\
&=\mathbb{E}\Big\{\int_0^T\frac{1}{2}\Big[\big\langle h_x^\top,b_{xx}(T,s)X_1^\varepsilon(s)^2\big\rangle+\big\langle\zeta(s),\sigma_{xx}(T,s)X_1^\varepsilon(s)^2\big\rangle\\
&\quad+\int_s^T\Big(\big\langle Y(t),b_{xx}(t,s)X_1^\varepsilon(s)^2\big\rangle+\big\langle Z(t,s),\sigma_{xx}(t,s)X_1^\varepsilon(s)^2\big\rangle\Big)dt\Big]ds\\
&\quad+\int_0^T\Big[\big\langle h_x^\top,\delta b(T,s)\big\rangle+\big\langle\zeta(s),\delta\sigma(T,s)\big\rangle\\
&\quad+\int_s^T\Big(\big\langle Y(t),\delta b(t,s)\big\rangle+\big\langle Z(t,s),\delta\sigma(t,s)\big\rangle\Big)dt\Big]\mathbf{1}_{[\tau,\tau+\varepsilon]}(s)ds\\
&\quad+\int_0^T\Big(\big\langle\zeta(s),\delta\sigma_x(T,s)X_1^\varepsilon(s)\big\rangle+\int_s^T\big\langle Z(t,s),\delta\sigma_x(t,s)X_1^\varepsilon(s)\big\rangle dt\Big)\mathbf{1}_{[\tau,\tau+\varepsilon]}(s)ds\Big\}.
\end{aligned}
$$

On the other hand,

$$
\begin{aligned}
&\mathbb{E}\Big|\int_0^T\Big(\big\langle\zeta(s),\delta\sigma_x(T,s)X_1^\varepsilon(s)\big\rangle+\int_s^T\big\langle Z(t,s),\delta\sigma_x(t,s)X_1^\varepsilon(s)\big\rangle dt\Big)\mathbf{1}_{[\tau,\tau+\varepsilon]}(s)ds\Big|\\
&\leqslant K\int_\tau^{\tau+\varepsilon}\mathbb{E}\Big[\Big(|\zeta(s)|+\int_s^T|Z(t,s)|dt\Big)|X_1^\varepsilon(s)|\Big]ds\\
&\leqslant K\Big[\int_\tau^{\tau+\varepsilon}\mathbb{E}\Big(|\zeta(s)|+\int_s^T|Z(t,s)|dt\Big)^2ds\Big]^{\frac{1}{2}}\Big[\int_\tau^{\tau+\varepsilon}\mathbb{E}|X_1^\varepsilon(s)|^2ds\Big]^{\frac{1}{2}}\\
&\leqslant K\varepsilon\Big[\int_\tau^{\tau+\varepsilon}\mathbb{E}\Big(|\zeta(s)|+\int_s^T|Z(t,s)|dt\Big)^2ds\Big]^{\frac{1}{2}}=o(\varepsilon).
\end{aligned}
$$

Hence, from (10.3.27),

$$
\begin{aligned}
&J\big(u^\varepsilon(\cdot)\big)-J\big(\bar u(\cdot)\big)\\
&=\mathbb{E}\Big[h_x\big[X_1^\varepsilon(T)+X_2^\varepsilon(T)\big]+\int_0^T g_x(s)\big[X_1^\varepsilon(s)+X_2^\varepsilon(s)\big]ds+\frac{1}{2}X_1^\varepsilon(T)^\top h_{xx}X_1^\varepsilon(T)\\
&\quad+\int_0^T\Big(\frac{1}{2}X_1^\varepsilon(s)^\top g_{xx}(s)X_1^\varepsilon(s)+\delta g(s)\mathbf{1}_{[\tau,\tau+\varepsilon]}(s)\Big)ds\Big]+o(\varepsilon)\\
&=\frac{1}{2}\mathbb{E}\Big\{h_{xx}\big(\bar X(T)\big)X_1^\varepsilon(T)^2+\int_0^T\Big[\big\langle h_x(\bar X(T))^\top,b_{xx}(T,s,\bar X(s),\bar u(s))X_1^\varepsilon(s)^2\big\rangle\\
&\quad+\big\langle\zeta(s),\sigma_{xx}(T,s,\bar X(s),\bar u(s))X_1^\varepsilon(s)^2\big\rangle+g_{xx}(s,\bar X(s),\bar u(s))X_1^\varepsilon(s)^2\\
&\quad+\int_s^T\Big(\big\langle Y(t),b_{xx}(t,s,\bar X(s),\bar u(s))X_1^\varepsilon(s)^2\big\rangle\\
&\quad+\big\langle Z(t,s),\sigma_{xx}(t,s,\bar X(s),\bar u(s))X_1^\varepsilon(s)^2\big\rangle\Big)dt\Big]ds\Big\}\\
&\quad+\mathbb{E}\int_\tau^{\tau+\varepsilon}\Big[\big\langle h_x(\bar X(T))^\top,b(T,s,\bar X(s),u)-b(T,s,\bar X(s),\bar u(s))\big\rangle\\
&\qquad\qquad+\big\langle\zeta(s),\sigma(T,s,\bar X(s),u)-\sigma(T,s,\bar X(s),\bar u(s))\big\rangle\\
&\qquad\qquad+g(s,\bar X(s),u)-g(s,\bar X(s),\bar u(s))\\
&\quad+\int_s^T\Big(\big\langle Y(t),b(t,s,\bar X(s),u)-b(t,s,\bar X(s),\bar u(s))\big\rangle\\
&\qquad\qquad+\big\langle Z(t,s),\sigma(t,s,\bar X(s),u)-\sigma(t,s,\bar X(s),\bar u(s))\big\rangle\Big)dt\Big]ds+o(\varepsilon)\\
&=\frac{1}{2}\mathscr{E}(t)+\mathbb{E}\int_\tau^{\tau+\varepsilon}\Delta H(t)dt+o(\varepsilon).
\end{aligned}
$$

This proves (10.3.22). □

We have seen that the above implies that if $(\bar X(\cdot),\bar u(\cdot))$ is an optimal pair, then the following variational inequality holds:

$$
\frac{1}{2}\mathscr{E}(t)+\mathbb{E}\int_\tau^{\tau+\varepsilon}\Delta H(t)dt\geqslant 0,\qquad \forall u\in U,\quad \varepsilon>0,\quad \tau\in[0,T].
$$

However, this is not very useful, since the quadratic form $\mathscr{E}(t)$ depends on $X_1^\varepsilon(\cdot)$, which is depending on the spike variation of the control. Our next goal is to use duality principle involving BSVIEs to get rid of it.

3.3 Representation of the Quadratic Form and the Second Order Adjoint Equation

Recalling (10.3.6) (suppressing the superscript ε in $X_1^\varepsilon(\cdot)$), we have

$$
\begin{aligned}
&X_1(t)=0, && t\in[0,\tau],\\
&X_1(t)=\int_\tau^t b_x(t,s)X_1(s)ds+\int_\tau^t\big[\sigma_x(t,s)X_1(s)+\delta\sigma^\varepsilon(t,s)\big]dW(s), && t\in[\tau,T],
\end{aligned}
$$

where

$$\delta\sigma^{\varepsilon}(t,s)\equiv\delta\sigma(t,s)\mathbf{1}_{[\tau,\tau+\varepsilon]}(s).$$

Thus, $X_1(\cdot)$ is the image of $\delta\sigma^{\varepsilon}(\cdot\,,\cdot)$ under a linear operator, i.e.,

$$X_1(t)=\Phi[\delta\sigma^{\varepsilon}(\cdot\,,\cdot)](t),\qquad t\in[\tau,T],$$

for some operator Φ. Let

$$X_1(T)=\Phi[\delta\sigma^{\varepsilon}(\cdot\,,\cdot)](T)\equiv\widehat{\Phi}[\delta\sigma^{\varepsilon}(\cdot\,,\cdot)].$$

Consequently, by misusing the notation a little, we roughly have

$$\mathscr{E}(\varepsilon)=\big\langle\big(\widehat{\Phi}^*h_{xx}\widehat{\Phi}+\Phi^*H_{xx}\Phi\big)\delta\sigma^{\varepsilon},\delta\sigma^{\varepsilon}\big\rangle. \tag{10.3.29}$$

In the above, the dependence of $\mathscr{E}(\varepsilon)$ on $\delta\sigma^{\varepsilon}(\cdot\,,\cdot)$ is not directly enough, due to the fact that the operator Φ is complicated. Therefore, (10.3.29) is not an integral of a bilinear form in $\mathbb{R}^n$. The purpose of this subsection is to obtain a more direct form of $\mathscr{E}(\cdot)$ in terms of $\delta\sigma^{\varepsilon}(\cdot\,,\cdot)$, whose procedure naturally leads to the second order adjoint equation. To express the main idea behind, let us first look at the following simple calculation. Such an idea comes from the standard study of LQ problems.

Consider the following linear SDE:

$$\begin{cases}dX(s)=\big[AX(s)+b(s)\big]ds+\big[CX(s)+\sigma(s)\big]dW(s), & s\in[t,T],\\ X(t)=x,\end{cases}$$

with suitable matrices A,C and processes $b(\cdot),\sigma(\cdot)$. Also, we are given the following quadratic functional

$$J=\mathbb{E}\Big[X(T)^{\top}GX(T)+\int_t^T\Big(X(s)^{\top}QX(s)+2q(s)^{\top}X(s)\Big)ds\Big], \tag{10.3.30}$$

with suitable G, Q, and $q(\cdot)$. We want to absorb the first term on the right-hand side of the above into the second one. To this end, we let $(P(\cdot),\Lambda(\cdot))$ be the adapted solution to the following BSDE (for symmetric matrix valued processes):

$$\begin{cases}dP(t)=-\Gamma(t)dt+\Lambda(t)dW(t), & t\in[0,T],\\ P(T)=G,\end{cases} \tag{10.3.31}$$

with $\Gamma(\cdot)$ undetermined (symmetric matrix valued process). Then by Itô's formula (suppressing s), we have

$$d(PX)=\Big(-\Gamma X+P(AX+b)+\Lambda(CX+\sigma)\Big)ds+\Big(\Lambda X+P(CX+\sigma)\Big)dW.$$

Therefore,

$$
\begin{aligned}
d(X^\top PX) = \Big[&(X^\top A^\top + b^\top)PX + X^\top(-\Gamma X + PAX + Pb + \Lambda CX + \Lambda\sigma) \\
&+(X^\top C^\top + \sigma^\top)(\Lambda X + PCX + P\sigma)\Big]ds + \{\cdots\}dW(s) \\
= \Big[&X^\top\Big(-\Gamma + PA + A^\top P + C^\top PC + \Lambda C + C^\top \Lambda\Big)X^\top \\
&+X^\top(Pb + C^\top P\sigma + \Lambda\sigma) + \Big(b^\top P + \sigma^\top PC + D^\top \Lambda\Big)X \\
&+\sigma^\top P\sigma\Big]ds + \Big\{\cdots\Big\}dW(s).
\end{aligned}
$$

Then

$$
\begin{aligned}
J &= \mathbb{E}\Big[X(T)^\top GX(T) + \int_t^T \Big(X(s)^\top QX(s) + 2q(s)^\top X(s)\Big)ds\Big] \\
&= \mathbb{E}\int_t^T \Big[X^\top\Big(-\Gamma + PA + A^\top P + C^\top PC + \Lambda C + C^\top \Lambda + Q\Big)X \\
&\qquad +2\Big(b^\top P + \sigma^\top PC + \sigma^\top \Lambda + q^\top\Big)X + \sigma^\top P\sigma\Big]ds + x^\top P(t)x.
\end{aligned}
$$

We see that the term $X(T)^\top GX(T)$ has been absorbed. Further, it is easy to see that to elimnate the quadratic term $X(\cdot)^\top(\cdots)X(\cdot)$ under the integral, we should choose

$$
\Gamma = PA + A^\top P + C^\top PC + \Lambda C + C^\top \Lambda + Q,
$$

so that one has

$$
J = x^\top P(t)x + \mathbb{E}\int_t^T \Big[2\Big(Pb + C^\top P\sigma + \Lambda\sigma + q\Big)^\top X + \sigma^\top P\sigma\Big]ds.
$$

Note that the first term in (10.3.30) is absorbed in the integral term and only the linear term under the integral appears. In the current case, (10.3.32) reads

$$
\begin{cases}
dP(t) = \big[P(t)A + A^\top P(t) + C^\top P(t)C + \Lambda(t)C + C^\top \Lambda(t) + Q\big]dt \\
\qquad\qquad + \Lambda(t)dW(t), \qquad t \in [0, T], \\
P(T) = G.
\end{cases}
\tag{10.3.32}
$$

Such an idea will be used several times below.

Note that the above could not be applied directly to $\mathscr{E}(\varepsilon)$ as $t \mapsto X_1(t)$ does not satisfy an SDE, and therefore, the Itô's formula could not be applied. To overcome this difficulty, we introduce an auxiliary process $\mathcal{X}_1(\cdot\,,\cdot)$ as follows

$$
\begin{aligned}
\mathcal{X}_1(t,r) = \int_\tau^r b_x(t,s)X_1(s)ds + \int_\tau^r \big[\sigma_x(t,s)X_1(s) + \delta\sigma^\varepsilon(t,s)\big]dW(s), \\
\tau \leqslant r \leqslant t \leqslant T.
\end{aligned}
\tag{10.3.33}
$$

Note that $r \mapsto \mathcal{X}_1(t, r)$ satisfies an SDE on $[\tau, t]$, with t being a parameter, and thus the Itô's formula can be used. Moreover, for any $t \in [\tau, T]$, by the second condition in (10.3.12), one has

$$\sup_{t\in[\tau,T]} \mathbb{E}\Big[\sup_{r\in[\tau,t]} |\mathcal{X}_1(t, r)|^k\Big] = O(\varepsilon^{\frac{k}{2}}), \quad k > 2. \tag{10.3.34}$$

Further, it is clear that

$$\mathcal{X}_1(r, r) = X_1(r), \qquad r \in [\tau, T].$$

Thus,

$$\mathscr{E}(\varepsilon) = \mathbb{E}\Big[\mathcal{X}_1(T, T)^\top h_{xx}\mathcal{X}_1(T, T) + \int_0^T \mathcal{X}_1(s, s)^\top H_{xx}(s)\mathcal{X}_1(s, s)ds\Big]. \tag{10.3.35}$$

We now treat the terms in $\mathscr{E}(\varepsilon)$. The procedure will be split into several steps.

Step 1. A general consideration.

Let $\Theta : \Delta_*[\tau, T] \times \Omega \to \mathbb{R}^{n\times n}$ be a process such that for each $t \in [\tau, T]$, $\Theta(t, \cdot) \in L^2_{\mathbb{F}}(\tau, t; \mathbb{R}^{n\times n})$. For each $(t, s) \in \Delta_*[\tau, T]$, applying the martingale representation theorem to the random variable $\Theta(t, s)$, one can find a unique $\Lambda(t, s, \cdot)$ such that

$$\begin{aligned}\Pi(t, s, r) \equiv \mathbb{E}_r[\Theta(t, s)] = \Theta(t, s) - \int_r^s \Lambda(t, s, \theta)dW(\theta),\\ \tau \leqslant r \leqslant s \leqslant t \leqslant T.\end{aligned} \tag{10.3.36}$$

Then, applying Itô's formula to the map $r \mapsto \Pi(t, s, r)\mathcal{X}_1(s, r)$, we get

$$\begin{aligned}&d\big[\Pi(t, s, r)\mathcal{X}_1(s, r)\big]\\ &= \Big[\Big(\Pi(t, s, r)b_x(s, r) + \Lambda(t, s, r)\sigma_x(s, r)\Big)X_1(r) + \Lambda(t, s, r)\delta\sigma^\varepsilon(s, r)\Big]dr\\ &\quad +\Big[\Lambda(t, s, r)\mathcal{X}_1(s, r) + \Pi(t, s, r)\Big(\sigma_x(s, r)X_1(r) + \delta\sigma^\varepsilon(s, r)\Big)\Big]dW(r).\end{aligned}$$

Further, applying the Itô's formula to the map $r \mapsto \mathcal{X}_1(t, r)^\top \Pi(t, s, r)\mathcal{X}_1(s, r)$ yields

$$\begin{aligned}&d\big[\mathcal{X}_1(t, r)^\top \Pi(t, s, r)\mathcal{X}_1(s, r)\big]\\ &= \Big[X_1(r)^\top b_x(t, r)^\top \Pi(t, s, r)\mathcal{X}_1(s, r)\end{aligned}$$

$$
\begin{aligned}
&+\mathcal{X}_1(t,r)^\top\Big(\Pi(t,s,r)b_x(s,r)+\Lambda(t,s,r)\sigma_x(s,r)\Big)X_1(r)\\
&+\mathcal{X}_1(r,s)^\top\Lambda(t,s,r)\delta\sigma^\varepsilon(s,r)+\Big(X_1(r)^\top\sigma_x(t,r)^\top+\delta\sigma^\varepsilon(t,r)^\top\Big)\\
&\quad\cdot\Big(\Lambda(t,s,r)\mathcal{X}_1(s,r)+\Pi(t,s,r)\sigma_x(s,r)X_1(r)+\Pi(t,s,r)\delta\sigma^\varepsilon(s,r)\Big)\Big]dr\\
&\qquad\qquad+\Gamma(t,s,r)dW(r)\\
=&\Big[X_1(r)^\top\Big(b_x(t,r)^\top\Pi(t,s,r)+\sigma_x(t,r)^\top\Lambda(t,s,r)\Big)\mathcal{X}_1(s,r)\\
&+\mathcal{X}_1(t,r)^\top\Big(\Pi(t,s,r)b_x(s,r)+\Lambda(t,s,r)\sigma_x(s,r)\Big)X_1(r)\\
&+X_1(r)^\top\sigma_x(t,r)^\top\Pi(t,s,r)\sigma_x(s,r)X_1(r)+\delta\sigma^\varepsilon(t,r)^\top\Pi(t,s,r)\delta\sigma^\varepsilon(s,r)\\
&+\delta\sigma^\varepsilon(t,r)^\top\Big(\Lambda(t,s,r)\mathcal{X}_1(s,r)+\Pi(t,s,r)\sigma_x(s,r)X_1(r)\Big)\\
&+\Big(\mathcal{X}_1(r,s)^\top\Lambda(t,s,r)+X_1(r)^\top\sigma_x(t,r)^\top\Pi(t,s,r)\Big)\delta\sigma^\varepsilon(s,r)\Big]dr\\
&+\Gamma(t,s,r)dW(r),
\end{aligned}
$$

where

$$
\begin{aligned}
\Gamma(t,s,r)=&\Big(X_1(r)^\top\sigma_x(t,r)^\top+\delta\sigma^\varepsilon(t,r)^\top\Big)\Pi(t,s,r)\mathcal{X}_1(s,r)\\
&+\mathcal{X}_1(t,r)^\top\Big[\Lambda(t,s,r)\mathcal{X}_1(s,r)+\Pi(t,s,r)\Big(\sigma_x(s,r)X_1(r)+\delta\sigma^\varepsilon(s,r)\Big)\Big].
\end{aligned}
$$

By the integrability of (Π, Λ), X_1, $\mathcal{X}_1$, and (H10.3.1), we see that

$$
\begin{aligned}
\mathbb{E}\int_0^T&\Big[\delta\sigma^\varepsilon(t,r)^\top\Big(\Lambda(t,s,r)\mathcal{X}_1(s,r)+\Pi(t,s,r)\sigma_x(s,r)X_1(r)\Big)\\
&+\Big(\mathcal{X}_1(r,s)^\top\Lambda(t,s,r)+X_1(r)^\top\sigma_x(t,r)^\top\Pi(t,s,r)\Big)\delta\sigma^\varepsilon(s,r)\Big]dr=o(\varepsilon).
\end{aligned}
\tag{10.3.37}
$$

Thus, (noting $\tau\leqslant r\leqslant s\leqslant t\leqslant T$, and $\mathcal{X}_1(t,r)$ and $\mathcal{X}_1(s,r)$ being $\mathcal{F}_r$-measurable)

$$
\begin{aligned}
&\mathbb{E}\big[\mathcal{X}_1(t,r)^\top\Theta(t,s)\mathcal{X}_1(s,r)\big]=\mathbb{E}\big[\mathcal{X}_1(t,r)^\top\Pi(t,s,r)\mathcal{X}_1(s,r)\big]\\
&=\mathbb{E}\int_\tau^r\Big[X_1(\theta)^\top\Big(b_x(t,\theta)^\top\Pi(t,s,\theta)+\sigma_x(t,\theta)^\top\Lambda(t,s,\theta)\Big)\mathcal{X}_1(s,\theta)\\
&\qquad+\mathcal{X}_1(t,\theta)^\top\Big(\Pi(t,s,\theta)b_x(s,\theta)+\Lambda(t,s,\theta)\sigma_x(s,\theta)\Big)X_1(\theta)\\
&\qquad+X_1(\theta)^\top\sigma_x(t,\theta)^\top\Pi(t,s,\theta)\sigma_x(s,\theta)X_1(\theta)\\
&\qquad+\delta\sigma^\varepsilon(t,\theta)^\top\Pi(t,s,\theta)\delta\sigma^\varepsilon(s,\theta)\Big]d\theta+o(\varepsilon)\\
&=\mathbb{E}\int_\tau^r\Big[X_1(\theta)^\top\Big(b_x(t,\theta)^\top\Theta(t,s)+\sigma_x(t,\theta)^\top\Lambda(t,s,\theta)\Big)\mathcal{X}_1(s,\theta)\\
&\qquad\quad+\mathcal{X}_1(t,\theta)^\top\Big(\Theta(t,s)b_x(s,\theta)+\Lambda(t,s,\theta)\sigma_x(s,\theta)\Big)X_1(\theta)\\
&\qquad\quad+X_1(\theta)^\top\sigma_x(t,\theta)^\top\Theta(t,s)\sigma_x(s,\theta)X_1(\theta)\\
&\qquad\quad+\delta\sigma^\varepsilon(t,\theta)^\top\Theta(t,s)\delta\sigma^\varepsilon(s,\theta)\Big]d\theta+o(\varepsilon).
\end{aligned}
\tag{10.3.38}
$$

In the above, to get rid of the Itô's integral of $\Gamma(t,s,\cdot)$ under the expectation, some standard localization arguments via stopping times have been used. Now, we are ready to look at the terms in $\mathscr{E}(\varepsilon)$ (see (10.3.24)).

Step 2. Treatment of the first term in $\mathscr{E}(\varepsilon)$.

Take $t=s=T$ and $\Theta(T,T)=h_{xx}$ in (10.3.36). Denote the adapted solution of (10.3.36) by

$$(\Pi(T,T,r),\Lambda(T,T,r))\equiv(P_1(r),Q_1(r)),\qquad r\in[\tau,T].$$

Then (10.3.36) reads

$$P_1(r)=h_{xx}-\int_r^T Q_1(\theta)dW(\theta),\qquad r\in[\tau,T]. \tag{10.3.39}$$

By the uniqueness, $P_1(\cdot)$ and $Q_1(\cdot)$ take values in $\mathbb{S}^n$. Making use of (10.3.12) and (H10.3.1), one has

$$\begin{aligned}&\mathbb{E}\int_\tau^{\tau+\varepsilon}|X_1(s)||P_1(s)||\delta\sigma(T,s)|ds\\&\leqslant\Big[\mathbb{E}\int_\tau^{\tau+\varepsilon}|X_1(s)|^4ds\Big]^{\frac14}\Big[\mathbb{E}\int_\tau^{\tau+\varepsilon}|\delta\sigma(T,s)|^4ds\Big]^{\frac14}\Big[\mathbb{E}\int_\tau^{\tau+\varepsilon}|P_1(s)|^2ds\Big]^{\frac12}=o(\varepsilon),\end{aligned}$$

and

$$\begin{aligned}&\mathbb{E}\int_\tau^{\tau+\varepsilon}|\mathcal{X}_1(T,s)||Q_1(s)||\delta\sigma(T,s)|ds\\&\leqslant\Big[\mathbb{E}\int_\tau^{\tau+\varepsilon}|\mathcal{X}_1(T,s)|^4ds\Big]^{\frac14}\Big[\mathbb{E}\int_\tau^{\tau+\varepsilon}|\delta\sigma(T,s)|^4ds\Big]^{\frac14}\Big[\mathbb{E}\int_\tau^{\tau+\varepsilon}|Q_1(s)|^2ds\Big]^{\frac12}=o(\varepsilon).\end{aligned}$$

Therefore, the corresponding (10.3.37) follows. On the other hand, in the current case, we have

$$\begin{aligned}\Gamma(T,T,r)=&\,\mathcal{X}_1(T,r)^\top Q_1(r)\mathcal{X}_1(T,r)\\&+\mathcal{X}_1(T,r)^\top P_1(r)\Big(\sigma_x(T,r)X_1(r)+\delta\sigma^\varepsilon(T,r)\Big)\\&+\Big(X_1(r)^\top\sigma_x(T,r)^\top+\delta\sigma^\varepsilon(T,r)^\top\Big)P_1(r)\mathcal{X}_1(T,r).\end{aligned}$$

Again, by a localization argument, we can get rid of the Itô's integral of $\Gamma(T,T,\cdot)$ under expectation to obtain the following, which is a particular case of (10.3.38)

$$
\begin{aligned}
&\mathbb{E}\big[X_1(T)^\top h_{xx}X_1(T)\big]=\mathbb{E}\big[\mathcal{X}_1(T,T)^\top P_1(T)\mathcal{X}_1(T,T)\big]\\
&=\mathbb{E}\int_\tau^T\Big[X_1(r)^\top\Big(b_x(T,r)^\top P_1(r)+\sigma_x(T,r)^\top Q_1(r)\Big)\mathcal{X}_1(T,r)\\
&\quad+\mathcal{X}_1(T,r)^\top\Big(P_1(r)b_x(T,r)+Q_1(r)\sigma_x(T,r)\Big)X_1(r)\\
&\quad+X_1(r)^\top\sigma_x(T,r)^\top P_1(r)\sigma_x(T,r)X_1(r)\\
&\quad+\delta\sigma^\varepsilon(T,r)^\top P_1(r)\delta\sigma^\varepsilon(T,r)\Big]dr+o(\varepsilon)\\
&\equiv\mathbb{E}\int_\tau^T\Big[X_1(r)^\top F_1(r)\mathcal{X}_1(T,r)+\mathcal{X}_1(T,r)^\top F_1(r)^\top X_1(r)\\
&\quad+X_1(r)^\top G_1(r)X_1(r)+\delta\sigma^\varepsilon(T,r)^\top P_1(r)\delta\sigma^\varepsilon(T,r)\Big]dr+o(\varepsilon).
\end{aligned}
$$

Here,

$$
\begin{cases}
F_1(r)\overset{\Delta}{=}b_x(T,r)^\top P_1(r)+\sigma_x(T,r)^\top Q_1(r),\\
G_1(r)\overset{\Delta}{=}\sigma_x(T,r)^\top P_1(r)\sigma_x(T,r).
\end{cases}
\tag{10.3.40}
$$

Note that both $F_1(r)$ and $G_1(r)$ are $\mathcal{F}_r$-measurable and independent of $u^\varepsilon(\cdot)$. Consequently, at the end of this step, one has

$$
\begin{aligned}
\mathcal{E}(\varepsilon)=\mathbb{E}\int_\tau^T&\Big[X_1(r)^\top F_1(r)\mathcal{X}_1(T,r)+\mathcal{X}_1(T,r)^\top F_1(r)^\top X_1(r)\\
&+X_1(r)^\top\Big(H_{xx}(r)+G_1(r)\Big)X_1(r)\\
&+\delta\sigma^\varepsilon(T,r)^\top P_1(r)\delta\sigma^\varepsilon(T,r)\Big]dr+o(\varepsilon).
\end{aligned}
\tag{10.3.41}
$$

We see that, with the above step, the term $\mathbb{E}\big[X_1(T)^\top h_{xx}X_1(T)\big]$ is absorbed. However, the new term of the form $X_1(r)^\top F_1(r)\mathcal{X}_1(T,r)$ and its transpose appear under integral. This will be handled by the next step.

Step 3. Treatment of the term $X_1(r)^\top F_1(r)\mathcal{X}_1(T,r)$ and its transpose.

Let $t=T$ in (10.3.36) and $\Theta(T,s)=\Theta_2(T,s)$ be undetermined (which is $\mathcal{F}_s$-measurable). By martingale representation theorem, one has

$$
\Theta_2(T,s)=\mathbb{E}_r[\Theta_2(T,s)]+\int_r^s\Lambda_2(T,s,\theta)dW(\theta),\qquad \tau\leqslant r\leqslant s\leqslant T.
\tag{10.3.42}
$$

In this case, (10.3.38) reads

$$
\begin{aligned}
&\mathbb{E}\big[\mathcal{X}_1(T,r)^\top \Theta_2(T,r)X_1(r)\big]\\
&=\mathbb{E}\int_\tau^r \Big[X_1(\theta)^\top\Big(b_x(T,\theta)^\top\Theta_2(T,r)+\sigma_x(T,\theta)^\top\Lambda_2(T,r,\theta)\Big)\mathcal{X}_1(r,\theta)\\
&\quad+\mathcal{X}_1(T,\theta)^\top\Big(\Theta_2(T,r)b_x(r,\theta)+\Lambda_2(T,r,\theta)\sigma_x(r,\theta)\Big)X_1(\theta)\\
&\quad+X_1(\theta)^\top\sigma_x(T,\theta)^\top\Theta_2(T,r)\sigma_x(r,\theta)X_1(\theta)\\
&\quad+\delta\sigma^\varepsilon(T,\theta)^\top\Theta_2(T,r)\delta\sigma^\varepsilon(r,\theta)\Big]d\theta+o(\varepsilon)\\
&\equiv\mathbb{E}\int_\tau^r\Big[X_1(\theta)^\top F_2(r,\theta)\mathcal{X}_1(r,\theta)+\mathcal{X}_1(T,\theta)^\top\widetilde{F}_2(r,\theta)^\top X_1(\theta)\\
&\quad+\frac{1}{2}X_1(\theta)^\top G_2(r,\theta)X_1(\theta)+\delta\sigma^\varepsilon(T,\theta)^\top\Theta_2(T,r)\delta\sigma^\varepsilon(r,\theta)\Big]d\theta+o(\varepsilon),
\end{aligned}
\tag{10.3.43}
$$

where

$$
\begin{cases}
F_2(r,\theta)\stackrel{\Delta}{=}b_x(T,\theta)^\top\Theta_2(T,r)+\sigma_x(T,\theta)^\top\Lambda_2(T,r,\theta),\\
\widetilde{F}_2(r,\theta)\stackrel{\Delta}{=}b_x(r,\theta)^\top\Theta_2(T,r)^\top+\sigma_x(r,\theta)^\top\Lambda_2(T,r,\theta)^\top,\\
G_2(r,\theta)\stackrel{\Delta}{=}\sigma_x(T,\theta)^\top\Theta_2(T,r)\sigma_x(r,\theta)+\sigma_x(r,\theta)^\top\Theta_2(T,r)^\top\sigma_x(T,\theta).
\end{cases}
\tag{10.3.44}
$$

Thus,

$$
\begin{aligned}
&\mathbb{E}\int_\tau^T\big[\mathcal{X}_1(T,r)^\top\Theta_2(T,r)X_1(r)+X_1(r)^\top\Theta_2(T,r)^\top\mathcal{X}_1(T,r)\big]dr\\
&=\mathbb{E}\int_\tau^T\int_\tau^r\Big[X_1(\theta)^\top F_2(r,\theta)\mathcal{X}_1(r,\theta)+\mathcal{X}_1(r,\theta)F_2(r,\theta)^\top X_1(\theta)\\
&\quad+\mathcal{X}_1(T,\theta)^\top\widetilde{F}_2(r,\theta)^\top X_1(\theta)+X_1(\theta)^\top\widetilde{F}_2(r,\theta)\mathcal{X}_1(T,\theta)\\
&\quad+X_1(\theta)^\top G_2(r,\theta)X_1(\theta)+\delta\sigma^\varepsilon(T,\theta)^\top\Theta_2(T,r)\delta\sigma^\varepsilon(r,\theta)\\
&\quad+\delta\sigma^\varepsilon(r,\theta)^\top\Theta_2(T,r)^\top\delta\sigma^\varepsilon(T,\theta)\Big]d\theta dr+o(\varepsilon)\\
&=\mathbb{E}\int_\tau^T\int_r^T\Big[X_1(r)^\top F_2(\theta,r)\mathcal{X}_1(\theta,r)+\mathcal{X}_1(\theta,r)F_2(\theta,r)^\top X_1(r)\\
&\quad+\mathcal{X}_1(T,r)^\top\widetilde{F}_2(\theta,r)^\top X_1(r)+X_1(r)^\top\widetilde{F}_2(\theta,r)\mathcal{X}_1(T,r)\\
&\quad+X_1(r)^\top G_2(\theta,r)X_1(r)+\delta\sigma^\varepsilon(T,r)^\top\Theta_2(T,\theta)\delta\sigma^\varepsilon(\theta,r)\\
&\quad+\delta\sigma^\varepsilon(\theta,r)^\top\Theta_2(T,\theta)^\top\delta\sigma^\varepsilon(T,r)\Big]d\theta dr+o(\varepsilon).
\end{aligned}
$$

Consequently, (10.3.41) becomes

$$
\begin{aligned}
\mathscr{E}(\varepsilon)&\equiv\mathbb{E}\Big[X_1(T)^\top h_{xx}X_1(T)+\int_\tau^T X_1(r)^\top H_{xx}(r)X_1(r)dr\Big]\\
&=\mathbb{E}\int_\tau^T\Big[X_1(r)^\top\Big(F_1(r)-\Theta_2(T,r)^\top+\int_r^T\widetilde{F}_2(\theta,r)d\theta\Big)\mathcal{X}_1(T,r)
\end{aligned}
$$

$$
\begin{aligned}
&+\mathcal{X}_1(T,r)^\top\Big(F_1(r)^\top-\Theta_2(T,r)+\int_r^T\widetilde{F}_2(\theta,r)^\top d\theta\Big)X_1(r)\\
&+\int_r^T\big[X_1(r)^\top F_2(\theta,r)\mathcal{X}_1(\theta,r)+\mathcal{X}_1(\theta,r)^\top F_2(\theta,r)^\top X_1(r)\big]d\theta\\
&+X_1(r)^\top\Big(H_{xx}(r)+G_1(r)+\int_r^T G_2(\theta,r)d\theta\Big)X_1(r)\\
&+\delta\sigma^\varepsilon(T,r)^\top P_1(r)\delta\sigma^\varepsilon(T,r)+\int_r^T\big[\delta\sigma^\varepsilon(T,r)^\top\Theta_2(T,\theta)\delta\sigma^\varepsilon(\theta,r)\\
&+\delta\sigma^\varepsilon(\theta,r)^\top\Theta_2(T,\theta)^\top\delta\sigma^\varepsilon(T,r)\big]d\theta\Big]dr+o(\varepsilon).
\end{aligned}
$$

Thus, in order to eliminate the term of the form $X_1(r)^\top(\cdots)\mathcal{X}_1(T,r)$ and its transpose, let

$$
\begin{aligned}
\Theta_2(T,r)&=F_1(r)^\top+\mathbb{E}_r\int_r^T\widetilde{F}_2(\theta,r)^\top d\theta\\
&=\mathbb{E}_r\Big[P_1(r)b_x(T,r)+Q_1(r)\sigma_x(T,r)+\int_r^T\Big(\Theta_2(T,\theta)b_x(\theta,r)\\
&\qquad+\Lambda_2(T,\theta,r)\sigma_x(\theta,r)\Big)d\theta\Big].
\end{aligned}
$$

Now, by denoting

$$
P_2(r)\stackrel{\Delta}{=}\Theta_2(T,r)^\top,\qquad Q_2(\theta,r)\stackrel{\Delta}{=}\Lambda_2(T,\theta,r)^\top,\qquad \tau\leqslant r\leqslant\theta\leqslant T,
$$

we have

$$
\begin{aligned}
P_2(r)&=\mathbb{E}_r\Big[b_x(T,r)^\top P_1(r)+\sigma_x(T,r)^\top Q_1(r)\\
&\qquad+\int_r^T\Big(b_x(\theta,r)^\top P_2(\theta)+\sigma_x(\theta,r)^\top Q_2(\theta,r)\Big)d\theta\Big]\\
&=b_x(T,r)^\top P_1(r)+\sigma_x(T,r)^\top Q_1(r)+\int_r^T\Big(b_x(\theta,r)^\top P_2(\theta)\\
&\qquad+\sigma_x(\theta,r)^\top Q_2(\theta,r)\Big)d\theta-\int_r^T\widetilde{Q}_2(r,\theta)dW(\theta),
\end{aligned}
$$

for some $\widetilde{Q}_2(r,\theta)$ defined on $\Delta^*[\tau,T]$, or for $\tau\leqslant r\leqslant\theta\leqslant T$, with a suitable measurability and integrability. At the same time, (10.3.42) reads

$$
P_2(s)=\mathbb{E}_r[P_2(s)]+\int_r^s Q_2(s,\theta)dW(\theta),\qquad \tau\leqslant r\leqslant s\leqslant T,
$$

with $Q_2(s,\theta)$ defined on $\Delta_*[t,T]$, or for $\tau\leqslant\theta\leqslant s\leqslant T$. We now extend $Q_2(s,\theta)$ from $\Delta_*[\tau,T]$ to $[\tau,T]^2$ by

$$
Q_2(s,\theta)\stackrel{\Delta}{=}\widetilde{Q}_2(s,\theta),\qquad \tau\leqslant s\leqslant\theta\leqslant T.
$$

Then $(P_2(\cdot), Q_2(\cdot\,,\cdot))$ is the unique adapted M-solution to the following Type-II BSVIE:

$$\begin{aligned}P_2(r)=&b_x(T,r)^\top P_1(r)+\sigma_x(T,r)^\top Q_1(r)+\int_r^T\Big(b_x(\theta,r)^\top P_2(\theta)\\&+\sigma_x(\theta,r)^\top Q_2(\theta,r)\Big)d\theta-\int_r^T Q_2(r,\theta)dW(\theta),\quad r\in[\tau,T],\end{aligned}\tag{10.3.45}$$

which is independent of spike variation $u^\varepsilon(\cdot)$ of $\bar u(\cdot)$. Having the above result, we obtain

$$\begin{aligned}\mathscr{E}(\varepsilon)=&\mathbb{E}\int_\tau^T\Big[X_1(r)^\top\Big(H_{xx}(r)+G_1(r)+\int_r^T G_2(\theta,r)d\theta\Big)X_1(r)\\&+\int_r^T\Big(X_1(r)^\top F_2(\theta,r)\mathcal{X}_1(\theta,r)+\mathcal{X}_1(\theta,r)^\top F_2(\theta,r)^\top X_1(r)\Big)d\theta\Big]dr\\&+\mathbb{E}\int_\tau^T\Big[\delta\sigma^\varepsilon(T,r)^\top P_1(r)\delta\sigma^\varepsilon(T,r)+\int_r^T\Big(\delta\sigma^\varepsilon(T,r)^\top P_2(\theta)^\top\delta\sigma^\varepsilon(\theta,r)\\&+\delta\sigma^\varepsilon(\theta,r)^\top P_2(\theta)\delta\sigma^\varepsilon(T,r)\Big)d\theta\Big]dr+o(\varepsilon).\end{aligned}\tag{10.3.46}$$

We see by (10.3.44) that

$$\begin{cases}F_2(r,\theta)\stackrel{\Delta}{=}b_x(T,\theta)^\top P_2(r)^\top+\sigma_x(T,\theta)^\top Q_2(r,\theta),\\ G_2(r,\theta)\stackrel{\Delta}{=}\sigma_x(T,\theta)^\top P_2(r)^\top\sigma_x(r,\theta)+\sigma_x(r,\theta)^\top P_2(r)\sigma_x(T,\theta).\end{cases}\tag{10.3.47}$$

Now, in $\mathscr{E}(\varepsilon)$ (see (10.3.46)), the terms of the forms $X_1(r)^\top(\cdots)X_1(r)$ and $X_1(r)^\top(\cdots)\mathcal{X}_1(\theta,r)$, together with its transpose, need to be handled. In the next step, we will get rid of those terms simultaneously.

Step 4. Treatment of the terms $X_1(r)^\top(\cdots)X_1(r)$, $\mathcal{X}_1(\theta,r)^\top(\cdots)X_1(r)$, and its transpose.

To treat $X_1(r)^\top(\cdots)X_1(r)$, we take $t=s$ in (10.3.36) with $\Theta(s,s)=\Theta_3(s)$ taking values in $\mathbb{S}^n$, undetermined. Again, by martingale representation theorem, we have a unique $\Lambda_3(s,s,\cdot)$ such that

$$\Theta_3(s)=\mathbb{E}_r[\Theta_3(s)]+\int_r^s\Lambda_3(s,s,\theta)dW(\theta),\qquad \tau\leqslant r\leqslant s\leqslant T.\tag{10.3.48}$$

Since $\Theta_3(s)$ is symmetric, so is $\Lambda_3(s,s,r)$. Now, (10.3.38) becomes

$$\begin{aligned}&\mathbb{E}\big[X_1(r)^\top\Theta_3(r)X_1(r)\big]\\&=\mathbb{E}\int_\tau^r\Big[X_1(\theta)^\top\Big(b_x(r,\theta)^\top\Theta_3(r)+\sigma_x(r,\theta)^\top\Lambda_3(r,r,\theta)\Big)\mathcal{X}_1(r,\theta)\\&\quad+\mathcal{X}_1(r,\theta)^\top\Big(\Theta_3(r)b_x(r,\theta)+\Lambda_3(r,r,\theta)\sigma_x(r,\theta)\Big)X_1(\theta)\end{aligned}$$

$$
+ X_1(\theta)^\top \sigma_x(r,\theta)^\top \Theta_3(r)\sigma_x(r,\theta)X_1(\theta) + \delta\sigma^\varepsilon(r,\theta)^\top \Theta_3(r)\delta\sigma^\varepsilon(r,\theta)\Big]d\theta + o(\varepsilon)
$$

$$
= \mathbb{E}\int_\tau^r \Big[X_1(\theta)^\top F_3(r,\theta)\mathcal{X}_1(r,\theta) + \mathcal{X}_1(r,\theta)^\top F_3(r,\theta)^\top X_1(\theta)
$$

$$
+ X_1(\theta)^\top G_3(r,\theta)X_1(\theta) + \delta\sigma^\varepsilon(r,\theta)^\top \Theta_3(r)\delta\sigma^\varepsilon(r,\theta)\Big]d\theta + o(\varepsilon),
$$

where

$$
\begin{cases} F_3(r,\theta) \triangleq b_x(r,\theta)^\top \Theta_3(r) + \sigma_x(r,\theta)^\top \Lambda_3(r,r,\theta), \\ G_3(r,\theta) \triangleq \sigma_x(r,\theta)^\top \Theta_3(r)\sigma_x(r,\theta). \end{cases} \tag{10.3.49}
$$

Then by integrating it over $[\tau, T]$, one has

$$
\begin{aligned}
&\mathbb{E}\int_\tau^T \big[X_1(r)^\top \Theta_3(r)X_1(r)\big]dr \\
&= \mathbb{E}\int_\tau^T\int_\tau^r \Big[X_1(\theta)^\top F_3(r,\theta)\mathcal{X}_1(r,\theta) + \mathcal{X}_1(r,\theta)^\top F_3(r,\theta)^\top X_1(\theta) \\
&\quad + X_1(\theta)^\top G_3(r,\theta)X_1(\theta) + \delta\sigma^\varepsilon(r,\theta)^\top \Theta_3(r)\delta\sigma^\varepsilon(r,\theta)\Big]d\theta dr + o(\varepsilon) \\
&= \mathbb{E}\int_\tau^T\int_r^T \Big[X_1(r)^\top F_3(\theta,r)\mathcal{X}_1(\theta,r) + \mathcal{X}_1(\theta,r)^\top F_3(\theta,r)^\top X_1(r) \\
&\quad + X_1(r)^\top G_3(\theta,r)X_1(r) + \delta\sigma^\varepsilon(\theta,r)^\top \Theta_3(\theta)\delta\sigma^\varepsilon(\theta,r)\Big]d\theta dr + o(\varepsilon).
\end{aligned} \tag{10.3.50}
$$

Next, for the term $\mathcal{X}_1(\theta,r)^\top(\cdots)X_1(r)$, let $\Theta(t,s) = \Theta_4(t,s)$ be undetermined. By martingale representation theorem,

$$
\Theta_4(t,s) = \mathbb{E}_r[\Theta_4(t,s)] + \int_r^s \Lambda_4(t,s,\theta)dW(\theta), \qquad \tau \leqslant r \leqslant s \leqslant t \leqslant T.
$$

Then, (10.3.38) reads

$$
\begin{aligned}
&\mathbb{E}\big[\mathcal{X}_1(\theta,r)^\top \Theta_4(\theta,r)X_1(r)\big] \\
&= \mathbb{E}\int_\tau^r \Big[X_1(\theta')^\top \Big(b_x(\theta,\theta')^\top \Theta_4(\theta,r) + \sigma_x(\theta,\theta')^\top \Lambda_4(\theta,r,\theta')\Big)\mathcal{X}_1(r,\theta') \\
&\quad + \mathcal{X}_1(\theta,\theta')^\top \Big(\Theta_4(\theta,r)b_x(r,\theta') + \Lambda_4(\theta,r,\theta')\sigma_x(r,\theta')\Big)X_1(\theta') \\
&\quad + X_1(\theta')^\top \sigma_x(\theta,\theta')^\top \Theta_4(\theta,r)\sigma_x(r,\theta')X_1(\theta') \\
&\quad + \delta\sigma^\varepsilon(\theta,\theta')^\top \Theta_4(\theta,r)\delta\sigma^\varepsilon(r,\theta')\Big]d\theta' + o(\varepsilon) \\
&\equiv \mathbb{E}\int_\tau^r \Big[X_1(\theta')^\top F_4(\theta,r,\theta')\mathcal{X}_1(r,\theta') + \mathcal{X}_1(\theta,\theta')^\top \widetilde{F}_4(\theta,r,\theta')^\top X_1(\theta') \\
&\quad + \frac{1}{2}X_1(\theta')^\top G_4(\theta,r,\theta')X_1(\theta') + \delta\sigma^\varepsilon(\theta,\theta')^\top \Theta_4(\theta,r)\delta\sigma^\varepsilon(r,\theta')\Big]d\theta' + o(\varepsilon),
\end{aligned}
$$

where (note that $\tau \leqslant \theta' \leqslant r \leqslant \theta \leqslant T$)

$$\begin{cases} F_4(\theta,r,\theta') \stackrel{\Delta}{=} b_x(\theta,\theta')^\top \Theta_4(\theta,r) + \sigma_x(\theta,\theta')^\top \Lambda_4(\theta,r,\theta'), \\ \widetilde{F}_4(\theta,r,\theta') \stackrel{\Delta}{=} b_x(r,\theta')^\top \Theta_4(\theta,r)^\top + \sigma_x(r,\theta')^\top \Lambda_4(\theta,r,\theta')^\top, \\ G_4(\theta,r,\theta') \stackrel{\Delta}{=} \sigma_x(\theta,\theta')^\top \Theta_4(\theta,r)\sigma_x(r,\theta') + \sigma_x(r,\theta')^\top \Theta_4(\theta,r)^\top \sigma_x(\theta,\theta'). \end{cases} \tag{10.3.51}$$

Therefore, we have

$$\begin{aligned}
&\mathbb{E}\int_\tau^T \int_r^T \mathcal{X}_1(\theta,r)^\top \Theta_4(\theta,r) X_1(r) d\theta dr \\
&= \mathbb{E}\int_\tau^T \int_r^T \int_\tau^r \Big[X_1(\theta')^\top F_4(\theta,r,\theta')\mathcal{X}_1(r,\theta') + \mathcal{X}_1(\theta,\theta')^\top \widetilde{F}_4(\theta,r,\theta')^\top X_1(\theta') \\
&\quad + \frac{1}{2} X_1(\theta')^\top G_4(\theta,r,\theta') X_1(\theta') + \delta\sigma^\varepsilon(\theta,\theta')^\top \Theta_4(\theta,r)\delta\sigma^\varepsilon(r,\theta')\Big] d\theta' d\theta dr + o(\varepsilon) \\
&= \mathbb{E}\int_\tau^T \int_{\theta'}^T \int_\tau^{\theta'} \Big[X_1(r)^\top F_4(\theta,\theta',r)\mathcal{X}_1(\theta',r) + \mathcal{X}_1(\theta,r)^\top \widetilde{F}_4(\theta,\theta',r)^\top X_1(r) \\
&\quad + \frac{1}{2} X_1(r)^\top G_4(\theta,\theta',r) X_1(r) + \delta\sigma^\varepsilon(\theta,r)^\top \Theta_4(\theta,\theta')\delta\sigma^\varepsilon(\theta',r)\Big] dr d\theta d\theta' + o(\varepsilon) \\
&= \mathbb{E}\int_\tau^T \int_\tau^{\theta'} \int_{\theta'}^T \Big[X_1(r)^\top F_4(\theta,\theta',r)\mathcal{X}_1(\theta',r) + \mathcal{X}_1(\theta,r)^\top \widetilde{F}_4(\theta,\theta',r)^\top X_1(r) \\
&\quad + \frac{1}{2} X_1(r)^\top G_4(\theta,\theta',r) X_1(r) + \delta\sigma^\varepsilon(\theta,r)^\top \Theta_4(\theta,\theta')\delta\sigma^\varepsilon(\theta',r)\Big] d\theta dr d\theta' + o(\varepsilon) \\
&= \mathbb{E}\int_\tau^T \int_r^T \int_{\theta'}^T \Big[X_1(r)^\top F_4(\theta,\theta',r)\mathcal{X}_1(\theta',r) + \mathcal{X}_1(\theta,r)^\top \widetilde{F}_4(\theta,\theta',r)^\top X_1(r) \\
&\quad + \frac{1}{2} X_1(r)^\top G_4(\theta,\theta',r) X_1(r) + \delta\sigma^\varepsilon(\theta,r)^\top \Theta_4(\theta,\theta')\delta\sigma^\varepsilon(\theta',r)\Big] d\theta d\theta' dr + o(\varepsilon) \\
&= \mathbb{E}\int_\tau^T \Big[\int_r^T \int_\theta^T X_1(r)^\top F_4(\theta',\theta,r)\mathcal{X}_1(\theta,r) d\theta' d\theta + \int_r^T \int_r^\theta \mathcal{X}_1(\theta,r)^\top \widetilde{F}_4(\theta,\theta',r)^\top X_1(r) d\theta' d\theta \\
&\quad + \int_r^T \int_\theta^T \Big(\frac{1}{2} X_1(r)^\top G_4(\theta',\theta,r) X_1(r) + \delta\sigma^\varepsilon(\theta',r)^\top \Theta_4(\theta',\theta)\delta\sigma^\varepsilon(\theta,r)\Big) d\theta' d\theta\Big] dr + o(\varepsilon).
\end{aligned}$$

Hence, combining (10.3.46) and (10.3.50) with the above, we have

$$\begin{aligned}
\mathscr{E}(\varepsilon) = \mathbb{E}\int_\tau^T \Big\{ & X_1(r)^\top \Big[H_{xx}(r) + G_1(r) \\
& + \int_r^T \Big(G_2(\theta,r) + G_3(\theta,r) + \int_\theta^T G_4(\theta',\theta,r) d\theta'\Big) d\theta - \Theta_3(r)\Big] X_1(r) \\
& + \int_r^T \Big[X_1(r)^\top \Big(F_2(\theta,r) + F_3(\theta,r) - \Theta_4(\theta,r)^\top \\
& + \int_\theta^T F_4(\theta',\theta,r) d\theta' + \int_r^\theta \widetilde{F}_4(\theta,\theta',r) d\theta'\Big) \mathcal{X}_1(\theta,r) \\
& + \mathcal{X}_1(\theta,r)^\top \Big(F_2(\theta,r) + F_3(\theta,r) - \Theta_4(\theta,r)^\top
\end{aligned}$$

$$
\begin{aligned}
&+\int_\theta^T F_4(\theta',\theta,r)d\theta'+\int_r^\theta \widetilde{F}_4(\theta,\theta',r)d\theta'\Big)^\top X_1(r)\Big]d\theta \\
&+\delta\sigma^\varepsilon(T,r)^\top P_1(r)\delta\sigma^\varepsilon(T,r) \\
&+\int_r^T\Big[\delta\sigma^\varepsilon(T,r)^\top P_2(\theta)^\top\delta\sigma^\varepsilon(\theta,r)+\delta\sigma^\varepsilon(\theta,r)^\top P_2(\theta)\delta\sigma^\varepsilon(T,r) \\
&+\delta\sigma^\varepsilon(\theta,r)^\top\Theta_3(\theta)\delta\sigma^\varepsilon(\theta,r)+\int_\theta^T\Big(\delta\sigma^\varepsilon(\theta',r)^\top\Theta_4(\theta',\theta)\delta\sigma^\varepsilon(\theta,r) \\
&+\delta\sigma^\varepsilon(\theta,r)^\top\Theta_4(\theta',\theta)^\top\delta\sigma^\varepsilon(\theta',r)\Big)d\theta'\Big]d\theta\Big\}dr+o(\varepsilon).
\end{aligned}
$$

Consequently, to eliminate the term $X_1(r)^\top(\cdots)X_1(r)$, we should let

$$
\begin{aligned}
\Theta_3(r)=&H_{xx}(r)+G_1(r)+\int_r^T\Big(G_2(\theta,r)+G_3(\theta,r)+\int_\theta^T G_4(\theta',\theta,r)d\theta'\Big)d\theta \\
&-\int_r^T\widetilde{Q}_3(r,\theta)dW(\theta) \\
=&H_{xx}(r)+\sigma_x(T,r)^\top P_1(r)\sigma_x(T,r)+\int_r^T\Big(\sigma_x(T,r)^\top P_2(\theta)^\top\sigma_x(\theta,r) \\
&+\sigma_x(\theta,r)^\top P_2(\theta)\sigma_x(T,r)+\sigma_x(\theta,r)^\top\Theta_3(\theta)\sigma_x(\theta,r)\Big)d\theta \\
&+\int_r^T\int_\theta^T\Big(\sigma_x(\theta',r)^\top\Theta_4(\theta',\theta)\sigma_x(\theta,r)+\sigma_x(\theta,r)^\top\Theta_4(\theta',\theta)^\top\sigma_x(\theta',r)\Big)d\theta'd\theta \\
&-\int_r^T\widetilde{Q}_3(r,\theta)dW(\theta),\qquad \tau\leqslant r\leqslant T,
\end{aligned}
$$

for some $\widetilde{Q}_3(r,\theta)$ defined for $\tau\leqslant r\leqslant\theta\leqslant T$. Similarly, to eliminate $X_1(r)^\top(\cdots)\mathcal{X}_1(\theta,r)$ and its transpose, we should let

$$
\begin{aligned}
\Theta_4(\theta,r)^\top=&F_2(\theta,r)+F_3(\theta,r)+\int_\theta^T F_4(\theta',\theta,r)d\theta'+\int_r^\theta\widetilde{F}_4(\theta,\theta',r)d\theta' \\
&-\int_r^T\widetilde{Q}_4(\theta,r,\theta')dW(\theta') \\
=&b_x(T,r)^\top P_2(\theta)^\top+\sigma_x(T,r)^\top Q_2(\theta,r)^\top+b_x(\theta,r)^\top\Theta_3(\theta)+\sigma_x(\theta,r)^\top\Lambda_3(\theta,\theta,r) \\
&+\int_\theta^T\Big(b_x(\theta',r)^\top\Theta_4(\theta',\theta)+\sigma_x(\theta',r)^\top\Lambda_4(\theta',\theta,r)\Big)d\theta' \\
&+\int_r^\theta\Big(b_x(\theta',r)^\top\Theta_4(\theta,\theta')^\top+\sigma_x(\theta',r)^\top\Lambda_4(\theta,\theta',r)^\top\Big)d\theta' \\
&-\int_r^T\widetilde{Q}_4(\theta,r,\theta')dW(\theta'),\qquad \tau\leqslant r\leqslant\theta\leqslant T,
\end{aligned}
$$

for some $\widetilde{Q}_4(\theta, r, \theta')$ defined for $\tau \leqslant r \leqslant \theta' \leqslant T$ and $\tau \leqslant r \leqslant \theta \leqslant T$. Let

$$P_3(\theta) \triangleq \Theta_3(\theta), \quad Q_3(\theta, r) \triangleq \begin{cases} \Lambda_3(\theta, \theta, r), & \tau \leqslant r \leqslant \theta \leqslant T, \\ \widetilde{Q}_3(\theta, r), & \tau \leqslant \theta \leqslant r \leqslant T, \end{cases}$$
$$P_4(\theta, \theta') \triangleq \Theta_4(\theta, \theta')^\top, \quad Q_4(\theta, r, \theta') \triangleq \begin{cases} \Lambda_4(\theta, r, \theta')^\top, & \tau \leqslant \theta' \leqslant r \leqslant \theta \leqslant T, \\ \widetilde{Q}_4(\theta, r, \theta')^\top, & \tau \leqslant r \leqslant \theta' \leqslant T. \end{cases}$$

Then,

$$\begin{aligned} P_3(r) &= H_{xx}(r) + \sigma_x(T, r)^\top P_1(r)\sigma_x(T, r) + \int_r^T \Big(\sigma_x(T, r)^\top P_2(\theta)^\top \sigma_x(\theta, r) \\ &\quad +\sigma_x(\theta, r)^\top P_2(\theta)\sigma_x(T, r) + \sigma_x(\theta, r)^\top P_3(\theta)\sigma_x(\theta, r)\Big)d\theta \\ &\quad +\int_r^T\!\!\int_\theta^T \Big(\sigma_x(\theta', r)^\top P_4(\theta', \theta)^\top \sigma_x(\theta, r) + \sigma_x(\theta, r)^\top P_4(\theta', \theta)\sigma_x(\theta', r)\Big)d\theta' d\theta \\ &\quad -\int_r^T Q_3(r, \theta)dW(\theta), \qquad \tau \leqslant r \leqslant T, \\ P_4(\theta, r) &= b_x(T, r)^\top P_2(\theta)^\top + \sigma_x(T, r)^\top Q_2(\theta, r)^\top \\ &\qquad +b_x(\theta, r)^\top P_3(\theta) + \sigma_x(\theta, r)^\top Q_3(\theta, r) \\ &\qquad +\int_\theta^T \Big(b_x(\theta', r)^\top P_4(\theta', \theta)^\top + \sigma_x(\theta', r)^\top Q_4(\theta', \theta, r)^\top\Big)d\theta' \\ &\qquad +\int_r^\theta \Big(b_x(\theta', r)^\top P_4(\theta, \theta') + \sigma_x(\theta', r)^\top Q_4(\theta, \theta', r)\Big)d\theta' \\ &\qquad -\int_r^T Q_4(\theta, r, \theta')dW(\theta'), \qquad \tau \leqslant r \leqslant \theta \leqslant T. \end{aligned}$$

Thus, we obtain the following system:

$$\begin{cases} P_1(t) = h_{xx} - \displaystyle\int_t^T Q_1(s)dW(s), \qquad \tau \leqslant t \leqslant T, \\ P_2(t) = b_x(T, t)^\top P_1(t) + \sigma_x(T, t)^\top Q_1(t) \\ \qquad + \displaystyle\int_t^T \Big(b_x(s, t)^\top P_2(s) + \sigma_x(s, t)^\top Q_2(s, t)\Big)ds \\ \qquad\qquad - \displaystyle\int_t^T Q_2(t, s)dW(s), \qquad \tau \leqslant t \leqslant T, \\ P_3(t) = H_{xx}(t) + \sigma_x(T, t)^\top P_1(t)\sigma_x(T, t) \\ \qquad +\displaystyle\int_t^T \Big[\sigma_x(T, t)^\top P_2(s)^\top \sigma_x(s, t) + \sigma_x(s, t)^\top P_2(s)\sigma_x(T, t) \\ \qquad +\sigma_x(s, t)^\top P_3(s)\sigma_x(s, t) + \displaystyle\int_t^T \Big(\sigma_x(\theta, t)^\top P_4(\theta, s)^\top \sigma_x(s, t) \\ \qquad\quad +\sigma_x(s, t)^\top P_4(\theta, s)\sigma_x(\theta, t)\Big)d\theta\Big]ds - \displaystyle\int_t^T Q_3(t, s)dW(s), \\ \qquad\qquad\qquad \tau \leqslant t \leqslant T, \end{cases} \tag{10.3.52}$$

$$
\begin{cases}
P_4(r,t)=b_x(T,t)^\top P_2(r)^\top+\sigma_x(T,t)^\top Q_2(r,t)^\top \\
\qquad\qquad +b_x(r,t)^\top P_3(r)+\sigma_x(r,t)^\top Q_3(r,t) \\
\qquad\qquad +\int_r^T \Big(b_x(s,t)^\top P_4(s,r)^\top+\sigma_x(s,t)^\top Q_4(s,r,t)^\top\Big)ds \\
\qquad\qquad +\int_t^r \Big(b_x(s,t)^\top P_4(r,s)+\sigma_x(s,t)^\top Q_4(r,s,t)\Big)ds \\
\qquad\qquad -\int_t^T Q_4(r,t,s)dW(s), \qquad \tau\leqslant t\leqslant r\leqslant T.
\end{cases}
$$

By saying (P_i, Q_i) to be the adapted M-solutions of the corresponding BSVIEs, the following constraints must hold:

$$
\begin{cases}
P_2(t)=\mathbb{E}_\theta[P_2(t)]+\int_\theta^t Q_2(t,s)dW(s), \qquad \tau\leqslant\theta\leqslant t\leqslant T, \\
P_3(t)=\mathbb{E}_\theta[P_3(t)]+\int_\theta^t Q_3(t,s)dW(s), \qquad \tau\leqslant\theta\leqslant t\leqslant T, \\
P_4(r,t)=\mathbb{E}_\theta[P_4(r,t)]+\int_\theta^t Q_4(r,t,s)dW(s), \quad \tau\leqslant\theta\leqslant t\leqslant r\leqslant T.
\end{cases}
\tag{10.3.53}
$$

The above is called the *second order adjoint equation*. With the above system, we eventually end up with the following representation theorem.

Theorem 10.3.4 *Let* (H10.3.1)–(H10.3.2) *hold and* $(\bar X(\cdot),\bar u(\cdot))$ *be a given state-control pair. Let* (10.3.52) *admit an adapted M-solution* (P_i,Q_i) $(1\leqslant i\leqslant 4)$ *such that*

$$
\begin{aligned}
&(P_1,Q_1)\in L^2_{\mathbb{F}}(\Omega;C([\tau,T];\mathbb{S}^n))\times L^2_{\mathbb{F}}(\tau,T;\mathbb{S}^n),\\
&(P_2,P_3,P_4)\in L^2_{\mathbb{F}}(\tau,T;\mathbb{R}^{n\times n})\times L^2_{\mathbb{F}}(\tau,T;\mathbb{S}^n)\times L^2_{\mathbb{F}}(\Omega;L^2(\Delta_*[\tau,T];\mathbb{R}^{n\times n})).
\end{aligned}
$$

Then $\mathscr{E}(\varepsilon)$ *admits the following representation.*

$$
\begin{aligned}
\mathscr{E}(\varepsilon)=\mathbb{E}\int_\tau^T\Big\{&\delta\sigma^\varepsilon(T,r)^\top P_1(r)\delta\sigma^\varepsilon(T,r)+\int_r^T\Big[\delta\sigma^\varepsilon(T,r)^\top P_2(\theta)^\top\delta\sigma^\varepsilon(\theta,r)\\
&+\delta\sigma^\varepsilon(\theta,r)^\top P_2(\theta)\delta\sigma^\varepsilon(T,r)+\delta\sigma^\varepsilon(\theta,r)^\top P_3(\theta)\delta\sigma^\varepsilon(\theta,r)\\
&+\int_\theta^T\Big(\delta\sigma^\varepsilon(\theta',r)^\top P_4(\theta',\theta)^\top\delta\sigma^\varepsilon(\theta,r)\\
&+\delta\sigma^\varepsilon(\theta,r)^\top P_4(\theta',\theta)\delta\sigma^\varepsilon(\theta',r)\Big)d\theta'\Big]d\theta\Big\}dr+o(\varepsilon).
\end{aligned}
$$

This will lead to the maximum principle and so on, whose precise statements will be given in Sect. 3.4.

3.4 Well-Posedness of the Second Order Adjoint Equations

In this section, we will establish the well-posedness of system (10.3.52). To begin with, let us make some observations. The existence and uniqueness of (P_1, Q_1) is easy to see, while the existence and uniqueness of (P_2, Q_2) follows from Theorem 3.4.3 in Chap. 3. The remaining solvability of (10.3.52) lies in (P_3, Q_3, P_4, Q_4). For the equation of (P_3, Q_3), it is a classical linear BSVIE with given (P_1, P_2, P_4). However, as to the equation of (P_4, Q_4), its well-posedness cannot be directly given by the current BSVIE theory. Therefore, we need to carefully treat it.

Note that by applying conditional expectation operator $\mathbb{E}_t$ on both sides of fourth equation in (10.3.52), we end up with the following system (noting the third equality in (10.3.53))

$$\begin{cases} P(r,t) = F(r,t) + \displaystyle\int_r^T \big[b_x(s,t)^\top \mathbb{E}_t P(s,r)^\top + \sigma_x(s,t)^\top Q(s,r,t)^\top\big] ds \\ \qquad\qquad + \displaystyle\int_t^r \big[b_x(s,t)^\top \mathbb{E}_t P(r,s) + \sigma_x(s,t)^\top Q(r,s,t)\big] ds, \quad r \geqslant t, \\ \mathbb{E}_\theta P(r,t) = P(r,t) - \displaystyle\int_\theta^t Q(r,t,s) dW(s), \quad \tau \leqslant \theta \leqslant t \leqslant r \leqslant T, \end{cases} \tag{10.3.54}$$

with $P(r,t) = P_4(r,t)$, $Q(r,t,s) = Q_4(r,t,s)$ and

$$\begin{aligned} F(r,t) = \mathbb{E}_t\Big[& b_x(T,t)^\top P_2(r)^\top + \sigma_x(T,t)^\top Q_2(r,t)^\top \\ & + b_x(r,t)^\top P_3(r) + \sigma_x(r,t)^\top Q_3(r,t)\Big]. \end{aligned} \tag{10.3.55}$$

The following gives the well-posedness for (10.3.54).

Lemma 10.3.5 *Let b_x, σ_x be bounded, $F(\cdot,\cdot) \in L^2_{\mathbb{F}}(\Omega; L^2(\Delta_*[\tau,T]; \mathbb{R}^{n\times n}))$. Then (10.3.54) admits a unique solution $P(\cdot,\cdot) \in L^2_{\mathbb{F}}(\Omega; L^2(\Delta_*[\tau,T]; \mathbb{R}^{n\times n}))$. In addition, for any $\rho \in [\tau, T]$, the following estimate holds:*

$$\mathbb{E}\int_\rho^T \int_\rho^r |P(r,t)|^2 dt dr \leqslant K \mathbb{E}\int_\rho^T \int_\rho^r |F(r,t)|^2 dt dr, \tag{10.3.56}$$

where K only depends on $\|b_x\|_\infty$, $\|\sigma_x\|_\infty$ and T, but not on ρ.

Proof Let $\rho \in [\tau, T]$ be fixed. For any $p(\cdot,\cdot) \in L^2_{\mathbb{F}}(\Omega; L^2(\Delta_*[\tau,T]; \mathbb{R}^{n\times n}))$, we look at the following system

$$
\begin{cases} P(r,t)=F(r,t)+\displaystyle\int_r^T\big[b_x(s,t)^\top\mathbb{E}_t p(s,r)^\top+\sigma_x(s,t)^\top q(s,r,t)^\top\big]ds \\ \qquad\qquad +\displaystyle\int_t^r\big[b_x(s,t)^\top\mathbb{E}_t p(r,s)+\sigma_x(s,t)^\top q(r,s,t)\big]ds, \quad r\geqslant t, \\ \mathbb{E}_\theta p(r,t)=p(r,t)-\displaystyle\int_\theta^t q(r,t,s)dW(s), \qquad \rho\leqslant\theta\leqslant t\leqslant r\leqslant T. \end{cases} \tag{10.3.57}
$$

By the first equality in (10.3.57), we have (for any $\beta>0$)

$$
\begin{aligned}
&\mathbb{E}\int_\rho^T\int_\rho^r e^{\beta(r+t)}|P(r,t)|^2dtdr\leqslant K\mathbb{E}\int_\rho^T\int_\rho^r e^{\beta(r+t)}|F(r,t)|^2dtdr\\
&+K\|b_x\|_\infty^2\mathbb{E}\int_\rho^T\int_\rho^r e^{\beta(r+t)}\Big\{\Big(\int_t^r|p(r,s)|ds\Big)^2+\Big(\int_r^T|p(s,r)|ds\Big)^2\Big\}dtdr\\
&+K\|\sigma_x\|_\infty^2\mathbb{E}\int_\rho^T\int_\rho^r e^{\beta(r+t)}\Big\{\Big(\int_t^r|q(r,s,t)|ds\Big)^2+\Big(\int_r^T|q(s,r,t)|ds\Big)^2\Big\}dtdr.
\end{aligned} \tag{10.3.58}
$$

We now estimate the right-hand side of (10.3.58). For the second term, by Fubini theorem, we have

$$
\begin{aligned}
&\mathbb{E}\int_\rho^T\int_\rho^r e^{\beta(r+t)}\Big\{\Big(\int_t^r|p(r,s)|ds\Big)^2+\Big(\int_r^T|p(s,r)|ds\Big)^2\Big\}dtdr\\
&\leqslant\mathbb{E}\int_\rho^T\int_\rho^r e^{\beta(r+t)}\Big\{\Big(\int_t^r e^{-\beta s}ds\Big)\Big(\int_t^r e^{\beta s}|p(r,s)|^2ds\Big)\\
&\qquad\qquad+\Big(\int_r^T e^{-\beta s}ds\Big)\Big(\int_r^T e^{\beta s}|p(s,r)|^2ds\Big)\Big\}dtdr\\
&=\mathbb{E}\int_\rho^T\int_\rho^r\int_t^r e^{\beta(r+t)}\frac{e^{-\beta t}-e^{-\beta r}}{\beta}e^{\beta s}|p(r,s)|^2dsdtdr\\
&\quad+\mathbb{E}\int_\rho^T\int_\rho^r\int_r^T e^{\beta(r+t)}\frac{e^{-\beta r}-e^{-\beta T}}{\beta}e^{\beta s}|p(s,r)|^2dsdtdr\\
&=\frac{1}{\beta}\mathbb{E}\int_\rho^T\int_\rho^r\int_\rho^s\Big(e^{\beta(r+s)}-e^{\beta(t+s)}\Big)|p(r,s)|^2dtdsdr\\
&\quad+\frac{1}{\beta}\mathbb{E}\int_\rho^T\int_r^T\int_\rho^r\Big(e^{\beta(t+s)}-e^{\beta(r+s+t-T)}\Big)|p(s,r)|^2dtdsdr\\
&=\frac{1}{\beta}\mathbb{E}\int_\rho^T\int_\rho^r\int_\rho^s\Big(e^{\beta(r+s)}-e^{\beta(t+s)}\Big)|p(r,s)|^2dtdsdr\\
&\quad+\frac{1}{\beta}\mathbb{E}\int_\rho^T\int_\rho^s\int_\rho^r\Big(e^{\beta(t+s)}-e^{\beta(r+s+t-T)}\Big)|p(s,r)|^2dtdrds\\
&\leqslant\frac{2T}{\beta}\mathbb{E}\int_\rho^T\int_\rho^r e^{\beta(r+s)}|p(r,s)|^2dsdr.
\end{aligned}
$$

For the third term on the right-hand side of (10.3.58), similarly, by Fubini theorem,

$$
\begin{aligned}
&\mathbb{E}\int_\rho^T\int_\rho^r e^{\beta(r+t)}\Big\{\Big(\int_t^r |q(r,s,t)|ds\Big)^2+\Big(\int_r^T |q(s,r,t)|ds\Big)^2\Big\}dtdr\\
&\leqslant \mathbb{E}\int_\rho^T\int_\rho^r e^{\beta(r+t)}\Big\{\Big(\int_t^r e^{-\beta s}ds\Big)\Big(\int_t^r e^{\beta s}|q(r,s,t)|^2ds\Big)\\
&\qquad\qquad +\Big(\int_r^T e^{-\beta s}ds\Big)\Big(\int_r^T e^{\beta s}|q(s,r,t)|^2ds\Big)\Big\}dtdr\\
&=\frac{1}{\beta}\mathbb{E}\int_\rho^T\int_\rho^r\int_\rho^s\Big(e^{\beta(r+s)}-e^{\beta(t+s)}\Big)|q(r,s,t)|^2dtdsdr\\
&\quad +\frac{1}{\beta}\mathbb{E}\int_\rho^T\int_\rho^s\int_\rho^r\Big(e^{\beta(t+s)}-e^{\beta(r+s+t-T)}\Big)|q(s,r,t)|^2dtdrds\\
&\leqslant \frac{2}{\beta}\mathbb{E}\int_\rho^T\int_\rho^r\int_\rho^s e^{\beta(r+s)}|q(r,s,t)|^2dtdsdr\\
&=\frac{2}{\beta}\int_\rho^T\int_\rho^r e^{\beta(r+s)}\mathbb{E}\Big|\int_\rho^s q(r,s,t)dW(t)\Big|^2dsdr\\
&\leqslant \frac{2}{\beta}\mathbb{E}\int_\rho^T\int_\rho^r e^{\beta(r+s)}|p(r,s)|^2dsdr.
\end{aligned}
$$

To sum up the above arguments, with any $\rho\in[\tau,T]$, we have

$$
\begin{aligned}
&\mathbb{E}\int_\rho^T\int_\rho^r e^{\beta(r+t)}|P(r,t)|^2dtdr\\
&\leqslant K\mathbb{E}\int_\rho^T\int_\rho^r e^{\beta(r+t)}|F(r,t)|^2dtdr+\frac{K}{\beta}\mathbb{E}\int_\rho^T\int_\rho^r e^{\beta(r+t)}|p(r,t)|^2dtdr<\infty.
\end{aligned}
$$

Hence $P(\cdot,\cdot)\in L^2_{\mathbb{F}}(\Omega;L^2(\Delta_*[\tau,T];\mathbb{R}^{n\times n}))$, and one can define a map Ξ from $L^2_{\mathbb{F}}(\Omega;L^2(\Delta_*[\tau,T];\mathbb{R}^{n\times n}))$ to itself as $\Xi(p)=P$.

Suppose $\bar p$, $\widetilde p$ are two elements in $L^2_{\mathbb{F}}(\Omega;L^2(\Delta_*[\tau,T];\mathbb{R}^{n\times n}))$, and $(\bar q,\widetilde q)$ is defined similarly as the second equality in (10.3.57). By the previous arguments, we have $\bar P=\Xi(\bar p)$, $\widetilde P=\Xi(\widetilde p)$, and define $\bar Q$, $\widetilde Q$ accordingly. Also let

$$
\widehat p\overset{\Delta}{=}p-\widetilde p,\quad \widehat q\overset{\Delta}{=}q-\widetilde q,\quad \widehat P\overset{\Delta}{=}P-\widetilde P,\quad \widehat Q\overset{\Delta}{=}Q-\widetilde Q.
$$

Then we have

$$
\begin{cases}
\widehat P(r,t)=\displaystyle\int_r^T\big[b_x(s,t)^\top\mathbb{E}_t\widehat p(s,r)^\top+\sigma_x(s,t)^\top\widehat q(s,r,t)^\top\big]ds\\
\qquad\qquad +\displaystyle\int_t^r\big[b_x(s,t)^\top\mathbb{E}_t\widehat p(r,s)+\sigma_x(s,t)^\top\widehat q(r,s,t)\big]ds,\quad r\geqslant t,\\
\mathbb{E}_\theta\widehat p(r,t)=\widehat p(r,t)-\displaystyle\int_\theta^t\widehat q(r,t,s)dW(s),\quad \rho\leqslant\theta\leqslant t\leqslant r\leqslant T.
\end{cases}
\tag{10.3.59}
$$

By the first equality in (10.3.59), for any $\rho \in [\tau, T]$,

$$
\begin{aligned}
&\mathbb{E}\int_\rho^T\int_\rho^r e^{\beta(r+t)}|\widehat{P}(r,t)|^2 dtdr\\
&\leqslant K\mathbb{E}\int_\rho^T\int_\rho^r e^{\beta(r+t)}\Big\{\Big(\int_t^r|\widehat{p}(r,s)|ds\Big)^2+\Big(\int_r^T|\widehat{p}(s,r)|ds\Big)^2\\
&\qquad+\Big(\int_t^r|\widehat{q}(r,s,t)|ds\Big)^2+\Big(\int_r^T|\widehat{q}(s,r,t)|ds\Big)^2\Big\}dtdr.
\end{aligned}
\tag{10.3.60}
$$

Similar as the estimation of (10.3.58), we have

$$
\mathbb{E}\int_\rho^T\int_\rho^r e^{\beta(r+t)}|\widehat{P}(r,t)|^2 dtdr \leqslant \frac{K}{\beta}\mathbb{E}\int_\rho^T\int_\rho^r e^{\beta(r+t)}|\widehat{p}(r,t)|^2 dtdr.
$$

Therefore, by choosing $\beta > 0$ large enough, we obtain the existence and uniqueness of $P(\cdot,\cdot)$ on $[\rho, T]$, as well as the above conclusion (10.3.56). □

Now we return to system (10.3.52), the second-order adjoint system in our scenario.

Theorem 10.3.6 *Suppose* (H10.3.1)–(H10.3.2) *hold. Then* (10.3.52) *admits a unique adapted M-solution* (P_i, Q_i), $1 \leqslant i \leqslant 4$, *such that* $(P_3, P_4) \in L^2_{\mathbb{F}}(\tau, T; \mathbb{R}^{n\times n}) \times L^2_{\mathbb{F}}(\Omega; L^2(\Delta_*[\tau, T]; \mathbb{R}^{n\times n}))$.

Proof By Theorem 3.1.4 in Chap. 3 and Theorem 5.2.2 in Chap. 5, we have the existence and uniqueness of

$$
\begin{aligned}
&(P_1, Q_1) \in L^2_{\mathbb{F}}\big(\Omega; C([\tau, T]; \mathbb{S}^n)\big) \times L^2_{\mathbb{F}}\big(\tau, T; \mathbb{S}^n\big),\\
&(P_2, Q_2) \in L^2_{\mathbb{F}}\big(\tau, T; \mathbb{R}^{n\times n}\big) \times L^2\big(\tau, T; L^2_{\mathbb{F}}(\tau, T; \mathbb{R}^{n\times n})\big).
\end{aligned}
$$

We now prove the remaining well-posedness of (P_3, P_4). Recall (10.3.35); we see that $H_{xx}(\cdot) \in L^2_{\mathbb{F}}(\tau, T; \mathbb{R}^{n\times n})$. By applying conditional expectation operator $\mathbb{E}_t$ on both sides of the equations of (P_3, P_4) in (10.3.52), we have

$$
\left\{
\begin{aligned}
P_3(t) &= \mathbf{F}_3(t) + \mathbb{E}_t\int_t^T\int_s^T\Big(\sigma_x(\theta,t)^\top P_4(\theta,s)^\top\sigma_x(s,t)\\
&\qquad+\sigma_x(s,t)^\top P_4(\theta,s)\sigma_x(\theta,t)\Big)d\theta ds\\
&\qquad+\mathbb{E}_t\int_t^T\sigma_x(s,t)^\top P_3(s)\sigma_x(s,t)ds, \qquad \tau\leqslant t\leqslant T,\\
P_4(r,t) &= \mathbf{F}_4(r,t) + b_x(r,t)^\top\mathbb{E}_t P_3(r) + \sigma_x(r,t)^\top Q_3(r,t)\\
&\qquad+\mathbb{E}_t\int_r^T\Big(b_x(s,t)^\top P_4(s,r)^\top + \sigma_x(s,t)^\top Q_4(s,r,t)^\top\Big)ds\\
&\qquad+\mathbb{E}_t\int_t^r\Big(b_x(s,t)^\top P_4(r,s) + \sigma_x(s,t)^\top Q_4(r,s,t)\Big)ds,\\
&\qquad\qquad \tau\leqslant t\leqslant r\leqslant T,
\end{aligned}
\right.
\tag{10.3.61}
$$

where $\mathbf{F}_3(\cdot)$ and $\mathbf{F}_4(\cdot\,,\cdot)$ are given by the following, depending on (P_1, Q_1) and (P_2, Q_2),

$$\begin{aligned}\mathbf{F}_3(t) &\triangleq H_{xx}(t) + \sigma_x(T,t)^\top P_1(t)\sigma_x(T,t)\\&\quad+\mathbb{E}_t\int_t^T \big[\sigma_x(s,t)^\top P_2(s)\sigma_x(T,t) + \sigma_x(T,t)^\top P_2(s)^\top \sigma_x(s,t)\big]ds,\\ \mathbf{F}_4(r,t) &\triangleq b_x(T,t)^\top \mathbb{E}_t P_2(r)^\top + \sigma_x(T,t)^\top Q_2(r,t)^\top.\end{aligned}$$

Thus,

$$\mathbb{E}\Big(\int_\tau^T |\mathbf{F}_3(t)|^2 dt + \int_\tau^T\int_\tau^r |\mathbf{F}_4(r,t)|^2 dt dr\Big) < \infty.$$

For a given $p_3(\cdot) \in L^2_{\mathbb{F}}(\tau,T) \equiv L^2_{\mathbb{F}}(\tau,T;\mathbb{R}^{n\times n})$, let the associated $q_3(\cdot\,,\cdot)$ be determined by martingale representation theorem:

$$p_3(t) = \mathbb{E}_r p_3(t) + \int_r^t q_3(t,s)dW(s), \qquad \tau \leqslant r \leqslant t \leqslant T.$$

Then

$$\mathbb{E}|p_3(t)|^2 = \mathbb{E}|\mathbb{E}_r p_3(t)|^2 + \mathbb{E}\int_r^t |q_3(t,s)|^2 ds \geqslant \mathbb{E}\int_r^t |q_3(t,s)|^2 ds. \tag{10.3.62}$$

Now, let us consider

$$\left\{\begin{aligned}P_3(t) &= \mathbf{F}_3(t)+\mathbb{E}_t\int_t^T\int_s^T\Big(\sigma_x(\theta,t)^\top P_4(\theta,s)^\top \sigma_x(s,t)\\&\qquad+\sigma_x(s,t)^\top P_4(\theta,s)\sigma_x(\theta,t)\Big)d\theta ds\\&\qquad+\mathbb{E}_t\int_t^T \sigma_x(s,t)^\top p_3(s)\sigma_x(s,t)ds, \quad \tau\leqslant t\leqslant T,\\ P_4(r,t) &= \mathbf{F}_4(r,t) + b_x(r,t)^\top \mathbb{E}_t p_3(r) + \sigma_x(r,t)^\top q_3(r,t)\\&\qquad+\mathbb{E}_t\int_r^T\Big(b_x(s,t)^\top P_4(s,r)^\top + \sigma_x(s,t)^\top Q_4(s,r,t)^\top\Big)ds\\&\qquad+\mathbb{E}_t\int_t^r\Big(b_x(s,t)^\top P_4(r,s) + \sigma_x(s,t)^\top Q_4(r,s,t)\Big)ds,\\&\qquad\qquad\tau\leqslant t\leqslant r\leqslant T.\end{aligned}\right. \tag{10.3.63}$$

By Lemma 10.3.5, the second equation in (10.3.63) is solvable and for any $\rho \in [\tau, T]$ (noting (10.3.62))

$$
\begin{aligned}
&\mathbb{E}\int_\rho^T\int_\rho^r |P_4(r,t)|^2 dtdr \\
&\leqslant K\mathbb{E}\Big(\int_\rho^T\int_\rho^r |\mathbf{F}_4(r,t)|^2 dtdr + \int_\rho^T |p_3(r)|^2 dr + \int_\rho^T\int_\rho^r |q_3(r,t)|^2 dtdr\Big) \qquad (10.3.64)\\
&\leqslant K\mathbb{E}\Big(\int_\rho^T\int_\rho^r |\mathbf{F}_4(r,t)|^2 dtdr + \int_\rho^T |p_3(r)|^2 dr\Big) < \infty.
\end{aligned}
$$

Therefore, for any given $\beta > 0$,

$$
\begin{aligned}
&\mathbb{E}\int_\tau^T e^{\beta t}\Big|\int_t^T\int_s^T \sigma_x(s,t)^\top P_4(\theta,s)\sigma_x(\theta,t)d\theta ds\Big|^2 dt \\
&\leqslant K\mathbb{E}\int_\tau^T e^{\beta t}\Big(\int_t^T\int_s^T |P_4(\theta,s)|^2 d\theta ds\Big)dt \\
&= K\mathbb{E}\int_\tau^T e^{\beta t}\Big(\int_t^T\int_t^\theta |P_4(\theta,s)|^2 dsd\theta\Big)dt \\
&\leqslant K\mathbb{E}\int_\tau^T e^{\beta t}\Big(\int_t^T\int_t^r |\mathbf{F}_4(r,t)|^2 dtdr + \int_t^T |p_3(r)|^2 dr\Big)dt
\end{aligned}
$$

$$
\begin{aligned}
&\leqslant \frac{K(e^{\beta T}-e^{\beta\tau})}{\beta}\mathbb{E}\int_\tau^T\int_\tau^r |\mathbf{F}_4(r,t)|^2 dtdr + K\mathbb{E}\int_\tau^T\int_\tau^r e^{\beta t}|p_3(r)|^2 dtdr \\
&\leqslant \frac{Ke^{\beta T}}{\beta}\mathbb{E}\int_\tau^T\int_\tau^r |\mathbf{F}_4(r,t)|^2 dtdr + \frac{K}{\beta}\mathbb{E}\int_\tau^T e^{\beta r}|p_3(r)|^2 dtdr < \infty.
\end{aligned}
$$

Similarly,

$$
\begin{aligned}
&\mathbb{E}\int_\tau^T e^{\beta t}\Big|\int_t^T\int_s^T \sigma_x(\theta,t)^\top P_4(\theta,s)^\top\sigma_x(s,t)d\theta ds\Big|^2 dt \\
&\leqslant \frac{Ke^{\beta T}}{\beta}\mathbb{E}\int_\tau^T\int_\tau^r |\mathbf{F}_4(r,t)|^2 dtdr + \frac{K}{\beta}\mathbb{E}\int_\tau^T e^{\beta r}|p_3(r)|^2 dtdr < \infty.
\end{aligned}
$$

Therefore, for the equality of $P_3(\cdot)$ in (10.3.63), for any given $\beta > 0$, we have (noting (10.3.62))

$$
\begin{aligned}
&\mathbb{E}\int_\tau^T e^{\beta t}|P_3(t)|^2 dt \\
&\leqslant K\mathbb{E}\Big[e^{\beta T}\int_\tau^T |\mathbf{F}_3(t)|^2 dt + \int_\tau^T e^{\beta t}\Big(\int_t^T\int_t^\theta |P_4(\theta,s)|^2 dsd\theta + \int_t^T |p_3(s)|^2 ds\Big)dt\Big] \\
&\leqslant K\mathbb{E}\Big[e^{\beta T}\int_\tau^T |\mathbf{F}_3(t)|^2 dt + \frac{e^{\beta T}}{\beta}\int_\tau^T\int_\tau^r |\mathbf{F}_4(r,s)|^2 dsdr + \int_\tau^T\int_\tau^s e^{\beta t}|p_3(s)|^2 dtds\Big] \\
&\leqslant K\mathbb{E}\Big[e^{\beta T}\int_\tau^T |\mathbf{F}_3(t)|^2 dt + \frac{e^{\beta T}}{\beta}\int_\tau^T\int_\tau^r \mathbf{F}_4(r,s)|^2 dsdr + \frac{1}{\beta}\int_\tau^T e^{\beta s}|p_3(s)|^2 dtds\Big].
\end{aligned}
\qquad (10.3.65)
$$

Hence, we have $P_3(\cdot) \in L^2_{\mathbb{F}}(\tau, T; \mathbb{R}^{n\times n})$, and the map $\Xi(p_3) = P_3$ in $L^2_{\mathbb{F}}(\tau, T; \mathbb{R}^{n\times n})$ is well-defined.

Now, suppose $\bar{p}_3$, $\widetilde{p}_3$ are two elements in $L^2_{\mathbb{F}}(\tau, T; \mathbb{R}^{n\times n})$. We then have $\Xi(\bar{p}_3) = \bar{P}_3$, $\Xi(\widetilde{p}_3) = \widetilde{P}_3$, and define

$$\widehat{p}_3 \overset{\Delta}{=} \bar{p}_3 - \widetilde{p}_3, \quad \widehat{P}_3 \overset{\Delta}{=} \bar{P}_3 - \widetilde{P}_3, \quad \widehat{P}_4 \overset{\Delta}{=} \bar{P}_2 - \widetilde{P}_4.$$

Then for any $t \in [\tau, T]$, by (10.3.64)–(10.3.65) (with the corresponding $\mathbf{F}_3(\cdot) = 0$ and $\mathbf{F}_4(\cdot\,,\cdot) = 0$),

$$\mathbb{E}\int_\tau^T e^{\beta r}|\widehat{P}_3(r)|^2 dr \leqslant \frac{K}{\beta}\mathbb{E}\int_\tau^T e^{\beta t}|\widehat{p}_3(t)|^2 dt.$$

By choosing β large, we see that the map Ξ is a contraction. Hence, it admits a unique fixed point $P_3(\cdot)$. The rest of the conclusion follows easily. □

Remark 10.3.7 In the above two results, there are two crucial points, i.e., the introduction of equivalent β-norm of $L^2_{\mathbb{F}}\big(\Omega; L^2(\Delta_*[\tau, T]; \mathbb{R}^{n\times n})\big)$ and $L^2_{\mathbb{F}}(\tau, T; \mathbb{R}^{n\times n})$, respectively, and the subtle use of Lemma 10.3.5 in Theorem 10.3.6.

3.5 *Maximum Principle for FSVIEs*

At this moment, we are ready to state the maximum principle of Problem (C). For convenience, let us recall the following first-order adjoint equations:

$$\begin{cases} \mathbb{E}_t[h_x^\top] = h_x^\top - \displaystyle\int_t^T \zeta(s)dW(s), \qquad t \in [0, T], \\ Y(t) = g_x(t)^\top + b_x(T, t)^\top h_x^\top + \sigma_x(T, t)^\top \zeta(t) + \displaystyle\int_t^T \Big(b_x(s, t)^\top Y(s) \\ \qquad + \sigma_x(s, t)^\top Z(s, t)\Big)ds - \displaystyle\int_t^T Z(t, s)dW(s), \quad t \in [0, T], \end{cases} \tag{10.3.66}$$

and the Hamiltonian

$$\begin{aligned} &H\big(s, x, u, \eta, \zeta, Y(\cdot), Z(\cdot\,, s)\big) \\ &= g(s, x, u) + \langle \eta, b(T, s, x, u)\rangle + \langle \zeta, \sigma(T, s, x, u)\rangle \\ &\quad + \int_s^T \Big(\langle Y(t), b(t, s, x, u)\rangle + \langle Z(t, s), \sigma(t, s, x, u)\rangle\Big)dt. \end{aligned} \tag{10.3.67}$$

Then the Pontryagin maximum principle for Problem (C) can be stated as follows.

Theorem 10.3.8 *Let* (H10.3.1)–(H10.3.2) *hold and* $(\bar{X}(\cdot), \bar{u}(\cdot))$ *be an optimal pair of Problem* (C). *Then*

$$
\begin{aligned}
&\mathbb{E}_\tau\Big[H\big(\tau, \bar{X}(\tau), u, \mathbb{E}_\tau[h_x^\top], \zeta(\tau), \bar{Y}(\cdot), \bar{Z}(\cdot\,,\tau)\big) \\
&\qquad -H\big(\tau, \bar{X}(\tau), \bar{u}(\tau), \mathbb{E}_\tau[h_x^\top], \zeta(\tau), \bar{Y}(\cdot), \bar{Z}(\cdot\,,\tau)\big)\Big] \\
&+\frac{1}{2}\Big[\delta\sigma(T,\tau)^\top P_1(\tau)\delta\sigma(T,\tau) + \mathbb{E}_\tau\int_\tau^T\Big(\delta\sigma(\theta,\tau)^\top P_3(\theta)\delta\sigma(\theta,\tau) \\
&\quad +\delta\sigma(T,\tau)^\top P_2(\theta)^\top\delta\sigma(\theta,\tau) + \delta\sigma(\theta,\tau)^\top P_2(\theta)\delta\sigma(T,\tau) \\
&\quad +\int_\theta^T\big[\delta\sigma(\theta',\tau)^\top P_4(\theta',\theta)^\top\delta\sigma(\theta,\tau) \\
&\qquad +\delta\sigma(\theta,\tau)^\top P_4(\theta',\theta)\delta\sigma(\theta',\tau)\big]d\theta'\Big)d\theta\Big] \geqslant 0, \\
&\qquad\qquad\qquad\qquad \text{a.s, a.e. } \tau\in[0,T],\ \forall u\in U,
\end{aligned}
\tag{10.3.68}
$$

where H *is defined in* (10.3.67), $\zeta(\cdot)$ *satisfies the first relation in* (10.3.66), $(\bar{Y}(\cdot), \bar{Z}(\cdot\,,\cdot)))$ *is the adapted M-solution to the Type-II BSVIE in* (10.3.66), *and* $(P_i(\cdot), Q_i(\cdot))$ $(i = 1, 2, 3, 4)$ *is the unique adapted M-solution of the second-order adjoint equation* (10.3.52).

Proof Recall that

$$
\delta\sigma^\varepsilon(t,s) \overset{\Delta}{=} \big[\sigma(t,s,\bar{X}(s),u) - \sigma(t,s,\bar{X}(s),\bar{u}(s))\big]\mathbf{1}_{[\tau,\tau+\varepsilon]}(s). \tag{10.3.69}
$$

Thanks to (10.3.69) and Lebesgue differentiability theorem, for almost $\tau \in [0, T]$, by Theorem 10.3.4 (of this chapter)

$$
\begin{aligned}
\lim_{\varepsilon\to 0}\frac{\mathscr{E}(\varepsilon)}{\varepsilon} &= \mathbb{E}\Big\{\delta\sigma(T,\tau)^\top P_1(\tau)\delta\sigma(T,\tau) + \mathbb{E}_\tau\int_\tau^T\Big[\delta\sigma(T,\tau)^\top P_2(\theta)^\top\delta\sigma(\theta,\tau) \\
&+\delta\sigma(\theta,\tau)^\top P_2(\theta)\delta\sigma(T,\tau) + \delta\sigma(\theta,\tau)^\top P_3(\theta)\delta\sigma(\theta,\tau)d\theta \\
&+\int_\theta^T\big[\delta\sigma(\theta',\tau)^\top P_4(\theta',\theta)^\top\delta\sigma(\theta,\tau)+\delta\sigma(\theta,\tau)^\top P_4(\theta',\theta)\delta\sigma(\theta',\tau)\big]d\theta'\Big]d\theta\Big\}.
\end{aligned}
$$

According to Theorems 10.3.2, Proposition 10.3.3, Theorem 10.3.4, Theorem 10.3.6, the optimality of $\bar{u}(\cdot)$, and the arbitrariness of $u \in U$, we obtain the above maximum condition (10.3.68) immediately. □

4 The FSDE Case

Now, we look at the above maximum principle for FSDEs. In this case, the following holds:

$$
b(t,s,x,u) \overset{\Delta}{=} b(s,x,u), \quad \sigma(t,s,x,u) \overset{\Delta}{=} \sigma(s,x,u), \quad \varphi(t) \overset{\Delta}{=} x. \tag{10.4.1}
$$

Let us look at the second-order adjoint system (10.3.52) under (10.4.1). The first equation is unchanged. The second equation becomes

$$\begin{aligned}P_2(t)&= b_x(t)^\top\Big(P_1(t)+\int_t^T P_2(s)ds\Big)+\sigma_x(t)^\top\Big(Q_1(t)+\int_t^T Q_2(s,t)ds\Big)\\&\quad-\int_t^T Q_2(t,s)dW(s)\\&\equiv b_x(t)^\top\mathcal{P}_2(t)+\sigma_x(t)^\top\mathcal{Q}_2(t)-\int_t^T Q_2(t,s)dW(s).\end{aligned}$$

The third equation becomes

$$\begin{aligned}P_3(t)&=H_{xx}(t)+\sigma_x(t)^\top P_1(t)\sigma_x(t)\\&\quad+\int_t^T\Big[\sigma_x(t)^\top P_2(s)^\top\sigma_x(t)+\sigma_x(t)^\top P_2(s)\sigma_x(t)\sigma_x(t)^\top P_3(s)\sigma_x(t)\\&\quad+\int_t^T\Big(\sigma_x(t)^\top P_4(\theta,s)^\top\sigma_x(t)+\sigma_x(t)^\top P_4(\theta,s)\sigma_x(\tau)\Big)d\theta\Big]ds\\&\qquad-\int_t^T Q_3(t,s)dW(s)\\&=H_{xx}(t)+\sigma_x(t)^\top\Big\{P_1(t)+\int_t^T\Big[P_2(s)^\top+P_2(s)+P_3(s)\\&\quad+\int_s^T\Big(P_4(\theta,s)^\top+P_4(\theta,s))\Big)d\theta\Big]ds\Big\}\sigma_x(t)-\int_t^T Q_3(t,s)dW(s)\\&\equiv H_{xx}(t)+\sigma_x(t)^\top\mathcal{P}_3(t)\sigma_x(t)-\int_t^T Q_3(t,s)dW(s).\end{aligned}$$

The fourth equation becomes

$$\begin{aligned}P_4(r,t)&=b_x(t)^\top P_2(r)^\top+\sigma_x(t)^\top Q_2(r,t)^\top+b_x(t)^\top P_3(r)+\sigma_x(t)^\top Q_3(r,t)\\&\quad+\int_r^T\Big(b_x(t)^\top P_4(s,r)^\top+\sigma_x(t)^\top Q_4(s,r,t)^\top\Big)ds\\&\quad+\int_t^r\Big(b_x(t)^\top P_4(r,s)+\sigma_x(t)^\top Q_4(r,s,t)\Big)ds-\int_t^T Q_4(r,t,s)dW(s)\\&=b_x(t)^\top\Big(P_2(r)^\top+P_3(r)+\int_r^T P_4(s,r)^\top ds+\int_t^r P_4(r,s)ds\Big)\\&\quad+\sigma_x(t)^\top\Big(Q_2(r,t)^\top+Q_3(r,t)+\int_r^T Q_4(s,r,t)^\top ds+\int_t^r Q_4(r,s,t)ds\Big)\\&\quad-\int_t^T Q_4(r,t,s)dW(s)\\&\equiv b_x(t)^\top\mathcal{P}_4(r,t)+\sigma_x(t)^\top\mathcal{Q}_4(r,t)-\int_t^T Q_4(r,t,s)dW(s),\end{aligned}$$

where

$$
\begin{cases}
\mathcal{P}_2(t) \triangleq P_1(t) + \int_t^T P_2(s)ds, \qquad \mathcal{Q}_2(t) \triangleq Q_1(t) + \int_t^T Q_2(s,t)ds, \\
\mathcal{P}_3(t) \triangleq P_1(t) + \int_t^T \Big[P_2(s)^\top + P_2(s) + P_3(s) + \int_s^T \Big(P_4(\theta,s)^\top + P_4(\theta,s)) \Big) d\theta \Big] ds, \\
\mathcal{P}_4(r,t) \triangleq P_2(r)^\top + P_3(r) + \int_r^T P_4(s,r)^\top ds + \int_t^r P_4(r,s)ds, \\
\mathcal{Q}_4(r,t) \triangleq Q_2(r,t)^\top + Q_3(r,t) + \int_r^T Q_4(s,r,t)^\top ds + \int_t^r Q_4(r,s,t)ds.
\end{cases}
$$

The Hamiltonian takes the form

$$
\begin{aligned}
& H\big(s, x, u, \mathbb{E}_s[h_x], \zeta(s), Y(\cdot), Z(\cdot\,, s)\big) \\
& = g(s,x,u) + \Big\langle \mathbb{E}_s[h_x^\top] + \mathbb{E}_s \int_s^T Y(t)dt, b(s,x,u) \Big\rangle \\
& \qquad\qquad + \Big\langle \zeta(s) + \mathbb{E}_s \int_s^T Z(t,s)dt, \sigma(s,x,u) \Big\rangle \\
& \equiv g(s,x,u) + \big\langle \mathcal{Y}(s), b(s,x,u) \big\rangle + \big\langle \mathcal{Z}(s), \sigma(s,x,u) \big\rangle \equiv \mathcal{H}\big(s,x,u,\mathcal{Y}(s),\mathcal{Z}(s)\big),
\end{aligned}
$$

where

$$
\mathcal{Y}(s) \triangleq \mathbb{E}_s[h_x^\top] + \mathbb{E}_s \int_s^T Y(t)dt, \qquad \mathcal{Z}(s) \triangleq \zeta(s) + \mathbb{E}_s \int_s^T Z(t,s)dt,
$$

and

$$
\mathcal{H}(s,x,u,\mathcal{Y},\mathcal{Z}) = g(s,x,u) + \langle \mathcal{Y}, b(s,x,u) \rangle + \langle \mathcal{Z}, \sigma(s,x,u) \rangle.
$$

The maximum condition becomes

$$
\begin{aligned}
0 & \leqslant \mathcal{H}(\tau, \bar{X}(\tau), u, \mathcal{Y}(\tau), \mathcal{Z}(\tau)) - \mathcal{H}(\tau, \bar{X}(\tau), \bar{u}(\tau), \mathcal{Y}(\tau), \mathcal{Z}(\tau)) \\
& \quad + \frac{1}{2} \delta\sigma(\tau)^\top \Big[P_1(\tau) + \mathbb{E}_\tau \int_\tau^T \Big(P_2(\theta)^\top + P_2(\theta) + P_3(\theta) \\
& \qquad\qquad + \int_\theta^T \big[P_4(\theta',\theta)^\top + P_4(\theta',\theta) \big] d\theta' \Big) d\theta \Big] \delta\sigma(\tau) \qquad (10.4.2) \\
& = \mathcal{H}(\tau, \bar{X}(\tau), u, \mathcal{Y}(\tau), \mathcal{Z}(\tau)) - \mathcal{H}(\tau, \bar{X}(\tau), \bar{u}(\tau), \mathcal{Y}(\tau), \mathcal{Z}(\tau)) \\
& \quad + \frac{1}{2} \delta\sigma(\tau)^\top \mathbb{E}_\tau \mathcal{P}_3(\tau) \delta\sigma(\tau).
\end{aligned}
$$

Next we check that $(\mathcal{Y}, \mathcal{Z})$ satisfies the following first-order adjoint equation

$$
\begin{cases}
d\mathcal{Y}(t) = -\big[g_x(t)^\top + b_x(t)^\top \mathcal{Y}(t) + \sigma_x(t)^\top \mathcal{Z}(t) \big] dt + \mathcal{Z}(t) dW(t), \\
\qquad\qquad\qquad\qquad\qquad\qquad\qquad\qquad t \in [0,T], \\
\mathcal{Y}(T) = h_x(\bar{X}(T))^\top.
\end{cases} \qquad (10.4.3)
$$

Note that in the current case, we have

$$\begin{aligned}Y(t) &= g_x(t)^\top + b_x(t)^\top h_x^\top + \sigma_x(t)^\top \zeta(t)\\&\quad+\int_t^T \Big(b_x(t)^\top Y(s) + \sigma_x(t)^\top Z(s,t)\Big)ds - \int_t^T Z(t,s)dW(s)\\&= g_x(t)^\top + b_x(t)^\top\Big(h_x^\top + \int_t^T Y(s)ds\Big) + \sigma_x(t)^\top\Big(\zeta(t) + \int_t^T Z(s,t)ds\Big)\\&\quad-\int_t^T Z(t,s)dW(s).\end{aligned}$$

Thus, applying $\mathbb{E}_t$ on both sides, we get

$$\begin{aligned}Y(t) &= g_x(t)^\top + b_x(t)^\top\Big(\eta(t) + \mathbb{E}_t\int_t^T Y(s)ds\Big)\\&\qquad+\sigma_x(t)^\top\Big(\zeta(t) + \int_t^T Z(s,t)ds\Big)\\&= g_x(t)^\top + b_x(t)^\top \mathcal{Y}(t) + \sigma_x(t)^\top \mathcal{Z}(t).\end{aligned}\tag{10.4.4}$$

By Fubini theorem,

$$\begin{aligned}\int_t^T Y(s)ds &= \mathbb{E}_t\int_t^T Y(s)ds + \int_t^T\int_t^s Z(s,r)dW(r)ds\\&= \mathbb{E}_t\int_t^T Y(s)ds + \int_t^T\int_r^T Z(s,r)dsdW(r).\end{aligned}$$

Hence

$$\begin{aligned}&h_x^\top + \int_t^T Y(s)ds\\&\quad= \mathbb{E}_t[h_x^\top] + \mathbb{E}_t\int_t^T Y(s)ds + \int_t^T\Big[\int_r^T Z(s,r)ds + \zeta(r)\Big]dW(r)\\&\quad= \mathcal{Y}(t) + \int_t^T \mathcal{Z}(r)dW(r).\end{aligned}$$

Plugging (10.4.4) into the above equation, we end up with BSDE (10.4.3), the first-order adjoint equation.

Finally, it suffices to prove that $\mathscr{M}(\cdot) \triangleq \mathbb{E}.\mathcal{P}_3(\cdot)$ satisfies the second-order adjoint equation

$$\begin{aligned}\mathscr{M}(r) &= h_{xx}(\bar{X}(T)) + \int_r^T\Big[b_x(t)^\top\mathscr{M}(t) + \mathscr{M}(t)b_x(t) + \sigma_x(t)^\top\mathscr{M}(t)\sigma_x(t)\\&\qquad+\sigma_x(t)^\top\mathscr{N}(t) + \mathscr{N}(t)\sigma_x(t) + H_{xx}(t)\Big]dt - \int_r^T \mathscr{N}(t)dW(t),\\&\qquad\qquad r\in[0,T],\end{aligned}\tag{10.4.5}$$

where

$$\mathscr{N}(t) \triangleq \mathcal{Q}_2(t) + \int_t^T \mathcal{Q}_4(s,t)ds. \tag{10.4.6}$$

To this end, we observe the following: For the equation of P_2, one has

$$\begin{aligned}&\int_r^T \big[P_2(s) + P_2(s)^\top\big]ds \\ &= \int_r^T \Big\{b_x(s)^\top \mathbb{E}_s \mathcal{P}_2(s) + \sigma_x(s)^\top \mathcal{Q}_2(s) + \mathbb{E}_s \mathcal{P}_2(s)^\top b_x(s) + \mathcal{Q}_2(s)^\top \sigma_x(s)\Big\}ds.\end{aligned}$$

For the equation of P_3, one has

$$\int_r^T P_3(t)dt = \int_r^T H_{xx}(t)dt + \int_r^T \sigma_x(t)^\top \mathbb{E}_t \mathcal{P}_3(t)\sigma_x(t)dt.$$

For the equation of P_4, we have

$$\int_r^T \int_t^T P_4(s,t)dsdt = \int_r^T \int_t^T \Big[b_x(t)^\top \mathbb{E}_s \mathcal{P}_4(s,t) + \sigma_x(t)^\top \mathcal{Q}_4(s,t)\Big]dsdt.$$

Therefore,

$$\begin{aligned}\mathcal{P}_3(r) = h_{xx}(\bar{X}(T)) + \int_r^T \Big[&b_x(t)^\top \mathscr{M}(t) + \sigma_x(t)^\top \mathscr{N}(t) + \mathscr{M}(t) b_x(t) \\ &+ \mathscr{N}(t)\sigma_x(t) + H_{xx}(t) + \sigma_x(t)^\top \mathscr{M}(t)\sigma_x(t)\Big]dt,\end{aligned} \tag{10.4.7}$$

where $\mathscr{N}(\cdot)$ is defined in (10.4.6).

By the definitions of Q_2, Q_3, and Fubini theorem, we see that

$$\begin{aligned}\int_r^T P_i(t)dt = \mathbb{E}_r \int_r^T P_i(t)dt + \int_r^T \int_s^T Q_i(t,s)dtdW(s), \quad i = 2, 3. \\ Q_4(s,t,\theta)dW(\theta)dsdt = \textstyle\int_r^T \int_\theta^T \int_t^T Q_4(s,t,\theta)dsdtdW(\theta).\end{aligned} \tag{10.4.8}$$

Therefore,

$$\begin{aligned}\int_r^T\int_t^T \big[P_4(s,t) + P_4(s,t)^\top\big]dsdt &= \mathbb{E}_r \int_r^T\int_t^T \big[P_4(s,t) + P_4(s,t)^\top\big]dsdt \\ &+ \int_r^T\int_\theta^T\int_t^T \big[Q_4(s,t,\theta) + Q_4(s,t,\theta)^\top\big]dsdtdW(\theta).\end{aligned} \tag{10.4.9}$$

Combining (10.4.8) and (10.4.9), we have

$$\mathcal{P}_3(r) = \mathscr{M}(r) + \int_r^T \mathscr{N}(\theta)dW(\theta).$$

Consequently, by means of (10.4.7), for $r \in [0, T]$, we have BSDE (10.4.5), the second-order adjoint equation.

To sum up, the maximum condition (10.4.2), the first-order adjoint equation (10.4.3) and the second-order adjoint equation (10.4.5) form the Pontryagin's maximum principle for optimal control of FSDEs. More precisely, we have the following:

Theorem 10.4.1 *Let* (H10.3.1)–(H10.3.2) *hold associated with* (10.4.1), $(\bar{X}(\cdot), \bar{u}(\cdot))$ *be optimal pair of Problem* (C), *and* $(\mathcal{Y}(\cdot), \mathcal{Z}(\cdot))$ *be the adapted solution of BSDE:*

$$\begin{aligned}\mathcal{Y}(t) = h_x(\bar{X}(T))^\top + \int_t^T \big[g_x(t)^\top + b_x(t)^\top \mathcal{Y}(t) + \sigma_x(t)^\top \mathcal{Z}(t)\big]dt \\ - \int_t^T \mathcal{Z}(t)dW(t), \qquad t \in [0, T],\end{aligned}$$

and $(\mathscr{M}(\cdot), \mathscr{N}(\cdot))$ *be the adapted solution of the following BSDE:*

$$\begin{aligned}\mathscr{M}(r) = h_{xx}(\bar{X}(T)) + \int_r^T \big[b_x(t)^\top \mathscr{M}(t) + \mathscr{M}(t) b_x(t) + \sigma_x(t)^\top \mathscr{M}(t)\sigma_x(t) \\ + \sigma_x(t)^\top \mathscr{N}(t) + \mathscr{N}(t)\sigma_x(t) + H_{xx}(t)\big]dt - \int_r^T \mathscr{N}(t)dW(t), \\ r \in [0, T].\end{aligned}$$

Then, the following maximum condition holds

$$\begin{aligned}\mathcal{H}\big(\tau, \bar{X}(\tau), u, \mathcal{Y}(\tau), \mathcal{Z}(\tau)\big) - \mathcal{H}\big(\tau, \bar{X}(\tau), \bar{u}(\tau), \mathcal{Y}(\tau), \mathcal{Z}(\tau)\big) \\ + \frac{1}{2}\delta\sigma(\tau)^\top \mathscr{M}(\tau)\delta\sigma(\tau) \geqslant 0, \ \forall u \in U, \ \text{a.e. } \tau \in [0, T], \ \text{a.s.}\end{aligned}$$

5 Bibliographical Remarks

One of the central topics of stochastic optimal control problems is to establish the necessary conditions for optimal controls. Pontryagin's maximum principle for (deterministic) controlled ordinary differential equations (ODEs) was formulated and proved by Boltyanskii-Gamkrelidze-Pontryagin in 1956 [32] (see also the classical book [167]). Since then, the research in this direction attracted many authors. Shortly after, people attempted to extend the results to stochastic differential equations (SDEs, for short). In 1965, Kushner firstly proved a maximum principle for SDEs with deterministic coefficients, for which the diffusion is independent of the state and control [129]. Therefore, the diffusion of the variational system vanishes so that the adjoint equation is an ODE, whose solution is the adjoint state process. In 1972, Kushner further considered the case that the diffusion of the state equation contains the state [130], in which the adjoint process was not successfully characterized. In 1973, Bismut [30] introduced the duality between an initial value problem of linear SDE (called FSDE) and a terminal value problem of a linear SDE (now called BSDE, a name coined by Pardoux-Peng in 1990 [158]). With this, he was able to prove the maximum principle for FSDEs with the diffusion containing state and control, provided the control domain is convex; the adjoint process was successfully characterized by the adapted solution of corresponding linear BSDE. There were a number of follow-up works. We mention those by Haussmann [97], Bensoussan [27], and the references cited therein. The maximum principle for general controlled SDEs of the following form

$$X(t) = x+\int_0^t b(s, X(s), u(s))ds+\int_0^t \sigma(s, X(s), u(s))dW(s), \qquad t \in [0, T], \tag{10.5.1}$$

without assuming the convexity of the control domain was proved by Peng in 1990 [161], with the development of spike variation technique for FSDEs. We refer to the monograph of Yong-Zhou [223] for a self-contained presentation.

For the optimal control theory of FVIEs, to our best knowledge, the earliest work was due to Friedman [79] in 1964. Later in 1967, Vinokurov [188] investigated the optimal control for FVIEs with various type constraints and closed control domains (described by level sets of smooth functions). There are quite a few literature concerning optimal control of FVIEs in the past several decades. We mention some of them here: Kamien-Muller [119], Medhin [144], Belbas [20, 21], Belbas-Schmidt [22], and Bonnans-de la Vega-Dupuis [34]. Recently, Lin-Yong [136] established a maximum principle for singular FVIEs with some state constraint by spike variation technique.

In contrast with the existing literatures for optimal control of (deterministic) FVIEs, there were much less literature on optimal control of FSVIEs, to our best knowledge. The simple reason is that the theory of adjoint equation for FSVIEs, i.e., the BSVIEs, was not available until the appearance of the work on Type-I BSVIEs by Lin [135] in 2002. Even then, the result was not connected to the maximum principle of optimal control for FSVIEs. In 2006, Yong introduced Type-II BSVIEs and established maximum principle for FSVIEs the first time [218], provided the control domain is convex (see also Yong [220]). There are some further extensions of maximum principle for FSVIEs that have appeared lately. Here are some of those results. Shi-Wang-Yong [177] studied the case of forward-backward SVIEs. Agram-Oksendal [5] and Agram-Oksendal-Yakhlef [6] presented some investigations by means of Malliavin calculus, when the control domain U is open. Wang-Shi-Zhu

[203] discussed the necessary conditions of optimal controls under partial information framework. Hamaguchi [91] extended the maximum principle with convex control domain to the case of FSVIEs with delay. Wang-Zhang [207] considered the case with U being a closed set with the necessary optimality conditions were obtained from those directions that approximate convex perturbations at the optimal control are admissible. In such a case, the spike variations in the current chapter are not necessary. For the study in infinite dimensional case, we refer to [12, 33, 209].

Wang [199] studied the general case without assuming the convexity of the control domain U and attempted to develop the spike variation techniques for FSVIEs. In doing so, the following form *quadratic functional*

$$\mathscr{E}(\varepsilon) = \mathbb{E}\Big[X_1^\varepsilon(T)^\top h_{xx} X_1^\varepsilon(T) + \int_0^T X_1^\varepsilon(t)^\top H_{xx}(t) X_1^\varepsilon(t)dt\Big], \tag{10.5.2}$$

appeared in the variation of the cost functional, where h_{xx} and $H_{xx}(\cdot)$ are $\mathbb{S}^n$ (the set of all real $(n \times n)$ symmetric matrices) valued functions only depend on the optimal pair $(\bar{X}(\cdot), \bar{u}(\cdot))$, and $X_1^\varepsilon(\cdot)$ is the solution of the first-order variational equation, depending on the spike variation of the control. In the case that the control domain is convex, convex variation of the control is allowed and by doing that the above quadratic form does not appear. Thus, it is not necessary for [5, 177, 207, 218, 220] to introduce techniques of handling the above. On the other hand, recall that similar quadratic form as $\mathscr{E}(\varepsilon)$ appears in the FSDE case, which can be handled by a duality between the second-order adjoint SDE and the variational SDE satisfied by $X_1^\varepsilon(\cdot)X_1^\varepsilon(\cdot)^\top$ (see Peng [161]). It is unlikely that for FSVIEs, one can obtain a good-looking linear FSVIE for $X_1^\varepsilon(\cdot)X_1^\varepsilon(\cdot)^\top$. Hence, the idea of Peng [161] cannot be directly borrowed here. In fact, in Wang [199], the author had to introduce two abstract and implicit operator-valued stochastic processes, which are regarded as a replacement of the second-order adjoint processes, in the statement of the maximum principle. Moreover, there was no second-order adjoint equation derived in [199].

In the current chapter, we have developed a spike variation technique for optimal controls of FSVIEs. The main contribution is the derivation of the second-order adjoint equation which is a little different from a standard BSVIE. Thus, the well-posedness of such a BSVIE is a part of novelty. We have used a trick used in the derivation of Riccati equation in (stochastic) linear-quadratic optimal control, together with the introduction of the auxiliary process so that the Itô's formula can be used. This part of the results is an adoption of Wang-Yong [206]. It is worthy mentioning that the second-order adjoint equation developed in this chapter was also used to treat the maximum principle for stochastic delay differential system; see Meng-Shi-Wang-Zhang [147].

Eventually, we mentioned that besides the Pontryagin's type maximum principle, linear quadratic optimal control problems, another important part of optimal control theory, for stochastic Volterra integral equations attracted more attention in recent years. For example, some causal feedback optimal control/strategy was introduced and discussed in, e.g., Gong-Wang [87], Hamaguchi-Wang [94, 95], and Wang-Yong-Zhou [196]. In addition, by borrowing the study in [94, 95], Meng-Wang-Zhang [146] studied linear quadratic problem for a quite large class of stochastic delay differential equation and obtained the closed optimal strategy/control via introduced Riccati system. In these papers, the BSVIEs play various roles in different frameworks. Some relevant papers also include Abi Jaber-Miller-Pham [2], Chen-Yong [43], Wang [198], and Wang-Shi [202].

Chapter 11
Some Applications

We have presented some general theory of BSVIEs in the previous chapters. In this chapter, we are going to present some interesting applications in mathematical finance.

1 Pricing Problem

1.1 European Type Contingent Claims

Let us begin with a European call option. It is a contract whose holder has the right to come to the writer at a given time moment to buy a share of certain stock. However, the holder does not have the obligation to do so. Such a contract is called a *European call option*. For example, suppose we are considering a European call option for stock A, and its price at time t is denoted by $X_A(t)$. The *maturity time* is T (and current time is $t < T$), at which the call option has to be exercised. The *strike price* is a priori set to be \$10, at which the writer has to sell one share of stock A to the holder at T, wherever the holder comes to buy. Now when time comes to T, there will be two situations: If the stock price $X_A(T) > 10$, then the holder will come to exercise the option, buying one share of stock A at the strike price \$10 from the writer, and then sell it in the market to get a profit $X_A(T) - 10$. On the other hand, if $X_A(T) \leqslant 10$, the holder will discard the contract (no obligation to exercise). Therefore, by holding such a contract, one will have a payoff $(X_A(T) - 10)^+$ at a future time T. Of course, such a contract is not free. To determine the price of such an option is called an *option pricing* problem.

In general, if a payment is made $t < T$ so that a random payoff, say, ξ at T will be expected, then it is called a European *contingent claim* with a payoff at some future maturity time.

T. Wang, J. Yong, *Backward Stochastic Volterra Integral Equations*,
Probability Theory and Stochastic Modelling 109,
https://doi.org/10.1007/978-3-032-20848-4_11

To be definite, let us look at a market in which there is one riskless bond and n risky stocks are continuously traded, whose price processes follow the following:

$$\begin{cases} dX_0(t) = r(t)X_0(t)dt, \\ dX_i(t) = X_i(r)\Big[b_i(t)dt + \sum_{j=1}^{d} \sigma_{ij}(t)dW_j(t)\Big], \quad 1 \leqslant i \leqslant n. \end{cases} \tag{11.1.1}$$

Therefore, $X_0(\cdot)$ is the price process of the bond and $X_i(\cdot)$ is the price process of the i-th stock. We call $r(\cdot)$ the *interest rate*, $b_i(\cdot)$ the *appreciation rate* of the i-th stock, and $\sigma(\cdot) = \big(\sigma_{ij}(\cdot)\big)$ the *volatility*. By investing in such a market (with *self-financing*, i.e., no fund is to be put into or withdraw from the market strictly between 0 and T), the wealth process $X(\cdot)$ for the investor (or agent) follows the following linear SDE:

$$\begin{cases} dX(t) = \big[r(t)X(t) + \langle b(t) - r(t)\mathbb{1}, u(t)\rangle\big]dt \\ \qquad\qquad + \langle u(t), \sigma(t)dW(t)\rangle, \quad t \in [0, T], \\ X(0) = x_0, \end{cases} \tag{11.1.2}$$

where

$$b(\cdot) = (b_1(\cdot), \cdots, b_n(\cdot))^\top, \qquad \mathbb{1} = (1, 1, \cdots, 1)^\top \in \mathbb{R}^n,$$

and $u(\cdot)$ is called a *portfolio*, y_0 is called *initial endowment*. The wealth process can also be noted $X(\cdot\,; x_0, u(\cdot))$. The market is said to be complete if for any $\xi \in L^{1+}_{\mathcal{F}_T}(\Omega; \mathbb{R}) \equiv \bigcup_{p>1} L^p_{\mathcal{F}_T}(\Omega; \mathbb{R})$, there is a $u(\cdot) \in L^{1+}_{\mathcal{F}_T}(\Omega; \mathbb{R}^n) \equiv \bigcup_{p>1} L^p_{\mathcal{F}_T}(\Omega; \mathbb{R}^n)$ such that (11.1.2) admits a solution satisfying

$$X(T; x_0, u(\cdot)) = \xi, \qquad \text{a.s.}$$

In a financial market, the pricing of a given contingent claim ξ is based on the following: if we start with the price of the claim (to be determined) as initial endowment and invest it in the bond-stock assets, the value of the portfolio at time maturity time T must be enough to reach ξ. From this viewpoint, the price of the European contingent claim at $t \in [0, T]$, denoted by $Y(t)$, can be determined by the following:

$$\begin{cases} dY(t) = \big[r(t)Y(t) + \langle b(t) - r(t)\mathbb{1}, u(t)\rangle\big]dt \\ \qquad\qquad + \langle u(t), \sigma(t)dW(t)\rangle, \quad t \in [0, T], \\ Y(T) = \xi. \end{cases} \tag{11.1.3}$$

Equivalently, we may rewrite the above as

$$Y(t)=\xi -\int_t^T\Big(r(s)Y(s)+\langle b(s)-r(s)\mathbb{1}, u(s)\rangle\Big)ds-\int_t^T\langle\sigma(s)^\top u(s), dW(s)\rangle. \tag{11.1.4}$$

To transform (11.1.3) (or the equivalent form (11.1.4)) into a conventional BSDE, let us make some preparations.

To this end, we assume that the following equation

$$\sigma(t)\theta(t) = b(t) - r(t)\mathbb{1}, \qquad t \in [0, T] \tag{11.1.5}$$

admits a solution $\theta(t)$, called the *risk premium*, or *instantaneous Sharpe ratio* (SR, for short), such that the so-called *Novikov condition* holds:

$$\mathbb{E}\Big[\exp\Big(\frac{1}{2}\int_0^T |\theta(t)|^2 dt\Big)\Big] < \infty. \tag{11.1.6}$$

In this case, if we define (see (2.3.23) of Chap. 2)

$$\begin{aligned}&\mathbb{P}^\theta(A) = \mathbb{E}\Big[\widehat{\mathcal{E}}\{\theta\}(T)\mathbb{1}_A\Big], \qquad \forall A \in \mathcal{F},\\ &\widehat{\mathcal{E}}\{\theta\}(t) = \exp\Big(-\int_0^t \langle\theta(s), dW(s)\rangle - \frac{1}{2}\int_0^t |\theta(s)|^2 ds\Big), \quad 0 \leqslant t \leqslant T,\end{aligned}$$

then, $\mathbb{P}^\theta$ is a probability measure which is equivalent to $\mathbb{P}$ with proper θ, and under which

$$\mathbb{P}^\theta(t) = W(t) + \int_0^t \theta(s)ds, \qquad t \in [0, T] \tag{11.1.7}$$

is a standard Brownian motion (Girsanov theorem, Chap. 2, Theorem 2.3.9). We refer to $\mathbb{P}^\theta$ as an *equivalent martingale measure* (EMM, for short). One can show that the existence of such an EMM implies that the market is arbitrage-free. Further, the converse is roughly true, i.e., if a market is arbitrage-free, then an EMM exists (for which the Novikov condition (11.1.6) might not be guaranteed). We prefer not to get into the technical generalities. The following condition

$$\sigma(t)\sigma(t)^\top \geqslant \delta I, \qquad t \in [0, T], \tag{11.1.8}$$

for some $\delta > 0$ ensures equation (11.1.5) to have a solution

$$\theta(t) = \sigma(t)^\top[\sigma(t)\sigma(t)^\top]^{-1}[b(t) - r(t)\mathbb{1}], \qquad t \in [0, T]. \tag{11.1.9}$$

Then, Novikov condition (11.1.6) will be true if some integrability condition holds for $b(\cdot) - r(\cdot)\mathbb{1}$. In this case, by setting $Z(\cdot) = \sigma(\cdot)^\top u(\cdot)$, we can show that the price $Y(\cdot)$ of the European contingent claim is determined by the following standard BSDE:

$$\begin{cases} dY(t) = \big[r(t)Y(t) + \langle\theta(t), Z(t)\rangle\big]dt + \langle Z(t), dW(t)\rangle, & t \in [0,T], \\ Y(T) = \xi, \end{cases} \tag{11.1.10}$$

or equivalently,

$$Y(t) = \xi - \int_t^T \Big(r(s)Y(s) + \langle\theta(s), Z(s)\rangle\Big)ds - \int_t^T \langle Z(s), dW(s)\rangle. \tag{11.1.11}$$

The above gives a solution of European type contingent claim pricing problem via BSDE. We refer to such an approach the *BSDE method*.

Because of the linear generator in (11.1.11), next we represent the solution Y in more explicit form. In what follows, we will assume (11.1.8), without assuming the completeness of the market and some other bounded conditions on b and σ. Thus, we have BSDE (11.1.11) with $\theta(\cdot)$ being a solution of (11.1.5), satisfying the Novikov condition (11.1.6). By (11.1.7), we may write (11.1.10) as

$$\begin{cases} dY(t) = r(t)Y(t)dt + \langle Z(t), dW^\theta(t)\rangle, & t \in [0,T], \\ Y(T) = \xi, \end{cases} \tag{11.1.12}$$

which is equivalent to

$$Y(t) = \xi - \int_t^T r(s)Y(s)ds - \int_t^T \langle Z(s), dW^\theta(s)\rangle, \quad t \in [0,T]. \tag{11.1.13}$$

On the one hand, by taking conditional expectation, we end up with

$$Y(t) = \mathbb{E}^{\mathbb{P}^\theta}\xi - \mathbb{E}^{\mathbb{P}^\theta}\int_t^T r(s)Y(s)ds, \quad t \in [0,T]. \tag{11.1.14}$$

On the other hand, (11.1.13) is equivalent with

$$Y(t) = e^{\int_t^T r(s)ds}\xi - \int_t^T e^{\int_t^s r(\tau)d\tau}\langle Z(s), dW^\theta(s)\rangle, \quad t \in [0,T].$$

Consequently,

$$\begin{aligned} Y(t) &\equiv Y(t;\xi) = \mathbb{E}_t^{\mathbb{P}^\theta}\Big[e^{\int_t^T r(s)ds}\xi\Big] \\ &= \mathbb{E}_t\Big[\frac{\widehat{\mathcal{E}}\{\theta\}(T)}{\widehat{\mathcal{E}}\{\theta\}(t)}e^{\int_t^T r(s)ds}\xi\Big] \equiv \mathbb{E}_t\Big[\frac{\varphi(T)}{\varphi(t)}\xi\Big]. \end{aligned} \tag{11.1.15}$$

In the above,

$$\begin{aligned}\varphi(t) &\overset{\Delta}{=} \widehat{\mathcal{E}}\{\theta\}(t)e^{\int_0^t r(s)ds}\\ &= \exp\Big[-\int_0^t\langle\theta(s), dW(s)\rangle+\int_0^t\Big(r(s)-\frac{1}{2}|\theta(s)|^2\Big)ds\Big], \quad 0\leqslant t\leqslant T,\end{aligned}$$

which is called the *state price*. We point out that in both (11.1.14) and (11.1.15), the price $Y(\cdot)$ is given (at least formally) without involving $Z(\cdot)$ when (11.1.5) has a solution $\theta(\cdot)$ satisfying the Novikov condition (11.1.6).

1.2 Pricing of Non-adapted Cash Flows

Let $\psi(\cdot)$ be a process, called an *anticipated payoff process*, which is merely $\mathcal{F}_T$-measurable and not necessarily $\mathbb{F}$-adapted. Process $\psi(\cdot)$ could be a collection of contingent claims with different maturities, some expected dividend payments of coupon bonds, the costs of future events, certain mortgage payments before the borrower's bankrupt, etc. To begin with, let us present some simple examples of non $\mathbb{F}$-adapted cash flow process.

Example 11.1.1 Suppose an agent is holding $N_0(t)$ shares of bond whose price follows $X_0(\cdot)$ and $N_1(t)$ shares of stock whose price follows $X_1(\cdot)$ at time t. Then his/her wealth at time t should be

$$\psi(t)=N_0(t)X_0(t)+N_1(t)X_1(t), \qquad t\in[0,T].$$

In the real world, at time t, the agent will not change the portfolio unless the price of risky stock fluctuates in a level, say, $[0.9S(t), 1.1S(t)]$. Therefore, the real wealth should be adjusted into

$$\psi(t)=N_0(t)X_0(t)+N_1(t)X_1(\tau(t)), \quad t\in[0,T],$$

where

$$\tau(t)=\inf\{s\in[t,T]\;\big|\;|X_1(s)-X_1(t)|>0.1X_1(t)\}, \quad \inf\varnothing=T.$$

We see that $\psi(\cdot)$ is $\mathcal{F}_T$-measurable, but not $\mathbb{F}$-adapted. □

Example 11.1.2 Let us describe an American type contingent claim (ACC, for short) as follows (compare with European type contingent claim in the previous subsection): An individual pays an amount $\delta>0$ to buy the following right: He/she could receive a cumulative income, denoted by $C(\cdot)$ (with $C(0)=0$), and a lump-sum settlement $L(\tau)$ at the exercise time $\tau\in\mathscr{T}[0,T]$, giving up the future income on $(\tau, T]$. Assume all the involved processes are càglàg. Thus, the total profit of the buyer is (for simplicity, we assume the riskless bond interest rate is 0)

$$\psi(t) = \max_{\tau\in\mathscr{T}[0,T]} \Big[C(t\wedge\tau) + L(\tau)\mathbf{1}_{\{t\geqslant\tau\}} - \delta\Big].$$

Clearly, $\psi(\cdot)$ is not $\mathbb{F}$-adapted but is only $\mathcal{F}_T$-measurable. Let us look at one special case of the above.

- *American call option.* For this case, $C(\cdot)\equiv 0$, and the holder entitles to buy one share of a given stock at any time prior to T at a specified strike price $q > 0$, which corresponds to $L(t) = (X_1(t) - q)^+$.

We point out the above is also applicable to explain *the prepayable mortgage.* This can be regarded as follows: The mortgage company lent (or pays) to the individual amount $\delta > 0$. Then the individual chooses the time of prepayment τ. Thus, $C(t)$ is the cumulative mortgage payment made up to time t, $L(t)$ is the remaining principal after cumulative payment by the time t, and T is the term of the mortgage. Thus, the cash flow of the individual is

$$\psi(t) = \min_{\tau\in\mathscr{T}[0,T]} \Big[C(t\wedge\tau) + L(\tau)\mathbf{1}_{\{t\geqslant\tau\}} - \delta\Big].$$

Of course, $\psi(\cdot)$ is a $\mathcal{F}_T$-measurable process. □

Now, we consider the following situation. Suppose the market is arbitrage-free so that the set $\mathcal{Q}$ of EMMs is not empty. However, the completeness (of the market) is not assumed so that $\mathcal{Q}$ might not be a singleton. Usually, the EMM is chosen at the time zero and is applied for later evaluation time of pricing for any maturity of the claim. Clearly, such a choice may not be the best, since here we are considering some (non-$\mathbb{F}$-adapted) time dependent cash flow to be priced, and the market is usually incomplete as time goes by. To better capture and reflect different situations, one had better to allow EMM to be time-dependent. Then, some new approach needs to be developed.

Let us now fix an arbitrage-free incomplete market with a non-empty set $\mathcal{Q}$ of EMMs. We define an *EMM string* as a $\mathcal{Q}$-valued map $\mathbb{Q}_\tau$, $\tau\in[0,T]$. It allows us to make a selection within the set $\mathcal{Q}$ in a dynamic manner. Under $\mathbb{Q}_\tau$, we define the instantaneous Sharpe ratio (SR) by $\theta(\tau,\cdot) \overset{\Delta}{=} \theta^{\mathbb{Q}_\tau}(\cdot)$ on $[0,T]$ and obtain the following relation

$$\frac{d\mathbb{Q}_\tau}{d\mathbb{P}}\Big|_{\mathcal{F}_t} = \widehat{\mathcal{E}}(\theta(\tau,\cdot))_t = \exp\Big(-\frac{1}{2}\int_0^t |\theta(\tau,s)|^2 ds + \int_0^t \langle\theta(\tau,s), dW(s)\rangle\Big), \quad t\in[0,T]. \tag{11.1.16}$$

(See (2.3.23) of Chap. 2). For given $\tau\in[0,T]$, $s\mapsto\theta(\tau,s)$ is $\mathbb{F}$-adapted on $[0,T]$. Further, we assume that $\theta(\tau,\cdot)$ satisfies the Novikov condition (11.1.6).

According to the above discussion, it should be an interesting topic to study pricing problem for (non-adapted) cash flows $\psi(\cdot)$ under dynamical $\mathbb{Q}_\cdot$. According to (11.1.14) and (11.1.15), the corresponding price at time t should be

$$Y_1(t) = \mathbb{E}^{\mathbb{Q}_t}\psi(t) - \mathbb{E}^{\mathbb{Q}_t}\int_t^T r(s)Y_1(s)ds, \quad t \in [0, T], \tag{11.1.17}$$

and

$$Y_2(t) = \mathbb{E}_t^{\mathbb{Q}_t}\Big[e^{\int_t^T r(s)ds}\psi(t)\Big], \quad t \in [0, T]. \tag{11.1.18}$$

At this moment, there are two questions coming up naturally. First, do we still have $Y_1 = Y_2$ which happens in the previous section? Second, to represent $Y_i(\cdot)$, what kind of backward stochastic system shall we need in this scenario? Mimicking pricing of (time-independent) European type contingent claim in a complete arbitrage-free market via BSDEs, one possibility is (compare with (11.1.11))

$$\begin{aligned} Y(t,\tau) = \psi(t) - \int_\tau^T \Big(r(s)Y(t,s) + \langle\theta(t,s), Z(t,s)\rangle\Big)ds \\ - \int_\tau^T \langle Z(t,s), dW(s)\rangle, \qquad \tau \in [t, T], \end{aligned} \tag{11.1.19}$$

with $t \in [0, T]$ being the parameter. Intuitively, a reasonable price for $\psi(\cdot)$ at t should be $Y(t,t)$. However, by setting $\tau = t$ in (11.1.19), we have the following:

$$\begin{aligned} Y(t,t) = \psi(t) - \int_t^T \Big(r(s)Y(t,s) + \langle\theta(t,s), Z(t,s)\rangle\Big)ds \\ - \int_t^T \langle Z(t,s), dW(s)\rangle, \qquad t \in [0, T]. \end{aligned} \tag{11.1.20}$$

Mathematically speaking, the above is not a qualified equation for the diagonal form $Y(t,t)$ since $Y(t,s)$ appears in the integral. Therefore, it is desirable to come up with one system only involving the diagonal expression. On the one hand, to build a backward system for the diagonal term, it is natural that the current value $Y(t,t)$ depends on the future values $Y(s,s)$, $t < s < T$. On the other hand, by observing (11.1.19), for any $t \in [0, T]$, the utility $Y(t,\cdot)$ on $[t, T]$ depends on t. This means the agents need to update their pricing for any given time *naively*. Hence, it is expected to obtain some *equilibrium* pricing in the sense that the previous instant dependence on t can be dropped. From this sense, another possible one is the following Type-I BSVIE:

$$\begin{aligned} Y(t) = \psi(t) - \int_t^T \Big(r(s)Y(s) + \langle\theta(t,s), Z(t,s)\rangle\Big)ds \\ - \int_t^T \langle Z(t,s), dW(s)\rangle, \qquad t \in [0, T], \end{aligned} \tag{11.1.21}$$

which admits a unique adapted solution $(Y(\cdot), Z(\cdot\,,\cdot))$, under proper condition for $\theta(\cdot\,,\cdot)$. We have the following result.

Theorem 11.1.3 *Suppose* $\theta(\cdot\,,\cdot)$ *is bounded and* $\psi(\cdot) \in L^2_{\mathcal{F}_T}(0, T; \mathbb{R})$. *Then* $Y_1(\cdot)$, $Y_2(\cdot)$ *satisfies* (11.1.21) *and* (11.1.20), *respectively.*

Proof We first look at the case of $Y_2(\cdot)$. For a given $t \in [0, T]$, we consider the following BSDE with adapted solution pair $(\eta(t,\cdot), \zeta(t,\cdot))$:

$$\begin{aligned}\eta(t,\tau) = \psi(t) - \int_\tau^T \Big(r(s)\eta(t,s) + \langle\theta(t,s), \zeta(t,s)\rangle\Big)ds \\ - \int_\tau^T \langle\zeta(t,s), dW(s)\rangle, \quad \tau \in [t,T].\end{aligned} \tag{11.1.22}$$

By the definition of $\mathbb{Q}_\tau$, the above can be rewritten as

$$\eta(t,\tau) = \psi(t) - \int_\tau^T r(s)\eta(t,s)ds - \int_\tau^T \langle\zeta(t,s), dW^{\mathbb{Q}_\tau}(s)\rangle, \quad \tau \in [t,T],$$

where $W^{\mathbb{Q}_\tau}(\cdot)$ is a standard Brownian motion under $\mathbb{Q}_\tau$ (see (11.1.16)),

$$W^{\mathbb{Q}_\tau}(t) \triangleq W(t) - \int_0^t \theta(\tau,s)ds, \quad t \in [0,T]. \tag{11.1.23}$$

This leads to

$$\eta(t,\tau) = \mathbb{E}_\tau^{\mathbb{Q}_\tau}\Big[e^{-\int_\tau^T r(s')ds'}\psi(t)\Big], \qquad \tau \in [t,T].$$

Let $\tau = t$; by uniqueness, we see that $Y_2(t) = \eta(t,t)$.

Next let us consider the result of $Y_1(\cdot)$. By the definition of $\mathbb{Q}_\tau$, we can rewrite (11.1.21) as

$$Y(t) = \psi(t) - \int_\tau^T r(s)Y(s)ds - \int_\tau^T \langle\zeta(t,s), dW^{\mathbb{Q}_\tau}(s)\rangle, \quad t \in [0,T].$$

Taking expectation on both sides, we have

$$Y(t) = \mathbb{E}^{\mathbb{Q}_t}\psi(t) - \mathbb{E}^{\mathbb{Q}_t}\int_\tau^T r(s)Y(s)ds, \quad t \in [0,T].$$

By the uniqueness, we obtain the result of $Y_1(\cdot)$. □

By the above result, we see that in general one has $Y_1(\cdot) \neq Y_2(\cdot)$. This is different from the discussion in the previous section. To conclude this subsection, we present an example.

Example 11.1.4 We look at an American contingent claim with payoff $C(\cdot)$ before the exercise (stopping time) τ and the settlement $L(\tau)$ at τ and maturity T in an arbitrage-free and complete market. In this case, the set of EMMs $\mathcal{Q}$ is a singleton $\{\theta(\cdot)\}$. Further, we assume $\theta(\cdot)$ satisfies the Novikov condition (11.1.6). Denote the total payoff at τ by

$$\xi(\tau) = \int_0^\tau C(s)ds + L(\tau), \qquad \tau \in [0, T].$$

Here we have assumed that the riskless interest rate $r(\cdot) = 0$, for simplicity. In addition, $\frac{d\mathbb{Q}}{d\mathbb{P}}\Big|_{\mathcal{F}_t} = \mathcal{E}(\theta(\cdot))_t$ is defined in (11.1.16) associated with $\theta(\cdot)$. The aim of the buyer of this American contingent claim, purchased at t, is to find a stopping time $\bar{\tau} \in \mathscr{T}[0, T]$ such that the discounted return is maximized:

$$Y(t) \stackrel{\Delta}{=} \mathbb{E}_t^{\mathbb{Q}}\big[\xi(\bar{\tau})\big] = \operatorname*{esssup}_{\tau \in \mathscr{T}[t,T]} \mathbb{E}_t^{\mathbb{Q}}\big[\xi(\tau)\big], \quad t \in [0, T). \tag{11.1.24}$$

Here and next we point out $\bar{\tau}$ in general depends on t. One can show that

$$\bar{\tau} \stackrel{\Delta}{=} \inf\big\{t' \in [t, T] \bigm| Y(t') = \xi(t')\big\}, \qquad \inf \varnothing \stackrel{\Delta}{=} T.$$

Moreover, there exists a square integrable continuous increasing process $\kappa(\cdot)$ and a square integrable portfolio process $\zeta(\cdot)$ such that

$$Y(t) = \xi(T) + \kappa(T) - \kappa(t) - \int_t^T \zeta(s)dW^{\mathbb{Q}}(s), \quad t \in [0, T],$$

where $W^{\mathbb{Q}}(\cdot)$ is that in (11.1.7), and $\big[\zeta(\cdot)^\top \sigma(\cdot)\big] \in L^2_{\mathbb{F}}(0, T; \mathbb{R}^d)$. In addition, $\kappa(\cdot)$ satisfies the so-called Skorokhod condition

$$\int_0^T \big[Y(s) - \xi(s)\big]d\kappa(s) = 0. \tag{11.1.25}$$

We can put the above into the framework of BSVIEs. In fact, by the Skorokhod condition and the continuity of $K(\cdot)$, for given $t \in [0, T)$, it is easy to see $\kappa(\bar{\tau}) - \kappa(t) = 0$. Then above BSDE can be rewritten as

$$\begin{aligned} Y(t) &= \xi(\bar{\tau}) - \int_t^{\bar{\tau}} \zeta(s)dW^{\mathbb{Q}}(s) \\ &= \xi(\bar{\tau}) - \int_t^{\bar{\tau}} \theta(s)\zeta(s)ds - \int_{t'}^{\bar{\tau}} \zeta(s)dW(s) \\ &= \psi(t) - \int_t^T \theta(s)Z(t, s)ds - \int_t^T Z(t, s)dW(s), \quad t \in [0, T], \end{aligned}$$

where $\psi(t) = \xi(\bar\tau(t))$, and

$$Z(t,s) = \mathbf{1}_{[t,\bar\tau]}\zeta(s).$$

Clearly, the last equation is a Type-I BSVIE with (11.1.24).

The above example shows that American contingent (in a complete market) pricing problem (at least for some special cases) can be put into the framework of BSVIEs.

1.3 Nonlinear Pricing via BSVIEs

Now, we consider more general situation. First, let us introduce the following definition.

Definition 11.1.5 A (*nonlinear*) *price system* Ψ maps a claim process $\psi(\cdot) \in C_{\mathcal{F}_T}([0,T]; L^p(\Omega;\mathbb{R}))$ onto a price process $\Psi(\cdot\,;\psi(\cdot)) \in C_{\mathbb{F}}([0,T]; L^p(\Omega;\mathbb{R}))$ such that

(i) the price process $\Psi(\cdot\,;0)$ for a null contingent claim process is the null process,
(ii) the price system is nondecreasing, i.e., $\psi_1(\cdot) \geqslant \psi_0(\cdot)$ implies that $\Psi(\cdot\,;\psi_1(\cdot)) \geqslant \Psi(\cdot\,;\psi_0(\cdot))$,
(iii) arbitrage-free, i.e., if $\psi_1(\cdot) \geqslant \psi_0(\cdot)$, and for any $t \in [0,T]$, and $A \in \mathcal{F}_t$ satisfying $\mathbb{P}(A) > 0$, $\Psi(t;\psi_1(\cdot)) = \Psi(t;\psi_0(\cdot))$ on A, one has $\psi_1(t) = \psi_0(t)$ on A, a.s..

We now present a class of price system by the following one-dimensional Type-I BSVIE

$$\begin{aligned} Y(t) = \psi(t) &+ \int_t^T g(t,s,Y(s),Z(t,s))ds \\ &- \int_t^T Z(t,s)dW(s), \qquad t \in [0,T], \end{aligned} \tag{11.1.26}$$

with the generator $g : \Omega \times \Delta^*[0,T] \times \mathbb{R} \times \mathbb{R} \to \mathbb{R}$. For simplicity, the dimension of Brownian motion is one.

Theorem 11.1.6 *Suppose* (H6.2.1) *holds with* $m = 1$, *and* $y \mapsto g(t,s,y,z)$ *is increasing. For any* $\psi(\cdot) \in C_{\mathcal{F}_T}([0,T]; L^p(\Omega;\mathbb{R}))$, *let* $(Y(\cdot), Z(\cdot\,,\cdot))$ *be the corresponding adapted solution of Type-I BSVIE* (11.1.26). *Then* $\Psi(\cdot\,;\psi(\cdot)) \stackrel{\Delta}{=} Y(\cdot)$ *is a nonlinear price system.*

Proof For any $\psi(\cdot) \in C_{\mathcal{F}_T}([0,T]; L^p(\Omega;\mathbb{R}))$, the existence and uniqueness of $(Y(\cdot), Z(\cdot\,,\cdot))$ with $Y(\cdot) \in C_{\mathbb{F}}([0,T]; L^p(\Omega;\mathbb{R}))$ is implied by Proposition 6.2.1 in Chap. 6. Also by Corollary 6.2.4 of Chap. 6, the first two properties in Definition 11.1.5 hold. Then to show Ψ is a nonlinear price system, we only verify the third property.

Given $\psi_i(\cdot)$, $i = 0, 1$, we consider

$$Y_i(t) = \psi_i(t) + \int_t^T g(t, s, Y_i(s), Z_i(t, s))ds - \int_t^T Z_i(t, s)dW(s), \quad t \in [0, T].$$

Since $\psi_1(\cdot) \geqslant \psi_0(\cdot)$, then the comparison theorem implies that $Y_1(\cdot) \geqslant Y_0(\cdot)$.

For $t, s \in [0, T]$, we define

$$\widehat{Y}(t) \triangleq Y_1(t) - Y_0(t), \quad \widehat{Z}(t, s) \triangleq Z_1(t, s) - Z_0(t, s), \quad \widehat{\psi}(t) \triangleq \psi_1(t) - \psi_0(t).$$

It is easy to see

$$\begin{aligned}\widehat{Y}(t) = \widehat{\psi}(t) &+ \int_t^T \widehat{g}(t, s, Y_1(s), Y_0(s), Z_1(t, s))ds \\ &+ \int_t^T L_z(t, s)\widehat{Z}(t, s)ds - \int_t^T \widehat{Z}(t, s)dW(s), \quad t \in [0, T],\end{aligned}$$

where

$$\begin{aligned}&\widehat{g}(t, s, Y_1(s), Y_0(s), Z_1(t, s)) \\ &\triangleq g(t, s, Y_1(s), Z_1(t, s)) - g(t, s, Y_0(s), Z_1(t, s)), \\ &L_z(t, s) \triangleq \frac{g(t, s, Y_0(s), Z_1(t, s)) - g(t, s, Y_0(s), Z_2(t, s))}{\widehat{Z}(t, s)} \mathbf{1}_{\{\widehat{Z}(t,s)\neq 0\}}.\end{aligned}$$

By extending $L_z(t, s)$ to be zero for $s \in [0, t)$, and defining for given t,

$$\widehat{W}(t, s) \triangleq W(s) - \int_0^s L_z(t, r)dr, \quad s \in [0, T],$$

we see that $\widehat{W}(t, \cdot)$ is a standard BM under probability $\mathbb{Q}^t$ defined as $\frac{d\mathbb{Q}^t}{d\mathbb{P}} = \mathcal{E}\big[L_z(t, \cdot)\big](T)$. Therefore,

$$\widehat{Y}(t) = \widehat{\psi}(t) + \int_t^T \widehat{g}(t, s, Y_1(s), Y_0(s), Z_1(t, s))ds - \int_t^T \widehat{Z}(t, s)d\widehat{W}(t, s),$$

and for any $A \in \mathcal{F}_t$ such that $\mathbb{P}(A) > 0$ and $\Psi(t; \psi_1) = \Psi(t; \psi_2)$ on A,

$$\widehat{Y}(t)\mathbf{1}_A(\omega) = \mathbb{E}_t^{\mathbb{Q}^t}\big[\widehat{\psi}(t)\big]\mathbf{1}_A(\omega) + \mathbb{E}_t^{\mathbb{Q}^t}\int_t^T \widehat{g}(t, s, Y_1(s), Y_0(s), Z_1(t, s))ds\mathbf{1}_A(\omega).$$

Since $y \mapsto g(t, s, y, z)$ is increasing, we then have

$$g_1(t, s, z) \stackrel{\Delta}{=} g(t, s, Y_1(s), z) \geqslant g_2(t, s, z) \stackrel{\Delta}{=} g(t, s, Y_0(s), z),$$

which then implies that

$$\widehat{g}(t, s, Y_1(s), Y_0(s), Z_1(t, s)) \geqslant 0.$$

Considering the definition of A, we have $\widehat{Y}(t)\mathbf{1}_A(\omega) = 0$ which then implies that $\widehat{\psi}(t) = 0$ on A. □

Hereafter, we call $Z(\cdot\,,\cdot)$ the *portfolio strategy process* associated with the generator $g(\cdot)$ and claim process $\psi(\cdot)$. Also $g(\cdot)$ is called a *standard increasing driver* if $(y, z) \mapsto g(t, s, y, z)$ is Lipschitz, $g(\cdot, \cdot, 0, 0)$ is bounded, $y \mapsto g(t, s, y, z)$ is increasing and for any $(t, s), (t', s) \in \Delta^*[0, T]$, $(y, z) \in \mathbb{R}^2$,

$$\big|g(t, s, y, z) - g(t', s, y, z)\big| \leqslant \rho(|t - t'|)\big[1 + |y| + |z|\big],$$

for some modulus of continuity ρ. The following is concerned with a property between the supremum of price systems and the supremum of the associated drivers. We now denote by $\mathcal{A}$ the set of parameter α.

Theorem 11.1.7 *For $\alpha \in \mathcal{A}$, let $g^\alpha : \Omega \times \Delta^*[0, T] \times \mathbb{R} \times \mathbb{R} \to \mathbb{R}$ be standard increasing drivers (with the same modulus of continuity ρ) and $\psi^\alpha(\cdot) \in C_{\mathcal{F}_T}([0, T]; L^p(\Omega; \mathbb{R}))$ be claim process. Suppose that there exists a parameter $\bar{\alpha} \in \mathcal{A}$ such that*

$$\begin{aligned}
&\operatorname*{esssup}_{\alpha\in\mathcal{A}} g^\alpha(t, s, y, z) = g^{\bar{\alpha}}(t, s, y, z), \qquad (t, s, y, z) \in \Delta^*[0, T] \times \mathbb{R}^2, \text{ a.s.}\\
&\operatorname*{esssup}_{\alpha\in\mathcal{A}} \psi^\alpha(t) = \psi^{\bar{\alpha}}(t), \qquad t \in [0, T], \text{ a.s.}
\end{aligned}$$

Then, the following holds:

$$\bar{Y}(t) = \operatorname*{esssup}_{\alpha\in\mathcal{A}} Y^\alpha(t), \qquad t \in [0, T], \text{ a.s.}$$

where $(Y^\alpha(\cdot), Z^\alpha(\cdot\,,\cdot))$, $(\bar{Y}(\cdot), \bar{Z}(\cdot\,,\cdot))$ is the adapted solution of the BSVIE associated with (ψ^α, g^α), $(\psi^{\bar{\alpha}}, g^{\bar{\alpha}})$, respectively.

Proof It is a direct consequence of the uniqueness of BSVIEs and comparison theorems (Corollary 6.2.4 of Chap. 6). □

One motivation of studying parameter dependent standard increasing drivers and pricing systems is pricing/hedging problem with constrained portfolios taking in some closed convex subset of Euclidean space. It is possible that the essential supremum or infimum may not be reachable. In this case, we introduce the following definition.

Definition 11.1.8 Given claim prices $\psi(\cdot)$, and a family of standard increasing drivers $\{g^{\alpha}(\cdot) \mid \alpha \in \mathcal{A}\}$, the associated *upper price* and *lower price* at time t of the contingent claim process $\psi(\cdot)$ is defined by the following, respectively,

$$\overline{\Psi}(t;\psi(\cdot)) = \operatorname*{esssup}_{\alpha\in\mathcal{A}} \Psi^{\alpha}(t;\psi(\cdot)); \qquad \underline{\Psi}(t;\psi(\cdot)) = \operatorname*{essinf}_{\alpha\in\mathcal{A}} \Psi^{\alpha}(t;\psi(\cdot)),$$

where $\Psi^{\alpha}(\cdot;\psi(\cdot))$ is the price system with respect to claim process $\psi(\cdot)$. The interval $[\underline{\Psi}(t;\psi(\cdot)), \overline{\Psi}(t;\psi(\cdot))]$ is the arbitrage price interval for the claim process $\psi(\cdot)$ at time t associated with the price driver family $\{g^{\alpha}(\cdot) \mid \alpha \in \mathcal{A}\}$.

We have the following result on the upper price and lower price.

Theorem 11.1.9 *Suppose $\{g^{\alpha}(\cdot) \mid \alpha \in \mathcal{A}\}$ is a family of standard increasing drivers (with the same modulus of continuity ρ). Then for $(t,s,y,z) \in \Delta^*[0,T] \times \mathbb{R}^2 \to \mathbb{R}$,*

$$\overline{g}(t,s,y,z) \triangleq \operatorname*{esssup}_{\alpha\in\mathcal{A}} g^{\alpha}(t,s,y,z), \quad \underline{g}(t,s,y,z) \triangleq \operatorname*{essinf}_{\alpha\in\mathcal{A}} g^{\alpha}(t,s,y,z),$$

are also standard increasing drivers. Let $\overline{\Psi}(\cdot\,;\psi(\cdot))$, $\underline{\Psi}(\cdot\,;\psi(\cdot))$ be the price process of $\psi(\cdot)$ associated with $\overline{g}(\cdot)$, $\underline{g}(\cdot)$. Then, $\overline{\Psi}(\cdot\,;\psi(\cdot))$, $\underline{\Psi}(\cdot;\psi(\cdot))$ is, respectively, the upper price and lower price of the claim process $\psi(\cdot)$.

Proof In the following, we only consider the driver $\overline{g}(\cdot)$. Given a denumerable dense family $\{(y_k, z_k) \in \mathbb{R}^2 \mid k \geqslant 1\}$, we define, for each k, a measurable process $\overline{g}(t,s,y_k,z_k)$ and a $d\mathbb{P}\otimes dt$-null set N such that for $(\omega,t) \in N^c$,

$$\overline{g}(\omega,t,s,y_k,z_k) = \operatorname*{esssup}_{\alpha\in\mathcal{A}} g(\omega,t,s,y_k,z_k), \quad \forall k \geqslant 1.$$

For $(\omega,t) \in N^c$, $\overline{g}(\omega,t,s,y,z)$ is defined as the limit of $\overline{g}(\omega,t,s,y_k,z_k)$ as (y_k,z_k) goes to (y,z). The increasing condition, Lipschitz condition, and the modulus continuity of $\overline{g}(\cdot)$ are easy to see. □

Next, we consider so-called convex driver $g(t,s,y,z)$ which is a standard increasing driver, and it is convex with respect to y,z. Now, for given $(t,s,\eta,\zeta) \in [0,T]^2 \times \mathbb{R}^2$, we define $G(t,s,\eta,\zeta)$ as follows:

$$G(t,s,\eta,\zeta) = \inf_{(y,z)\in\mathbb{R}^2} \big[g(t,s,y,z) + \eta y + \zeta z\big].$$

The *domain* of $G(\cdot)$ is

$$\mathscr{D}_G \triangleq \big\{(t,s,\eta,\zeta) \in \Omega \times [0,T]^2 \times \mathbb{R}^2,\ G(t,s,\eta,\zeta) > -\infty\big\}.$$

Now, given the above $G(\cdot)$, we denote by $\mathcal{A}_0$ the set of admissible control parameters $(\beta,\gamma)\ :\ [0,T]^2\times\Omega\ \to\ [-L,L]^2$ satisfying joint measurability, $s\mapsto(\beta,\gamma)(t,s)$ being $\mathbb{F}$-adapted, and

$$\mathbb{E}\int_0^T\int_t^T G(t,s,\beta(t,s),\gamma(t,s))^2dsdt<\infty.$$

For any $(\beta,\gamma)\in\mathcal{A}_0$, we define the convex driver $g^{\beta,\gamma}(t,s,\cdot,\cdot)$ as follows

$$g^{\beta,\gamma}(t,s,y,z)\stackrel{\Delta}{=}G(t,s,\beta(t,s),\gamma(t,s))-\beta(t,s)y-\gamma(t,s)z,\qquad y,z\in\mathbb{R}.$$

Obviously, $(y,z)\ \mapsto\ g^{\beta,\gamma}(t,s,y,z)$ is linear. The following result exhibits the relation between price systems determined by g and $g^{\beta,\gamma}$.

Theorem 11.1.10 *Suppose $\psi(\cdot)$ is claim process, $g(\cdot)$ is a convex driver, and $\{g^{\beta,\gamma}(\cdot)\ \big|\ (\beta,\gamma)\in\mathcal{A}_0\}$ is a family of convex drivers. Then the price process of $\psi(\cdot)$ determined by $g(\cdot)$ is the upper-price process of $\psi(\cdot)$ determined by family $\{g^{\beta,\gamma}(\cdot)\ \big|\ (\beta,\gamma)\in\mathcal{A}_0\}$. Moreover, there exists $(\bar\beta,\bar\gamma)\in\mathcal{A}_0$ such that the nonlinear price process of $\psi(\cdot)$ determined by $g(\cdot)$ coincides with the price process determined by $g^{\bar\alpha,\bar\beta}(\cdot)$.*

Proof Since $(y,z)\ \mapsto\ g(t,s,y,z)$ is convex and continuous, the above $g(\cdot)$ and $G(\cdot)$ satisfy the conjugate relation

$$g(t,s,y,z)=\sup_{(\eta,\zeta)\in\mathscr{D}_G^{(t,s,\omega)}}\big[G(t,s,\eta,\zeta)-\eta y-\zeta z\big],$$

where

$$\mathscr{D}_G^{(t,s,\omega)}=\{(\eta,\zeta)\in[-L,L]^2\ \big|\ G(t,s,\eta,\zeta)>-\infty\}.$$

Recall the definition of $g^{\beta,\gamma}$; it is easy to see

$$g(t,s,y,z)\geqslant g^{\beta,\gamma}(t,s,y,z),\qquad\forall(\beta,\gamma)\in\mathcal{A}_0.$$

Next we prove that there exists $(\bar\beta,\bar\gamma)\in\mathcal{A}_0$ such that

$$g(t,s,Y(s),Z(t,s))=g^{\bar\beta,\bar\gamma}(t,s,Y(s),Z(t,s)),\quad d\mathbb{P}\otimes ds,\quad\text{a.s.}\qquad(11.1.27)$$

where $(Y(\cdot),Z(\cdot\,,\cdot))$ is the adapted solution of BSVIE

$$Y(t)=\psi(t)+\int_t^T g(t,s,Y(s),Z(t,s))ds-\int_t^T Z(t,s)dW(s),\quad t\in[0,T].$$

Therefore, our conclusions hold naturally. In fact, for fixed (t, s, ω), by the definition of supremum, there exists a bounded sequence $(\eta^k, \zeta^k)_{k\in\mathbb{N}} \in \mathscr{D}_G^{(t,s,\omega)}$ such that

$$g(t, s, Y(s), Z(t, s)) = \lim_{k\to\infty} \big[G(t, s, \eta^k, \zeta^k) - \eta^k Y(s) - \zeta^k Z(t, s)\big].$$

Since the sequence $\{\eta^k, \zeta^k\}_{k\geqslant 1}$ is bounded, there exists a subsequence still denoted by $\{\eta^k, \zeta^k\}_{k\geqslant 1}$ which converges in $[-L, L]$ to $(\bar\beta, \bar\gamma)$ depending on (t, s, ω). Therefore,

$$\begin{aligned} &g(t, s, Y(s), Z(t, s)) = \lim_{k\to\infty} \big[G(t, s, \eta^k, \zeta^k) - \eta^k Y(s) - \zeta^k Z(t, s)\big] \\ &= G(t, s, \bar\beta(t, s), \bar\gamma(t, s)) - \bar\beta(t, s)Y(s) - \bar\gamma(t, s)Z(t, s) = g^{\bar\beta,\bar\gamma}(t, s, Y(s), Z(t, s)). \end{aligned}$$

Therefore, (11.1.27) holds. Since $g(\cdot\,,\cdot\,, Y(\cdot), Z(\cdot\,,\cdot))$, $Y(\cdot)$ and $Z(\cdot\,,\cdot)$ are measurable processes, by a measurable selection theorem, we see that $(\bar\beta, \bar\gamma)$ are measurable (bounded) processes in $[-L, L]$. In addition, $s \mapsto (\bar\beta(t, s), \bar\gamma(t, s))$ is $\mathbb{F}$-adapted. Eventually, by the integrability of $g(\cdot\,,\cdot\,, Y(\cdot), Z(\cdot\,,\cdot))$, $Y(\cdot)$, $Z(\cdot\,,\cdot)$, and the boundedness of $(\bar\beta, \bar\gamma)$, we see that

$$\mathbb{E}\int_0^T\int_t^T G(t, s, \bar\beta(t, s), \bar\gamma(t, s))^2 ds dt < \infty.$$

Then Theorem 11.1.9 applies. □

2 Equilibrium Recursive Utility Processes

In this section, we would like to apply the BSVIE theory developed in the previous chapters to some recursive utility models.

2.1 *Recursive Utility by BSDEs*

We begin with the following situation. Suppose $x \in [0, \infty)$ is the quantity of certain commodity (say, bottles of water). Then the satisfaction level it brings from $x = 0$ to $x = 1$ is huge since it is lifesaving. However, the satisfaction level it brings from $x = 100$ to $x = 101$ is very tiny. Thus, satisfaction level can be described by a nondecreasing concave function $h : (a, \infty) \to \mathbb{R}$ for some $a \in [-\infty, \infty)$. We call $h(x)$ the *utility* of x. Next, let ξ be a random variable representing a payoff in a future time. Then the *expected utility* reads $\mathbb{E}\big[h(\xi)\big]$, which is nothing but the expected satisfaction level.

Now, we extend the above notions. First of all, we define

$$Y(t) = \mathbb{E}_t\big[h(\xi)\big], \qquad t \in [0, T],$$

then $Y(0) = \mathbb{E}[h(\xi)]$. We call the above defined $Y(\cdot)$ an *expected utility process*. Further, if a discount rate $\lambda > 0$ exists, then the *discounted expected utility process* is given by the following:

$$Y(t) = \mathbb{E}_t\Big[e^{-\lambda(T-t)}h(\xi)\Big], \qquad t \in [0, T].$$

On the other hand, one can define the so-called expected recursive utility process, through a *recursive* way, i.e., the current expected utility depends on the future expected utility, as well as certain consumption process $c(\cdot)$. Mathematically, this can be described via the following way:

$$Y(t) = \mathbb{E}_t\Big[h(\xi) + \int_t^T g(s, c(s), Y(s))ds\Big], \qquad t \in [0, T],$$

where $Y(t)$ is the expected utility at t. Combining the above two cases, we could define the *discounted expected recursive utility process* as the solution $Y(\cdot)$ to the following equation:

$$Y(t) = \mathbb{E}_t\Big[e^{-\lambda(T-t)}h(\xi) + \int_t^T e^{-\lambda(s-t)}g(s, c(s), Y(s))ds\Big], \quad t \in [0, T]. \tag{11.2.1}$$

One can show that the above $Y(\cdot)$, together with some $Z(\cdot)$, will be the adapted solution to the following BSDE:

$$Y(t) = h(\xi) + \int_t^T \big(\lambda Y(s) + g(s, c(s), Y(s))\big)ds - \int_t^T Z(s)dW(s), \quad t \in [0, T]. \tag{11.2.2}$$

Of course, to keep the existence of $Z(\cdot)$, one has to require proper conditions on the terminal value $h(\xi)$ and consumption process $c(\cdot)$. Such an observation inspires us to introduce the *generalized stochastic recursive utility process* to be $Y(\cdot)$, which together with some $Z(\cdot)$ satisfies the following BSDE:

$$Y(t) = h(\xi) + \int_t^T g(s, c(s), Y(s), Z(s))ds - \int_t^T Z(s)dW(s), \quad t \in [0, T]. \tag{11.2.3}$$

In other words, when $(Y(\cdot), Z(\cdot))$ is the adapted solution to the above BSDE, we may define $Y(\cdot)$ as a generalized stochastic recursive utility process of $h(\xi)$ and $c(\cdot)$.

2.2 Equilibrium Recursive Utility by BSVIEs

In contrast with the recursive utility given by BSDEs, let us show the analogue theory via BSVIEs.

2.2.1 Three Motivational Cases

To begin with we show the advantages and improvements of BSVIEs from the standpoints of three cases.

Case I: We look at the following example which shows that the generator/driver of BSDEs can also be improved in some sense.

Example 11.2.1 We consider the following quadratic BSDE

$$Y(t) = \int_t^T \Big[u(c(s)) - \frac{1}{2\gamma}|Z(s)|^2\Big]ds - \int_t^T Z(s)dW(s), \qquad t \in [0, T].$$

By supposing utility function $u(\cdot)$, consumption $c(\cdot)$ to satisfy proper boundedness/integrability, one can obtain the well-posedness of the above system. Then $Y(\cdot)$ can be seen as a recursive utility process where $\frac{1}{2\gamma}|z|^2$ can be considered as a risk penalization of this recursive utility. Another interpretation of γ can also be given from the perspective of information. A positive γ implies the monotonicity of utility in the underlying filtration, i.e., more and earlier information increases the agent's utility.

We take a closer look at the parameter γ which indicates the agent time preference in the time duration $[0, T]$. In the practical world, instead of the constant assumption, it is more natural that people keep changing their risk tolerance parameter as time goes by. Therefore, it is better for us to use the map $t \mapsto \gamma(t)$. From this sense, we have the following BSVIE for utility process $Y(\cdot)$,

$$Y(t) = \int_t^T \big[u(c(s)) - \frac{1}{2\gamma(t)}|Z(t, s)|^2\big]ds - \int_t^T Z(t, s)dW(s), \quad t \in [0, T].$$

The above example shows a class of BSDEs/BSVIEs arising from mathematical finance when the generator is quadratic growth. More importantly, by allowing some financial parameters to be time-dependent in proper way, one then ends up with certain BSVIEs instead of BSDEs.

Case II: Next we discuss the second case to use BSVIEs rather than BSDEs.

Given terminal payoff ξ, and consumption process $c(\cdot)$, one way to measure their utility is (11.2.1) which can be rewritten as

$$Y(t) = e^{-\lambda(T-t)}h(\xi) + \int_t^T e^{-\lambda(s-t)}g(s, c(s), Y(s))ds - \int_t^T e^{\lambda t}Z(s)dW(s).$$

Here $h(\cdot)$, $g(\cdot)$ are chosen properly to ensure the existence and uniqueness of (Y, Z), λ is called the discount rate, and $e^{-\lambda(s-t)}$ in the running cost is exponential functions with parameter λ. We observe that the constant assumption on the discounting rate is too ideal. In fact, people usually overdiscount on the utility for the outcome of immediate future events. In some sense, people are impatient about choices in the short term but are patient among the long-term alternatives. Mathematically, such a situation could be described by the so-called nonexponential discounting, meaning that the discounting terms $e^{-\lambda(T-t)}$ and $e^{-\lambda(s-t)}$ by some more general functions $\mu(T, t)$ and $\nu(s, t)$, respectively. For given $\lambda_1, \lambda_2 > 0$, suppose

$$\mu(t, T) \stackrel{\triangle}{=} \frac{1}{1 + \lambda_1(T - t)}, \quad \nu(t, r) = \frac{1}{1 + \lambda_2(r - t)}.$$

Then we end up with a utility process with hyperbolic discounting function. Suppose

$$\mu(t, T) \stackrel{\triangle}{=} e^{-\lambda_1(T-t)}, \quad \nu(t, r) = e^{-\lambda_2(r-t)}, \quad \lambda_1 \neq \lambda_2,$$

then we end up with a utility process with heterogeneous discounting function. Suppose

$$\begin{aligned} &\mu(t, T) \stackrel{\triangle}{=} \alpha e^{-\lambda_1(T-t)} + (1-\alpha)e^{-\lambda_2(T-t)}, \\ &\nu(t, r) = \alpha e^{-\lambda_1(r-t)} + (1-\alpha)e^{-\lambda_2(r-t)}, \quad \lambda_1 \neq \lambda_2, \ \alpha \in (0, 1), \end{aligned}$$

then we end up with a utility process with convex combination of two exponential discounting function. Suppose $\lambda > 0, \alpha > 0$,

$$\mu(t, T) \stackrel{\triangle}{=} (1 + \alpha(T - t))e^{-\lambda(T-t)}, \quad \nu(t, r) = (1 + \alpha(r - t))e^{-\lambda(r-t)},$$

then we end up with a utility process with quasi-exponential discounting function.

From the standpoint of nonexponential discounting function, the above BSDE can be replaced by BSVIE of

$$Y(t) = \mu(T, t)h(\xi) + \int_t^T \nu(s, t)g(s, c(s), Y(s))ds - \int_t^T Z(t, s)dW(s), \tag{11.2.4}$$

which is more closer to the real world.

Case III: Now we give the third way of transforming BSDEs into general BSVIEs. Suppose there exists the underlying economy which is assumed to contain one representative agent with fixed belief probability $\mathbb{P}$. Under $\mathbb{P}$, according to the previous discussion, the stochastic differential utility $Y(\cdot) \triangleq U(\cdot; c(\cdot))$ can be expressed as

$$Y(t) = \mathbb{E}_t^{\mathbb{P}} \int_t^T g(c(s), Y(s))ds, \quad t \in [0, T] \tag{11.2.5}$$

(for proper function g) which can be rewritten as a BSDE of

$$Y(t) = \int_t^T g(c(s), Y(s))ds - \int_t^T Z(s)dW(s), \quad t \in [0, T].$$

In real world, for the agent, his/her subjective beliefs may be changeable over time. In this sense, at time t, one special case of the believe system is $\mathbb{P}_t$ which is mutually absolute continuous w.r.t. $\mathbb{P}$ with $\frac{d\mathbb{P}_t}{d\mathbb{P}} = \mathcal{E}(\theta(t, \cdot))_T$ and some bounded map $\theta(t, \cdot)$. To give more details, let us look at the following example.

Consider an agent with a pessimistic belief $\mathbb{P}_0 \triangleq \bar{\mathbb{P}}$ at time 0 with $\frac{d\bar{\mathbb{P}}}{d\mathbb{P}} = \mathcal{E}(\bar{\theta})_T$ and a rather optimistic belief $\mathbb{P}_T \triangleq \widetilde{\mathbb{P}}$ at time T with $\frac{d\widetilde{\mathbb{P}}}{d\mathbb{P}} = \mathcal{E}(\widetilde{\theta})_T$. The degree of pessimism $\alpha(\cdot) : [0, T] \to [0, 1]$ starts at 0 and is of bounded variation and continuous. This yields the following flow of beliefs

$$\mathbb{P}_t = (1 - \alpha(t))\bar{\mathbb{P}} + \alpha(t)\widetilde{\mathbb{P}}, \quad t \in [0, T].$$

Notice that

$$\frac{d\mathbb{P}_t}{d\mathbb{P}} = (1 - \alpha(t))\frac{d\bar{\mathbb{P}}}{d\mathbb{P}} + \alpha(t)\frac{d\widetilde{\mathbb{P}}}{d\mathbb{P}}, \quad t \in [0, T],$$

and for $t, r \in [0, T]$,

$$\begin{aligned} &X_1(t) \triangleq \frac{d\bar{\mathbb{P}}}{d\mathbb{P}}\Big|_{\mathcal{F}_t}, \quad X_2(t) \triangleq \frac{d\widetilde{\mathbb{P}}}{d\mathbb{P}}\Big|_{\mathcal{F}_t}, \\ &X(t, r) \triangleq (1 - \alpha(t))X_1(r) + \alpha(t)X_2(r), \end{aligned}$$

we thus have

$$\begin{aligned} &X_1(r) = 1 + \int_0^r \bar{\theta}(s)X_1(s)dW(s), \quad r \in [0, T], \\ &X_2(r) = 1 + \int_0^r \widetilde{\theta}(s)X_2(s)dW(s), \quad r \in [0, T], \\ &X(t, r) = 1 + \int_0^r \Big[(1 - \alpha(t))\bar{\theta}(s)X_1(s) + \alpha(t)\widetilde{\theta}(s)X_2(s)\Big]dW(s) \\ &= 1 + \int_0^r \theta(t, s)X(t, s)dW(s), \end{aligned}$$

where

$$\theta(t,s) \triangleq \frac{(1-\alpha(t))\bar{\theta}(s)X_1(s)+\alpha(t)\widetilde{\theta}(s)X_2(s)}{(1-\alpha(t))X_1(s)+\alpha(t)X_2(s)}. \tag{11.2.6}$$

Since $X_i(\cdot) > 0$, $0 \leqslant \alpha(t) \leqslant 1$, $\bar{\theta}$, $\widetilde{\theta}$ are bounded, then it is easy to see that $\theta(\cdot,\cdot)$ is bounded and $\frac{d\mathbb{P}_t}{d\mathbb{P}} = \mathcal{E}(\theta(t,\cdot))_T$.

Under the probability $\mathbb{P}_t$, we can improve the stochastic differential utility (11.2.5) as follows:

$$Y(t) \triangleq \mathbb{E}_t^{\mathbb{P}_t}\int_t^T u(c(s),Y(s))ds, \tag{11.2.7}$$

which satisfies a linear BSVIE of the form

$$Y(t) = \int_t^T \big[u(c(s),Y(s)) + \theta(t,s)Z(t,s)\big]ds - \int_t^T Z(t,s)dW(s). \tag{11.2.8}$$

Of course, it is particular case of BSVIE.

2.2.2 Equilibrium Recursive Utility by BSVIEs

For simplicity, in the current part, we omit the terminal ξ and consider the generalized recursive utility of merely having the running consumption process $c(\cdot)$. Inspired by the above three cases, we should consider the following:

$$Y(t) = \int_t^T g(t,s,c(s),Y(s),Z(t,s))ds - \int_t^T Z(t,s)dW(s), \qquad t \in [0,T], \tag{11.2.9}$$

where $g : \Delta^*[0,T] \times \mathbb{R} \times \mathbb{R} \times \mathbb{R} \to \mathbb{R}$ is called the intertemporal aggregator. In the following, we make the following assumptions on $c(\cdot)$ and g.

(H11.2.1) Suppose $(y,z) \mapsto g(t,s,c,y,z)$ is Lipschitz, $g(\cdot,\cdot,0,0)$ is bounded, $y \mapsto g(t,s,c,y,z)$ is increasing and for any $(t,s),(t',s) \in \Delta^*[0,T]$, $(c,y,z) \in \mathbb{R}^3$,

$$\big|g(t,s,c,y,z) - g(t',s,c,y,z)\big| \leqslant \rho(|t-t'|)\big[1+|y|+|z|\big],$$

for some modulus of continuity ρ. In addition, $c(\cdot)$ is a bounded adapted process.

Notice that the boundedness of consumption $c(\cdot)$ can be relaxed in a proper manner. We prefer not get into further details along this line. The following result reveals the connection between consumption and utility process.

Proposition 11.2.2 *Suppose* (H11.2.1) *holds.*

(i) *If* $c \mapsto g(t, s, c, y, z)$ *is nondecreasing, then* $Y(\cdot)$ *in* (11.2.9) *is nondecreasing in* $c(\cdot)$ *in the sense that* $c_0(\cdot) \leqslant c_1(\cdot)$ *implies that* $Y_0(t) \leqslant Y_1(t)$ *for any* $t \in [0, T]$.

(ii) *If* $(c, y, z) \mapsto g(t, s, c, y, z)$ *is concave, then* $Y(\cdot)$ *in* (11.2.9) *is concave with respect to consumption* $c(\cdot)$ *in the sense that for given* $c_0(\cdot)$, $c_1(\cdot)$, $c_\lambda(\cdot) \equiv (1-\lambda)c_0(\cdot)+\lambda c_1(\cdot)$, *one has* $Y_\lambda(t) \geqslant (1-\lambda)Y_0(t)+\lambda Y_1(t)$ *for any* $t \in [0, T]$.

Proof (i) Consider a pair of consumption $c_0(\cdot)$, $c_1(\cdot)$ with $c_0(\cdot) \leqslant c_1(\cdot)$. Then by assumption on $g(\cdot)$ we have, for any $(y, z) \in \mathbb{R}^2$,

$$g(t, s, c_0(s), y, z) \leqslant g(t, s, c_1(s), y, z).$$

By the comparison theorem of BSVIE (e.g., Corollary 6.2.4 in Chap. 6), we have $Y_0(\cdot) \leqslant Y_1(\cdot)$, which implies the conclusion naturally.

(ii) For $i = 0, 1$, let $(Y_i(\cdot), Z_i(\cdot\,,\cdot))$ be the adapted solution of BSVIE (11.2.9) associated with $c_i(\cdot)$, i.e.,

$$Y_i(t) = \int_t^T g(t, s, c_i(s), Y_i(s), Z_i(t, s))ds - \int_t^T Z_i(t, s)dW(s), \qquad t \in [0, T]. \tag{11.2.10}$$

Now, for $\lambda \in (0, 1)$, let $(Y_\lambda(\cdot), Z_\lambda(\cdot\,,\cdot)))$ be the adapted solution of the following BSVIE:

$$\begin{aligned} Y_\lambda(t) = &\int_t^T g(t, s, (1-\lambda)c_0(s) + \lambda c_1(s), Y_\lambda(s), Z_\lambda(t, s))ds \\ &- \int_t^T Z_\lambda(t, s)dW(s), \qquad t \in [0, T]. \end{aligned} \tag{11.2.11}$$

We need to show that

$$Y_\lambda(t) \geqslant (1-\lambda)Y_0(t) + \lambda Y_1(t), \qquad t \in [0, T]. \tag{11.2.12}$$

In fact, from (11.2.10), we see that

$$\begin{aligned} &(1-\lambda)Y_0(t) + \lambda Y_1(t) \\ &= \int_t^T \Big[g\big(t, s, (1-\lambda)c_0(s) + \lambda c_1(s), (1-\lambda)Y_0(s) + \lambda Y_1(s), \\ &\qquad\qquad (1-\lambda)Z_0(t, s) + \lambda Z_1(t, s)\big) - \delta g(t, s)\Big]ds \\ &\qquad - \int_t^T \Big[(1-\lambda)Z_0(t, s) + \lambda Z_1(t, s)\Big]dW(s), \end{aligned}$$

where

$$\delta g(t,s) = g\big(t,s,(1-\lambda)c_0(s)+\lambda c_1(s),(1-\lambda)Y_0(s)+\lambda Y_1(s),\\ (1-\lambda)Z_0(t,s)+\lambda Z_1(t,s)\big)\\ -(1-\lambda)g\big(t,s,c_0(s),Y_0(s),Z_0(t,s)\big)-\lambda g\big(t,s,c_1(s),Y_1(s),Z_1(t,s)\big)\geqslant 0,$$

by the concavity condition of $g(\cdot)$. Then, (11.2.12) follows from Corollary (6.2.4) in Chap. 6. □

Now, for any bounded consumption process $c(\cdot)$, let us denote $(Y^{c(\cdot)}(\cdot), Z^{c(\cdot)}(\cdot\,,\cdot))$ the adapted solution of the BSVIE (11.2.9). Then there is an interesting comparison between $Y^{c(\cdot)}(\cdot)$ and $Y^{\bar c(\cdot)}(\cdot)$, where $\bar c(\cdot) = \mathbb{E}[c(\cdot)]$, i.e., the group people under consideration do not prefer the random consumption to the average of that. In another word, *risk-averse* people more likely prefer the latter (i.e., $\bar c(\cdot)$) to the former (i.e., $c(\cdot)$). This leads to the following notion.

Definition 11.2.3 A utility process $Y^{c(\cdot)}(\cdot)$ determined through (11.2.9) is said to be *risk-averse* if

$$Y^{\bar c(\cdot)}(t) \geqslant Y^{c(\cdot)}(t), \qquad t\in[0,T].$$

We have the following result, which presents the existence of risk-averse utility process.

Proposition 11.2.4 *Suppose* (H11.2.1) *holds such that* $c \mapsto g(t,s,c,y,z)$ *is concave. Moreover,*

$$g(t,s,c,y,z)\leqslant g(t,s,c,y,0),\quad 0\leqslant t\leqslant s\leqslant T,\ (c,y,z)\in\mathbb{R}^3. \tag{11.2.13}$$

Then the utility process $Y^{c(\cdot)}(\cdot)$ *via* (11.2.9) *is risk-averse.*

Proof First, since $g(\cdot)$ is Markovian (e.g., ω-independent), for deterministic function $\bar c(\cdot)=\mathbb{E}[c(\cdot)]$, $(Y^{\bar c(\cdot)}(\cdot),0)$ is deterministic and satisfies the following equation:

$$Y^{\bar c(\cdot)}(t) = \int_t^T g(t,s,\bar c(s),Y^{\bar c(\cdot)}(s),0)ds,\quad t\in[0,T]. \tag{11.2.14}$$

Next, let $(Y_0^{c(\cdot)}(\cdot), Z_0^{c(\cdot)}(\cdot\,,\cdot))$ be the adapted solution of BSVIE

$$Y_0^{c(\cdot)}(t) = \int_t^T g(t,s,c(s),Y_0^{c(\cdot)}(s),0)ds - \int_t^T Z_0^{c(\cdot)}(t,s)dW(s),\quad t\in[0,T].$$

Then by (11.2.13), (11.2.9), together with comparison theorem for BSVIEs (Corollary 6.2.4 in Chap. 6), one has

$$Y^{c(\cdot)}(t)\leqslant Y_0^{c(\cdot)}(t),\quad \text{a.s.},\ \forall t\in[0,T]. \tag{11.2.15}$$

Since $c \mapsto g(t,s,c,y,0)$ is concave, we have

$$\begin{aligned} Y^{\bar{c}(\cdot)}(t) &= \int_t^T g(t,s,\bar{c}(s),Y^{\bar{c}(\cdot)}(s),0)ds = \mathbb{E}_t \int_t^T g(t,s,\bar{c}(s),Y^{\bar{c}(\cdot)}(s),0)ds \\ &\geqslant \mathbb{E}_t \int_t^T g(t,s,c(s),Y^{\bar{c}(\cdot)}(s),0)ds \equiv \mathcal{Y}_0^{c(\cdot)}(t). \end{aligned} \tag{11.2.16}$$

Since $y \mapsto g(t,s,c,y,z)$ is increasing, we then have

$$\begin{aligned} \mathcal{Y}_0^{c(\cdot)}(t) &= \mathbb{E}_t \int_t^T g(t,s,c(s),Y^{\bar{c}(\cdot)}(s),0)ds \\ &\geqslant \mathbb{E}_t \int_t^T g(t,s,c(s),\mathcal{Y}_0^{c(\cdot)}(s),0)ds \equiv \mathcal{Y}_1^{c(\cdot)}(t). \end{aligned}$$

Inductively, for any $n \geqslant 1$, we have

$$\begin{aligned} \mathcal{Y}_n^{c(\cdot)}(t) &= \mathbb{E}_t \int_t^T g(t,s,c(s),\mathcal{Y}_{n-1}^{c(\cdot)}(s),0)ds \\ &\geqslant \mathbb{E}_t \int_t^T g(t,s,c(s),\mathcal{Y}_n^{c(\cdot)}(s),0)ds \equiv \mathcal{Y}_{n+1}^{c(\cdot)}(t). \end{aligned}$$

On the other hand, we observe that

$$Y_0^{c(\cdot)}(t) = \mathbb{E}_t \int_t^T g(t,s,c(s),Y_0^{c(\cdot)}(s),0)ds, \qquad t \in [0,T],$$

from which we see that

$$\mathbb{E}\big|\mathcal{Y}_1^{c(\cdot)}(t) - Y_0^{c(\cdot)}(t)\big| \leqslant L \int_t^T \mathbb{E}|\mathcal{Y}_0^{c(\cdot)}(s) - Y_0^{c(\cdot)}(s)|ds,$$

and

$$\begin{aligned} &\mathbb{E}\big|\mathcal{Y}_n^{c(\cdot)}(t) - Y_0^{c(\cdot)}(t)\big| \leqslant L \int_t^T \mathbb{E}|\mathcal{Y}_{n-1}^{c(\cdot)}(s) - Y_0^{c(\cdot)}(s)|ds \\ &\leqslant L^2 \int_t^T\!\!\int_s^T \mathbb{E}|\mathcal{Y}_{n-2}^{c(\cdot)}(\tau) - Y_0^{c(\cdot)}(\tau)|d\tau ds \\ &= L^2 \int_t^T\!\!\int_t^\tau \mathbb{E}|\mathcal{Y}_{n-2}^{c(\cdot)}(\tau) - Y_0^{c(\cdot)}(\tau)|ds d\tau \\ &= L^2 \int_t^T (s-t)\mathbb{E}||\mathcal{Y}_{n-2}^{c(\cdot)}(s) - Y_0^{c(\cdot)}(s)|ds = \cdots \\ &\leqslant \frac{L^n}{(n-1)!} \int_t^T (s-t)^{n-1}\mathbb{E}|\mathcal{Y}_0^{c(\cdot)}(s) - Y_0^{c(\cdot)}(s)|ds. \end{aligned}$$

Therefore, we see that

$$\begin{aligned}&\mathbb{E}\big|\mathcal{Y}_n^{c(\cdot)}(t) - Y_0^{c(\cdot)}(t)\big|\\&\leqslant \frac{L^n}{(n-1)!}(T-t)^{n-1}\int_t^T \mathbb{E}|\mathcal{Y}_0^{c(\cdot)}(s) - Y_0^{c(\cdot)}(s)|ds \to 0,\end{aligned}$$

as $n \to \infty$. Therefore, by (11.2.16) and the above discussions, we have

$$Y^{\bar{c}(\cdot)}(t) \geqslant \mathcal{Y}_n^{c(\cdot)}(t) \to Y_0^{c(\cdot)}(t) \geqslant Y^{c(\cdot)}(t), \quad \forall t \in [0,T].$$

The last inequality follows from (11.2.15). Thus, $Y^{c(\cdot)}(\cdot)$ is risk-averse. □

Next, we compare the *risk aversion* between different utility processes. To this end, we introduce the following definition.

Definition 11.2.5 Given two utility processes $Y_0^{c(\cdot)}(\cdot)$, $Y_1^{c(\cdot)}(\cdot)$ determined by BSVIEs

$$\begin{aligned}Y_i^{c(\cdot)}(t) = &\int_t^T g_i(t,s,c(s),Y_i^{c(\cdot)}(s),Z_i^{c(\cdot)}(t,s))ds\\&-\int_t^T Z_i^{c(\cdot)}(t,s)dW(s), \quad t \in [0,T],\end{aligned} \tag{11.2.17}$$

with $i = 0, 1$. We say utility process $Y_0^{c(\cdot)}$ is more risk-averse than $Y_1^{c(\cdot)}$ if for each consumption process $c(\cdot)$ and $\bar{c}(\cdot) = \mathbb{E}_t[c(\cdot)]$,

$$Y_0^{c(\cdot)}(t) \leqslant Y_0^{\bar{c}(\cdot)}(t) \quad \Rightarrow \quad Y_1^{c(\cdot)}(t) \leqslant Y_1^{\bar{c}(\cdot)}(t), \quad t \in [0,T].$$

The result next provides a comparison between two different utility processes via BSVIEs.

Proposition 11.2.6 *Suppose* $g_0, g_1 : \Delta^*[0,T] \times \mathbb{R}^3 \to \mathbb{R}$ *satisfy* (H11.2.1). *Further,*

$$g_0(t,s,c,y,0) = g_1(t,s,c,y,0), \quad g_0(t,s,c,y,z) \geqslant g_1(t,s,c,y,z),$$

with $0 \leqslant t \leqslant s \leqslant T$, $c, y, z \in \mathbb{R}$. *Then we have*

$$Y_0^{c(\cdot)}(t) \leqslant Y_0^{\bar{c}(\cdot)}(t) \quad \Rightarrow \quad Y_1^{c(\cdot)}(t) \leqslant Y_1^{\bar{c}(\cdot)}(t), \quad t \in [0,T].$$

Proof Since

$$g_0(t,s,c,y,z) \geqslant g_1(t,s,c,y,z), \quad \forall (t,s,y,z) \in \Delta^*[0,T] \times \mathbb{R} \times \mathbb{R} \times \mathbb{R},$$

it follows from the comparison theorem for BSVIEs (Corollary 6.2.4 in Chap. 6) that

$$Y_0^{c(\cdot)}(t) \geqslant Y_1^{c(\cdot)}(t), \qquad t \in [0, T].$$

On the other hand, if consumption process takes $\bar{c}(s) = \mathbb{E}[c(s)]$ with $s \in [t, T]$, then

$$Y_i^{\bar{c}(\cdot)}(t) = \int_t^T g_i(t, s, \bar{c}(s), Y_i^{\bar{c}(\cdot)}(s), 0), \quad t \in [0, T].$$

By assumption above and the uniqueness of BSVIEs, for any $t \in [0, T]$, we have $Y_0^{\bar{c}(\cdot)}(t) = Y_1^{\bar{c}(\cdot)}(t)$. Then our conclusion holds naturally. □

3 Equilibrium Dynamic Risk Measures

3.1 Static Risk Measures

In various situations of real life, people frequently face risks. It is desirable to have some ways to measure the risks so that one can control/manage them. Let us consider such a case. Suppose an investment decision is made at time $t = 0$ and the corresponding *return* at some future time T is denoted by ξ which is a random variable, and thus it could be positive or negative. For simplicity, we assume that there is a one-to-one correspondence between the return ξ at T and the decision made at $t = 0$. Hence, one may identify ξ with the decision. Let Ξ be the set of all possible returns at time T. We assume that Ξ is a convex cone, i.e.,

$$\xi, \eta \in \Xi \quad \Rightarrow \quad \lambda\xi + \mu\eta \in \Xi, \qquad \forall \lambda, \mu \in [0, \infty).$$

This means that if ξ, η are two returns, then so is the nonnegative linear combination $\lambda\xi + \mu\eta$, for $\lambda, \mu \geqslant 0$. Now, let us introduce the following definition.

Definition 11.3.1

(i) A map $\rho : \Xi \to \mathbb{R}$ is called a *static risk measure* of Ξ with reference return r if the following conditions are satisfied:

- (Monotonicity): For any $\xi, \bar{\xi} \in \Xi$ with $\xi \leqslant \bar{\xi}$ almost surely, it holds

$$\rho(\xi) \geqslant \rho(\eta);$$

- (Relevance): For any $\xi \in \Xi \setminus \{0\}$ with $\xi \leqslant 0$ almost surely, it holds

$$\rho(\xi) > 0;$$

- (Translation invariance): For any $\xi \in \Xi$ and $\alpha \geqslant 0$, it holds

$$\rho(\xi + \alpha r) = \rho(\xi) - \alpha,$$

for some $r > 0$.

(ii) A static risk measure $\rho : \Xi \to \mathbb{R}$ is said to be *convex* if the map $\xi \mapsto \rho(\xi)$ is convex, i.e.,

$$\rho\big(\lambda\xi + (1-\lambda)\eta\big) \leqslant \lambda\rho(\xi) + (1-\lambda)\rho(\eta), \qquad \forall \xi, \eta \in \Xi,\ \lambda \in [0, 1].$$

(iii) A static risk measure $\rho : \Xi \to \mathbb{R}$ is said to be *coherent* if the following hold:

- (Positive homogeneity):

$$\rho(\lambda\xi) = \lambda\rho(\xi), \qquad \forall \xi \in \Xi,\ \lambda \geqslant 0;$$

- (Subadditivity):

$$\rho(\xi + \eta) \leqslant \rho(\xi) + \rho(\eta), \qquad \forall \xi, \eta \in \Xi.$$

Let us make some brief comments on the above definition. First of all, as a measure of risk, it is natural to require that the larger the payoff ξ is, the smaller the risk is, which gives the monotonicity. Second, as long as the return $\xi \leqslant 0$ almost surely and has a positive probability that $\xi < 0$, the risk $\rho(\xi)$ should be positive, which gives the relevance. Third, by adding αr to the return ξ, it makes the original risk reduced by α, which is the translation invariance. Observe that if we choose $\alpha = \rho(\xi)$ in the translation invariance property, one has

$$\rho(\xi + \rho(\xi)r) = \rho(\xi) - \rho(\xi) = 0.$$

It means that $\rho(\xi)$ equals the *risk premium*, by paying which, the risk can be taken care of. Fourth, subadditivity and convexity means that diversifying the investment should reduce the total risk, and positive homogeneity means that a multiple of the same type of investment will proportionally change the total risk, either increased if $\lambda > 1$, meaning to enlarge the investment, or decreased if $\lambda < 1$, meaning to shrink the investment.

Now, we present some popular static risk measures.

- Value-at-risk. Suppose X is a random variable, representing the value of certain asset (or a portfolio) at some future time. Then the value-at-risk of X at level $\alpha \in (0, 1)$, denoted by $\mathrm{VaR}_\alpha(X)$, is defined by the following:

$$\mathrm{VaR}_\alpha(X) = \inf\{x \in \mathbb{R} \bigm| F_X(x) > \alpha\},$$

where $F_X(\cdot)$ is the distribution function of X. For example, with $\alpha = 95\%$ and X being the market value of an asset (or a portfolio), then $\mathrm{VaR}_\alpha(X)$ is the least value that X can drop with the 95% confidence. It is known that value-at-risk is not coherent because the subadditivity fails.

- Average value-at-risk. It is also called *conditional value-at-risk*, *expected shortfall*, *expected tail loss*, or *superquantile*. For given random variable X, its *average value-at-risk* is defined by

$$\mathrm{AVaR}_\varepsilon(X) = \frac{1}{\varepsilon}\int_0^\varepsilon \mathrm{VaR}_\alpha(X)d\alpha.$$

  One can show that this risk measure is coherent.
- Entropic value-at-risk: Let X be a random variable such that its moment generating function $M_X(z) = \mathbb{E}[e^{zX}]$ is well defined for all $z > 0$. Then the *entropic value-at-risk* is defined by

$$\mathrm{EVaR}_\alpha(X) = \inf_{z>0}\Big[z^{-1}\ln\Big(\frac{M_X(z)}{1-\alpha}\Big)\Big], \qquad \alpha \in (0,1).$$

  This is also coherent.

3.2 *Dynamic Risk Measures by BSDEs*

Next, let $\xi \in L^p_{\mathcal{F}_T}(\Omega)$ be a future payoff at T, and we are holding such a contingent claim. We not only want to know the risk we are facing at T but also want to know the risk any time between 0 and T. Namely, we want to measure the dynamic risk associated. To study this, the following notion is needed.

Definition 11.3.2 A map $\rho : [0,T] \times L^\infty_{\mathcal{F}_T}(\Omega) \to \mathbb{R}$ is called a *dynamic risk measure* if the following are satisfied:

(i) (Adaptiveness) For any $\xi \in L^\infty_{\mathcal{F}_T}(\Omega)$, $t \mapsto \rho(t;\xi)$ is $\mathbb{F}$-adapted;
(ii) (Monotonicity) For any $\xi, \bar\xi \in L^\infty_{\mathcal{F}_T}(\Omega)$ with $\xi \leqslant \bar\xi$, one has $\rho(t;\xi) \geqslant \rho(t;\bar\xi)$, for all $t \in [0,T]$;
(iii) (Translation Invariant) For any $\sigma \in L^\infty_{\mathcal{F}_T}(\Omega)$ and $c \in \mathbb{R}$, $\rho(t;\xi+c) = \rho(t;\xi) - c$. Further, ρ is said to be *convex* if the following holds:
(iv) (Convexity) $\xi \mapsto \rho(t;\xi)$ is convex; and ρ is said to be *coherent* if the following are satisfied:
(v) (Positive Homogeneity): For any $\xi \in L^\infty_{\mathcal{F}_T}(\Omega)$ and $\lambda \geqslant 0$, $\rho(t;\lambda\xi) = \lambda\rho(t;\xi)$;
(vi) (Subadditivity): For any $\xi, \bar\xi \in L^\infty_{\mathcal{F}_T}(\Omega)$, $\rho(t;\xi+\bar\xi) \leqslant \rho(t;\xi) + \rho(t;\bar\xi)$.

Each item in the above definition will have similar explanation as static risk measures. For example, (ii) means that between two gains, the dominantly larger one has a smaller risk; (vi) means that combining two investments will have smaller risk. The following is known.

Proposition 11.3.3 *Let $g : [0, T] \times \mathbb{R} \to \mathbb{R}$ be measurable such that $z \mapsto g(t, z)$ is convex and grows at most quadratically. Then for any $\xi \in L^\infty_{\mathcal{F}_T}(\Omega)$, the following BSDE*

$$Y(t) = -\xi + \int_t^T g(s, Z(s))ds - \int_t^T Z(s)dW(s), \quad t \in [0, T], \tag{11.3.1}$$

admits a unique adapted solution $(Y(\cdot), Z(\cdot)) \equiv (Y(\cdot\,; \xi), Z(\cdot\,; \xi))$. Let $\rho : [0, T] \times L^\infty_{\mathcal{F}_T}(\Omega) \to \mathbb{R}$ be defined by the following:

$$\rho(t, \xi) = Y(t; \xi), \quad (t, \xi) \in [0, T] \times L^\infty_{\mathcal{F}_T}(\Omega).$$

Then ρ is a dynamic convex risk measure.

One of the most interesting examples is the following.

$$Y(t) = -\xi + \int_t^T \frac{1}{2\gamma}|Z(s)|^2 ds - \int_t^T Z(s)dW(s), \quad t \in [0, T].$$

The above admits a unique adapted solution: $(Y(\cdot), Z(\cdot))$, and

$$\rho(t, \xi) \equiv Y(t) = \gamma \ln \mathbb{E}\Big[e^{-\frac{\xi}{\gamma}} \,\Big|\, \mathcal{F}_t\Big] \equiv e_{\gamma,t}(\xi), \quad t \in [0, T],$$

is called a *dynamic entropic risk measure* for ξ.

3.3 Equilibrium Dynamic Risk Measures by BSVIEs

In this section, we will use the following Type-I BSVIE to derive a class of dynamic risk measure:

$$Y(t) = -\psi(t) + \int_t^T g(t, s, Y(s), Z(t, s))ds - \int_t^T Z(t, s)dW(s), \quad t \in [0, T]. \tag{11.3.2}$$

In contrast with the above BSDE study, let us show some advantages by two cases.

Case I: Suppose at time t, a financial unit (e.g., a bank) has $N_i(t)$ shares of the i-th asset, with the price $p_i(t)$, $i = 1, 2, \cdots, k$. For definiteness, we assume that each $p_i(\cdot)$ is $\mathbb{F}$-adapted processes with $(\Omega, \mathcal{F}, \mathbb{F}, \mathbb{P})$ being a filtered probability space and

$\mathbb{F} = \{\mathcal{F}_t\}_{t\geqslant 0}$ being a filtration. Therefore, the (nominal) market value of the wealth is

$$\bar{\psi}(t) = \sum_{i=1}^{k} p_i(t)N_i(t), \quad \forall t \in [0, T].$$

The unit will hold the i-th asset for some period, say, until some $\tau_i(t) > t$, before selling it. One way to determine $\tau_i(t)$ could be the following:

$$\tau_i(t) = \inf\{s > t \mid p_i(s) > \alpha_i p_i(t) \text{ or } p_i(s) < \beta_i p_i(t)\},$$

for some $\alpha_i > 0$ and $\beta_i \in (0, 1)$. Therefore, at time t, instead of $\bar{\psi}(t)$, the unit should regard its wealth as

$$\psi(t) = \sum_{i=1}^{k} p_i(\tau_i(t))N_i(t), \quad \forall t \in [0, T].$$

Since in general $\tau_i(t) > t$ with t being the current time, the process $\psi(\cdot)$ is not $\mathbb{F}$-adapted. Now, the question is how one can measure the risk of the *position process* $\psi(\cdot)$. Of course, it goes beyond the random variable framework in the above sections.

Mimicking the above BSDE case, the corresponding dynamic risk should be measured via

$$Y(t, r) = -\psi(t) + \int_r^T g(s, Y(t, s), Z(t, s))ds - \int_r^T Z(t, s)dW(s), \quad (r, t) \in \Delta^*[0, T],$$

and the current dynamic risk should be $Y(t; t)$. But, simply taking $r = t$ in the above leads to the following:

$$Y(t, t) = -\psi(t) + \int_t^T g(s, Y(t, s), Z(t, s))ds - \int_t^T Z(t, s)dW(s), \quad t \in [0, T],$$

which is not a closed form equation for the pair $(Y(t, t), Z(t, s))$. To figure out the above issue, we propose the following modified equation, a special case of (11.3.2),

$$Y(t) = \psi(t) + \int_t^T g(s, Y(s), Z(t, s))ds - \int_t^T Z(t, s)dW(s). \tag{11.3.3}$$

Note that the above modification is simply to force $Y(t; s) = Y(s; s)$ and then rename $Y(t; t)$ to be $Y(t)$. The above (11.3.3) is special case of BSVIE with driver g independent of t. In fact, there is indeed some convincing mathematical justification for the model (11.3.3). On the one hand, it has been shown in Chap. 7

that the BSVIEs can be seen as a proper limit for BSDEs from which dynamic risk measure can be obtained. On the other hand, one expects that the dynamic risk measure should be equilibrium in the sense the value of the risk today (for a process $\psi(\cdot)$) should match the one expected yesterday. Therefore, it is natural to use BSVIEs (11.3.3) to describe/measure the dynamic risk of the process $\psi(\cdot)$.

Case II: Let us look at the following example which shows that the generator/driver of BSDEs can also be improved in some sense.

Example 11.3.4 In the above section, we have mentioned the notion of *entropic risk measure* where $\gamma > 0$ is called a *risk aversion parameter*. We take a closer look at the parameter γ which indicates the agent time preference in the time duration $[0, T]$. In the practical world, instead of the constant assumption, it is more natural that people keep changing their risk tolerance parameter as time goes by. Therefore, it is better for us to use the map $t \mapsto \gamma(t)$. From this sense, a new general form of entropic risk measure could be

$$Y(t) = -\gamma(t) \log \mathbb{E}_t\Big[\exp\Big(\frac{\xi}{\gamma(t)}\Big)\Big], \qquad t \in [0, T].$$

It corresponds to the solution $Y(\cdot)$ for the following BSVIE

$$Y(t) = -k_1\xi + \int_t^T \frac{1}{2\gamma(t)}|Z(t,s)|^2 ds - \int_t^T Z(t,s)dW(s), \qquad t \in [0, T].$$

The above shows BSDE/BSVIE arising from mathematical finance when the generator is quadratic growth, see Chap. 8 for more details. Next let us look at another way of transforming BSDEs into general BSVIEs. Suppose there exists the underlying economy which is assumed to contain one representative agent with changing subjective beliefs over time. For example, at time t, the believe system is $\mathbb{P}_t$ and is mutually absolute continuous w.r.t. $\mathbb{P}$ with $\frac{d\mathbb{P}_t}{d\mathbb{P}} = \mathcal{E}(\theta(t,\cdot))_T$ with some bounded map $\theta(t,\cdot)$. We refer to Case III in the above Sect. 2.2.1 for a detailed example. Under the probability $\mathbb{P}_t$, given contingent claim ξ, the dynamic risk measure $Y(\cdot) \triangleq \mathbb{E}_t^{\mathbb{P}_t}\xi$ satisfies a linear BSVIE of the form

$$Y(t) = \xi + \int_t^T \theta(t,s)Z(t,s)ds - \int_t^T Z(t,s)dW(s), \quad t \in [0, T]. \tag{11.3.4}$$

Of course, it is particular case of BSVIE (11.3.2).

Based on the above discussions, in the following, we define dynamic risk measure for *position process* $\psi(\cdot) \in C_{\mathcal{F}_T}([0, T]; L^2(\Omega; \mathbb{R}))$. Now, mimicking Definition 11.3.2, we introduce the following.

Definition 11.3.5 A map $\rho : [0,T]\times C_{\mathcal{F}_T}([0,T];L^2(\Omega;\mathbb{R}))\to C_{\mathbb{F}}([0,T];L^2(\Omega;\mathbb{R}))$ is called an *equilibrium dynamic risk measure* if the following hold:

(i) (Past Independence) For any $\psi_0(\cdot),\psi_1(\cdot)\in C_{\mathcal{F}_T}([0,T];L^2(\Omega;\mathbb{R}))$, if

$$\psi_0(s)=\psi_1(s),\qquad \text{a.s.},\quad \text{a.e. } s\in[t,T],$$

for some $t\in[0,T)$, then

$$\rho(t;\psi_0(\cdot))=\rho(t;\psi_1(\cdot)),\qquad \text{a.s.}$$

(ii) (Monotonicity) For any $\psi_0(\cdot),\psi_1(\cdot)\in C_{\mathcal{F}_T}([0,T];L^2(\Omega;\mathbb{R}))$, if

$$\psi_0(s)\leqslant\psi_1(s),\qquad \text{a.s.},\quad \text{a.e. } s\in[t,T],$$

for some $t\in[0,T)$, then

$$\rho(s;\psi_0(\cdot))\geqslant\rho(s;\psi_1(\cdot)),\qquad \text{a.s.},\quad s\in[t,T].$$

(iii) (Translation Invariance) There exists a deterministic integrable function $r(\cdot)$ such that for any $\psi(\cdot)\in C_{\mathcal{F}_T}([0,T];L^2(\Omega;\mathbb{R}))$,

$$\rho(t;\psi(\cdot)+c)=\rho(t;\psi(\cdot))-ce^{\int_t^T r(s)ds},\qquad \text{a.s.},\quad t\in[0,T]. \tag{11.3.5}$$

Further, ρ is said to be *convex* if the following holds:

(iv) (Convexity) For any $\psi_0(\cdot),\psi_1(\cdot)\in C_{\mathcal{F}_T}([0,T];L^2(\Omega;\mathbb{R}))$ and $\lambda\in[0,1]$,

$$\begin{aligned}\rho(t;(1-\lambda)\psi_0(\cdot)+\lambda\psi_1(\cdot))\leqslant(1-\lambda)\rho(t;\psi_0(\cdot))+\lambda\rho(t;\psi_1(\cdot)),\\ \text{a.s.},\quad t\in[0,T].\end{aligned} \tag{11.3.6}$$

And ρ is said to be *coherent* if the following are satisfied:

(v) (Positive Homogeneity) For any $\psi(\cdot)\in C_{\mathcal{F}_T}([0,T];L^2(\Omega;\mathbb{R}))$ and $\lambda>0$,

$$\rho(t;\lambda\psi(\cdot))=\lambda\rho(t;\psi(\cdot)),\qquad \text{a.s.},\quad t\in[0,T].$$

(vi) (Subadditivity) For any $\psi_0(\cdot),\psi_1(\cdot)\in C_{\mathcal{F}_T}([0,T];L^2(\Omega;\mathbb{R}))$,

$$\rho(t;\psi_0(\cdot)+\psi_1(\cdot))\leqslant\rho(t;\psi_0(\cdot))+\rho(t;\psi_1(\cdot)),\qquad \text{a.s.},\quad t\in[0,T].$$

The word "equilibrium" indicates the value of the risk today (for a process $\psi(\cdot)$) should match the one expected yesterday, and it is some kind of modification of the naive one by family of BSDEs aforementioned. In the following, we will assume that proper conditions stated in Chap. 4, Sect. 2 and Chap. 6, Sect. 2, are satisfied. Then, one has the following result.

Proposition 11.3.6 *Let the generator be given by*

$$g(t,s,y,z) \equiv r(s)y + g_0(t,s,z), \quad (t,s,y,z) \in \Delta^*[0,T] \times \mathbb{R} \times \mathbb{R}, \tag{11.3.7}$$

where $r(\cdot)$ is a nonnegative deterministic function. Then the following are true:

(i) *The map $\psi(\cdot) \mapsto \rho(t;\psi(\cdot))$ is past independent.*
(ii) *The map $\psi(\cdot) \mapsto \rho(t;\psi(\cdot))$ is monotone.*
(iii) *The map $\psi(\cdot) \mapsto \rho(t;\psi(\cdot))$ is translation invariant.*
(iv) *Suppose $z \mapsto g_0(t,s,z)$ is convex, so is $\psi(\cdot) \mapsto \rho(t;\psi(\cdot))$.*
(v) *Suppose $z \mapsto g_0(t,s,z)$ is positively homogeneous, so is $\psi(\cdot) \mapsto \rho(t;\psi(\cdot))$.*
(vi) *Suppose $z \mapsto g_0(t,s,z)$ is subadditive, so is $\psi(\cdot) \mapsto \rho(t;\psi(\cdot))$.*

Proof (i) and (ii) are clear. (iii) Let $(Y(\cdot).Z(\cdot\,,\cdot))$ be the adapted solution of BSVIE with generator $g(\cdot)$ given by (11.3.7). Then

$$\begin{aligned} Y(t) - ce^{\int_t^T r(s)ds} &= -\psi(t) + \int_t^T \Big[r(s)Y(s) + g_0(t,s,Z(t,s))\Big]ds \\ &\quad - \int_t^T Z(t,s)dW(s) - ce^{\int_t^T r(\tau)d\tau} \\ &= -\big[\psi(t)+c\big] + \int_t^T \Big[r(s)\big[Y(s) - ce^{\int_s^T r(\tau)d\tau}\big] + g_0(t,s,Z(t,s))\Big]ds \\ &\quad - \int_t^T Z(t,s)dW(s), \qquad t\in[0,T]. \end{aligned}$$

This means $(Y(\cdot) - ce^{\int_\cdot^T r(s)ds}, Z(\cdot\,,\cdot))$ is the adapted solution to BSVIE (11.3.2) with free term $-\big[\psi(\cdot)+c\big]$. Hence, (11.3.5) holds.

(iv) Let $z \mapsto g_0(t,s,z)$ be convex. Then, for any $\lambda \in (0,1)$ and $\psi_0(\cdot), \psi_1(\cdot) \in C_{\mathcal{F}_T}\big([0,T]; L^2(\Omega;\mathbb{R})\big)$. Let $(Y_i(\cdot), Z_i(\cdot\,,\cdot))$ be the adapted solution corresponding to $\psi_i(\cdot)$ $(i=0,1)$. Then for $t\in[0,T]$,

$$\begin{aligned} (1-\lambda)Y_0(t) + \lambda Y_1(t) &= -[(1-\lambda)\psi_0(t) + \lambda\psi_1(t)] \\ &\quad + \int_t^T \Big[r(s)\big[(1-\lambda)Y_0(s) + \lambda Y_1(s)\big] \\ &\quad + (1-\lambda)g_0(t,s,Z_0(t,s)) + \lambda g_0(t,s,Z_1(t,s))\Big]ds \\ &\quad - \int_t^T \Big[(1-\lambda)Z_0(t,s) + \lambda Z_1(t,s)\Big]dW(s). \end{aligned}$$

Now, let $(Y(\cdot), Z(\cdot\,,\cdot))$ be the adapted solution of BSVIE (11.3.2) corresponding to the free $-[(1-\lambda)\psi_0(\cdot) + \lambda\psi_1(\cdot)]$. By the convexity condition of $z \mapsto g_0(t,s,z)$, and by the comparison theorem in Chap. 6, one has

$$Y(t) \leqslant (1-\lambda)Y_0(t) + \lambda Y_1(t), \qquad \text{a.s. } \forall t \in [0,T].$$

This proves (iv),

(v), and (vi). The proof is similar to that of (iv). □

By Proposition 11.3.6, we can construct a large class of equilibrium dynamic risk measures by choosing suitable generator $g(\cdot)$ of BSVIE (11.3.2). More precisely, we have the following result.

Theorem 11.3.7 *Let* (11.3.7) *hold. If* $z \mapsto g_0(t,s,z)$ *is convex, then* $\psi(\cdot) \mapsto \rho(t;\psi(\cdot))$ *is an equilibrium dynamic convex risk measure. If* $z \mapsto g_0(t,s,z)$ *is positively homogeneous and subadditive, then* $\psi(\cdot) \mapsto \rho(t;\psi(\cdot))$ *is an equilibrium dynamic coherent risk measure.*

According to the above results, we can have some examples of equilibrium dynamic risk measures by the choices of $g_0(t,s,z)$: If

$$g_0(t,s,z) = \bar{g}(t,s)|z|, \qquad \bar{g}(t,s) \geqslant 0,$$

then, it is subadditive and positively homogeneous in z. The corresponding equilibrium dynamic risk measure is coherent. If

$$g_0(t,s,z) = \bar{g}(t,s)\sqrt{1+|z|^2}, \qquad \bar{g}(t,s) \geqslant 0,$$

then, it is convex in z. The corresponding equilibrium dynamic risk measure is convex. If

$$g_0(t,s,z) = \bar{g}(t,s)|z|^2, \qquad \bar{g}(t,s) \geqslant 0,$$

then one has an entropy type equilibrium dynamic risk measure.

To conclude this part, we discuss a class of risk minimization problem where the dynamic risk measure is given by BSVIEs. For simplicity, suppose there are two financial assets in the market, i.e., one non-risky asset (e.g., bond) used as numeraire and one risky asset (e.g., stock) described as

$$\begin{cases} dS(r) = \alpha(r)S(r)dr + \beta(r)S(r)dW(r), & r \in [0,T], \\ S(0) = s_0, \end{cases}$$

where $\alpha(\cdot)$, $\beta(\cdot)$ are bounded adapted processes, $\epsilon \leqslant |\beta(\cdot)| \leq K$, $0 < \epsilon < K$. Let $\pi(\cdot) \in L^2_{\mathbb{F}}(0,T;\mathbb{R})$ be dollar amount of the agent invested in the risky asset. Given initial wealth x and initial time t, the wealth process is,

$$X(s) = x + \int_t^s \alpha(u)\pi(u)du + \int_t^s \beta(u)\pi(u)dW(u), \qquad s \in [t,T]. \tag{11.3.8}$$

For given (11.3.8), we want to find $\bar{\pi}$ to minimize risk measure $\rho(t; X(T)) = Y(t)$, where $Y(t)$ satisfies a BSVIE

$$Y(t) = -X(T) + \int_t^T g(t, s, Z(t, s))ds - \int_t^T Z(t, s)dW(s), \quad t \in [0, T], \tag{11.3.9}$$

Here g is measurable such that $g(t, s, 0) = 0$, $z \mapsto g(t, s, z)$ is Lipschitz continuous and convex, and $t \mapsto g(t, s, z)$ is continuous.

By taking $p(\cdot) \equiv \pi(\cdot)\beta(\cdot)$, $\theta(\cdot) \equiv \alpha(\cdot)\beta^{-1}(\cdot)$, for $t \in [0, T]$, we can rewrite (11.3.8), (11.3.9) as

$$\begin{cases} X^p(s) = x + \int_t^s \theta(u)p(u)du + \int_t^s p(u)dW(u), \quad s \geqslant t, \\ Y^p(t) = -X^p(T) + \int_t^T g(t, s, Z^p(t, s))ds - \int_t^T Z^p(t, s)dW(s). \end{cases}$$

For the above backward equation at time t, it is easy to see

$$Y^p(t) = -x + \int_t^T g_1(t, s, \mathcal{Z}^p(t, s), p(s))ds - \int_t^T \mathcal{Z}^p(t, s)dW(s), \tag{11.3.10}$$

where for $t, s \in [0, T]$,

$$\mathcal{Z}^p(t, s) \equiv Z^p(t, s) + p(s), \quad g_1(t, s, z, p(s)) \equiv g(t, s, z - p(s)) - p(s)\theta(s).$$

By the convexity of g, for $t, s \in [0, T]$, we define the Fenchel-Legendre transform as

$$\begin{aligned} G(t, s, \theta(s)) &\equiv \sup_{r \in \mathbb{R}} \widetilde{g}(t, s, \theta(s), r), \quad \text{a.s.}, \\ \widetilde{g}(t, s, \theta(s), r) &\equiv -\theta(s) \cdot r - g(t, s, r). \end{aligned} \tag{11.3.11}$$

Given L the Lipschitz constant of g, $G < \infty$ is lead by $|\theta(\cdot)| < L$.

Suppose for given $\theta(\cdot)$, there exists measurable mapping $\bar{p} : [0, T] \times [0, T] \times \Omega \to \mathbb{R}$ such that (11.3.10) admits a unique pair of $(\bar{Y}(\cdot), \bar{\mathcal{Z}}(\cdot, \cdot))$, and for $s \in [0, T]$,

$$G(t, s, \theta(s)) = -\theta(s)[\bar{\mathcal{Z}}(t, s) - \bar{p}(t, s)] - g(t, s, \bar{\mathcal{Z}}(t, s) - \bar{p}(t, s)). \text{ a.s.} \tag{11.3.12}$$

From (11.3.12), for any measurable $p(\cdot, \cdot)$ satisfying $p(t, \cdot) \in L^2_{\mathbb{F}}(t, T; \mathbb{R})$ for any t, it is easy to see

$$-G(t, s, \theta(s)) - \theta(s)\bar{\mathcal{Z}}(t, s) = g_1(t, s, \bar{\mathcal{Z}}(t, s), \bar{p}(t, s)) \leqslant g_1(t, s, \bar{\mathcal{Z}}(t, s), p(t, s)).$$

By the comparison theorem of BSVIEs,

$$\bar{\pi}(t,s) = \bar{p}(t,s)\beta^{-1}(s), \quad s \in [0,T], \tag{11.3.13}$$

is an optimal strategy in the sense that $Y^{\bar{\pi}}(t)$ is the minimized risk measure. We will provide two explicit examples to verify assumption (11.3.12).

Example 11.3.8 Given bounded continuous function $k : [0,T] \times [0,T] \to \mathbb{R}^+$, $\alpha(\cdot)$, $\beta(\cdot)$ in (11.3.8), let us consider

$$g(t,s,z) \equiv k(t,s)\Big[\sqrt{1+|z|^2} - 1\Big], \quad t,s \in [0,T], \quad z \in \mathbb{R}.$$

For constants a_i $(i = 1,2,3,4)$ satisfying $0 < a_1 < a_2 < 1$, suppose $a_1 < \frac{|\theta(\cdot)|}{k(\cdot,x)} \leqslant a_2$. From (11.3.11), we have

$$\widetilde{g}(t,s,\theta(s),r) = -\theta(s)r - k(t,s)\big[\sqrt{1+|r|^2} - 1\big].$$

Given s, t, we obtain $r \mapsto \widetilde{g}(t,s,\theta(s),r)$ is concave in $\mathbb{R}$ with

$$\widetilde{g}_r(t,s,x,\theta(s),r) = -\theta(s) - \frac{k(t,s)r}{\sqrt{1+r^2}}, \quad \widetilde{g}_{rr}(t,s,\theta(s),r) = -\frac{k(t,s)}{\big(\sqrt{1+r^2}\big)^{\frac{3}{2}}}.$$

Consequently, the supremum in (11.3.11) is reachable at $r = \frac{-\theta(s)}{\sqrt{k^2(t,s)-\theta^2(s)}}$. Putting it back to $\widetilde{g}(t,s,\theta(s),r)$, we have

$$G(t,s,\theta(s)) = -\Big[\sqrt{k^2(t,s)-\theta^2(s)} - k(t,s)\Big].$$

Therefore, we obtain a BSVIE for $(\bar{Y},\bar{Z})$ as follows:

$$\begin{aligned}\bar{Y}(t) = -x + &\int_t^T \Big[\sqrt{k^2(t,s)-\theta^2(s)} - k(t,s) - \bar{Z}(t,s)\theta(s)\Big]ds \\ &- \int_t^T \bar{Z}(t,s)dW(s), \quad t \in [0,T].\end{aligned} \tag{11.3.14}$$

With $\bar{Z}(\cdot,\cdot)$ satisfying (11.3.14), we define $\bar{p}(\cdot,\cdot)$ as

$$\bar{p}(t,s) \equiv \left[\frac{\theta(s)}{\sqrt{k^2(t,s)-\theta^2(s)}} + \bar{Z}(t,s)\right], \quad t \leqslant s.$$

Thus (11.3.12) is fulfilled. According to (11.3.13), the following

$$\bar{\pi}(t,s) = \beta^{-1}(s)\left[\frac{\theta(s)}{\sqrt{k^2(t,s)-\theta^2(s)}} + \bar{Z}(t,s)\right], \quad t \leqslant s, \tag{11.3.15}$$

is an optimal strategy.

As to coefficient k, we consider it as a subjective *risk aversion parameter* representing investors' reluctance toward risk. For example, suppose two investors A, B assess the risk of ξ as $\rho^A(\xi)$, $\rho^B(\xi)$. If $k_A \leq k_B$, by comparison theorems of BSVIEs in Chap. 6, the risk on behalf of A is less than B. In other words, A is is more risk tolerant (less risk-averse) than B.

If $\theta(\cdot)$ is deterministic, we present some conclusions on optimal strategy, which fit the common economic knowledge in the real life. First, in this case, the term $\bar{Z}(\cdot,\cdot)$, arising from the randomness of θ, becomes null and

$$\bar{\pi}(t,s) = \frac{\theta(s)}{\sqrt{k^2(t,s)-\theta^2(s)}}\beta^{-1}(s), \quad t \leqslant s. \tag{11.3.16}$$

Second, from (11.3.16), there is a negative monotonic link between k and $|\bar{\pi}|$. With a higher risk aversion parameter, people are so conservative such that they put less money into risky assets. If $\beta > 0$, $\bar{\pi}$ has positive monotonicity with market price of risk θ.

Example 11.3.9 Given $\ell : [0;T] \times [0,T] \to \mathbb{R}_+$, we consider

$$g(t,s,z) = \ell(t,s)\ln\frac{1+e^{-z}}{2}, \qquad s,t \in [0,T],\ z \in \mathbb{R}_+.$$

For constants a_i, $i = 1,2$ satisfying $1 < a_1 < a_2$, suppose that $\frac{2a_3-1}{2a_3} < \frac{\theta(\cdot)}{l(\cdot,x)} < \frac{2a_4-1}{2a_4}$. First, by (11.3.11),

$$\widetilde{g}(t,s,\theta(s),r) = -\theta(s)r - \ell(t,s)\ln\frac{1+e^{-r}}{2}.$$

Given t,s, we obtain $r \mapsto \widetilde{g}(t,s,\theta(s),r)$ is concave in $\mathbb{R}$ with

$$\widetilde{g}_r(t,s,\theta(s),r) = -\theta(s) + \frac{\ell(t,s)}{1+e^r}, \quad \widetilde{g}_{rr}(t,s,\theta(s),r) = -\frac{\ell(t,s)e^r}{(1+e^r)^2}.$$

Consequently, the supremum is reached at $r = \ln(\frac{\ell(t,s)}{\theta(s)} - 1)$. Putting it back to $\widetilde{g}(t,s,\theta(s),r)$, we have

$$G(t,s,\theta) = \theta(s)\ln\frac{\theta(s)}{\ell(t,s)-\theta(s)} - \ell(t,s)\ln\frac{\ell(t,s)}{2[\ell(t,s)-\theta(s)]}.$$

With $\bar{Z}(\cdot,\cdot)$ satisfying the following,

$$\bar{Y}(t) = -x + \int_t^T \Big[\ell(t,s)\ln\frac{\ell(t,s)}{2[\ell(t,s)-\theta(s)]} - \theta(s)\ln\frac{\theta(s)}{\ell(t,s)-\theta(s)} \\ -\theta(s)\bar{Z}(t,s)\Big]ds - \int_t^T \bar{Z}(t,s)dW(s), \tag{11.3.17}$$

we define $\bar{p}(\cdot,\cdot)$ as

$$\bar{p}(t,s) = \Big[\bar{Z}(t,s) - \ln\Big(\frac{\ell(t,s)}{\theta(s)} - 1\Big)\Big], \quad s,t \in [0,T]. \tag{11.3.18}$$

Thus (11.3.12) is fulfilled. According to (11.3.13), the following $\bar{\pi}$ is an optimal strategy

$$\bar{\pi}(t,s) = \beta(s)^{-1}\ln\Big[\frac{\theta(s)}{\ell(t,s)-\theta(s)}\Big] + \beta(s)^{-1}\bar{Z}(t,s), \quad s \in [t,T]. \tag{11.3.19}$$

In the above, $\ell(\cdot,\cdot)$ can still be identified as a subjective risk aversion parameter.

If $\theta(\cdot)$ is deterministic, $\beta(\cdot) > 0$, $\alpha(\cdot) > 0$, optimal strategy has positive correlation with market price of risk, initial wealth, while it keeps negative monotonicity with respect to risk aversion parameter, the conclusions of which are consistent with those in the real world.

4 Dynamic Capital Allocation

In this section, we are going to use BSVIEs further to discuss the so-called dynamic capital allocation problem.

To begin with, let us give a brief introduction to the capital allocation problem. Suppose we are given a time duration T, risk measure ρ, as well as financial position ξ (with maturity T) that can be separated into subunits $\xi_1, \cdots, \xi_n$, i.e., $\xi = \sum_{i=1}^{n} \xi_n$. For example, ξ can be seen as the profit or loss of a portfolio composed by different asset, while ξ_i as the profit and losses asset i. The capital allocation problem deals with the issue of sharing in the proper way the margin (or deposit) $\rho(\xi)$ required to hedge the riskiness of financial position ξ among the different subunits $\xi_1, \cdots, \xi_n$. Since the margin is usually represented by risk measure, hence the capital allocation problem is inherently related to the latter. The aim of the financial problem is to find $k_1, \cdots, k_n$ such that $\rho(\xi) = k_1 + \cdots + k_n$, where k_i represents the capital to be allocated to ξ_i or, in other words, the risk contribution of ξ_i to the total risk capital $\rho(\xi)$ of ξ. To arrive such a destination, one of the most studied and applied tool is the gradient allocation, denoted by $\nabla_{\xi_i}\rho(\xi)$, which is the Gâteaux-derivative of risk measure $\rho(\xi)$ in the direction of subunit ξ_i. In other words,

$$\nabla_{\xi_i}\rho(\xi) \equiv \lim_{h\to 0}\frac{\rho(\xi+h\xi_i)-\rho(\xi)}{h} = \frac{d}{du}\rho(\xi+u\xi_i)\Big|_{u=0}.$$

Another important notion is Aumann-Shapley allocation which is defined as follows:

$$\nabla^{AS}_{\xi_i}\rho(\xi) \equiv \int_0^1 \nabla_{\xi_i}\rho(\gamma\xi)d\gamma, \quad i=1,\cdots,n.$$

Next we would like to apply the above idea to position process (rather than random variable) by means of BSVIEs. Suppose $\psi(\cdot)$ is a position process, the risk measure of which is given by $\rho(\cdot;\psi) \equiv Y^\psi(\cdot)$,

$$\begin{aligned} Y^\psi(t) = -\psi(t) + \int_t^T \big[r(s)Y^\psi(s) + g_0(t,s,Z^\psi(t,s))\big]ds \\ -\int_t^T Z^\psi(t,s)dW(s), \qquad t\in[0,T], \end{aligned} \tag{11.4.1}$$

see Proposition 11.3.6. In addition, $\psi(\cdot) = \sum\limits_{i=1}^{n}\psi_n(\cdot)$ for subunits $\psi_1(\cdot),\cdots,\psi_n(\cdot)$. The corresponding capital allocation problem is to find $k_1(\cdot),\cdots,k_n(\cdot)$ defined on $[0,T]$ such that $\rho\big(\cdot;\psi\big) = k_1(\cdot)+\cdots+k_n(\cdot)$, where $k_i(\cdot)$ represents the risk contribution of $\psi_i(\cdot)$ to the total risk capital $\rho(\cdot;\psi)$ of $\psi(\cdot)$. To study this problem, we need some differentiability results for BSVIEs. The following result in some sense has the similar spirit of Proposition 3.2.2 in Chap. 3.

(H11.4.1) The map $g : \Delta^*[0,T]\times\mathbb{R}^m\times\mathbb{R}^m \to \mathbb{R}^m$ is measurable such that $(y,z)\mapsto g(t,s,y,z)$ is differentiable, with bounded and continuous partial derivatives g_y, g_z. In addition, $g(\cdot,\cdot,0,0)=0$, and

$$\begin{aligned} &|g(t,s,y,z)-g(\bar t,s,y,z)| \leqslant \rho(|t-\bar t|)(1+|y|+|z|), \\ &|f(t,s,y,z)-f(\bar t,s,y,z)| \leqslant \rho(|t-\bar t|), \quad f\equiv g_y,\ g_z, \end{aligned}$$

for some modulus of continuity $\rho:\mathbb{R}^+\to\mathbb{R}^+$.

Notice that the above assumption can be slightly relaxed properly. We have the following result.

Proposition 11.4.1 *Let* (H11.4.1) *hold. Let* $\psi(\cdot),\varphi(\cdot)\in C_{\mathcal{F}_T}\big([0,T];L^2(\Omega;\mathbb{R}^m)\big)$. *For* $\tau\geqslant 0$, *let* $(Y^\tau(\cdot),Z^\tau(\cdot\,,\cdot))$ *be the adapted solution of the following BSVIE:*

$$\begin{aligned} Y^\tau(t) = -\Big[\psi(t)+\tau\varphi(t)\Big] + \int_t^T g(t,s,Y^\tau(s),Z^\tau(t,s))ds \\ -\int_t^T Z^\tau(t,s)dW(s), \qquad t\in[0,T]. \end{aligned} \tag{11.4.2}$$

Then $\tau \mapsto (Y^\tau(\cdot), Z^\tau(\cdot\,,\cdot))$ *is differentiable at* $\tau = 0$ *with the derivative* $(\nabla Y(\cdot), \nabla Z(\cdot\,,\cdot))$ *being the adapted solution of the following BSVIE:*

$$\begin{aligned}\nabla Y(t) = -\varphi(t) + \int_t^T \Big(& g_y(t,s,Y^0(s),Z^0(t,s))\nabla Y(s) \\ &+ g_z(t,s,Y^0(s),Z^0(t,s))\nabla Z(t,s)\Big)ds \\ &- \int_t^T \nabla Z(t,s)dW(s), \qquad t \in [0,T].\end{aligned} \tag{11.4.3}$$

Proof First, by Proposition 6.2.1 in Chap. 6, we see that BSVIEs (11.4.2) and (11.4.3) admit unique adapted solution $(Y^\tau(\cdot), Z^\tau(\cdot\,,\cdot))$, $(\nabla Y(\cdot), \nabla Z(\cdot\,,\cdot))$ such that

$$\begin{aligned}&Y^\tau(\cdot),\ \ \nabla Y(\cdot) \in C_{\mathbb{F}}([0,T]; L^2(\Omega;\mathbb{R}^m)),\\ &Z^\tau(\cdot,\cdot),\ \ \nabla Z(\cdot,\cdot) \in C\big([0,T]; L^2_{\mathbb{F}}(\Omega; L^2(\cdot,T;\mathbb{R}^m))\big).\end{aligned}$$

We look at the difference quotient:

$$\begin{aligned}\frac{1}{\tau}\Big[Y^\tau(t) - Y^0(t)\Big] &= -\varphi(t)\\ &+ \int_t^T \frac{1}{\tau}\Big[g(t,s,Y^\tau(s),Z^\tau(t,s)) - g(t,s,Y^0(s),Z^0(t,s))\Big]ds\\ &- \int_t^T \frac{1}{\tau}\Big[Z^\tau(t,s) - Z^0(t,s)\Big]dW(s), \qquad t \in [0,T].\end{aligned}$$

Consequently,

$$\begin{aligned}&\frac{Y^\tau(t) - Y^0(t)}{\tau} - \nabla Y(t)\\ &= \int_t^T \Big\{\frac{1}{\tau}\Big[g(t,s,Y^\tau(s),Z^\tau(t,s)) - g(t,s,Y^0(s),Z^0(t,s))\Big]\\ &\qquad - g_y(t,s,Y^0(s),Z^0(t,s))\nabla Y(s) - g_z(t,s,Y^0(s),Z^0(t,s))\nabla Z(t,s)\Big\}ds\\ &\quad - \int_t^T \Big\{\frac{1}{\tau}\Big[Z^\tau(t,s) - Z^0(\cdot)(t,s)\Big] - \nabla Z(t,s)\Big\}dW(s)\\ &= \int_t^T \Big[\widetilde{g}_y(t,s)\frac{Y^\tau(s) - Y^0(s)}{\tau} - g_y(t,s,Y^0(s),Z^0(t,s))\nabla Y(s)\\ &\qquad + \widetilde{g}_z(t,s)\frac{Z^\tau(t,s) - Z^0(t,s)}{\tau} - g_y(t,s,Y^0(s),Z^0(t,s))\nabla Z(t,s)\Big]ds\\ &\quad - \int_t^T \Big[\frac{Z^\tau(t,s) - Z^0(t,s)}{\tau} - \nabla Z(t,s)\Big]dW(s),\end{aligned}$$

where

$$\begin{aligned}\widetilde{g}_y(t,s)=\int_0^1 & g_y\big(t,s,Y^0(s)+\alpha[Y^\tau(s)-Y^0(s)],\\ & Z^0(t,s)+\alpha[Z^\tau(t,s)-Z^0(t,s)]\big)d\alpha,\\ \widetilde{g}_z(t,s)=\int_0^1 & g_z\big(t,s,Y^0(s)+\alpha[Y^\tau(s)-Y^0(s)],\\ & Z^0(t,s)+\alpha[Z^\tau(t,s)-Z^0(t,s)]\big)d\alpha.\end{aligned}$$

Then

$$\begin{aligned}\frac{Y^\tau(t)-Y^0(t)}{\tau}-\nabla Y(t)=\int_t^T & \Big[\widetilde{g}_y(t,s)\Big(\frac{Y^\tau(s)-Y^0(s)}{\tau}-\nabla Y(s)\Big)\\ & +\Big(\widetilde{g}_y(t,s)-g_y(t,s,Y^0(s),z^0(t,s)\Big)\nabla Y(s)\\ & +\widetilde{g}_z(t,s)\Big(\frac{Z^\tau(t,s)-Z^0(t,s)}{\tau}-\nabla Z(t,s)\Big)\\ & +\Big(\widetilde{g}_z(t,s)-g_z(t,s,Y^0(s),z^0(t,s)\Big)\nabla Z(t,s)\Big]ds\\ & -\int_t^T\Big[\frac{Z^\tau(t,s)-Z^0(t,s)}{\tau}-\nabla Z(t,s)\Big]dW(s).\end{aligned}$$

Hence, by the stability estimate for BSVIE (e.g., Proposition 6.2.1 in Chap. 6), one has

$$\begin{aligned}&\Big\|\frac{Y^\tau(\cdot)-Y^0(\cdot)}{\tau}-\nabla Y(\cdot)\Big\|^2_{C_{\mathbb{F}}([0,T];L^2(\Omega;\mathbb{R}^m))}\\ &\quad+\Big\|\frac{Z^\tau(\cdot\,,\cdot)-Z^0(\cdot\,,\cdot)}{\tau}-\nabla Z(\cdot\,,\cdot)\Big\|^2_{C([0,T];L^2_{\mathbb{F}}(\Omega;L^2(\cdot,T;\mathbb{R}^m)))}\\ &\leqslant K\sup_{t\in[0,T]}\mathbb{E}\Big[\Big(\int_t^T|\widetilde{g}_y(t,s)-g_y(t,s,Y^0(s),Z^0(t,s))|\,|\nabla Y(s)|ds\Big)^2\\ &\quad+\Big(\int_t^T|\widetilde{g}_z(t,s)-g_z(t,s,Y^0(s),Z^0(t,s))|\,|\nabla Z(t,s)|ds\Big)^2\Big].\end{aligned}$$

Thus, by the continuity of $g_y(\cdot)$ and $g_z(\cdot)$, together with the bounded convergence theorem, we get the convergence. □

Let us return back the asset allocation problem. To this end, we use the gradient allocation approach, denoted by $\nabla_{\psi_i}\rho(\cdot;\psi)$, which is the Gâteaux-derivative of risk measure in the direction of subunit $\psi_i(\cdot)$. In other words,

$$\nabla_{\psi_i}\rho(\cdot;\psi)\equiv\lim_{h\to0}\frac{\rho(\cdot;\psi+h\psi_i)-\rho(\cdot;\psi)}{h}=\frac{d}{du}\rho(\cdot;\psi+u\psi_i)\Big|_{u=0}.$$

Now, let $m=1$ and consider BSVIE (11.4.1) with $r(\cdot)$ being nonnegative and deterministic. By Proposition 11.3.6, $\psi(\cdot)\mapsto\rho(t;\psi(\cdot))$ defines a dynamic risk measure with proper generator g. If $z \mapsto g_0(t, s, z)$ is differentiable, then according to Proposition 11.4.1, $\psi(\cdot) \mapsto \rho(t; \psi(\cdot))$ is Gâteaux differentiable in the direction of $\varphi(\cdot)$, with the derivative $(\nabla_\varphi Y^\psi(\cdot), \nabla_\varphi Z^\psi(\cdot\,,\cdot))$ being the unique adapted solution to the following BSVIE:

$$\begin{aligned}\nabla_\varphi Y^\psi(t) = -\varphi(t) + \int_t^T \Big(r(s)\nabla_\varphi Y^\psi(s) + g_{0,z}(t, s, Z^\psi(t, s))\nabla_\varphi Z^\psi(t, s)\Big)ds \\ - \int_t^T \nabla_\varphi Z^\psi(t, s)dW(s), \qquad t \in [0, T].\end{aligned}$$

By taking $\varphi(\cdot) \equiv \psi_i(\cdot)$, we then obtain gradient allocation at $\psi(\cdot)$ in the direction of $\psi_i(\cdot)$. In addition, one can define the analogue Aumann-Shapley allocation in the scenario,

$$\nabla^{AS}_{\psi_i}\rho(\cdot; \psi) \equiv \int_0^1 \nabla_{\psi_i}\rho(\cdot; \gamma\psi)d\gamma, \quad i = 1, \cdots, n.$$

Notice that

$$\begin{aligned}\sum_{i=1}^n \nabla_{\psi_i} Y^\psi(t) = -\sum_{i=1}^n \psi_i(t) + \int_t^T \Big(r(s)\sum_{i=1}^n \nabla_{\psi_i} Y^\psi(s) \\ + g_{0,z}(t, s, Z^\psi(t, s))\sum_{i=1}^n \nabla_{\psi_i} Z^\psi(t, s)\Big)ds \\ - \int_t^T \sum_{i=1}^n \nabla_{\psi_i} Z^\psi(t, s)dW(s), \qquad t \in [0, T].\end{aligned}$$

Recall that $\psi(\cdot) = \sum_{i=1}^n \psi_i(\cdot)$, by the uniqueness of BSVIEs, we have

$$\nabla_\psi\rho(t; \psi) = \sum_{i=1}^n \nabla_{\psi_i}\rho(t; \psi) = \sum_{i=1}^n \nabla_{\psi_i} Y^\psi(t). \tag{11.4.4}$$

If the risk measure $\psi \mapsto \rho(\cdot; \psi)$ is positively homogeneous (see, e.g., Definition 11.3.5 in the current chapter), then we claim that the sum of risk contributions for $\psi_i(\cdot)$ adds up to the risk of the portfolio $\psi(\cdot)$, i.e.,

$$\rho(\cdot; \psi) = \sum_{i=1}^n \nabla_{\psi_i}\rho(\cdot; \psi). \tag{11.4.5}$$

Such a result is named the *full allocation property* in the literature. From this result, we can take the above $k_i(\cdot) \equiv \nabla_{\psi_i}\rho(\cdot; \psi)$.

To explain this result, for any $\gamma \in (0, 1)$, let us first observe that

$$\begin{aligned}\nabla_\psi \rho(\cdot; \gamma\psi) &\equiv \lim_{h\to 0} \frac{\rho(\cdot; \gamma\psi + h\psi) - \rho(\cdot; \gamma\psi)}{h} \\ &= \lim_{h\to 0} \frac{\rho(\cdot; \psi + \frac{h}{\gamma}\psi) - \rho(\cdot; \psi)}{\frac{h}{\gamma}} = \nabla_\psi \rho(\cdot; \psi).\end{aligned} \tag{11.4.6}$$

In the first equality of the second line, we use the positively homogeneous property of $\psi \mapsto \rho(\cdot; \psi)$. On the other hand,

$$\begin{aligned}&\rho(\cdot; \psi) = \rho(\cdot; \psi) - \rho(\cdot; 0) = \int_0^1 \frac{d}{d\gamma}(\rho(\cdot; \gamma\psi))d\gamma \\ &= \int_0^1 \lim_{\varepsilon\to 0} \frac{\rho(\cdot; [\gamma+\varepsilon]\psi) - \rho(\cdot; \gamma\psi)}{\varepsilon} d\gamma \\ &= \int_0^1 \lim_{\varepsilon\to 0} \frac{\rho(\cdot; \gamma\psi + \varepsilon\psi) - \rho(\cdot; \gamma\psi)}{\varepsilon} d\gamma = \int_0^1 \nabla_\psi \rho(\cdot; \gamma\psi)d\gamma.\end{aligned} \tag{11.4.7}$$

By the above (11.4.4), (11.4.6), (11.4.7), we then obtain (11.4.5).

5 Rough Heston Model

In the previous sections, we directly applied the tool of BSVIEs to financial models. In this part, we discuss another financial model where BSVIEs can also be used potentially and indirectly.

Let us revisit with the geometric Brownian motion describing the asset price:

$$dS(t) = bS(t)dt + \sigma S(t)dW(t), \qquad t \geqslant 0. \tag{11.5.1}$$

Here $S(0) = s_0 \in \mathbb{R}$. In the above, b is called the appreciation rate of the asset, and σ is called the volatility of the asset. Such a model is oversimplified. Many real financial data show that both the drift term and diffusion term are nonlinear in $S(\cdot)$. Therefore, one could accept that asset price process $S(\cdot)$ follows general SDEs:

$$dS(t) = b(t, S(t))dt + \sigma(t, S(t))dW(t), \qquad t \geqslant 0. \tag{11.5.2}$$

This will make relevant problems fall into the classical stochastic optimal control area. Because such a model is too general, the results are very general and coarse. We hope to get more finer model so that some important feature of the asset price process could be captured.

There are three features of the asset process we should pay enough attention:

- *Leverage effect:* The negative correlation between price returns and volatility increments, i.e., when the prices of the asset is increasing, its volatility drops, while when it decreases, the volatility tends to become larger.
- *Time-varying:* The volatility should depend on the time t.
- *Fat tails:* As the random variable, the distribution of the price process at each time t is of fat tail.

To this end, let us look at the following so-called classical Heston model:

$$\begin{cases} dS(t) = bS(t)dt + \sigma\sqrt{V(t)}S(t)dW(t), \\ dV(t) = \kappa[\theta - V(t)]dt + \gamma\sqrt{V(t)}dB(t), \end{cases} \tag{11.5.3}$$

where $W(\cdot)$ and $B(\cdot)$ are Brownian motions satisfying

$$\langle dW(t), dB(t)\rangle = \rho dt,$$

and $b, \sigma, \kappa, \theta, \gamma > 0$ are constants. Such a model captures the abovementioned three features.

Note that in the classical Heston model, the volatility is a Brownian motion semimartingale. However, when one considers high frequency trading, it is found that the behavior of the volatility is much rougher than that driven by a Brownian motion. In fact, the dynamics of log-volatility is typically modeled by a fractional Brownian motion with Hurst parameter $H \approx 0.1$.

In order to allow for a rougher behavior of the volatility in a Heston model, intuitively, one could consider the following system, instead of (11.5.3).

$$\begin{cases} S(t) = S(0) + \int_0^t bS(s)ds + \int_0^t \sigma\sqrt{V(s)}S(s)dW(s), \\ V(t) = V(0) + \kappa\int_0^t (t-s)^{\alpha-1}[\theta - V(s)]ds \\ \qquad +\gamma\int_0^t (t-s)^{\alpha-1}\sqrt{V(s)}dB(s), \\ \langle dW(t), dB(t)\rangle = \rho dt, \end{cases} \tag{11.5.4}$$

with $\alpha, \rho \in (0, 1)$. This is an FSVIE having a singular kernel. In the above, each $W(\cdot)$ and $B(\cdot)$ is standard Brownian motion. Now, let

$$\widetilde{W}(t) = \frac{-\rho}{\sqrt{1-\rho^2}}W(t) + \frac{1}{\sqrt{1-\rho^2}}B(t), \qquad t \geqslant 0.$$

Then $(W(\cdot), \widetilde{W}(\cdot))$ is a two-dimensional standard Brownian motion, and the above can be equivalently written as

$$
\begin{cases}
S(t) = S(0) + \int_0^t bS(s)ds + \int_0^t \sigma\sqrt{V(s)}S(s)dW(s),\\
V(t) = V(0) + \kappa \int_0^t (t-s)^{\alpha-1}[\theta - V(s)]ds \\
\qquad +\gamma \int_0^t (t-s)^{\alpha-1}\sqrt{V(s)}\Big[\rho dW(s) + \sqrt{1-\rho^2}d\widetilde{W}(s)].
\end{cases}
\tag{11.5.5}
$$

From such a model, we can formulate various (optimal control) problems. Then BSVIEs are naturally involved.

To conclude this section, we consider a situation. Let the market consist of one bond and n stocks whose price processes $S_0(\cdot)$ and $S(\cdot) = (S_1(\cdot), \cdots, S_n(\cdot))$ follow the following system:

$$
\begin{cases}
S_0(t) = S_0(0) + \int_0^t r(s)S_0(s)ds,\\
S_k(t) = S_k(0) + \int_0^t b_k(s)S_k(s)ds + \int_0^t \sum_{\ell=1}^n \sigma_{k\ell}(s)\sqrt{V_k(s)}S_k(s)dW_\ell(s),\\
V_k(t) = V_k(0) + \int_0^t \sum_{\ell=1}^n \kappa_{k\ell}(t-s)^{\alpha_\ell-1}[\theta_{k\ell} - V_\ell(s)]ds\\
\qquad + \int_0^t \sum_{\ell=1}^n \gamma_{k\ell}\rho_{k\ell}(t-s)^{\alpha_\ell-1}\sqrt{V_\ell(s)}dW_\ell(s)\\
\qquad + \int_0^t \sum_{\ell=1}^n \gamma_{k\ell}\sqrt{1-\rho_{k\ell}^2}(t-s)^{\alpha_\ell-1}\sqrt{V_\ell(s)}d\widetilde{W}_\ell(s),\\
\qquad\qquad k = 1, 2, \cdots, n.
\end{cases}
\tag{11.5.6}
$$

Here, we assume $(W_1(\cdot), \cdots, W_n(\cdot), \widetilde{W}_1(\cdot), \cdots, \widetilde{W}_n(\cdot))$ to be a $2n$-dimensional standard Brownian motion. Under self-financing condition, the total wealth process, denoted by $Y(\cdot)$, satisfies

$$
\begin{cases}
Y(t) = Y(0) + \int_0^t \Big[r(s)Y(s) + \langle b - r\mathbb{1}, u(s)\rangle\Big]ds\\
\qquad + \int_0^t \langle u(s), \sigma(s, V(s))dW(s)\rangle + \int_0^t \langle u(s), \widetilde{\sigma}(s, V(s))d\widetilde{W}(s)\rangle,\\
V_k(t) = V_k(0) + \int_0^t \sum_{\ell=1}^n \kappa_{k\ell}(t-s)^{\alpha_\ell-1}[\theta_{k\ell} - V_\ell(s)]ds\\
\qquad + \int_0^t \sum_{\ell=1}^n \gamma_{k\ell}\rho_{k\ell}(t-s)^{\alpha_\ell-1}\sqrt{V_\ell(s)}dW_\ell(s)\\
\qquad + \int_0^t \sum_{\ell=1}^n \gamma_{k\ell}\sqrt{1-\rho_{k\ell}^2}(t-s)^{\alpha_\ell-1}\sqrt{V_\ell(s)}d\widetilde{W}_\ell(s),\\
\qquad\qquad k = 1, 2, \cdots, n.
\end{cases}
\tag{11.5.7}
$$

Here

$$\sigma(t, V(t)) = \begin{pmatrix} \sigma_{11}(t)\sqrt{V_1(t)} & \sigma_{12}(t)\sqrt{V_1(t)} & \cdots & \sigma_{1n}(t)\sqrt{V_1(t)} \\ \sigma_{21}(t)\sqrt{V_2(t)} & \sigma_{22}(t)\sqrt{V_2(t)} & \cdots & \sigma_{n2}(t)\sqrt{V_2(t)} \\ \vdots & \vdots & \ddots & \vdots \\ \sigma_{n1}(t)\sqrt{V_n(t)} & \sigma_{n2}(t)\sqrt{V_n(t)} & \cdots & \sigma_{nn}(t)\sqrt{V_n(t)} \end{pmatrix}$$

and

$$\widetilde{\sigma}(t, V(t)) = \begin{pmatrix} \widetilde{\sigma}_{11}(t)\sqrt{V_1(t)} & \widetilde{\sigma}_{12}(t)\sqrt{V_1(t)} & \cdots & \widetilde{\sigma}_{1n}(t)\sqrt{V_1(t)} \\ \widetilde{\sigma}_{21}(t)\sqrt{V_2(t)} & \widetilde{\sigma}_{22}(t)\sqrt{V_2(t)} & \cdots & \widetilde{\sigma}_{n2}(t)\sqrt{V_2(t)} \\ \vdots & \vdots & \ddots & \vdots \\ \widetilde{\sigma}_{n1}(t)\sqrt{V_n(t)} & \widetilde{\sigma}_{n2}(t)\sqrt{V_n(t)} & \cdots & \widetilde{\sigma}_{nn}(t)\sqrt{V_n(t)} \end{pmatrix}.$$

Thus, (11.5.7) is a controlled FSVIE. We then can formulate proper problems where BSVIEs can be applied.

6 Bibliographical Remarks

This chapter has a brief presentation of applications of BSVIEs in mathematical finance. Section 1 is a little modification or extension of classical results since middle of the 1990s, from Duffie [62, Chapter 10], El Kouri–Quenez [69, Section 3], Karatzas–Shreve [121, Chapter 2], and Beissner-Rosazza Gianin [19], in terms of BSVIEs.

Let us look at Sect. 2. In 1992, Duffie–Epstein [63] introduced stochastic differential utility (see also Duffie–Lions [64]). It was realized by El Karoui–Peng–Quenez [68] that stochastic differential utility is just the $Y(\cdot)$ part of an adapted solution of a BSDE. Then, the notion was extended to general BSDEs by Lazrak–Quenez [134] in 2003 and by Lazrak [133] in 2004. Note that these abovementioned works can only treat the case of European type continent claim (as the terminal state of the BSDE), but cannot treat the case of "time-dependent terminal state." The notion of equilibrium recursive utility was introduced by Wang–Sun–Yong [192]. By using BSVIEs, the case of "time-dependent terminal state" (which is just the free term in BSVIEs) can be successfully treated. The word "equilibrium" was borrowed from the study of time-inconsistent problems (see Yong [221], Wang–Yong [193]). The Case II and Case III in Sect. 2.2.1 are inspired by [193] and [19]. The contents in Sect. 2.2.2 seem to be new and extend the counterparts in [133, 134] to the BSVIE setting.

Static coherent risk measure was introduced by Artzner–Delbaen–Eber–Heath [13] in 1999 and was extended to convex risk measures by Föllmer–Schied [75] in 2002. In 2004, Riedel [171] studied discrete time dynamic coherent risk measures and was extended by Detlefsen–Scandolo [58] to discrete dynamic convex risk measures in 2005. Here, dynamic means the "terminal state" could be time-varying. We realized that, for continuous time case, this was perfectly fitted in the framework of BSVIE. So we have the main part of Sect. 3. Again, the word "equilibrium" was borrowed from the study of time-inconsistent problems. The results in Sect. 3 were modification of contents in Yong [219], Wang–Yong [204], and Liu–Wang [139]. Relevant papers on dynamic risk measure and BSVIEs can be found in, e.g., Agram [3] and Miao–Chen–Xiao–Hu [149].

The main results in Sect. 4 are some brief modifications/improvements of Kromer–Overbeck [127]. It essentially says that BSVIE is Gâteux differentiable with respect to the free term. Recently, Rosazza Gianin–Zullino had some further works along this line [172].

As to fractional stochastic volatility model with the Hurst parameter $H > 1/2$, we refer to the work by Comte–Renault in 1998 [51]. The rough volatility model was proposed by Gatheral–Jaisson–Rosenbaum in 2018 [85], for the case the Hurst parameter $H < 1/2$. The model presented in Sect. 5 is a modification of that in the work of El Euch–Rosenbaum [67]. See also El Euch–Rosenbaum [66] and El Euch–Fukasawa-Rosenbaum [65]. Recently, Chen–Shue–Li–Yong built a modal suitable for high-frequency trading with self-exciting volatility [44]. It should be natural to obtain optimal control problems for FSVIEs. We leave the details to the interested readers.

Chapter 12
Appendices

1 Malliavin Calculus

In this section, we briefly recall some relevant results of Malliavin calculus. It requires some knowledge of functional analysis/operator theory. The first time readers could skip this section and may also skip the later results involving Malliavin calculus. By doing that, the development of the main theory will not be affected much.

Let Ξ be the set of all (scalar) $\mathcal{F}_T$-measurable random variables ξ of form

$$\xi = f\Big(\int_0^T h(s)dW(s)\Big), \tag{12.1.1}$$

where $f \in C_b^1(\mathbb{R}^k)$ (the set of all scalar bounded continuously differentiable functions with bounded gradient) for some $k \geqslant 1$, and

$$h(\cdot) \equiv \big(h_1(\cdot), \cdots, h_d(\cdot)\big), \qquad h_i(\cdot) \in L^\infty(0, T; \mathbb{R}^k), \quad 1 \leqslant i \leqslant d.$$

Since $h(\cdot)$ is $\mathbb{R}^{k\times d}$-valued bounded and deterministic, the Itô's integral involved in (12.1.1) is well-defined, $\mathbb{R}^k$-valued random variable. One can show that Ξ is dense in $L^p_{\mathcal{F}_T}(\Omega; \mathbb{R})$ for any $p \in [1, \infty)$. For any $\xi \in \Xi$ of form (12.1.1), we define

$$D_\tau^{(i)}\xi = \langle f_x\Big(\int_0^T h(s)dW(s)\Big), h_i(\tau)\rangle, \qquad \text{a.e. } \tau \in [0, T],\ 1 \leqslant i \leqslant d, \ \text{ a.s.} \tag{12.1.2}$$

In particular, for $f(x) = x \in \mathbb{R}^k$, with $k = 1$,

$$D_\tau^{(i)}\xi \equiv D_\tau^{(i)}\int_0^T h(s)dW(s) = h_i(\tau), \qquad \text{a.e. } \tau \in [0, T],\ 1 \leqslant i \leqslant d.$$

T. Wang, J. Yong, *Backward Stochastic Volterra Integral Equations*,
Probability Theory and Stochastic Modelling 109,
https://doi.org/10.1007/978-3-032-20848-4_12

Thus, by taking $h(s) = h_j(s)e_j$, with $h_j(s) \in \mathbb{R}$ and $e_j \in \mathbb{R}^d$ being the vector whose j-th entry is 1 and all others are zero, we have

$$D_\tau^{(i)}\xi \equiv D_\tau^{(i)}\int_0^T h_j(s)dW_j(s) = \delta_{ij}h_i(\tau), \qquad \text{a.e. } \tau \in [0,T],\ 1 \leqslant i,j \leqslant d,$$

with δ_{ij} being the Kronecker delta. We call $D_\tau^{(i)}\xi$ the *Malliavin derivative* of (such a) ξ with respect to $W_i(\cdot)$. Denote

$$D_\tau\xi = \big(D_\tau^{(1)}\xi, D_\tau^{(2)}\xi, \cdots, D_\tau^{(d)}\xi\big), \qquad \forall \xi \in \Xi.$$

Clearly, for any $\xi \in \Xi$ of form (12.1.1), $D\xi$ is $\mathbb{R}^{1\times d}$-valued $\mathcal{F}_T$-measurable (not necessarily $\mathcal{F}_\tau$-measurable) random process, and

$$|D_\tau^{(i)}\xi| \leqslant \|f_x(\cdot)\|_\infty |h_i(\tau)|, \qquad \text{a.e. } \tau \in [0,T],\ 1 \leqslant i \leqslant d, \ \text{a.s.},$$

where $\|f_x(\cdot)\|_\infty$ is the L^∞-norm of $f_x(\cdot)$. Hence, for any $\xi \in \Xi$, we may define

$$\begin{aligned} \|\xi\|_{\mathbb{D}_{1,p}} &\triangleq \Big[\mathbb{E}|\xi|^p + \mathbb{E}\Big(\int_0^T |D_\tau\xi|^2 d\tau\Big)^{\frac{p}{2}}\Big]^{\frac{1}{p}} \\ &\equiv \Big(\|\xi\|^p_{L^p_{\mathcal{F}_T}(\Omega;\mathbb{R})} + \|D.\xi\|^p_{L^p_{\mathcal{F}_T}(\Omega;L^2(0,T;\mathbb{R}^d))}\Big)^{\frac{1}{p}} \\ &\leqslant K\Big(1 + \|h(\cdot)\|_{L^2(0,T;\mathbb{R}^{k\times d})}\Big) < \infty, \qquad 1 \leqslant p < \infty. \end{aligned} \tag{12.1.3}$$

In the above, the constant $K > 0$ depends on the bounds of $f(\cdot)$ and $f_x(\cdot)$. Clearly, $\|\cdot\|_{\mathbb{D}_{1,p}}$ is a norm on Ξ. Let $\mathbb{D}_{1,p}$ be the completion of Ξ under the norm $\|\cdot\|_{\mathbb{D}_{1,p}}$, which is a Banach space and is densely imbedded in $L^p_{\mathcal{F}_T}(\Omega;\mathbb{R})$. Further, it is known that operator $D \equiv (D^{(1)}, D^{(2)}, \cdots, D^{(d)})$ admits a closed extension, still denoted by itself, with the domain $\mathscr{D}(D) = \mathbb{D}_{1,p}$. Namely, if $\xi_k \in \mathbb{D}_{1,p}$ with

$$\lim_{k\to\infty}\|\xi_k - \xi\|_{\mathbb{D}_{1,p}} = 0, \qquad \lim_{k\to\infty}\|D.\xi_k - \eta(\cdot)\|_{L^p_{\mathcal{F}_T}(\Omega;L^2(0,T;\mathbb{R}^{1\times d}))} = 0,$$

for some $\xi \in L^p_{\mathcal{F}_T}(\Omega;\mathbb{R})$ and $\eta(\cdot) \in L^p_{\mathcal{F}_T}(\Omega;L^2(0,T;\mathbb{R}^{1\times d}))$, then $D.\xi = \eta(\cdot)$. Thus, $D : \mathbb{D}_{1,p} \subseteq L^p_{\mathcal{F}_T}(\Omega;\mathbb{R}) \to L^p_{\mathcal{F}_T}(\Omega;L^2(0,T;\mathbb{R}^{1\times d}))$ is a linear (unbounded) closed operator. We let $\mathbb{D}^m_{1,p}$ and $\mathbb{D}^{m\times n}_{1,p}$ be the set of all $\mathbb{R}^m$ and $\mathbb{R}^{m\times n}$-valued random variables with each component belonging to $\mathbb{D}_{1,p}$. Thus, if $\xi \in \mathbb{D}^m_{1,p}$, then $\xi \in L^p_{\mathcal{F}_T}(\Omega;\mathbb{R}^m)$ and $D.\xi \in L^p_{\mathcal{F}_T}(\Omega;L^2(0,T;\mathbb{R}^{m\times d}))$; and if $\xi \in \mathbb{D}^{m\times n}_{1,p}$, then for each $i = 1,2,\cdots,d$, $D^{(i)}_{\cdot}\xi \in L^p_{\mathcal{F}_T}(\Omega;L^2(0,T;\mathbb{R}^{m\times n}))$.

Now, let $\mathbb{W}^m_{1,2}(0,T)$ be the set of all $\mathbb{R}^m$-valued random variables ξ which is Malliavin differentiable such that

$$\|\xi\|_{\mathbb{W}^m_{1,2}(0,T)} \equiv |\xi| + \Big(\int_0^T \sum_{i=1}^d |D^{(i)}_\tau \xi|^2 d\tau\Big)^{\frac{1}{2}} < \infty, \qquad \text{a.s.} \tag{12.1.4}$$

The right-hand side of the above is a scalar valued nonnegative random variable. Naturally, we let $L^p(\Omega; \mathbb{W}^m_{1,2}(0,T))$ be the set of all ξ such that

$$\mathbb{E}\|\xi\|^p_{\mathbb{W}^m_{1,2}(0,T)} < \infty.$$

Then the following holds:

$$\begin{aligned}
\Big[\mathbb{E}\|\xi\|^p_{\mathbb{W}^m_{1,2}(0,T)}\Big]^{\frac{1}{p}} &= \Big\{\mathbb{E}\Big[|\xi| + \Big(\int_0^T \sum_{i=1}^d |D^{(i)}_\tau \xi|^2 d\tau\Big)^{\frac{1}{2}}\Big]^p\Big\}^{\frac{1}{p}} \\
&\leqslant K\Big[\mathbb{E}|\xi|^p + \mathbb{E}\Big(\int_0^T \sum_{i=1}^d |D^{(i)}_\tau \xi|^2 d\tau\Big)^{\frac{p}{2}}\Big]^{\frac{1}{p}} \equiv K\|\xi\|_{\mathbb{D}^m_{1,p}} \\
&\leqslant K\Big\{\mathbb{E}\Big[|\xi| + \Big(\int_0^T \sum_{i=1}^d |D^{(i)}_\tau \xi|^2 d\tau\Big)^{\frac{1}{2}}\Big]^p\Big\}^{\frac{1}{p}} = K\Big[\mathbb{E}\|\xi\|^p_{\mathbb{W}^m_{1,2}(0,T)}\Big]^{\frac{1}{p}}.
\end{aligned} \tag{12.1.5}$$

Thus, $\|\cdot\|_{\mathbb{D}^m_{1,p}}$ is equivalent to $\big[\mathbb{E}\|\cdot\|^p_{\mathbb{W}^m_{1,2}(0,T)}\big]^{\frac{1}{p}}$ and

$$\mathbb{D}^m_{1,p} = L^p(\Omega; \mathbb{W}^m_{1,2}(0,T)). \tag{12.1.6}$$

Such an equivalence will be useful later.

The following collects some basic properties of the Malliavin differential operator D.

Proposition 12.1.1 *Let $p \in (1,\infty)$.*

(i) *Let $A \in \mathcal{B}[0,T]$ and*

$$\mathcal{F}_A = \sigma\Big\{W(B) \equiv \int_B dW(s) \mid B \in \mathcal{B}[0,T],\ B \subseteq A\Big\}.$$

Then, for any $\xi \in \mathbb{D}_{1,p}$, $\mathbb{E}[\xi \mid \mathcal{F}_A] \in \mathbb{D}_{1,p}$ and

$$D_\tau\big(\mathbb{E}[\xi \mid \mathcal{F}_A]\big) = \mathbb{E}[D_\tau \xi \mid \mathcal{F}_A]\mathbf{1}_A(\tau), \qquad \text{a.e. } \tau \in A,\ \text{a.s.} \tag{12.1.7}$$

In particular, for any $t \in [0,T]$, $\Phi_t \equiv \mathcal{F}_{[0,t]}$,

$$D_\tau\big(\mathbb{E}[\xi \mid \mathcal{F}_t]\big) = \mathbb{E}[D_\tau \xi \mid \mathcal{F}_t]\mathbf{1}_{[0,t]}(\tau), \qquad \text{a.e. } \tau \in [0,t],\ \text{a.s.} \tag{12.1.8}$$

Consequently, for any $\xi \in \mathbb{D}^m_{1,p} \subseteq L^p_{\mathcal{F}_T}(\Omega;\mathbb{R}^m)$,

$$\xi \text{ is } \mathcal{F}_t\text{-measurable} \quad \Rightarrow \quad D_\tau \xi = 0, \quad \forall \tau \in (t,T], \tag{12.1.9}$$

(ii) *If* $\xi \in \mathbb{D}_{1,1}$ *such that* $D\xi = 0$*, then* $\xi = \mathbb{E}\xi$*, i.e.,* ξ *must be deterministic.*

(iii) *Let* $A \in \mathcal{F}$*. Then* $I_A \in \mathbb{D}_{1,1}$ *if and only if* $\mathbb{P}(A) = 0, 1$*. Consequently, by choosing* $A \in \mathcal{F}$ *with* $\mathbb{P}(A) \neq 0, 1$*, it is seen that*

$$\mathbb{D}_{1,p} \neq L^p_{\mathcal{F}_T}(\Omega;\mathbb{R}). \tag{12.1.10}$$

(iv) ***(Clark-Ocone's formula)*** *For any* $\xi \in \mathbb{D}^m_{1,p}$*, it holds*

$$\xi = \mathbb{E}\xi + \int_0^T \big[\mathbb{E}_t(D_t\xi)\big]dW(t). \tag{12.1.11}$$

(v) *For any* $\xi \in \mathbb{D}^m_{1,p}$*,* $\eta \in \mathbb{D}_{1,p}$*, and* $h(\cdot) \in L^2(0,T;\mathbb{R}^{m\times d})$*,*

$$\mathbb{E}\Big[\eta\,\xi^\top\int_0^T h(t)dW(t)\Big] = \mathbb{E}\Big[\eta\int_0^T tr\big[(D_t\xi)h(t)\big]dt + \xi^\top\int_0^T h(t)(D_t\eta)^\top dt\Big].$$

In particular, by taking $\eta = 1$*, it holds*

$$\mathbb{E}\Big[\xi^\top\int_0^T h(t)dW(t)\Big] = \mathbb{E}\int_0^T tr\Big[(D_t\xi)h(t)\Big]dt. \tag{12.1.12}$$

(vi) *If* $z(\cdot) \in L^1_{\mathcal{F}_T}(0,T;\mathbb{R}^m)$ *such that for almost all* $t \in [0,T]$*,* $z(t)$ *is Malliavin differentiable and for almost every* $\tau \in [0,T]$*,* $D_\tau z(\cdot) \in L^1(0.T;\mathbb{R}^{m\times d})$*, then*

$$D_\tau\int_0^T z(t)dt = \int_0^T D_\tau z(t)dt, \qquad \text{a.e. } \tau \in [0,T]. \tag{12.1.13}$$

There are some remarks in order.

- In general, if ξ is merely $\mathcal{F}_T$-measurable, for each $\tau \in [0,T)$, $D_\tau\xi$ is (still just) $\mathcal{F}_T$-measurable, not necessarily $\mathcal{F}_\tau$-measurable. Consequently, if $z(\cdot)$ is $\mathbb{F}$-adapted and for each $t \in [0,T]$, $z(t)$ is Malliavin differentiable, then for each $\tau \in [0,T]$, $t \mapsto D_\tau z(t)$ is $\mathbb{F}$-adapted.
- By martingale representation theorem (see Theorem 2.3.8 in Chap. 2), for any $\xi \in L^p_{\mathcal{F}_T}(\Omega;\mathbb{R}^m)$, there exists a unique $\zeta(\cdot) \in L^p_{\mathbb{F}}(\Omega;L^2(0,T;\mathbb{R}^{m\times d}))$ such that

$$\xi = \mathbb{E}\xi + \int_0^T \zeta(t)dW(t). \tag{12.1.14}$$

Comparing (12.1.11) and (12.1.14), we see that in the case $\xi \in \mathbb{D}^m_{1,p}$, one has

$$\zeta(t) = \mathbb{E}_t[D_t\xi], \qquad \text{a.e. } t \in [0, T], \;\; \text{a.s.}$$

However, having representation (12.1.14) does not mean that $\xi \in \mathbb{D}^m_{1,p}$ which is a proper subspace of $L^p_{\mathcal{F}_T}(\Omega; \mathbb{R}^m)$. We also point out that since $t \mapsto \mathbb{E}_t[D_t\xi]$ has a version that is $\mathbb{F}$-progressively measurable, as long as

$$\mathbb{E}\int_0^T |\mathbb{E}_t[D_t\xi]|^2 dt < \infty,$$

the Itô's integral in (12.1.11) is well-defined.

- Relation (12.1.12) is a consequence of Clark-Ocone's formula and a property of Itô's integral.
- (12.1.13) follows from the closeness of D.

In what follows, for any k-dimensional random vector η and any random field f defined on $[0, T] \times \mathbb{R}^k \times \Omega$, we will distinguish $D_\tau[f(t, \eta(\omega), \omega)]$ from

$$D_\tau f(t, x, \omega)\Big|_{x=\eta(\omega)} \triangleq [D_\tau f](t, \eta(\omega), \omega).$$

Recall that Itô's integral is defined for $\mathbb{F}$-progressively measurable processes. We now would like to extend Itô's integral so that for some non $\mathbb{F}$-progressively measurable processes, one can also define its stochastic integral. This will be useful later.

Let $p \in (1, \infty)$. Since $D : \mathscr{D}(D) = \mathbb{D}_{1,p} \subseteq L^p_{\mathcal{F}_T}(\Omega; \mathbb{R}) \to L^p_{\mathcal{F}_T}(\Omega; L^2(0, T; \mathbb{R}^{1\times d}))$ is a densely defined closed linear operator, it admits the adjoint operator $D^* : \mathscr{D}(D^*) \subseteq L^{p'}_{\mathcal{F}_T}(\Omega; L^2(0, T; \mathbb{R}^{1\times d})) \to L^{p'}_{\mathcal{F}_T}(\Omega; \mathbb{R})$, with $p' = \frac{p}{p-1}$. Here, $\mathscr{D}(D^*)$ is the set of all $\zeta(\cdot) \in L^{p'}_{\mathcal{F}_T}(\Omega; L^2(0, T; \mathbb{R}^{1\times d}))$ such that

$$\Big|\mathbb{E}\int_0^T \zeta(t)(D_t\xi)^\top dt\Big| \leqslant C\big[\mathbb{E}|\xi|^p\big]^{\frac{1}{p}}, \qquad \forall \xi \in \mathbb{D}_{1,p}\,.$$

Now, for any $\zeta(\cdot) \in \mathscr{D}(D^*)$, we define $D^*\zeta(\cdot) \in L^{p'}_{\mathcal{F}_T}(\Omega; \mathbb{R})$ by the following:

$$\mathbb{E}\big\{[D^*\zeta(\cdot)]\xi\big\} = \mathbb{E}\int_0^T \zeta(t)(D_t\xi)^\top dt, \qquad \forall \xi \in \mathbb{D}_{1,p}, \tag{12.1.15}$$

and call it the *Skorokhod integral* of the process $\zeta(\cdot)$ with respect to $W(\cdot)$. Note that, in general, $\zeta(\cdot) \in \mathscr{D}(D^*) \subseteq L^{p'}_{\mathcal{F}_T}(\Omega; L^2(0, T; \mathbb{R}^{1\times d}))$ is not necessarily $\mathbb{F}$-adapted. Consequently, the Itô's integral $\int_0^T \zeta(t)dW(t)$ might not be defined. In the Skorokhod integral, the integrand is allowed to be not non-anticipating. Skorokhod integral is also referred to as an *anticipating integral*. For convenience, for any $\zeta(\cdot) \in \mathscr{D}(D^*)$, we denote[1]

$$D^*\zeta(\cdot) = \int_0^T \zeta(t)dW(t).$$

With this notation, (12.1.15) can be rewritten as

$$\mathbb{E}\Big[\Big(\int_0^T \zeta(t)dW(t)\Big)\xi\Big] = \mathbb{E}\int_0^T \zeta(t)(D_t\xi)^\top dt, \qquad \forall \xi \in \mathbb{D}_{1,p}. \tag{12.1.16}$$

By taking $\xi \equiv 1$ in the above, we see that

$$\mathbb{E}\int_0^T \zeta(t)dW(t) = 0, \tag{12.1.17}$$

which is a same property that Itô's integral has. Note that in the above definition of Skorokhod integral, $p > 1$ is involved, for example, $\mathscr{D}(D^*)$ is actually depending on p. But, such a dependence can easily be identified from the contexts, and we do not bother to indicate that below. On the other hand, we let $\mathscr{D}(D^*)^m$ be the set of all $z : [0, T] \times \Omega \to \mathbb{R}^{m\times d}$ such that each row is in $\mathscr{D}(D^*)$, and therefore, $\int_0^T \zeta(s)dW(s)$ is an $\mathbb{R}^m$-valued random variable. Then it is natural to let $L^p_{\mathcal{F}_T}(\mathscr{D}(D^*)^m)$ be the set of all $\zeta(\cdot)$ such that the Skorokhod integral exists and

$$\mathbb{E}\Big|\int_0^T \zeta(s)dW(s)\Big|^p < \infty. \tag{12.1.18}$$

Thus, $L^p_{\mathcal{F}_T}(\mathscr{D}(D^*))$ is a Banach space. Further, it is known that any $\zeta(\cdot) \in L^p_{\mathbb{F}}(\Omega; L^2(0, T; \mathbb{R}^{m\times d}))$ admits the Skorokhod integral which coincides with the Itô's integral of $\zeta(\cdot)$. By Burkholder-Davis-Gundy's inequalities, we have that

$$L^p_{\mathbb{F}}(\Omega; L^2(0, T; \mathbb{R}^m)) \subseteq L^p_{\mathcal{F}_T}(\mathscr{D}(D^*)^m). \tag{12.1.19}$$

Next, for $p, q \in [1, \infty)$, we define $L^q_{\mathcal{F}_T}(0, T; \mathbb{D}^{m\times d}_{1,p})$ to be the set of all $\mathcal{F}_T$-measurable processes $z \equiv (z_1, \cdots, z_d) : [0, T] \times \Omega \to \mathbb{R}^{m\times d}$ such that for almost all $t \in [0, T]$, $z(t) \in \mathbb{D}^{m\times d}_{1,p}$, and

$$\begin{aligned}\|z(\cdot)\|^q_{L^q_{\mathcal{F}_T}(0,T;\mathbb{D}^{m\times d}_{1,p})} &\equiv \int_0^T \|z(t)\|^q_{\mathbb{D}^{m\times d}_{1,p}}dt \equiv \sum_{j=1}^d \int_0^T \|z_j(t)\|^q_{\mathbb{D}^m_{1,p}} dt \\ &\equiv \sum_{j=1}^d \int_0^T \Big[\mathbb{E}|z_j(t)|^p + \mathbb{E}\Big(\int_0^T \sum_{i=1}^d |D^{(i)}_\tau z_j(t)|^2 d\tau\Big)^{\frac{p}{2}}\Big]^{\frac{q}{p}} dt < \infty.\end{aligned} \tag{12.1.20}$$

[1] In some literature, people use $\delta(\zeta(\cdot))$ instead of $D^*\zeta(\cdot)$.

By (12.1.6), we see that the norm defined through the above is equivalent to that through the following:

$$\|z(\cdot)\|^q_{L^q(0,T;L^p(\Omega;\mathbb{W}^m_{1,2}(0,T)))^d} = \sum_{j=1}^d \int_0^T \Big(\mathbb{E}\|z_j(t)\|^p_{\mathbb{W}^m_{1,2}(0,T)}\Big)^{\frac{q}{p}} dt. \tag{12.1.21}$$

Thus,

$$L^q_{\mathcal{F}_T}(0,T;\mathbb{D}^{m\times d}_{1,p}) = L^q_{\mathcal{F}_T}(0,T;L^p(\Omega;\mathbb{W}^m_{1,2}(0,T)))^d. \tag{12.1.22}$$

We can define the similar spaces with additional $\mathbb{F}$-progressive measurability requirement. More precisely, let $L^q_{\mathbb{F}}(0,T;\mathbb{D}^{m\times d}_{1,p})$ be the subset of all $z(\cdot) \in L^q_{\mathcal{F}_T}(0,T;\mathbb{D}^{m\times d}_{1,p})$ such that for almost all $\tau \in [0,T]$, $(t,\omega) \mapsto D_\tau z(t,\omega)$ (with the image in $\mathbb{R}^{m\times d}$) admits an $\mathbb{F}$-progressively measurable version (still denote it by itself). Similar to (12.1.22), we have

$$L^q_{\mathbb{F}}(0,T;\mathbb{D}^{m\times d}_{1,p}) = L^q_{\mathbb{F}}(0,T;L^p(\Omega;\mathbb{W}^m_{1,2}(0,T)))^d. \tag{12.1.23}$$

The spaces $L^\infty_{\mathbb{F}}(0,T;\mathbb{D}^m_{1,p})$, $C_{\mathbb{F}}([0,T];\mathbb{D}^{m\times d}_{1,p})$ and $C^\#_{\mathbb{F}}([0,T];\mathbb{D}^{m\times d}_{1,p})$ can be defined similarly. In the above, we may replace $[0,T]$ by any $[S,T] \subseteq [0,T]$, with $\mathbb{F}$ replaced by $\mathbb{F}^{[S,T]}$ (see (2.2.3) in Chapter 2 for the definition), to define corresponding spaces.

Now, let $\mathbb{L}^{m\times d}_{1,p}$ be the set of $\mathbb{R}^{m\times d}$-valued $\mathbb{F}$-progressively measurable processes $z(\cdot)$ such that

(i) For almost all $t \in [0,T]$, $\omega \mapsto z(t,\omega)$ is Malliavin differentiable and

$$\int_0^T |D^{(i)}_\tau z(t)|^2 d\tau = \int_0^t |D^{(i)}_\tau z(t)|^2 d\tau < \infty, \qquad \text{a.s., } 1 \leqslant i \leqslant d;$$

(ii) For almost all $\tau \in [0,T]$, $(t,\omega) \mapsto D^{(i)}_\tau z(t,\omega)$ admits an $\mathbb{F}$-progressively measurable version, $1 \leqslant i \leqslant d$;

(iii) The following holds:

$$\begin{aligned}\|z(\cdot)\|^p_{\mathbb{L}^{m\times d}_{1,p}} = \sum_{j=1}^d \mathbb{E}\Big[\Big(\int_0^T |z_j(t)|^2 dt\Big)^{\frac{p}{2}} \\ +\Big(\int_0^T\int_0^T \sum_{i=1}^d |D^{(i)}_\tau z_j(t)|^2 d\tau dt\Big)^{\frac{p}{2}}\Big] < \infty.\end{aligned} \tag{12.1.24}$$

Note that the norm defined through (12.1.20) (with $q = p$) is not equivalent to that defined through the above. This can also be seen from the following simple fact. For any $z(\cdot) \in \mathbb{L}_{1,p}^{m\times d}$, the following Itô's integrals exist:

$$\int_0^T z(t)dW(t), \qquad \int_0^T D_\tau^{(i)} z(t)dW(t), \qquad \text{a.e. } \tau \in [0,T],\ 1 \leqslant i \leqslant d. \tag{12.1.25}$$

Whereas, if $p \in [1,2)$ and $z(\cdot) \in L_{\mathbb{F}}^p(0,T;\mathbb{D}_{1,p}^{m\times d})$, the first Itô's integral might not exist.

Next, we would like to find relation between $\mathbb{L}_{1,p}^{m\times d}$ and $L_{\mathbb{F}}^p(0,T;\mathbb{D}_{1,p}^{m\times d})$. We first have (see (12.1.5))

$$\begin{aligned}
\|z(\cdot)\|_{\mathbb{L}_{1,p}^{m\times d}}^p &\equiv \sum_{j=1}^d \mathbb{E}\Big[\Big(\int_0^T |z_j(t)|^2 dt\Big)^{\frac{p}{2}} + \Big(\int_0^T\int_0^T \sum_{i=1}^d |D_\tau^{(i)} z_j(t)|^2 d\tau dt\Big)^{\frac{p}{2}}\Big]\\
&\leqslant K\mathbb{E}\Big[\int_0^T \Big(|z(t)|^2 + \int_0^T \sum_{i=1}^d |D_\tau^{(i)} z(t)|^2 d\tau\Big) dt\Big]^{\frac{p}{2}}\\
&\leqslant K\sum_{j=1}^d \mathbb{E}\Big\{\int_0^T \Big[|z_j(t)| + \Big(\int_0^T \sum_{i=1}^d |D_\tau^{(i)} z_j(t)|^2 d\tau\Big)^{\frac12}\Big]^2 dt\Big\}^{\frac{p}{2}}\\
&= K\sum_{j=1}^d \mathbb{E}\Big[\int_0^T \|z_j(t)\|_{\mathbb{W}_{1,2}^m(0,T)}^2 dt\Big]^{\frac{p}{2}} \equiv K\|z(\cdot)\|_{L^p(\Omega;L^2(0,T;\mathbb{W}_{1,2}^m(0,T)))^d},
\end{aligned}$$

and

$$\begin{aligned}
\|z(\cdot)\|_{L^p(\Omega;L^2(0,T;\mathbb{W}_{1,2}^m(0,T)))^d}^p &\equiv \sum_{j=1}^d \mathbb{E}\Big[\int_0^T \|z_j(t)\|_{\mathbb{W}_{1,2}^m(0,T)}^2\Big]^{\frac{p}{2}}\\
&= \sum_{j=1}^d \mathbb{E}\Big\{\int_0^T \Big[|z_j(t)| + \Big(\int_0^T \sum_{i=1}^d |D_\tau^{(i)} z_j(t)|^2 d\tau\Big)^{\frac12}\Big]^2 dt\Big\}^{\frac{p}{2}}\\
&\leqslant K\sum_{j=1}^d \mathbb{E}\Big\{\int_0^T \Big[|z_j(t)|^2 + \Big(\int_0^T \sum_{i=1}^d |D_\tau^{(i)} z_j(t)|^2 d\tau\Big)\Big] dt\Big\}^{\frac{p}{2}} \equiv K\|z(\cdot)\|_{\mathbb{L}_{1,p}^{m\times d}}^p.
\end{aligned}$$

This means that the following holds

$$\mathbb{L}_{1,p}^{m\times d} = L_{\mathbb{F}}^p(\Omega;L^2(0,T;\mathbb{W}_{1,2}^m(0,T)))^d, \tag{12.1.26}$$

and their norms are equivalent. Notationally, the left-hand side is simpler. However, the right-hand side contains accurate information of the space. Recall (12.1.23) with $q = p \in [1,\infty)$, one has

$$L_{\mathbb{F}}^p(0,T;\mathbb{D}_{1,p}^{m\times d}) = L_{\mathbb{F}}^p(0,T;L^p(\Omega;\mathbb{W}_{1,2}^m(0,T)))^d. \tag{12.1.27}$$

Therefore, for $p = 2$,

$$\mathbb{L}_{1,2}^{m\times d} = L_{\mathbb{F}}^2(0,T;L^2(\Omega;\mathbb{W}_{1,2}^m(0,T)))^d = L_{\mathbb{F}}^2(0,T;\mathbb{D}_{1,2}^{m\times d}).$$

Now, we can make a comparison in the spirit of Lemma 2.4.2 of Chap. 2. For $p \leqslant 2$,

$$\begin{aligned}&\sum_{j=1}^d \|z_j(\cdot)\|^p_{L^p_{\mathbb{F}}(0,T;\mathbb{D}^m_{1,p})} \equiv \sum_{j=1}^d \mathbb{E}\Big\{\int_0^T |z_j(t)|^p dt + \int_0^T \Big(\int_0^T |D_\tau z_j(t)|^2 d\tau\Big)^{\frac{p}{2}} dt\Big\}\\ &\leqslant K\sum_{j=1}^d \mathbb{E}\Big\{\Big(\int_0^T |z_j(t)|^2 dt\Big)^{\frac{p}{2}} + \Big(\int_0^T\int_0^T |D_\tau z_j(t)|^2 d\tau dt\Big)^{\frac{p}{2}}\Big\} = K\|z_j(\cdot)\|^p_{\mathbb{L}^{m\times d}_{1,p}},\end{aligned}$$

and for $p \geqslant 2$,

$$\begin{aligned}&\|z(\cdot)\|^p_{\mathbb{L}^{m\times d}_{1,p}} = \sum_{j=1}^d \mathbb{E}\Big[\Big(\int_0^T |z_j(t)|^2 dt\Big)^{\frac{p}{2}} + \Big(\int_0^T\int_0^T \sum_{i=1}^d |D_\tau^{(i)} z_j(t)|^2 d\tau dt\Big)^{\frac{p}{2}}\Big]\\ &\leqslant K\sum_{j=1}^d \mathbb{E}\Big[\int_0^T |z_j(t)|^p dt + \int_0^T\Big(\int_0^T |D_\tau z_j(t)|^2 d\tau\Big)^{\frac{p}{2}} dt\Big] \leqslant K\|z(\cdot)\|^p_{L^p_{\mathbb{F}}(0,T;\mathbb{D}^{m\times d}_{1,p})}.\end{aligned}$$

Hence,

$$\begin{cases}\mathbb{L}_{1,p}^{m\times d} = L_{\mathbb{F}}^p(\Omega;L^2(0,T;\mathbb{W}_{1,2}^m(0,T)))^d \\ \qquad \subseteq L_{\mathbb{F}}^p(0,T;L^p(\Omega;\mathbb{W}_{1,2}^m(0,T)))^d = L_{\mathbb{F}}^p(0,T;\mathbb{D}_{1,p}^{m\times d}), \quad \text{if } 1 \leqslant p \leqslant 2,\\ \mathbb{L}_{1,p}^{m\times d} = L_{\mathbb{F}}^p(\Omega;L^2(0,T;\mathbb{W}_{1,2}^m(0,T)))^d \\ \qquad \supseteq L_{\mathbb{F}}^p(0,T;L^p(\Omega;\mathbb{W}_{1,2}^m(0,T)))^d = L_{\mathbb{F}}^p(0,T;\mathbb{D}_{1,p}^{m\times d}), \quad \text{if } 2 \leqslant p < \infty.\end{cases}$$

Consequently, for any $z(\cdot) \in L_{\mathbb{F}}^p(0,T;\mathbb{D}_{1,p}^{m\times d})$, in order to keep Itô's integrals (12.1.25) meaningful, we had better assume that $2 \leqslant p < \infty$. We also see that

$$L_{\mathbb{F}}^p(\Omega;L^2(0,T;\mathbb{R}^{m\times d})) \cap L_{\mathbb{F}}^p(0,T;\mathbb{D}_{1,p}^m) \subseteq \mathbb{L}_{1,p}^{m\times d}, \qquad \forall 1 \leqslant p < \infty.$$

The following collects some important properties of Skorokhod integrals.

Proposition 12.1.2

(i) *Let $z(\cdot) \in \mathscr{D}(D^*)^m$, $\xi \in \mathbb{D}_{1,2}^m$ such that*

$$\mathbb{E}\Big[|\xi|^2\int_0^T |z(s)|^2 ds\Big] < \infty.$$

Then

$$\xi^\top \int_0^T z(s)dW(s) = \int_0^T \xi^\top z(s)dW(s) + \int_0^T tr\Big([D_t\xi]z(t)\Big)dt. \tag{12.1.28}$$

Consequently, if $\mathcal{S}$ is the set of all $z(\cdot)$ of the following form:

$$z(t) = \sum_{i=1}^{k} \xi_i^\top h_i(t), \tag{12.1.29}$$

with $\xi_i \in \mathbb{D}_{1,2}^m$ and $h_i(\cdot) \in L^2(0,T;\mathbb{R}^{m\times d})$, then $\mathcal{S} \subseteq \mathscr{D}(D^*)$, and for any $z(\cdot) \in \mathcal{S}$,

$$\int_0^T z(t)dW(t) = \sum_{i=1}^{k} \xi_i^\top \int_0^T h_i(t)dW(t) - \sum_{i=1}^{k} \int_0^T tr\Big[(D_t\xi_i)^\top h_i(t)\Big]dt. \tag{12.1.30}$$

In particular, for $\xi \in \mathbb{D}_{1,2}$ and $h(\cdot) \in L^2(0,T;\mathbb{R}^d)$,

$$\int_0^T \xi h(t)dW(t) = \xi \int_0^T h(t)dW(t) - \int_0^T (D_t\xi)h(t)dt. \tag{12.1.31}$$

(ii) *Let $z(\cdot),\ \zeta(\cdot) \in L^2_{\mathcal{F}_T}(0,T;\mathbb{D}_{1,2}^{1\times d})$. Then*

$$\begin{aligned}&\mathbb{E}\Big[\Big(\int_0^T z(t)dW(t)\Big)\Big(\int_0^T \zeta(t)dW(t)\Big)\Big]\\ &= \mathbb{E}\int_0^T z(t)\zeta(t)^\top dt + \mathbb{E}\int_0^T\int_0^T tr\Big[[D_t z(s)]^\top D_s\zeta(t)\Big]dtds.\end{aligned} \tag{12.1.32}$$

In particular,

$$\begin{aligned}&\mathbb{E}\Big(\int_0^T z(t)dW(t)\Big)^2\\ &= \mathbb{E}\Big[\int_0^T |z(t)|^2dt + \int_0^T\int_0^T tr\Big([D_t z(s)]^\top[D_s z(t)]\Big)dtds\Big].\end{aligned} \tag{12.1.33}$$

(iii) *Let $z(\cdot) \in L^p_{\mathbb{F}}(\Omega; L^2(0,T;\mathbb{R}^{m\times d}))$. Then $z(\cdot) \in \mathbb{L}_{1,p}^{m\times d}$ if and only if $\int_0^T z(s)dW(s) \in \mathbb{D}_{1,p}^m$. In this case, the process $t \mapsto \int_0^t z(s)dW(s)$ is in $\mathbb{L}_{1,p}^{m\times d}$ and for each $i = 1, 2, \cdots, d$, for almost all $\tau \in [0,T]$, $D_\tau^{(i)} z(\cdot) \in \mathscr{D}(D^*)$ and there exists a version $\int_0^T [D_\cdot^{(i)} z(s)]dW(s) \in L^p((0,T)\times\Omega;\mathbb{R}^m)$. Moreover,* *for any* $S \in [0, T)$,

$$D_\tau^{(i)} \int_S^T z(s)dW(s) = z_i(\tau)\mathbf{1}_{(S,T]}(\tau) + \int_{\tau \vee S}^T \big[D_\tau^{(i)} z(s)\big]dW(s), \quad \text{a.e. } \tau \in [0, T]. \tag{12.1.34}$$

Consequently, the above is true if $z(\cdot) \in L^p_{\mathbb{F}}(0, T; \mathbb{D}^m_{1,p})$ *with* $2 \leqslant p < \infty$.

Now, let us present Fubini type theorems for Skorokhod integrals, which will be useful below. To this end, let us first introduce the second order Malliavin derivative. For any $\xi \in \Xi$ of form (12.1.1), we define

$$D_\tau^{(i)} D_\theta^{(j)} \xi = \langle f_{xx}\Big(\int_0^T h(s)dW(s)\Big)h_i(\theta), h_j(\tau)\rangle, \quad \theta, \tau \in [0, T], \ 1 \leqslant i, j \leqslant d,$$

and denote

$$D^2\xi \equiv D\big\{D\xi\big\} = \big(D_\cdot^{(i)}[D_\cdot^{(j)}\xi]\big),$$

valued in $\mathbb{S}^d$ (the set of all $d \times d$ symmetric matrices). Let

$$\|\xi\|_{\mathbb{D}_{2,p}} = \Big\{\mathbb{E}\Big[|\xi|^2 + \Big(\int_0^T |D_\theta \xi|^2\Big)^{\frac{p}{2}} + \Big(\int_0^T \int_0^T |D_\tau\{D_\theta \xi\}|^2 d\theta d\tau\Big)^{\frac{p}{2}}\Big]\Big\}^{\frac{1}{p}}.$$

Let $\mathbb{D}_{2,p}$ be the closure of Ξ under the above norm. We also need a few results as follows.

Proposition 12.1.3 *The space* $\mathbb{L}_{1,2}$ *is included in* $\mathscr{D}(D^*)$. *In addition, for* $u, v \in \mathbb{L}_{1,2}$,

$$\mathbb{E}\Big[D^*(u)D^*(v)\Big] = \mathbb{E}\int_0^T u(t)v(t)dt + \mathbb{E}\int_0^T \int_0^T D_s u(t) D_t v(s) ds dt.$$

(H12.1.1) Suppose $u(\cdot) \in \mathbb{L}^{1,2}$, for almost t, $D_t u(\cdot)$ is Skorokhod integrable, and there is a version of $\int_0^T D_\cdot u(s)dW(s)$ belonging to $L^2_{\mathcal{F}_T}\big((0, T) \times \Omega; \mathbb{R}\big)$.

Proposition 12.1.4 *Suppose (H12.1.1) holds. Then* $D^*u \in \mathbb{D}^{1,2}$, *and*

$$D_t\big(D^*u\big) = u(t) + \int_0^T D_t u(s)dW(s), \quad t \in [0, T].$$

(H12.1.2) Given $t \in [0, T)$, suppose F is a random variable satisfying

$$\mathbb{E}|F|^2 + \mathbb{E}\int_0^t \big|D_r F\big|^2 dr < \infty.$$

Let $u \in L^2_{\mathcal{F}_T}(0,T;\mathbb{R})$ such that $u\mathbf{1}_{[0,t]} \in \mathscr{D}(D^*)$, $Fu\mathbf{1}_{[0,t]} \in L^2_{\mathcal{F}_T}(0,T;\mathbb{R})$.

Proposition 12.1.5 *Suppose (H12.1.2) holds. Then* $Fu\mathbf{1}_{[0,t]} \in \mathscr{D}(D^*)$, *and*

$$D^*\big(Fu\mathbf{1}_{[0,t]}\big) = FD^*\big(u\mathbf{1}_{[0,t]}\big) - \int_0^t D_r Fu(r)dr,$$

provided the right-hand terms are square integrable.

Proposition 12.1.6 *Suppose* F_1, F_2 *are two elements in* $\mathbb{D}^{1,2}$ *such that* F_1, $\|DF_1\|_{L^2(0,T;\mathbb{R})}$ *are bounded. Then* $F_1F_2 \in \mathbb{D}^{1,2}$ *and* $D(F_1F_2) = F_1DF_2 + F_2DF_1$.

In the following, given proper function H and $t, s \in [0,T]$, we define by

$$\bar{H}_1(t) \equiv \int_0^T H(t,s)dW(s),\ \ \bar{H}_2(s) \equiv \int_0^T H(t,s)dW(t),\ \ \bar{H}_3(s) \equiv \int_0^T H(t,s)dt.$$

Theorem 12.1.7 *Suppose* $H(\cdot,\cdot) \in L^2_{\mathcal{F}_T}([0,T]^2 \times \Omega;\mathbb{R}^{n\times d})$.

(i) For almost all $t \in [0,T]$, $H(t,\cdot) \in \mathscr{D}(D^*)$, $\bar{H}_1(\cdot) \in \mathscr{D}(D^*)$, *and for almost all* $s \in [0,T]$, $H(\cdot,s) \in \mathscr{D}(D^*)$, $\bar{H}_2(\cdot) \in \mathscr{D}(D^*)$. *Then* $D^*\big(\bar{H}_1\big) = D^*\big(\bar{H}_2\big)$.

(ii) For almost all $t \in [0,T]$, $H(t,\cdot) \in \mathscr{D}(D^*)$, $H_1(\cdot) \in L^1(0,T;\mathbb{R})$ *and for almost all* $s \in [0,T]$, $H(\cdot,s) \in L^1(0,T;\mathbb{R})$, $\bar{H}_3(\cdot) \in \mathscr{D}(D^*)$. *Then*

$$\int_0^T \bar{H}_1(t)dt = \int_0^T \bar{H}_3(s)ds.$$

Proof For simplicity, let $n = d = 1$.

(i) For any $F \in \mathbb{D}^{2,2}$, we have

$$\begin{aligned}
\mathbb{E}\Big[F\int_0^T \bar{H}_1(t)dW(t)\Big] &= \mathbb{E}\int_0^T D_tF H_1(t)dt,\\
\mathbb{E}\Big[D_tF\int_0^T H(t,s)dW(s)\Big] &= \mathbb{E}\int_0^T \big[D_sD_tF\big]H(t,s)ds,
\end{aligned}$$

which implies that

$$\mathbb{E}\Big[F\int_0^T \bar{H}_1(t)dW(t)\Big] = \mathbb{E}\int_0^T\int_0^T \big[D_sD_tF\big]H(t,s)dsdt.$$

Similarly

$$\mathbb{E}\Big[F\int_0^T \bar{H}_2(s)dW(s)\Big] = \mathbb{E}\int_0^T\int_0^T \big[D_tD_sF\big]H(t,s)dtds.$$

Thanks to $D_s D_t F = D_t D_s F$, we then have

$$\mathbb{E}\Big[F\Big\{\int_0^T \bar{H}_1(t)dW(t) - \int_0^T \bar{H}_2(s)dW(s)\Big\}\Big] = 0,$$

from which we obtain the conclusion.

(ii) Notice that

$$\mathbb{E}\Big[F\int_0^T \bar{H}_1(t)dt\Big] = \mathbb{E}\int_0^T F\bar{H}_1(t)dt,$$
$$\mathbb{E}\Big[F\int_0^T H(t,s)dW(s)\Big] = \mathbb{E}\int_0^T \big[D_s F\big]H(t,s)ds,$$

from which we see that

$$\mathbb{E}\Big[F\int_0^T \bar{H}_1(t)dt\Big] = \mathbb{E}\int_0^T\int_0^T \big[D_s F\big]H(t,s)dsdt.$$

Also

$$\mathbb{E}\Big[F\int_0^T \bar{H}_3(s)dW(s)\Big] = \mathbb{E}\int_0^T D_s F H_3(s)ds,$$
$$\mathbb{E}\Big[D_s F\int_0^T H(t,s)dt\Big] = \mathbb{E}\int_0^T \big[D_s F\big]H(t,s)dt,$$

from which we see that

$$\mathbb{E}\Big[F\int_0^T \bar{H}_3(s)dW(s)\Big] = \mathbb{E}\int_0^T\int_0^T \big[D_s F\big]H(t,s)dtds.$$

By Fubini theorem,

$$\mathbb{E}\int_0^T\int_0^T \big[D_s F\big]H(t,s)dtds = \mathbb{E}\int_0^T\int_0^T \big[D_s F\big]H(t,s)dsdt,$$

from which we have

$$\mathbb{E}\Big[F\int_0^T \bar{H}_1(t)dt\Big] = \mathbb{E}\Big[F\int_0^T \bar{H}_3(s)dW(s)\Big].$$

Then it is easy to see the conclusion. □

2 Garsia-Rodemich-Rumsey Inequality and Its Consequence

Theorem 12.2.1 *Let* $\Psi : [0,\infty) \to [0,\infty)$ *and* $\Phi : [0,T] \to [0,\infty)$ *be continuous and strictly increasing with* $\Psi(0) = \Phi(0) = 0$, $\Psi(\infty) = \infty$. *Suppose* $f : [0,T] \to \mathbb{R}$ *is continuous such that*

$$\int_0^T \int_0^T \Psi\Big(\frac{|f(t) - f(s)|}{\Phi(|t-s|)}\Big) ds dt = F < \infty. \tag{12.2.1}$$

Then the following is true:

$$|f(t) - f(s)| \leqslant 8 \int_0^{|t-s|} \Psi^{-1}\Big(\frac{F}{r^2}\Big) d\Phi(r), \qquad \forall t, s \in [0,T].$$

Proof We first prove the conclusion for the case $f : [0,1] \to \mathbb{R}$. Let us define

$$I(t) = \int_0^1 \Psi\Big(\frac{f(t) - f(s)|}{\Phi(|t-s|)}\Big) ds, \quad t \in [0,1].$$

Pick $t_0 \in (0,1)$, so that

$$I(t_0) \leqslant F.$$

Define

$$d_0 = \Phi^{-1}\Big(\frac{1}{2}\Phi(t_0)\Big) < t_0.$$

Choose $t_1 \in (0, d_0)$ so that

$$I(t_1) \leqslant \frac{2F}{d_0}, \qquad \Psi\Big(\frac{|f(t_1) - f(t_0)|}{\Phi(|t_1 - t_0|)}\Big) \leqslant \frac{2I(t_0)}{d_0}. \tag{12.2.2}$$

Such a t_1 can always be done. In fact, if the first fails for all $t_1 \in (0, d_0)$, then

$$\frac{2F}{d_0} < I(t_1) = \int_0^1 \Psi\Big(\frac{|f(t_1) - f(s)|}{\Phi(|t_1 - s|)}\Big) ds = F,$$

which leads to $d_0 > 2$, a contradiction. Now, if the second fails for all $t_1 \in (0, d_0)$, then letting

$$\Gamma = \Big\{t \in (0, d_0) \mid \Psi\Big(\frac{|f(t_1) - f(s)|}{\Phi(|t_1 - s|)}\Big) > \frac{2I(t_0)}{d_0}\Big\},$$

we have

$$I(t_0) = \int_0^1 \Psi\Big(\frac{|f(s) - f(t_0)|}{\Phi(|s - t_0)}\Big)ds \geqslant \int_\Gamma \frac{2I(t_0)}{d_0}ds = \frac{2I(t_0)}{d_0}|\Gamma|.$$

This leads to $|\Gamma| \leqslant \frac{d_0}{2}$. This means that the above $t_1 \in (0, d_0)$ can be chosen so that (12.2.2) hold. By induction, suppose t_k is taken. Define

$$d_k = \Phi^{-1}\Big(\frac{1}{2}\Phi(t_k)\Big), \tag{12.2.3}$$

and choose $t_{k+1} \in (0, d_k]$ so that

$$I(t_{k+1}) \leqslant \frac{2F}{d_k}, \qquad \Psi\Big(\frac{|f(t_{k+1}) - f(t_k)|}{\Phi(|t_{k+1} - t_k|)}\Big) \leqslant \frac{2I(t_k)}{d_k}. \tag{12.2.4}$$

Then by (12.2.3), the first of (12.2.4) and the strict monotonicity of $\Phi(\cdot)$, we have

$$0 < t_{k+1} \leqslant d_k < t_k \leqslant d_{k-1}, \qquad k \geqslant 1.$$

Hence, we claim that $t_k \downarrow 0$. Otherwise, say $t_k \downarrow t^* > 0$, then from

$$\Phi(t_{k+1}) \leqslant \Phi(d_k) = \frac{1}{2}\Phi(t_k),$$

by sending $k \to \infty$, we end up a contradiction: $\Phi(t^*) \leqslant \frac{1}{2}\Phi(t^*)$. Now, the first of (12.2.4) implies

$$I(t_k) \leqslant \frac{2F}{d_k}, \qquad k \geqslant 1.$$

The above is trivially true for $k = 0$. Consequently, the second of (12.2.4), together with the strict monotonicity of $\Psi(\cdot)$, gives

$$|f(t_{k+1}) - f(t_k)| \leqslant \Phi\big(|t_{k+1} - t_k|\big)\Psi^{-1}\Big(\frac{2I(t_k)}{d_k}\Big) \leqslant \Phi\big(|t_{k+1} - t_k|\big)\Psi^{-1}\Big(\frac{4F}{d_k^2}\Big).$$

By (12.2.3) and monotonicity of $\Phi(\cdot)$,

$$\Phi(t_k - t_{k+1}) \leqslant \Phi(t_k) = 2\Phi(d_k).$$

On the other hand,

$$\Phi(d_{k+1}) = \frac{1}{2}\Phi(t_{k+1}) \leqslant \frac{1}{2}\Phi(d_k),$$

which leads to

$$2\Phi(d_k) \leqslant 4[\Phi(d_k) - \Phi(d_{k+1})].$$

Hence,

$$\Phi(t_k - t_{k+1}) \leqslant 4[\Phi(d_k) - \Phi(d_{k+1})].$$

Then

$$\begin{aligned}|f(t_0) - f(0)| &\leqslant 4\sum_{k=1}^{\infty}[\Phi(d_{k-1}) - \Phi(d_k)]\Psi^{-1}\Big(\frac{4F}{d_{k-1}^2}\Big)\\ &\leqslant 4\sum_{k=1}^{\infty}\int_{d_k}^{d_{k-1}}\Psi^{-1}\Big(\frac{4F}{r^2}\Big)d\Phi(r) \leqslant 4\int_0^1 \Psi^{-1}\Big(\frac{4F}{r^2}\Big)d\Phi(r).\end{aligned}$$

Applying the same argument to $f(1-t)$, we obtain the same bound for $|f(t_0) - f(1)|$. Hence,

$$|f(1) - f(0)| \leqslant 8\int_0^1 \Psi^{-1}\Big(\frac{4F}{r^2}\Big)d\Phi(r).$$

Now, for general $0 \leqslant s < t \leqslant 1$, let

$$\bar{f}(\tau) = f(s + \tau(t-s)), \qquad \tau \in [0, 1],$$

and

$$\bar{\Phi}(r) = \Phi(r(t-s)), \qquad r \in [0, 1].$$

Then

$$\begin{aligned}&\int_0^1\int_0^1 \Psi\Big(\frac{|\bar{f}(t') - \bar{f}(s')|}{\bar{\Phi}(|t'-s'|)}\Big)ds'dt'\\ &= \int_0^1\int_0^1 \Psi\Big(\frac{|f(s+t'(t-s)) - f(s+s'(t-s))|}{\Phi(|t'-s'|(t-s))}\Big)ds'dt'\\ &= \int_s^t\int_s^t \Psi\Big(\frac{|f(t'') - f(s'')|}{\Phi(|t''-s''|)}\Big)\frac{ds''dt''}{(t-s)^2}\\ &\leqslant \frac{1}{|t-s|^2}\int_0^1\int_0^1 \Psi\Big(\frac{|f(t'') - f(s'')|}{\Phi(|t''-s''|)}\Big)ds''dt'' = \frac{F}{(t-s)^2}.\end{aligned}$$

Hence,

$$\begin{aligned}&|f(t) - f(s)| = |\bar{f}(1) - \bar{f}(0)|\\ &\leqslant 8\int_0^1 \Psi^{-1}\Big(\frac{4F}{r^2(t-s)^2}\Big)d\Phi(r(t-s)) = 8\int_0^{|t-s|}\Psi^{-1}\Big(\frac{4F}{r^2}\Big)d\Phi(r).\end{aligned}$$

This proves the theorem for the case $T = 1$. We now look at the general $T > 0$. Let

$$\bar{f}(\tau) = f(T\tau), \qquad \tau \in [0, 1],$$

and

$$\bar{\Phi}(r) = \Phi(Tr), \qquad r \in [0, 1].$$

Then

$$\int_0^1 \int_0^1 \Psi\Big(\frac{|\bar{f}(t') - \bar{f}(s')|}{\bar{\Phi}(|t' - s'|)}\Big) ds' dt' = \int_0^1 \int_0^1 \Psi\Big(\frac{|f(Tt') - f(Ts')|}{\Phi(T|t' - s'|)}\Big) ds' dt'$$
$$= \frac{1}{T^2} \int_0^T \int_0^T \Psi\Big(\frac{|f(t'') - f(s'')|}{\Phi(|t'' - s''|)}\Big) ds'' dt'' = \frac{F}{T^2}.$$

Hence, for $t, s \in [0, T]$,

$$|f(t) - f(s)| = |\bar{f}(T^{-1}t) - \bar{f}(T^{-1}s)|$$
$$\leqslant 8 \int_0^{T^{-1}|t-s|} \Psi^{-1}\Big(\frac{4F}{r^2 T^2}\Big) d\Phi(rT) = 8 \int_0^{|t-s|} \Psi^{-1}\Big(\frac{4F}{r^2}\Big) d\Phi(r).$$

This proves the theorem □

Corollary 12.2.2 *Let $X(\cdot)$ be an $\mathbb{F}$-adapted stochastic process and the following holds*

$$\mathbb{E}|X(t) - X(s)|^p \leqslant K_0 |t - s|^{1+\beta}, \qquad t, s \in [0, T],$$

for some $p > 0$, $K_0, \beta > 0$. Then $X(\cdot)$ has a continuous version. More precisely, an $\mathcal{F}_T$-measurable random variable ξ defined by

$$\xi = \Big[\int_0^T \int_0^T \frac{|X(t) - X(s)|^p}{|t - s|^{2+\beta}\Big(\log \frac{\gamma}{|t-s|}\Big)^2} ds dt\Big]^{\frac{1}{p}}$$

is integrable, with $\gamma > e^{\frac{2}{2+\beta}}$, and

$$|X(t) - X(s)| \leqslant K\xi |t - s|^{\frac{\beta}{p} - \varepsilon}, \qquad \forall t, s \in [0, T]. \tag{12.2.5}$$

Consequently,

$$\mathbb{E}\Big[\sup_{0 \leqslant s \neq t \leqslant T} \frac{|X(t) - X(s)|}{|t - s|^{\frac{\beta}{p} - \varepsilon}}\Big] \leqslant K\mathbb{E}[\xi]. \tag{12.2.6}$$

Furthermore, it holds

$$\mathbb{E}\Big[\sup_{t\in[0,T]}|X(t)^p\Big]\leqslant K(1+|x|^p). \tag{12.2.7}$$

Proof Choose

$$\Psi(r)=r^p,\quad \Phi(r)=r^{\frac{2+\beta}{p}}\Big(\log\frac{\gamma}{r}\Big)^{\frac{2}{p}}.$$

Then

$$\begin{aligned}\Phi'(r)&=\frac{2+\beta}{p}r^{\frac{2+\beta-p}{p}}\Big(\log\frac{\gamma}{r}\Big)^{\frac{2}{p}}-\frac{2}{p}r^{\frac{2+\beta-p}{p}}\Big(\log\frac{\gamma}{r}\Big)^{\frac{2-p}{p}}\\&=r^{\frac{2+\beta-p}{p}}\frac{1}{p}\Big(\log\frac{\gamma}{r}\Big)^{\frac{2-p}{p}}\Big[(2+\beta)\Big(\log\frac{\gamma}{r}\Big)-2\Big].\end{aligned}$$

Thus, in order

$$(2+\beta)\Big(\log\frac{\gamma}{r}\Big)-2>0.\qquad \forall r\in[0,1],$$

we need

$$\gamma>e^{\frac{2}{2+\beta}}.$$

For such a case, $\Phi(\cdot)$ is strictly increasing on $[0,1]$. On the other hand, by defining F as (12.2.1) with $f(\cdot)=X(\cdot)$, we have

$$\begin{aligned}&\mathbb{E}[F]=\mathbb{E}\int_0^T\int_0^T\Psi\Big(\frac{|X(t)-X(s)|}{\Phi(|t-s|)}\Big)dsdt\\&=\int_0^T\int_0^T\frac{\mathbb{E}|X(t)-X(s)|^p}{|t-s|^{2+\beta}\Big(\log\frac{\gamma}{|t-s|}\Big)^2}dsdt\leqslant K_0\int_0^T\int_0^T\frac{dsdt}{|t-s|\Big(\log\frac{\gamma}{|t-s|}\Big)^2}<\infty,\end{aligned}$$

and

$$\begin{aligned}&|X(t)-X(s)|\leqslant 8\int_0^{|t-s|}\Psi^{-1}\Big(\frac{F}{r^2}\Big)d\Phi(r)=8\int_0^{|t-s|}\frac{F^{\frac{1}{p}}}{r^{\frac{2}{p}}}d\Big[r^{\frac{2+\beta}{p}}\Big(\log\frac{\gamma}{r}\Big)^{\frac{2}{p}}\Big]\\&=8F^{\frac{1}{p}}\Big[r^{\frac{\beta}{p}}\Big(\log\frac{\gamma}{r}\Big)^{\frac{2}{p}}\Big]\Big|_0^{|t-s|}+\frac{16F^{\frac{1}{p}}}{p}\int_0^{|t-s|}r^{\frac{\beta-p}{p}}\Big(\log\frac{\gamma}{r}\Big)^{\frac{2}{p}}dr.\end{aligned}$$

For any small $\varepsilon>0$,

$$r^{\varepsilon}\Big(\log\frac{\gamma}{r}\Big)^{\frac{2}{p}}\leqslant K,\qquad r\in(0,1],$$

for some $K > 0$ that only depends on ε, p, γ. Then we could continue the above estimation:

$$
\begin{aligned}
|X(t) - X(s)| &\leqslant 8F^{\frac{1}{p}}|t-s|^{\frac{\beta}{p}}\Big(\log\frac{\gamma}{|t-s|}\Big)^{\frac{2}{p}} + KF^{\frac{1}{p}}\int_0^{|t-s|} r^{\frac{\beta-p}{p}-\varepsilon}dr \\
&\leqslant 8F^{\frac{1}{p}}|t-s|^{\frac{\beta}{p}}\Big(\log\frac{\gamma}{|t-s|}\Big)^{\frac{2}{p}} + KF^{\frac{1}{p}}|t-s|^{\frac{\beta}{p}-\varepsilon} \leqslant KF^{\frac{1}{p}}|t-s|^{\frac{\beta}{p}-\varepsilon}.
\end{aligned}
$$

Let $\xi = F^{\frac{1}{p}}$. Then $\mathbb{E}[\xi] < \infty$. Thus, (12.2.6) follows. Finally,

$$
|X(t)|^p \leqslant K\Big[|x|^p + |X(t) - X(0)|^p\Big] \leqslant K\Big[|x|^p + \xi^p|t|^{\beta-\varepsilon p}\Big].
$$

Hence, (12.2.7) follows. This proves the theorem. □

3 Some Standard Results of PDEs

In this section, we recall some standard results on parabolic PDEs. First of all, for any suitable function $\varphi : [S,T] \times \mathbb{R}^n \to \mathbb{R}$, with $\alpha \in (0,1)$ and $S \in [0,T)$, let

$$
\begin{aligned}
&|\varphi|^{(0)} = \|\varphi\|_{L^\infty([S,T]\times\mathbb{R}^n)}, \qquad |\varphi|^{(1)} = |\varphi|^{(0)} + |\varphi_x|^{(0)}, \\
&|\varphi|^{(2)} = |\varphi|^{(1)} + |\varphi_s|^{(0)} + |\varphi_{xx}|^{(0)}, \\
&\langle\varphi\rangle_s^{(\frac{\alpha}{2})} = \sup_{\substack{s,s'\in[S,T],x\in\mathbb{R}^n\\ s\neq s'}} \frac{|\varphi(s,x) - \varphi(s',x)|}{|s-s'|^{\frac{\alpha}{2}}}, \\
&\langle\varphi\rangle_x^{(\alpha)} = \sup_{\substack{s\in[S,T],x,x'\in\mathbb{R}^n\\ 0<|x-x'|\leqslant 1}} \frac{|\varphi(s,x) - \varphi(s,x')|}{|x-x'|^{\alpha}}, \\
&\langle\varphi\rangle^{(\alpha)} = \langle\varphi\rangle_s^{(\frac{\alpha}{2})} + \langle\varphi\rangle_x^{(\alpha)}, \qquad |\varphi|^{(\alpha)} = |\varphi|^{(0)} + \langle\varphi\rangle^{(\alpha)}, \\
&|\varphi|^{(1+\alpha)} = |\varphi|^{(1)} + \langle\varphi_x\rangle^{(\alpha)} + \langle\varphi\rangle_s^{(\frac{1+\alpha}{2})}, \\
&|\varphi|^{(2+\alpha)} = |\varphi|^{(2)} + \langle\varphi_s\rangle^{(\alpha)} + \langle\varphi_{xx}\rangle^{(\alpha)} + \langle\varphi_x\rangle_s^{(\frac{1+\alpha}{2})}.
\end{aligned}
$$

When $[S,T]\times\mathbb{R}^n$ needs to be emphasized, we use, say, $|\varphi|^{(1)}_{[S,T]\times\mathbb{R}^n}$, etc. We denote

$$
C^{\frac{\alpha}{2},\alpha}([S,T]\times\mathbb{R}^n) = \Big\{\varphi : [S,T]\times\mathbb{R}^n \to \mathbb{R} \;\Big|\; |\varphi|^{(\alpha)}_{[S,T]\times\mathbb{R}^n} < \infty\Big\}.
$$

Clearly, $\varphi(\cdot\,,\cdot) \in C^{\frac{\alpha}{2},\alpha}([S,T]\times\mathbb{R}^n)$ if and only if $\varphi(\cdot\,,\cdot) \in L^\infty([S,T]\times\mathbb{R}^n)$ and $\forall s, s' \in [S,T],\ x, x' \in \mathbb{R}^n,\ |x-x'| \leqslant 1$,

$$
|\varphi(s,x) - \varphi(s',x')| \leqslant K\Big(|s-s'|^{\frac{\alpha}{2}} + |x-x'|^{\alpha}\Big).
$$

Also, we denote

$$C^{\frac{1+\alpha}{2},1+\alpha}([S,T]\times\mathbb{R}^n)=\big\{\varphi:[S,T]\to\mathbb{R}^n \bigm| |\varphi|^{(1+\alpha)}_{[S,T]\times\mathbb{R}^n}<\infty\big\},$$
$$C^{1+\frac{\alpha}{2},2+\alpha}([S,T]\times\mathbb{R}^n)=\big\{\varphi:[S,T]\to\mathbb{R}^n \bigm| |\varphi|^{(2+\alpha)}_{[S,T]\times\mathbb{R}^n}<\infty\big\}.$$

Let us introduce the following differential operator:

$$(\mathcal{L}v)(s,x)=v_s(s,x)+\sum_{i,j=1}^n a_{ij}(s,x)v_{x_ix_j}(s,x)+\sum_{i=1}^n b_i(s,x)v_{x_i}(s,x), \tag{12.3.1}$$

and for given $f(\cdot)$ and $h(\cdot)$, consider the following Cauchy problem:

$$\begin{cases}(\mathcal{L}v)(s,x)+f(s,x)=0, & (s,x)\in[t,T]\times\mathbb{R}^n,\\ v(T,x)=h(x), & x\in\mathbb{R}^n.\end{cases} \tag{12.3.2}$$

We introduce the following hypotheses:

(H12.3.1) There exist constant $\bar{\lambda}_0>0$ such that

$$\sum_{i,j=1}^n a_{ij}(s,x)\xi_i\xi_j\equiv\langle a(s,x)\xi,\xi\rangle\geqslant\bar{\lambda}_0|\xi|^2,\quad \forall\xi\in\mathbb{R}^n,\ (s,x)\in[0,T]\times\mathbb{R}^n.$$

(H12.3.2) Functions $a_{ij}(\cdot,\cdot)$, $b_i(\cdot,\cdot)$ are continuous and bounded, and for some $\alpha\in(0,1)$,

$$\begin{cases}|a_{ij}(s,x)-a_{ij}(s',x')|\leqslant K\big(|s-s'|^{\frac{\alpha}{2}}+|x-x'|^\alpha\big),\\ |b_i(s,x)-b_i(s,x')|\leqslant K|x-x'|^\alpha,\\ \qquad\qquad (s,x),(s',x')\in[0,T]\times\mathbb{R}^n,\ |x-x'|\leqslant 1.\end{cases}$$

When (H12.3.1) holds, $\mathcal{L}$ defined by (12.3.1) is called a *parabolic differential operator*. The following result is concerned with the Cauchy problem (12.3.2).

Proposition 12.3.1 *Let* (H12.3.1)–(H12.3.2) *hold. Assume that*

$$f(\cdot\,,\cdot)\in C^{\frac{\alpha}{2},\alpha}([0,T]\times\mathbb{R}^n),\quad h(\cdot)\in C^{2+\alpha}(\mathbb{R}^n),$$

for some $\alpha\in(0,1)$. *Then Cauchy problem* (12.3.2) *admits a unique classical solution* $v(\cdot\,,\cdot)\in C^{1+\frac{\alpha}{2},2+\alpha}([0,T]\times\mathbb{R}^n)$. *Moreover,* $v(\cdot\,,\cdot)$ *is represented as follows: for any* $(s,x)\in[0,T]\times\mathbb{R}^n$,

$$v(s,x)=\int_{\mathbb{R}^n}G(s,x;T,\eta)h(\eta)d\eta+\int_s^T\int_{\mathbb{R}^n}G(s,x;\tau,\eta)f(\tau,\eta)d\eta d\tau.$$

Here $G(s,x;\tau,\eta)$ is called the fundamental solution of the parabolic differential operator $\mathcal{L}$, which satisfies the following: There exists a $\lambda>0$ such that for any $x,\eta\in\mathbb{R}^n$ and $0\leqslant s<\tau\leqslant T$,

$$\begin{cases}|G(s,x;\tau,\eta)|\leqslant\dfrac{K}{(\tau-s)^{\frac{n}{2}}}e^{-\lambda\frac{|\eta-x|^2}{\tau-s}},\\ |G_x(s,x;\tau,\eta)|\leqslant\dfrac{K}{(\tau-s)^{\frac{n+1}{2}}}e^{\frac{-\lambda|\eta-x|^2}{\tau-s}},\\ |G_s(s,x;\tau,\eta)|+|G_{xx}(s,x;\tau,\eta)|\leqslant\dfrac{K}{(\tau-s)^{\frac{n+2}{2}}}e^{\frac{-\lambda|\eta-x|^2}{\tau-s}}.\end{cases}\tag{12.3.3}$$

Moreover,

$$|v|^{(2+\alpha)}_{[0,T]\times\mathbb{R}^n}\leqslant K\big(|f|^{(\alpha)}_{[0,T]\times\mathbb{R}^n}+|h|^{(2+\alpha)}_{\mathbb{R}^n}\big),\tag{12.3.4}$$

and for any $S\in[0,T)$,

$$|v|^{(1+\alpha)}_{[S,T]\times\mathbb{R}^n}\leqslant|h|^{(1+\alpha)}_{\mathbb{R}^n}+K(T-S)^{\frac{\alpha}{2}}\big(|f|^{(\alpha)}_{[S,T]\times\mathbb{R}^n}+|h|^{(2+\alpha)}_{\mathbb{R}^n}\big).\tag{12.3.5}$$

Proof The proof up to (12.3.4) is standard. Let us look at (12.3.5) which will play a crucial role below. Note that by defining

$$\begin{aligned}&\widetilde{v}(s,x)=v(s,x)-h(x),\\ &\widetilde{f}(s,x)=f(s,x)+\frac{1}{2}\sigma(s,x)^\top h_{xx}(x)+h_x(x)b(s,x),\qquad (s,x)\in[0,T]\times\mathbb{R}^n,\end{aligned}$$

we see that $\widetilde{v}(\cdot\,,\cdot)$ is the solution to the following:

$$\begin{cases}\widetilde{v}_s(s,x)+\dfrac{1}{2}\sigma(s,x)^\top\widetilde{v}_{xx}(s,x)\sigma(s,x)+\widetilde{v}_x(s,x)b(s,x)+\widetilde{f}(s,x)=0,\\ \qquad\qquad\qquad\qquad\qquad\qquad\qquad\qquad (s,x)\in[0,T]\times\mathbb{R}^n,\\ \widetilde{v}(T,x)=0,\qquad x\in\mathbb{R}^n.\end{cases}\tag{12.3.6}$$

Thus, the following representation holds:

$$\widetilde{v}(s,x)=\int_s^T\int_{\mathbb{R}^n}G(s,x;\tau,\eta)\widetilde{f}(\tau,\eta)d\eta d\tau,\qquad (s,x)\in[0,T]\times\mathbb{R}^n.$$

Similar as inequality (12.3.4), one can establish the following useful estimates:

$$\begin{cases}|\widetilde{v}|^{(0)}_{[S,T]\times\mathbb{R}^n}\leqslant K(T-S)|\widetilde{f}|^{(\alpha)}_{[S,T]\times\mathbb{R}^n},\\ |\widetilde{v}_x|^{(0)}_{[S,T]\times\mathbb{R}^n}\leqslant K(T-S)^{\frac{1+\alpha}{2}}|\widetilde{f}|^{(\alpha)}_{[S,T]\times\mathbb{R}^n},\\ |\widetilde{v}_s|^{(0)}_{[S,T]\times\mathbb{R}^n}+|\widetilde{v}_{xx}|^{(0)}_{[S,T]\times\mathbb{R}^n}\leqslant K(T-S)^{\frac{\alpha}{2}}|\widetilde{f}|^{(\alpha)}_{[S,T]\times\mathbb{R}^n}.\end{cases}\tag{12.3.7}$$

Also, for any $s, s' \in [S, T]$, $x, x' \in \mathbb{R}^n$,

$$\begin{cases} |\widetilde{v}_x(s, x) - \widetilde{v}_x(s', x)| \leqslant K|s - s'|^{\frac{1+\alpha}{2}} |\widetilde{f}|^{(\alpha)}_{[S,T]\times\mathbb{R}^n}, \\ |\widetilde{v}_{xx}(s, x) - \widetilde{v}_{xx}(s', x)| \leqslant K|s - s'|^{\frac{\alpha}{2}} |\widetilde{f}|^{(\alpha)}_{[S,T]\times\mathbb{R}^n}, \\ \qquad\qquad (s, x), (s', x) \in [S, T] \times \mathbb{R}^n. \end{cases} \tag{12.3.8}$$

From (12.3.7), for any $s, s' \in [S, T]$, $x, x' \in \mathbb{R}^n$, $|x - x'| \leqslant 1$, we further have

$$\begin{aligned} &|\widetilde{v}(s, x) - \widetilde{v}(s', x)| \\ &\leqslant \int_0^1 |\widetilde{v}_s(s' + \mu(s - s'), x)| d\mu\, |s - s'| \leqslant K(T - S)^{\frac{\alpha}{2}} |\widetilde{f}|^{(\alpha)}_{[S,T]\times\mathbb{R}^n} |s - s'| \\ &\leqslant K|\widetilde{f}|^{(\alpha)}_{[S,T]\times\mathbb{R}^n} \Big[(T - S)|s - s'|^{\frac{\alpha}{2}}\Big] \wedge \Big[(T - S)^{\frac{1}{2}} |s - s'|^{\frac{1+\alpha}{2}}\Big], \end{aligned}$$

and

$$\begin{aligned} |\widetilde{v}(s, x) - \widetilde{v}(s, x')| &\leqslant \int_0^1 |\widetilde{v}_x(s, x' + \mu(x - x'))| d\mu\, |x - x'| \\ &\leqslant K(T - S)^{\frac{1+\alpha}{2}} |\widetilde{f}|^{(\alpha)}_{[S,T]\times\mathbb{R}^n} |x - x'| \leqslant K(T - S)^{\frac{1+\alpha}{2}} |\widetilde{f}|^{(\alpha)}_{[S,T]\times\mathbb{R}^n} |x - x'|^{\alpha}, \end{aligned}$$

which leads to

$$\langle\widetilde{v}\rangle^{(\alpha)}_{[S,T]\times\mathbb{R}^n} \equiv \langle\widetilde{v}\rangle^{(\frac{\alpha}{2})}_{s,[S,T]\times\mathbb{R}^n} + \langle\widetilde{v}\rangle^{(\alpha)}_{x,[S,T]\times\mathbb{R}^n} \leqslant K(T - S)^{\frac{1+\alpha}{2}} |\widetilde{f}|^{(\alpha)}_{[S,T]\times\mathbb{R}^n},$$

as well as

$$\langle\widetilde{v}\rangle^{(\frac{1+\alpha}{2})}_{s,[S,T]\times\mathbb{R}^n} \leqslant K(T - S)^{\frac{1}{2}} |\widetilde{f}|^{(\alpha)}_{[S,T]\times\mathbb{R}^n}.$$

Next, the first inequality in (12.3.8) implies that

$$\begin{aligned} |\widetilde{v}_x(s, x) - \widetilde{v}_x(s', x)| &\leqslant K|\widetilde{f}|^{(\alpha)}_{[S,T]\times\mathbb{R}^n} |s - s'|^{\frac{1+\alpha}{2}} \\ &\leqslant K(T - S)^{\frac{1}{2}} |\widetilde{f}|^{(\alpha)}_{[S,T]\times\mathbb{R}^n} |s - s'|^{\frac{\alpha}{2}}. \end{aligned}$$

Similar to the above, for any $s \in [S, T]$, $x, x' \in \mathbb{R}^n$, $|x - x'| \leqslant 1$, making use of the third inequality in (12.3.7), we have

$$\begin{aligned} |\widetilde{v}_x(s, x) - \widetilde{v}_x(s, x')| &\leqslant \int_0^1 |\bar{v}_{xx}(s, x' + \mu(x - x')) d\mu|\, |x - x'| \\ &\leqslant K(T - S)^{\frac{\alpha}{2}} |\widetilde{f}|^{(\alpha)}_{[S,T]\times\mathbb{R}^n} |x - x'|, \end{aligned}$$

which leads to

$$\langle\widetilde{v}_x\rangle^{(\alpha)}_{[S,T]\times\mathbb{R}^n} \leqslant K(T - S)^{\frac{\alpha}{2}} |\widetilde{f}|^{(\alpha)}_{[S,T]\times\mathbb{R}^n}.$$

Hence, combining the above, we end up with

$$\begin{aligned}|\widetilde{v}|^{(1+\alpha)}_{[S,T]\times\mathbb{R}^n} &\equiv |\widetilde{v}|^{(0)}_{[S,T]\times\mathbb{R}^n} + |\widetilde{v}_x|^{(0)}_{[S,T]\times\mathbb{R}^n} + \langle\widetilde{v}\rangle^{(\frac{1+\alpha}{2})}_{s,[S,T]\times\mathbb{R}^n} + \langle\widetilde{v}_x\rangle^{(\alpha)}_{[S,T]\times\mathbb{R}^n}\\ &\leqslant K(T-S)^{\frac{\alpha}{2}}|\widetilde{f}|^{(\alpha)}_{[S,T]\times\mathbb{R}^n}.\end{aligned}$$

This implies (12.3.5). □

4 Bibliographical Remarks

Mallivan calculus materials are basically taken from [150], Chapter 1, and Section 5 of [68], with some small modifications. In particular, Theorem 12.1.7, i.e., the Fubini theorem for Skorokhod integral is adjusted from the analogue result in [151]. In 1970, Garsia-Rodemich-Rumsey proved an interesting inequality [84] which strengthened the well-knwon Kolmogorov continuity theorem. Later, the inequality was extended to the case of multidimensional times by Garsia [83]. As this result is not very standard, for reader's convenience, we provide a proof here, which is taken from [84], and the proof of the Corollary 12.2.2 is taken from [190], with some modifications. The PDE material provided in Sect. 3 is standard, which can be found in, say, [78].

References

1. N. H. Abel, *Solution de quelques problèmes à l.aide d'integrales défines,* In: Sylow L, Lie S, eds. Oeuvres Complètes de Niels Henrik Abel: Nouvelle Edition. Cambridge Library Collection - Mathematics. 2012: 11–27.
2. E. Abi Jaber, E. Miller, and H. Pham, *Linear-quadratic control for a class of stochastic Volterra equations: solvability and approximation,* Ann. Appl. Probab. 31 (2021), 2244–2274.
3. N. Agram, *Dynamic risk measure for BSVIE with jumps and semimartingale issues,* Stoch. Anal. Appl. 37 (2019), 361–376.
4. N. Agram, and B. Djehiche, *On a class of reflected backward stochastic Voletrra integral equations and related time-inconsistet optiml stopping problems,* Syst. Control Lett., 155 (2021), 104989;
5. N. Agram and B. Oksendal, *Malliavin calculus and optimal control of stochastic Volterra integral equations,* J. Optim. Theory Appl., 167 (2015), 1070–1094.
6. N. Agram, B. Oksendal and S. Yakhlef, *New approach to optimal control of stochastic Volterra integral equations,* Stochastics, 91 (2019), 873–894.
7. A. Ahmadova and N. Mahmudov, *Picard approximation of a singular backward stochastic nonlinear Volterra integral equation,* Qual. Theory Dyn. Syst., 23 (2024), 192.
8. N. Ahmed and X. Ding, *A semilinear McKean-Vlasov stochastic evolution equation in Hilbert space,* Stochastic Process. Appl., 60 (1995), 65–85.
9. A. Aman and M. N'Zi, *Backward stochastic nonlinear Volterra integral equations with local Lipschitz drift,* Probab. Math. Stat., 25 (2005), 105–127.
10. V. A. Ambarzumian, *Diffuse reflection of light by a foggy medium,* C. R. Acad. Sci. SSSR, 38 (1943), 229–232 (in Russian).
11. V. Anh and J. Yong, *backward stochastic Volterra integral equations in Hilbert spaces,* Differential and Difference Equations and Applications, Hindawi, New York, (2006), 57–66.
12. V. Anh, W. Grecksch and J. Yong, *Regularity of backward stochastic Volterra integral equations in Hilbert spaces,* Stoch. Anal. Appl., 29 (2011), 146–168.
13. P. Artzner, F. Delbaen, J. M. Eber, and D. Heath, *Coherent measures of risk,* Math. Finance, **9** (1999), 203–228.
14. G. Barles, R. Buckdahn, and E. Pardoux, *Backward stochastic differential equations and integral partial differential equations,* Stochastics, 60 (1997), 57–83.
15. P. Barrieu and N. El Karoui, *Monotone stability of quadratic semimartingales with applications to unbounded general quadratic BSDEs.* Ann. Probab., 41 (2013), 1831–1863.
16. R. Bass, *The measurability of hitting times,* Elect. Comm. Probab., 10 (2010), 99–105.

T. Wang, J. Yong, *Backward Stochastic Volterra Integral Equations*,
Probability Theory and Stochastic Modelling 109,
https://doi.org/10.1007/978-3-032-20848-4

17. C. Bayer, P. K. Friz, P. Gassiat, J. Martin, and B. Stemper, *A regularity structure for rough volatility,* Math. Finance, 30 (2020), 782–832.
18. P. R. Beesack, *Comparison theorems and integral inequalities for Volterra integral equations,* Proc. Amer. Math. Soc., 20 (1969), 61–66.
19. P. Beissner and E. R. Gianin, *The term structure of Sharpe ratios and arbitrage free asset pricing in continuous time,* Probab. Uncertain. Quant. Risk, 6 (2021), 23–52.
20. S. A. Belbas, *Iterative schemes for optimal control of Volterra integral equations,* Nonlinear Anal., 37 (1999), 57–79.
21. S. A. Belbas, *A reduction method for optimal control of Volterra integral equations,* Appl. Math. Comput., 197 (2008), 80–890.
22. S. A. Belbas and W. H. Schmidt, *Optimal control of Volterra equations with impulses,* Appl. Math. Comput., 166 (2005), 696–723.
23. R. Bellman, *On the theory of dynamic programming,* Proc. Nat, Acad. Sci. USA, 38 (1952), 716–719.
24. R. E. Bellman and R. E. Kalaba, *On the principle of invariant imbedding and propagation through inhomogeneous media,* Proc. Natl. Acad. Sci. USA, 42 (1956), 629–632.
25. R. E. Bellman and G. M. Wing, *An Introduction to Invariant Imbedding,* Wiley, New York, 1975.
26. C. Bender and S. Pokalyuk, *Discretization of backward stochastic Volterra integral equations,* Recent developments in computational finance, Interdiscip. Math. Sci., 14 (2013), 245–278.
27. A. Bensoussan, *Lecture on Stochastic Control,* Lecture Notes in Math., Vol.972, Springer-Verlag, 1981.
28. M. Berger and V. Mizel, *Volterra equations with Itô integrals, I, II,* J. Integral Equation., 2 (1980), 187–245; 319–337.
29. J.-M. Bismut, *Conjugate convex functions in optimal stochastic control,* J. Math. Anal. Appl., 44 (1973), 384–404.
30. J.-M. Bismut, *An introductory approach to duality in optimal stochastic control,* SIAM Rev., 20 (1978), 62–78.
31. M. Bôcher, *An Introduction to the Study on Integral Equations,* Cambridge Univ. Press, 1909.
32. V. G. Boltyanskii, R. V. Gamkreldze, and L. S. Pontryagin, *On the theory of optimal processes,* Dokl. Akad. Nauk SSSR (N.S.), 110 (1956), 7–10. (in Russian)
33. S. Bonaccorsi, F. Confortola and E. Mastrogiacomo, *Optimal control for stochastic Volterra equations with completely monotone kernels,* SIAM J. Control Optim. 50 (2012), 748–789.
34. J. F. Bonnans, C. de la Vega, and X. Dupuis, *First and second-order optimality conditions for optimal control problems of state constrained intergral equations,* J. Optim Theory Appl., 159 (2013), 1–40.
35. V. Borkar and K. Kumar, *McKean-Vlasov limit in portfolio optimization,* Stoch. Anal. Appl., 28 (2010), 884–906.
36. P. Briand and Y. Hu, *BSDE with quadratic growth and unbounded terminal value.* Probab. Theory Relat. Fields 136 (2006), 604–618.
37. P. Briand and Y. Hu, *Quadratic BSDEs with convex generators and unbounded terminal conditions.* Probab. Theory Relat. Fields 141 (2008), 543–567.
38. R. Buckdahn, B. Djehiche, J. Li and S. Peng, *Mean-field backward stochastic differential equations: a limit approach,* Ann. Probab., 37 (2009), 1524–1565.
39. R. Buckdahn, J. Li and S. Peng, *Mean-field backward stochastic differential equations and related partial differential equations,* Stochastic. Process. Appl., 119 (2009), 3133–3154.
40. Z. Cao and J. Yan, A comparison theorem for solutions of stochastic differential equations, *Adv. Math. (China)* 28 (1999), 304–308.
41. T. Chan, *Dynamics of the McKean–Vlasov equation,* Ann. Probab., 22 (1994), 431–441.
42. S. Chandrasekhar, *Radiative Transfer,* Dover, New York, 1960.
43. S. Chen and J. Yong, *A linear quadratic optimal control problems for stochastic Volterra integral equations,* in Control Theory and Related Topics: In Memory of Prof. Xunjing Li, edited by S. Tang and J. Yong, Word Scientific Publishing Company, 2007, 44–66.

44. Y. Chen-Shue, Y, Li and J. Yong, *A limit order book model for high frequency trading with rough volatility,* Numer. Algebra, Control Optim., 15 (2025), 1192–1243.
45. P. Cheridito, H. M. Soner, N. Touzi, and N. Victoir, *Second order backward stochastic differentail equations and fully nonlinear parabolic PDEs,* Comm. Pure Appl. Math., 60 (2007), 1081–1110.
46. P. Cheridito, F. Delbaen, and M. Kupper, *Coherent and covex monetary risk measures for bounded càdlàg processes,* Stochastic Process. Appl. 112 (2004), 1–22.
47. P. Cheridito, F. Delbaen, and M. Kupper, *Coherent and covex monetary risk measures for unbounded processes,* Finance Stoch. 10 (2006), 427–448.
48. P. Cheridito, F. Delbaen, and M. Kupper, *Dynamic monetary risk measures for bounded discrete time processes,* Electron. J. Probab., 11 (2006), 57–106.
49. J. A. Cochran, Analysis of Linear Integral Equations, McGraw-Hill, New York, 1972.
50. S. N. Cohen, R. J. Elliott, and C. E. M. Pearce, *A general comparison theorem for backward stochastic differential equations,* Adv. Appl. Prob., 42 (2010), 878–898.
51. F. Comte and E. Renault, *Long memory in continuous-time stochastic volatility models.* Math. Finance, 8 (1998), 291–323.
52. H. Coulibaly and A. Aman, *Backward stochastic Volterra integral equations with non-Lipschitz time delayed generators,* J. Math. Research, 11 (2019), 94–102.
53. M. H. A. Davis, *Martingale representation and all that,* Advances in Control, Communication Networks, and Tranportation Systems, E. H. Abed Editor, Birkhäuser, 2005, 57–68.
54. D. Dawson, *Critical dynamics and uctuations for a mean-field model of cooperative behavior,* J. Statist. Phys., 31 (1983), 29–85.
55. F. Delbaen, Y. Hu and X. Bao, *Backward SDEs with superquadratic growth,* Probab. Theory Relat. Fields, 150 (2010), 145–192.
56. F. Delbaen, Y. Hu and A. Richou, *On the uniqueness of solutions to quadratic BSDEs with convex generators and unbounded terminal conditions.* Ann. Inst. Henri Poincaré Probab. Stat. 47 (2011), 559–574.
57. C. Dellacherie and P.-A. Meyer, Probabilities and Potential, North-hollend, Amsterdam, 1978.
58. K. Detlefsen and G. Scandolo, *Conditional and dynamic convex risk measures,* Finance Stoch., 9 (2005), 539–561.
59. J. Djordjević and S. Janković, *On a class of backward stochastic Volterra integral equations,* Appl. Math. Lett., 26 (2013), 1192–1197.
60. J. Djordjević and S. Janković, *Backward stochastic Volterra integral equations with additive perturbations,* Appl. Math. Computation, 265 (2015), 903–910.
61. C. Doléans-Dade, *Variation quadratique des martingales continues à droite,* Ann. Math. Star., 40 (1969), 284–289.
62. D. Duffie, Dynamic Asset Pricing Theory, Princeton University Press, 1996.
63. D. Duffie and L. G. Epstein, *Stochastic differential utility,* Econometrica 60 (1992), 353–394.
64. D. Duffie and P.-L. Lions, *PDE solutions of stochastic differential utility,* J. Math. Econom., 21 (1992), 577–606.
65. O. El Euch, M. Fukasawa, and M. Rosenbaum, *The microstructural foundations of leverage effect and rough volatility,* Finance Stoch., 22 (2018), 241–280.
66. O. El Euch and M. Rosenbaum, *Perfect hedging in rough Heston models,* Ann. Appl. Probab., 28 (2018), 3813–3856.
67. O. El Euch and M. Rosenbaum, *The characteristic function of rough Heston models,* Math. Finance, 29 (2019), 3–38.
68. N. El Karoui, S. Peng and M. C. Quenez, *Backward stochastic differential equations in finance,* Math. Finance, 7 (1997), 1–71.
69. N. El Karoui and M. Quenez, *Non-linear pricing theory and backward stochastic differential equations.* Runggaldier, W.J. (eds) Financial Mathematics. Lecture Notes in Mathematics, vol *1656.* Springer, Berlin, Heidelberg, 1997.

70. S. Fan, Y. Hu and S. Tang, *Multi-dimensional backward stochastic differential equations of diagonally quadratic generators: the general result,* J. Differential Equation, 368 (2023), 105–140.
71. S. Fan, T. Wang and J. Yong, *Multi-dimensional super-linear backward stochastic Volterra integral equations,* J. Differential Equation. 437 (2025), 113350.
72. H. O. Fattorini, Infinite Dimensional Optimization and Control Theory, Cambridge Univ. Press, 1999.
73. G. Ferreyra and P. Sundar, *Comparison of stochastic Volterra equations,* Bernoulli 6 (2000), 1001–1006.
74. R. P. Feynman, *Space-time approach to nonrelativistic quantum mechanics,* Rev. Modern Phy., 20 (1948), 367–387.
75. H. Föllmer and A. Schied, *Convex measures of risk and trading constraints,* Finance Stoch., 6 (2002), 429–447.
76. I. Fredholm, *Sur une d'équations fonctionnelles,* Acta Math., 27 (1903), 365–390.
77. A. Friedman, On integral equations of Volterra type, J. Analyse Math. 11, 1963, 381–413.
78. A. Friedman, Partial Differential Equations of Parabolic Type, Prentice-Hall, Englewood Cliffs, N.J.,1964.
79. A. Friedman, *Optimal control for hereditary processes,* Arch. Rat. Mech. Anal. **15** (1964) 396–416.
80. C. Frei and G. Dos Reis, *A financial market with interacting investors: does an equilibrium exist? Math. Financ. Econ.* 4 (2011), 161–182.
81. M. Frittelli and G. Scandolo, *Risk measures and capital requirements for processes,* Math. Finance 16 (2005), 589–621.
82. L. I. Gal'cuk and M. H. A. Davis, *A note on a comparison theorem for equations with different diffusions,* Stochastics, 6 (1982) 147–149.
83. A. M. Garsia, *Continuity properties of Gaussian processes with multidimensional time parameter,* Proc. 6th Berkeley Symposium, V. *II*, 1972, 369–374.
84. A. M. Garsia, E. Rodemich and H. Rumsey, Jr., *A real variable lemma and the continuity of paths of some Gaussian proceses,* Indiana Math. J., 20 (1970), 65–578.
85. J. Gatheral, T. Jaisson, and M. Rosenbaum, *Volatility is rough,* Quant. Finance, 18 (2018), 933 -949.
86. C. Geiβ and R. Manthey, *Comparison theorems for stochastic differential equations in finite and infinite dimensions,* Stochastic Process. Appl., 53 (1994), 23–35.
87. J. Gong, and T. Wang, *Causal feedback strategies for controlled stochastic Volterra systems: a unified treatment, SIAM J.* Control Optim. 63 (2025), 1950–1980.
88. G. Gripenberg, S. O. Londen and O. Staffans, Volterra Integral and Functional Equations, Cambridge Univ. Press, 1990.
89. Y. Hamaguchi, *Extended backward stochastic Volterra integral equations and their applications to time-inconsistent stochastic recursive control problems,* Math. Control Relat. Fields, 11 (2021), 433–478.
90. Y. Hamaguchi, *Infinite horizon backward stochastic Volterra integral equations and discounted control problems,* ESAIM Control Opim. Calc. Var., 27 (2021), 101.
91. Y. Hamaguchi *On the maximum principle for optimal control problems of stochastic Volterra integral equations with delay,* Appl. Math. Optim., 87 (2023), 42.
92. Y. Hamaguchi, *Variation of constants formulae for forward and backward stochastic Volterrra integral equations, J.* Differential Equation., 343 (2023), 332–389.
93. Y. Hamaguchi and T. Dai, *Approximations for adapted M-solutions of type-II backward stochastic Volterra integral equations,* ESAIM Probab. Stat., 27 (2023), 19–79.
94. Y. Hamaguchi and T. Wang, *Linear-quadratic stochastic Volterra controls I: Causal feedback strategies,*Stochastic. Process. Appl., 16 (2024), 104449.
95. Y. Hamaguchi and T. Wang, *Linear-quadratic stochastic Volterra controls II: Optimal strategies and Riccati-Volterra equations,* ESAIM Control Optim. Calc Var., 30 (2024), 48.
96. R. A. Handelsman and W. E. Olmstead, *Asymptotic solution to a class of nonlieaner Volterra integral equations, SIAM J.* Appl. Math., 22 (1972), 373–384.

97. U. G. Haussmann, *General necessary conditions for optimal control of stochastic systems,* Math. Prog. Study. 6 (1976), 34–48.
98. S. He, J. Wang and J. Yan, Semimartingale Theory and Stochastic Calculus, Science Press and CRC Press, Beijing, 1992.
99. C. Hernández, *On quadratic multidimensional type-I BSVIEs, infinite families of BSDEs and their applications,* Stochastic. Process. Appl. 162 (2023), 249–298.
100. C. Hernández and D. Possamai, *A unified approach to well-posedness of type-I backward stochastic Volterra integral equations,* Electron J. Probab., 25 (2021), 89.
101. H. W. Hethcote and D. W. Tudor, *Integral equation models for endemic infectious diseases,* J. Math. Biology, 9 (1980), 37–47.
102. E. Hille and R. S. Phillips, Functional Analysis and Semi-Groups, AMS, R.I., 1957.
103. Y. Hu and B. Oksendal, *Linear Volterra backward stochastic integral equations,* Stochastic Process. Appl. 129 (2019), 626–633.
104. Y. Hu and S. Peng, *On the comparison theorem for multidimensional BSDEs,* C. R. Math. Acad. Sci. Paris, 343 (2006), 135–140.
105. Y. Hu and S. Tang, *Multi-dimensional backward stochastic differential equations of diagonally quadratic generators,* Stochastic. Process. Appl., 126 (2016), 1066–1086.
106. J. Huang, X. Li and T. Wang, *Mean-field linear-quadratic-Gaussian (LQG) games for stochastic integral systems,* IEEE Trans. Automat. Control, 61 (2016), 2670–2675
107. M. Huang, R. Malhamé and P. Caines, *Large population stochastic dynamic games: closed-loop McKean-Vlasov systems and the Nash certainty equivalence principle,* Comm. Inform. Systems, 6 (2006), 221–252.
108. Z. Huang, *A comparison theorem for solutions of stochastic differential equations and its applications,* Proc. Amer. Math. Soc, 91 (1984), 611–617.
109. N. Ikeda and S. Watanabe, *A comparison theorem for solutioins of stochastic differential equations and its applications,* Osaka J. Math., 14 (1977), 619–633.
110. I. Itô, *On the existence and uniqueness of solutions of stochastic integral equations of the volterra type,* Kodai Math. J. 2 (1979), 158–170.
111. K. Itô, *Stochastic integral,* Proc. Imp. Acad. Tokyo, 20 (1944), 519–524.
112. K. Itô, *On a stochastic integral equation,* Proc. Japan Acad., 22 (1946), 32–35.
113. K. Itô, *On a formla concerning stochastic differentials,* Nagoya Math. J., 3 (1951), 55–65.
114. A. Jamneshan, M. Kupper and P. Luo, *Multidimensional quadratic BSDEs with separated generators.* Electron. Commun. Probab. 22 (2017), 1–10.
115. A. J. Jerri, Introduction to Integral Equations with Applications, John Wiley & Sons, Inc., New York, 1999.
116. P. Jusselin and M. Rosenbaum, *No-arbitrage implies power-law market impact and rough volatility,* Math. Finance, 30 (2020), 1309–1336.
117. M. Kac, *On distributions of certain Wiener functionals,* Trans. Amer. Math. Soc., 65 (1949), 1–13.
118. R. E. Kalman, *Contributions to the theory of optimal control,* Bol. Soc. Mat. Mexicana, 5 (1960), 102–119.
119. M. I. Kamien and E. Muller, *Optimal control with integral state equations,* The Review Econ. Studies, 43 (1976), 469–473.
120. I. Karatzas and S. E. Shreve, Brownian Motion and Stochastic Calculus, 2nd Edition, Springer-Verlag, 1991.
121. I. Karatzas, S. E. Shreve, Methods of Mathematical Finance, Springer, New York, 1998.
122. N. Kazamaki, Continuous Exponential Martingales and BMO, Lecture Notes in Mathematics, *1579,* Springer-Verlag, Berlin, 1994.
123. J. B. Keller and W. E. Olmstead, *Temperature of a nonlinearly radiating semi-infinite solid,* Quarterly Appl. Math., 29 (1972), 559–566,
124. M. Kobylanski, *Backward stochastic differential equations and partial differential equations with quadratic growth.* Ann. Probab., 28 (2000), 558–602.
125. J. Kondo, Integral Equations, Kodansha, Tokyo, 1991.

126. P. Kotelenez and T. Kurtz, *Macroscopic limit for stochastic partial differential equations of McKean-Vlasov type,* Probab. Theory Relat. Fields, 146 (2010), 189–222.
127. E. Kromer and L. Overbeck, *Differentiability of BSVIEs and dynamic capital allocations,* Int. J. Theor. Appl. Finance, 20 (2017), 1750047.
128. H. Kushner, *Optimal stochastic control,* IEEE Trans. Automat. Control, 7 (1962), 120–122.
129. H. Kushner, *A maximum principle for stochastic control systems,* J. Math. Anal. Appl., 11 (1965), 78–92.
130. H. Kushner, *Necessary conditions for continuous parameter stochastic optimization problems,* SIAM J. Control, 10 (1972), 550–565.
131. H. Kushner and F. Schweppe, *A maximum principle for stochastic control systems,* J. Math. Anal. Appl., 8 (1964), 287–302.
132. J. Lasry and P. L. Lions, *Mean field games,* Japan J. Math., 2 (2007), 229–260.
133. A. Lazrak, *Generalized stochastic differential utility and preference for information,* Ann. Appl. Probab., 14 (2004), 2149–2175.
134. A. Lazrak and M. C. Quenez, *A generalized stochastic differential utility,* Math. Oper. Res. 28 (2003) 154–180.
135. J. Lin, *Adapted solution of a backward stochastic nonlinear Volterra integral equation,* Stochastic Anal. Appl., 20 (2002), 165–183.
136. P. Lin and J. Yong, *Controlled singular Volterra integral equations and Pontryagin maximum principle,* SIAM J. Control Optim. 58 (2020), 136–164.
137. J. Liouville, *Second Mémoire sur le développement des fonctions ou parties de fonctions en séries dont les divers terms sont assujétis à satisfaire à une même équation différentielle du second ordre, contenant un paramètre variable,* J. Math. Pure Appl., 2 (1837), 16–35.
138. J. Liu and J. Ren, *Comparison theorem for solutions of backward stochastic differential equations with continuous coefficient,* Statist. Probab. Lett. 56 (2002), 93–100
139. Z. Liu and T. Wang, *A class of stochastic Fredholm-algebraic equations and applications in finance,* Discrete. Contin. Dyn. Syst. Ser. B. 26 (2021), 3879–3903.
140. Q. Lü, J. Yong and X. Zhang, *Representation of Itô integrsls by Lebesgue/ Bochner integrsls,* J. Eur. Math. Soc., 14 (2012), 1795–1823.
141. J. Ma, P. Protter and J. Yong, *Solving forward-backward stochastic differential equations explicitly–a four step scheme,* Probab. Theory Related Fields, 98 (1994), 339–359.
142. J. Ma, Z. Wu, D. Zhang, and J. Zhang, *On well-posedness of forward-backward SDEs—a unified approach,* Ann. Appl. Probab., 25 (2015), 2168–2214.
143. J. Ma and J. Yong, Forward-backward stochastic differential equations and their applications, Lecture Notes in Mathematics, vol. 1702, Springer-Verlag, Berlin, 1999.
144. N. G. Medhin, *Optimal processes governed by integral equations,* J. Math. Anal. Appl., 120 (1986), 1–12.
145. A. V. Melnikov, *On the theory of stochastic equations in components of semimartingales,* Math. Ussr Sbornik, 38 (1981), 381–394.
146. W. Meng, T. Wang, and J. Zhang, *Closed-loop solvability of delayed control problems: A stochastic Volterra system approach,* arXiv:2510.02674v1.
147. W. Meng, J. Shi, T. Wang, and J. Zhang, *A general maximum principle for optimal control of stochastic differential delay systems,* SIAM J. Control Optim. 63 (2025), 175–205.
148. R. K. Miller, *Nonlinear Volterra Integral Equations,* Menlo Park, CA: Benjamin, 1971.
149. L. Miao, Y. Chen. X. Xiao, and Y. Hu, *Anticipated backward stochastic Volterra integral equationa with jumps and applications to dynamic risk measures,*Acta Math. Scientia, 43 (2023), 1365–1381.
150. D. Nualart, *The Malliavin Calculus and Related Topics,* Springer, 2006.
151. D. Nualart and M. Zakai, *Generalized multiple stochastic integrals and tbe representation of Wiener functionals,* Stochastic, 23 (1988), 311–330.
152. G. L. O'Brien, *A new comparison theorem for solution of stochastic differential equations,* Stochastics 3 (1980), 245–249.
153. B. Oksendal, *Stochastic Differential Equations: An Introduction with Applications, 5th Edition,* Springer-Verlag, 1998.

154. W. E. Olmstead and R. A. Handelsman, *Asymptotic solution to a class of nonlinear Volterra integral equations, II, SIAM J. Appl. Math.,* 30 (1976), 180–189.
155. Y. Ouknine and M. Rutkowski, *Strong comparison of solutions of one-dimensional stochastic differential equations,* Stochastic. Process. Appl., 36 (1990), 217–230.
156. L. Overbeck and J. A. L. Röder, *Path–dependent backward stochastic Volterra integral equations with jumps, differentiability and duality principle,* Probab. Uncerain quant. Risk, 3 (2018), 4.
157. E. Pardoux, *Generalized discontinuous backward stochastic differential equations,* In: N. El Karoui, L. Mazliak (Eds.), Backward Stochastic Differential Equations, in: Pitman Res. Notes Math., vol. 364, Longman, Harlow, 1997, 207–219.
158. E. Pardoux and S. Peng, *Adapted solution of a backward stochastic differential equation, Syst.* Control Lett., 14 (1990), 55–61.
159. E. Pardoux and S. Peng, *Backward stochastic differential equations and quasilinear parabolic partial differential equations,* Lecture Notes in Control and Inform. Sci., vol. 176, Springer, Berlin, 1992, 200–217.
160. E. Pardoux and P. Protter, *Stochastic Volterra equations with anticipating coefficients,* Ann. Probab., 18 (1990), 1635–1655.
161. S. Peng, *A general stochastic maximum principle for optimal control problems,* SIAM J. Control Optim., 28 (1990), 966–979.
162. S. Peng, *A nonlinear Feynman-Kac formula and applications,* Control theory, stochastic analysis and applications (Hangzhou, 1991), 173–184, World Sci. Publ., River Edge, NJ, 1991.
163. S. Peng, *A generalized dynamic programming principle and Hamilton-Jacobi-Bellman equation,* Stochastics Stochastics Rep. 38 (1992) 119–134.
164. S. Peng, *Backward stochastic differential equations, nonlinear expectation and their applications,* Proc. ICM, Vol 1, 393–432, Hindustn Book Agency, New Delhi, 2010.
165. S. Peng, Nonlinear Expectations and Stochastic Calculus under Uncertainty, with Robust CLT and G-Brownian Motion, Springer, 2019.
166. S. Peng and X. Zhu, *Necessary and sufficient condition for comparison theorem of 1-dimensional stochastic differential equations,* Stochastic Process Appl. 116 (2006) 370–380.
167. L. S. Pontryagin, V. G. Boltyanskii, R. V. Gamkrelidze, E. F. Mishchenko, The Mathematical Theory of Optimal Processes, Interscience Publishers, John Wiley & Sons, Inc., New York, 1962.
168. A. Popier, *Backward stochastic Volterra integral equations with jumps in a general filtration,* ESAIM: Probab, Stat. 25 (2021), 133–203.
169. P. Protter, *On the existence, uniqueness, convergence and explorsions of solutions of sysmtes of stochastic integral equations,* Ann. Probab., 5 (1977), 243–261.
170. Y. Ren, *On solutions of backward stochastic Volterra integralequations with jumps in Hilbert spaces,* J. Optim. Theory Appl., 144 (2010), 319–333.
171. F. Riedel, *Dynamic coherent risk measures,* Stochastic Process. Appl., 112 (2004), 185–200.
172. E. Rosazza Gianin and M. Zullino, *Capital allocation for cash-subadditive risk measures: from BSDEs to BSVIEs,* Probab. Uncertain. Quant. Risk, 9 (2024), 339–370.
173. T. Roubicek, Nonlinear Partial Differential Equations with Applications, Birkhäuser, 2005.
174. M. R. Scott, Invariant Imbedding and ts Applications to Ordinary Differential Equations, Addison-Wesley, Reading, Mass., 1973.
175. Y. Shi and T. Wang, *Solvability of general backward stochastic Volterra integral equations, J.* Korean Math. Soc., 49 (2012), 1301–1321.
176. Y. Shi, T. Wang, and J. Yong, *Mean-field backward stochastic Volterra integral equations,* Discrete Contin. Dyn. Syst. Ser. B, 18 (2013), 1929–1967.
177. Y. Shi, T. Wang and J. Yong, *Optimal control of forward-backward stochastic Volterra integral equations and applications,* Math. Control Related Fields, 5 (2015), 613–649.
178. Y. Shi, J. Wen and J. Xiong, *Backward doubly stochastic Volterra integral equations and applications.* J. Differential Equations. 269 (2020), 6492–6528.

179. A. V. Skorohod, *Studies in the theory of random processes*, Izdat. Kiev. Univ., Kiev, 1961; English transl., Addison-Wesley, 1965
180. H. M. Soner, N. Touzi, and J. Zhang, *Wellposednes of second order backward SDEs*, Probab. Theory Relat. Fields, 153 (2012), 149–190.
181. A. Sznitman, *Topics in Propagation of Chaos, Ecôle de Probabilites de Saint Flour, XIX-1989. Lecture Notes in Math, 1464*, Springer, Berlin, 1989, 165–251.
182. R. Tevzadze, *Solvability of backward stochastic differential equations with quadradic growth. Stochastic Process. Appl.* 118 (2008), 503–515.
183. F. G. Tricomi, *Integral Equations*, Interscience Publishers, Inc., New York, 1957.
184. C. P. Tsokos, *On a stochastic integral equation of the Volterra type, Math, System Theory*, 3 (1969), 222–231.
185. C. P. Tsokos and W. J. Padgett, *Random Integral Equations with Applications to Life Sciences and Engineering*, Academic Press, New York, 1974.
186. C. Tudor, *A comparison theorem for stochastic equations with Volterra drifts, Ann. Probab.*, 17 (1989), 1541–1545.
187. F. Viens and J. Zhang, *A martingale approach for fractional Brownian motions and related path dependent PDEs, Ann. Appl. Probab.*, 29 (2019), 3489–3540.
188. V. R. Vinokurov, *Optimal control of processes described by integral equations, I, II, III*, Izv. *Vysš. Učebn. Zaved. Matematika*, (1967), no.7 (62), 21–33; no.8 (63), 16–23; no.9 (64), 16–25; (in Russian). English transl. in *SIAM J. Control*, 7 (1969), 324–336, 337–345, 346–355.
189. V. Volterra, *Sulle inversione degl integrali definite*, Nota I, *Atti R. Accad. Sci. Torino*, 31 (1896), 311–323.
190. J. B. Walsh. *An Introduction to Stochastic Partial Differential Equations*, Lecture Notes in Math. vol. 1180, Springer-Varlag. 1984.
191. H. Wang, *Extended Backward Stochastic Volterra Integral Equations, Quasilinear Parabolic Equations, and Feynman-Kac Formula, Stoch. Dyn.*, 21 (2021), 2150003.
192. H. Wang, J. Sun and J. Yong, *Recursive utility processes, dynamic risk measures and quadratic backward stochastic Volterra integral equations, Appl. Math. Optim.*, 84 (2021), 145–190.
193. H. Wang, and J. Yong, *Time-inconsistent stochastic optimal control problems and backward stochastic Volterra integral equatioins, ESAIM: Control, Optim. Cal. Var.* 27 (2021), 22 https://doi.org/10.1051/cocv/2021027.
194. H. Wang, J. Yong and J. Zhang, *Path dependent Feynman–Kac formula for forward-backward stochastic Volterra integral equations, Ann, Inst, Henri Poincare Probab.*, 58 (2022), 603–638.
195. H. Wang, J. Yong, and C. Zhou. *Backward stochastic differential equations and backward stochastic Volterra integral equations with anticipating generators*, *Probab. Uncertain. Quant. Risk*, 7 (2022), 301–332.
196. H. Wang, J. Yong, and C. Zhou. *Linear-quadratic optimal controls for stochastic Volterra integral equations: causal state feedback and path-dependent Riccati equations, SIAM J. Control Optim.* 61 (2023) 2595–2629.
197. T. Wang, *L^p solutions of backward stochastic Volterra integral equations, Acta Math. Sinica*, 28 (2012), 1875–1882.
198. T. Wang, *Linear quadratic control problems of stochastic Volterra integral equations, ESAIM: Control, Optim. Cal. Var.* 24 (2018), 1849–1879.
199. T. Wang, *Necessary conditions of Pontraygin's type for general controlled stochastic Volterra integral equations, ESAIM: Control, Optim. Calc. Var.* 26 (2020) paper No.41, 34 pp.
200. T. Wang, *Backward stochastic Volterra intego-differential equations and applications in optimal control problems, SIAM J. Control Optim.* 60 (2022), 2393–2419.
201. T. Wang and Y. Shi, *Symmetrical solutions of backward stochastic Volterra integral equations and their applications, Discrete Contin. Dyn. Syst. Ser. B*, 14 (2010), 251–274.
202. T. Wang and Y. Shi, *Linear quadratic stochastic integral games and related topics, Sci. China, Math.* 58 (2015), 2405–2420.

203. T. Wang, Y. Shi and Q. Zhu, *Necessary and sufficient conditions of optimality for stochastic integral systems with partial information,* Proceedings of the 30th Chinese Control Conference, July 22–24, 2011, Yantai, China, 1950–1955.
204. T. Wang and J. Yong, *Comparison theorems for some backward stochastic Volterra integral equations,* Stochastic Process. Appl. 125 (2015), 1756–1798.
205. T. Wang and J. Yong, *Backward stochastic Volterra integral equations–representation of adapted solutions,* Stochastic. Process. Appl., 129 (2019), 4926–4964.
206. T. Wang and J. Yong, *Spike variations for Backward stochastic Volterra integral equations,* SIAM J. Control Optim., 61 (2023), 3608–3634.
207. T. Wang and H. Zhang, *Optimal control problems of forward-backward stochastic Volterra integral equations with closed control regions,* SIAM J. Control Optim. 55 (2017), 2574–2602.
208. T. Wang and H. Zheng, *Closed-loop equilibrium strategies for general time-inconsistent optimal control problems,* SIAM J. Control Optim. 59 (2021), 3152–3178.
209. T. Wang and M. Zheng, *Singular backward stochastic Volterra integral equations in infinite dimensional spaces,* J. Differential Equations. 407 (2024), 1–56.
210. Z. Wang, *Existence and uniqueness of solutions to stochastic Volterra equations with singular kernels and non-Lipschitz coefficients,* Stat. Probab. Lett., 78 (2008), 1062–1071.
211. Z. Wang and X. Zhang, *Non-Lipschitz backward stochastic Volterra type equations with jumps,* Stoch. Dyn., 7 (2007), 479–496.
212. A.-M. Wazwaz, Linear and Nonlinear Integral Equations—Methods and Applications, Higher Education Press, Beijing, 2011.
213. J. Wei, *Backward stochastic Volterra integral equations on Markov chains,* Stochastics, 90 (2018), 605–639.
214. J. Wen and Y. Shi, *Solvability of anticipated backward stochastic Volterra integral equations,* Statist. Probab. Lett. 156 (2020), 108599.
215. J. Wen and Y. Shi, *Symmetrical martingale solutions of backward doubly stochastic Volterra integral equations,* Comput. Math. Appl. 79 (2020), 1435–1446.
216. T. Yamada, *On a comparison theorem for solutions of stochastic differential equations and its applications,* J. Math. Kyoto Univ. 13 (1973), 497–512.
217. J. Yan, *A comparison theorem for semimartingales and its applications,*Séminaire de probabilités, 20 (1986), 349–351.
218. J. Yong, *Backward stochastic Volterra integral equations and some related problems,* Stochastic. Process. Appl., 116 (2006), 779–795.
219. J. Yong, *Continuous-time dynamic risk measure by backward stochastic Volterra integral equations,* Appl. Anal., 86 (2007), 1429–1442.
220. J. Yong, *Well-posedness and regularity of backward stochastic Volterra integral equations,* Probab. Theory Relat. Fields, 142 (2008), 21–77.
221. J. Yong, *Backward stochastic Volterra integral equations—a brief survey,* Appl. Math. J. Chinese Univ., 28 (2013), 383–394.
222. J. Yong, *Representation of adapted solutions to backward stochastic Volterra integral equations,* Sci. Sinica Math. 47 (2017), 305–345 (in Chinese).
223. J. Yong and X. Y. Zhou, Stochastic Controls: Hamiltonian Systems and HJB Equations, Springer-Verlag, New York, 1999.
224. S. M. Zemyan, The Classical Theory of Integral Equations, A Concise Treatment, Springer Science Business Media, LLC, 2012.
225. J. Zhang, *Representation of solutions to BSDEs associated with a degenerate FSDE, Ann. Appl. Probab.,* 15 (2005), 1798–1831.
226. J. Zhang, Backward Stochastic Differential Equations: From Linear to Fully Nonlinear Theory, Springer-Verlag, New York, 2017.
227. X. Zhang, *Stochastic Volterra equations in Banach spaces and stochastic partial differential equation,* J. Funct. Anal., 258 (2010), 1361–1425.

Index

T. Wang, J. Yong, *Backward Stochastic Volterra Integral Equations*,
Probability Theory and Stochastic Modelling 109,
https://doi.org/10.1007/978-3-032-20848-4